Biopolymers from Natural Resources

Biopolymers from Natural Resources

Editors

Rafael Antonio Balart Gimeno
Marina Patricia Arrieta Dillon
Daniel García García
Luís Jesús Quiles Carrillo
Vicent Fombuena Borrás

MDPI • Basel • Beijing • Wuhan • Barcelona • Belgrade • Manchester • Tokyo • Cluj • Tianjin

Editors
Rafael Antonio Balart Gimeno
Universitat Politècnica de València (UPV)
Spain

Marina Patricia Arrieta Dillon
Technical University of Madrid
Spain

Daniel García García
Universitat Politècnica de València (UPV)
Spain

Luís Jesús Quiles Carrillo
Universitat Politècnica de València (UPV)
Spain

Vicent Fombuena Borrás
Universitat Politècnica de València (UPV)
Spain

Editorial Office
MDPI
St. Alban-Anlage 66
4052 Basel, Switzerland

This is a reprint of articles from the Special Issue published online in the open access journal *Polymers* (ISSN 2073-4360) (available at: https://www.mdpi.com/journal/polymers/special_issues/Biopolymers_Natural_Resources).

For citation purposes, cite each article independently as indicated on the article page online and as indicated below:

LastName, A.A.; LastName, B.B.; LastName, C.C. Article Title. *Journal Name* **Year**, *Volume Number*, Page Range.

ISBN 978-3-0365-3967-6 (Hbk)
ISBN 978-3-0365-3968-3 (PDF)

Contents

About the Editors . . . . **ix**

Preface to "Biopolymers from Natural Resources" . . . . **xi**

Rafael Balart, Daniel Garcia-Garcia, Vicent Fombuena, Luis Quiles-Carrillo and Marina P. Arrieta
Biopolymers from Natural Resources
Reprinted from: *Polymers* **2021**, *13*, 2532, doi:10.3390/polym13152532 . . . . **1**

Doina Crucean, Bruno Pontoire, Gervaise Debucquet, Alain Le-Bail and Patricia Le-Bail
Influence of the Presence of Choline Chloride on the Classical Mechanism of "Gelatinization" of Starch
Reprinted from: *Polymers* **2021**, *13*, 1509, doi:10.3390/polym13091509 . . . . **11**

Ilze Irbe, Inese Filipova, Marite Skute, Anna Zajakina, Karina Spunde and Talis Juhna
Characterization of Novel Biopolymer Blend Mycocel from Plant Cellulose and Fungal Fibers
Reprinted from: *Polymers* **2021**, *13*, 1086, doi:10.3390/polym13071086 . . . . **23**

Maria Jorda-Reolid, Jaume Gomez-Caturla, Juan Ivorra-Martinez, Pablo Marcelo Stefani, Sandra Rojas-Lema and Luis Quiles-Carrillo
Upgrading Argan Shell Wastes in Wood Plastic Composites with Biobased Polyethylene Matrix and Different Compatibilizers
Reprinted from: *Polymers* **2021**, *13*, 922, doi:10.3390/polym13060922 . . . . **37**

David Alejandro González Martínez, Enrique Vigueras Santiago and Susana Hernández López
Yield and Selectivity Improvement in the Synthesis of Carbonated Linseed Oil by Catalytic Conversion of Carbon Dioxide
Reprinted from: *Polymers* **2021**, *13*, 852, doi:10.3390/polym13060852 . . . . **61**

Marcela Ferrándiz, Eduardo Fages, Sandra Rojas-Lema, Juan Ivorra-Martinez, Jaume Gomez-Caturla and Sergio Torres-Giner
Development and Characterization of Weft-Knitted Fabrics of Naturally Occurring Polymer Fibers for Sustainable and Functional Textiles
Reprinted from: *Polymers* **2021**, *13*, 665, doi:10.3390/polym13040665 . . . . **77**

Chiraz Ammar, Fahad M. Alminderej, Yassine EL-Ghoul, Mahjoub Jabli and Md. Shafiquzzaman
Preparation and Characterization of a New Polymeric Multi-Layered Material Based K-Carrageenan and Alginate for Efficient Bio-Sorption of Methylene Blue Dye
Reprinted from: *Polymers* **2021**, *13*, 411, doi:10.3390/polym13030411 . . . . **95**

Johannes Adamcyk, Stefan Beisl, Samaneh Amini, Thomas Jung, Florian Zikeli, Jalel Labidi and Anton Friedl
Production and Properties of Lignin Nanoparticles from Ethanol Organosolv Liquors—Influence of Origin and Pretreatment Conditions
Reprinted from: *Polymers* **2021**, *13*, 384, doi:10.3390/polym13030384 . . . . **107**

Pornchai Rachtanapun, Pensak Jantrawut, Warinporn Klunklin, Kittisak Jantanasakulwong, Yuthana Phimolsiripol, Noppol Leksawasdi, Phisit Seesuriyachan, Thanongsak Chaiyaso, Chayatip Insomphun, Suphat Phongthai, Sarana Rose Sommano, Winita Punyodom, Alissara Reungsang and Thi Minh Phuong Ngo
Carboxymethyl Bacterial Cellulose from Nata de Coco: Effects of NaOH
Reprinted from: *Polymers* **2021**, *13*, 348, doi:10.3390/polym13030348 . . . . **121**

Azlin Fazlina Osman, Lilian Siah, Awad A. Alrashdi, Anwar Ul-Hamid and Ismail Ibrahim
Improving the Tensile and Tear Properties of Thermoplastic Starch/Dolomite Biocomposite Film through Sonication Process
Reprinted from: *Polymers* **2021**, *13*, 274, doi:10.3390/polym13020274 . . . . **139**

Warinporn Klunklin, Kittisak Jantanasakulwong, Yuthana Phimolsiripol, Noppol Leksawasdi, Phisit Seesuriyachan, Thanongsak Chaiyaso, Chayatip Insomphun, Suphat Phongthai, Pensak Jantrawut, Sarana Rose Sommano, Winita Punyodom, Alissara Reungsang, Thi Minh Phuong Ngo and Pornchai Rachtanapun
Synthesis, Characterization, and Application of Carboxymethyl Cellulose from Asparagus Stalk End
Reprinted from: *Polymers* **2021**, *13*, 81, doi:10.3390/polym13010081 . . . . **159**

David Marset, Celia Dolza, Eduardo Fages, Eloi Gonga, Oscar Gutiérrez, Jaume Gomez-Caturla, Juan Ivorra-Martinez, Lourdes Sanchez-Nacher and Luis Quiles-Carrillo
The Effect of Halloysite Nanotubes on the Fire Retardancy Properties of Partially Biobased Polyamide 610
Reprinted from: *Polymers* **2020**, *12*, 3050, doi:10.3390/polym12123050 . . . . **175**

Pattarapom Panraksa, Suruk Udomsom, Pornchai Rachtanapun, Chuda Chittasupho, Warintorn Ruksiriwanich and Pensak Jantrawut
Hydroxypropyl Methylcellulose E15: A Hydrophilic Polymer for Fabrication of Orodispersible Film Using Syringe Extrusion 3D Printer
Reprinted from: *Polymers* **2020**, *12*, 2666, doi:10.3390/polym12112666 . . . . **197**

Salmah Husseinsyah, Nur Liyana Izyan Zailuddin, Azlin Fazlina Osman, Chew Li Li, Awad A. Alrashdi and Abdulkader Alakrach
Methyl Methacrylate (MMA) Treatment of Empty Fruit Bunch (EFB) to Improve the Properties of Regenerated Cellulose Biocomposite Films
Reprinted from: *Polymers* **2020**, *12*, 2618, doi:10.3390/polym12112618 . . . . **211**

Xian Zhang, Zhuangzhuang Teng and Runzhou Huang
Biodegradable Starch/Chitosan Foam via Microwave Assisted Preparation: Morphology and Performance Properties
Reprinted from: *Polymers* **2020**, *12*, 2612, doi:10.3390/polym12112612 . . . . **227**

Mohd Ibnu Haikal Ahmad Sohaimy and Mohd Ikmar Nizam bin Mohamad Isa
Natural Inspired Carboxymethyl Cellulose (CMC) Doped with Ammonium Carbonate (AC) as Biopolymer Electrolyte
Reprinted from: *Polymers* **2020**, *12*, 2487, doi:10.3390/polym12112487 . . . . **245**

Franco Dominici, Francesca Luzi, Paolo Benincasa, Luigi Torre and Debora Puglia
Biocomposites Based on Plasticized Wheat Flours: Effect of Bran Content on Thermomechanical Behavior
Reprinted from: *Polymers* **2020**, *12*, 2248, doi:10.3390/polym12102248 . . . . **259**

María Carolina Otálora, Robinson Camelo, Andrea Wilches-Torres, Agobardo Cárdenas-Chaparro and Jovanny A. Gómez Castaño
Encapsulation Effect on the In Vitro Bioaccessibility of Sacha Inchi Oil (*Plukenetia volubilis* L.) by Soft Capsules Composed of Gelatin and Cactus Mucilage Biopolymers
Reprinted from: *Polymers* **2020**, *12*, 1995, doi:10.3390/polym12091995 . . . . **275**

Mayra Beatriz Gómez-Patiño, Rosa Estrada-Reyes, María Elena Vargas-Diaz and Daniel Arrieta-Baez
Cutin from *Solanum Myriacanthum* Dunal and *Solanum Aculeatissimum* Jacq. as a Potential Raw Material for Biopolymers
Reprinted from: *Polymers* **2020**, *12*, 1945, doi:10.3390/polym12091945 . . . . **285**

Marina Ramos, Elena Fortunati, Ana Beltrán, Mercedes Peltzer, Francesco Cristofaro, Livia Visai, Artur J.M. Valente, Alfonso Jiménez, José María Kenny and María Carmen Garrigós
Controlled Release, Disintegration, Antioxidant, and Antimicrobial Properties of Poly (Lactic Acid) /Thymol/Nanoclay Composites
Reprinted from: *Polymers* **2020**, *12*, 1878, doi:10.3390/polym12091878 . . . . **297**

Yurong Chen, Yanxia Liu, Yidan Chen, Yagang Zhang and Xingjie Zan
Design and Preparation of Polysulfide Flexible Polymers Based on Cottonseed Oil and Its Derivatives
Reprinted from: *Polymers* **2020**, *12*, 1858, doi:10.3390/polym12091858 . . . . **315**

Freddys R. Beltrán, Marina P. Arrieta, Gerald Gaspar, María U. de la Orden and Joaquín Martínez Urreaga
Effect of lignocellulosic Nanoparticles Extracted from Yerba Mate (*Ilex paraguariensis*) on the Structural, Thermal, Optical and Barrier Properties of Mechanically Recycled Poly(lactic acid)
Reprinted from: *Polymers* **2020**, *12*, 1690, doi:10.3390/polym12081690 . . . . **325**

Marina Patricia Arrieta, Luan Garrido, Simón Faba, Abel Guarda, María José Galotto and Carol López de Dicastillo
Cucumis metuliferus Fruit Extract Loaded Acetate Cellulose Coatings for Antioxidant Active Packaging
Reprinted from: *Polymers* **2020**, *12*, 1248, doi:10.3390/polym12061248 . . . . **345**

Juan Ivorra-Martinez, Isabel Verdu, Octavio Fenollar, Lourdes Sanchez-Nacher, Rafael Balart and Luis Quiles-Carrillo
Manufacturing and Properties of Binary Blend from Bacterial Polyester Poly(3-hydroxybutyrate-*co*-3- hydroxyhexanoate) and Poly(caprolactone) with Improved Toughness
Reprinted from: *Polymers* **2020**, *12*, 1118, doi:10.3390/polym12051118 . . . . **365**

Juan Ivorra-Martinez, Jose Manuel-Mañogil, Teodomiro Boronat, Lourdes Sanchez-Nacher, Rafael Balart and Luis Quiles-Carrillo
Development and Characterization of Sustainable Composites from Bacterial Polyester Poly(3-Hydroxybutyrate-*co*-3-hydroxyhexanoate) and Almond Shell Flour by Reactive Extrusion with Oligomers of Lactic Acid
Reprinted from: *Polymers* **2020**, *12*, 1097, doi:10.3390/polym12051097 . . . . **385**

About the Editors

Rafael Antonio Balart Gimeno (h-index 41, Scopus) is a faculty staff member in the Department of Mechanical and Materials Engineering, in the area of Materials Science and Metallurgical Engineering. He has been a Full Professor at Universitat Politècnica de València since 2015. His research activity has been developed in the Institute of Materials Technology at UPV in the "RESEARCH GROUP ON POLYMERS AND GREEN COMPOSITES - GiPCEco", a group that he has led since its formation. This group focuses its activity on the following research lines, all of them with a marked environmental concern: (1) the potential of polymers of renewable origin and their transfer to industrial sectors; (2) minimizing the impact of additives and active ingredients of natural origin; (3) the development of materials under the "bio-refinery" concept; (4) the use of waste from the agro-food industry and agro-forestry sector through their incorporation into polymer and composite formulations; and (5) circular economy.

Marina Patricia Arrieta Dillon (h-index 30, Scopus) works at the Department of Industrial and Environmental Chemical Engineering at the School of Technical Industrial Engineering of the Technical University of Madrid, (ETSII-UPM) Madrid, Spain. She holds a Biochemistry BS from the National University of Córdoba (Argentina), a MS in Food Technology from the Catholic University of Cordoba (Argentina), a MS in Polymer Science and Technology from UNED (Spain), and an international PhD in Science, Technology and Food Management from the Polytechnic University of Valencia (Spain), awarded with the Extraordinary PhD Thesis Award. She has been an AECID and Santiago Grisolía predoctoral fellow, as well as a Juan de la Cierva (formación and Incorporación) and UCM postdoctoral fellow. Dr. Arrieta has experience on the synthesis, processing (melt-blending, extrusion, injection moulding, electrospinning, etc.), recycling, and characterization of multifunctional sustainable polymers and their nanocomposites with active and multifunctional properties for sustainable food packaging or agricultural applications.

Daniel García García (h-index 19, Scopus) is an Associate Professor of Materials Science and Engineering at the Department of Mechanical and Materials Engineering (UPV Campus d'Alcoi). He is a Technical Engineer in Industrial Design and Materials Engineering. He studied a Master's degree in Engineering, Processing and Characterisation of Materials and obtained an international Ph.D. in Engineering and Industrial Production with a national FPU grant at the UPV. He was awarded with the National Award for End of Degree in University Education by the Spanish Government and with the Extraordinary Doctoral Thesis Award from the UPV. He joined the Technological Institute of Materials (ITM) in 2009 and specializes on the development and formulation of sustainable polymers. His areas of interest include the extraction of active compounds from agroforestry residues for use as additives in biopolymers and the development and characterization of bionanocomposites.

Luís Jesús Quiles Carrillo (h-index 18, Scopus) works at the Department of Mechanical and Materials Engineering of the Escuela Politécnica Superior de Alcoy (EPSA) of the Universitat Politècnica de València, (UPV) Alcoy, Spain. He holds a degree in Mechanical Engineering, a Master's degree in Engineering, Processing and Characterisation of Materials, and obtained an international PhD in engineering and industrial production with a national FPU grant at the UPV. He was awarded the Best Doctoral Thesis in Polymers Award by the Royal Spanish Society of Chemistry and Physics (RSEQ and RSEF), granted by the group specialised in polymers (GEP).

His research work has focused on the search for new formulations of polymers with high environmental performance from waste and natural compounds. He has extensive experience in the characterisation of polymeric and composite materials, the reuse of waste and by-products for the creation of highly eco-efficient packaging, and works extensively with techniques such as electrospinning and polymer injection moulding.

Vicent Fombuena Borrás (h-index 17, Scopus) works at the Department of Nuclear and Chemical Engineering of the Escuela Politécnica Superior de Alcoy (EPSA) of the Universitat Politècnica de València, (UPV) Alcoy, Spain. He holds degrees in Technical Engineering in Industrial Chemical Engineering and Materials Engineering. He joined the Technological Institute of Materials (ITM) in 2008 as a research technician and obtained an international PhD in Engineering and Industrial Production in 2012. His research work has focused on the implementation of a circular economy model in the polymer industry. He has focused his efforts on obtaining multiple active compounds from seeds and agroforestry residues to be applied in thermoplastic and thermosetting polymers. He has extensive experience in techniques of the chemical, thermal and mechanical characterization of materials.

Preface to "Biopolymers from Natural Resources"

The increase in the production and application of biobased polymers has been posed as an extremely promising method of sustainable development goals, offering the option of replacing traditional petroleum polymers in several industrial sectors. In this context, there is great interest in biobased building blocks, including: the revalorization of agri-food wastes; biopolymers extracted directly from biomass, such as polysaccharides and proteins, as well as those produced by yeast biomass or by bacterial fermentation; medical devices; food packaging; agricultural films; membrane process applications, and so on. Despite the successful development of biobased polymers to an industrial scale, huge interest in their optimization still exists in order to extend their industrial exploitation. Therefore, significant attention has been given to the improvement of these sustainable plastics' overall performance, with the hope of further expanding their advantageous properties. As a versatile product, biobased polymers can be modified through chemical synthesis, copolymerization, or surface modification, as well as by the use of different additives, i.e., micro and nanoparticles, plasticizers, and active agents, among others.

This work covers all aspects related to recent, original, and cutting-edge research works focused on enhancing the performance of biopolymers from natural resources, and emphasising their potential use in industrial sectors, in line with the worldwide trend of building a circular economy. Each chapter is concerned with a given polymer's sustainable origin, in addition to their prospective obtainment, production, design, and processing at an industrial level, as well as possible improvements for specific industrial applications.

We acknowledge each author for their substantial contribution to the research field of biopolymers from natural resources.

Rafael Antonio Balart Gimeno, Marina Patricia Arrieta Dillon, Daniel García García, Luís Jesús Quiles Carrillo, and Vicent Fombuena Borrás

Editors

polymers

Editorial
Biopolymers from Natural Resources

Rafael Balart [1,*], Daniel Garcia-Garcia [1,*], Vicent Fombuena [1,*], Luis Quiles-Carrillo [1,*] and Marina P. Arrieta [2,3,*]

1 Technological Institute of Materials (ITM), Universitat Politècnica de València (UPV), Plaza Ferrándiz y Carbonell 1, 03801 Alcoy, Spain
2 Departamento de Ingeniería Química y del Medio Ambiente, Escuela Técnica Superior de Ingenieros Industriales, Universidad Politécnica de Madrid (ETSII-UPM), Calle José Gutiérrez Abascal 2, 28006 Madrid, Spain
3 Grupo de Investigación: Polímeros, Caracterización y Aplicaciones (POLCA), 28006 Madrid, Spain
* Correspondence: rbalart@mcm.upv.es (R.B.); dagarga4@epsa.upv.es (D.G.-G.); vifombor@upv.es (V.F.); luiquic1@epsa.upv.es (L.Q.-C.); m.arrieta@upm.es (M.P.A.)

Citation: Balart, R.; Garcia-Garcia, D.; Fombuena, V.; Quiles-Carrillo, L.; Arrieta, M.P. Biopolymers from Natural Resources. *Polymers* **2021**, *13*, 2532. https://doi.org/10.3390/polym13152532

Received: 30 June 2021
Accepted: 20 July 2021
Published: 30 July 2021

During the last decades, the increasing ecology in the reduction of environmental impact caused by traditional plastics is contributing to the growth of more sustainable plastics with the aim to reduce the consumption of non-renewable resources for their production. Thus, in the last few years, the increased bio-based polymer production and application have positioned biopolymers as one of the most promising ways to meet the sustainable development goal of replacing traditional petroleum polymers with more sustainable materials in several industrial sectors.

In this context, the revalorization of agro-food wastes, biopolymers extracted directly from biomass such as polysaccharides, proteins and lipids, as well as those produced by yeast biomass, algae or by bacterial fermentation, have attracted a significant amount of interest, especially for medical devices, food packaging, agricultural films, membrane process applications, sustainable clothes and so on—interest which, despite successful recent developments in bio-based polymers up to the industrial scale, extends to their optimization for industrial exploitation. Therefore, significant attention has been paid to the improvement of those sustainable plastics directly derived from biomass overall performance to provide more than one advantageous property attributed to their versatility to be modified through chemical synthesis, copolymerization, and surface modification, as well as through the use of different additives (i.e., micro- and nanoparticles, plasticizers, and active agents, among others). Moreover, nowadays, the starting monomers or building blocks to obtain traditional plastics can be obtained from biomass instead of petrochemical resources, known as drop-in bioplastics. Drop-in bioplastics present the same advantages of traditional plastics in terms of performance and are more sustainable, while at the same time they can be directly transformed into the desired products by means of the same technology already available at the plastic industrial sector. Figure 1 summarizes several biobased polymers classified on the basis of their origin and obtainment process.

The present Special Issue gathers a series of twenty-four articles focused on the manufacturing and characterization of biopolymers and building blocks extracted from natural resources, as well as their potential scalability to the industrial level for several industrial applications. The Special Issue paid special focus on the improvement of the overall performance of biobased polymers by chemical modification process, the incorporation of sustainable additives such as biobased oligomers, by blending with another biobased or biodegrable polymeric matrix, as well as by the development of composites and nanocomposites through the incorporation of naturally occurring particles and nanoparticles.

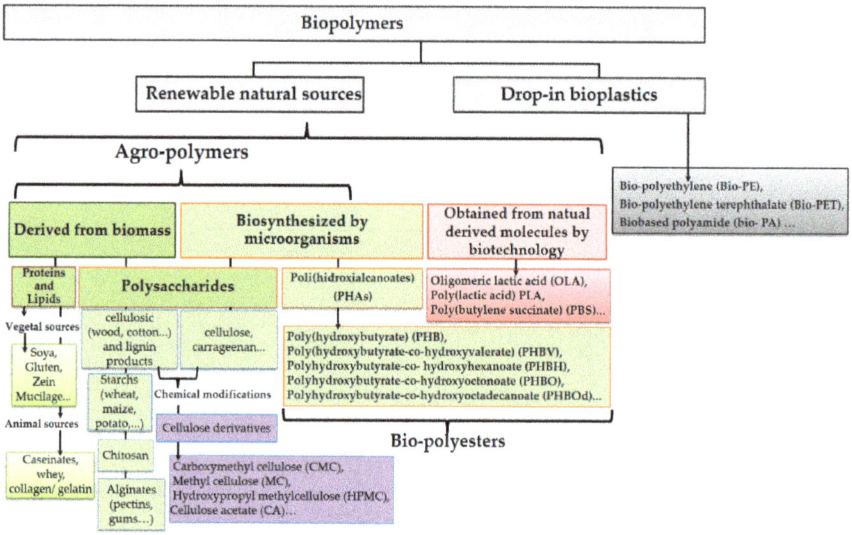

Figure 1. Classification of biobased polymers.

Among other naturally derived biopolymers, polysaccharides are nontoxic and biodegradable, which increases their potential application in several fields. Polysaccharides, with cellulose as the most important representative since it is the most abundant polymer in nature and is found all around the world, are widely investigated. Although the most-used biopolysaccharides are obtained from plant origin (e.g., cellulose), they can be obtained from animal origin (e.g., chitin/chitosan) and also from microbial origin (e.g., bacterial cellulose). In this sense, the production of biobased polymers by using microorganisms has gained considerably attention during the last decade as a sustainable production method. In this regard, Irbe et al. [1] developed novel blends based on biopolysaccharides derived from biomass, softwood cellulose fibers such as softwood Kraft (KF) or hemp fibers (HF) and fungal fibers (hyphae). Regarding the fungal fibers, they concluded that among several fungal fibers, the fibers from screened basidiomycota fungi *Ganoderma applanatum* (Ga), *Fomes fomentarius* (Ff), *Agaricus bisporus* (Ab), and *Trametes versicolor* (Tv) were good for blending with softwood cellulose fibers. Thus, the fungal fibers were blended with KF and HF fiber pulp in different mass ratios (i.e., 50:50 and 33:33:33). The materials were characterized in terms of mechanical properties, air permeability and virus filtration efficiency. The overall performance of the hyphal polysaccharide/cellulose blends was highly affected by the microstructural features of each cellulosic material, while the ability of functional groups of hyphal polysaccharides to establish hydrogen bonding interactions with cellulose fibers was one of the key elements of network formation which determined the final blend properties and the potential applications. In this context, highly fibrillated hemp fibers showed the highest mechanical strength, while the blends containing Kraft fibers increased air permeability and showed a high virus reduction capacity. Thus, depending on the intended use the different microstructure of Hemp fibers or Kraft fibers will allow obtaining materials with interesting properties. For instance, materials with a low air permeability are interesting for food packaging applications where high gas barrier properties are of fundamental importance. Meanwhile, materials with a higher air permeability and virus filtration efficiency have potential for being used as gas permeable membranes, for example, as biobased filter layers in face masks. Despite its outstanding characteristics, cellulose in its native form is hard and provides some limitations for the plastic industry in terms of cellulose processing. Nevertheless, cellulose derivatives can overcome such problems. Thus, cellulose chemical modifications have been widely studied and nowadays several cellulose derivatives can be found commercially such as carboxymethyl cellulose (CMC), methyl cellulose (MC), hydroxyethyl cellulose (HEC), hydroxypropyl methylcellulose (HPMC), and cellulose acetate (CA).

Among others, CMC possesses good film-forming abilities because of the presence of a hydrophilic carboxyl group (–CH_2COONa), which allows its dissolution in water. In this context, Rachtanapun et al. [2] revalorized coconut juice, a waste product from the coconut milk industry, into bacterial cellulose using *Acetobacter xylinum* to further obtain carboxymethyl cellulose (CMC) through chemical synthesis with different degrees of substitution. With this propose in mind, CMC was obtained from bacterial cellulose in the presence of monochloroacetic acid (MCA) via a carboxymethylation reaction and by simply varying the sodium hydroxide content (NaOH from 20% to 60%), the degree of substitution was controlled. The degree of substitution significantly increased with NaOH concentration up to 30% and then progressively decreased with further NaOH concentrations. As CMC is a water soluble cellulose derivative, the higher the degree of substitution, the higher the water vapor permeability. The flexibility of the material also increased with the degree of substitution. Thus, the developed method to modify cellulose from nata de coco allows tuning of the water permeability and the mechanical properties of the final material according to the need of the intended application by simple varying the NaOH content. CMC with different degrees of substitutions has been also obtained by Kluklin et al. [3] from the *Asparagus officinalis* stalk end, a typical herbal medicine used in Asia, by simply varying the NaOH content (from 20% to 60%) too, and it was also found that there is a higher degree of substitution for the NaOH 30% (w/v). CMC has been also used in the development of solid polymer electrolyte (SPE) film for sustainable energy storage systems due to its solubility in water and its naturally high degree of an amorphous phase that allows easier transport of conducting ions like lithium (Li^+) and proton (H^+). In this context, Sohaimy et al. [4] tuck this advantageous properties of CMC to develop CMC-based films for electrochemical applications. However, the inherent solid properties of the polymeric film impede ionic mobility leading to a low ionic conductivity, especially at room temperature. Thus, ammonium carbonate (AC), with two groups of ammonium ions, was used as a dopant in amounts from 1 wt.% to 11 wt.% with the main objective of injecting more H^+ into the system and the CMC-AC thin electrolyte film was obtained by a solvent casting method. Both CMC and AC salts dissolved completely and had a low degree of crystallinity (X_c) as it was demonstrated by X-ray diffraction (XRD). Although, the ionic conductivity for a biopolymer electrolyte intended for commercial energy storage technology needs to be at least a minimum value of ~10^{-4} Scm^{-1}, they obtained promising results since the highest ionic conductivity obtained for CMC doped with 7 wt.% was 7.71×10^{-6} Scm^{-1}, resulting in higher results than other diammonium salts tested (i.e., ammonium adipate and ammonium sulfate), showing that further enhancement is still needed to increase the ionic conductivity for suitable electrochemical applications (~10^{-4} Scm^{-1}) and the CMC-AC films show promising prospects for electrochemical applications.

Panraksa et al. [5] developed orodispersible films (ODF) based on hydroxypropyl methylcellulose (HPMC: E5 and E15) by extrusion-3D printer technology to enhance the solubility and dissolution of poorly water-soluble drugs (i.e., phenytoin). ODFs are required to rapidly disintegrate in the buccal cavity (i.e., within a minute), without requiring water, thus HPMC, which is a well-known hydrophilic and film-forming polymer, is an ideal candidate for this propose. HPMC E5 and E15 were used as the film-forming polymers, while propylene glycol and glycerin were used as plasticizers. They optimized the processing conditions and the results showed that the phenytoin-loaded HPMC E15 (at a concentration of 10% *w/v* of HPMC E15) was the most suitable formulation of HPMC, since it exhibited a good physical appearance, good mechanical strength, low Young's modulus and high elongation at break, rapid in vitro disintegration time (within 5 s) as well as a rapid drug release (up to 80% within 10 min), showing the improvement of solubility and dissolution rate of phenytoin at the same time as allowing ease of handling and application.

Another widely used cellulose derivative is cellulose acetate (CA). For instance, it has been used by Arrieta et al. [6] as a carrier of *Cucumis metuliferus* fruit extract to provide to corona-treated low-density polyethylene (LDPE) films with antioxidant activity by coating a CA-based polymeric solution, containing different amounts of *Cucumis metuliferus* fruit extract (1, 3 and 5 wt.%), onto the LDPE film. The materials showed antioxidant activity (between 0.3 and 0.6 mg Trolox/dm^2) after performing migration studies in a fatty food simulant and also showed their ability to reduce the browning effect of fresh-cut apples in direct contact. Additionally, the coating not only provided an antioxidant active functionality, but also improved the oxygen barrier performance to the final bilayer material, showing their use in active food packaging applications.

Another interesting approach to obtain a solution of insoluble cellulose is regenerated cellulose (RC). RC refers to the chemical dissolution of insoluble natural cellulose followed by the recovery of the material from the solution by means of different methods, such as cellulose carbamate processes, lyocell processes, and viscose processes. In this context, Husseinsyah et al. [7] used an ionic liquid (1-butyl-3-methylimidazolium chloride) as a novel method to obtain RC from empty fruit bunches (EFB) to develop a self-reinforced composite for biodegradable packaging applications. They studied the possibility of improving the interphase bonding between the cellulosic matrix and the filler in the self-reinforced composite by means of EFB chemical treatment with a methyl methacrylate (MMA) chemical treatment. They observed that the substitution of the –OH group of the EFB cellulose with the ester group of the MMA allowed the dissolution process of the EFB during the regeneration process owing to weakening of the hydrogen bonding of the cellulose. This allowed us to obtain a more crystalline and homogeneous EFB RC biocomposite film, with an improved thermal stability and mechanical performance. Another very important cellulosic material is cotton, which is widely distributed worldwide. It has a higher yield compared to other commercial crops, and thus several products are obtained from cotton seed, such as oil. In this context, Chen et al. [8] used bio-based cottonseed oil (CO) and two of its derivatives, fatty acid of cottonseed oil (COF) and sodium soap of cottonseed oil (COS), to tune the density and mechanical strength of polysulfide polymers by means of a free radical addition mechanism. COF reacts with sulfur to generate serials of polysulfide-derived polymers with a lesser viscosity and tensile strength. Whereas COS was not involved in the reaction with sulfur, as a consequence of the high melting point of sodium linoleate and sodium oleate. Nevertheless, COS it was able to increase the density and tensile strength of polysulfide-derived polymers when it was used as a filler. Moreover, the developed polysulfide-based polymers showed good reprocessability and recyclability, showing their potential as bio-based functional supplementary additives.

Ammar et al. [9] developed a novel adsorbent material for cationic dyes based on a polyelectrolyte multi-layered (PEM) system cross-linked to a cellulosic-based material. The PEM was formed by an alternation of layers of two polysaccharides, sodium alginate polyanion and reticulated citric acid with k-carrageenan, thus providing many carboxylate and sulfonate groups grafted on the cellulosic surface. The developed system provides outstanding adsorption capacities for cationic dyes (e.g., with a capacity above 522.4 mg/g for methylene blue).

The processing and revalorization of lignocellulosic materials obtained from agro-food wastes, as well as the extraction through sustainable approaches, have gained considerable interest. Gómez-Patiño et al. [10] extracted cellulosic cutins from of *S. aculeatissimum* and *S. myriacanthum* fruit peels, using enzymatic (i.e., *A. niger pectinase*, *A. niger cellulose* and *A. niger hemicellulose*) and chemical (trifluoroacetic acid hydrolysis) methods. They selected these fruits since they are not for human or animal consumption and are thus interesting as a potential raw material to produce sustainable materials. The hydrolyzed cutins showed mainly 10,16-dihydroxyhexadecanoic acid (10,16-DHPA) monomers composition and the obtained films showed a good homogeneity.

Lignocellulosic derivatives have also gain considerable interest as micro and nanofillers. In this context, Adamcyk et al. [11] studied the extraction of lignin from wheat straw, spruce, and beech using ethanol organosolv pretreatment at temperatures from 160–220 °C, and further precipitated by solvent-shifting obtaining lignin micro- and nanoparticles (with mean hydrodynamic diameters from 67.8 nm to 1156.4 nm). They observed that higher pretreatment temperatures increased the delignification of the raw materials but also favor depolymerization and structural alteration of the extracted lignin, increasing the particle sizes as well as agglomeration at pretreatment temperatures of over 200 °C. The obtained particles were then purified by dialysis and the showed interesting antioxidant activity (i.e., from 19.1 and 50.4 mg Lignin/mg ascorbic acid equivalents).

On the other side, lignocellulosic nanoparticles have been obtained from yerba mate (*Ilex paraguariensis*) waste, a typical infusion widely consumed in Latin America (i.e., Argentina, Brazil, Uruguay and Paraguay), by means of an aqueous extraction procedure (with mean average size of 495 nm). The obtained nanoparticles were used by Beltrán et al. [12] to reinforce mechanically recycled bioplastic (i.e., polylactic acid). Poly(lactic acid) (PLA) is chemically synthesized starting from simple sugars obtained from biomass fermented to lactic acid. It is one of the most promising biopolyesters for massive industrial applications due to its availability in the market at a competitive cost, and it can be recycled following the mechanical recycling process used for other traditional plastics (e.g., polyethylene terephthalate, PET). Low amounts of yerba mate nanoparticles (i.e., 1 wt.%) were able to increase the intrinsic viscosity and to improve the oxygen barrier properties of mechanically recycled PLA, showing the interest of such nanocomposites in the food packaging field. Ramos et al. [13] also studied PLA-based nanocomposites for food packaging applications. In this case, PLA was loaded with commercial montmorillonite (D43B) at two different concentrations (2.5 and 5 wt.%) and was further incorporated with thymol as an antioxidant and antimicrobial agent for active food packaging applications. The results showed that the addition of 2.5 wt.% D43B and 8 wt.% thymol leads to a material with a good balance of properties, with antibacterial activity against *Escherichia coli* and *Staphylococcus aureus*.

Another interesting source for sustainable additives are seeds fruits, peel and/or nut's shells, from which can be obtained flours and natural particles and/or nanoparticles, as well as vegetable oils. In this regard, Jorda-Reolid et al. [14] obtained micronized Argan shell (MAS) particles from residues of the *Argania spinosa* plant, a tropical plant whose fruit is used in Morocco to prepare oil. The authors used MAS particles (with mean average long of 70 μm and mean average wide of 45 μm) as reinforcing fillers of high-density polyethylene obtained from sugarcane (Bio-HDPE). Meanwhile, to improve the low compatibility between MAS and Bio-HDPE, they studied two compatibilizing agents, polyethylene-grafted maleic anhydride (PE-g-MA) and maleinized linseed oil (MLO); as well as halloysite nanotubes (HNTs), a second reinforcing filler. The obtained materials showed improved stiffness, high ductility, and good thermal stability, as well as a visual appearance very similar to that of reddish-color woods, showing their interest as wood plastic composites with the additional advantage of being chipper rather than neat Bio-HDPE. González Martínez et al. [15] optimized the carbonation reaction of epoxidized linseed oil (ELO) with carbon dioxide (CO_2) in the presence of tetrabutylammonium bromide (TBAB) as catalysts, to obtain carbonated vegetable oils (CVOs), which have potential applications as interesting starting materials for the formation of polymers (e.g., monomers, additives, lubricants and plasticizers), particularly in the synthesis of non-isocyanate polyurethanes (NIPUs). They were able to obtain large conversion (96%), carbonation (95%), and selectivity (99%) at low temperature (90 °C), and thus the high carbonate content obtained for carbonated ELO showed their promising performance for the synthesis of NIPU with the required properties and more sustainable characteristics. On the other hand, Dominici et al. [16] studied the possibility of obtaining bioplastics from wheat flour particles as a novel source and an energetically and economically cheap alternative to other thermoplastics. The refined flours, with different contents of bran, were first plasticized with glycerol and then the authors attempted to find how different contents of grinded bran could affect the deformability of the flour by blending with low melting polymeric fractions (such as poly(ε-caprolactone) (PCL), and polybutylene-adipate-*co*-terephthalate (PBAT)). They studied different processing parameters and they were able to obtain a novel thermoplastic wheat flour (TPWF), which showed an improved performance when glycerol was partially replaced by water and when it was blended with PCL, as well as adding citric acid as a compatibilizer.

Ferrándiz et al. [17] studied the potential use of soy protein (SP) fibers functionalized with an undisclosed antimicrobial agent as well chitin (with inherent antimicrobial properties) fibers as sustainable active textiles to replace synthetic polymers widely used in this sector. The thermal resistance of both weft-knitted fabrics was similar to that of cotton, whereas their air permeability was higher, particularly in the case of chitin due to its higher fineness, which makes these natural fibers very promising for summer clothes.

After cellulose, starch is the second most abundant polymer in nature and, thus, has been widely studied for several sustainable industrial applications. Zhang et al. [18] developed potato starch-based foams by means of a microwave treatment. They blended potato starch with chitosan as a reinforcing phase and it was found that starch-based foams with a larger proportion of starch showed a small pore size and a low density with higher compressive strength ascribed to the good compatibility between both polymeric matrices due to the formation of hydrogen bond interactions between amino groups of chitosan and hydroxyl groups of starch. Those interactions were also responsible to maintained intact the morphological structure of the foam in water for 10 days, while the starch-based foams completely degraded in water after 30 days. Thus, the developed starch/chitosan foams resulted in interesting results for their use as active materials in biomedical applications as well as in drug and food packaging.

For several industrial applications, starch is particularly used in its thermoplastic form, known as thermoplastic starch (TPS). To obtain TPS, native starch requires disruption of the granule organization by means of a combination of high water content and heat, leading to the starch granule swelling and consequently starch gelatinization. In this regard, Crucean et al. [19] studied the gelatinization process of three native starches: (i) wheat, (ii) potato, and (iii) waxy corn starch, as well as starch from a wheat flour in mixtures of water and choline chloride (an ionic compound, which has a "structure making" effect). They demonstrated that choline chloride/water system exhibits an allotropic change at low water concentrations and solubilization for water contents greater than 30%. They observed a stabilization, or a better organization, of the ordered regions of the starch in the presence of choline chloride as it was demonstrated by an X-ray diffraction analysis in a heating cell that the crystalline rearrangement of the structure of the starch grain takes place simultaneously with the solubilization phenomenon of amylose.

Osman et al. [20] studied thermoplastic starch (TPS)-based biocomposites using dolomite (DOL) as a filler in its pristine form (DOL(P)), as well as dolomites, after a simple and scalable sonication process (DOL(U)). TPS-DOL biocomposites intended for packaging applications were prepared at different loadings (i.e., 1, 2, 3, 4 and 5 wt.%). The TPS-based biocomposites with a high dolomite loading (i.e., 4 and 5 wt.%) showed better mechanical performance, showing greater tensile and tear properties, particularly in the case of biocomposites loaded with a sonicated process assisted by dolomite due to a reduction in particle size allowing their better dispersion within the TPS matrix. Thus, the high abundancy and low cost of dolomite, in combination with the simple, scalable and environmental friendly method of sonication, showed their interest to be of use as reinforcing fillers for TPS to produce a sustainable biocomposite for packaging applications. Although biomass-derived biopolymers are mainly extracted from plants, they can also be obtained from animal resources. For instance, gelatin has been traditionally used in the development of soft capsules (softgels) due to its biodegradable nature, as well as its ability to form thermo-responsive hydrogels, inspiring interest in the biomedical sector and food industry. In recent years, new materials have been explored in the food industry to partially or completely replace gelatin with other non-animal natural hydrocolloids. In this context, Otálora et al. [21] developed gelatin/cactus mucilage softgel capsules to encapsulate oil extracted from sacha inchi (*Plukenetia volubilis L.*) seeds, a rich source of polyunsaturated fatty acids (PUFAs) that are beneficial to human health. They submitted the softgel capsules to an in vitro digestion process to simulate gastric conditions and the protective capacity of the gelatin/cactus mucilage-based softgel against digestive processes was evaluated. Although the study revealed a reduction in the content of polyunsaturated fatty acids (PUFAs) after the digestion process and, thus, a reduction of the nutritional value, they concluded that gelatin/cactus mucilage microcapsules can act as interesting bioactive delivery systems for acidic food (e.g.,: fruit juices or dairy drinks) before being subjected to digestive processes.

As has already been mentioned, the biobased polymer obtainment, by means of microorganisms, has gained a lot of interest. In this context, the family of polyhydroxyalkanoates (PHAs) are polyesters biologically synthesized by controlled bacterial fermentation in response to nutrient limitation as an intracellular storage of food and energy. The homopolymer, poly(hydroxybutyrate) (PHB), is the simplest and most common representative of the PHA family, but PHAs also comprise many copolyesters, polyhydroxybutyrate-*co*-hydroxyalkanoates, that have gained high industrial interest in the bioplastic sector, such as polyhydroxybutyrate-*co*-hydroxyvalerates (PHBV), or polyhydroxybutyrate-*co*-hydroxyhexanoate (PHBH), polyhydroxybutyrate-*co*-hydroxyoctonoate (PHBO), and polyhydroxybutyrate-*co*-hydroxyoctadecanoate (PHBOd). For instance, Ivorra-Martinez et al. [22] developed PHBH/PCL blends with different compositions ranging from 0 to 40 of PCL wt.% by means of extrusion process followed by injection molding, with the main objective of reducing the inherent embrittlement of PHBH produced due to the aging process (i.e., secondary crystallization). Although they mainly found a lack of miscibility between both polymeric matrices, as revealed by the thermogravimetric analysis, they concluded that the addition of high amounts of PCL (i.e., 40 wt.%) contributed to an increase in ductility and to a decrease in the typical brittleness of PHA. PCL considerably improved the toughness, as well as the impact resistance, of neat PHBH. Therefore, PHBH/PCL blends showed their suitability for the packaging industry. Then, Ivorra-Martinez et al. [23] developed Wood Plastic Composites (WPCs) based on PHBH loaded with lignocellulosic particles of almond shell flour (ASF), a by-product from the agro-food industry, by means of extrusion process followed by injection molding. Since the addition of ASF (with a maximum particle size of 150 μm) leads to an embrittlement and reduced toughness of the WPC, they used oligomeric lactic acid (OLA) to provide improved properties to the final PHBH-ASF/OLA material. Interestingly, a remarkable increase in impact strength with 20 phr OLA addition was achieved. In fact, OLA provides PHBH polymer chains of a high mobility due to a plasticizing effect, leading to an improvement in toughness, even on composites with 30 wt.% ASF. Additionally, OLA decreased the water absorption capacity of WPC, thus broadening potential industrial applications of PHBH-ASF/OLA in high humidity environments.

As it was previously mentioned, nowadays many traditional plastics, which were obtained for decades through the classic petrochemical production routes, have recently found "green" routes for their production (e.g., Bio-PE, Bio-PET, etc.). In this regard, polyamide 6 (PA6) can be replaced with a biobased variant that exhibits a high renewable content and with very similar properties known as PA610. In this context, Marset et al. [24] used PA610 partially biobased polyamide (Bio-PA) to develop nanocomposites reinforced with 10%, 20%, and 30% halloysite nanotubes (HNTs) as a natural flame retardancy filler by means of an extrusion process, followed by injection molding. The results showed that HNTs promote a significant reduction in the optical density and in the number of toxic gases (i.e., CO_2) emitted during combustion, especially when PA610 was loaded with 30% HNTs. Thus, the nanocomposites showed good flame retardancy properties.

In summary, the Special Issue Biopolymers from Natural resources published in *Polymers*, compiles the recent research works in biopolymers obtained from natural resources and the strategies to improve their overall performance for sustainable industrial applications. From the above, improvements of biobased polymers was successfully achieved by means of copolymerization, blending, the use of compatibilizers, or plasticizers (vegetable oils, oligomers, etc.), as well as with naturally occurring or naturally derived particles on the micro- and nano-scale. Special interest has been shown to the revalorization of agro-food wastes for the extraction/production of additives or polymers with an interest in the industrial sector. The resulting, more sustainable materials can find potential applications in several industrial sectors, such as in the controlled release of active compounds for active food packaging, bioactive delivery systems, or biomedical devices (e.g., soft capsules, etc.), textiles (e.g., sustainable clothes) or as wood plastic composites (e.g., furniture and outdoor applications).

Future research efforts on biobased polymer extraction and production methods, as well as the optimization of their final production into co-polymers, blends, composites, and nanocomposites, are still needed to properly transfer the proposed developments, as well as future ones from the laboratory scale level to the industrial sector to reach more environmentally friendly materials within a circular economy approach.

Author Contributions: All the guest editors read the twenty-four articles of the Special Issue and wrote this editorial letter. All authors have then reviewed the final version of the manuscript and agreed its publication. All authors have read and agreed to the published version of the manuscript.

Acknowledgments: The guest editors thank all the authors for submitting their work to this Special Issue and for its successful completion. We also acknowledge all the reviewers participating in the peer-review process of the submitted manuscripts for enhancing their quality and impact. We are also grateful to Chris Chen and the editorial assistants of *Polymers* who made the entire Special Issue creation a smooth and efficient process.

Conflicts of Interest: The authors declare no conflict of interest.

References

1. Irbe, I.; Filipova, I.; Skute, M.; Zajakina, A.; Spunde, K.; Juhna, T. Characterization of Novel Biopolymer Blend Mycocel from Plant Cellulose and Fungal Fibers. *Polymers* **2021**, *13*, 1086. [CrossRef]
2. Rachtanapun, P.; Jantrawut, P.; Klunklin, W.; Jantanasakulwong, K.; Phimolsiripol, Y.; Leksawasdi, N.; Seesuriyachan, P.; Chaiyaso, T.; Insomphun, C.; Phongthai, S. Carboxymethyl Bacterial Cellulose from Nata de Coco: Effects of NaOH. *Polymers* **2021**, *13*, 348. [CrossRef] [PubMed]
3. Klunklin, W.; Jantanasakulwong, K.; Phimolsiripol, Y.; Leksawasdi, N.; Seesuriyachan, P.; Chaiyaso, T.; Insomphun, C.; Phongthai, S.; Jantrawut, P.; Sommano, S.R. Synthesis, Characterization, and Application of Carboxymethyl Cellulose from Asparagus Stalk End. *Polymers* **2021**, *13*, 81. [CrossRef]
4. Sohaimy, M.I.H.A.; Isa, M.I.N.M. Natural Inspired Carboxymethyl Cellulose (CMC) Doped with Ammonium Carbonate (AC) as Biopolymer Electrolyte. *Polymers* **2020**, *12*, 2487. [CrossRef]
5. Panraksa, P.; Udomsom, S.; Rachtanapun, P.; Chittasupho, C.; Ruksiriwanich, W.; Jantrawut, P. Hydroxypropyl Methylcellulose E15: A Hydrophilic Polymer for Fabrication of Orodispersible Film Using Syringe Extrusion 3D Printer. *Polymers* **2020**, *12*, 2666. [CrossRef] [PubMed]
6. Arrieta, M.P.; Garrido, L.; Faba, S.; Guarda, A.; Galotto, M.J.; López de Dicastillo, C. Cucumis metuliferus Fruit Extract Loaded Acetate Cellulose Coatings for Antioxidant Active Packaging. *Polymers* **2020**, *12*, 1248. [CrossRef]
7. Husseinsyah, S.; Zailuddin, N.L.I.; Osman, A.F.; Li Li, C.; Alrashdi, A.A.; Alakrach, A. Methyl Methacrylate (MMA) Treatment of Empty Fruit Bunch (EFB) to Improve the Properties of Regenerated Cellulose Biocomposite Films. *Polymers* **2020**, *12*, 2618. [CrossRef]
8. Chen, Y.; Liu, Y.; Chen, Y.; Zhang, Y.; Zan, X. Design and Preparation of Polysulfide Flexible Polymers Based on Cottonseed Oil and Its Derivatives. *Polymers* **2020**, *12*, 1858. [CrossRef] [PubMed]
9. Ammar, C.; Alminderej, F.M.; EL-Ghoul, Y.; Jabli, M.; Shafiquzzaman, M. Preparation and Characterization of a New Polymeric Multi-Layered Material Based K-Carrageenan and Alginate for Efficient Bio-Sorption of Methylene Blue Dye. *Polymers* **2021**, *13*, 411. [CrossRef] [PubMed]
10. Gómez-Patiño, M.B.; Estrada-Reyes, R.; Vargas-Diaz, M.E.; Arrieta-Baez, D. Cutin from Solanum Myriacanthum Dunal and Solanum Aculeatissimum Jacq. as a Potential Raw Material for Biopolymers. *Polymers* **2020**, *12*, 1945. [CrossRef]

11. Adamcyk, J.; Beisl, S.; Amini, S.; Jung, T.; Zikeli, F.; Labidi, J.; Friedl, A. Production and Properties of Lignin Nanoparticles from Ethanol Organosolv Liquors—Influence of Origin and Pretreatment Conditions. *Polymers* **2021**, *13*, 384. [CrossRef]
12. Beltrán, F.R.; Arrieta, M.P.; Gaspar, G.; de la Orden, M.U.; Martínez Urreaga, J. Effect of lignocellulosic Nanoparticles Extracted from Yerba Mate (Ilex paraguariensis) on the Structural, Thermal, Optical and Barrier Properties of Mechanically Recycled Poly(lactic acid). *Polymers* **2020**, *12*, 1690. [CrossRef]
13. Ramos, M.; Fortunati, E.; Beltrán, A.; Peltzer, M.; Cristofaro, F.; Visai, L.; Valente, A.J.M.; Jiménez, A.; Kenny, J.M.; Garrigós, M.C. Controlled Release, Disintegration, Antioxidant, and Antimicrobial Properties of Poly (Lactic Acid)/Thymol/Nanoclay Composites. *Polymers* **2020**, *12*, 1878. [CrossRef] [PubMed]
14. Jorda-Reolid, M.; Gomez-Caturla, J.; Ivorra-Martinez, J.; Stefani, P.M.; Rojas-Lema, S.; Quiles-Carrillo, L. Upgrading Argan Shell Wastes in Wood Plastic Composites with Biobased Polyethylene Matrix and Different Compatibilizers. *Polymers* **2021**, *13*, 922. [CrossRef]
15. González Martínez, D.A.; Vigueras Santiago, E.; Hernández López, S. Yield and Selectivity Improvement in the Synthesis of Carbonated Linseed Oil by Catalytic Conversion of Carbon Dioxide. *Polymers* **2021**, *13*, 852. [CrossRef] [PubMed]
16. Dominici, F.; Luzi, F.; Benincasa, P.; Torre, L.; Puglia, D. Biocomposites Based on Plasticized Wheat Flours: Effect of Bran Content on Thermomechanical Behavior. *Polymers* **2020**, *12*, 2248. [CrossRef]
17. Ferrándiz, M.; Fages, E.; Rojas-Lema, S.; Ivorra-Martinez, J.; Gomez-Caturla, J.; Torres-Giner, S. Development and Characterization of Weft-Knitted Fabrics of Naturally Occurring Polymer Fibers for Sustainable and Functional Textiles. *Polymers* **2021**, *13*, 665. [CrossRef] [PubMed]
18. Zhang, X.; Teng, Z.; Huang, R. Biodegradable Starch/Chitosan Foam via Microwave Assisted Preparation: Morphology and Performance Properties. *Polymers* **2020**, *12*, 2612. [CrossRef] [PubMed]
19. Crucean, D.; Pontoire, B.; Debucquet, G.; Le-Bail, A.; Le-Bail, P. Influence of the Presence of Choline Chloride on the Classical Mechanism of "Gelatinization" of Starch. *Polymers* **2021**, *13*, 1509. [CrossRef]
20. Osman, A.F.; Siah, L.; Alrashdi, A.A.; Ul-Hamid, A.; Ibrahim, I. Improving the Tensile and Tear Properties of Thermoplastic Starch/Dolomite Biocomposite Film through Sonication Process. *Polymers* **2021**, *13*, 274. [CrossRef]
21. Otálora, M.C.; Camelo, R.; Wilches-Torres, A.; Cárdenas-Chaparro, A.; Gómez Castaño, J.A. Encapsulation Effect on the In Vitro Bioaccessibility of Sacha Inchi Oil (*Plukenetia volubilis* L.) by Soft Capsules Composed of Gelatin and Cactus Mucilage Biopolymers. *Polymers* **2020**, *12*, 1995. [CrossRef]
22. Ivorra-Martinez, J.; Verdu, I.; Fenollar, O.; Sanchez-Nacher, L.; Balart, R.; Quiles-Carrillo, L. Manufacturing and Properties of Binary Blend from Bacterial Polyester Poly(3-hydroxybutyrate-co-3-hydroxyhexanoate) and Poly(caprolactone) with Improved Toughness. *Polymers* **2020**, *12*, 1118. [CrossRef] [PubMed]
23. Ivorra-Martinez, J.; Manuel-Mañogil, J.; Boronat, T.; Sanchez-Nacher, L.; Balart, R.; Quiles-Carrillo, L. Development and Characterization of Sustainable Composites from Bacterial Polyester Poly(3-Hydroxybutyrate-co-3-hydroxyhexanoate) and Almond Shell Flour by Reactive Extrusion with Oligomers of Lactic Acid. *Polymers* **2020**, *12*, 1097. [CrossRef] [PubMed]
24. Marset, D.; Dolza, C.; Fages, E.; Gonga, E.; Gutiérrez, O.; Gomez-Caturla, J.; Ivorra-Martinez, J.; Sanchez-Nacher, L.; Quiles-Carrillo, L. The Effect of Halloysite Nanotubes on the Fire Retardancy Properties of Partially Biobased Polyamide 610. *Polymers* **2020**, *12*, 3050. [CrossRef] [PubMed]

polymers

MDPI

Article

Influence of the Presence of Choline Chloride on the Classical Mechanism of "Gelatinization" of Starch

Doina Crucean [1,2,3], Bruno Pontoire [2,3], Gervaise Debucquet [4], Alain Le-Bail [1,3] and Patricia Le-Bail [2,3,*]

[1] ONIRIS, UMR 6144 GEPEA CNRS, F-44322 Nantes, France; doina.crucean@oniris-nantes.fr (D.C.); alain.lebail@oniris-nantes.fr (A.L.-B.)
[2] INRAe, UR1268 Biopolymers Interactions Assemblies, BP 71627, F-44316 Nantes, France; bruno.pontoire@inrae.fr
[3] SFR IBSM 4202, BP 71627, F-44316 Nantes, France
[4] Audencia Business School, Human and Social Sciences, 8 Route de la Jonelière, BP 31222, F-44312 Nantes, France; gdebucquet@audencia.com
* Correspondence: patricia.le-bail@inrae.fr

Abstract: The aim of this research is to contribute to a better understanding the destructuration of three native starches and a wheat flour in mixtures of water and choline chloride. Model systems have thus been defined to allow a better approach to hydrothermic transformations related to the interactions between choline chloride and starch. We have observed that choline chloride has an impact on the gelatinization of starch which corresponds to the stabilizing salts phenomenon. The depolymerization and dissolution of the starch have also been demonstrated and can there dominate the gelatinization. However, the results obtained in X-ray diffraction by heating cell have shown that the exotherm which appeared was not only related to the depolymerization of the starch, but that a stage of crystalline rearrangement of the starch coexisted with this phenomenon.

Keywords: allotropic transition; choline chloride; plasticizer; starch dissolution

Citation: Crucean, D.; Pontoire, B.; Debucquet, G.; Le-Bail, A.; Le-Bail, P. Influence of the Presence of Choline Chloride on the Classical Mechanism of "Gelatinization" of Starch. *Polymers* **2021**, *13*, 1509. https://doi.org/10.3390/polym13091509

Academic Editors: Rafael Antonio Balart Gimeno, Daniel García García, Vicent Fombuena Borrás, Luís Jesús Quiles Carrillo and Marina Patricia Arrieta Dillon

Received: 21 March 2021
Accepted: 28 April 2021
Published: 7 May 2021

1. Introduction

Choline chloride, because of its unique nutrient functionality and its ability to enhance flavor, provides a new option as a partial substitute for sodium chloride in reformulations of processed foods to reduce their sodium content. To date, very few studies have explored the potential of choline chloride to act as a substitute for salt. Locke and Fielding were the first to identify the properties of choline chloride to improve the salty taste of food products [1–3].

Starch is a natural polymer with particular properties unlike those of traditional polymers. As a heterogeneous material, it has macromolecular structures bound in a granular superstructure.

When native starch granules are heated in water (excess of water conditions, around 60% water wb), their semicrystalline nature and three-dimensional architecture are gradually disrupted, resulting in a phase transition from the ordered granular structure to a disordered state in water, which is known as "gelatinization" [4,5]. Gelatinization is an irreversible process that includes several steps, such as granule swelling, native crystalline melting (loss of birefringence), and leaching of fractions of polymers or of polymers in particular amylose chains resulting in a molecular solubilization [6]. The gelatinization process is essential in the processing of foods. For improving the properties of starch and adding new functionalities, it is common to carry out the hydrothermal transformation of starch in environments containing substances other than water. Various plasticizers and additives for starch processing have been used, including polyols (glycerol, glycol, sorbitol, etc.) and nitrogen-containing compounds (urea, ammonium derived, and amines) [7,8]. An alternative class of materials known as ionic liquids, now commonly defined as salts, that melt below 100 °C has recently attracted much interest for the processing of biopolymers such as starch.

Choline chloride solubilized in water gives an ionic liquid. Several studies have shown the effects of ionic liquids (solvents and plasticizers) based on the effects of choline chloride on starch [9–14]. A study conducted by Decaen et al. highlights the plasticization of starch by ionic liquids based on an examination of choline [15]. The destructuration of native maize starch in mixtures of water and ionic liquids containing choline cations was studied in dynamic heating conditions, combining calorimetry, rheology, microscopy, and chromatographic techniques by Sciarini et al. [16].

This work presents the mechanism of choline chloride penetration into the starch grain and its impact on gelatinization and the structural evolution of starch during heating kinetics.

2. Materials and Methods

2.1. Materials

Wheat flour (type 65) was supplied by Evelia, La Varenne, France. Wheat starch was purchased from Loryma GmbH (Zwingenberg, Germany), and waxy corn starch was purchased from Roquette Italia SpA (Cassano Spinola, Italy). Potato starch and choline chloride ($C_5H_{14}ClNO$) were supplied by Sigma-Aldrich (France).

2.2. Sample Preparations

The measurements were carried out on wheat flour (F_W) and wheat, potato, and waxy corn starch (S_W, S_P, and S_{WC}, respectively) solubilized in an aqueous solution containing different concentrations of choline chloride.

The model systems studied being composed of three ingredients (flour, chlorine choline, and water) and were presented under the form of Equation (1) with F_x, Cc_y, and W_z being the flour (or starch), choline chloride, and water content, respectively.

$$F_x Cc_y W_z \tag{1}$$

A mixing design has been used (1). The factors are the concentrations of each component of the mixture. Responses are expressed as a function of these concentrations with the sum of the mass fraction of each component being equal to 100 ($x + y + z = 100$). For each analysis, 5 g of mixture was prepared.

The mode of preparation (quoted ILS = "Ionic Liquid + Starch" for the rest of the paper) consisted in adding the choline chloride under the form of ionic liquid [15] (previous dissolution of Cc and water for one hour before starting the experiments); in that case, the script that was used was "(Cc + W)", and the ionic liquid was then added to the flour. These systems were written with the following script: $F_x[Cc_y W_z]$.

All analyses were repeated in triplicate. A statistical analysis was performed with variance analysis ($p < 0.005$) to detect significant differences.

2.3. Methods

2.3.1. Differential Scanning Calorimetry Measurements

Samples weighing 500 mg were placed in stainless steel pans, and a reference cell was prepared by adding an amount of water equal to the volume of water in the sample cells. The cells were then sealed with a heat-resistant seal. The cell containing the sample was agitated (50 rpm at room temperature) for one hour and then placed in the oven of the appliance. The pans were heated at a rate of 1 °C/min from 10 to 120 °C by using the SETARAM microcalorimeter (μDSC) III (France). Scans were run on the following starch suspensions: $S_{P20}[Cc_x W_y]$, $S_{W20}[Cc_x W_y]$, $S_{WC20}[Cc_x W_y]$, and $F_{W20}[Cc_x W_y]$. All measurements were made at least in triplicate.

Thermal transitions were defined as T_{Ge} (peak temperature gelatinization), and ΔH_{Ge} denotes the transition enthalpies for different endotherms. The enthalpies were determined by endotherm integration, and all traces were normalized to 1 mg of dried sample. The partial enthalpies were calculated from the onset of the endotherm to the end (every 1 °C).

2.3.2. X-ray Diffraction

The samples containing flour and starch suspended in $[Cc_{56}W_{24}]$ were prepared as follows: $S_{WC20}[Cc_{56}W_{24}]$, $S_{P20}[Cc_{56}W_{24}]$, $S_{W20}[Cc_{56}W_{24}]$, and $F_{W20}[Cc_{56}W_{24}]$. The appropriate amounts of choline chloride, water (when required), and starch were weighed. The same formulations without choline chloride were studied: $S_{WC45}W_{55}$, $S_{P45}W_{55}$, $S_{W45}W_{55}$, and $F_{W45}W_{55}$, corresponding to the same F/W ratio as the previous mixtures.

The samples were examined by wide-angle (WAX) X-ray diffraction. The measurements were performed using a D8 Discover spectrometer from Bruker-AXS (Karlsruhe, Germany). Cu $K\alpha_1$ radiation, produced in a sealed tube at 40 kV and 40 mA, was selected and parallelized using a double Gobël mirror parallel optics system and collimated to produce a 500 µm beam diameter with sample alignment by a microscopic video and laser. The data were monitored by a VANTEC 500 2D detector (Bruker, Karlsruhe, Germany) for 10 min and normalized.

The sample was placed in a capillary with a diameter of 1.5 mm, and a second capillary was introduced into the first to avoid loss of water during the rise in temperature; the two capillaries were sealed. The as-prepared capillary was placed in a heating stage HFS91 (Linkam, Tadworth, UK). The detector was positioned at a focusing distance of 8.6 cm from the sample surface. It was in direct beam position [17]. The heating kinetic applied to the sample was 1 °C/min from 20 to 120 °C. Every 10 °C, a plateau of 10 min was introduced to enable the acquisition of the diffraction spectrum.

3. Results and Discussion

3.1. Gelatinization of Different Starches

The thermograms of wheat flour (F_W), wheat starch (S_W), potato starch (S_P), and waxy-corn starch (S_{WC}) in excess water (W = 80%) should serve as a reference and allow comparison with the $F_{20}[Cc_xW_y]$ systems, i.e., when choline chloride is added to the aqueous phase. All the samples exhibited an endotherm on their thermograms, which corresponded to the gelatinization of the starch at the following temperatures: F_W 61.7 ± 0.2 °C, S_W 58.1 ± 0.1 °C, S_{WC} 70.8 ± 0.1 °C, and S_P 65.7 ± 0.1 °C (Table 1). The gelatinization temperatures of the different samples studied increased in the following order: T_{Ge} (S_W) < T_{Ge} (F_W) < T_{Ge} (S_P) < T_{Ge} (S_{WC}), whereas the gelatinization enthalpies increased in the following order: ΔH_{Ge} (F_W) < ΔH_{Ge} (S_W) < ΔH_{Ge} (S_{WC}) < ΔH_{Ge} (S_P).

Table 1. Enthalpies (ΔHGe, J/g) and gelatinization temperatures (TGe, °C) of different flour/starch and plasticizer model systems. Enthalpies were calculated on flour or starch total dry basis.

x,y	Waxy-Corn Starch $S_{WC20}[Cc_xW_y]$		Potato Starch $S_{P20}[Cc_xW_y]$		Wheat Starch $S_{W20}[Cc_xW_y]$		Wheat Flour $F_{W20}[Cc_xW_y]$	
	ΔH_{Ge}, J/g	T_{Ge}, °C	ΔH_{Ge}, J/g	T_{Ge}, °C	ΔH_{Ge}, J/g	T_{Ge}, °C	ΔH_{Ge}, J/g	T_{Ge}, °C
x = 0 y = 80	10.8 ± 0.2	70.8 ± 0.1	12.0 ± 0.1	65.6 ± 0.1	7.2 ± 0.1	58.1 ± 0.1	5.2 ± 0.1	61.7 ± 0.2
x = 8 y = 72	12.4 ± 0.1	82.9 ± 0.1	12.2 ± 0.3	71.4 ± 0.1	8.4 ± 0.4	71.0 ± 0.1	6.3 ± 0.3	73.4 ± 0.4
x = 16 y = 64	13.3 ± 0.1	90.1 ± 0.2	12.8 ± 0.1	77.2 ± 0.1	8.8 ± 0.9	78.6 ± 0.1	6.7 ± 0.6	80.8 ± 0.3
x = 24 y = 56	14.9 ± 0.4	96.1 ± 0.2	13.4 ± 0.4	84.2 ± 0.1	10.3 ± 0.2	85.2 ± 0.1	7.7 ± 0.6	86.8 ± 0.7
x = 32 y = 48	14.9 ± 0.3	100.0 ± 0.1	13.6 ± 0.1	91.7 ± 0.1	11.4 ± 0.5	89.9 ± 0.1	8.6 ± 0.2	91.9 ± 0.2
x = 40 y = 40	14.9 ± 0.3	101.3 ± 0.1	14.2 ± 0.1	99.1 ± 0.1	12.2 ± 0.3	92.7 ± 0.1	9.2 ± 0.9	94.8 ± 0.1
x = 48 y = 32	16.8 ± 0.5	98.6 ± 0.1	14.0 ± 0.2	104.5 ± 0.1	13.0 ± 0.2	92.2 ± 0.1	9.7 ± 0.2	93.8 ± 0.1
x = 56 y = 24	16.1 ± 0.1	90.7 ± 0.1	13.5 ± 0.3	102.1 ± 0.1	12.0 ± 0.2	87.1 ± 0.3	9.2 ± 0.6	87.4 ± 0.2
x = 64 y = 16	N/A	79.9 ± 0.2	N/A	95.5 ± 0.2	N/A	81.0 ± 0.6	N/A	84.5 ± 0.5

The thermograms of the F_W and S_W samples showed a second endotherm characterized by the fusion of the amylose-endogenous lipid complexes formed during gelatinization [18,19]. The melting temperatures of the amylose-lipid complexes for S_W and F_W were 97 ± 1 °C and 92.3 ± 0.2 °C, respectively.

3.2. *Influence of Choline Chloride on Starch Destructuration*

Since choline chloride behaves as an ionic liquid when it is in the presence of water, its behavior during the heating of a starch granule suspension may strongly influence gelatinization. A study comparing the previously discussed aqueous-phase suspensions containing solubilized choline chloride was carried out.

Micro-DSC was used to study the physicochemical characteristics of the following suspensions: $S_{WC20}[Cc_xW_y]$, $S_{P20}[Cc_xW_y]$, $S_{W20}[Cc_xW_y]$, and $F_{W20}[Cc_xW_y]$.

The thermograms of the suspensions are shown in Figure 1, and their enthalpies ΔH_{Ge} and gelatinization temperatures T_{Ge} are shown in Table 1.

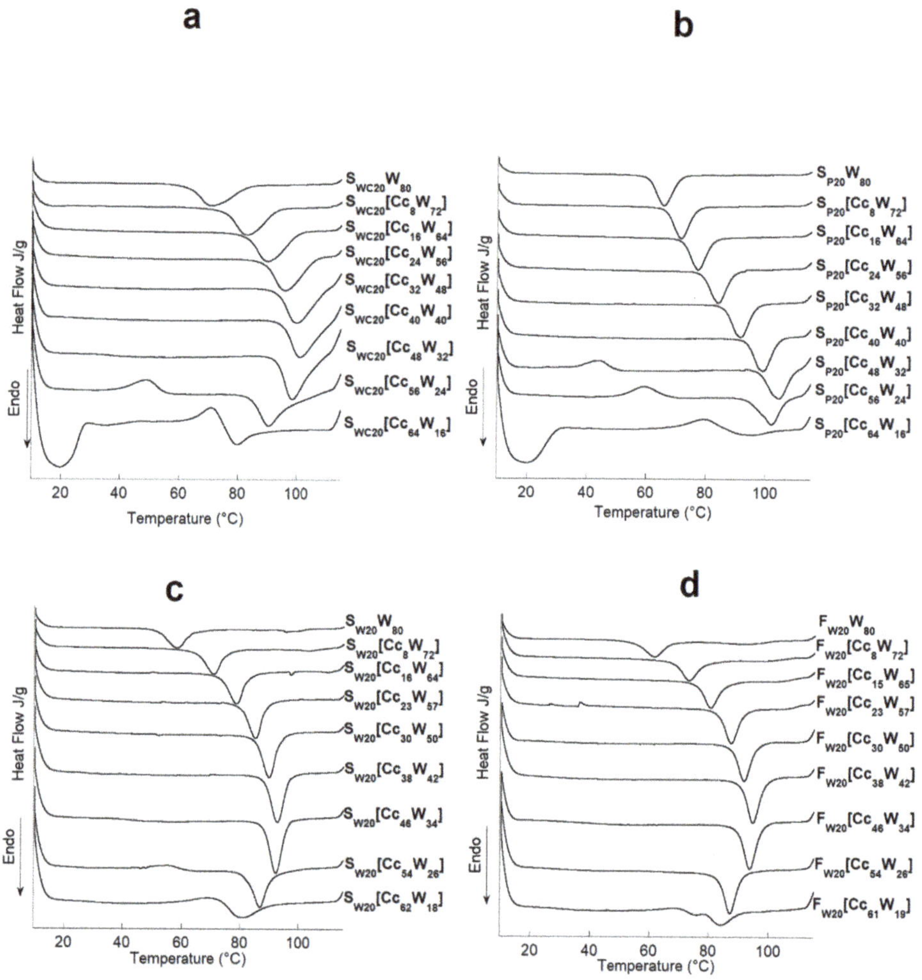

Figure 1. µDSC heating curves, 10–120 °C at 1 °C/min, of: (**a**) waxy-corn starch, (**b**) potato starch, (**c**) wheat starch, and (**d**) flour at different water and chlorine chloride contents.

At low concentrations of Cc, the starch undergoes a typical gelatinization, represented by an endothermic transition. At the same concentration, the endothermic transition attributed to the fusion of endogenous amylose-lipid complexes is also observed for F_W and S_W. When the concentration of Cc increases from 0 to 40% for S_{WC}, S_W, and F_W and from 0 to 48% for S_P, the peak of gelatinization moves to a higher temperature and then decreases for all three starches and flour. The observed gelatinization temperature is always higher than that observed in pure water. However, beyond 40% for S_{WC}, S_P, and F_W and beyond 48% for S_W, the gelatinization of the starch decreases with an increase in the B4 concentration. The same trend is observed for the enthalpies of gelatinization.

The increase in gelatinization temperature caused by Cc and the subsequent decrease with an increasing Cc level agree with the results of Chiotelli et al. for wheat and potato starches, Chungcharoen and Lund for rice starch, Jane for corn starch, and those of Ahmad and William and Ghani et al. for sargo starch [20–24].

The shift of the gelatinization temperature toward higher temperatures as well as the increase in the total enthalpy of gelatinization in the presence of Cc may be due to the reduction of the water activity in the starch/plasticizer solution, which causes an increase in the energy required for chemical and physical reactions involving water [25] Jane [22] claimed that the effect of salts on starch gelatinization follows the "Hofmeister" series, the increase in the gelatinization temperature follows the charge density order of the ions (LiCl > NaCl > KCl > RbCl), that is, the order of the "structure making" effect. However, at concentrations greater than 4 M, these salts decrease the onset temperature of gelatinization. Choline reacts as an "ion structure maker".

For high Cc concentrations, exothermic transitions were observed. The exotherms started at higher temperatures as the water content decreased, and the corresponding heat released (ΔH) increased as well. There was a critical concentration, which depends on the type of starch used, for which both exothermic and endothermic transitions took place. For this concentration, both phenomena seem to happen at very close temperatures, at which they overlapped, resulting in a neutralization with no visible thermal effect. The same behavior was found by Sciarini et al. and Mateyawa et al. working with choline acetate and EMIMAc and by Koganti et al. using N-methyl morpholine N-oxide (NMMO) [12,16,26]. These authors attributed the exothermic transition to starch dissolution in these solvents. For low water contents (16%), the thermograms of S_P and S_{WC} show wide endotherms at low temperatures (19.7 °C). This phenomenon was not visible for S_W and F_W. A DSC study was then carried out on the $S_{WC20}[Cc_xW_y]$ systems (with x included between 56 and 70 and y between 24 and 10) to understand this phenomenon (Figure 2).

The thermogram of pure Cc (data not shown) shows an endotherm near 73.6 ± 0.1 °C. With the addition of water, this endotherm moves to lower temperatures and undergoes a strong shift from the baseline until disappearing. In the presence of waxy maize starch (20%), the concentration of Cc was increased in steps of 1% (from 56 to 70%), and water was decreased likewise (from 24 to 10%), which produced an endotherm at low temperatures for the sample with 17% water.

During these investigations, the Cc was solubilized in water to form the plasticizer, and then the starch was added to the plasticizer. When the amount of water was sufficient, complete solubilization of the Cc was achieved. When the water concentration was less than 17%, the solubilization of Cc was partial, and we observed a solubilization endotherm that broadened with a decrease in the water content. For the 14% water sample, there was enough unsolubilized choline chloride for the allotropic change in choline chloride at 68 °C to be observed.

It was assumed that the solution (water + Cc) became more viscous with increasing Cc concentrations, resulting in more difficulty for the Cc solution to penetrate inside the starch granule; in turn, a shift to higher gelatinization temperatures was observed.

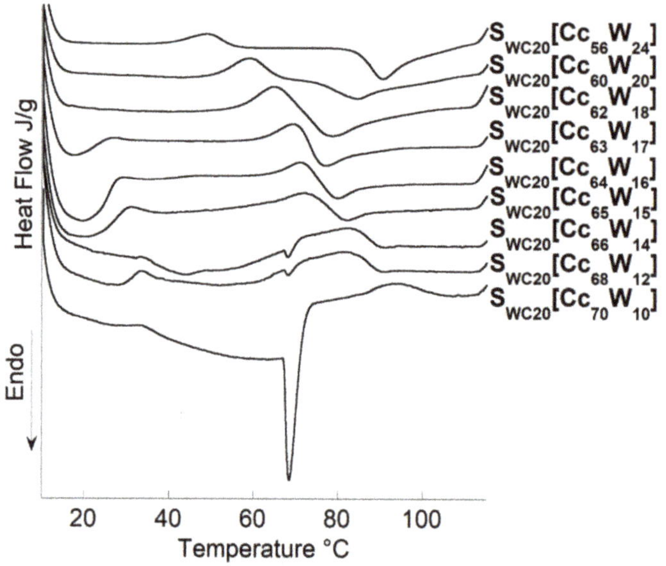

Figure 2. Microcalorimeter heating curves, 10–120 °C at 1 °C/min, of waxy-corn starch at different water, and choline chloride contents.

The effect of chlorine choline (Cc) on the gelatinization of starch is very different from that of NaCl observed in the literature. Indeed, if the gelatinization temperature follows the same trend when adding Cc or NaCl, the enthalpy of gelatinization shows the opposite behavior. The decrease in the total enthalpy of starch gelatinization at high NaCl concentrations suggests a destabilization of the ordered regions of the starch in the presence of NaCl. As explained by Chiotelli et al., sodium chloride may be hindered by polymer–polymer interactions in favor of water–polymer interactions, resulting in a lower enthalpy for fusion of organized regions. The addition of Cc causes a reorganization of the internal structure of the starch grain [20].

3.3. *Penetration of the Plasticizer into the Starch Grain*

To understand the penetration kinetics of the ionic liquid (Cc + water) in the starch granule, the cumulative enthalpy curves were calculated and reported as a function of temperature.

The curves of the partial enthalpy of gelatinization for all the plasticizer contents studied are shown in Figure 3. This study presents the kinetics of the loss of ordered structure in starch for temperatures between 40 and 110 °C. An offset of the melting toward higher temperatures was observed when the Cc content increased; this was observed for the three starches and the flour.

The shapes of the curves shown in Figure 3 are rather different for S_P and S_W: the melting of the potato starch ordered zones was more abrupt than it was for the wheat starch and the wheat flour. These observations are in good agreement with the results of Waigh et al. on the two-stage destructuration of S_W (crystallinity loss and helices destruction) [27]. Fannon, Hauber, and Bemiller showed that the S_P granules appear smooth on the surface by scanning electron microscopy, whereas cereal grains such as S_W have pores of about 100 nm at their surfaces, which are generated at the time of granule biosynthesis and cause a greater hydration sensitivity [28]. These observations may partly explain the differences in the plasticizer penetration into the granules for wheat starch and potato starch.

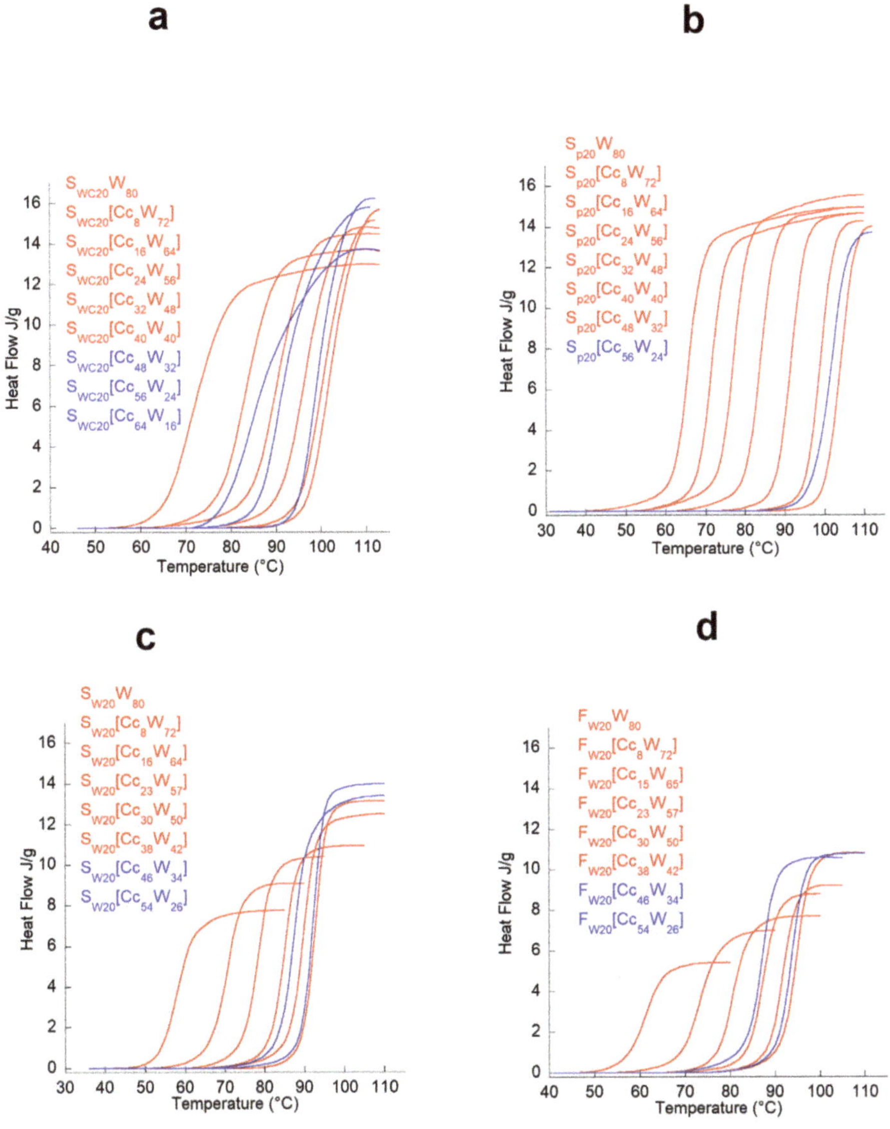

Figure 3. Curves of the partial enthalpy of gelatinization for all the plasticizer contents studied. (**a**) $[S_{WCx}Cc_y]W_z$ systems; (**b**) $[S_{Px}Cc_y]W_z$ system$_s$; (**c**) $[S_{Wx}Cc_y]W_z$ systems; (**d**) $[F_{Wx}Cc_y]W_z$ systems.

For potato starch, the total gelatinization enthalpy (ΔH) in the presence of Cc remained nearly constant at low concentrations of Cc but decreased slightly for higher concentrations. For wheat starch, the ΔH increased significantly as the concentration of Cc increased, to a greater extent than that for the potato starch. This considerable increase in the total enthalpy of gelatinization of wheat starch at high Cc concentrations suggests a better organization of the ordered regions of starch in the presence of Cc; on the other hand, when the concentration reached 40% for the Cc, the total enthalpy of the starch due to the fusion decreased perceptibly. The effect of high Cc concentrations on the gelatinization process was different for potato starch compared to waxy maize and wheat starch and wheat flour; it is assumed that this was due to the differences in the polymorphic structures.

The effect of Cc on the gelatinization of S_P is more pronounced in the first step of the gelatinization process (retardation of the loss of crystalline order), whereas for S_W, F_W, and S_{WC}, Cc also affects the second stage (delay in the loss of molecular order) and the overall enthalpy of the transition. These results are again opposite to those found in the literature for potato starch and wheat starch in the presence of high salt concentrations (NaCl) [20].

3.4. *Impact of the Formulation on Starch Gelatinization*

As previously described, choline chloride in the absence of water is not considered as an ionic liquid. Therefore, it seemed interesting to observe the action of Cc on starch when Cc is no longer acting as a plasticizer, but when it enters into the formulation as dry matter, Cc is placed directly in competition with water in front of starch. The relevant suspensions were formulated as follows: the starch (or flour) was mixed with the choline chloride (dry powder), and the water was added to the mixture. The obtained mixes were quoted using the following script: $[F_xCc_y]W_z$. These mixtures were studied by μDSC.

The temperatures and enthalpies of starch gelatinization did not vary with the formulation. However, there was a marked increase in the solubilization/rearrangement exotherm when the choline chloride was used as an ionic liquid (data not shown). This observation is explained by the fact that the solubilization of Cc before deposition on the starch allowed better penetration into the grain.

Another comparative study was carried out on the $F_{W20}Cc_{62}W_{18}$ system, with three different formulations (Figure 4):

- Formulation 1: the wheat flour was mixed with water before the introduction of choline chloride for $[F_{W20}W_{18}]Cc_{62}$.
- Formulation 2: choline chloride was mixed with water before being applied to the wheat flour for $F_{W20}[Cc_{62}W_{18}]$.
- Formulation 3: the wheat flour was mixed with choline chloride before the introduction of water for $[F_{W20}Cc_{62}]W_{18}$.

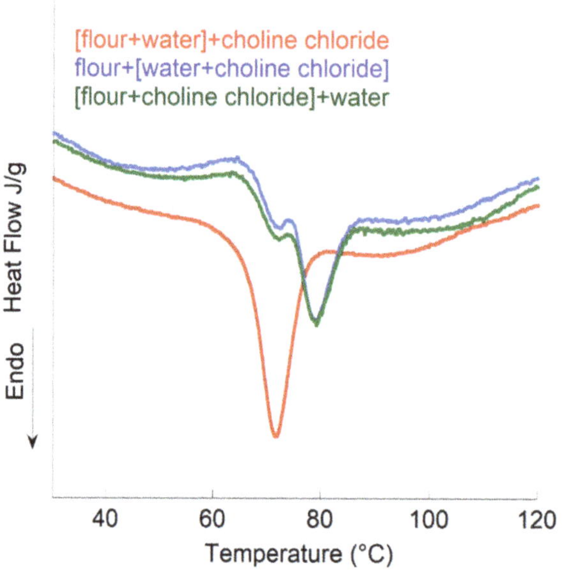

Figure 4. Comparative study of the thermograms of the systems $[F_{W20}Cc_{64}]W_{16}$, $F_{W20}[Cc_{64}W_{16}]$, and $[F_{W20}W_{16}]Cc_{64}$.

The results clearly showed that when the water (in small quantities) is premixed with the flour, it becomes inaccessible to the choline chloride, causing gelatinization at 90 °C corresponding to a hydration of approximately 50%, which agrees with the starting hydration (following formula).

$F_{20}W_{18}Cc_{62}$ → $F_{52}W_{48}$

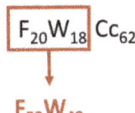

Black formula: amount of water in relation to the amount of F and Cc
Red formula: amount of water in relation to the amount of F

There is also a broad endotherm of allotropic change in Cc, suggesting that choline chloride has access to only a very small amount of water.

For the other two formulations, very similar results were obtained: namely, no change in the gelatinization temperature, a slightly more prominent exotherm for formulation 2, and a small endotherm associated with the allotropic change in Cc.

In this part, we have shown that the order of incorporation of water into the system can influence the gelatinization of the studied starches. Indeed, if the water is added to the starch first, it binds enough to the starch to become unavailable for Cc. On the other hand, if the water is first added to the choline chloride, the latter is solubilized before entering as an ionic liquid in the starch.

3.5. Evolution of the Structure of Different Starches in the Presence of Choline Chloride during Heating Kinetics

To understand the phenomena underlying the exothermic transitions observed with the microcalorimeter, measurements by an XRD heating cell were made.

Two types of systems were compared with and without Cc. The starch/flour and water concentrations were the same in both systems. The X-ray diffraction spectra of each of the suspensions of waxy corn starch and potato starch are collated in Figure 5.

S_{WC}, S_W, and F_W possess A-type polymorphic structures, and S_P possesses a B-type polymorphic structure. In the diffraction spectra of the $S_{W20}W_{24}$ (not shown), $F_{W20}W_{24}$ (not shown), and $S_{WC20}W_{24}$ systems, characteristic peaks of the A-type structure were identified. Spectra of the $S_{P20}W_{24}$ system showed characteristic peaks of the B-type structure.

A notable difference is seen in the evolution of the spectra during heating between the samples without Cc (Figure 5a,c) and those containing Cc (Figure 5b,d). Indeed, the peaks are more well defined in the systems without Cc, as with the addition of Cc at 20 °C. This confirms the previous conclusion: Cc causes a loss of crystalline order for all types of starch at room temperature. For the $S_{P45}W_{55}$ system, the type-B structure disappears at a temperature of 90 °C. However, for type-A systems, the structure collapses near 90 °C; this is consistent with Waigh's observations [29]. When Cc was added to the systems, a collapse of structure A or B was observed at much higher temperatures, which agrees with the results obtained by DSC.

For the $S_{P20}[Cc_{56}W_{24}]$ system, an increase in the peak intensity occurred at 5.6° in 2Θ at a temperature of 60 °C, followed by the onset of collapse at 100 °C and the total disappearance of the structure at 110 °C. This phenomenon corresponds to the appearance of the exotherm on the thermogram from the DSC study.

For systems $S_{WC20}[Cc_{56}W_{24}]$ and $S_{W20}[Cc_{56}W_{24}]$, the transition to an increase in crystallinity before melting is much less visible, and it is even absent for the $F_{W20}[Cc_{56}W_{24}]$ system. This explains the absence of the exotherm for the $F_{W20}[Cc_{56}W_{24}]$ system. In the literature, the exotherm is related to the solubilization of the starch in the plasticizer. However, during the X-ray diffraction study, we showed that the exotherm also corresponded to the increase in the intensity of the diffraction peaks and therefore to the increase in the crystallinity of the starch before its solubilization and then its melting.

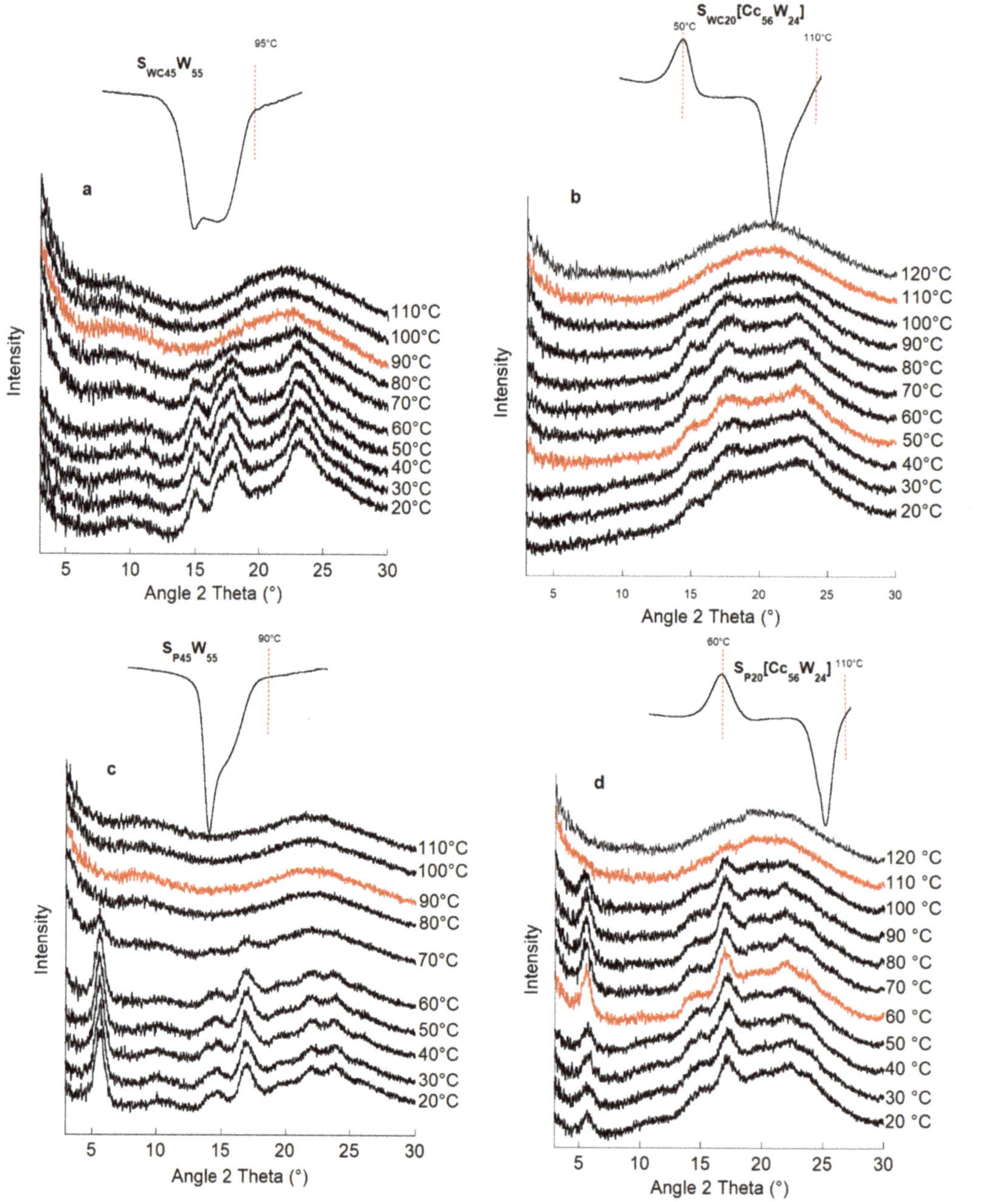

Figure 5. Evolution of the X-ray diffraction spectra during heating associated with the thermogram: waxy maize starch without Cc: $S_{WC45}W_{55}$ (**a**) and with Cc: $S_{WC20}[Cc_{56}W_{24}]$ (**b**); potato starch without Cc: $S_{P45}W_{55}$ (**c**) and with Cc: $S_{P20}[Cc_{56}W_{24}]$ (**d**).

To complete this work, a μDSC study was conducted on the $S_{P20}[Cc_{56}W_{24}]$ system at low rates (0.1 °C/min). Two peaks were clearly observed (Figure 6), indicating that more than one phenomenon is responsible for the exothermic transition observed.

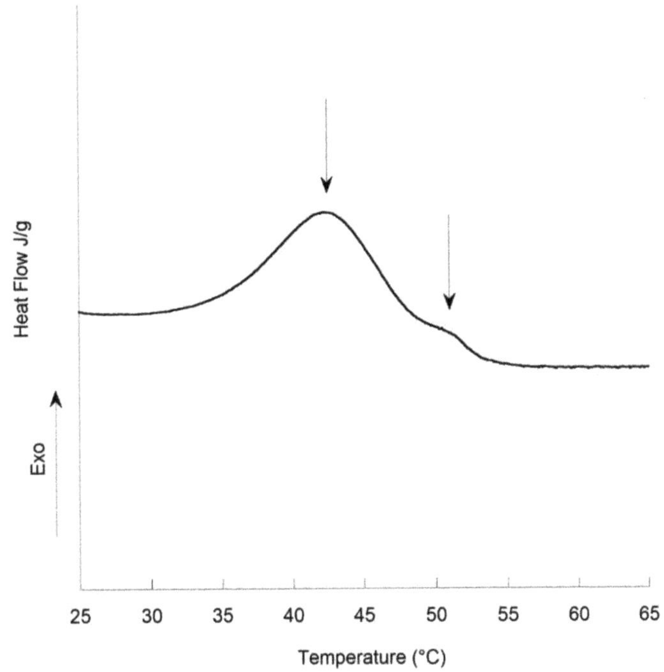

Figure 6. μDSC heating curves, 20–120 °C at 0.1 °C/min, for $S_{P20}[C_{C56}W_{24}]$.

4. Conclusions

The starch/water system shows a gelatinization accompanied or unaccompanied by a fusion, according to the water content of the matrix. The choline chloride/water system possesses the characteristics of an ionic liquid and exhibits an allotropic change at low water concentrations and solubilization for water contents greater than 30%.

Choline chloride is an ionic compound, which, like NaCl, has a "structure making" effect: that is, it increases the viscosity of the aqueous solution while decreasing the water fraction with high mobility. As a result, the gelatinization temperatures are displaced to higher temperatures. In contrast to the addition of NaCl, the addition of choline chloride entails a significant increase in the gelatinization enthalpy, which suggests stabilization or better organization of the ordered regions of the starch in the presence of choline chloride.

An X-ray diffraction study in a heating cell demonstrated the crystalline rearrangement of the structure of the starch grain, which takes place simultaneously with the solubilization phenomenon of amylose.

Author Contributions: Conceptualization, A.L.-B. and P.L.-B.; methodology, D.C. and B.P.; formal analysis, D.C.; data curation, D.C., P.L.-B.; writing—original draft preparation, D.C., A.L.-B. et P.L.-B.; writing—review and editing, D.C., G.D. and P.L.-B.; supervision, P.L.-B.; project administration and funding acquisition, A.L.-B. All authors have read and agreed to the published version of the manuscript.

Funding: Co fundings by French Ministry of Agriculture (ONIRIS-ID for FOODS programme), by ONIRIS-GEPEA and by INRA-BIA.

Institutional Review Board Statement: Nothing to report.

Informed Consent Statement: Not applicable.

Data Availability Statement: Nothing to report.

Acknowledgments: This project was supported by the French Ministry of Agriculture (ONIRIS-ID for FOODS programme), by ONIRIS-GEPEA and by INRA-BIA co fundings.

Conflicts of Interest: I certify that there is no conflict of interest.

References

1. Locke, K.W.; Fielding, S. Enhancement of salt intake by choline chloride. *Physiol. Behav.* **1994**, *55*, 1039. [CrossRef]
2. Yamaguchi, M.; Kainuma, K.; French, D. Electron microscopic observations of waxy maize starch. *J. Ultrasructure Res.* **1979**, *69*, 249. [CrossRef]
3. Oostergetel, G.T.; van Bruggen, F.J. On the Origin of a Low Angle Spacing in Starch. *Starch Stärke* **1989**, *41*, 331. [CrossRef]
4. Atwell, W.A.; Hood, L.F.; Lineback, D.R.; Marston, E.V.; Zobel, H.F. The terminology and methodology associated with basic starch phenomenon. *Cereal Foods World* **1988**, *33*, 306.
5. Lelievre, J. Starch gelatinization. *J. Appl. Polym. Sci.* **1974**, *18*, 293. [CrossRef]
6. Nikolic, M.A.L.; O'Sullivan, C.; Rounsefell, B.; Halley, P.J.; Truss, R.; Clarke, W.P. The anaerobic degradability of thermoplastic starch: Polyvinyl alcohol blends: Potential biodegradable food packaging materials. *Bioresour. Technol.* **2009**, *100*, 1705.
7. Liu, H.; Xie, F.; Yu, L.; Chen, L.; Li, L. Thermal processing of starch-based polymers. *Prog. Polym. Sci.* **2009**, *34*, 1348. [CrossRef]
8. Xie, F.; Halley, P.J.; Avérous, L. Rheology to understand and optimize processibility, structures and properties of starch polymeric materials. *Prog. Polym. Sci.* **2012**, *37*, 595. [CrossRef]
9. Biswas, A.; Shogren, R.L.; Stevenson, G.D.; Willett, J.L.; Bhowmik, P.K. Ionic liquids as solvents for biopolymers: Acylation of starch and zein protein. *Carbohydr. Polym.* **2006**, *66*, 546. [CrossRef]
10. Lappalainen, K.; Kärkkäinen, J.; Lajunen, M. Dissolution and depolymerization of barley starch in selected ionic liquids. *Carbohydr. Polym.* **2013**, *93*, 89. [CrossRef]
11. Liu, W.; Budtova, T. Dissolution of unmodified waxy starch in ionic liquid and solution rheological properties. *Carbohydr. Polym.* **2013**, *93*, 199. [CrossRef]
12. Mateyawa, S.; Xie, D.F.; Truss, R.W.; Halley, P.J.; Nicholson, T.M.; Shamshina, J.L.; Rogers, R.D.; Boehm, M.W.; McNally, T. Effect of the ionic liquid 1-ethyl-3-methylimidazolium acetate on the phase transition of starch: Dissolution or gelatinization? *Carbohydr. Polym.* **2013**, *94*, 520. [CrossRef]
13. Stevenson, D.G.; Biswas, A.; Jane, J.-L.; Inglett, G.E. Changes in structure and properties of starch of four botanical sources dispersed in the ionic liquid, 1-butyl-3-methylimidazolium chloride. *Carbohydr. Polym.* **2007**, *67*, 21. [CrossRef]
14. Zdanowicz, M.; Spychaj, T. Ionic liquids as starch plasticizers or solvents. *Polim. Polym.* **2011**, *56*, 861.
15. Decaen, P.; Rolland-Sabaté, A.; Guilois, S.; Jury, V.; Allanic, N.; Colomines, G.; Lourdin, D.; Leroy, E. Choline chloride vs choline ionic liquids for starch thermoplasticization. *Carbohydr. Polym.* **2017**, *177*, 424. [CrossRef]
16. Sciarini, L.S.; Rolland-Sabaté, A.; Guilois, S.; Decaen, P.; Leroy, E.; Le Bail, P. Understanding the destructuration of starch in water–ionic liquid mixtures. *Green Chem.* **2015**, *17*, 291. [CrossRef]
17. Le-Bail, P.; Houinsou-Houssou, B.; Kosta, M.; Pontoire, B.; Gore, E.; Le-Bail, A. Molecular encapsulation of linoleic and linolenic acids by amylose using hydrothermal and high-pressure treatments. *Food Res. Int.* **2015**, *67*, 223. [CrossRef]
18. Biliaderis, C.G.; Page, C.M.; Slade, L.; Sirett, R.R. Thermal behaviour of amylose-lipid complexes. *Carbohydr. Polym.* **1985**, *47*, 73.
19. Kugimiya, M.; Donova, J.W.; Wong, R.Y. Phase Transitions of Amylose-Lipid Complexes in Starches: A Calorimetric Study. *Starch Stärke* **1980**, *32*, 265. [CrossRef]
20. Chiotelli, E.; Pilosio, G.; Le Meste, M. Effect of sodium chloride on the gelatinization of starch: A multimeasurement study. *Biopolymers* **2002**, *63*, 41. [CrossRef]
21. Chungcharoen, A.; Lund, D.B. Influence of Solutes and Water on Rice Starch Gelatinization. *Cereal Chem.* **1987**, *64*, 240.
22. Jane, J.-L. Mechanism of Starch Gelatinization in Neutral Salt Solutions. *Starch Stärke* **1993**, *45*, 161. [CrossRef]
23. Ahmad, F.B.; William, P.A. Effect of Salts on the Gelatinization and Rheological Properties of Sago Starch. *J. Agric. Food Chem.* **1999**, *47*, 3359. [CrossRef] [PubMed]
24. Ghani, M.B.A.; Che Man, Y.B.; Ali, A.B.; Hashim, D.B.M. Differential scanning calorimetry: Gelatinisation of sago starch in the presence of sucrose and sodium chloride. *J. Sci. Food Agric.* **1999**, *79*, 2001. [CrossRef]
25. Seetharaman, K.; Yao, N.; Rout, M.K. Role of Water in Pretzel Dough Development and Final Product Quality. *Cereal Chem.* **2004**, *81*, 336. [CrossRef]
26. Koganti, N.; Mitchell, J.R.; Ibbett, R.N.; Foster, T.J. Solvent Effects on Starch Dissolution and Gelatinization. *Biomacromolecus* **2011**, *12*, 2888. [CrossRef]
27. Waigh, T.A.; Kato, K.L.; Donald, A.M.; Gidley, M.J.; Clarke, C.J.; Riekel, C. Side-Chain Liquid-Crystalline Model for Starch. *Starch Stärke* **2000**, *52*, 450. [CrossRef]
28. Fannon, J.E.; Hauber, R.J.; BeMiller, J.N. Surface Pores of Starch Granules. *Cereal Chem.* **1992**, *69*, 284.
29. Waigh, T.A.; Gidley, M.J.; Komanshek, B.U.; Donald, A.M. The phase transformations in starch during gelatinisation: A liquid crystalline approach. *Carbohydr. Polym.* **2000**, *328*, 165. [CrossRef]

polymers MDPI

Article

Characterization of Novel Biopolymer Blend Mycocel from Plant Cellulose and Fungal Fibers

Ilze Irbe [1,*], Inese Filipova [1], Marite Skute [1], Anna Zajakina [2,*], Karina Spunde [2] and Talis Juhna [3]

1 Latvian State Institute of Wood Chemistry, Dzerbenes 27, LV-1006 Riga, Latvia; inese.filipova@inbox.lv (I.F.); polarlapsa@inbox.lv (M.S.)
2 Latvian Biomedical Research and Study Centre, Ratsupites 1 k. 1, LV-1067 Riga, Latvia; karina.spunde@biomed.lu.lv
3 Water Research and Environmental Biotechnology Laboratory, Riga Technical University, P. Valdena 1-303, LV-1048 Riga, Latvia; talis.juhna@rtu.lv
* Correspondence: ilzeirbe@edi.lv (I.I.); anna@biomed.lu.lv (A.Z.)

Abstract: In this study unique blended biopolymer mycocel from naturally derived biomass was developed. Softwood Kraft (KF) or hemp (HF) cellulose fibers were mixed with fungal fibers (FF) in different ratios and the obtained materials were characterized regarding microstructure, air permeability, mechanical properties, and virus filtration efficiency. The fibers from screened Basidiomycota fungi *Ganoderma applanatum* (Ga), *Fomes fomentarius* (Ff), *Agaricus bisporus* (Ab), and *Trametes versicolor* (Tv) were applicable for blending with cellulose fibers. Fungi with trimitic hyphal system (Ga, Ff) in combinations with KF formed a microporous membrane with increased air permeability (>8820 mL/min) and limited mechanical strength (tensile index 9–14 Nm/g). HF combination with trimitic fungal hyphae formed a dense fibrillary net with low air permeability (77–115 mL/min) and higher strength 31–36 Nm/g. The hyphal bundles of monomitic fibers of Tv mycelium and Ab stipes made a tight structure with KF with increased strength (26–43 Nm/g) and limited air permeability (14–1630 mL/min). The blends KF FF (Ga) and KF FF (Tv) revealed relatively high virus filtration capacity: the $\log_{10}$ virus titer reduction values (LRV) corresponded to 4.54 LRV and 2.12 LRV, respectively. Mycocel biopolymers are biodegradable and have potential to be used in water microfiltration, food packaging, and virus filtration membranes.

Keywords: air permeability; fungal fibers; hemp fibers; microstructure; mechanical properties; mycocel; softwood fibers; virus membrane filtration

Citation: Irbe, I.; Filipova, I.; Skute, M.; Zajakina, A.; Spunde, K.; Juhna, T. Characterization of Novel Biopolymer Blend Mycocel from Plant Cellulose and Fungal Fibers. *Polymers* **2021**, *13*, 1086. https://doi.org/10.3390/polym13071086

Academic Editor: Rafael Antonio Balart Gimeno

Received: 3 March 2021
Accepted: 25 March 2021
Published: 30 March 2021

1. Introduction

The importance of biobased polymers is well known, and much research and development activities concern the use of biobased polymers in science, engineering, and industry. Biopolymers from renewable resources are used in multiple fields, namely health, food, energy, and the environment, due to their intrinsic features, versatility, biocompatibility, and degradability. Besides, the widespread use of biopolymers also addresses concerns about environmental sustainability [1]. As long as biopolymers stand as a sustainable, biodegradable compound, new biopolymer products and materials filled, blended, or reinforced with natural fibers, will predominate with the promise of sustainability benefits [2].

Generally, biobased polymers are classified into three classes: (1) naturally derived biomass polymers such as cellulose, cellulose acetate, starches, chitin, modified starch, etc.; (2) bio-engineered polymers bio-synthesized by using microorganisms and plants such as poly(hydroxy alkanoates (PHAs), poly(glutamic acid), etc.; (3) synthetic polymers produced from naturally derived molecules or by the breakdown of naturally derived macromolecules through the combination of chemical and biochemical processes such as polylactide (PLA), poly(butylene succinate) (PBS), bio-polyolefins, bio-poly(ethylene terephtalic acid) (bio-PET) [3]. The first and second class polymers are biodegradable, they allow for more efficient production, which can produce desired functionalities and physical properties, but chemical structure designs have limited flexibility. The third class polymers such as bio-polyolefins and bio-PET are not biodegradable and the only contribution for reducing environmental impact comes from reducing the carbon footprint. The origin of a polymer does not determine its biodegradability; this condition depends on the chemical structure of the polymer [3].

The naturally derived biopolymers, among others, include polysaccharides of plant and fungal origin. Polysaccharides are nontoxic and biodegradable, which increases their potential application in biopolymers. The most used biopolysaccharides are obtained from plant origin (e.g., cellulose), microbial origin (e.g., bacterial cellulose), and animal origin (e.g., chitin/chitosan). Moreover, these natural derivatives present a considerable number of reactive functional groups (e.g., hydroxyl, carboxyl, and amino groups), which significantly increase their applicability through chemical modification or physical blend [1]. Cellulose is the most abundant polysaccharide of natural origin in the world, and is mostly produced by plants. The cellulose chains structure leads to areas of high crystallinity within the polymer and to high stability structures, which as a consequence promote considerable strength, remarkable inertness, and insolubility in water and common organic solvents [4]. In its turn, fungal cell walls share a common chemical structure composed of homo- and heteropolysaccharides, protein, protein–polysaccharide complexes, lipids, melanin, and polysaccharide chains of chitin. Chitin is a biopolymer of N-acetylglucosamine with some glucosamine, which is the main component of the cell walls of fungi and is considered the second most abundant natural polymer after cellulose [5].

Current research in membrane science is now focusing more on biopolymers from natural raw materials with a well-defined structure to develop new membrane materials. In fact, the combination of polysaccharides and proteins is a method frequently used to design blended materials with improved performance regarding swelling, mechanical resistance, and biocompatibility, among other features. The attention has been focused also on membranes based on chitosan blended with other biomacromolecules such as alginate, cellulose, collagen, gelatin, keratin, sericin, and soy protein [6]. Chitin-related materials from fungal sources with focus on nanocomposites and nanopapers have been suggested as greener alternative to synthetic polymers [7]. Fungal chitin–glucan nanopapers have manufactured from chitin nanofibrils, which is a native composite material (chitin–glucan) combining the strength of chitin and the toughness of glucan. These nanopapers showed distinct physico-chemical surface properties, being more hydrophobic than crustacean chitin [8]. Nanopapers from chitin nanofibrils have exhibited tunable mechanical and surface properties with potential use in coatings, membranes, packaging [9] and ultra-filtration of organic solvents and water [10]. Fungal mycelium–nanocellulose has been produced by the agitated liquid culture of a white-rot fungus with nanocellulose as part of the culture media. The obtained biomaterials are suggested for diverse applications, including packaging, filtration, and hygiene products [11].

The literature survey demonstrates that fungal chitin nanofibrils have been used to manufacture nanopapers. Additionally, fungal mycelium has been combined with nanocellulose to obtain a blended biomaterial. However, there are no studies on blended biopolymers from plant cellulose and fungal hyphae. The aim of this study was to develop a novel blended biopolymer from naturally derived biomass that is made of softwood cellulose fibers and fungal fibers (hyphae) and to test it for air permeability, mechanical properties, and virus filtration efficiency. More detailed investigation of specific properties of biomaterial blend containing fungal fibers from basidiomycete *Ganoderma applanatum* is published in our previous paper [12]. The cellulose fibers from softwood and hemp shives are easily available natural resource. The basidiomycetes selected for this study have shown a high importance in medical and nutritional applications. For example, *Ganoderma applanatum* has been reported for its phytochemical properties for potential application in nanotechological engineering for clinical use [13]. *Fomes fomentarius* has been used in a traditional medicine for centuries and is still economically important as a source of medicinal and neutraceutical products [14]. *Trametes versicolor* is known for its general health-promoting effects and is widely employed in traditional medicine [15]. *Agaricus bisporus* represents the leading position among edible cultivated mushrooms [16].

2. Materials and Methods

2.1. Synthesis of Biopolymers

The screening of several basidiomycetes was carried out to select the fungal candidates forming homogenous hyphal biomass convenient for biopolymer material development. The fungal biomass was obtained from (1) fruiting bodies of the forest growing species *Ganoderma applanatum* (Pers.) Pat., *Fomes fomentarius* (L.) Fr., *Lentinus lepideus* (Fr.) Fr., *Polyporus squamosus* (Huds.) Fr., *Fomitopsis betulina* (Bull.) B.K. Cui, (2) commercially cultivated mushroom stipes of *Agaricus bisporus* (J.E. Lange) Imbach and (3) mycelium of pure culture strain *Trametes versicolor* CTB 863 A.

Fungal biomass was kept in 4% NaOH solution for 24 h at a room temperature in order to extract proteins and alkali soluble polysaccharides. Then the samples were washed in tap water and mechanically disintegrated using Blendtec 725 (Orem, UT, USA) at 360 W for 30 s. Obtained fungal fibers (FF) were dried at room temperature and kept in a dry state until used.

Bleached softwood Kraft fibers (KF) were provided by Metsä Fibre (Äänekoski, Finland) as pressed sheets and used without specific pre-treatment.

Hemp fibers (HF) were obtained from industrial hemp *Cannabis sativa* (USO-31). After the decortication fibers were treated in 4% NaOH solution at 165 °C for 75 min, then washed with tap water to neutral and refined using Blendtec 725 (Orem, UT, USA) at 179 W for 7 min at 1.5% consistency, dried at room temperature and kept in dry state until used.

For biopolymer blend development, a certain amount (g) of FF in combinations with KF and HF fiber pulp in different mass ratios (50:50 and 33:33:33) was placed in a glass baker and soaked in 1–2 L of distilled water for 8 h, then disintegrated using 75,000 revolutions in the disintegrator (Frank PTI, Laakirchen, Austria). Material sheets were produced according to ISO 5269-2:2004 with a Rapid Köthen paper machine (Frank PTI, Laakirchen, Austria). Grammage (weight per unit area in g/m^2) of samples was calculated by dividing the mass with area according to ISO 536:2019. At least five parallel samples at grammage 50 ± 15 g/m^2 of each composition were prepared.

2.2. Micromorphology

The micromorphology was examined by a light microscope (LM) Leica DMLB (Leica Microsystems GmbH, Wetzlar, Germany) at a magnification of 200×. Lactophenol blue solution (Fluka) was used to observe cyanophilic reaction of fungal hyphae. The images were captured by a video camera Leica DFC490 using calibrated image analysis software Image-Pro plus 6.3 (Media Cybernetics, Inc., Rockville, MD, USA).

For scanning electron microscopy (SEM), the surface of samples was coated with gold plasma using a K550X sputter coater (Emitech, Ashford, UK) and examined with Vega TC (Tescan, Brno-Kohoutovice, Czech Republic) with accelerating voltage of 15 kV, software 2.9.9.21.

2.3. Air Permeability

Air permeability was tested by Bendsen method according to ISO 5636-3:2013 using an air permeability tester 266 (Lorentzen and Wettre, Stockholm, Sweden). Samples were clamped between a metal ring and a rubber gasket and the air flow rate was measured through sample area of 10 cm^2 under 1.47 kPa air pressure for 5 s. Commercially available disposable face masks consisting of three layers made of polypropylene spunbond-meltblown-spunbond nonwoven fabrics were tested for air permeability as comparison for developed biopolymer samples.

2.4. Mechanical Properties

For evaluation of tensile properties samples with a width of 1 cm were prepared with a strip cutter (Frank-PTI, Laakirchen, Austria) and tested according to ISO 1924-1:1992 using a tensile tester vertical F81838 (Frank-PTI, Laakirchen, Austria). Tensile index (Nm/g) was calculated by dividing measured maximum tensile strength (N/m) by grammage (g/m^2) of test sample. Breaking length (km) and stretch (%) was used as calculated by software of equipment. Burst index (kPa m^2/g) was calculated dividing the measured burst strength (kPa) by grammage (g/m^2), where burst strength was measured as hydrostatic pressure necessary to cause rupture in a circular area of a 4 cm diameter of sample according to ISO 2758:2014 using burst tester (Frank-PTI, Laakirchen, Austria).

2.5. Virus Preparation and Membrane Filtration Procedure

The recombinant Semliki forest virus (SFV) pSFVenh/Luc, encoding firefly luciferase gene, was produced as previously described [17,18]. The virus-containing cell medium was harvested and concentrated by ultracentrifugation through two sucrose cushions, as previously described [19]. The virus was rapidly frozen and subsequently used as a virus stock. The virus titre expressed in infectious units per ml (i.u./mL) was quantified by immunostaining with rabbit polyclonal antibodies specific to the nsp1 subunit of SFV replicase, as previously described [20].

The ability of developed biopolymer materials to retain/remove virus particles was tested using centrifugal filtration test. EMA/CPMP/BWP/268 guidelines were considered for development of virus filtration procedures and calculation of virus reduction values. Two-layer cellulose hygienic paper (AB Grigeo, Vilnius, Lithuania), three-layer cotton mask with antibacterial *Silverplus* coating (SIA P.E.M.T., Riga, Latvia) and a standard surgical mask type II EN 14,683 (Matopat, Toruń, Poland) were used as reference materials.

A 1.5 × 1.5 cm sample was cut out from each material, folded in a form of a conical funnel, placed into truncated 1 mL plastic pipette tip (for better fixation), and subsequently placed into an Eppendorf 1.5 mL tube as presented in Figure 1.

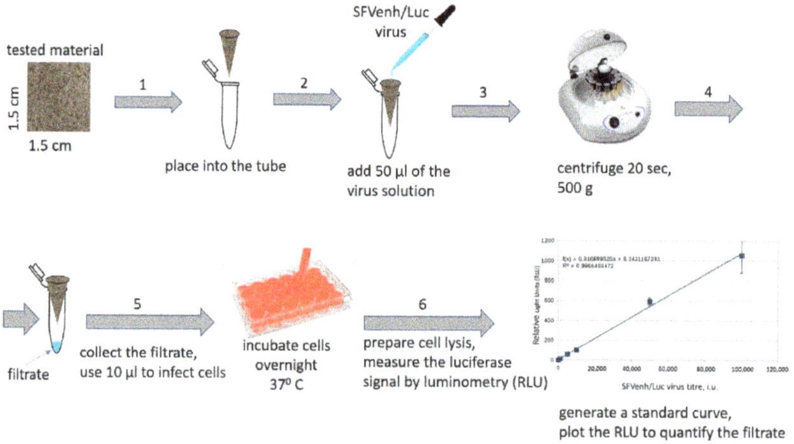

Figure 1. Schematic representation of the virus filtration test (technical details are provided in text). (1) 1.5 cm × 1.5 cm filter sample is cut out from each material, folded into the form of a conical funnel, and placed into a 1.5 mL tube; (2) 50 µL of recombinant Semliki forest virus (SFV)-enh/Luc virus solution (107 i.u./mL) is added into the cone; (3) the tube is centrifuged to allow the virus to pass through the material; (4) the filtrate (indicated by arrow) is collected; (5) the filtrated is diluted and used for cell infection in a 24-well cell culture plate; (6) after overnight incubation of the plate the cell lysates are prepared and the virus infection is measured by detection of the luciferase activity in infected cells (luminometry). The cell infection with the standard dilutions of the virus is used to generate a standard curve and to calculate the amount of virus in the filtrate.

50 μL of recombinant SFVenh/Luc virus solution with a titer of 10^7 infectious units per milliliter (i.u./mL) were poured into each cone containing the test material. The tubes were then centrifuged for 20 s at 500× *g* (3500 rpm) by FVL-2400N Combi-Spin, Mini-Centrifuge/Vortex (Biosan, Riga, Latvia) to allow the liquid to pass (filtrate) through the material under low pressure conditions. The filtrated virus samples were collected (at least 20 μL each) and used for BHK-21 cell infection in 24-well plate as previously described [18]. Briefly, 10 μL of the virus sample in duplicate was mixed with 190 μL PBS (containing M^{2+} and Ca^{2+}) and incubated with BHK-21 cells for 1 h at 37 °C, 5% CO_2, then 800 μL of BHK medium (1% fetal bovine serum) was added. In parallel, standard dilutions of the SFVenh/Luc virus in a range 1×10^3–5×10^5 i.u./mL per well were generated and used for BHK-21 cell infection in duplicate in the 24-well plate. The cells were incubated overnight at 37 °C, 5% CO_2 to allow complete cell infection to complete and expression of firefly luciferase gene.

To quantify the virus titer in filtered samples the relative luminescence units (RLU) were measured in cell lysates and the respective values were plotted on the standard curve with serial SFVenh/Luc virus dilutions. The RLUs were measured by the Luciferase assay (Promega, Madison, WI, USA), as recommended by manufacturer. Briefly, the cell medium was removed, and the cells (24-well) were lysed in 100 μL of the Cell Culture Lysis buffer (Promega, Madison, WI, USA), centrifuged at 600 cfr for 5 min, and 1 μL of the cell lysate was used immediately to measure the luciferase enzymatic activity by luminometer Luminoskan Ascent (Thermo Scientific, Loughborough, UK). The cell infection was done in duplicate in each independent experiment. Two independent experiments were performed with each sample (repeats). The virus titer standard curve was generated in each experiment and the negative control signal (RLU of uninfected cells) was subtracted from all values. The $\log_{10}$ reduction value (LRV) represents the difference between loaded and eluted virus infectious units per ml, and respectively was calculated according to the following equation:

$$\text{LRV} = [\log_{10} (\text{virus titer before filtration}) - \log_{10} (\text{virus titer after filtration})]. \quad (1)$$

Statistical analysis of obtained results was performed using an Excel 2016 MSO data statistical analysis tool.

3. Results and Discussion

The fruiting bodies of *L. lepideus*, *P. squamosus*, and *F. betulina* formed gelatinous biomass in NaOH solution which was unusable for further experiments. FF were successfully separated from the biomass of fungi *G. applanatum*, *F. fomentarius*, *A. bisporus*, and *T. versicolor* and integrated in biopolymer compositions with softwood and hemp cellulose fibers.

3.1. Morphological Characterization

The microscopic structure of mycocel biopolymer blends and macrostructure of developed materials is shown in Figure 2.

Kraft fibers, hemp fibers, and fungal fibers were most likely physically bound together making a net of the biopolymer material. The fungal fibers separately or in bundles were randomly distributed within the net of cellulose fibers. The structure of materials was affected by individual properties of fibers. The size of Kraft fibers was average 2 mm in length and 30 μm in width. Hemp fibers were 1 mm long and 10–20 μm wide with a net of microfibers (2–5 μm). Fungal fibers from fruiting bodies of poroid species (*G. applanatum*, *F. fomentarius*) (Figure 2a,b) were 2–7 μm across. The fibers of *T. versicolor* mycelium were 2–3 μm wide (Figure 2c). The stipes of mushroom *A. bisporus* were composed of swollen hyphae, ascending 7–20 μm across (Figure 2d).

(**a**)

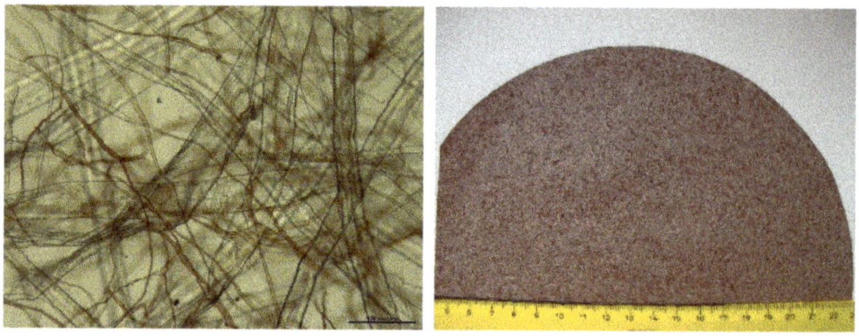

(**b**)

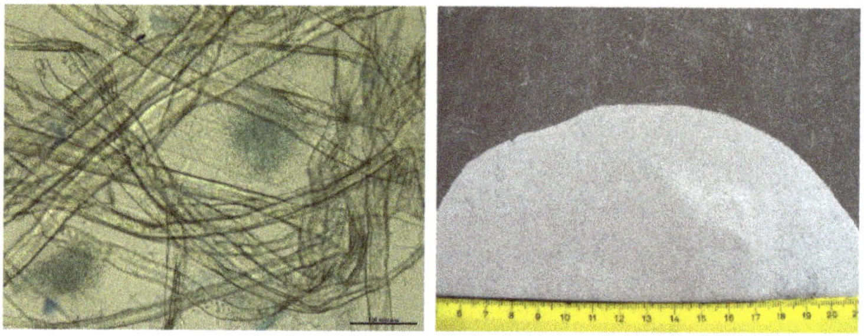

(**c**)

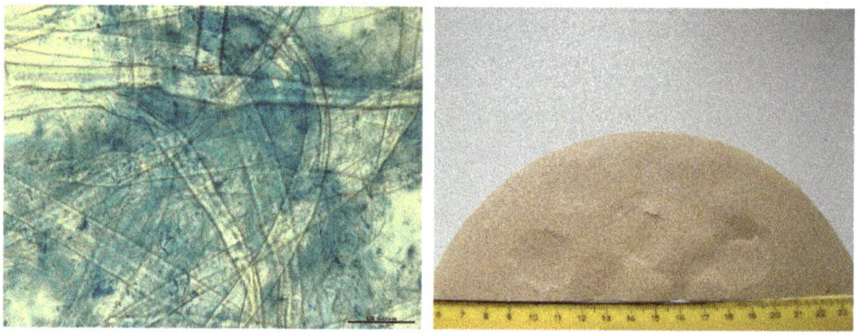

(**d**)

Figure 2. *Cont.*

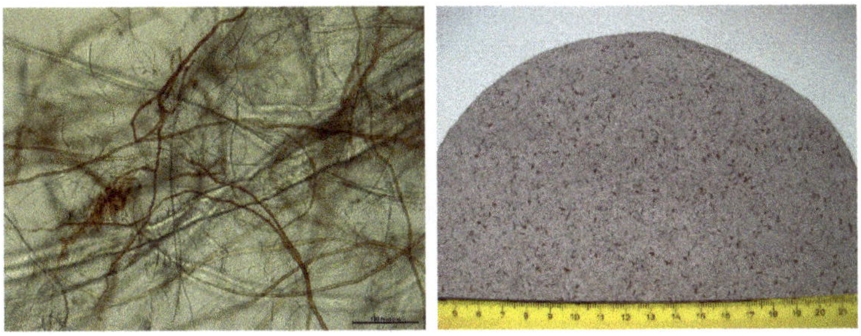

(e)

Figure 2. Microstructure (left column) and macrostructure (right column) of mycocel biopolymer compositions made of Kraft fibers (KF), hemp fibers (HF), and fungal fibers (FF): (**a**) KF FF (Ff); (**b**) KF FF (Ga); (**c**) KF FF (Tv); (**d**) KF FF (Ab); (**e**) KF HF FF (Ga). Microimages show flat hemp and softwood fibers (10–30 μm) (**a**–**e**) and narrow fungal fibers (2–7 μm) of polypores (**a**–**c**,**e**) and swollen hyphae (7–20 μm) of agaric (**d**). LM, 200×. Bar = 100 μm.

The hyphae of basidiomycetes (primarily poroid species) are divided in three main types: generative, binding, and skeletal hyphae with key differences in cell wall thickness, internal structure, and branching characteristics. Monomitic species comprise only generative hyphae, dimitic species comprise two hyphal types (usually generative and skeletal) and trimitic species contain all three hyphal types [21].

The fungal fruiting bodies under this study, namely, *G. applanatum* and *F. fomentarius*, had trimitic hyphal system. Depending on the species, the generative hyphae were 2–5 μm, skeletal hyphae 3–7 μm and binding hyphae 2–4 μm across (Figure 2a,b). In the case of poroid species *T. versicolor*, the mycelium of pure culture strain consisted of thin-walled generative hyphae, while hyphae of agaric *A. bisporus* were flat and swollen. The fibers of two later species formed dense hyphal bundles among the cellulose fibers. These can be observed in Figure 2c,d where cyanophilic reaction turned hyaline hyphae of Tv and Ab blue after staining. The trimitic species Ga and Ff presented brown pigmented hyphal network (Figure 2a,b,e). The pigmented layer within the hyphal cell wall is likely cross-linked to polysaccharides and has a mesh-like structure with pores, through which small and large molecules can penetrate into cells. Pigment melanin is believed to enhance the strength of the cell wall and has antioxidant properties and propensity to bind to a variety of substances [22].

SEM micrographs (Figure 3) support the findings of LM and display a detailed ultrastructure of the raw materials and mycocel biopolymer blends. Fiber orientation appeared randomly distributed, entangled fibers were physically bound together with H bonds, forming non-oriented, multi-layered net. Kraft fibers (Figure 3a) were flattened and made a microporous network. Similarly, the fungal fibers Ga (Figure 3c) formed a network of microporous structures. Hemp material (Figure 3b) displayed a dense structure composed of flat fibers and thin microfibers.

The microstructure of mycocel biopolymers was affected by individual properties of raw materials. The structure of KF FF (Ga) (Figure 3d) was determined by individual properties of Kraft fibers and fungal fibers. Distribution FF among the flattened KF favoured formation a loose microporous structure of the material. The structure of biopolymer HF FF (Ga) was affected by a dense network of variable diameter hemp fibers (Figure 3e). The FF were incorporated in a tight HF network of microfibers forming a compact structure with decreased porosity. The surface of KF HF FF (Ga) material (Figure 3f) displayed dense areas of hemp fibers and Kraft fibers with few areas of loose fungal fibers. The porosity and the tortuosity fractal dimension are two critical parameters to determine the permeability. It is reported [23] that increase in the tortuosity fractal dimension leads to decrease in the dimensionless permeability and absolute permeability; an increase in the porosity increases the dimensionless permeability; increase in the fiber diameter yields an increase in the absolute permeability.

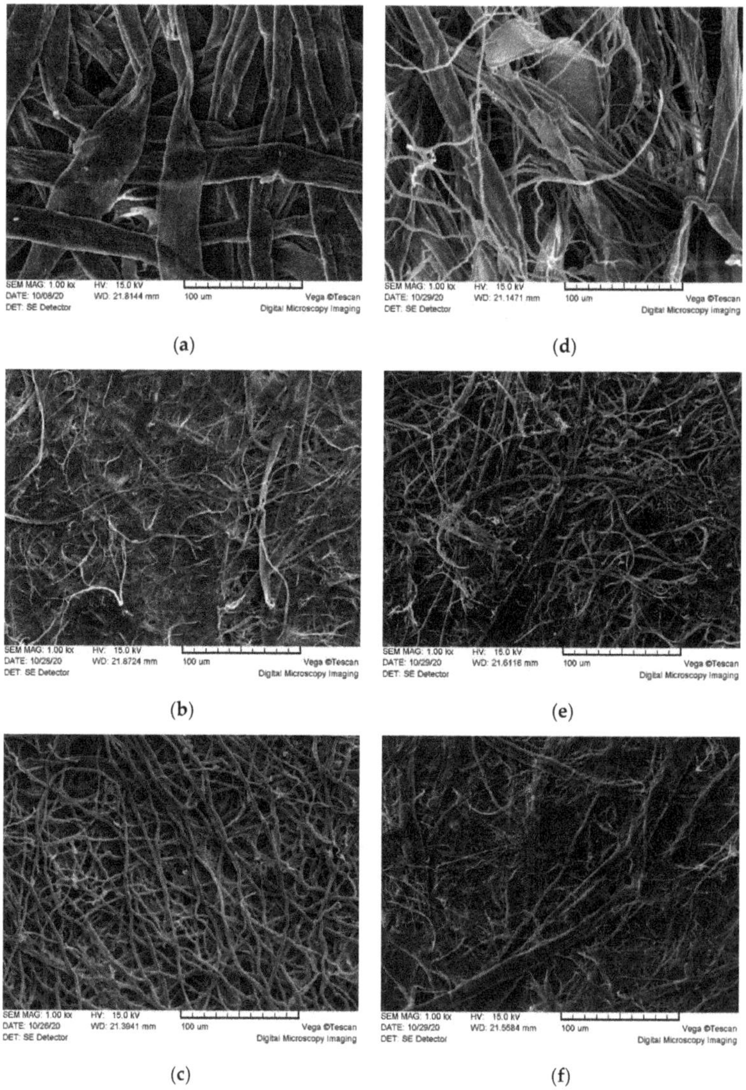

Figure 3. Ultrastructure of raw materials (left column) from (**a**) Kraft fibers (KF), (**b**) hemp fibers (HF), and (**c**) fungal fibers (FF Ga), and mycocel biopolymer compositions (right column): (**d**) KF FF (Ga); (**e**) HF FF (Ga); (**f**) KF HF FF (Ga). SEM, 1000×. Bar = 100 μm.

3.2. Air Permeability

The air permeability results were supported by microscopy findings. Mycocel biopolymer compositions had significant variations in air permeability properties starting from almost non-air permeable ones to materials with air permeability above measuring limit of device, which was 8820 mL/min (Table 1). The fungal type of hyphae had a significant effect on results. Biopolymer blends having *F. fomentarius* (Ff) or *G. applanatum* (Ga) trimitic hyphal system in combination with Kraft fibers (KF FF (Ff) and KF FF (Ga)) had the highest air permeability >8820 mL/min. Biopolymers consisting of cellulose fibers (KF and HF) and *A. bisporus* (Ab) hyphae had the lowest air permeability 14.1 mL/min, <1 mL/min and 7.5 mL/min for KF FF (Ab), HF FF (Ab), and KF HF FF (Ab), respectively.

Table 1. Mechanical and air permeability properties of mycocel biopolymer materials (KF = Kraft fibers; HF = hemp fibers; FF = fungal fibers; Ga = *G. applanatum*; Ab = *A. bisporus*; Tv = *T. versicolor*; Ff = *F. fomentarius*). The marked (*) samples have been described previously [12].

Sample	Tensile Index, Dry Nm/g	Tensile Index, Wet Nm/g	Burst Index kPa m^2/g	Breaking Length km	Stretch %	Air Permeability mL/min
KF *	16.9	0.8	1.0	1.7	0.9	8275
HF *	60.4	10.9	4.6	6.1	3.4	32
FF (Ga) *	8.2	-	0.9	0.8	2.0	6935
KF FF (Ga) *	13.9	1.3	1.0	1.4	2.0	>8820
HF FF (Ga) *	30.8	-	2.1	3.1	3.4	77
HF FF (Ab)	32.5	5.3	2.0	3.3	0.7	<1
KF HF FF (Ga) *	35.9	1.9	2.8	3.7	3.9	115
KF FF (Tv)	26.5	2.4	2.0	2.7	2.0	1630
KF FF (Ff)	8.9	-	1.5	0.9	1.1	>8820
KF FF (Ab)	43.5	3.0	1.9	4.4	1.0	14.1
KF HF FF (Ab)	46.0	3.0	2.4	4.7	1.6	7.5

It should be noted that materials with HF in their composition had the lowest air permeability numbers because of high Shopper Riegler freeness of hemp cellulose fiber (91.5 °SR) [12], which led to high bonding and tight networks of fibers (Figure 3b). The bonding of cellulose fibers is based primarily on hydrogen bonding between hydroxyl functional groups during close contacting of fibers [24]. The same bonding theory can be attributed to investigated mycocel biopolymer materials, since cellulose and chitin—the main constituent of hyphae—are biopolymers and have similar polysaccharide chain structure with the main difference being the replacement of one of the three hydroxyl groups with acetyl amine group in chitin monomeric unit [25]. The ability of functional groups of hyphal polysaccharides to make hydrogen bonding with cellulose fibers is one of the key elements of network formation in investigated mycocel materials. Furthermore, better bonding leads to a more compact packing of fibers and lower free volumes associated with lower air permeability [26], which can be seen in the cases of biopolymer blends with Tv mycelial and Ab stipe fibers. The higher numbers of air permeability showed lower ability of Ga and Ff hyphae to get involved in the network of cellulose fibers through hydrogen bonding. It can be explained by presence of non-polysaccharidic substances, such as pigments, which were indicated by the dark color of fruiting bodies and isolated hyphae from Ga and Ff.

Mycocel blends with higher air permeability have potential for using as gas permeable membranes, for example, as biobased filter layer in face masks. Disposal medical face masks, used as reference material, correspondingly showed high air permeability above 8820 mL/min, therefore only KF FF (Ga) and KF FF (Ff) are appropriate for this application. Blends with very low air permeability, such as those containing Ab hyphae can be used in applications, where high air or gas barrier properties are important, for example, food packaging materials [27].

3.3. Mechanical Properties

Comparison of cellulose and fungal fiber biopolymers showed significant differences among different fungal species and hyphal types regarding their effects on mechanical properties of investigated materials (Table 1). It was possible to produce a pure fiber material from *G. applanatum* (Ga) and *A. bisporus* (Ab) biomass, however FF (Ga) material had a rather low tensile index 8.2 Nm/g, but FF (Ab) material was too fragile for handling and it was not possible to measure mechanical properties. When compare two-component blends of Kraft fibers and fungal fibers, the best results were obtained in the case of KF FF

(Ab) and KF FF (Tv), 32.5 and 26.5 Nm/g, respectively, which is by 92% and 57% more than tensile index measured for material containing only KF. However, addition of *G. applanatum* (Ga) or *F. fomentarius* (Ff) fibers to cellulose fibers decreased mechanical strength by 18% for KF FF (Ga) and by 48% for KF FF (Ff) if compare with pure KF. This can be explained by specific character of Ab and Tv hyphae which formed a dense microstructure with cellulose fibers increasing the mechanical strength of material. It is reported [28] that generative hyphae alone (monomitic hyphal system), which are hollow and contain cytoplasm, are suggested to provide limited mechanical performance, with binding hyphae (dimitic and trimitic hyphal systems) responsible for material strength. Contrary, our results showed that the trimitic hyphal system of fungi Ga and Ff did not favor an improvement of mechanical properties of mycocel compositions. Lower mechanical performance of Ga and Ff hyphae containing materials can be explained also by the aspects mentioned in description of air permeability properties and are related to lower ability for bonding with cellulose caused by presence of non-polysaccharidic substances and lower amount of hydrogen bonds formed. It is known that fiber to fiber bonding or bonding strength is directly related to mechanical strength of cellulose fiber-based materials and can be evaluated using results of mechanical tests [29].

Tensile index was almost threefold higher in wet stage when compare KF and KF FF (Ab) showing the ability of Ab hyphae to significantly improve the strength of fiber material in wet conditions. Two-component blends consisting of hemp fibers and fungal fibers HF FF were produced using hemp cellulose and Ga or Ab fibers. It must be noted that highly fibrillated hemp fibers with freeness 91.5 °SR used in the research, had the most significant effect on the mechanical strength of all investigated materials. Furthermore, pure HF material had the highest tensile index among all the tested materials by reaching 60.4 Nm/g. Addition of hyphae significantly decreased tensile index reaching 30.8 Nm/g in the case of HF FF (Ga) and 32.5 Nm/g in the case of HF FF (Ab). When compare three-component blends KF HF FF, here also Ab hyphae showed the highest effect on mechanical properties of composites by KF HF FF (Ab) reaching tensile index 46 Nm/g, while KF HF FF (Ga) reached 35.9 Nm/g. Overall, *A. bisporus* (Ab) and *T. versicolor* (Tv) fibers showed higher potential as component of biopolymer for improving the mechanical properties of cellulose fiber-based materials, whether they consist of wood or hemp cellulose fibers. Biopolymers containing Ga hyphae demonstrated the highest flexibility measured as percentage elongation or stretch. HF FF (Ga) and KF HF FF (Ga) samples reached 3.4% and 3.9% stretch, while other composites had numbers equal or below 2.0%.

Mechanical performance of materials was evaluated also by measuring burst index, which is the hydrostatic pressure necessary to cause rupture in a circular area of a given diameter and tells how much pressure material can tolerate before rupture. Three component blends showed the highest burst index 2.8 kPa m^2/g and 2.4 kPa m^2/g for KF FF FF (Ga) and KF HF FF (Ab) respectively, indicating the prevalence of fiber diversity presented in mechanical blend. If two-component blends are compared, Tv and Ab containing KF FF materials have higher results than Ga and Ff containing ones. Breaking length is used to characterize inherent strength of material and is defined as the length of imaginary material strip, if suspended vertically from one end, would break by its own weight. Numeric values of breaking length of biopolymer blends were in direct correlation with tensile strength values showing the predominance of Ab and Tv containing materials and lower results in Ga and Ff containing materials. Mechanical properties of disposal medical face masks layers were not possible to measure according to the standard and device used for biopolymer blends testing, however apparently they are less strong than materials of biopolymer blends.

3.4. Virus Filtration Properties

In order to evaluate the filtration properties of the experimental biomaterials, the recombinant Semliki forest virus (SFV) was applied, which belongs to the Togaviridae family of enveloped RNA viruses (60–70 nm in diameter) structurally similar to human pathogenic viruses such as influenza and coronaviruses. The test system was based on a safe replication deficient SFV vector (pSFVenh/Luc), allowing to perform one round cell infection to be performed with precise quantification of the virus filtration rates using rapid measurement of luciferase activity in infected cells.

The efficiency of the SFV/enhLuc virus filtration through the tested materials are summarized in Table 2. The efficient virus retention properties were observed both for raw materials (Kraft fibers, hemp fibers, and fungal fibers Ga) and mycocel blends KF FF (Ga) and KF FF (Tv), which revealed very low virus permeability rates (<2%). Cellulose hygiene paper also showed the ability to retain the virus, albeit with lower efficiency (the permeability <30%) comparing to cellulose containing biopolymers. The properties of cellulose materials are determined by production technology and additives which can result in lower mechanical strength, density, and virus filtering efficiency as observed in the case of hygiene paper. Regarding the reference materials, i.e., face masks, the highest virus permeability was observed for the tested surgical mask (>92%) and hydrophobic outer layer of the cotton mask (>70%), whereas the *Silverplus* layer retained the virus significantly (the permeability <1%).

Table 2. SFV1enh/Luc recombinant virus titer change after pressure filtration through experimental materials (KF = Kraft fibers; HF = hemp fibers; FF = fungal fibers; Ga = *G. applanatum*; Tv = *T. versicolor*).

Sample	Virus Titer i.u./mL ± SD	Virus Titer [Log_{10}]	Log_{10} Reduction Value LRV	Virus Amount after Filtration Relative to Nonfiltered Control %
Non-filtered virus	$(1 \pm 0.066) \times 10^7$	7.00	-	100
Surgical mask (all layers)	$(8.28 \pm 0.066) \times 10^6$	6.92	0.08	92.63
Cellulose hygienic paper	$(2.66 \pm 0.076) \times 10^6$	6.42	0.58	27.05
KF	$(1.43 \pm 0.413) \times 10^4$	4.16	2.84	0.16
KF FF (Ga)	$(2.86 \pm 0.076) \times 10^2$	2.46	4.54	0.00
KF FF (Tv)	$(7.66 \pm 0.791) \times 10^4$	4.88	2.12	0.78
FF (Ga)	$(2.25 \pm 1.460) \times 10^4$	4.35	2.65	0.26
HF	$(6.27 \pm 2.120) \times 10^3$	3.80	3.20	0.08
Cotton outer layer (hydrophobic)	$(7.21 \pm 0.0330) \times 10^6$	6.86	0.14	73.44
Cotton Silverplus middle layer	$(2.52 \pm 3.350) \times 10^3$	3.40	3.60	0.03
Cotton inner layer	$(7.31 \pm 1.300) \times 10^5$	5.86	1.14	7.44

According to EMA guidelines (CPMP/BWP/268/95, European Medicines Agency) the $\log_{10}$ reduction value (LRV) is an important parameter to quantify the virus reduction capacity. Only the LRV > 4 is considered as a "very high" virus reduction potential. In this study, the biopolymer KF FF (Ga) demonstrated sufficiently high virus reduction capacity (4.54 LRV). Remarkably, the silver containing cotton layer of the commercial mask also showed efficient virus removal properties (3.6 LRV), which can be related to the viricidal capacity of the immobilized silver nanoparticles [30].

Due to safety aspects, handling of human viruses is a labor-intensive and time-consuming process. In this study, we have established a protocol for a relatively simple virus filtration method, which is based on replication deficient SFV vector encoding firefly luciferase used for rapid virus quantification. The proposed method can be efficiently applied for primary screening of virus filtration properties of the membranes.

The mechanisms underlying the membrane filtration of viruses include size exclusion and/or adsorptive interactions (e.g., hydrophobic/hydrophilic and electrostatic interactions) between virus envelope and membrane compounds [31]. The diameter of the virus particles is much smaller than the pore size of the tested membranes. Therefore, the adsorptive interactions can be considered as the main mechanism of filtration in the tested system. A surgical mask, which is made of hydrophobic polypropylene layers, did not adsorb aqueous virus containing solution, resulting in low virus retention values (Table 2). Practically, the surgical masks serve as a barrier to aqueous aerosol and are designed to block direct fluid entry into the wearer's respiratory tract and mostly act as a repellent of the water-based liquids [32]. In contrast to surgical mask and tested cotton mask, the row fiber materials and mycocel blends exhibited hygroscopic properties and showed the ability to binding, or adsorption of SFV particles. A high virus filtration capacity of mycocel biopolymer blends might be attributed to the structural properties of the fungal cell wall as an insoluble polysaccharide-based sorbent. The fungal polysaccharides possess direct virus inactivation properties as shown for several highly pathogenic human viruses, including human immunodeficiency virus, herpes simplex virus, etc. [33]. The application of mycocel-based filters alone or in combination with polypropylene-based layers can represent an advanced individual respiratory protective device against airborne pathogens. Furthermore, chitin, one of the main polymers of fungal cell walls, is widely used for controlled drug delivery systems, protein and enzyme carriers, and packaging materials, based on its natural antimicrobial activity. Chitin and its derivatives (e.g., chitosan) have many useful properties that make them suitable for a wide variety of biomedical applications. Their products are known to be antibacterial, antifungal, antiviral, nontoxic, and nonallergic [34]. Therefore, the proposed mycocel-based biopolymers are promising multifunctional materials for biomedical and bioengineering applications.

4. Conclusions

A novel mycocel biopolymer from naturally derived biomass of plant cellulose fibers and fungal fibers (hyphae) was developed and characterized regarding its air permeability, mechanical properties, and virus filtration efficiency.

Air permeability and mechanical properties of mycocel biopolymer blends were affected by microstructural features of raw materials. Highly fibrillated hemp fibers had the most significant effect on the mechanical strength while Kraft fibers revealed increased air permeability. The ability of functional groups of hyphal polysaccharides to make hydrogen bonding with cellulose fibers was one of the key elements of network formation which determined biopolymer properties. The loose fiber net of trimitic fungal species *G. applanatum* and *F. fomentarius* with incorporated Kraft fibers formed a microporous structure of mycocel blends with improved air permeability (>8820 m/min) and limited mechanical properties in comparison with individual raw fibers. *A. bisporus* and *T. versicolor* fibers showed higher potential as components of biopolymer for improving the mechanical properties of cellulose fiber-based materials, whether they consist of wood or hemp cellulose fibers.

Virus testing provided promising results regarding virus filtration efficiency of biopolymer blends. Mycocel biopolymer KF FF (Ga) demonstrated sufficiently high virus reduction capacity (4.54 LRV) than surgical mask and outer and inner layers of commercial face mask. Feasibly, the adsorptive interactions can be considered as the main mechanism of filtration in tested system.

The natural, biodegradable mycocel blends have a potential for use in biomaterial membranes depending on the target application. Blends with higher air permeability and virus filtration efficiency have potential for being used as gas permeable membranes, for example, as biobased filter layer in face masks. Blends with low air permeability can be used in areas where high air or gas barrier properties are important, for example, food packaging materials. The water microfiltration and ultrafiltration also are considered for future application.

Author Contributions: Conceptualization, I.I. and I.F.; methodology, M.S., K.S., A.Z.; validation, I.I., I.F. and A.Z.; formal analysis, K.S., A.Z.; writing—original draft preparation, I.I.; writing—review and editing, I.F., A.Z. and T.J.; project administration, I.F. and T.J.; funding acquisition, T.J. All authors have read and agreed to the published version of the manuscript.

Funding: This research is funded by the Ministry of Education and Science, Republic of Latvia, Project "Integration of reliable technologies for protection against Covid-19 in healthcare and high-risk areas",

Project No. VPP-COVID-2020/1-0004. The Article Processing Charge was kindly provided by the Latvian State Institute of Wood Chemistry Base financing source.

Institutional Review Board Statement: Not applicable.

Informed Consent Statement: Not applicable.

Data Availability Statement: Not applicable.

Conflicts of Interest: The authors declare no conflict of interest.

References

1. Silva, S.S.; Rodrigues, L.C.; Fernandes, E.M.; Reis, R.L. Fundamentals on biopolymers and global demand. In *Biopolymer Membranes and Films*; de Moraes, M.A., da Silva, C.F., Vieira, R.S., Eds.; Elsevier: Amsterdam, The Netherlands, 2020; pp. 3–34. [CrossRef]
2. Sukumaran, N.P.; Gopi, S. Overview of biopolymers: Resources, demands, sustainability, and life cycle assessment modeling and simulation. In *Biopolymers and Their Industrial Applications*; Thomas, S., Gopi, S., Amalraj, A., Eds.; Elsevier: Amsterdam, The Netherlands, 2021; pp. 1–19. [CrossRef]
3. Nakajima, H.; Dijkstra, P.; Loos, K. The Recent Developments in Biobased Polymers toward General and Engineering Applications: Polymers that are Upgraded from Biodegradable Polymers, Analogous to Petroleum-Derived Polymers, and Newly Developed. *Polymers* **2017**, *9*, 523. [CrossRef]
4. Rudnik, E. 10—Compostable Polymer Materials: Definitions, Structures, and Methods of Preparation. In *Handbook of Biopolymers and Biodegradable Plastics*; Ebnesajjad, S., Ed.; William Andrew Publishing: Boston, MA, USA, 2013; Volume 10, pp. 189–211. [CrossRef]
5. El-Enshasy, H.A. Filamentous Fungal Cultures—Process Characteristics, Products, and Applications. In *Bioprocessing for Value-Added Products from Renewable Resources*; Yang, S.-T., Ed.; Elsevier: Amsterdam, The Netherlands, 2007; pp. 225–261. [CrossRef]
6. De Moraes, M.A.; Da Silva, C.F.; Vieira, R.S. *Biopolymer Membranes and Films: Health, Food, Environment, and Energy Applications*, 1st ed.; Elsevier: Amsterdam, The Netherlands, 2020; p. 654.
7. Nawawi, W.M.; Jones, M.; Murphy, R.J.; Lee, K.-Y.; Kontturi, E.; Bismarck, A. Nanomaterials Derived from Fungal Sources—Is It the New Hype? *Biomacromolecules* **2020**, *21*, 30–55. [CrossRef] [PubMed]
8. Nawawi, W.M.; Lee, K.-Y.; Kontturi, E.; Bismarck, A.; Mautner, A. Surface properties of chitin-glucan nanopapers from Agaricus bisporus. *Int. J. Biol. Macromol.* **2020**, *148*, 677–687. [CrossRef]
9. Jones, M.; Weiland, K.; Kujundzic, M.; Theiner, J.; Kählig, H.; Kontturi, E.; John, S.; Bismarck, A.; Mautner, A. Waste-Derived Low-Cost Mycelium Nanopapers with Tunable Mechanical and Surface Properties. *Biomacromolecules* **2019**, *20*, 3513–3523. [CrossRef] [PubMed]
10. Yousefi, N.; Jones, M.; Bismarck, A.; Mautner, A. Fungal chitin-glucan nanopapers with heavy metal adsorption properties for ultrafiltration of organic solvents and water. *Carbohydr. Polym.* **2021**, *253*, 117273. [CrossRef] [PubMed]
11. Attias, N.; Reid, M.; Mijowska, S.C.; Dobryden, I.; Isaksson, M.; Pokroy, B.; Grobman, Y.J.; Abitbol, T. Biofabrication of Nanocellulose–Mycelium Hybrid Materials. *Adv. Sustain. Syst.* **2021**, *5*. [CrossRef]
12. Filipova, I.; Irbe, I.; Spade, M.; Skute, M.; Dāboliņa, I.; Baltiņa, I.; Vecbiskena, L. Mechanical and Air Permeability Performance of Novel Biobased Materials from Fungal Hyphae and Cellulose Fibers. *Materials* **2021**, *14*, 136. [CrossRef] [PubMed]
13. Manasseh, A.T.; Godwin, J.T.A.; Emanghe, E.U.; Borisde, O.O. Phytochemical properties of Ganoderma applanatum as potential agents in the application of nanotechnology in modern day medical practice. *Asian Pac. J. Trop. Biomed.* **2012**, *2*, S580–S583. [CrossRef]
14. Elkhateeb, W.; Daba, G.; Elnahas, M.; Thomas, P. Fomes fomentarius and Polyporus squamosus Models of Marvel Medicinal Mushrooms. *Biomed. Res. Rev.* **2020**, *3*. [CrossRef]
15. Habtemariam, S. Trametes versicolor (Synn. Coriolus versicolor) Polysaccharides in Cancer Therapy: Targets and Efficacy. *Biomedicines* **2020**, *8*, 135. [CrossRef]
16. Holban, A.M.; Grumezescu, A.M. Therapeutic Food. In *Handbook of Food Bioengineering*; Academic Press: Cambridge, MA, USA, 2018; Volume 8, p. 538.
17. Zajakina, A.; Vasilevska, J.; Zhulenkovs, D.; Skrastina, D.; Spaks, A.; Plotniece, A.; Kozlovska, T. High efficiency of alphaviral gene transfer in combination with 5-fluorouracil in a mouse mammary tumor model. *BMC Cancer* **2014**, *14*, 460. [CrossRef]
18. Kurena, B.; Vežāne, A.; Skrastiņa, D.; Trofimova, O.; Zajakina, A. Magnetic nanoparticles for efficient cell transduction with Semliki Forest virus. *J. Virol. Methods* **2017**, *245*, 28–34. [CrossRef] [PubMed]
19. Hutornojs, V.N.-O.B.; Kozlovska, T.; Zajakina, A. Comparison of ultracentrifugation methods for concentration of recombinant alphaviruses: Sucrose and iodixanol cushions. *Environ. Exp. Biol.* **2012**, *10*, 117–123.
20. Vasilevska, J.; Skrastina, D.; Spunde, K.; Garoff, H.; Kozlovska, T.; Zajakina, A. Semliki Forest virus biodistribution in tumor-free and 4T1 mammary tumor-bearing mice: A comparison of transgene delivery by recombinant virus particles and naked RNA replicon. *Cancer Gene Ther.* **2012**, *19*, 579–587. [CrossRef] [PubMed]
21. Ryvarden, L.M.I. *Poroid Fungi of Europe*, 2nd ed.; Fungiflora: Oslo, Norway, 2014; Volume 31, p. 455.

22. Eisenman, H.C.; Casadevall, A. Synthesis and assembly of fungal melanin. *Appl. Microbiol. Biotechnol.* **2012**, *93*, 931–940. [CrossRef] [PubMed]
23. Xiao, B.; Wang, W.; Zhang, X.; Long, G.; Fan, J.; Chen, H.; Deng, L. A novel fractal solution for permeability and Kozeny-Carman constant of fibrous porous media made up of solid particles and porous fibers. *Powder Technol.* **2019**, *349*, 92–98. [CrossRef]
24. Gardner, D.J.; Oporto, G.S.; Mills, R.; Samir, M.A.S.A. Adhesion and Surface Issues in Cellulose and Nanocellulose. *J. Adhes. Sci. Technol.* **2008**, *22*, 545–567. [CrossRef]
25. Ling, S.; Chen, W.; Fan, Y.; Zheng, K.; Jin, K.; Yu, H.; Buehler, M.J.; Kaplan, D.L. Biopolymer nanofibrils: Structure, modeling, preparation, and applications. *Prog. Polym. Sci.* **2018**, *85*, 1–56. [CrossRef]
26. Aulin, C.; Gällstedt, M.; Lindström, T. Oxygen and oil barrier properties of microfibrillated cellulose films and coatings. *Cellulose* **2010**, *17*, 559–574. [CrossRef]
27. Priyadarshi, R.; Rhim, J.-W. Chitosan-based biodegradable functional films for food packaging applications. *Innov. Food Sci. Emerg. Technol.* **2020**, *62*, 102346. [CrossRef]
28. Jones, M.; Mautner, A.; Luenco, S.; Bismarck, A.; John, S. Engineered mycelium composite construction materials from fungal biorefineries: A critical review. *Mater. Des.* **2020**, *187*, 108397. [CrossRef]
29. Koubaa, A.; Koran, Z. Measure of the internal bond strength of paper/board. *TAPPI J.* **1995**, *78*, 103–111.
30. Naik, K.; Kowshik, M. The silver lining: Towards the responsible and limited usage of silver. *J. Appl. Microbiol.* **2017**, *123*, 1068–1087. [CrossRef] [PubMed]
31. Shirasaki, N.; Matsushita, T.; Matsui, Y.; Murai, K. Assessment of the efficacy of membrane filtration processes to remove human enteric viruses and the suitability of bacteriophages and a plant virus as surrogates for those viruses. *Water Res.* **2017**, *115*, 29–39. [CrossRef]
32. Centers for Disease Control and Prevention, NIOSH. Use of Respirators and Surgical Masks for Protection Against Healthcare Hazards. Available online: https://www.cdc.gov/niosh/topics/healthcarehsps/respiratory.html (accessed on 15 March 2021).
33. Martinez, M.J.A.; Olmo, L.M.B.D.; Benito, P.B. Antiviral Activities of Polysaccharides from Natural Sources. *Stud. Nat. Prod. Chem.* **2005**, *30*, 393–418. [CrossRef]
34. Shamshina, J.L.; Berton, P.; Rogers, R.D. Advances in Functional Chitin Materials: A Review. *ACS Sustain. Chem. Eng.* **2019**, *7*, 6444–6457. [CrossRef]

polymers

MDPI

Article

Upgrading Argan Shell Wastes in Wood Plastic Composites with Biobased Polyethylene Matrix and Different Compatibilizers

Maria Jorda-Reolid [1], Jaume Gomez-Caturla [2], Juan Ivorra-Martinez [2,*], Pablo Marcelo Stefani [3], Sandra Rojas-Lema [1] and Luis Quiles-Carrillo [1,*]

1 Departamento de Materiales y Tecnologías, Asociación de Investigación de la Industria del Juguete, Conexas y Afines (AIJU), Av. de la Industria, 23, 03440 Ibi, Spain; mariajorda@aiju.es (M.J.-R.); sanrole@epsa.upv.es (S.R.-L.)
2 Instituto de Tecnología de Materiales (ITM), Universitat Politècnica de València (UPV), Plaza Ferrándiz y Carbonell 1, 03801 Alcoy, Spain; jaugoca@epsa.upv.es
3 Instituto de Investigaciones en Ciencia y Tecnología de Materiales (INTEMA), Consejo Nacional de Investigaciones Científicas y Técnicas (CONICET), Universidad Nacional de Mar del Plata (UNMdP), Av. Colón 10850, Mar del Plata 7600, Argentina; pmstefan@fi.mdp.edu.ar
* Correspondence: juaivmar@doctor.upv.es (J.I.-M.); luiquic1@epsa.upv.es (L.Q.-C.)

Abstract: The present study reports on the development of wood plastic composites (WPC) based on micronized argan shell (MAS) as a filler and high-density polyethylene obtained from sugarcane (Bio-HDPE), following the principles proposed by the circular economy in which the aim is to achieve zero waste by the introduction of residues of argan as a filler. The blends were prepared by extrusion and injection molding processes. In order to improve compatibility between the argan particles and the green polyolefin, different compatibilizers and additional filler were used, namely polyethylene grafted maleic anhydride (PE-g-MA 3 wt.-%), maleinized linseed oil (MLO 7.5 phr), halloysite nanotubes (HNTs 7.5 phr), and a combination of MLO and HNTs (3.75 phr each). The mechanical, morphological, thermal, thermomechanical, colorimetric, and wettability properties of each blend were analyzed. The results show that MAS acts as a reinforcing filler, increasing the stiffness of the Bio-HDPE, and that HNTs further increases this reinforcing effect. MLO and PE-g-MA, altogether with HNTs, improve the compatibility between MAS and Bio-HDPE, particularly due to bonds formed between oxygen-based groups present in each compound. Thermal stability was also improved provided by the addition of MAS and HNTs. All in all, reddish-like brown wood plastic composites with improved stiffness, good thermal stability, enhanced compatibility, and good wettability properties were obtained.

Keywords: argan shell particles; wood plastic composite; polyethylene; mechanical properties; compatibilization

Citation: Jorda-Reolid, M.; Gomez-Caturla, J.; Ivorra-Martinez, J.; Stefani, P.M.; Rojas-Lema, S.; Quiles-Carrillo, L. Upgrading Argan Shell Wastes in Wood Plastic Composites with Biobased Polyethylene Matrix and Different Compatibilizers. *Polymers* **2021**, *13*, 922. https://doi.org/10.3390/polym13060922

Academic Editor: Gianluca Tondi

Received: 27 February 2021
Accepted: 15 March 2021
Published: 17 March 2021

1. Introduction

In the last decades, an increasing concern about environmental issues related to the great use of petrochemical-derived plastics has been rising, as well as the necessity to reduce the carbon footprint associated with the production processes of those polymers. This trend is driving the industry toward the use of polymeric materials derived from natural resources, with the objective of moving away from petroleum. Moreover, this fact has propitiated the use of biodegradable polymers, some of which are capable of decomposing in composting conditions. At the same time, they exhibit very similar properties to those of their petrochemical counterparts [1]. One of the main drawbacks of plastics derived from natural sources is that they are more expensive than those from petrochemical origin. This leads to a direct rejection by industry. For this reason, incorporation of fillers and creation of new more economic materials is gaining increasing relevance.

Among those materials, "Wood plastic composites" (WPC) are becoming increasingly popular. This technology implies the addition of natural organic fillers obtained from wastes into polymer matrices, with the objective of decreasing their processing cost. At the same time, those fillers enhance their environmental value and vary their mechanical, physical and chemical properties [2]. Initially, only wood flour or sawdust were used as bio-fillers [3–6], but over time fillers from agroforestry waste or food industry have started to be used, such as orange peel [7,8], almond shell [9], pomegranate peel [10], or argan seed shell [11]. Those fillers are introduced in the structure of polymers like poly (lactic acid) (PLA), high density polyethylene (HDPE) or polyester [12–14]. Nowadays, WPC can be employed as a substitute of wood due to its appearance for the fabrication of indoor and outdoor furniture, decks pergolas, and so on [15,16].

In this sense, high density bio-polyethylene (Bio-HDPE) is one of the most interesting polymer matrices, as it combines its natural origin with the ease of processability of polyolefins. This polymer is produced by conventional polymerization of ethylene obtained from catalytic dehydration of bio-ethanol [17], which is extracted from natural sources, such as sugarcane [18]. Bio-HDPE possesses the same physical properties than its fossil-derived counterpart; particularly, good mechanical resistance and high ductility [19]. Injection-molded Bio-HDPE pieces have great application in packaging [20]. However, one of the main problems that limits the use of bio-fillers as reinforcing agents is their hydrophilic nature, which causes a poor interfacial adhesion with a hydrophobic (non-polar) matrix, such as Bio-HDPE. This provokes a deterioration in the mechanical properties of the material and a bad compatibility between filler and matrix [21,22].

Several studies have been carried out to improve the compatibilization between those fillers and polymeric matrices. On the one hand, some of the methods that have been applied to improve this compatibility include the modification of the surface of the filler particles from the waste, such as mercerization, benzoylation, acetylation, silanation, or graft copolymerization [23,24]. On the other hand, compatibilizing agents have also been used in order to improve the adhesion and affinity between filler and matrix. Additionally, they have also been utilized to increase the dispersion of the reinforcing particles within the matrix and give additional properties to the blend [25–27]. In this context, polyethylene-grafted maleic anhydride (PE-g-MA) is one of the most efficient compatibilizers. It acts as a chemical connection between ligno-cellulosic particles and polymeric chains, owing to its double functionality. First, the polyethylene fraction of PE-g-MA interacts with polymeric chains due to its chemical affinity, while the anhydride groups can react with the hydroxyl groups from the ligno-cellulosic structure in the organic particles by esterification. Thus, leading to an increase in the matrix–filler interaction and a positive effect in the dispersion of the particles and the stiffness of the blend [28–30]. However, ductile properties tend not to get better, because of the fragility given by those fillers when they form stress-concentrator aggregates in the matrix.

In recent years, vegetable oils have generated a great deal of interest as compatibilizing additives. Modified vegetable oils have been used as plasticizers, stabilizers, crosslinkers, compatibilizers, and so on. The use of epoxidized vegetable oils (EVOS) should be remarked, such as epoxidized soybean oil (ESBO) [31,32], epoxidized linseed oil (ELO) [33,34], or epoxidized palm oil (EPO) [35]. All of them have been tested in polymeric blends and composites, achieving positive compatibilization/plasticization effects, owing to the reactivity of the oxirane group. Maleinization is an interesting alternative chemical modification for vegetable oils. Maleinized linseed oil (MLO) has been widely used in WPC, for example, in a PBS matrix with almond shell flour [36]. Furthermore, the excellent ductile properties that this oil is capable of giving to composites have also been reported [13]. In addition, inorganic particles can also be introduced as reinforcing agents in polymeric matrices, such as carbon nanotubes (CNTs), nanoclays, or metallic oxide nanoparticles [37–39]. In recent years, halloysite nanotubes (HNTs) have gained great importance in this sense. They are cheaper and more abundant than, for example, carbon nanotubes (CNTs). In this sense, halloysite is a natural mineral clay with nanotube morphology, similar to kaolinite ($Al_2[Si_2O_5(OH)_4].nH_2O$), where n is 2 for hydrated HNTs, and 0 for dehydrated HNTs [40]. Frost et al. [41] reported two types of hydroxyl groups, inner and outer hydroxyl groups, placed between the nanotubes layers and in the surface of HNTs [41]. Because of their multilayer structure, most of the hydroxyl groups are in the inner side, while a few of them are in the surface of HNTs. As a result of this, HNTs dispersion in polymer matrices is simpler than other inorganic nanoparticles [40]. Numerous studies have reported the ability of HNTs to improve thermal stability, fire resistance, mechanical properties, and crystalline behavior of polymers such as polypropylene [42–46], low density polyethylene (LLDPE) [47], polyamide 6 (PA6) [48], epoxy resin [49], polylactide (PLA) [50,51], and so on. These are the reasons for utilizing HNTs in this work. As it is the case with organic particles, the challenge here is to achieve a good compatibility and adhesion between HNTs and the polymer matrix, especially in non-polar matrices, such as polyolefins.

The objective of this work is the reuse of micronized argan shell (MAS) as a reinforcing organic filler in a Bio-HDPE matrix derived from sugarcane, with the objective of developing a wood plastic composite with enhanced properties compared to the original matrix following the trends proposed by the circular economy, which seeks to reuse or recycling the waste and to reintroduce them into the industry [52]. Argan (*Argania spinosa*) is a tropical plant that belongs to the *Sapotaceae* family, whose fruit is used in morocco to prepare oil [53]. This oil has great application in cosmetics owing to its high E vitamin content [53], while fruit residues and seeds are generally used to feed cattle [54]. The main drawback of micronized argan shell (MAS) is its low compatibility with Bio-HDPE. In particular, the challenge of this investigation is to improve the affinity between both elements through the use of compatibilizing agents. Several formulations have been developed to meet this end, using the compatibilizers PE-g-MA and MLO; and halloysite nanotubes (HNTs) as additional reinforcing fillers; altogether with Bio-HDPE as the polymer matrix and argan shell particles as the main reinforcing filler; these elements have been used both individually and in combination. Additionally, MLO and HNTs concentrations have been varied (7.5 parts per hundred resin (phr) when used individually and 3.75 phr when used in combination). In order to evaluate the effects of these compatibilizers and fillers over each blend, mechanical, morphological, thermal, thermomechanical, colorimetric, and wettability properties are presented.

2. Materials and Methods

2.1. Materials

Bio-HDPE, SHA7260 grade, was supplied in pellets form by FKuR Kunststoff GmbH (Willich, Germany) and manufactured by Braskem (São Paulo, Brazil). This green polyethylene has a density of 0.955 g cm^{-3} and a melt flow index (MFI) of 20 g/10 min, measured with a load of 2.16 kg and a temperature of 190 °C. Micronized argan shell (MAS) was supplied by MICRONIZADOS VEGETALES S.L. (Benamejí (Córdoba), Spain) company. Halloysite nanotubes were supplied by Sigma Aldrich (Madrid, Spain) with CAS number 1332-58-7.

Interlocking is one of the most important mechanisms in polymer composites. This mechanism depends on the filler shape. Figure 1a shows the morphology of the argan particles observed by SEM. Most of the particles exhibit a rough surface, which can be ascribed to the milling process due to the high hardness of this filler. A closer observation of the particles shows some porosity on their surface, which can produce a good interaction with the polymer matrix acting as mechanical anchoring points. A similar structure was observed by Laaziz et al. [55] when studying argan nut shell particles with different treatments. Additionally, Figure 1b and c show histograms for the length and diameter of the particles, respectively, determined from the SEM image. Particles were average 70 µm long and 45 µm wide. These parameters are also important to consider when talking about mechanical properties. Too large particles could lead to a great mechanical impairment, increasing the heterogeneity of the blends. Crespo et al. [56] observed this effect for almond shell particles superior to 150 μm diameter.

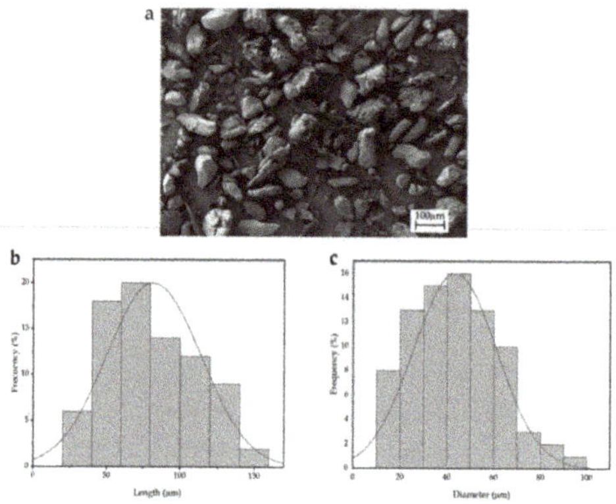

Figure 1. (**a**) Scanning electron microscope image (SEM) of micronized argan shell (MAS). Image was taken with a magnification of 100× and a scale marker of 100 μm; (**b**) histogram of the argan particles length; (**c**) histogram of the argan particles diameter.

In relation to the additives, polyethylene-grafted maleic anhydride (PE-g-MA) with CAS Number 9006-26-2 and MFI values of 5 g/10 min (190 °C/2.16 kg), was obtained from Sigma-Aldrich S.A. (Madrid, Spain). This PE-based copolymer was selected due to its functionality. The maleinized linseed oil—MLO, VEOMER LIN was supplied by Vandeputte (Mouscron, Belgium). This modified vegetable oil is characterized by a viscosity of 10 dPa s at 20 °C and an acid value comprised in the 105–130 mg KOH g^{-1} range.

2.2. Preparation of Bio-HDPE Blends

Bio-HDPE, PE-g-MA, MAS y HNTs were initially dried at 40 °C for 48 h in a dehumidifying dryer MDEO (Barcelona, Spain) to remove any residual moisture prior to processing to avoid the possibility of hydrolysis due to the moisture. Then, the corresponding wt.% of each component (see Table 1) were mixed and pre-homogenized in a zipper bag. The corresponding formulations were compounded in a twin-screw co-rotating extruder from Construcciones Mecánicas Dupra, S.L. (Alicante, Spain). This extruder has a 25 mm diameter with a length-to-diameter ratio (L/D) of 24 to allow a correct blending process. The extrusion process was carried out at a rate of 22 rpm, using the following temperature profile (from the hopper to the die): 140–145–150–155 °C. The compounded materials were pelletized using an air-knife unit. In all cases, residence time was approximately 1 min. Table 1 shows the compositions of the formulations developed in this work.

To transform the pellets into standard samples, a Meteor 270/75 injector from Mateu & Solé (Barcelona, Spain) was used. The temperature profile in the injection molding unit was 135 °C (hopper), 140 °C, 150 °C, and 160 °C (injection nozzle). A clamping force of 75 tons was applied while the cavity filling and cooling times were set to 1 and 10 s, respectively. Standard samples for mechanical and thermal characterization with an average thickness of 4 mm were obtained.

Table 1. Summary of compositions according to the weight content (wt.%) of Bio-HDPE/MAS and different compatibilizers and additives.

Code	BIO-HDPE (wt.%)	PE-g-MA (wt.%)	MAS (wt.%)	HNTs (phr)	MLO (phr)
Bio-HDPE	100	0	0	0	0
Bio-HDPE /PE-g-MA/MAS	67	3	30	0	0
Bio-HDPE /PE-g-MA/MAS/HNT	67	3	30	7.5	0
Bio-HDPE /PE-g-MA/MAS/MLO	67	3	30	0	7.5
Bio-HDPE /PE-g-MA/MAS/HNT/MLO	67	3	30	3.75	3.75

2.3. *Characterization of Bio-HDPE Blends*

2.3.1. Mechanical Characterization

Mechanical properties were obtained with different tests like tensile test, shore hardness, and Charpy impact test. Tensile properties of Bio-HDPE/PE-g-MA/MAS blends were obtained in a universal testing machine ELIB 50 from S.A.E. Ibertest (Madrid, Spain) as recommended by ISO 527-1:2012 with dog bone samples 1B specification. A 5-kN load cell was used and the cross-head speed was set to 5 mm/min. Shore hardness was measured in a 676-D durometer from J. Bot Instruments (Barcelona, Spain), using the D-scale, on rectangular samples with dimensions 80 × 10 × 4 mm^3, according to ISO 868:2003. The impact strength was also studied on injection-molded rectangular samples with dimensions of 80 × 10 × 4 mm^3 in a Charpy pendulum (1-J) from Metrotec S.A. (San Sebastián, Spain) on notched samples (0.25 mm radius V-notch), following the specifications of ISO 179-1:2010. All mechanical tests were performed at room temperature, and at least six samples of each material were tested and the corresponding values were averaged.

2.3.2. Morphology Characterization

The morphology of fractured samples from Charpy tests, obtained from the impact tests, were studied by field emission scanning electron microscopy (FESEM) in a ZEISS ULTRA 55 microscope from Oxford Instruments (Abingdon, UK). Before placing the samples in the vacuum chamber, they were sputtered with a gold-palladium alloy in an EMITECH sputter coating SC7620 model from Quorum Technologies, Ltd. (East Sussex, UK). The FESEM was operated at an acceleration voltage of 2 kV.

2.3.3. Thermal Analysis

The most relevant thermal transitions of Bio-HDPE/PE-g-MA/MAS blends were obtained by differential scanning calorimetry (DSC) in a Mettler-Toledo 821 calorimeter (Schwerzenbach, Switzerland). Samples with an average weight of 6–7 mg, were subjected to a thermal program divided into three stages: a first heating from 25 °C to 160 °C followed by a cooling to 0 °C, and a second heating to 250 °C. Both heating and cooling rates were set to 10 °C/min. All tests were run in nitrogen atmosphere with a flow rate of 66 mL/min using standard sealed aluminum crucibles with a capacity of 40 µL.

The thermal degradation of the Bio-HDPE/PE-g-MA/MAS blends was assessed by thermogravimetric analysis (TGA). TGA tests were performed in a LINSEIS TGA 1000 (Selb, Germany). Samples with a weight of 15–17 mg were placed in 70 µL alumina crucibles and subjected to a dynamic heating program from 40 °C to 700 °C at a heating rate of 10 °C/min in air atmosphere. The first derivative thermogravimetric (DTG) curves were also determined. All tests were carried out at least three times in order to obtain reliable results.

2.3.4. Thermomechanical Properties

In order to obtain the thermomechanical properties of the Bio-HDPE composites a dynamical mechanical thermal analysis (DMTA) was carried out in a DMA1 dynamic analyzer from Mettler-Toledo (Schwerzenbach, Switzerland), working in single cantilever flexural conditions. Rectangular samples with dimensions 20 × 6 × 2.7 mm^3 were subjected to a dynamic temperature sweep from −150 °C to 120 °C at a constant heating rate of 2 °C/min. The selected frequency was 1 Hz and the maximum flexural deformation or cantilever deflection was set to 10 μm.

2.3.5. Color and Wetting Characterization

The colorimetric modifications of the samples were measured with a Konica CM-3600d Colorflex-DIFF2, from Hunter Associates Laboratory, Inc. (Reston, VA, USA) was used. Color coordinates (L*a*b*) were measured according to the following criteria: L* = 0, darkness; L* = 100, lightness; a* represents the green (a* < 0) to red (a* > 0); b* stands for the blue (b* < 0) to yellow (b* > 0) coordinate.

Hydrophilicity/ hydrophobicity of each blend was assessed by contact angle measurements through time were carried out with an EasyDrop Standard goniometer model FM140 (KRÜSS GmbH, Hamburg, Deutschland) which is equipped with a video capture kit and analysis software (Drop Shape Analysis SW21; DSA1). Double distilled water was used as test liquid, a drop of it was put in each sample and contact angle measurements were taken at 0, 5, 10, 15, 20, and 30 min after the administration of the water. For color and wetting measurement, traction samples were used to facilitate the measures.

2.3.6. Water Absorption Test

The water absorption capacity of the Bio-HDPE/PE-g-MA/MAS blends was evaluated by the water uptake method. Impact specimens (80 × 10 × 4 mm^3) from each blend were first weighted in a balance and then put inside a beaker filled with distilled water, all of them wrapped with tiny pieces of a metal grid so they could sink. After that, the weight of all samples was measured in intervals of several hours the first day, and then measured each week for 14 weeks in order to evaluate the amount of taken up water. In every measurement, the moisture in the surface of the samples was removed with tissue paper. During the immersion time, the temperature was controlled to 23 °C.

2.3.7. Infrared Spectroscopy

In order to obtain the interactions between the different elements, a chemical analysis of the Bio-HDPE/PE-g-MA/MAS blends was carried out by attenuated total reflection-Fourier transform infrared (ATR-FTIR) spectroscopy. Spectra were recorded using a Bruker S.A Vector 22 (Madrid, Spain) coupled to a PIKE MIRacleTM single reflection diamond ATR accessory (Madison, WI, USA). Data were collected as the average of 10 scans between 4000 and 500 cm^{-1} with a spectral resolution of 2 cm^{-1}. Samples from impact specimens (80 × 10 × 4 mm^3) were employed in this test and the measurement was performed at room temperature.

2.3.8. Statistical Analysis

The significant differences among the samples were evaluated at 95% confidence level ($p \leq 0.05$) by one-way analysis of variance (ANOVA) following Tukey's test. Software employed for this propose was the open source R software V4.0.3 (http://www.r-project.org (accessed on 12 March 2021))

3. Results

3.1. Mechanical Properties

The results concerning the mechanical properties of Bio-HDPE/MAS blends with different compatibilizers are shown in Table 2. Those results are of great interest to evaluate the effectivity of the different agents used in terms of resistance improvement of Bio-HDPE and compatibility between Bio-HDPE and MAS.

Table 2. Summary of mechanical properties of the injection-molded samples of Bio-HDPE blends. Tensile modulus (E), maximum tensile strength (σ_{max}) elongation at break (ε_b), Shore D hardness, and impact (Charpy) strength.

Code	E (MPa)	σ_{max} (MPa)	ε_b (%)	Shore D Hardness	Impact Strength (kJ/m^2)
Bio-HDPE	750 ± 47 [a]	14.48 ± 0.78 [a]	Nb [a]	56.2 ± 1.3 [a]	2.7 ± 0.2 [a]
Bio-HDPE/PE-g-MA/MAS	846 ± 36 [b]	7.57 ± 0.81 [b]	20.7 ± 2.0 [b]	59.2 ± 0.8 [a]	1.4 ± 0.1 [b]
Bio-HDPE/PE-g-MA/MAS/HNTs	1126 ± 65 [c]	7.57 ± 0.90 [b]	6.0 ± 0.9 [c]	60.6 ± 1.3 [a]	1.7 ± 0.1 [c]
Bio-HDPE/PE-g-MA/MAS/MLO	442 ± 33 [d]	6.66 ± 0.39 [c]	41.5 ± 1.7 [d]	53.2 ± 0.8 [b]	2.2 ± 0.3 [d]
Bio-HDPE/PE-g-MA/MAS/HNTs/MLO	523 ± 26 [e]	6.98 ± 0.59 [c]	33.9 ± 3.5 [e]	54.6 ± 0.5 [b]	2.1 ± 0.3 [d]

[a–e] Different letters in the same column indicate a significant difference among the samples ($p < 0.05$).

As it can be observed in Figure 2, neat Bio-HDPE presents a Young modulus (E) and tensile strength of 750 and 14.48 MPa, respectively. Elongation at break (ε_b) could not be determined because the tensile test machine reached its maximum elongation without breaking the sample. Those values are indicative of certain stiffness and high ductility for Bio-HDPE, as it has been also reported by other authors [57]. Incorporation of MAS (30 wt.%) with PE-g-MA (3 wt.%) into the Bio-HDPE matrix increases the Young modulus up to 846 MPa, in this case the statical analysis showed a significant difference by the incorporation of the filler. This can be directly related to a proper distribution of MAS particles in the polymer matrix, leading to a good adhesion of the filler with the polymer and obtaining a more rigid material [26,27]. However, there is a clear reduction in tensile strength and elongation at break (Figure 2), this can be related to an excess of MAS utilized, increasing the stiffness of the material. In this context Essabir et al. [11] observed a similar behavior when treating HDPE with argan particles, in a way that Young modulus increased with MAS concentration, but tensile strength started to decrease with a MAS particle content superior to 5%. Addition of HNTs in the blend showed an additional enhancement of stiffness; the incorporation of only 7.5 phr HNTs produced a significant difference against the previous sample and a 33% of improvement. In this context, Bio-HDPE/PE-g-MA/MAS/HNTs sample presented the highest Young modulus of all blends (1126 MPa). This can be attributed to the reinforcing effect of HNTs, which seem to be well-dispersed within the matrix owing to the presence of the compatibilizer PE-g-MA. Similar behavior was reported by Pratap et. al. [58]. On the other hand, HNTs do not produce changes in tensile strength (no significant differences could be appreciated between them) and reduce elongation at break. Therefore, the material becomes less ductile, as a result of the combination of MAS and HNTs reinforcing effects. MLO-compatibilized blends saw their Young modulus and tensile strength reduced in comparison with the rest of the blends. Nevertheless, elongation at break increased 100% referring to the Bio-HDPE/PE-g-MA/MAS blend and the Tukey test showed a significant difference between the samples. In this sense, the plasticizing effect of MLO can be verified, highly increasing the ductile properties and compatibility between components in the blend. Similar results were observed by Quiles-Carrillo et. al. [59] when treating a PA1010/Bio-HDPE blend with MLO. As to Bio-HDPE/PE-g-MA/MAS/HNTs/MLO blend, it shows intermediate values between the sample with HNTs and the sample with MLO individually. Although the plasticizing effect of MLO seems to prevail over the reinforcing effect of HNTs, due to the very high elongation at break and the relatively low Young modulus and tensile strength.

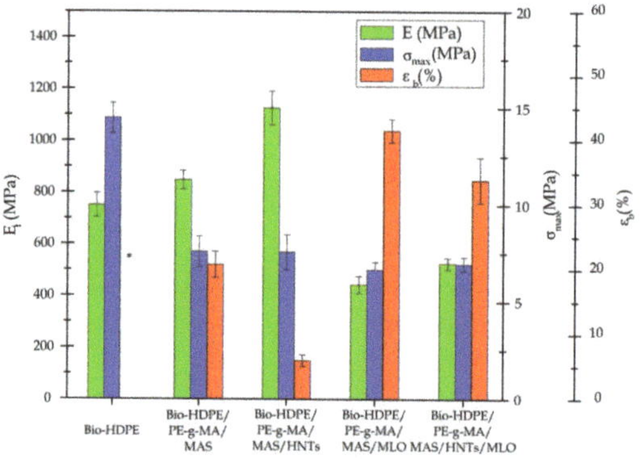

Figure 2. Mechanical properties of the injection-molded samples of Bio-HDPE blends. Tensile modulus (E), maximum tensile strength (σ_{max}) elongation at break (ε_b). * Elongation at break for the Bio-HDPE could not be assessed because breakage is not achieved.

Regarding the impact strength results, neat Bio-HDPE presents a value of 2.7 kJ/m^2, which is a ductile behavior indicator. Bio-HDPE/PE-g-MA/MAS blend shows a reduction of impact strength down to 1.4 kJ/m^2, confirming the great stiffness of the blend, probably due to the quantity of MAS utilized [11]. The presence of HNTs in the polymer matrix increases the impact strength to 0.3 kJ/m^2, which means a significant difference between the mentioned samples. This effect is probably associated with an interaction between argan particles and HNTs. Moreover, it is possible that HNTs presence leads to a good dispersion of the argan particles, as well as HNTs themselves, boosting the reinforcing effect of the fillers [58]. As it was expected, MLO-compatibilized system shows a great impact strength due to the plasticizing effect of MLO. Several studies have reported the effectivity of MLO as a compatibilizing agent to increase the impact strength of fragile materials, such as PLA or PLA/almond shell flour blends [33]. Impact strength of HNTs/MLO combinated system is very similar to that of MLO blend, remarking the plasticizing effect of this oil. In this sense, the addition of HNTs to the composites with MLO did not provide a significant difference.

With regard to Shore D hardness, a similar trend to that observed in Young modulus (E) can be observed. HNTs sample shows the highest hardness (60.6), giving evidence of the reinforcing effect of nanotubes and MAS. A very similar value is displayed by the ternary blend Bio-HDPE/PE-g-MA/MAS (59.2), reflecting the hardening effect of MAS, composites without MLO did not show significant differences between them. As expected, MLO samples exhibit the lowest hardness (53.2), due to the great ductility this component bestows upon the blend. Ferri et. al. reported this same effect in PLA/TPS blends with different proportions of MLO [60]. The combination of HNTs and MLO leads to a slightly superior hardness to that of the MLO sample, this provoked by the hardening effect of HNTs and the reduction in MLO concentration (going from 7.5 phr to 3.75 phr). The ANOVA analysis showed that MLO had a clear effect on the hardness.

3.2. Morphology of Bio-HDPE/MAS Blends

Internal morphology of these materials is directly related to their mechanical properties. Figure 3 shows the field emission scanning electron microscopy (FESEM) images at 1000× of argan particles and the surface of fractured impact samples of each one of the blends. First, Figure 3a shows the morphology of an argan particle, which presents a rough surface, emphasized by the presence of holes on its structure (arrow). Those holes play an important role in the adhesion of the particles into the matrix according to Crespo et al. [56,61], who reported a very similar morphology in almond shell flour particles. Figure 3b corresponds to neat Bio-HDPE, which exhibits the typical irregular, rough, and cavernous surface of a ductile polymer, as it was also reported by Quiles-Carrillo et. al. [57].

Regarding the incorporation of argan particles, they show great cohesion within the matrix, as it is shown in Figure 3c–f. The gap between the perimeter of the particles and the polymer matrix is really narrow (arrows), which implies good interaction between both elements. This behavior can be well-related to the action of PE-g-MA as a compatibilizer and justifies the increase in stiffness, in terms of elastic modulus, of the samples that do not have MLO in their structure. Samples in Figure 3e,f show wire-shaped long structures in the polymer matrix (circles). They indicate a more ductile fracture and they confirm the plasticizing effect of MLO. This is directly related to the high elongation at break displayed in mechanical properties. Moreover, argan particles seem to be more immersed in the matrix in samples with MLO, revealing a contribution of the oil to compatibilization. This was also observed by Quiles-Carrillo et. al. [13], when adding MLO to PLA and almond shell flour blends. Additionally, with the presence of HNTs and MLO in the structure, a positive decrease in the number of gaps inside the matrix can be observed. Those gaps are originated by the fall of argan particles after fracture, thus being indicative of a worse interaction between MAS and Bio-HDPE.

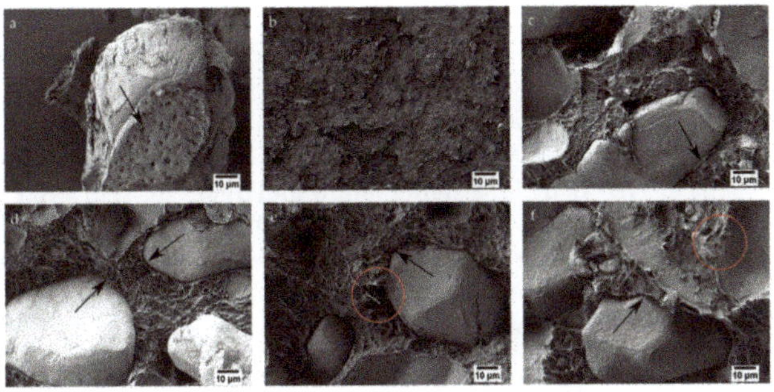

Figure 3. Field emission scanning electron microscopy (FESEM) images at 1000× of the fractured surfaces of: (**a**) MAS; (**b**) neat Bio-HDPE; (**c**) Bio-HDPE/PE-g-MA/MAS; (**d**) Bio-HDPE/PE-g-MA/MAS/HNTs; (**e**) Bio-HDPE/PE-g-MA/MAS/MLO; (**f**) Bio-HDPE/PE-g-MA/MAS/HNTs/MLO.

Presence of HNTs in the blends can be observed in argan particles in Figure 3d,f. Figure 4a,b better illustrates them, where 5000× FESEM images of Bio-HDPE/PE-g-MA/MAS/HNTs and Bio-HDPE/PE-g-MA/MAS/HNTs/MLO samples are shown, respectively. In this context, a correct distribution of nanotubes can be observed in the surface of argan particles, which are indicated in the images by circles. This behavior could be attributed to a synergy between hydroxyl groups of HNTs and hydroxyl and anhydride groups present in PE-g-MA and lignocellulosic compounds of MAS [62]. As expected, nanotubes density is superior in the 7.5 phr HNTs sample (Figure 4a), where nanotubes cover almost all the argan particle surface. While in the 3.75 phr HNTs sample (Figure 4b), nanotubes are disposed in several agglomerates in the surface of MAS. This fact could be responsible for the low increase in resistance in 3.75 phr HNTs sample, altogether with the plasticizing effect of MLO.

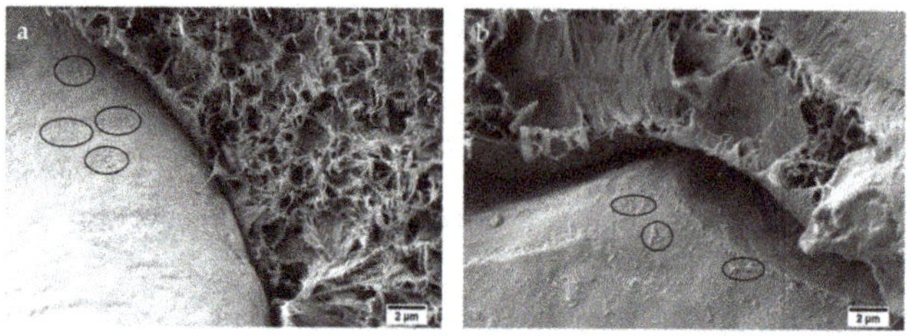

Figure 4. Field emission scanning electron microscopy (FESEM) images at 5000× of the fractured surfaces of samples compatibilized with HNTs: (**a**) Bio-HDPE/PE-g-MA/MAS/HNTs (7.5 phr HNTs); (**b**) Bio-HDPE/PE-g-MA/MAS/HNTs/MLO (3.75 phr HNTs).

3.3. Thermal Properties of Bio-HDPE/MAS Blends

Figure 5 gathers the DSC thermograms corresponding to the second heating cycle of the studied samples; Table 3 gathers melting temperature (T_m) and crystallinity (X_c) values of every Bio-HDPE blend. Regarding thermograms, in this case only melting temperature can be observed, because glass transition temperature is really low for Bio-HDPE (−100 °C) [63]. In the case of neat Bio-HDPE, it shows a melting temperature of 131 °C and a crystallinity X_C of 66.3%. Similar values were reported by Quiles-Carrillo et al. [64]. After the incorporation of argan particles alongside PE-g-MA into the Bio-HDPE, melting temperature undergoes a slight decrease down to 130.4 °C. On the other hand, crystallinity highly decreases to 49.7%. This could be due to the fact that, although argan particles can induce heterogeneous nucleation over the crystallization process of the polymer, high MAS concentrations (over 10% MAS) can compromise nucleation because of the particle–particle contact. Therefore, a limitation in space for crystal formation and growth is produced. This phenomenon was observed by Essabir et al. [65]. HNTs increase T_m up to 133.1 °C. This effect was also observed by Erding et al. [66] after studying the melting temperature of polyethylene and halloysite nanotubes composites (PE/HNTs). This can be ascribed to the fact that HNTs have some insulating properties against heat transfer, which is common in clay-based materials. Crystallinity suffers a great decrease in this case (37.6%), probably due to an excess of nucleating agents in the blend. This excess is originated by the combined presence of HNTs and argan particles. In this context, several studies have demonstrated the ability of HNTs to favor crystallization of polymer matrices, as long as there is no overload (+20 phr), to avoid the formation of aggregates that hinder the crystal nucleation and growth process [67]. Regarding MLO, it does not vary T_m, but reduces, along with argan, crystallinity down to 39.8%. MLO improves compatibilization and, therefore, argan particle dispersion in the polymer matrix, thus contributing to reducing polymer–polymer interactions and favoring crystal formation [68]. However, in this case there are excess of argan particles, as it was aforementioned, that diminishes the crystallinity of the matrix. HNTs, MAS, and MLO sample hardly presents any change referring to the melting temperature and it reduces crystallinity in a very similar manner to that of MLO sample, due to obstruction of heterogeneous nucleation, as it has been previously explained. Regarding the statistical analysis of the variance, the melting temperatures did not showed modifications. A different trend could be observed when talking about the melting enthalpy. The first significant difference could be observed when the filler was introduced and as a result a dilution effect was produced on the blends. Depending on the compatibilization strategy of each material, the degree of crystallinity showed minor modifications as commented above.

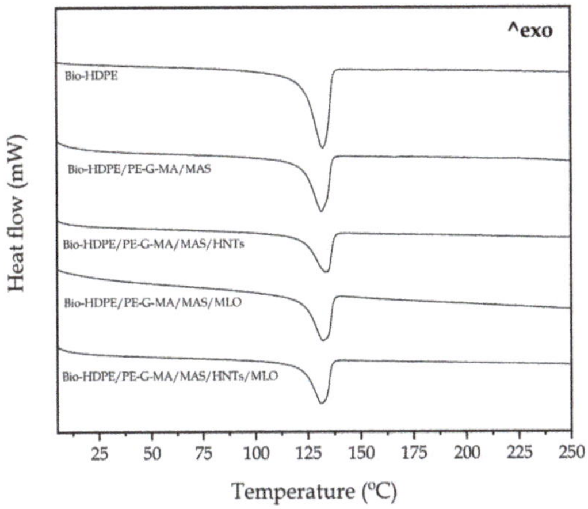

Figure 5. Differential scanning calorimetry (DSC) thermograms of Bio-HDPE/PE-g-MA/MAS blends with different compatibilizers.

Table 3. Melting temperature (T_m), melting enthalpy (ΔH_m) and crystallinity (X_C) of Bio-HDPE/PE-g-MA/MAS blends with different compatibilizers, obtained by differential scanning calorimetry (DSC).

Code	T_m (°C)	ΔH_m ($J \cdot g^{-1}$)	X_C (%)
Bio-HDPE	131.0 ± 0.5 [a]	194.2 ± 1.5 [a]	66.3 ± 0.9 [a]
Bio-HDPE/PE-g-MA/MAS	130.4 ± 0.3 [a]	145.7 ± 1.2 [b]	49.7 ± 0.7 [b]
Bio-HDPE/PE-g-MA/MAS/HNTs	133.1 ± 0.4 [a]	110.3 ± 1.0 [c]	37.6 ± 0.6 [c]
Bio-HDPE/PE-g-MA/MAS/MLO	131.2 ± 0.3 [a]	116.7 ± 1.2 [d]	39.8 ± 0.9 [c]
Bio-HDPE/PE-g-MA/MAS/HNT/MLO	130.4 ± 0.2 [a]	118.8 ± 2.0 [d]	40.5 ± 0.7 [c]

[a–d] Different letters in the same column indicate a significant difference among the samples ($p < 0.05$).

Thermal stability has also been studied. Figure 6 shows both TGA curves and their first derivative (DTG), while Table 4 gathers the main quantitative parameters related to thermal degradation. Neat Bio-HDPE presented a temperature of approximately 345 °C for a mass loss of 5 wt.-% ($T_{5\%}$), while its maximum degradation rate temperature was (T_{deg}) 447 °C. Additionally, after a single-phase degradation, its residual mass at 700 °C was 0.3%. Montanes et. al. [69] observed a similar thermal profile for neat Bio-HDPE. Addition of argan fillers and compatibilizers seem to have decreased the initial degradation temperature to values between 275 and 290 °C, the Tukey analysis showed a significant difference in these cases. This is mainly attributed to the presence of MAS in the polymer structure, which reduces in certain manner its thermal stability. This was also reported by Essabir et. al. [26] when they studied the properties of polypropylene and almond shell flour blends. Interestingly, it can be seen as all other samples undergo several degradation stages. Particularly, Bio-HDPE/PE-g-MA/MAS sample shows a three-phase degradation, attributed to the hemicellulose and pectin degradations (280–340 °C); cellulose degradation (340–448 °C) and lignin degradation (448–477 °C) which are present in argan particles [26,70]. These different phases appear in all samples except for neat Bio-HDPE. MLO samples seem to provide some more thermal stability to the blend, due to its capacity to link polymer chains (crosslinking), which is a positive effect toward stability. With regard to the maximum degradation rate temperature, it has increased in all cases except the MLO sample. This is highly related to the compatibilizing effect of PE-g-MA altogether with argan particles and HNTs [58], leading to a good particle dispersion and retarding the maximum degradation peak to 481.3 and 458.3 °C for Bio-HDPE/PE-g-

MA/MAS and Bio-HDPE/PE-g-MA/MAS/HNTs samples, respectively. Referring to HNTs, their tubular hollow structure, which is capable of holding degradation products inside, is responsible for slowing the mass transport mechanism, as it was reported by Pratap et al. [58].

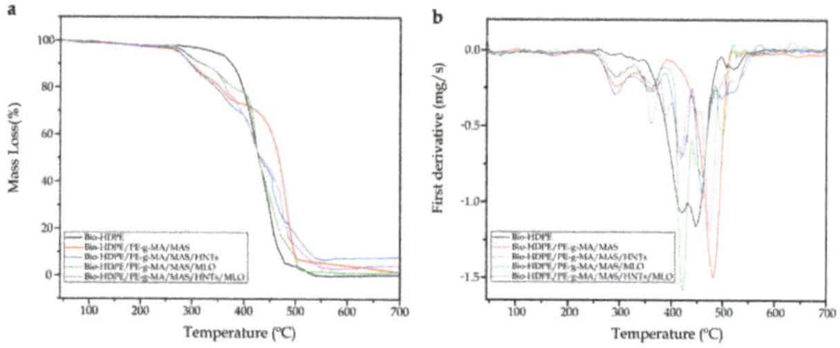

Figure 6. (**a**) Thermogravimetric analysis (TGA) curves and (**b**) first derivative (DTG) of Bio-HDPE/PE-g-MA/MAS blends with different compatibilizers.

Table 4. Main thermal degradation parameters of the Bio-HDPE/MAS blends with different compatibilizers in terms of the onset degradation temperature at a mass loss of 5 wt.% ($T_{5\%}$), maximum degradation rate (peak) temperature (T_{deg}), and residual mass at 700 °C.

Code	$T_{5\%}$ (°C)	T_{deg} (°C)	Residual Weight (%)
Bio-HDPE	342.8 ± 0.9 [a]	447.3 ± 2.2 [a]	0.3 ± 0.1 [a]
Bio-HDPE/PE-g-MA/MAS	275.8 ± 1.2 [b]	481.3 ± 1.7 [b]	1.5 ± 0.1 [b]
Bio-HDPE/PE-g-MA/MAS/HNT	276.8 ± 1.3 [b]	458.3 ± 1.5 [c]	7.8 ± 0.2 [c]
Bio-HDPE/PE-g-MA/MAS/MLO	285.3 ± 0.8 [b]	421.8 ± 1.8 [d]	1.5 ± 0.2 [d]
Bio-HDPE/PE-g-MA/MAS/HNT/MLO	287.8 ± 1.1 [b]	475.3 ± 0.9 [e]	4.3 ± 0.1 [e]

[a–e] Different letters in the same column indicate a significant difference among the samples ($p < 0.05$).

According to residual mass results, all samples have more residual mass than neat Bio-HDPE. This is especially logic with HNTs, as they are mostly inorganic and possess a very high degradation temperature [58]. In this case 7.5 phr HNTs sample retains a residual mass of 7.8 wt.-% while the sample with 3.75 phr HNTs maintains a 4.3 wt.-% residual mass. The introduction of the lignocellulosic fillers also provided an increasement of the residual mass due to the introduction of lignocellulosic compounds that cannot be degraded at 700 °C as Liminana et al. reported for the Almon Shell Flour [36]. Regarding the residual mass, all the samples show significant differences between the previous blend, this is manly by the incorporation of the HNT and the MAS that provide different residual mass as mentioned before.

3.4. Dynamic-Mechanical Behavior of Bio-HDPE/PE-g-MA/MAS Blends

Thermomechanical properties of blends were studied by DMTA technique. Figure 7a shows the evolution of the storage modulus (G′) from −150 to 120 °C. In relation to neat Bio-HDPE, an initial decrease in G′ can be observed until −100 °C approximately. This is directly connected with the glass transition of the material. Then, the storage modulus further decreased due to softening of the polymer matrix. Quiles-Carrillo et. al. [57] observed a very similar profile for this polyolefin. Addition of MAS and PE-g-MA into the matrix slightly increases G′ modulus along all the temperature range. Specifically, a rise is produced at −140 °C (see Table 5) from 2513 MPa (Bio-HDPE) to 2523 MPa (Bio-HDPE/PE-g-MA/MAS). This effect emphasizes as temperature decreases, going from 1309 to 1413 MPa at −25 °C. This is caused by a stiffening of the polymer due to the reinforcing effect of argan particles [65]. As expected, HNTs presence highly augments the storage modulus, providing values of 3111 and 1898 MPa at −140 and −25 °C (see Table 5), respectively, in comparison to the values obtained previously for neat Bio-HDPE. This is indicative of an increase in the stiffness of the material, confirming the results commented in the mechanical properties section, associated with a high dispersion of HNTs in the matrix. Thus provoking a decrease in polymer chain mobility due to physical interactions between nanotubes and adjacent chains [71]. In the case of Bio-HDPE/PE-g-MA/MAS/MLO sample, superior values of storage modulus were observed until 0 °C in comparison to Bio-HDPE and Bio-HDPE/PE-g-MA/MAS samples, but the modulus was inferior in the last section, till 120 °C. This reduction can be associated with the plasticizing effect of MLO. Although the difference is not significant and, given the modulus increase in the initial section of the diagram, it can be asserted that MLO acts as a dispersing element of the argan particles, in combination with PE-g-MA. Quiles-Carrillo et. al. [57] observed a similar behavior by studying the effect of MLO to compatibilize Bio-HDPE/PLA blends. Finally, the combined effect of HNTs (3.75 phr) with MLO (3.75 phr) enhances the stiffness of Bio-HDPE in all the temperature range, but to a lesser extent than that of 7.5 phr HNTs sample, as the concentration is lower in this case. Nonetheless, a good compatibilization and dispersion effect of MLO toward HNTs and MAS can be seen. The statistical analysis shows a similar trend of the storage modulus with respect to the elastic modulus.

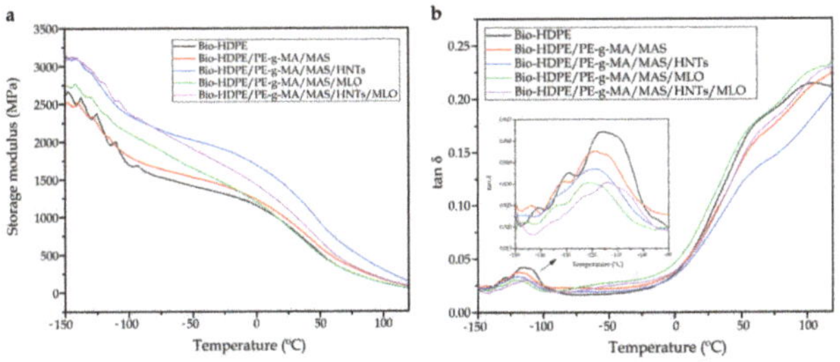

Figure 7. Plot evolution of (**a**) the storage modulus (G′) and (**b**) the dynamic damping factor (tan δ) of the injection-molded samples of Bio-HDPE/PE-g-MA/MAS blends with different compatibilizers.

Table 5. Dynamic-mechanical properties of injection-molded samples of Bio-HDPE/PE-g-MA/MAS blends with different compatibilizers, at different temperatures.

Parts	E′ (MPa) at −140 °C	E′ (MPa) at −25 °C	E′ (MPa) at 100 °C	$T_{g\ BIO\text{-}HDPE}$ (°C) *
Bio-HDPE	2513 ± 30 [a]	1309 ± 14 [a]	124 ± 2 [a]	−115.0 ± 1.1 [a]
Bio-HDPE/PE-g-MA/MAS	2523 ± 25 [a]	1413 ± 16 [b]	170 ± 4 [b]	−118.7 ± 1.3 [a]
Bio-HDPE/PE-g-MA/MAS/HNTs	3111 ± 39 [b]	1898 ± 20 [c]	299 ± 8 [c]	−119.0 ± 2.0 [a]
Bio-HDPE/PE-g-MA/MAS/MLO	2730 ± 31 [c]	1449 ± 17 [d]	127 ± 2 [d]	−120.5 ± 3.2 [a]
Bio-HDPE/PE-g-MA/MAS/HNT/MLO	3090 ± 37 [d]	1685 ± 23 [e]	176 ± 5 [e]	−114.3 ± 1.2 [b]

* The T_g has been measured using the tan δ peak maximum criterion. [a–e] Different letters in the same column indicate a significant difference among the samples ($p < 0.05$).

Figure 7b shows the evolution of the damping coefficient (tan δ) with temperature for every studied blend. The peak observed at −115 °C for Bio-HDPE corresponds to γ-relaxation of the green polyolefin, which is directly related to its glass transition temperature T_g [72], which is considered as the maximum peak value. This value is very similar to those observed in other studies [57]. A second relaxation, which is called α-relaxation, can be located between 50 and 120 °C and is associated with an interlaminar shear process. α-relaxation is often separated into two processes (α y α′) due to an inhomogeneity in crystalline regions [73]. The T_g change given by the addition of MAS with different compatibilizers is not so significant. In the case of Bio-HDPE/PE-g-MA/MAS and Bio-HDPE/PE-g-MA/MAS/HNTs samples, a reduction of about 3–4 °C is produced, while MLO reduces it even further (5 °C). It is logic that the 7.5 phr MLO sample possesses the lowest T_g due to the plasticizing effect previously mentioned. Regarding the combination of HNTs (3.75 phr) and MLO (3.75 phr), they seem to exert a synergetic effect that moves the glass transition peak +1 °C approximately. This can be associated with a good dispersion of HNTs into the polymer matrix, which immobilize Bio-HDPE chains to some extent. Only with the superposition of all the effects commented before, a significant difference could be observed, HNT/MLO blend reduced the T_g to the lowest temperature (−114.3 °C).

3.5. Color Measurement of BIO-HDPE/PE-g-MA/MAS Blends

Color, luminance, and transparency are essential issues to be considered in materials that try to imitate wood or WPC, in this case reddish-like, as those parameters determine how similar in appearance are these materials to wood. Table 6 gathers color coordinates values of Bio-HDPE/PE-g-MA/MAS blends, while Figure 8 shows the visual appearance of tensile test samples. All samples are opaque, mainly due to the semicrystalline nature of Bio-HDPE, which does not allow light to pass through [74].

Table 6. Luminance and color coordinates (L*a*b*) of the Bio-HDPE/PE-g-MA/MAS blends with different compatibilizers.

Code	L *	a *	b *
Bio-HDPE	72.7 ± 0.3^{a}	-2.29 ± 0.01^{a}	-5.35 ± 0.07^{a}
Bio-HDPE/PE-g-MA/MAS	37.6 ± 0.7^{b}	6.21 ± 0.48^{b}	4.95 ± 0.61^{b}
Bio-HDPE/PE-g-MA/MAS/HNT	39.2 ± 2.4^{b}	5.76 ± 0.37^{c}	5.05 ± 1.12^{b}
Bio-HDPE/PE-g-MA/MAS/MLO	36.1 ± 0.2^{b}	5.53 ± 0.31^{c}	4.23 ± 0.06^{c}
Bio-HDPE/PE-g-MA/MAS/HNT/MLO	37.0 ± 0.1^{b}	5.21 ± 0.16^{c}	4.39 ± 0.24^{c}

a–c Different letters in the same column indicate a significant difference among the samples ($p < 0.05$).

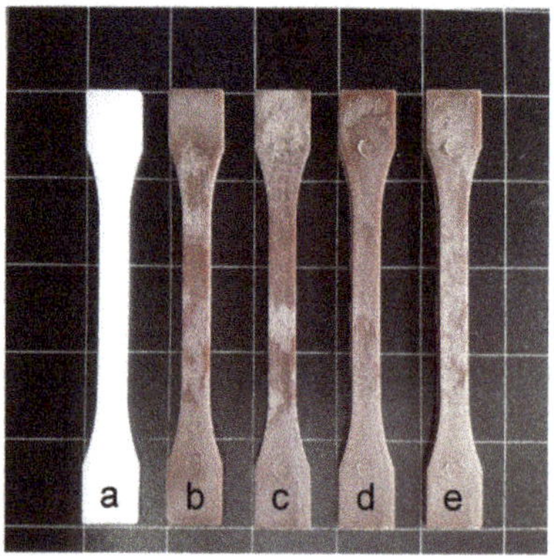

Figure 8. Visual appearance of the samples: (**a**) Bio-HDPE; (**b**) Bio-HDPE/PE-g-MA/MAS; (**c**) Bio-HDPE/PE-g-MA/MAS/HNT; (**d**) Bio-HDPE/PE-g-MA/MAS/MLO; (**e**) Bio-HDPE/PE-g-MA/MAS/HNT/MLO.

L*a*b* color coordinates were measured on injection-molded samples in order to analyze the color variations and luminance. Luminance (L*) is indicative of lightness. Neat Bio-HDPE showed high brightness as a consequence of its characteristic white color, in contrast with the brown-like color of the other samples, which present low luminance as a result of the addition of fillers and compatibilizers. Referring to color coordinate a*, it is indicative of the color change between green (negative) and red (positive). Particularly, neat Bio-HDPE presents a negative value but close to 0, −2.29 in this case, because of the proximity of the sample to white color. Similar value was reported by Rojas-Lema et al. [75] in neat Bio-HDPE films. The rest of materials have positive and quite similar values each due to the characteristic brown color of micronized argan shell [76]. However, a slight change in brown tonality can be observed between those samples with and without MLO. MLO samples display a darker brown color, as a result they present a lower a* value than samples lacking MLO, which have a clearer brown color as well as closer to pure red. This can be due to the trend of MLO to take a* coordinate to more negative values, as it was observed by Quiles-Carrillo et. al. [77]. Regarding b* coordinate color, it defines blue (negative) and yellow (positive) colors. Neat Bio-HDPE has a negative value of −5.35, which is similar to that reported by Rojas-Lema et. al. [75]. While the other materials present similar values between 4 and 5, due to approaching the intrinsic yellow-like color of the argan particles and MLO. It should be noted that yellow tonality of MLO is less intense than that of argan particles, therefore reaching lower b* values. On the other hand, HNTs do not affect color significantly but seem to reduce color coordinate a* and slightly increase coordinate b*. The statistical results show that the biggest color difference was assessed by the introduction of the filler, the compatibilization strategies followed slightly provided modifications between them.

Colors showed by the materials studied here make them perfect candidates for wood plastic composites [78], presenting reddish-like brown colors that allow them to be used in wood-based products fabrication, where quality and aesthetic are vital and determine the success or failure of the product [79].

3.6. Wetting Properties and Water Absorption of Bio-HDPE/PE-g-MA/MAS Blends

One of the main disadvantages of green composites or any material based on hydrophilic fillers is their tendency to absorb water. With the objective of evaluating the behavior of Bio-HDPE/PE-g-MA/MAS blends against water, contact angle was measured at different times after applying a water drop over each sample surface. A great contact angle is indicative of a poor affinity toward water. Table 7 shows different contact angles for the blends at 0, 5, 10, 15, 20, and 30 min after the administration of the water drop. It can be seen as initially, all samples are hydrophobic, as their contact angles are far superior to 65°, which is the hydrophilic threshold according to Vogler [80]. This can be ascribed to the fact that the polymer matrix (Bio-HDPE) is a completely nonpolar compound, formed only by C-H bonds. This bond is formed by atoms with practically the same electronegativity. It is this reason that makes neat Bio-HDPE barely reduce its contact angle over time. When PE-g-MA and MAS are introduced into the polymer, the contact angle suffers a quick reduction (56.7° at 30 min). This can be attributed to the oxygen-based groups in PE-g-MA (anhydride groups) and to cellulose, hemicellulose, and lignin present in argan particles (hydroxyl and carbonyl groups). These groups give polarity to the material and can form hydrogen bonds with water (polar solvent), providing hydrophilicity over time [81]. Addition of HNTs into the blend provokes a faster decrease in contact angle than the previously mentioned samples. However, the contact angle at 30 min is very similar and slightly superior to that of Bio-HDPE/PE-g-MA/MAS. This could be attributed, on the one hand, to polar functional groups (hydroxyl) present in HNTs structure and, on the other hand, to their ability to retain different molecules within their tubular hollow structure, water being one of those molecules [82]. As expected, Bio-HDPE/PE-g-MA/MAS/MLO sample shows the lowest contact angle over time, reaching a considerably lower value than the rest of the samples (22.1°). This is ascribed to the high polarity of the MLO molecule, which has a great amount of anhydride and carbonyl groups, through which water can form hydrogen bonds as long as water and sample remain in contact [83]. Figure 9 perfectly reflects this behavior on the MLO sample, where the drop of water flattens over time. Finally, MLO and HNTs sample greatly reduces the hydrophobicity of the material due to the aforementioned reasons, verifying the effect of MLO over wettability.

Table 7. Contact angle (θ_w) of different Bio-HDPE/PE-g-MA/MAS blends with different compatibilizers at several times of exposure to water: 0, 5, 10, 15, 20 and 30 min.

Code/Time	0 min	5 min	10 min	15 min	20 min	30 min
Bio-HDPE	$90.1 \pm 3.2°$ ᵃ	$88.1 \pm 2.5°$ ᵃ	$88.2 \pm 1.3°$ ᵃ	$84.5 \pm 4.2°$ ᵃ	$83.3 \pm 1.0°$ ᵃ	$81.1 \pm 2.1°$ ᵃ
Bio-HDPE/PE-g-MA/MAS	$89.9 \pm 2.4°$ ᵃ	$80.0 \pm 0.8°$ ᵇ	$78.6 \pm 2.0°$ ᵇ	$76.1 \pm 1.7°$ ᵇ	$72.2 \pm 0.6°$ ᵇ	$56.7 \pm 0.5°$ ᵇ
Bio-HDPE/PE-g-MA/MAS/HNTs	$87.6 \pm 1.3°$ ᵃ	$77.9 \pm 0.9°$ ᵇ	$70.1 \pm 1.8°$ ᶜ	$67.7 \pm 1.3°$ ᶜ	$61.0 \pm 0.9°$ ᶜ	$59.4 \pm 0.7°$ ᶜ
Bio-HDPE/PE-g-MA/MAS/MLO	$84.7 \pm 0.9°$ ᵇ	$73.0 \pm 1.1°$ ᶜ	$67.1 \pm 1.7°$ ᶜ	$59.1 \pm 0.8°$ ᵈ	$52.2 \pm 0.6°$ ᵈ	$22.1 \pm 0.4°$ ᵈ
Bio-HDPE/PE-g-MA/MAS/HNTs/MLO	$92.5 \pm 3.1°$ ᶜ	$81.4 \pm 1.9°$ ᵈ	$78.8 \pm 1.2°$ ᵈ	$73.8 \pm 1.3°$ ᵉ	$66.4 \pm 0.8°$ ᵉ	$47.7 \pm 0.8°$ ᵉ

a–e Different letters in the same column indicate a significant difference among the samples ($p < 0.05$).

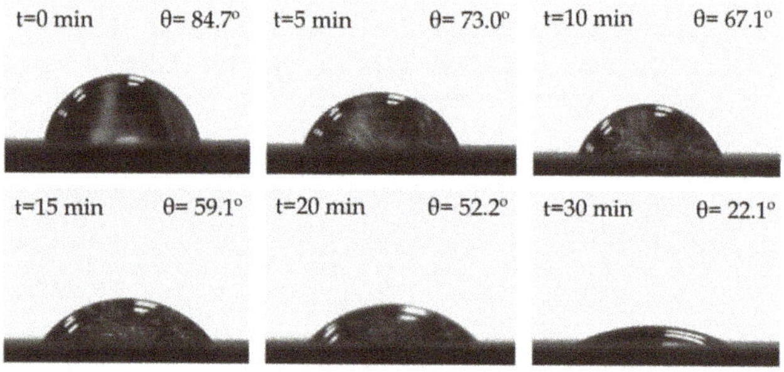

Figure 9. Water contact angle of the sample Bio-HDPE/PE-g-MA/MAS/MLO over time: 0, 5, 10, 15, 20, and 30 min.

From all these results it can be concluded that all fillers and compatibilizers used do not significantly vary hydrophilicity of Bio-HDPE at first, but they do increase water absorption in case of exposure for long periods of time. This could be corroborated by the statistical study where only some significant differences appeared at the initial time, but these were clearly accentuated over the time due to the different behaviors of the blends.

Additionally, water absorption capabilities under long time exposure were studied for each material by means of water absorption test. Figure 10 shows the water absorption of each sample after 14 weeks of immersion in distilled water. It can be observed as neat Bio-HDPE barely absorbed water, presenting an asymptotic value of 0.05 wt.-%. This remarks the hard hydrophobic nature of Bio-HDPE, as it has been previously said during the contact angle analysis. With the addition of argan particles and PE-g-MA, water absorption increases up to 0.98 wt.-%, after a period of 14 weeks. This means an increase of 94% in relation to neat Bio-HDPE. This is closely related to hydroxyl groups in lignocellulosic compounds of argan particles, which increase the facility of the material toward capturing humidity [84]. The incorporation of HNTs (7.5 phr) in the blend causes an increment in water absorption up to 1.27 wt.-%, which can be ascribed to the hollow tubular morphology of nanotubes. Their structure helps them to trap water molecules with ease, along with the effect of polar hydroxyl groups. Lastly, MLO sample achieves the highest water absorption (1.55 wt.-%) due to the plasticizing effect that it exerts over Bio-HDPE matrix, increasing its free volume and making the diffusion of water within its structure easier, as it was observed by Quiles-Carrillo et. al. [85]. As expected, 3.75 phr HNTs and 3.75 phr MLO sample presents intermediate results between those of individual HNTs and MLO samples (7.5 phr each). It should be also noted that addition of HNTs and MLO increases the water absorption speed during the first week, denoted by a more pronounced slope in the initial section showed by all three samples, in comparison with Bio-HDPE/MAS/PE-g-MA sample.

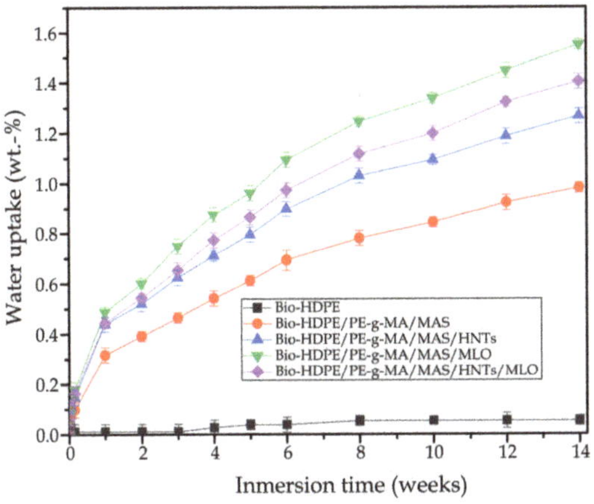

Figure 10. Water uptake of the injection-molded pieces of neat Bio-HDPE and its green composites with argan shell micronized (MAS) compatibilized with PE-g-MA, MLO, and with HNTs as additional filler.

According to these results, although hydrophilicity provided by argan particles (MAS) could prove a disadvantage in some ambits, it could certainly give applicability to these materials in some other fields. One of the applications could be flowerpot fabrication, so that when the plant is irrigated, water excess is absorbed by the pot material over time.

3.7. Infrared Spectroscopy

Chemical composition of the injection-molded samples was analyzed by means of Fourier transformed infrared spectroscopy (FTIR). Figure 11a gathers the individual spectrums of the compatibilizers and fillers used (PE-g-MA, MAS, HNTs y MLO) from 4000 to 600 cm^{-1}. First, PE-g-MA spectrum is very similar to that of Bio-HDPE, as it is a polyethylene-based (PE) compound. Absorption bands at 2840 and 2910 cm^{-1} are associated with the stretching vibration for -CH-, $-CH_2-$, or $-CH_3$ [86]. The peak at 1468 cm^{-1} corresponds to the deformation vibration of $-CH_2-$ or $-CH_3$ [86]. On the other hand, the peak at 720 cm^{-1} is due to $(CH_2)_n$ rock when $n \geq 4$ [86]. In the case of MAS, two bands stand out at 1030 and 900 cm^{-1}, which are related to the stretching vibration of C-O and C-OH in polysaccharide rings in cellulose [87,88]. A peak at 1114 cm^{-1} can also be appreciated due to the symmetric glycosidic stretching of C-O-C bond in polysaccharide compounds of cellulose. Moreover, little absorption bands in the 1500 and 1130 cm^{-1} region can be observed, which indicate the presence of characteristic groups of cellulose, hemicellulose, and lignin. Some examples of those groups are the peak at 1227 cm^{-1}, related to the stretching vibration C=O of the acetyl group in lignins [89]; and the band at 1420 cm^{-1}, which corresponds to the vibration of aromatic rings in lignin [87]. The low-intensity band at 1720 cm^{-1}, is ascribed to non-cellulosic compounds (pectin, lignin, and hemicellulose) [89]. Another weak peak can be found at 2900 cm^{-1}, which is related to stretching vibration of C-H bonds in CH and CH_2 groups, present in cellulose and hemicellulose [65]. A final little band is found at 3300 cm^{-1}, associated to the stretching of O-H bonds in carbohydrates (cellulose and hemicellulose) [55]. Regarding the halloysite nanotubes (HNTs) spectrum, peaks at 3695 and 3619 cm^{-1} are due to the stretching vibration of Al_2OH- (each OH is linked to two Al atoms) [90]. Bands at 3670 and 3650 cm^{-1} are ascribed to the O-H stretching of inner-surface hydroxyl groups and the out-of-phase vibration of the inner surface hydroxyl groups, respectively [90–93]. The low intensity peak at 1649 cm^{-1} is attributed to bending vibrations due to absorbed water [94], the band at 1117 cm^{-1} is due to apical Si-O bonds, while 1022 and 682 cm^{-1} bands are associated with the perpendicular stretching of Si-O-Si bonds [93,94]. Peaks at 937 and 909 cm^{-1} are consequence of the O-H deformation of inner-surface hydroxyl groups and

O-H deformation of inner hydroxyl groups, respectively [93,94]. Two last bands analyzed are located at 790 and 749 cm^{-1}, and can be ascribed to the O-H translation vibrations of halloysite O-H units [93]. With regard to MLO spectrum, a first peak is located at 3006 cm^{-1}, corresponding to =C-H stretching of double carbon-carbon bonds, and those at 2925 and 2850 cm^{-1} refer to the antisymmetric and symmetric stretching vibration C-H of saturated carbon-carbon bonds (C-C), respectively [95]. Other bonds can be found at 1740 and 1709 cm^{-1}, indicative of the C=O carbonyl stretching vibration of the ester and maleic anhydride, respectively; another band at 1158 cm^{-1} is due to the stretching vibration C-O-C, C-O, and C-C in ester groups; the peak at 720 cm^{-1} is normally related to the out-of-plane C-H stretching vibration of saturated C-C bonds [95]. Finally, the band at 1462 cm^{-1} is due to C-H bending vibration, while the absorption peaks at 1862 and 1784 cm^{-1} are attributed to anhydride groups.

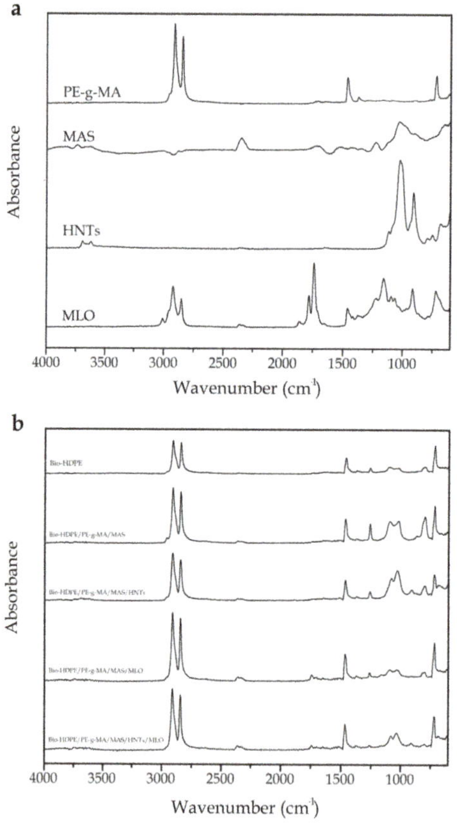

Figure 11. (**a**) Fourier transform infrared (FTIR) spectra, from bottom to top, of maleinized linseed oil (MLO), halloysite nanotubes (HNTs), micronized argan shell (MAS), Polyethylene-graft-maleic anhydride (PE-g-MA) (**b**) FTIR spectra of the different blends, from top to bottom Bio-HDPE, Bio-HDPE(67 wt.%)/PE-g-MA(3 wt.%)/MAS(30 wt.%),Bio-HDPE(67 wt.%)/PE-g-MA(3 wt.%)/MAS(30 wt.%)/HNTs(7.5 phr), Bio-HDPE(67 wt.%)/PE-g-MA(3 wt.%)/MAS(30 wt.%)/MLO(7.5 phr), Bio-HDPE(67 wt.%)/PE-g-MA (3 wt.%)/MAS(30 wt.%)/HNTs(3.75 phr)/MLO(3.75 phr).

Figure 11b gathers infrared spectrums for neat Bio-HDPE as well as all the developed blends. The main bands of neat Bio-HDPE are located at 2914, 2846, 1466, and 718 cm^{-1}, and they are related to the stretching vibration, bending deformations, and rocking of methylene groups (CH_2) [96]. These peaks are found in all samples, but with higher intensity, due to the presence of PE-g-MA, which is a PE-based material and as such presents the same nature as Bio-HDPE. Thus, their spectrums are very similar between each other. The weak absorption bands between 1370 and 1350 cm^{-1} are associated with wagging and symmetric deformations of CH_2 and CH_3 groups, respectively. The addition of MAS compatibilized with PE-g-MA makes a band in the range 1025–1091 cm^{-1} to appear, which is mainly attributed to the formation of PE-g-MA dimers or oligomers, identified with the stretching of C-C and C-O bonds [59]. Moreover, this band could be combined with another one at 1030 cm^{-1}, relative to the presence of C-O and C-OH bonds in MAS particles cellulose. The presence of HNTs in the blend deforms and increases the intensity of the peak at 1030 cm^{-1}, due to the perpendicular Si-O-Si stretching, as it has been previously mentioned in the individual spectra of HNTs. Additionally, in HNTs samples a weaker band at 900 cm^{-1} can be observed, ascribed to the deformation of O-H groups [90]. Moreover, little peaks can be located in the 3650 cm^{-1} region, which have been previously identified in the individual analysis of HNTs, associated with the stretching vibration of inner-surface O-H groups in the nanotubes. Regarding the MLO presence in the blend, it produces changes found at 1690 cm^{-1} approximately, related to the C=O stretching in hydrolyzed anhydride groups in MLO [13]. This peak has already been observed in MLO individual spectrum, as it has been aforementioned, and it does not appear in those samples without MLO on their structure. The appearance of low-intensity bands at 3700 and 3600 cm^{-1} can be ascribed to the absorption of water, catalyzed by MLO and argan particles, which are polar compounds with great amount of functional oxygen-based groups and great affinity toward water. HNTs and MLO blend shows the same characteristic peaks described for each individual sample, but with less absorption intensity, as their concentration reduces from 7.5 phr to 3.75 phr.

These results seem to reveal certain compatibilization between elements. This fact is denoted by a general increase in the intensity of peaks related to oxygen-based groups, indicating a possible bonding between anhydride groups in PE-g-MA and MLO with hydroxyl groups in MAS and HNTs, increasing the affinity reciprocally within the polymer matrix.

4. Discussion

The present work shows the incorporation of argan shell wastes in wood plastic composites with biobased polyethylene matrix and different compatibilizers. The incorporation of this type of loads can be effectively used as new reinforcement elements in order to create parts prepared by conventional industrial processes for thermoplastic materials, in special injection molding. In relation to the mechanical properties, the incorporation of MAS (30 wt.%) with PE-g-MA (3 wt.%) into the Bio-HDPE matrix increases the Young modulus up to 846 MPa. This can be directly related to a proper distribution of MAS particles in the polymer matrix, leading to a good adhesion of the filler with the polymer and obtaining a more rigid material. Addition of HNTs in the blend showed an additional enhancement of stiffness. In this context, Bio-HDPE/PE-g-MA/MAS/HNTs sample presented the highest Young modulus of all blends (1126 MPa). The incorporation of MLO in the green composites increased 100% of the elongation at break of the Bio-HDPE/PE-g-MA/MAS blend. From a morphological point of view, the incorporation of argan particles in the BioHDPE matrix, showed great cohesion, which implies good interaction between both elements. Furthermore, the addition of MLO verified the results obtained in the mechanical properties, showing a more ductile fracture and confirming the plasticizing effect of MLO. However, the incorporation of the lignocellulosic filler leads to a reduction in thermal properties and a decrease in the crystallinity of the compound. In relation to the thermomechanical properties, the incorporation of HNTs highly augmented the storage modulus, providing values of 3111 and 1898 MPa. This is indicative of an increase in the stiffness of the material, confirming the results commented in the mechanical properties section, associated with a high dispersion of HNTs in the matrix. In general, the incorporation of argan particles and additives provided the materials with reddish-brown colors that allow their use in the manufacturing of wood-based products. In relation to one of the main drawbacks of green composites, it can be seen how the incorporation of lignocellulosic particles greatly increases the water absorption of the composites, generating certain disadvantages. In addition, the incorporation of the MLO increases the water absorption capacity of the composites, due to the plasticization effect it exerts over the Bio-HDPE matrix.

5. Conclusions

The results obtained in this work indicate that it is possible to obtain WPC with high renewable content with Bio-HDPE, lignocellulosic fillers and natural additives. This type of green composites can greatly favor the generation of circular economies focused on giving added value to agri-food waste from the Mediterranean basin, also favoring the creation of highly efficient polymers at very competitive costs. The materials obtained in this work have proved to possess excellent properties at a reduced cost in comparison with the neat Bio-HDPE. High ductile properties, relatively high stiffness and good thermal stability were reported, as well as a visual appearance very similar to that of reddish-color woods, which is essential in a wood plastic composite. The samples showed certain hydrophilicity over time, which can prove to be a disadvantage, although it is unusual for these materials to be immersed in water for long periods of time. However, if it were the case, they have some applications too, as it is the fabrication of plant pots. All in all, the affinity between argan particles and Bio-HDPE has been successfully increased in this study with the use of compatibilizers, and their properties vastly displayed and demonstrated. This work opens a new line of research regarding the development of new materials formed by polar fillers and non-polar polymeric matrices.

Author Contributions: Conceptualization, L.Q.-C. and P.M.S.; methodology, S.R.-L.; validation, J.I.-M., J.G.-C., and L.Q.-C.; formal analysis, M.J.-R.; investigation, M.J.-R.; data curation, J.I.-M.; writing—original draft preparation, M.J.-R.; writing—review and editing, J.G.-C. and J.I.-M.; supervision, P.M.S.; project administration, L.Q.-C.; All authors have read and agreed to the published version of the manuscript.

Funding: This research was funded by the Ministry of Science, Innovation, and Universities (MICIU) project number MAT2017-84909-C2-2-R.

Institutional Review Board Statement: Not applicable.

Informed Consent Statement: Not applicable.

Data Availability Statement: Not applicable.

Acknowledgments: L. Q.-C. wants to thank Universitat Politècnica de València for his post-doctoral PAID-10-20 grant from (SP20200073). J. I.-M. wants to thank the Spanish Ministry of Science, Innovation and Universities for his FPU grant (FPU19/01759). J. G.-C. wants to thank Universitat Politècnica de València for his FPI grant from (SP20200080). S.R.-L. for her Santiago Grisolia grant from Generalitat Valenciana (GVA) (GRISOLIAP/2019/132).

Conflicts of Interest: The authors declare no conflict of interest.

References

1. Tahir, N.; Bhatti, H.N.; Iqbal, M.; Noreen, S. Biopolymers composites with peanut hull waste biomass and application for Crystal Violet adsorption. *Int. J. Biol. Macromol.* **2017**, *94*, 210–220. [CrossRef]
2. Wechsler, A.; Hiziroglu, S. Some of the properties of wood–plastic composites. *Build. Environ.* **2007**, *42*, 2637–2644. [CrossRef]
3. Ballerini, A.; Reyes, N.; Núñez, M.; Wechsler, A.J.V.U.C. *Development of Chemical Additives from Tall Oil to Improve the Compatibility of Wood Plastic Composites*; FONDEF: Santiago, Chile, 2005.
4. Charrier, J.-M. *Polymeric Materials and Processing: Plastics, Elastomers, and Composites*; CRC Press: Boca Raton, FL, USA, 1991.
5. Otto, G.P.; Moisés, M.P.; Carvalho, G.; Rinaldi, A.W.; Garcia, J.C.; Radovanovic, E.; Fávaro, S.L. Mechanical properties of a polyurethane hybrid composite with natural lignocellulosic fibers. *Compos. Part B Eng.* **2017**, *110*, 459–465. [CrossRef]
6. Simonsen, J. The mechanical properties of wood fiber plastic composites: Theoretical vs. experimental. *For. Prod. Soc. Proc.* **1995**, 47–55.
7. Younes, U.E.; Perry, J.H.; Rosthauser, J.W. Long-Fiber Thermoset Composite with Low Orange Peel. U.S. Patent 20120003454A1, 5 January 2012.

8. Quiles-Carrillo, L.; Montanes, N.; Lagaron, J.M.; Balart, R.; Torres-Giner, S. On the use of acrylated epoxidized soybean oil as a reactive compatibilizer in injection-molded compostable pieces consisting of polylactide filled with orange peel flour. *Polym. Int.* **2018**, *67*, 1341–1351. [CrossRef]
9. Pirayesh, H.; Khazaeian, A. Using almond (Prunus amygdalus L.) shell as a bio-waste resource in wood based composite. *Compos. Part B Eng.* **2012**, *43*, 1475–1479. [CrossRef]
10. Jasim, F.A.; Hashim, A.; Hadi, A.G.; Lafta, F.; Salman, S.R.; Ahmed, H. Preparation of (pomegranate peel-polystyrene) composites and study their optical properties. *Res. J. Appl. Sci.* **2013**, *8*, 439–441.
11. Essabir, H.; El Achaby, M.; Hilali, E.M.; Bouhfid, R.; Qaiss, A.; Hilali, E.M. Morphological, Structural, Thermal and Tensile Properties of High Density Polyethylene Composites Reinforced with Treated Argan Nut Shell Particles. *J. Bionic Eng.* **2015**, *12*, 129–141. [CrossRef]
12. Abass, R.U. *Mechanical Behavior of Natural Material (Orange Peel) Reinforced Polyester Composite*; CiteSeer: Princeton, NJ, USA, 2015.
13. Quiles-Carrillo, L.; Montanes, N.; Sammon, C.; Balart, R.; Torres-Giner, S. Compatibilization of highly sustainable polylactide/almond shell flour composites by reactive extrusion with maleinized linseed oil. *Ind. Crop. Prod.* **2018**, *111*, 878–888. [CrossRef]
14. Victor, A.; Atuanya, C.; Igogori, E.; Ihom, P. Development of high-density polyethylene/orange peels particulate bio-composite. *Gazi Univ. J. Sci.* **2013**, *26*, 107–117.
15. Ivorra-Martinez, J.; Manuel-Mañogil, J.; Boronat, T.; Sanchez-Nacher, L.; Balart, R.; Quiles-Carrillo, L. Development and Characterization of Sustainable Composites from Bacterial Polyester Poly(3-Hydroxybutyrate-co-3-hydroxyhexanoate) and Almond Shell Flour by Reactive Extrusion with Oligomers of Lactic Acid. *Polymers* **2020**, *12*, 1097. [CrossRef] [PubMed]
16. Montanes, N.; Quiles-Carrillo, L.; Ferrandiz, S.; Fenollar, O.; Boronat, T. Effects of Lignocellulosic Fillers from Waste Thyme on Melt Flow Behavior and Processability of Wood Plastic Composites (WPC) with Biobased Poly(ethylene) by Injection Molding. *J. Polym. Environ.* **2019**, *27*, 747–756. [CrossRef]
17. Chen, G.; Li, S.; Jiao, F.; Yuan, Q. Catalytic dehydration of bioethanol to ethylene over TiO2/γ-Al2O3 catalysts in microchannel reactors. *Catal. Today* **2007**, *125*, 111–119. [CrossRef]
18. Goldemberg, J.; Coelho, S.T.; Guardabassi, P. The sustainability of ethanol production from sugarcane. *Energy Policy* **2008**, *36*, 2086–2097. [CrossRef]
19. Babu, R.P.; O'Connor, K.; Seeram, R. Current progress on bio-based polymers and their future trends. *Prog. Biomater.* **2013**, *2*, 1–16. [CrossRef] [PubMed]
20. Siracusa, V.; Blanco, I. Bio-Polyethylene (Bio-PE), Bio-Polypropylene (Bio-PP) and Bio-Poly (ethylene tereph-thalate)(Bio-PET): Recent developments in bio-based polymers analogous to petroleum-derived ones for packaging and engineering applications. *Polymers* **2020**, *12*, 1641. [CrossRef] [PubMed]
21. Boujmal, R.; Essabir, H.; Nekhlaoui, S.; Bensalah, M.; Bouhfid, R.; Qaiss, A. Bioenergy. Composite from poly-propylene and henna fiber: Structural, mechanical and thermal properties. *J. Biobased Mater. Bioenergy* **2014**, *8*, 246–252. [CrossRef]
22. Salleh, F.M.; Hassan, A.; Yahya, R.; Azzahari, A.D. Effects of extrusion temperature on the rheological, dynamic mechanical and tensile properties of kenaf fiber/HDPE composites. *Compos. Part B Eng.* **2014**, *58*, 259–266. [CrossRef]
23. Arrakhiz, F.; El Achaby, M.; Kakou, A.; Vaudreuil, S.; Benmoussa, K.; Bouhfid, R.; Fassi-Fehri, O.; Qaiss, A. Mechanical properties of high density polyethylene reinforced with chemically modified coir fibers: Impact of chemical treatments. *Mater. Des.* **2012**, *37*, 379–383. [CrossRef]
24. Arrakhiz, F.; El Achaby, M.; Malha, M.; Bensalah, M.; Fassi-Fehri, O.; Bouhfid, R.; Benmoussa, K.; Qaiss, A. Mechanical and thermal properties of natural fibers reinforced polymer composites: Doum/low density polyethylene. *Mater. Des.* **2013**, *43*, 200–205. [CrossRef]
25. Arrakhiz, F.; El Achaby, M.; Benmoussa, K.; Bouhfid, R.; Essassi, E.; Qaiss, A. Evaluation of mechanical and thermal properties of Pine cone fibers reinforced compatibilized polypropylene. *Mater. Des.* **2012**, *40*, 528–535. [CrossRef]
26. Essabir, H.; Hilali, E.; Elgharad, A.; El Minor, H.; Imad, A.; ElAmraoui, A.; Al Gaoudi, O. Mechanical and thermal properties of bio-composites based on polypropylene reinforced with Nutshells of Argan particles. *Mater. Des.* **2013**, *49*, 442–448. [CrossRef]
27. Essabir, H.; Nekhlaoui, S.; Malha, M.; Bensalah, M.; Arrakhiz, F.; Qaiss, A.; Bouhfid, R. Bio-composites based on polypropylene reinforced with Almond Shells particles: Mechanical and thermal properties. *Mater. Des.* **2013**, *51*, 225–230. [CrossRef]
28. Ayrilmis, N.; Kaymakci, A. Fast growing biomass as reinforcing filler in thermoplastic composites: Paulownia elongata wood. *Ind. Crop. Prod.* **2013**, *43*, 457–464. [CrossRef]
29. Dányádi, L.; Móczó, J.; Pukánszky, B. Effect of various surface modifications of wood flour on the properties of PP/wood composites. *Compos. Part A Appl. Sci. Manuf.* **2010**, *41*, 199–206. [CrossRef]
30. Garcia-Garcia, D.; Carbonell-Verdu, A.; Jordá-Vilaplana, A.; Balart, R.; Garcia-Sanoguera, D. Development and characterization of green composites from bio-based polyethylene and peanut shell. *J. Appl. Polym. Sci.* **2016**, *133*. [CrossRef]
31. Fombuena, V.; Samper, M.D. Study of the properties of thermoset materials derived from epoxidized soybean oil and protein fillers. *J. Am. Oil Chem. Soc.* **2013**, *90*, 449–457. [CrossRef]
32. Quiles-Carrillo, L.; Duart, S.; Montanes, N.; Torres-Giner, S.; Balart, R. Enhancement of the mechanical and thermal properties of injection-molded polylactide parts by the addition of acrylated epoxidized soybean oil. *Mater. Des.* **2018**, *140*, 54–63. [CrossRef]

33. Balart, J.; Fombuena, V.; Fenollar, O.; Boronat, T.; Sánchez-Nacher, L. Processing and characterization of high environmental efficiency composites based on PLA and hazelnut shell flour (HSF) with biobased plasticizers derived from epoxidized linseed oil (ELO). *Compos. Part B Eng.* **2016**, *86*, 168–177. [CrossRef]
34. Garcia-Garcia, D.; Ferri, J.M.; Montanes, N.; Lopez-Martinez, J.; Balart, R. Plasticization effects of epoxidized vegetable oils on mechanical properties of poly(3-hydroxybutyrate). *Polym. Int.* **2016**, *65*, 1157–1164. [CrossRef]
35. Sarwono, A.; Man, Z.; Bustam, M.A. Blending of Epoxidised Palm Oil with Epoxy Resin: The Effect on Morphology, Thermal and Mechanical Properties. *J. Polym. Environ.* **2012**, *20*, 540–549. [CrossRef]
36. Liminana, P.; Quiles-Carrillo, L.; Boronat, T.; Balart, R.; Montanes, N. The Effect of Varying Almond Shell Flour (ASF) Loading in Composites with Poly(Butylene Succinate (PBS) Matrix Compatibilized with Maleinized Linseed Oil (MLO). *Materials* **2018**, *11*, 2179. [CrossRef]
37. Osman, M.A.; Atallah, A. Effect of the particle size on the viscoelastic properties of filled polyethylene. *Polymer* **2006**, *47*, 2357–2368. [CrossRef]
38. Tang, Y.; Yang, C.; Gao, P.; Ye, L.; Zhao, C.; Lin, W. Rheological study on high-density polyeth-ylene/organoclay composites. *Polym. Eng. Sci.* **2011**, *51*, 133–142. [CrossRef]
39. Vega, J.F.; Martinez-Salazar, J.; Trujillo, M.; Arnal, M.L.; Muller, A.J.; Bredeau, S.; Dubois, P. Rheology, Processing, Tensile Properties, and Crystallization of Polyethylene/Carbon Nanotube Nanocomposites. *Macromolecules* **2009**, *42*, 4719–4727. [CrossRef]
40. Du, M.; Guo, B.; Jia, D. Newly emerging applications of halloysite nanotubes: A review. *Polym. Int.* **2010**, *59*, 574–582. [CrossRef]
41. Frost, R.; Shurvell, H.F. Minerals, C. Raman microprobe spectroscopy of halloysite. *Clays Clay Miner.* **1997**, *45*, 68–72. [CrossRef]
42. Du, M.; Guo, B.; Jia, D. Thermal stability and flame retardant effects of halloysite nanotubes on poly(propylene). *Eur. Polym. J.* **2006**, *42*, 1362–1369. [CrossRef]
43. Du, M.; Guo, B.; Wan, J.; Zou, Q.; Jia, D. Effects of halloysite nanotubes on kinetics and activation energy of non-isothermal crystallization of polypropylene. *J. Polym. Res.* **2009**, *17*, 109–118. [CrossRef]
44. Lecouvet, B.; Bourbigot, S.; Sclavons, M.; Bailly, C. Kinetics of the thermal and thermo-oxidative degradation of polypropylene/halloysite nanocomposites. *Polym. Degrad. Stab.* **2012**, *97*, 1745–1754. [CrossRef]
45. Lecouvet, B.; Sclavons, M.; Bourbigot, S.; Devaux, J.; Bailly, C. Water-assisted extrusion as a novel processing route to prepare polypropylene/halloysite nanotube nanocomposites: Structure and properties. *Polymer* **2011**, *52*, 4284–4295. [CrossRef]
46. Ning, N.-Y.; Yin, Q.-J.; Luo, F.; Zhang, Q.; Du, R.; Fu, Q. Crystallization behavior and mechanical properties of polypropylene/halloysite composites. *Polymer* **2007**, *48*, 7374–7384. [CrossRef]
47. Jia, Z.; Luo, Y.; Guo, B.; Yang, B.; Du, M.; Jia, D. Reinforcing and Flame-Retardant Effects of Halloysite Nanotubes on LLDPE. *Polym. Technol. Eng.* **2009**, *48*, 607–613. [CrossRef]
48. Handge, U.A.; Hedicke-Höchstötter, K.; Altstädt, V. Composites of polyamide 6 and silicate nanotubes of the mineral halloysite: Influence of molecular weight on thermal, mechanical and rheological properties. *Polymer* **2010**, *51*, 2690–2699. [CrossRef]
49. Liu, M.; Guo, B.; Du, M.; Cai, X.; Jia, D. Properties of halloysite nanotube–epoxy resin hybrids and the interfacial reactions in the systems. *Nanotechnology* **2007**, *18*, 455703. [CrossRef]
50. Liu, M.; Zhang, Y.; Zhou, C. Nanocomposites of halloysite and polylactide. *Appl. Clay Sci.* **2013**, *75–76*, 52–59. [CrossRef]
51. Prashantha, K.; Lecouvet, B.; Sclavons, M.; Lacrampe, M.F.; Krawczak, P. Poly(lactic acid)/halloysite nanotubes nanocomposites: Structure, thermal, and mechanical properties as a function of halloysite treatment. *J. Appl. Polym. Sci.* **2012**, *128*, 1895–1903. [CrossRef]
52. Garcia, D.; Balart, R.; Sánchez, L.; López, J. Compatibility of recycled PVC/ABS blends. Effect of previous degradation. *Polym. Eng. Sci.* **2007**, *47*, 789–796. [CrossRef]
53. Charrouf, Z.; Guillaume, D. Technology. Argan oil: Occurrence, composition and impact on human health. *Eur. J. Lipid Sci. Technol.* **2008**, *110*, 632–636. [CrossRef]
54. Martínez-Gómez, P.; Correa, D.; Sánchez-Blanco, M.; Majourhat, K.; Rubio, M.; Martínez-García, P.J. Posibilidades del cultivo del argán [*Argania spinosa* (L.) Skeels] en el Sureste español. *Rev. Fruticul.* **2018**, *66*, 26–41.
55. Laaziz, S.A.; Raji, M.; Hilali, E.; Essabir, H.; Rodrigue, D.; Bouhfid, R.; Qaiss, A.E.K. Bio-composites based on polylactic acid and argan nut shell: Production and properties. *Int. J. Biol. Macromol.* **2017**, *104*, 30–42. [CrossRef] [PubMed]
56. Crespo, J.; Balart, R.; Sánchez, L.; Lopez, J. Mechanical behaviour of vinyl plastisols with cellulosic fillers. Analysis of the interface between particles and matrices. *Int. J. Adhes. Adhes.* **2007**, *27*, 422–428. [CrossRef]
57. Quiles-Carrillo, L.; Montanes, N.; Jorda-Vilaplana, A.; Balart, R.; Torres-Giner, S. A comparative study on the effect of different reactive compatibilizers on injection-molded pieces of bio-based high-density polyethylene/polylactide blends. *J. Appl. Polym. Sci.* **2019**, *136*, 47396. [CrossRef]
58. Singh, V.P.; Vimal, K.; Kapur, G.; Sharma, S.; Choudhary, V. High-density polyethylene/halloysite nanocompo-sites: Morphology and rheological behaviour under extensional and shear flow. *J. Polym. Res.* **2016**, *23*, 43. [CrossRef]
59. Quiles-Carrillo, L.; Montanes, N.; Fombuena, V.; Balart, R.; Torres-Giner, S. Enhancement of the processing window and performance of polyamide 1010/bio-based high-density polyethylene blends by melt mixing with natural additives. *Polym. Int.* **2020**, *69*, 61–71. [CrossRef]
60. Ferri, J.; Garcia-Garcia, D.; Sánchez-Nacher, L.; Fenollar, O.; Balart, R. The effect of maleinized linseed oil (MLO) on mechanical performance of poly(lactic acid)-thermoplastic starch (PLA-TPS) blends. *Carbohydr. Polym.* **2016**, *147*, 60–68. [CrossRef] [PubMed]

61. Crespo, J.; Sanchez, L.; Parres, F.; López, J. Mechanical and morphological characterization of PVC plastisol composites with almond husk fillers. *Polym. Compos.* **2007**, *28*, 71–77. [CrossRef]
62. Lommerse, J.P.M.; Price, S.L.; Taylor, R. Hydrogen bonding of carbonyl, ether, and ester oxygen atoms with alkanol hydroxyl groups. *J. Comput. Chem.* **1997**, *18*, 757–774. [CrossRef]
63. Khonakdar, H.; Morshedian, J.; Wagenknecht, U.; Jafari, S. An investigation of chemical crosslinking effect on properties of high-density polyethylene. *Polymer* **2003**, *44*, 4301–4309. [CrossRef]
64. Quiles-Carrillo, L.; Montava-Jordà, S.; Boronat, T.; Sammon, C.; Balart, R.; Torres-Giner, S. On the Use of Gallic Acid as a Potential Natural Antioxidant and Ultraviolet Light Stabilizer in Cast-Extruded Bio-Based High-Density Polyethylene Films. *Polymers* **2019**, *12*, 31. [CrossRef] [PubMed]
65. Essabir, H.; Bensalah, M.O.; Rodrigue, D.; Bouhfid, R.; Qaiss, A.E.K. Biocomposites based on Argan nut shell and a polymer matrix: Effect of filler content and coupling agent. *Carbohydr. Polym.* **2016**, *143*, 70–83. [CrossRef] [PubMed]
66. Tas, C.E.; Hendessi, S.; Baysal, M.; Unal, S.; Cebeci, F.C.; Menceloglu, Y.Z.; Unal, H. Halloysite nano-tubes/polyethylene nanocomposites for active food packaging materials with ethylene scavenging and gas barrier properties. *Food Bioprocess. Technol.* **2017**, *10*, 789–798. [CrossRef]
67. Liu, M.; Guo, B.; Du, M.; Chen, F.; Jia, D. Halloysite nanotubes as a novel β-nucleating agent for isotactic polypropylene. *Polymer* **2009**, *50*, 3022–3030. [CrossRef]
68. Chieng, B.W.; Ibrahim, N.A.; Then, Y.Y.; Loo, Y.Y. Epoxidized Vegetable Oils Plasticized Poly(lactic acid) Biocomposites: Mechanical, Thermal and Morphology Properties. *Molecules* **2014**, *19*, 16024–16038. [CrossRef]
69. Montanes, N.; Garcia-Sanoguera, D.; Segui, V.; Fenollar, O.; Boronat, T. Processing and charac-terization of environmentally friendly composites from biobased polyethylene and natural fillers from thyme herbs. *J. Polym. Environ.* **2018**, *26*, 1218–1230. [CrossRef]
70. Ouajai, S.; Shanks, R. Composition, structure and thermal degradation of hemp cellulose after chemical treatments. *Polym. Degrad. Stab.* **2005**, *89*, 327–335. [CrossRef]
71. Berahman, R.; Raiati, M.; Mazidi, M.M.; Paran, S.M.R. Preparation and characterization of vulcanized silicone rubber/halloysite nanotube nanocomposites: Effect of matrix hardness and HNT content. *Mater. Des.* **2016**, *104*, 333–345. [CrossRef]
72. Castro, D.O.; Ruvolofilho, A.; Frollini, E. Materials prepared from biopolyethylene and curaua fibers: Composites from biomass. *Polym. Test.* **2012**, *31*, 880–888. [CrossRef]
73. Pegoretti, A.; Ashkar, M.; Migliaresi, C.; Marom, G. Relaxation processes in polyethylene fibre-reinforced polyethylene composites. *Compos. Sci. Technol.* **2000**, *60*, 1181–1189. [CrossRef]
74. Shen, L.; Nickmans, K.; Severn, J.; Bastiaansen, C.W.M. Improving the Transparency of Ultra-Drawn Melt-Crystallized Polyethylenes: Toward High-Modulus/High-Strength Window Application. *ACS Appl. Mater. Interfaces* **2016**, *8*, 17549–17554. [CrossRef]
75. Rojas-Lema, S.; Torres-Giner, S.; Quiles-Carrillo, L.; Gomez-Caturla, J.; Garcia-Garcia, D.; Balart, R. On the Use of Phenolic Compounds Present in Citrus Fruits and Grapes as Natural Antioxidants for Thermo-Compressed Bio-Based High-Density Polyethylene Films. *Antioxidants* **2020**, *10*, 14. [CrossRef]
76. González-Fernández, M.J.; Manzano-Agugliaro, F.; Zapata-Sierra, A.; Belarbi, E.H.; Guil-Guerrero, J.L. Green argan oil extraction from roasted and unroasted seeds by using various polarity solvents allowed by the EU legislation. *J. Clean. Prod.* **2020**, *276*, 123081. [CrossRef]
77. Quiles-Carrillo, L.; Fenollar, O.; Balart, R.; Torres-Giner, S.; Rallini, M.; Dominici, F.; Torre, L. A comparative study on the reactive compatibilization of melt-processed polyamide 1010/polylactide blends by multi-functionalized additives derived from linseed oil and petroleum. *Express Polym. Lett.* **2020**, *14*, 583–604. [CrossRef]
78. Klyosov, A.A. *Wood-Plastic Composites*; John Wiley & Sons: Hoboken, NJ, USA, 2007.
79. Fabiyi, J.S.; McDonald, A.G.; Wolcott, M.P.; Griffiths, P.R. Wood plastic composites weathering: Visual appearance and chemical changes. *Polym. Degrad. Stab.* **2008**, *93*, 1405–1414. [CrossRef]
80. Vogler, E.A. Structure and reactivity of water at biomaterial surfaces. *Adv. Colloid Interface Sci.* **1998**, *74*, 69–117. [CrossRef]
81. Lee, J.H.; Park, J.W.; Lee, H.B. Cell adhesion and growth on polymer surfaces with hydroxyl groups prepared by water vapour plasma treatment. *Biomaterials* **1991**, *12*, 443–448. [CrossRef]
82. Sadeh, P.; Najafipour, I.; Gholami, M. Adsorption kinetics of halloysite nanotube and modified halloysite at the Palm oil-water interface and Pickering emulsion stabilized by halloysite nanotube and modified halloysite nanotube. *Colloids Surfaces A Physicochem. Eng. Asp.* **2019**, *577*, 231–239. [CrossRef]
83. Ellison, A.H.; Zisman, W.A. Wettability Studies on Nylon, Polyethylene Terephthalate and Polystyrene. *J. Phys. Chem.* **1954**, *58*, 503–506. [CrossRef]
84. Kuciel, S.; Jakubowska, P.; Kuźniar, P. A study on the mechanical properties and the influence of water uptake and temperature on biocomposites based on polyethylene from renewable sources. *Compos. Part B Eng.* **2014**, *64*, 72–77. [CrossRef]
85. Quiles-Carrillo, L.; Montanes, N.; Garcia-Garcia, D.; Carbonell-Verdu, A.; Balart, R.; Torres-Giner, S. Effect of different compatibilizers on injection-molded green composite pieces based on polylactide filled with almond shell flour. *Compos. Part B Eng.* **2018**, *147*, 76–85. [CrossRef]
86. Chen, W.; Qu, B. Structural characteristics and thermal properties of PE-g-MA/MgAl-LDH exfoliation nanocom-posites synthesized by solution intercalation. *Chem. Mater.* **2003**, *15*, 3208–3213. [CrossRef]

87. Essabir, H.; Elkhaoulani, A.; Benmoussa, K.; Bouhfid, R.; Arrakhiz, F.; Qaiss, A. Dynamic mechanical thermal behavior analysis of doum fibers reinforced polypropylene composites. *Mater. Des.* **2013**, *51*, 780–788. [CrossRef]
88. Liu, D.; Han, G.; Huang, J.; Zhang, Y. Composition and structure study of natural Nelumbo nucifera fiber. *Carbohydr. Polym.* **2009**, *75*, 39–43. [CrossRef]
89. El Mechtali, F.Z.; Essabir, H.; Nekhlaoui, S.; Bensalah, M.O.; Jawaid, M.; Bouhfid, R.; Qaiss, A. Mechanical and thermal properties of polypropylene reinforced with almond shells particles: Impact of chemical treatments. *J. Bionic Eng.* **2015**, *12*, 483–494. [CrossRef]
90. Szczepanik, B.; Słomkiewicz, P.; Garnuszek, M.; Czech, K.; Banaś, D.; Kubala-Kukuś, A.; Stabrawa, I. The effect of chemical modification on the physico-chemical characteristics of halloysite: FTIR, XRF, and XRD studies. *J. Mol. Struct.* **2015**, *1084*, 16–22. [CrossRef]
91. Mellouk, S.; Cherifi, S.; Sassi, M.; Marouf-Khelifa, K.; Bengueddach, A.; Schott, J.; Khelifa, A. Intercalation of halloysite from Djebel Debagh (Algeria) and adsorption of copper ions. *Appl. Clay Sci.* **2009**, *44*, 230–236. [CrossRef]
92. Joussein, E.; Petit, S.; Delvaux, B. Behavior of halloysite clay under formamide treatment. *Appl. Clay Sci.* **2007**, *35*, 17–24. [CrossRef]
93. Frost, R.L.; Kristof, J.; Schmidt, J.M.; Kloprogge, J. Raman spectroscopy of potassium acetate-intercalated kaolinites at liquid nitrogen temperature. *Spectrochim. Acta Part A Mol. Biomol. Spectrosc.* **2001**, *57*, 603–609. [CrossRef]
94. Cheng, H.; Frost, R.L.; Yang, J.; Liu, Q.; He, J. Infrared and infrared emission spectroscopic study of typical Chinese kaolinite and halloysite. *Spectrochim. Acta Part A Mol. Biomol. Spectrosc.* **2010**, *77*, 1014–1020. [CrossRef]
95. Gomez, N.A.; Abonía, R.; Cadavid, H.; Vargas, I.H. Chemical and spectroscopic characterization of a vegetable oil used as dielectric coolant in distribution transformers. *J. Braz. Chem. Soc.* **2011**, *22*, 2292–2303. [CrossRef]
96. Gulmine, J.; Janissek, P.; Heise, H.; Akcelrud, L. Polyethylene characterization by FTIR. *Polym. Test.* **2002**, *21*, 557–563. [CrossRef]

polymers

MDPI

Article

Yield and Selectivity Improvement in the Synthesis of Carbonated Linseed Oil by Catalytic Conversion of Carbon Dioxide

David Alejandro González Martínez, Enrique Vigueras Santiago and Susana Hernández López *

Laboratorio de Investigación y Desarrollo de Materiales Avanzados, Facultad de Química, Universidad Autónoma del Estado de México, Campus Rosedal, Toluca 50200, Mexico; dgonzalezmartinez31@gmail.com (D.A.G.M.); eviguerass@uaemex.mx (E.V.S.)
* Correspondence: shernandezl@uaemex.mx

Abstract: Carbonation of epoxidized linseed oil (CELO) containing five-membered cyclic carbonate (CC5) groups has been optimized to 95% by reacting epoxidized linseed oil (ELO) with carbon dioxide (CO_2) and tetrabutylammonium bromide (TBAB) as catalysts. The effect of reaction variables (temperature, CO_2 pressure, and catalyst concentration) on the reaction parameters (conversion, carbonation and selectivity) in an autoclave system was investigated. The reactions were monitored, and the products were characterized by Fourier Transform Infrared Spectroscopy (FT-IR), carbon-13 nuclear magnetic resonance (^{13}C-NMR) and proton nuclear magnetic resonance (^{1}H-NMR) spectroscopies. The results showed that when carrying out the reaction at high temperature (from 90 °C to 120 °C) and CO_2 pressure (60–120 psi), the reaction's conversion improves; however, the selectivity of the reaction decreases due to the promotion of side reactions. Regarding the catalyst, increasing the TBAB concentration from 2.0 to 5.0 w/w% favors selectivity. The presence of a secondary mechanism is based on the formation of a carboxylate ion, which was formed due to the interaction of CO_2 with the catalyst and was demonstrated through ^{13}C-NMR and FT-IR. The combination of these factors makes it possible to obtain the largest conversion (96%), carbonation (95%), and selectivity (99%) values reported until now, which are obtained at low temperature (90 °C), low pressure (60 psi) and high catalyst concentration (5.0% TBAB).

Keywords: carbonation reaction; selectivity optimization; carbonated epoxidized linseed oil; non-isocyanate polyurethane

Citation: González Martínez, D.A.; Vigueras Santiago, E.; Hernández López, S. Yield and Selectivity Improvement in the Synthesis of Carbonated Linseed Oil by Catalytic Conversion of Carbon Dioxide. *Polymers* **2021**, *13*, 852. https://doi.org/10.3390/polym13060852

Academic Editor: Rafael Antonio Balart Gimeno

Received: 14 January 2021
Accepted: 5 March 2021
Published: 10 March 2021

1. Introduction

Obtaining cyclic carbonates (CCs) has received significant attention because CCs have attractive properties, such as low toxicity, high solubility, and boiling points, can be used as solvents, and have high reactivity with amines [1,2]. Although 5-membered cyclic carbonates (CC5) are less reactive than 6-, 7-, and 8-element cyclic carbonates (CC6, CC7, and CC8), respectively, the synthesis of CC5 has been more studied because CC5 synthesis does not involve the use of toxic precursors such as CS_2, phosgene, and ethyl chloroformate, among other environmentally harmful solvents [3,4]. The most common synthesis method of CC5 remains the insertion of carbon dioxide (CO_2) in cyclic ethers because it is considered a safe process, and the atom economy is close to 100% [4]. This synthesis route has been extensively studied with small molecules such as ethylene and propylene oxide [2,5], while long-chain molecules, such as vegetable oils (VOs), have been less studied. Carbonated vegetable oils (CVOs) have various applications as solvents, lubricants, additives, plasticizers, and monomers for the formation of polymers [6–9]. Currently, the main application is in the synthesis of non-isocyanate polyurethanes (NIPUs) through the aminolysis reaction of CVO because NIPU does not require highly toxic materials such as isocyanates [10–16].

Although CC has been synthesized since 1933 by Carothers et al. [17], it was not until 2004 that Tamami et al. obtained CC5 from epoxidized soybean oil (ESO) and CO_2 for the first time in the presence of TBAB as a catalyst (Scheme 1) [18]. A significant amount of research has focused on developing catalysts or co-catalysts to enhance carbonation kinetics. Generally, catalysts are composed of a Lewis acid for the oxirane's electrophilic activation and a Lewis base that acts as a nucleophile. Some of these catalysts are quaternary phosphate compounds, phosphine complexes, metal complexes, alkali metal halides, quaternary ammonium salts, and ionic liquids, among others [19–25]. Additionally, it has been observed that the catalytic activity increases when the acidity of the cation and the nucleophilicity of the halide increase [19]. Therefore, co-catalyst systems have been proposed, including $CaCl_2$, SiO_2-I, palladium doped with $H_3PW_{12}O_{40}/ZrO_2$, TBAB+$SnCl_4$, and TBAB+H_2O [10,18,19,26]. Among the catalyst systems studied, TBAB is the most commonly used catalyst in the carbonation reaction due to bromine's effectiveness as a leaving group [27]. However, one disadvantage of TBAB is that at high temperatures (150–190 °C), it shows decomposition to volatile components (Hofmann reaction) (Scheme 2) [6].

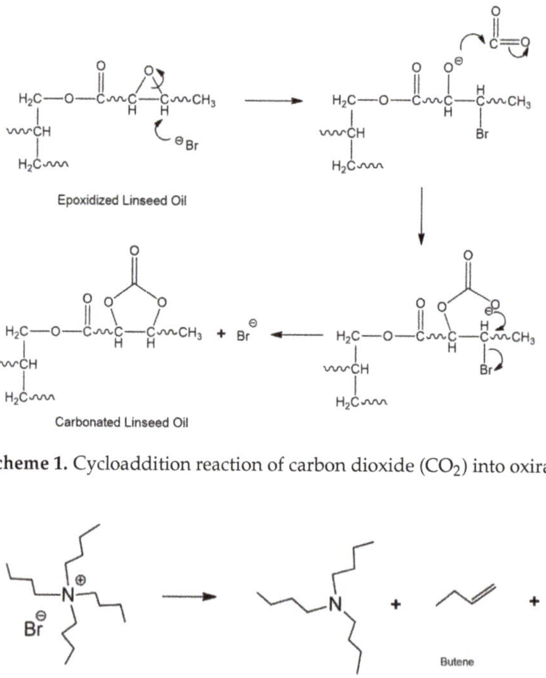

Scheme 1. Cycloaddition reaction of carbon dioxide (CO_2) into oxirane rings [18].

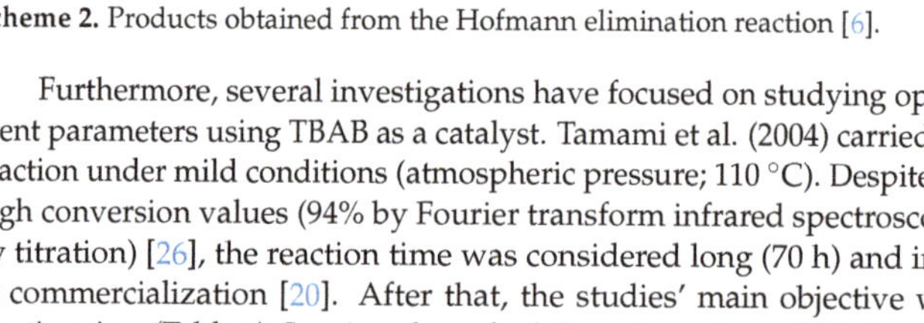

Scheme 2. Products obtained from the Hofmann elimination reaction [6].

Furthermore, several investigations have focused on studying operational and equipment parameters using TBAB as a catalyst. Tamami et al. (2004) carried out the carbonation reaction under mild conditions (atmospheric pressure; 110 °C). Despite obtaining relatively high conversion values (94% by Fourier transform infrared spectroscopy (FT-IR) and 78% by titration) [26], the reaction time was considered long (70 h) and involved an obstacle to commercialization [20]. After that, the studies' main objective was to optimize the reaction time (Table 1). Javni et al. studied the carbonation of ESBO at higher CO_2 pressure (56.5 bar), demonstrating that reaction time and conversion are a function of temperature, pressure, and catalyst concentration [2,18]. Doll et al. intensified the carbonation conditions using CO_2 in the supercritical state (103 MPa, 100 °C) and reduced the reaction time to 40 h and 100% conversion. Similar results were obtained by Mann et al.; however, using supercritical conditions is a disadvantage due to high energy consumption [6,28]. Regarding the efforts to develop equipment to improve carbonation, Mazo et al. used microwave heating (1 atm, 120 °C) and a water ratio (1:3; H_2O/poxy), reaching 87% conversion in 40 h [29]. Zheng et al. proposed using a continuous-flow microwave reactor, obtaining 73% conversion in 7 h (6 bar, 120 °C). Nevertheless, homogeneity and microwave penetration are still problems to scale to the industry level [30].

Despite the progress made in the investigations of carbonation reactions in vegetable oils, there are still some drawbacks. As shown in Table 1, the main objective of these studies is to achieve high conversion degrees, that is, ensuring that the epoxide groups react mostly. However, the carbonation levels that have been reached are not high, reaching no more than 78% [26,31]. Furthermore, this variable is not measured in most of the reported works. The carbonation level allows us to know if effectively the epoxides are converted into cyclic carbonates or if there is the generation of reaction byproducts that may impact the selectivity of the carbonation reaction.

The conversion kinetics of the reaction improves with increasing temperature and pressure. Nevertheless, the impact of these variables on the degree of carbonation and selectivity has not been studied in detail. Therefore, there are differences in the published results, so it has not been possible to establish the reaction conditions that render a higher percent of carbonation and selectivity, not just conversion (Table 1). This is because we consider it relevant to systematically study the influence of temperature, CO_2 pressure and TBAB concentration on the conversion, carbonation and selectivity degree to find a balance of those variables for obtaining the highest reaction parameters.

Table 1. Literature Reports of Carbonation Reaction in Epoxidized Vegetable Oils (EVOs) Using Tetrabutylammonium Bromide (TBAB).

Oil Type	Catalyst Type	Reaction Conditions (Carbonation)				Reaction Results (Carbonation)			Equipment/ Process Characteristics	Reference
		Pressure	Temperature	Time	% Catalyst	% Conversion	% Carbonation	% Selectivity		
ESO	TBAB	1 atm	120 °C	70 h	5%	87%	77%	89%	1:3 (H_2O/Epoxy)	[8]
ELO	TBAB	10 bar	140 °C	96 h	---	91%	26.7%	---	---	[18]
ESO	TBAB	1 atm	110 °C	70 h	5%	94%	---	---	Constant CO_2 flow	[20]
ESO	TBAB	1 atm 57 bar	110 °C 140 °C	89 h 20 h	2.5% 2.5%	63% 100%	--- ---	--- ---	--- ---	[2]
ESO	TBAB	10 bar	120 °C	20 h	3%	71.3%	---	---	Autoclave	[28]
ECSO	TBAB	30 bar	140 °C	24 h	3.75%	99.9%	---	---	Autoclave	[32]
ESO	TBAB	103 bar	100 °C	40 h	5%	100%	---	---	Supercritical CO_2	[7]
ECO	TBAB	5 bar	130 °C	8 h	5%	93.4%	57.7%	61.7%	Oxirane esterification	[33]
ECSO	TBAB	6 bar	120 °C	7 h	8%	73%	---	---	Continuous Flow microwave reactor	[30]
EVNO	TBAB	59 bar	100 °C	46 h	---	95.3%	---	---	Supercritical CO_2 with stirring (150 rpm)	[34]
ESO	TBAB	1 atm	120 °C	40 h	5%	86.7%	77.4%	88.6%	1:3 (H_2O/Epoxy) + Microwave	[29]

ESO = Epoxidized soybean oil; ELO = Epoxidized linseed oil; ECSO = Epoxidized cottonseed oil; ECO = Epoxidized castor oil; EVNO = Epoxidized vernonia oil.

2. Materials and Methods

2.1. Materials

Linseed oil (LO) was purchased commercially through a local distributor (Jalisco, Mexico). LO has a clear yellow oil appearance. The molecular weight (M_W), the number of double bonds (DB), and iodine value (IV) were determined by proton nuclear magnetic resonance (^{1}H-NMR) as described in [9,35,36], rendering 920.5 g/mol, 6.86 (DB), and 189.16 (IV), respectively. IR120 (AIR-120H) Amberlite catalyst (1.8 meq/mL by wetted bed volume), TBAB (≥98.0% assay), solvents such as ethyl acetate (≥99.5% assay) and toluene (≥99.5% assay) were purchased from Sigma-Aldrich Química, S.L. (Toluca, EdoMex, México) Acetic acid (≥99.7%), and hydrogen peroxide (50% concentration) were obtained from Fermont (Monterrey, N.L, México). Industrial grade CO_2 gas (≥99.5%) was purchased from Praxair (Toluca, EdoMex, México). All reagents were used as received except the LO, which was passed through a chromatographic column filled with α-alumina.

2.2. Characterization

The structural analysis of products and the monitoring of epoxidation and carbonation reactions were studied using FT-IR, Carbon-13 nuclear magnetic resonance (^{13}C-NMR), and proton magnetic resonance (^{1}H-NMR) spectroscopies. FT-IR spectra were obtained on an FT-IR Prestige 21 spectrometer, Shimadzu Scientific Instruments, Inc. in México (Tultitlán, EdoMex, México) equipped with a diamond crystal and a horizontal attenuated total reflectance (HART) module. The infrared spectra were obtained in absorbance mode with 64 scans and a resolution of 4 cm^{-1} in the range of 560–4000 cm^{-1}. All FT-IR spectra were normalized to the signal at 1736 cm^{-1} [37], corresponding to the triglyceride ester group's carbonyl vibration. An Avance III spectrometer (Bruker Mexicana, S.A. de C.V, Cd. de México, México) was used for ^{13}C-NMR and ^{1}H-NMR analysis. The analysis was performed at 300 MHz, with a spectrum width of 3689.22 Hz, a pulse width of 4.75 μs, 32 scans at 293 K, 90 pulse width of 9.5 μs. CDC13 was used as the solvent, and tetramethylsilane was used as the internal standard. Through ^{1}H-NMR, the number of epoxide and carbonate groups was quantified, as well as conversion, epoxidation, carbonation, and selectivity values of the reactions [8,13,35].

2.3. Synthesis of Epoxidized Linseed Oil (ELO)

ELO was obtained according to the Prileschajew reaction (industrial method), where DB of LO reacts in situ with a percarboxylic acid (peracetic acid) in the presence of a heterogeneous catalyst (Scheme 3), such as an ion-exchange resin (Amberlite IR120) [38–40]. The conditions of the epoxidation reaction were obtained from a recently optimized methodology [35]. The general procedure consists of placing LO (0.054 mol), toluene (25 mL), acetic acid (0.53 mol/DB), and Amberlite IR120 (12.5 g) inside a three-necked flask equipped with a thermometer, magnetic stirring, and condenser with reflux. The initial mixture was heated to 50 °C, and H_2O_2 (50 wt%) (1.54 mol/DB) was added. Subsequently, the reaction temperature was adjusted to 80°C and maintained under these conditions for 90 min. Immediately, the reaction mixture was cooled to room temperature, and the product was purified. Then, 150 mL of ethyl acetate was added to the mixture and filtered under vacuum to remove the catalyst. The organic phase was washed with 500 mL of a 10% sodium bicarbonate solution to neutralize acetic acid. Subsequently, the organic phase was dried with anhydrous magnesium sulfate and filtered. The solvent is removed using a rotary evaporator and then placed in a vacuum desiccator [37]. Finally, the product obtained was characterized by both FT-IR and ^{1}H-NMR.

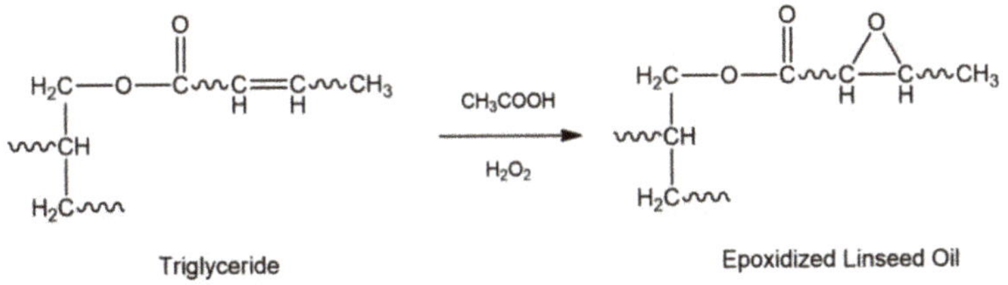

Scheme 3. Epoxidation reaction of vegetable oils (VOs).

2.4. Synthesis of Carbonated Epoxidized Linseed Oil (CELO)

As shown in Table 1, various authors have systematically investigated the effect of different variables, both in the process and in operation, on the epoxidation degree of vegetable oils [8,26,32]. From the results obtained, it has been shown that high epoxidation values can be obtained and are spectroscopically similar to each other (e.g., soybean oil, cottonseed oil, linseed oil). However, in most of the studies, a low carbonation degree was obtained. Therefore, it was necessary to study different reaction conditions to determine the best possible parameters to obtain carbonated oils with the highest carbonation degree and reaction selectivity. The carbonation reaction was monitored as a function of time, temperature, CO_2 pressure, and catalyst concentration, quantifying by ^{1}H-NMR the content of oxirane and carbonate groups to determine the change in the epoxidation, carbonation, and selectivity percentage. The synthesis was carried out in 500 mL Teflon vessel, inserted into a stainless steel reactor equipped with a temperature controller from room temperature to 400 °C and inlets necessary to flow gas and pressurize the reaction in a range from 15 to 200 psi. All the settings of the experimental conditions and the reaction parameters are shown in Table 2. The first variable that was studied was temperature. The reaction temperature was modified from 90 to 120 °C, keeping the pressure (90 psi) and the catalyst concentration (2.5%) constant and setting the temperature that presented the best carbonation reaction results. Subsequently, CO_2 pressure was modified in an interval from 60 to 120 psi, keeping the temperature (90 °C) and catalyst concentration (3.5%) constant. Finally, once the impact of both the temperature and the CO_2 pressure on the reaction performance was defined, the catalyst concentration (from 2.5% to 5.0%) was varied at temperature (90 °C) and constant pressure (60 psi). Finally, once the effect of both temperature and CO_2 pressure on the reaction yield was defined, the catalyst concentration (from 2.5% to 5.0%) was varied at constant temperature (90 °C) and pressure (60 psi). The general methodology consisted of placing 5 g of ELO (M_W = 1007.3 g/mol; 6.26 epoxide content, EC, values obtained by ^{1}H-NMR) and the amount of TBAB (2.5–5.0% mol with respect to EC) inside the reactor. The mixture was stirred manually for 10 min until complete homogeneity was achieved and the reactor was closed. Prior to initiating the reaction, an oxygen-free atmosphere is generated by performing three CO_2 purges. Afterward, the system is adjusted to reaction conditions by holding both temperature (90–120 °C) and CO_2 pressure (60–120 psi) constant for a defined time (12–92 h). After the reaction time was increased, the system was cooled to room temperature and depressurized. ELO purification is initiated by solving the reaction mixture with ethyl acetate. Subsequently, at least 3 or 5 washes with hot water (50 °C) are carried out to remove the catalyst. The organic phase is dried with anhydrous magnesium sulfate. The highest amount of ethyl acetate is removed by distillation at reduced pressure, and the residual solvent is eliminated in a vacuum desiccator for 24 h. Finally, the dry samples were characterized by FT-IR and ^{13}C-NMR. The amount of carbonate groups formed was quantified by ^{1}H-NMR. The runs were performed in triplicate and the results plotted are the average value of each test.

Table 2. Experimental Settings and Average Values of the Reaction Parameters: Conversion, Carbonation and Selectivity Percentages.

Run	Temperature (°C)	Pressure (psi)	Catalyst (%)	Time (h)	Conversion (%)	Carbonation (%)	Selectivity (%)
1	90	90	2.5	24	55.0 ± 0.51	51	92.7
2	100	90	2.5	24	63.2 ± 0.40	53.2	85.8
3	110	90	2.5	24	74.3 ± 0.48	56.5	76.1
4	120	90	2.5	24	85.3 ± 0.59	57.1	66.9
5	90	60	3.5	68	74.0 ± 0.59	66.7	90.1
6	90	90	3.5	68	83.9 ± 0.55	73.6	87.6
7	90	120	3.5	68	88.3 ± 0.61	77.2	87.2
8	90	120	2.5	86	80.8 ± 0.58	70.9	87.8
9	90	120	3.5	86	94.1 ± 0.51	83.2	88.5
10	90	120	5.0	86	96.1 ± 0.56	95.8	99.8

3. Results

3.1. Characterization of Epoxidated Linseed Oil (ELO)

FT-IR spectra were obtained for both the raw material (LO) and the corresponding epoxide (ELO), consistent with those reported by different authors [35,37,41–43]. The most representative LO vibration signals (Figure 1a) correspond to the ester carbonyl group at 1736 cm^{-1} (C=O) and double-bound signals at 3021 cm^{-1} (=C–H), 1652 cm^{-1} (C=C) and 720 cm^{-1} ($HC{=}CH_{cis}$). The other vibration bands correspond to ester carbonyl (1736 and 1159 cm^{-1}), methyl (2922, 1456, and 1377 cm^{-1}), and methylene (2852 and 719 cm^{-1}) groups. In the ELO spectrum (Figure 1b), the epoxy ring's vibration signals are identified at 1250 and 823 cm^{-1} (C–O–C), the last being the most representative. The FT-IR technique allowed us to qualitatively verify the formation of ELO from LO through the disappearance of DB signals and the appearance of bands corresponding to the epoxy rings. Moreover, it is important to mention that there is no evidence of hydrolyzed chains or interruption of ester bonds, which generally occur at 3200–3550 cm^{-1} (hydroxyl zones) and at 1650 cm^{-1} (carboxylic acids), respectively [37,44].

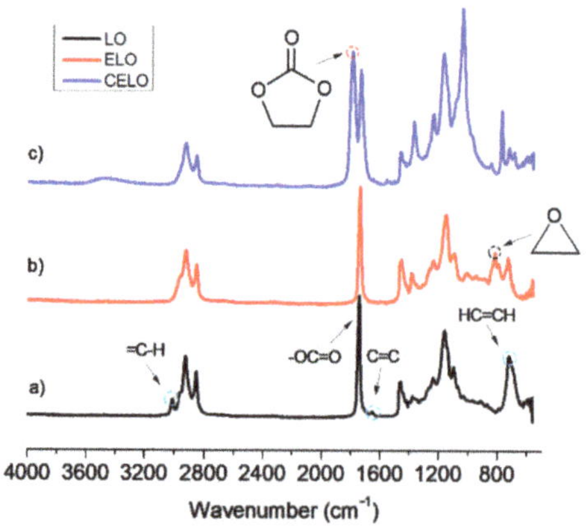

Figure 1. Fourier transform infrared spectroscopy (FT-IR) spectrum of (**a**) linseed oil (LO), (**b**) epoxidized linseed oil (ELO), and (**c**) carbonated epoxidized linseed oil (CELO).

The signals of representative ^{1}H-NMR spectra for LO and ELO are shown in Figure 2, which are consistent with those reported in the literature [35,37,45]. In the ^{1}H-NMR spectrum of LO (Figure 2a), the most characteristic signal belongs to DB (vinyl hydrogens) at 5.25–5.45 ppm (L), which overlaps with the central hydrogen of glycerol (K). The rest of the hydrogen signals are found at 4.10–4.34 ppm (J, glycerol methylene), 2.74–2.80 ppm

(G, internal allylic), 2.27–2.35 ppm (F, α-carboxylic), 1.97–2.13 ppm (E, external allylic), 1.55–1.68 ppm (D, β-carboxylic), 1.23–1.40 ppm (C, aliphatic methyl), 0.94–1.01 ppm (B, fatty acid methylene) and 0.84–0.92 ppm (A, methyl). The ^{1}H-NMR spectrum of ELO (Figure 2b) is similar to that of LO; hence, the presence of an oxirane ring was corroborated mainly with the appearance of the signal in the region of 2.86–3.23 ppm (I, –CHOCH–). Furthermore, a significant decrease and signal shift corresponding to the unreacted vinyl hydrogens (5.6 ppm region) is observed. The epoxide group content present in ELO, as well as the molecular weight of oil, was determined by ^{1}H-NMR, giving values of 1007.3 g/mol and 6.26 epoxide groups, respectively [35].

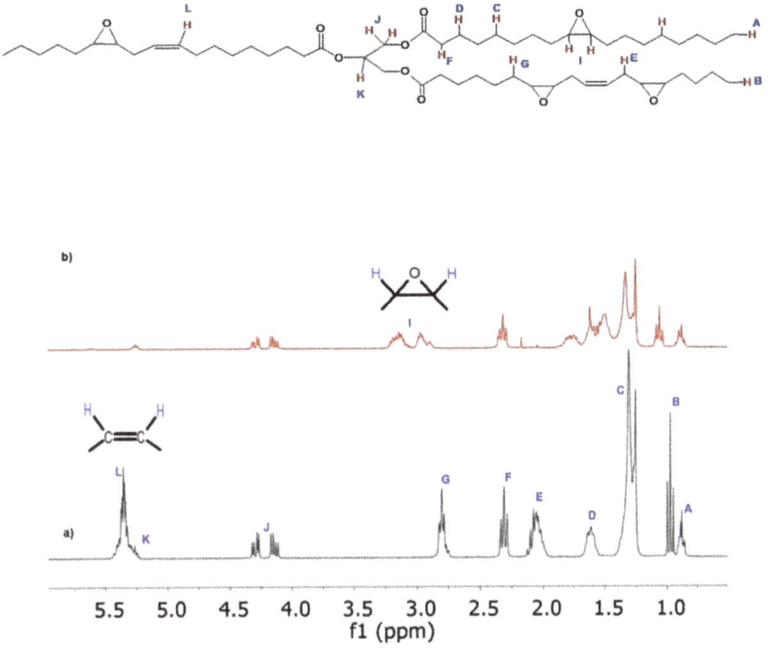

Figure 2. Proton nuclear magnetic resonance (^{1}H-NMR) spectra of (**a**) LO and (**b**) ELO.

3.2. *Characterization of Carbonated Epoxidized Linseed Oil (CELO)*

CELO was obtained by reacting the oxirane groups of ELO obtained above with CO_2 in the presence of TBAB. The reaction mechanism that has been adopted involves a nucleophilic attack by the bromide ion on the epoxy ring. Alkoxide generation is promoted, which in turn carries out a nucleophilic attack on CO_2. Finally, the oxyanion displaces the bromine, generating the corresponding CC5 (Scheme 1) [8,18].

The structural characterization of CELO was carried out using FT-IR, ^{13}C-NMR, and ^{1}H-NMR spectroscopic techniques that are consistent with the results reported in the literature [6,17,20,31,46–48]. To corroborate the formation of the carbonate group, the infrared spectra of ELSO and CELO were compared. In the CELO spectrum (Figure 1c), the carbonate group's vibration signals appear at 1798 and 1045 cm^{-1}. The most notable signal is at 1798 cm^{-1}, which corresponds to the carbonyl of the cyclic carbonate (C=O), while the signal at 1045 cm^{-1} is assigned to the C-O bond. Also, the disappearance of epoxide group signals (823 cm^{-1}) is observed.

As additional qualitative evidence, ^{13}C-NMR spectra were useful to identify the carbonate group's presence in the CELO structure, as reported by Doley and Dolui, Zheng et al. and Liu and Lu [10,47,48]. The ^{13}C-NMR spectrum of CELO (Figure 3b) showed the following carbon signals: end methyl (9.7–13.3 ppm), triglyceride chain methylene (20.2–33.4 ppm), remaining epoxy rings (53.2–57.4 ppm), central methine (61.5 ppm), methylene of glycerol (68.2 ppm), and carbonyls of fatty acids (172.5–173.0 ppm). Carbon signals at 79.2 ppm (C–O), 81.4 ppm (C–O), and 154.1 ppm (C=O) confirm the formation of CELO.

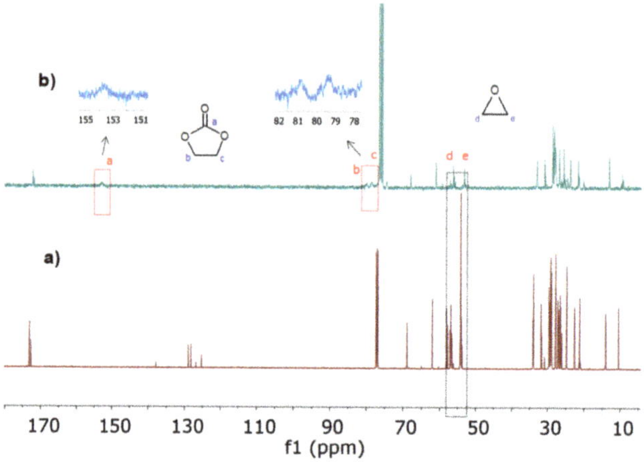

Figure 3. Carbon-13 nuclear magnetic resonance (^{13}C-NMR) spectra of (**a**) ELO and (**b**) CELO.

Figure 4 shows a representative ^{1}H-NMR spectrum of CELO, which was useful for qualitative characterization of CELO and quantifying the carbonation reaction progress. In the ^{1}H-NMR spectrum of CELO (Figure 4), the signal corresponding to the hydrogen of the central carbon of glycerol at 5.25 ppm (K) is observed. In the 2.85–3.28 ppm (I) region, the hydrogen is associated with the unreacted epoxide groups. To corroborate CELO formation, signals associated with cyclic carbonate hydrogens appear in the interval of 4.45–5.10 ppm (M).

The ^{1}H-NMR technique is widely used to characterize materials' chemical structure, both qualitatively and quantitatively [49]. ^{1}H-NMR has been shown to be an effective technique, compared to traditional methods, to determine the chemical composition of triglycerides and their derivatives and reaction parameters [36,50]. The performance of the carbonation reaction was evaluated through conversion (% C), carbonation, or yield (% Y), and selectivity (% S) parameters. The numbers of epoxide and carbonate groups present in CELO were calculated from the integrals of the ^{1}H-NMR spectrum (Figure 4). The central carbon (K) signal was taken as the spectrum normalization factor. The number of epoxide groups (E_m) was calculated with the signals corresponding to epoxy rings (I) (Equation (1)). Similarly, through Equation (2), the number of carbonate groups (C_m) was calculated from the associated signals (M). Regarding the reaction parameters, the conversion, carbonation, and selectivity values were obtained using Equations (3)–(5) [8,29]. The runs were performed in triplicate and the results plotted are the average value of each test:

$$E_m = \frac{I}{2K} \quad (1)$$

$$C_m = \frac{M}{2K} \quad (2)$$

$$\%C = \left(\frac{E_{mi} - E_{mf}}{E_{mi}}\right) \times 100 \quad (3)$$

$$\%Y = \left(\frac{C_m}{E_{mi}}\right) \times 100 \quad (4)$$

$$\%S = \left(\frac{\%Y}{\%C}\right) \times 100 \quad (5)$$

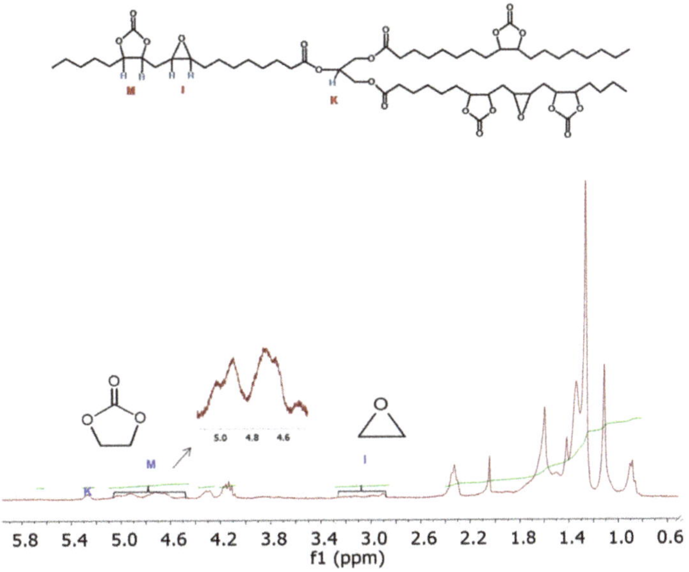

Figure 4. Values obtained by ^{1}H-NMR in the carbonation reaction at 90 °C, 60 psi, 3.5% TBAB, and 68 h.

3.2.1. Effect of Temperature

To determine the optimal conditions for obtaining CELO that contains the highest degree of carbonation (high content of carbonate groups) and selectivity (least amount of byproducts), the carbonation reaction was monitored based on temperature, CO_2 pressure, and the amount of catalyst. In general, in the CELO structure, the epoxide groups show steric hindrance because the epoxides' positions are located in the middle of the molecular chain (25). To obtain high levels of conversion, it is necessary to use a slightly elevated temperature (Scheme 2), avoiding reaching the decomposition temperature of the catalyst (determined as 150–190 °C by Doll et al. [6]). Therefore, temperature is a critical parameter in the carbonation reaction.

Figure 5 shows the effect of reaction temperatures (90, 100, 110, and 120 °C) on the conversion, carbonation, and selectivity degree while keeping the pressure (90 psi), catalyst concentration (2.5% TBAB), and time (24 h) constant. At 90 °C, conversion reaches 55%. When the temperature increases from 90 to 120 °C, the conversion increases from 55 to 85.3%. Similar behavior is observed in the carbonation percentage. By increasing the temperature to 120 °C, carbonate formation increases slightly from 51 to 57.1%. It is observed that the growth rate in carbonation is slower than that in conversion. The opposite behavior is observed in selectivity, whereby increasing the temperature to 120 °C, the selectivity decreases from 92.7 to 66.9%. Through FT-IR, it was detected that as the reaction temperature increased from 90 to 120 °C, the signal corresponding to the hydroxyl zone also increased (3200–3600 cm^{-1}). Usually, in chemical reactions, it is desirable to have high selectivity values. Low values indicate that some of the epoxide groups are not being converted to the desired product, i.e., they are not being formed to cyclic carbonates, which indicates that some alternative reactions.

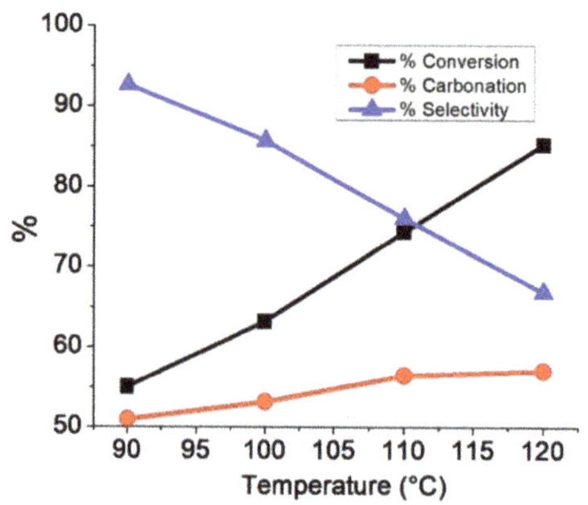

Figure 5. Effect of temperature on the conversion, carbonation, and selectivity degree.

Although the carbonation reaction was carried out below the decomposition temperature of TBAB, evidence of the formation of byproducts at higher temperatures is presented in Figure 6. To verify this statement, the interaction between TBAB and CO_2 was studied in the absence of ELO at 120 °C and 90 psi for 48 h. TBAB decomposition is similar to the Hoffman reaction (Scheme 3). However, due to the presence of CO_2, the first step was the addition of the bromide to the CO_2 molecule to form the carboxylate ion, which extracts beta hydrogen from a butyl substitute of tetrabutylammonium, decomposing into CO_2 and hydrogen bromide, and generating butene and tributylamine as final products (Scheme 4). Figure 7 shows the ^{13}C-NMR spectra of the catalyst before (pure TBAB) and after being subjected to heat treatment in the presence of CO_2 (Figure 7b). The carbon signals of pure TBAB (Figure 7a) are found at 12.2 ppm (*d*, terminal methyl), 18.5 ppm (*e*, methylene), 22.8 ppm (*f*, methylene), and 57.8 ppm (*g*, quaternary amine). In Figure 7b, the formation of tributylamine is confirmed by the appearance of signals at 11.5 ppm (*h*, –CH3), 18.2 ppm (*i*, –CH2–), 23.2 ppm (*j*, –CH2–), and 50.4 ppm (*k*, tertiary amine). Because the spectrum was only obtained from the solid sample, the presence of butene and carboxylate was determined indirectly. However, it is necessary to continue with the byproducts characterization.

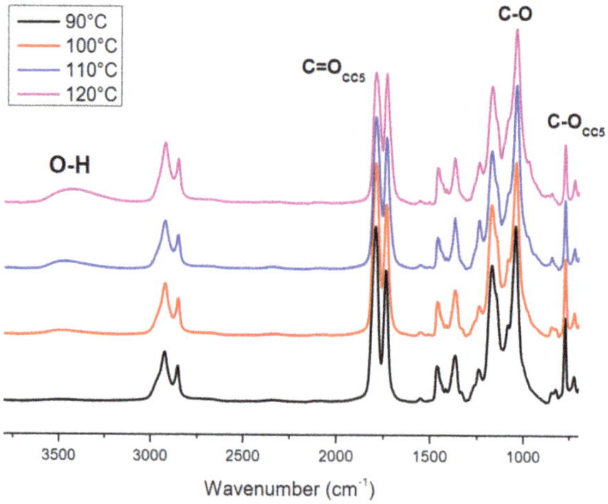

Figure 6. FT-IR spectrum of CELO at different temperatures.

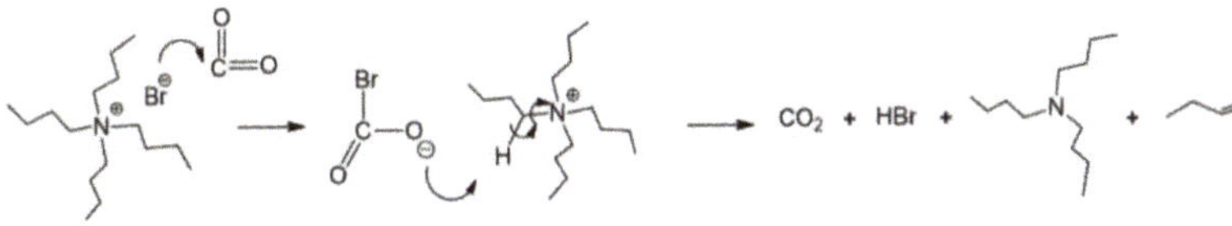

Scheme 4. Byproducts resulting from the interaction between TBAB and CO_2 at 120 °C.

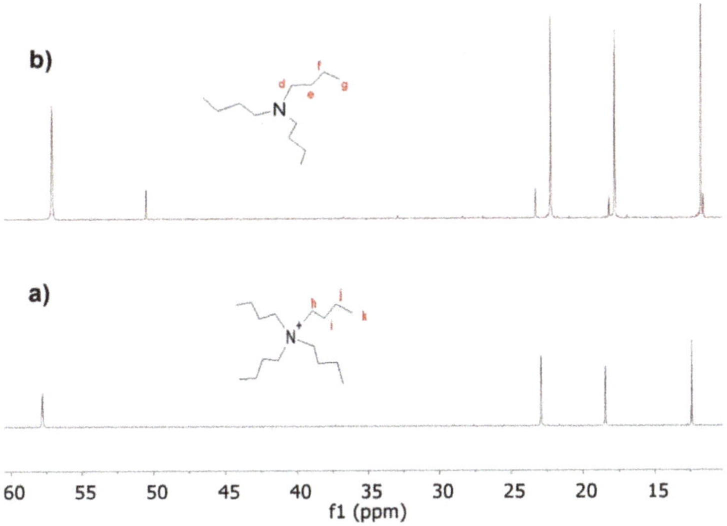

Figure 7. ^{13}C-NMR spectrum of (**a**) TBAB, and (**b**) trimethylamine as a byproduct or the TBAB+CO_2 reaction in absence of ELO at 120 °C, 90 psi, and 48 h.

The formation of carboxylate ions from metal catalysts and CO_2 at elevated pressures is well known [51]. The presence of another additional reaction mechanism to the one reported (Scheme 1) is possible at elevated temperatures and in the presence of the carboxylate ion. Hence, the interaction between epoxide and CO_2 may also depend on the reaction temperature. Similar results were reported by Kiara et al., 1993, who proposed the influence of CO_2 pressure on the reaction mechanisms [52]. At a lower temperature, the bromide ion generates the alkoxide that is subsequently transformed into the corresponding cycle (species 5a). However, when the carbonation reaction is carried out at a higher temperature, there may also be the formation of other species (oligomers) that negatively impact selectivity. A possible mechanism is shown in Scheme 5 (pathway 1). HBr promotes the formation of carboxylate ions, which by nucleophilic addition, interact with epoxide cations to form carboxylates. However, this species is not stable, so it reacts with another carboxylate molecule to form the final macrocyclic dimers (species 5b) or oligomer. However, analyzing closely the spectra in Figure 6 for runs 1 through 4 (Table 2), new bands corresponding to O–H bonds in 2200–3600 cm^{-1} region were observed for experiments at 100, 110 and 120 °C, as well as a broadening of the band from 990–1070 cm^{-1} due to the overlap of new C–O bonds. This broadening was determined from the area under the curve in the 990–1122 cm^{-1} region, from the normalized spectrum. These facts suggest that formation of hydroxyl groups. One explanation is that HBr generated at temperatures above 100 ° C carried out an epoxy ring opening reaction by the mechanism shown in Scheme 5 (pathway 2) forming a bromohydrin, instead of a CC5 (a) or a dicarbonate (c). Table 3 summarizes signals observed in the ^{13}C– and ^{1}H-NMR spectra for run 4 (120 °C), which would correspond to a bromohydrin (c). These assignments are supported by the work of Eren et al. [53]. These signals confirm that an increase in temperature enables the epoxy rings to react in alternative ways, decreasing the selectivity towards the carbonation reaction.

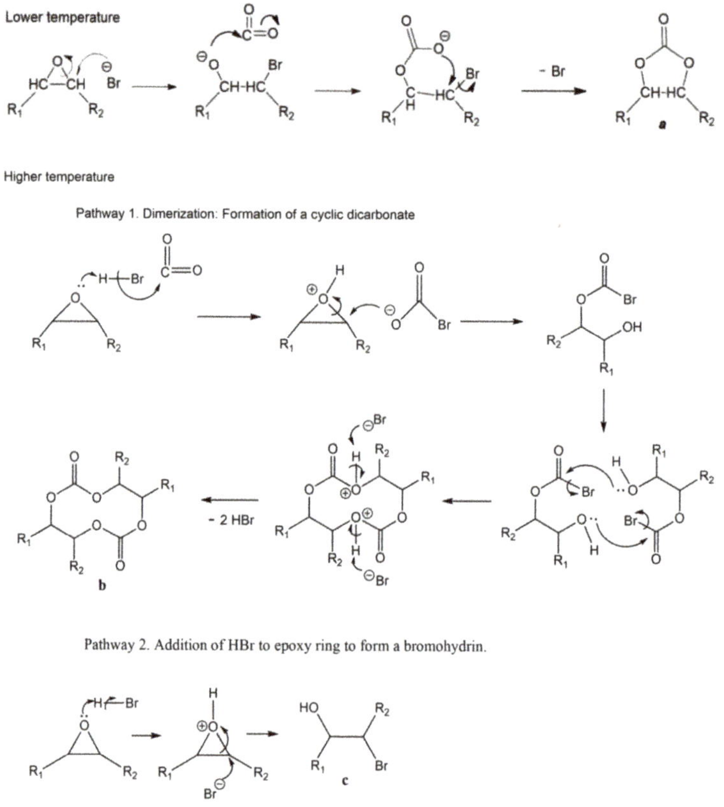

Scheme 5. Possible reaction pathways between CO_2 and epoxides at low and high temperatures.

Table 3. Signals that Corroborate the Formation of a Bromohydrin as a Side Product of the Carbonation Reaction of ELO (Run 4, 120 °C).

NMR Spectra	Chemical Bond	Signals (ppm)	
		Eren et al., 2004 [53]	Experimental
^{13}C-NMR	–CHBr–	64.8	64.4
	–CHOH–	75.3	74.5
^{1}H-NMR	–CHBr–	4.0	4.2
	–CHOH–	3.4	3.7

3.2.2. Effect of Pressure and Catalyst Concentration

Like temperature, pressure is a key parameter in the carbonation reaction, not only because it increases the solubility of CO_2 in oils but it also favors the carbonation reaction since it is the volume reduction reaction and it promotes the interaction between the oil and the catalyst as observed and studied by other authors [28,32,54,55]. Their studies show that the solubility of a gas (CO_2) in the liquid phase (ELO) is a function of the pressure of the system and is related by the Henry coefficient. In turn, this coefficient depends on both the properties of the liquid and the temperature [55]. Results published by Zhang et al. [32] have shown that the solubility of CO_2 increases with pressure, which can favor the conversion of epoxidized vegetable oil (opening of the oxirane ring) due to the higher concentration of CO_2 in the system. Regarding the glyceride series, they showed that the difference in polarity and molecular weight of the tested molecules affects the solubility in CO_2. Monoglyceride is logically more soluble in CO_2 than diglyceride and triglyceride, due to its lower number of carbons, evidence that the solubility of these derivates in carbon dioxide is governed mainly by their molecular weight. A terminal epoxy fatty acid diester was found to be more soluble and more reactive in CO_2 than an internal epoxy fatty acid diester [56,57]. The reactivity centers, i.e., epoxide group of epoxidized fatty esters is more accessible than the ones of epoxidized vegetable oils. Considering temperature and pressure, the higher they are, usually the greater the solubility of the epoxy ester in CO_2. They explain this effect by the fact that the increase in temperature leads to a decrease in the cohesiveness of the oil and, therefore, to an increase in its solubility in CO_2. Furthermore, due to the high molecular weight, the steric hindrance of ELO structure with limitedly accessible internal epoxy groups; it is believed that pressure and temperature can play an important role in increasing yield and selectivity in this particular epoxidized triglyceride. Therefore, in order to increase the degree of carbonation of ELO, it was established to evaluate the carbonation reaction at 90 °C, 3.5% TBAB, 68 h, and under different moderate pressures of CO_2 (60, 90 and 120 psi). The results obtained from the reaction parameters (% V, % C and% S) are presented in Figure 8a. A considerable increase in conversion from 60 to 120 psi is observed (74% and 88.3%, respectively). Cyclic carbonate formation is also enhanced with increasing CO_2 pressure, reaching a maximum carbonation degree of 77.2% at 120 psi. However, the growth rate of carbonation is less than the conversion rate. Hence, the selectivity of the reaction decreases from 90.1 to 87.2% with increasing CO_2 pressure. According to Ochiai and Endo, the mechanism in reactions between oxirane rings and CO_2 depends on the system pressure [46]. Therefore, at high pressures, the generation of carboxylate ions is favored, reducing the reaction's selectivity.

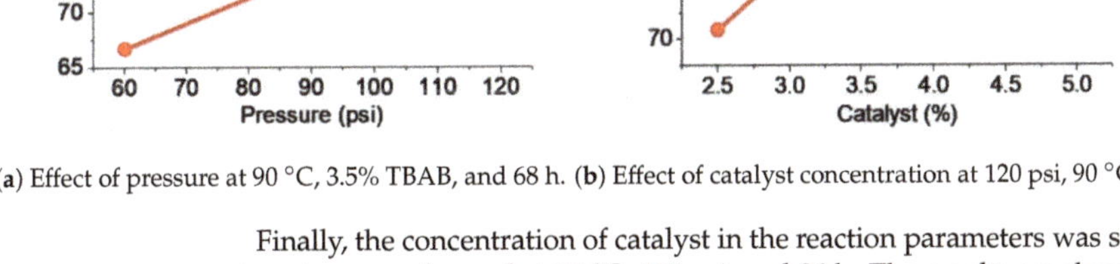

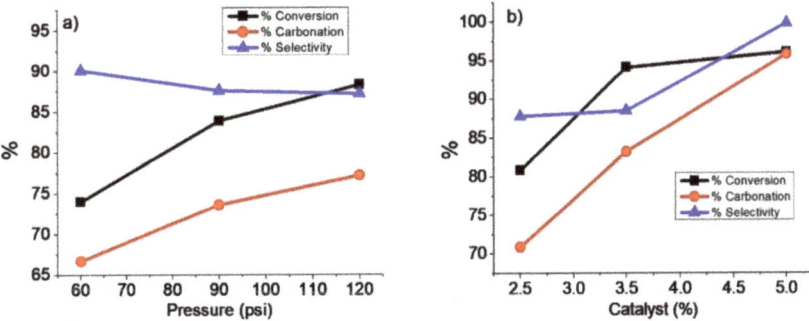

Figure 8. (**a**) Effect of pressure at 90 °C, 3.5% TBAB, and 68 h. (**b**) Effect of catalyst concentration at 120 psi, 90 °C, and 86 h.

Finally, the concentration of catalyst in the reaction parameters was studied. The experiment was performed at 90 °C, 120 psi, and 86 h. The results are shown in Figure 8b. The conversion degrees achieved were 80.8, 94.1, and 96.1% at 2.5, 3.5, and 5.0% TBAB, respectively. The increase in conversion using 2.5 to 3.5% TBAB is greater than 5.0% because as the reaction progresses, the material's viscosity is higher, decreasing the absorption of CO_2. The carbonation degree improves from 70.9 to 95.2% using 2.5–5.0% TBAB, indicating a better interaction between the catalyst, CO_2, and ELO at 5.0%. Similar behavior occurs with selectivity because as the concentration of TBAB increases, selectivity increases, obtaining maximum values of 99%.

The systematic study of the reaction variables made it possible to find the optimal conditions to obtain CELO with a high content of cyclic carbonates and a low content of residual epoxide groups. The temperature was found to be the most critical variable in the reaction. The selectivity value was lower with increasing reaction temperature. In the literature (Table 1), the carbonation reaction was carried out at 120–140 °C to accelerate it. However, they fail to achieve high carbonation percentages. Therefore, in our work, the temperature of 90 °C was determined as the optimum temperature value because at this temperature, the highest selectivity percentage was obtained. Subsequently, the CO_2 pressure was modified, and when the pressure increased, both the conversion and carbonation percentage increased. Although the selectivity decreases slightly as the pressure increases, the pressure at 120 psi has a higher reaction rate than that at 60 psi. Hence, 120 psi was considered acceptable and was selected as the optimum pressure value. Finally, at 90 °C and 120 psi, the catalyst concentration was changed from 2.5 to 5.0%. It was observed that as the catalyst concentration increased, both the conversion, carbonation, and selectivity percentage increased considerably. Therefore, 5% was set as the optimal catalyst concentration value because values of 96.1, 95.2, and 99% conversion, carbonation, and selectivity, respectively, were obtained. Therefore, 5% was set as the optimal catalyst concentration value because, at this concentration, the highest conversion (96.1%), carbonation (95.2%), and selectivity (99%) values were obtained. Although conversion values equal to 100% have been obtained in the literature, carbonation values above 77% and selectivity values above 89% have not been reported until now. Obtaining vegetable oils with a high carbonate content is essential to improve the quality of subsequent polymers such as NIPU with the required properties.

4. Conclusions

The carbonation reaction between epoxidized linseed oil and CO_2 in the presence of TBAB as a catalyst was studied by modifying the parameters of temperature (90–120 °C), CO_2 pressure (60–120 psi), and amount of catalyst (2.5–5.0% TBAB). The presence of CELO was corroborated using FT-IR, ^{13}C-NMR, and ^{1}H-NMR techniques by the appearance of signals corresponding to cyclic carbonates and a decrease in signals belonging to epoxide groups. The different reaction conditions have a significant impact on the conversion, carbonation, and selectivity degree. It was observed that by increasing the reaction temperature from 90 to 120 °C, both the conversion rate and the carbonation percentage are favored; however, selectivity is negatively impacted. An additional reaction mechanism is promoted, based on the generation of carboxyl ions as the result of the interaction between CO_2 and TBAB. Pressure exhibits similar behavior, and high CO_2 pressures (120 psi) also favor side reactions. Contrary to the catalyst concentration, selectivity is favored by increasing the percentage of TBAB in the system. From quantification of the ^{1}H-NMR spectra, it was established that at 90 °C, 60 psi, and 5.0% catalyst, it is possible to obtain the highest values reported up to the moment of conversion equal to 96%, 95% carbonation, and selectivity values of 99%.

Author Contributions: Conceptualization, S.H.L., D.A.G.M., and E.V.S.; validation, S.H.L. and E.V.S.; writing—original draft preparation, D.A.G.M.; writing—review and editing, S.H.L., D.A.G.M., and E.V.S.; visualization, S.H.L., and E.V.S.; supervision, E.V.S. and S.H.L.; funding acquisition, S.H.L. and E.V.S. All authors have read and agreed to the published version of the manuscript.

Funding: This research was funded by Project SIEA-UAEM No. 4965/2020CIB.

Data Availability Statement: The data is not yet publicly available as it is necessary to first present the dissertation paper. Once submitted, the data is made public in the Institutional Repository of the Universidad Autónoma del Estado de México (http://ri.uaemex.mx/ accessed on 1 March 2021).

Acknowledgments: Authors thank CONACYT for scholarship provided to the student (701376). To M.C. Nieves Zavala for the technical support in ^{1}H-NMR measurements (Project SHL2018).

Conflicts of Interest: The authors declare no conflict of interest.

References

1. Kreye, O.; Mutlu, H.; Meier, M.A. Sustainable routes to polyurethane precursors. *Green Chem.* **2013**, *15*, 1431–1455. [CrossRef]
2. Javni, I.; Hong, D.P.; Petrović, Z.S. Soy-based polyurethanes by nonisocyanate route. *J. Appl. Polym. Sci.* **2008**, *108*, 3867–3875. [CrossRef]
3. Rokicki, G.; Parzuchowski, P.G.; Mazurek, M. Non-isocyanate polyurethanes: Synthesis, properties, and applications. *Polym. Adv. Technol.* **2015**, *26*, 707–761. [CrossRef]
4. Cornille, A.; Auvergne, R.; Figovsky, O.; Boutevin, B.; Caillol, S. A perspective approach to sustainable routes for non-isocyanate polyurethanes. *Eur. Polym. J.* **2017**, *87*, 535–552. [CrossRef]
5. Aissa, K.A.; Zheng, J.L.; Estel, L.; Leveneur, S. Thermal stability of epoxidized and carbonated vegetable oils. *Org. Process Res. Dev.* **2016**, *20*, 948–953. [CrossRef]
6. Doll, K.M.; Erhan, S.Z. The improved synthesis of carbonated soybean oil using supercritical carbon dioxide at a reduced reaction time. *Green Chem.* **2005**, *7*, 849–854. [CrossRef]
7. Büttner, H.; Longwitz, L.; Steinbauer, J.; Wulf, C.; Werner, T. Recent developments in the synthesis of cyclic carbonates from epoxides and CO_2. *Top. Curr. Chem.* **2017**, *375*, 1–56. [CrossRef]
8. Mazo, P.; Rios, L. Carbonation of epoxidized soybean oil improved by the addition of water. *J. Am. Oil Chem. Soc.* **2013**, *90*, 725–730. [CrossRef]
9. Dehonor-Márquez, E.; Vigueras-Santiago, E.; Hernández-López, S. Thermal Study of Aluminum Trifluoromethyl Sulfonate as Effective Catalyst for the Polymerization of Epoxidized Linseed Oil. *Phys. Chem.* **2019**, *9*, 1–7. [CrossRef]
10. Zheng, J.L.; Burel, F.; Salmi, T.; Taouk, B.; Leveneur, S. Carbonation of Vegetable Oils: Influence of Mass Transfer on Reaction Kinetics. *Ind. Eng. Chem. Res.* **2015**, *54*, 10935–10944. [CrossRef]
11. Pérez-Sena, W.Y.; Cai, X.; Kebir, N.; Vernières-Hassimi, L.; Serra, C.; Salmi, T.; Leveneur, S. Aminolysis of cyclic-carbonate vegetable oils as a non-isocyanate route for the synthesis of polyurethane: A kinetic and thermal study. *Chem. Eng. J.* **2018**, *346*, 271–280. [CrossRef]
12. Błażek, K.; Datta, J. Renewable natural resources as green alternative substrates to obtain bio-based non-isocyanate polyurethanes-review. *Crit. Rev. Environ. Sci. Technol.* **2019**, *49*, 173–211. [CrossRef]
13. Błażek, K.; Kasprzyk, P.; Datta, J. Diamine derivatives of dimerized fatty acids and bio-based polyether polyol as sustainable platforms for the synthesis of non-isocyanate polyurethanes. *Polymer* **2020**, *205*, 122768. [CrossRef]
14. Suryawanshi, Y.; Sanap, P.; Wani, V. Advances in the synthesis of non-isocyanate polyurethanes. *Polym. Bull.* **2019**, *76*, 3233–3246. [CrossRef]
15. Ke, J.; Li, X.; Jiang, S.; Liang, C.; Wang, J.; Kang, M.; Li, Q.; Zhao, Y. Promising approaches to improve the performances of hybrid non-isocyanate polyurethane. *Polym. Int.* **2019**, *68*, 651–660. [CrossRef]
16. Ghasemlou, M.; Daver, F.; Ivanova, E.P.; Adhikari, B. Bio-based routes to synthesize cyclic carbonates and polyamines precursors of non-isocyanate polyurethanes: A review. *Eur. Polym. J.* **2019**, *118*, 668–684. [CrossRef]
17. Besse, V.; Camara, F.; Voirin, C.; Auvergne, R.; Caillol, S.; Boutevin, B. Synthesis and applications of unsaturated cyclocarbonates. *Polym. Chem.* **2013**, *4*, 4545–4561. [CrossRef]
18. Bähr, M.; Mülhaupt, R. Linseed and soybean oil-based polyurethanes prepared via the non-isocyanate route and catalytic carbon dioxide conversion. *Green Chem.* **2012**, *14*, 483–489. [CrossRef]
19. Kathalewar, M.S.; Joshi, P.B.; Sabnis, A.S.; Malshe, V.C. Non-isocyanate polyurethanes: From chemistry to applications. *RSC Adv.* **2013**, *3*, 4110–4129. [CrossRef]
20. Tamami, B.; Sohn, S.; Wilkes, G.L. Incorporation of carbon dioxide into soybean oil and subsequent preparation and studies of nonisocyanate polyurethane networks. *J. Appl. Polym. Sci.* **2004**, *92*, 883–891. [CrossRef]
21. Yue, S.; Wang, P.; Hao, X. Synthesis of cyclic carbonate from CO_2 and epoxide using bifunctional imidazolium ionic liquid under mild conditions. *Fuel* **2019**, *251*, 233–241. [CrossRef]
22. Saptal, V.B.; Bhanage, B.M. Bifunctional Ionic Liquids Derived from Biorenewable Sources as Sustainable Catalysts for Fixation of Carbon Dioxide. *ChemSusChem* **2017**, *10*, 1145–1151. [CrossRef]
23. Yang, H.; Wang, X.; Ma, Y.; Wang, L.; Zhang, J. Quaternary ammonium-based ionic liquids bearing different numbers of hydroxyl groups as highly efficient catalysts for the fixation of CO_2: A theoretical study by QM and MD. *Catal. Sci. Technol.* **2016**, *6*, 7773–7782. [CrossRef]
24. Monfared, A.; Mohammadi, R.; Hosseinian, A.; Sarhandi, S.; Nezhad, P.D. Cycloaddition of atmospheric CO_2 to epoxides under solvent-free conditions: A straightforward route to carbonates by green chemistry metrics. *RSC Adv.* **2019**, *9*, 3884–3899. [CrossRef]
25. Sun, J.; Zhang, S.; Cheng, W.; Ren, J. Hydroxyl-functionalized ionic liquid: A novel efficient catalyst for chemical fixation of CO_2 to cyclic carbonate. *Tetrahedron Lett.* **2008**, *49*, 3588–3591. [CrossRef]
26. Miloslavskiy, D.; Gotlib, E.; Figovsky, O.; Pashin, D. Cyclic Carbonates Based on Vegetable Oils. *Int. Lett. Chem. Phys. Astron.* **2014**, *27*, 20–29. [CrossRef]
27. Maltby, K.A.; Hutchby, M.; Plucinski, P.; Davidson, M.G.; Hintermair, U. Selective catalytic synthesis of 1,2- and 8,9-cyclic limonene carbonates as versatile building blocks for novel hydroxyurethanes. *Chem. Eur. J.* **2020**, *26*, 7405–7415. [CrossRef] [PubMed]

28. Li, Z.; Zhao, Y.; Yan, S.; Wang, X.; Kang, M.; Wang, J.; Xiang, H. Catalytic synthesis of carbonated soybean oil. *Catal. Lett.* **2008**, *123*, 246–251. [CrossRef]
29. Mazo, P.C.; Rios, L.A. Improved synthesis of carbonated vegetable oils using microwaves. *Chem. Eng. J.* **2012**, *210*, 333–338. [CrossRef]
30. Zheng, J.L.; Tolvanen, P.; Taouk, B.; Eränen, K.; Leveneur, S.; Salmi, T. Synthesis of carbonated vegetable oils: Investigation of microwave effect in a pressurized continuous-flow recycle batch reactor. *Chem. Eng. Res. Des.* **2018**, *132*, 9–18. [CrossRef]
31. Mahendran, A.R.; Aust, N.; Wuzella, G.; Müller, U.; Kandelbauer, A. Bio-Based Non-Isocyanate Urethane Derived from Plant Oil. *J. Polym. Environ.* **2012**, *20*, 926–931. [CrossRef]
32. Zhang, L.; Luo, Y.; Hou, Z.; He, Z.; Eli, W. Synthesis of carbonated cotton seed oil and its application as lubricating base oil. *J. Am. Oil Chem. Soc.* **2014**, *91*, 143–150. [CrossRef]
33. Guzmán, A.F.; Echeverri, D.A.; Rios, L.A. Carbonation of epoxidized castor oil: A new bio-based building block for the chemical industry. *J. Chem. Technol. Biotechnol.* **2017**, *92*, 1104–1110. [CrossRef]
34. Mann, N.; Mendon, S.K.; Rawlins, J.W.; Thames, S.F. Synthesis of carbonated vernonia oil. *J. Am. Oil Chem. Soc.* **2008**, *85*, 791–796. [CrossRef]
35. Dehonor-Márquez, E.; Nieto-Alarcón, J.F.; Vigueras-Santiago, E.; Hernández-López, S. Effective and Fast Epoxidation Reaction of Linseed Oil Using 50 wt% Hydrogen Peroxyde. *Am. J. Chem.* **2018**, *8*, 99–106. [CrossRef]
36. Farias, M.; Martinelli, M.; Bottega, D.P. Epoxidation of soybean oil using a homogeneous catalytic system based on a molybdenum (VI) complex. *Appl. Catal. A* **2010**, *384*, 213–219. [CrossRef]
37. López-Téllez, G.; Vigueras-Santiago, E.; Hernández-López, S. Characterization of linseed oil epoxidized at different percentages. *Superf. Vacío* **2009**, *22*, 5–10.
38. Goud, V.V.; Patwardhan, A.V.; Dinda, S.; Pradhan, N.C. Epoxidation of karanja (*Pongamia glabra*) oil catalysed by acidic ion exchange resin. *Eur. J. Lipid Sci. Technol.* **2007**, *109*, 575–584. [CrossRef]
39. Mungroo, R.; Pradhan, N.C.; Goud, V.V.; Dalai, A.K. Epoxidation of canola oil with hydrogen peroxide catalyzed by acidic ion exchange resin. *J. Am. Oil Chem. Soc.* **2008**, *85*, 887–896. [CrossRef]
40. Janković, M.R.; Govedarica, O.M.; Sinadinović-Fišer, S.V. The epoxidation of linseed oil with In Situ formed peracetic acid: A model with included influence of the oil fatty acid composition. *Ind. Crops Prod.* **2020**, *143*, 111881. [CrossRef]
41. Schönemann, A.; Edwards, H.G. Raman and FTIR microspectroscopic study of the alteration of Chinese tung oil and related drying oils during ageing. *Anal. Bioanal. Chem.* **2011**, *400*, 1173–1180. [CrossRef]
42. Audic, J.L.; Lemiègre, L.; Corre, Y.M. Thermal and mechanical properties of a polyhydroxyalkanoate plasticized with biobased epoxidized broccoli oil. *J. Appl. Polym. Sci.* **2014**, *131*, 39983. [CrossRef]
43. Cheng, W.; Liu, G.; Wang, X.; Liu, X.; Jing, L. Kinetics of the epoxidation of soybean oil with H_2O_2 catalyzed by phosphotungstic heteropoly acid in the presence of polyethylene glycol. *Eur. J. Lipid Sci. Technol.* **2015**, *117*, 1185–1191. [CrossRef]
44. Sharma, B.K.; Adhvaryu, A.; Erhan, S.Z. Synthesis of hydroxy thio-ether derivatives of vegetable oil. *J. Agric. Food Chem.* **2006**, *54*, 9866–9872. [CrossRef] [PubMed]
45. Albarrán-Preza, E.; Corona-Becerril, D.; Vigueras-Santiago, E.; Hernández-López, S. Sweet polymers: Synthesis and characterization of xylitol-based epoxidized linseed oil resins. *Eur. Polym. J.* **2016**, *75*, 539–551. [CrossRef]
46. Alves, M.; Grignard, B.; Gennen, S.; Detrembleur, C.; Jerome, C.; Tassaing, T. Organocatalytic synthesis of bio-based cyclic carbonates from CO_2 and vegetable oils. *RSC Adv.* **2015**, *5*, 53629–53636. [CrossRef]
47. Doley, S.; Dolui, S.K. Solvent and catalyst-free synthesis of sunflower oil based polyurethane through non-isocyanate route and its coatings properties. *Eur. Polym. J.* **2018**, *102*, 161–168. [CrossRef]
48. Liu, W.; Lu, G. Carbonation of epoxidized methyl soyates in tetrabutylammonium bromide-based deep eutectic solvents. *J. Oleo Sci.* **2018**, *67*, 609–616. [CrossRef]
49. Li, Y.; Sun, X.S. Synthesis and characterization of acrylic polyols and polymers from soybean oils for pressure-sensitive adhesives. *RSC Adv.* **2015**, *5*, 44009–44017. [CrossRef]
50. Miyake, Y.; Yokomizo, K.; Matsuzaki, N. Rapid Determination of Iodine Value by 1 H Nuclear Magnetic Resonance Spectroscopy. *J. Am. Oil Chem. Soc.* **1998**, *75*, 15–19. [CrossRef]
51. Ochiai, B.; Endo, T. Carbon dioxide and carbon disulfide as resources for functional polymers. *Prog. Polym. Sci.* **2005**, *30*, 183–215. [CrossRef]
52. Kihara, N.; Hara, N.; Endo, T. Catalytic Activity of Various Salts in the Reaction of 2,3-Epoxypropyl Phenyl Ether and Carbon Dioxide under Atmospheric Pressure. *J. Org. Chem.* **1993**, *58*, 6198–6202. [CrossRef]
53. Eren, T.; Küsefoğlu, S. One step hydroxybromination of fatty acid derivatives. *Eur. J. Lipid Sci. Technol.* **2004**, *106*, 27–34. [CrossRef]
54. Cai, X.; Matos, M.; Leveneur, S. Structure–reactivity: Comparison between the carbonation of epoxidized vegetable oils and the corresponding epoxidized fatty acid methyl Ester. *Ind. Eng. Chem. Res.* **2019**, *58*, 1548–1560. [CrossRef]
55. Cai, X.; Zheng, J.; Wärnå, J.; Salmi, T.; Taouk, B.; Leveneur, S. Influence of gas-liquid mass transfer on kinetic modeling: Carbonation of epoxidized vegetable oils. *Chem. Eng. J.* **2017**, *313*, 1168–1183. [CrossRef]
56. Boyer, A.; Cloutet, E.; Tassaing, T.; Gadenne, B.; Alfos, C.; Cramail, H. Solubility in CO_2 and carbonation studies of epoxidized fatty acid diesters: Towards novel precursors for polyurethane synthesis. *Green Chem.* **2010**, *12*, 2205–2213. [CrossRef]
57. Foltran, S.; Maisonneuve, L.; Cloutet, E.; Gadenne, B.; Alfos, C.; Tassaing, T.; Cramail, H. Solubility in CO_2 and swelling studies by in situ IR spectroscopy of vegetable-based epoxidized oils as polyurethane precursors. *Polym. Chem.* **2012**, *3*, 525–532. [CrossRef]

polymers

MDPI

Article

Development and Characterization of Weft-Knitted Fabrics of Naturally Occurring Polymer Fibers for Sustainable and Functional Textiles

Marcela Ferrándiz [1], Eduardo Fages [2], Sandra Rojas-Lema [2,*], Juan Ivorra-Martinez [2], Jaume Gomez-Caturla [2] and Sergio Torres-Giner [3,*]

1 Textile Industry Research Association (AITEX), Plaza Emilio Sala 1, E-03801 Alcoy, Spain; marcela@gmail.com
2 Technological Institute of Materials (ITM), Universitat Politècnica de València (UPV), Plaza Ferrándiz y Carbonell 1, 03801 Alcoy, Spain; efages@aitex.es (E.F.); juaivmar@doctor.upv.es (J.I.-M.); jaugoca@epsa.upv.es (J.G.-C.)
3 Research Institute of Food Engineering for Development (IIAD), Universitat Politècnica de València (UPV), Camino de Vera s/n, 46022 Valencia, Spain
* Correspondence: sanrole@epsa.upv.es (S.R.-L.); storresginer@upv.es (S.T.-G.)

Abstract: This study focuses on the potential uses in textiles of fibers of soy protein (SP) and chitin, which are naturally occurring polymers that can be obtained from agricultural and food processing by-products and wastes. The as-received natural fibers were first subjected to a three-step manufacturing process to develop yarns that were, thereafter, converted into fabrics by weft knitting. Different characterizations in terms of physical properties and comfort parameters were carried out on the natural fibers and compared to waste derived fibers of coir and also conventional cotton and cotton-based fibers, which are widely used in the textile industry. The evaluation of the geometry and mechanical properties revealed that both SP and chitin fibers showed similar fineness and tenacity values than cotton, whereas coir did not achieve the expected properties to develop fabrics. In relation to the moisture content, it was found that the SP fibers outperformed the other natural fibers, which could successfully avoid variations in the mechanical performance of their fabrics as well as impair the growth of microorganisms. In addition, the antimicrobial activity of the natural fibers was assessed against different bacteria and fungi that are typically found on the skin. The obtained results indicated that the fibers of chitin and also SP, being the latter functionalized with biocides during the fiber-formation process, showed a high antimicrobial activity. In particular, reductions of up to 100% and 60% were attained for the bacteria and fungi strains, respectively. Finally, textile comfort was evaluated on the weft-knitted fabrics of the chitin and SP fibers by means of thermal and tactile tests. The comfort analysis indicated that the thermal resistance of both fabrics was similar to that of cotton, whereas their air permeability was higher, particularly for chitin due to its higher fineness, which makes these natural fibers very promising for summer clothes. Both the SP and chitin fabrics also presented relatively similar values of fullness and softness than the pure cotton fabric in terms of body feeling and richness. However, the cotton/polyester fabric was the only one that achieved a good range for uses in winter-autumn cloths. Therefore, the results of this work demonstrate that non-conventional chitin and SP fibers can be considered as potential candidates to replace cotton fibers in fabrics for the textile industry due to their high comfort and improved sustainability. Furthermore, these natural fibers can also serve to develop novel functional textiles with antimicrobial properties.

Keywords: natural fibers; soy protein; chitin; coir; comfort; functional textiles; Circular Bioeconomy

Citation: Ferrándiz, M.; Fages, E.; Rojas-Lema, S.; Ivorra-Martinez, J.; Gomez-Caturla, J.; Torres-Giner, S. Development and Characterization of Weft-Knitted Fabrics of Naturally Occurring Polymer Fibers for Sustainable and Functional Textiles. *Polymers* **2021**, *13*, 665. https://doi.org/10.3390/polym13040665

Academic Editor: Hong Hu

Received: 3 February 2021
Accepted: 18 February 2021
Published: 23 February 2021

1. Introduction

Over the last several years, valorization of natural resources has become an important topic. Therefore, the idea of using agricultural and food processing by-products [1], or

even wastes of the agricultural and food industries can be very interesting due to it may contribute to the development of the Circular Bioeconomy [2]. In addition, it provides the possibility of develop fiber-based products with new characteristics and applications. More recently, fibrous porous media have also been elucidated with fractal models [3,4].

In this context, an interesting natural and renewable source for textiles fibers is represented by the antimicrobial polysaccharides. Among these, chitin offers, together with cellulose, show the highest worldwide availability. Chitin is a β-(1,4)-N-acetyl-D-glucosamine found in the shells of crabs, lobsters, and shrimps, which are marine fishery by-products. It is biodegradable and can be additionally blended with cellulose or silk [5]. Furthermore, chitin and chitosan, which is obtained from chitin deacetylation, are carbohydrates that exhibit several active and bioactive properties. For instance, some novel functional applications of chitin fibers are focused on the medical sector such as bandages, sutures, etc. since it shows non-toxicity to humans [6]. Chitin also offers antimicrobial protection due to its structure prevents the growth of microorganisms such as bacteria, fungi, and viruses [7–9]. However, despite some chitin fibers have been reported in literature, chitin is not widely used at industrial scale in textile applications. In particular, chitin has a semi-crystalline structure with highly extended hydrogen bonds, which cause some difficulty in its solubilization. In addition, its degree of acetylation (DA), which is normally 0.9, indicates the presence of some amino groups and it also affects its solubility. However, these groups are susceptible to modifications in order to improve this characteristic [10]. Therefore, artificially converting it to fibrous architectures has emerged as a novel approach to modulate their surface activity, structure, and molecular flexibility/rigidity for expanded textile applications. Therefore, various strategies have been adopted to fabricate fibers from chitin, including electrospinning as well as dry and wet spinning [11]. Fibers characteristics vary according to the solvent and carbohydrate contents, showing values of tenacity from ~1.59 to 3.2 g/d and elongation from ~3 to 20% [12].

Proteins represent another novel source for natural fibers derived from agricultural and industrial by-products and wastes [13]. Their properties and functionalities are greatly influenced by their amino acid sequences that initiate the self-assembly into higher order structures with unique physicochemical properties [14]. Among protein-based plants, soy proteins (SPs) are widely processed and consumed all over the world. In particular, different types of SP resins can obtained from soybean harvesting and processing as a by-product [15]. In this process, soy flour is purified to obtain soy protein powder. For that, soybean meal is first centrifuged followed by acidification and neutralization processes [16]. SPs can be classified into four categories, namely 2S, 7S, 11S, and 15S, according to their sedimentation coefficients, in which 7S and 11S surpass 80% of the total amount. β-conglycinin is the most prevalent 7S globulin and composed of three major subunits, namely α (~67 kDa), α′ (~71 kDa), and β (~50 kDa), jointed through noncovalent interactions. Native glycinin (11S globulin, hexamer) is made up of six subunits, owing a molecular weight (M_W) of 300–380 kDa. These subunits are either acidic or basic and form a hollow cylinder by electrostatic interaction, hydrogen, and disulfide bonds [17]. The quaternary structure of SPs is susceptible to changes of external conditions such as pH, temperature, and ionic strength. Moreover, its high β-sheet content (≥40%) provides the peptides necessary for self-assembled fibrillation [18]. During fabrication of amyloid-like fibrils, proteins dissociate and are hydrolyzed into peptides, which then assemble into protofilaments that intertwine in the final stage. This sequential process can be achieved by heating the proteins in an acid environment [19–21], where conditions of 80–95 °C and pH 1.6–2 are commonly used. In general, SP fibers exhibit a worm-like and periodic structure, typically with diameter of a few nm and length of several μm. SP can be molded into a wide variety of shapes, which include fibers, films, hydrogels, nanofibers, and solid parts [22,23]. Moreover, SP can be mixed with other components in order to improve their properties or achieve new ones, such as antioxidant capacity or antimicrobial activity [24]. SP fibers also present a luxurious appearance, good comfort, good chromaticity, which makes them brilliant, and improves dyeing fastness compared to other protein-based fibers, such as silk [25].

Nevertheless, one of the main drawbacks of natural fibers is that they are prone to the biological attack of microorganisms and insects [26]. Microorganism growth depends on their chemical structure, specific surface area, thread thickness, and capacity to retain moisture, oxygen, and nutrients [27–29]. Some of the problems caused by microorganisms are odor emissions, discoloring, degradation, and risk of causing infections. In addition, microorganisms can release enzymes and other organic compounds, which can trigger degradation. Additionally, it is necessary to take into account that the microbial growth can result in a deterioration of the mechanical properties with a loss in tensile strength, elongation, and visual changes [29]. As a result, the use of antimicrobial materials has recently gained importance in a wide variety of applications related to medical clothes and implements, housing, decorative, construction as well as textiles [30]. In this regard, *Escherichia coli* (*E. coli*, gram-negative) and *Staphylococcus aureus* (*S. aureus*, gram-positive) are commonly used to test the antimicrobial properties since both are human pathogens that show high resistance to common antimicrobial agents [31,32]. Furthermore, *Candida albicans* (*C. albicans*), *Trichophyton mentagrophytes* (*T. mentagrophytes*), and *Epidermophyton stockdaleae* (*E. stockdaleae*) are among the fungi tested since they can be located on the skin as well as in mucous membranes and are responsible for systemic infections with high mortality rate [33,34]. All of these microorganism have been associated with several infections due to their growth on the surface of textile fabrics, especially in hospitals or health care institutions [35].

This study originally assesses the potential of natural fibers of chitin, SP, and coir, which can be obtained from agricultural and food wastes, to produce functional fabrics produced by weft knitting for uses in textile applications. To this end, it involves the physical characterization and antimicrobial properties of the natural fibers against the most commonly tested bacteria and fungi. Finally, the thermal and tactile comforts of the natural fiber fabrics were also determined and compared to those of cotton and cotton-based fibers. The tactile comfort was related to mechanical properties and the interaction of the clothes with the human body, while thermal comfort characterization involved thermal resistance, water vapor resistance, and air permeability properties.

2. Materials and Methods

2.1. Materials

Chitin fibers were supplied by Tec Service S.r.L. (Villorba, Italy), whereas the SP fibers were provided by Swicofil (Emmen, Switzerland). According to the supplier, SP fibers are functionalized with undisclosed antimicrobial agents during the spinning process, which can restrain the growth of colon bacillus, impetico bacteria, and sporothric. Coir fibers, obtained from Swarna Trades (Kochi, India), were also tested for comparison purposes. Table 1 summarizes the main physical properties of these natural fibers indicated by their suppliers.

Table 1. Main physical properties of soy protein (SP), chitin, and coir fibers.

Properties	SP Fiber	Chitin Fiber	Coir Fiber
Density (g/cm^3)	1.29	1.43	1.40
Fiber length (mm)	38	37–39	50–150
Linear density (dtex)	0.9–3.0	1.57–1.77	-
Tenacity (g/tex)	38.8–40.8 *	23.4–27.5	10.0
Elongation at break (%)	18–21 *	18–22	30
Moisture content (%)	8.6	< 20	8

* dry properties.

2.2. Yarn Processing

The three types of natural fibers were selected for yarn processing and subsequent weaving. This process started with the opening of the fiber bales, which consists of the extraction of the pressed fiber from its packaging by an automatic machine UNIfloc A 12 from Rieter China Textile Instruments Co., Ltd. (Changzhou, China). The second step was the separation of the flock into homogenous small flakes of fibers to facilitate their disaggregation and individualization in subsequent processes. Finally, the third step was the carded process within the opening and cleaning of the fibers. This last operation completed the individualization of the fibers to obtain a regular tape, which was folded on a coil in a cylinder card. Thereafter, the resultant folded tape was stretched, in order to obtain a wick, which was subjected to a certain twist using a ring twisting machine form MEERA Industries limited (Gujarat, India) to obtain the yarn.

2.3. Fabric Manufacturing

In order to perform the comfort tests, it was necessary to obtain the fabrics from the fibers. Knitting is, after weaving, the second most frequently used method to produce fabrics and its popularity has recently grown due to the adaptability of the new fibers, high versatility, and the consumer demand for stretchable, wrinkle resistant, and snug-fitting fabrics. Knitting to shape can be realized either by weft knitting or by warp knitting. In general, the weft-knitting process is more flexible due to the needle selection capability available and the variety of structural designs possible. In this technique, the same thread feeds an entire mesh pass, progressively feeding all the needles. Therefore, during this process, fabrics of several consecutive rows of intermeshed loops were made in a horizontal way from a single yarn and each consecutive rows of loop built upon the prior loops consecutively. To this end, the small-diameter circular machine Galan Ratera (Manresa, Spain) was used with a speed of 350 rpm. The resultant fabrics are shown in Figure 1.

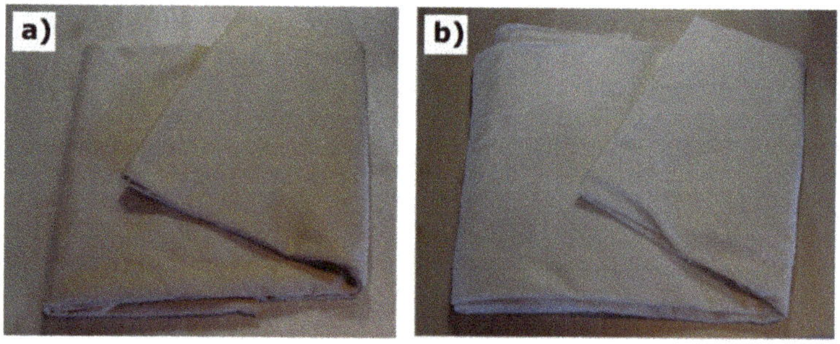

Figure 1. Weft-knitted fabrics of (**a**) soy protein (SP) (**b**) chitin fibers.

2.4. Evaluation of Fiber Geometry and Mechanical Tests

The evaluation of fiber geometry, which is related to the fiber fineness, was carried out according to the classification of Sekhri [36], as shown in Table 2.

Table 2. Fineness classification of fibers.

Type of Fiber	Fiber Fineness (dtex)
Thick fibers	>7
Semi-fine fibers	7–2.4
Fine fibers	2.4–1
Microfibers	1–0.3
Super microfibers	< 0.3

The mechanical properties of the fibers were studied according to the UNE EN ISO 5079-96 and UNE EN ISO 1973:1996 analyzing tensile strength, elongation at break, tenacity, and fineness. The UNE 40152:1984 was followed for determining the length of the fibers. All the mechanical properties were obtained by averaging five experimental values, and all of the samples were stored and tested under room conditions.

2.5. Moisture Content Determination

The ASTM D2495-07 standard test method, applied for moisture determination in cotton, was used to the study moisture content by oven drying. To this end, fibers were previously conditioned at 20 ± 2 °C and $65 \pm 2\%$ RH for 48 h, then, they were weighed and placed for drying at 105 ± 2 °C in an oven for 24 h. After this time, samples were cooled in a desiccator and weighed again. The moisture content was calculated following Equation (1),

$$MC\ (\%) = [(M - D)/M] \times 100 \quad (1)$$

where *MC* represents the moisture content (%), *M* is the mass of the specimen, and *D* is the oven-dry mass of the specimen. Samples were tested in triplicate.

2.6. Antimicrobial Tests

To ascertain the antibacterial and antifungal activities of the fibers, these were subjected to the presence of *E. coli* (ATCC 10536) and *S. aureus* (ATCC 6538) bacteria and *C. albicans* (CECT 1394), *T. mentagrophytes* (CECT 2958), and *E. stockdaleae* (CECT 2988) fungi. All bacteria and fungi were obtained from the Leibniz Institute DSMZ-German Collection of Microorganisms and Cell Cultures (Braunschweig, Germany). The evaluation of the antimicrobial efficacy of the fibers was carried out by following the principles of the AATCC Test Method 100-2004 standard. To this end, fiber samples of approximately 1 ± 0.1 g were introduced in the broth with the bacterial or fungal inoculum. Then, it was plated out by triplicate in agar plates after serial dilutions. Finally, samples were incubated for 18 h at 37 °C. After this time, the bacteria colony-forming units (CFU/mL) were determined for each plate. The microbial reduction or growth inhibition was quantified based on the number of CFU/mL identified after the incubation time for each type of fiber in relation to the number of bacteria initially counted. The same procedure was performed to analyze the antifungal activity.

2.7. Comfort Assessment

The comfort parameter is complex and subjective feature. It depends on both psychological and physiological perceptions of the body and the own fabric properties. It can be divided into tactile comfort and thermal comfort [37]. The comfort generated by a fabric depends, among other factors, on sensory parameters such as perception through the sense of touch, which is much more demanding and subjective than the rest. The aim of this measurements was to ascertain whether the comfort characteristics of the natural fiber fabrics could serve to replace textiles based on cotton. To this end, the KES-FB or Kawabata Evaluation System (Kato Tekko, Kyoto, Japan) was used to analyze the fabric low-stress mechanical and surface properties. It is based on different experimentation modulus: KES-FB1—automated module designed to measure the tension and shear properties of fabrics (tension energy, tension force, and stiffness and shear hysteresis), which applies opposing and parallel forces to the fabric until reach a maximum offset angle; KES-FB2—automated module to measure the flexion of a fabric or non-woven, where measures the force that a given fabric requires to be bended; KES-FB3—automated module to determine the compression properties of fabrics, in which a fabric of ~2 cm^2 is subdued to a compress between two plates and increasing the pressure at a maximum value of 50 gf/cm^2; and KES-FB4—automated module for measuring the surface of fabrics by determining the friction coefficient and geometric roughness with a displacement speed of the sample 1 mm/s [38]. Therefore, through the Kawabata Evaluation System, the three primary hand parameters shown in Table 3 were determined, which are habitually considered for analyzing the applications of fabrics in winter and autumn suiting [39].

Table 3. Primary hand parameters for fabrics following the Kawabata Evaluation System.

Test	Primary Hand Parameters
Koshi	Stiffness and stiffness/elasticity when folded
Numeri	Smoothness
Fukurami	Fullness and softness

The results obtained of tactile comfort in terms of primary hand values (PHVs) and total hand values (THVs) of the fabrics were qualified according to the evaluations described in Tables 4 and 5, respectively [40].

Table 4. Primary Hand Values (PHVs) for fabrics according to the feeling grade.

PHV	Feeling Grade
10	Extremely strong
9	Strong
8	-
7	-
6	-
5	Average
4	-
3	-
2	Weaker
1	Extremely weak
0	No feeling

Table 5. Total Hand Values (THVs) for fabrics.

THV	Grade
5	Excellent
4	Good
3	Average
2	Fair
1	Poor
0	Not useful

The thermal comfort is defined as the condition in which the user feels satisfied with the thermal environment. Therefore, this concept is related to the fabric heat transmission behavior. In this analysis, three different tests were performed: Water vapor resistance, thermal resistance, and air permeability [41]. All these measurements were performed through Skin Model of Weiss Umwelttechnik (Lindenstruth, Germany), whose device allows quantifying the breathability of a given fabric and, therefore, the heat flow by varying the environmental conditions (%RH and temperature).

To compare the comfort results of the natural fabrics with conventional ones, the same tests were carried out on pure cotton and blends of cotton/polyester (35/65) and cotton/acrylic (40/60) fabrics obtained from Tela Prat S.A (Sabadell, Barcelona), which have the same stitch per weft. All the measurements were carried out at 65 ± 3% of RH at 20 ± 1 °C, averaging five experimental values.

2.8. Statistical Analysis

Mean values and standard deviations were calculated from the experimental data obtained by analysis of variance (ANOVA) with 95% confidence interval level ($p \leq 0.05$). For this purpose, a multiple comparison test (Tukey) was followed using the software OriginPro8 (OriginLab Corporation, Northampton, MA, USA).

3. Results

3.1. Geometry and Mechanical Properties of the Natural Fibers

In the clothing industry, it is important to develop textiles with flexibility and elasticity that also offer functional properties that enjoy the growing demand for workwear and sportswear, thereby, improving the wearer's comfort and protection [42,43]. Therefore, mechanical parameters of the fibers have a great importance both in terms of performance and also during the spinning process since they influence on the machine preparation. In Table 6, one can observe the different mechanical values of the natural fibers tested herein and obtained from the test. In this regard, it is important to consider that the characteristics of the natural fibers regarding mechanical properties and chemical composition can vary considerably according to the cultivation place, environmental conditions, extraction and processing methods, the part of the plant from which the fibers are obtained (roots, stem, leaves, etc.), maturity of the plant, among others [44,45]. Fineness, also called linear density, determines the quality and price of the fibers. In general terms, fibers are not constant or regular. Therefore, if a given fiber is thick, it will be stiff, firmer, and with wrinkle resistance. As opposite, if the fiber is fine, it will present softness, flexibility, and good fall [46]. In this sense, Hari [47] reported that fineness for cotton is between 1 and 3 dtex. According to the present results, the SP fiber value was within this range. On the contrary, the chitin fiber showed a higher value, being statistically different, but it was close to the upper limit. According to the evaluation reported in previous Table 2, both natural fibers can be classified as "semi-fine" fibers. These results in fineness are promising because SP and chitin are within the same range of cotton.

Table 6. Mechanical properties of soy protein (SP), chitin, and coir fibers.

Properties	Fineness (dtex)	Length (mm)	Tensile Strength (cN)	Elongation at Break (%)	Tenacity (cN/tex)
SP fiber	2.45 ± 0.88^{a}	108.12 ± 0.74^{a}	7.87 ± 0.63^{a}	16.21 ± 0.98^{a}	32.12 ± 1.10^{a}
Chitin fiber	3.25 ± 1.21^{b}	87.26 ± 0.96^{b}	8.04 ± 0.74^{b}	17.87 ± 0.93^{a}	24.73 ± 0.79^{b}
Coir fiber *	-	98.40	-	30.00	9.80

* Obtained from the technical datasheet of the product; ^{a,b} Different letters in the same column indicate a significant difference among the samples ($p < 0.05$).

Fiber length is another very important feature due to it affects the processability that, in turn, also influences the physical properties. In the case of cotton fibers, their length ranges from 10 to 40 mm [48], being the most used worldwide those with a length between 20 and 35 mm [46]. Besides, it is known that the diameter of the cotton fiber, in most cases, is not more than 1/100 of the fiber length. The results indicate that the here-tested natural fibers are approximately 2 to 4 times longer than those of cotton, which is an indispensable factor to consider when choosing the spinning process and their application in textiles. In addition, another positive aspect of longer fibers is that they will require less twist than shorter fibers for yarn strengthen [47]. In this case, the SP fibers presented the highest value followed by coir and chitin, respectively.

The mechanical properties of the fibers also depend on their constitution, the amount of cellulose, and the crystallinity [49]. In this context, fibers obtained from fruits or seeds are habitually weaker than those obtained from steams of hemp or ramie. In particular, the higher tensile strength and modulus derive from their low microfibrillar angle and high cellulose content. However, it is also important to consider that the fibers strength can be more affected by their own defects than their structure [49]. In this context, SP and chitin, both natural fibers, showed similar values of elongation at break, that is, 16.21, and 17.87%, respectively. The ductility of these natural fibers is higher than that of cotton fibers, which has been reported to show an elongation-at-break value of approximately 7% [50]. For chitin, Fan et al. [7] indicated that blend fibers of chitin at 70 wt% with alginate yielded an elongation at break of 16.4%, which is close to the value attained in the present study. In addition, as it was mentioned above, the mechanical properties of the coir fibers were not analyzes due to their high brittleness, although other studies have reported values of elongation at break between 15 and 30% [51]. For instance, in the study performed by Sumi et al. [52], the elongation at break of coir fibers achieved a value of 16.67%, while Tomczak et al. [45] reported for 25-mm long fibers a value of approximately 25%.

In terms of tenacity, Jackman [53] described that the minimum value for most fiber fabrics is 22.12 cN/tex, though some fibers with lower values can be used in textiles due to they also show high elasticity and resilience. For instance, McKenna [50] reported that tenacity of cotton fiber is around 30 cN/tex, whereas Singh [54] indicated that the cotton fiber tenacity ranges between 25 to 40 cN/tex. Therefore, the previously reported values are close to the tenacity value of 32.12 cN/tex obtained herein for the SP fibers. The chitin fibers achieved a tenacity value of 24.73 cN/tex, which is similar to that reported previously by Pang et al. [55] for chitin/cellulose blended fibers, that is, 23 cN/tex. Therefore, one can consider that both SP and chitin fibers are above the minimum tenacity values required for being used in textiles. Considering that this property is related to their strength, which is the force per unit linear density necessary to break the fiber [36,56], these results make these natural fibers very promising to produce sustainable textiles.

3.2. Moisture Content of the Natural Fibers

Moisture content is one of the most important characteristics to be considered as it may affect the physical properties of fabrics [57]. For instance, it has been reported that moisture absorption for cotton can produce a significant increase in flexibility, toughness, elongation, length, tensile strength, and color, besides affecting the electrical properties [58,59]. Therefore, in textiles, controlling moisture content is necessary as an inadequate level of moisture can lead to fiber breakage, which causes a decrease in the quality of the fiber during harvesting and ginning. Additionally, moisture content of the fiber may also influence the fiber mass, which is an important factor in the textile market [60]. For instance, the increase in fiber strength in cotton is related to an increase in moisture content due to the formation of hydrogen bonding, while other fibers, such as rayon, become weaker when wet [36,56]. Another consequence of water absorption is swelling, which causes the increase in fiber diameter, and consequently, the contraction in the length of twisted ropes. In addition, a low moisture content will require excessive energy at the bale press, whereas a high moisture content can produce a deterioration of the fiber quality during storage due to microorganisms' attack. Furthermore, the control of moisture is also important to facilitate cleaning. Therefore, a good control of this parameter can guarantee the adequate quality of the fibers all over the production and supplying chain [59].

In textiles, this characteristic is normally expressed as the wet percentage material. One can observe in Figure 2 that the value obtained for SP fiber was 5.8%, whereas in the case of coir and chitin fibers, these values were 9.5%, and 9.2%, respectively. These moisture contents were also compared to those of cotton because this is the most popular natural fiber, particularly in clothing [48]. According to the study performed by Higgins et al. [61], the cotton fabric moisture content is around 8%, measured at 20 ± 2 °C with 65 ± 2% of RH. Similar results were attained for cotton fibers and cotton fabrics, that is, 8.5% in both cases [62]. According to Netravali [49], Tomczak et al. [45], and Mohanty et al. [63], the moisture content of coir fibers can be as high as 8%, while Cunha [64] indicated that chitin fiber is relatively hydrophobic. It is important to mention that, in general, plant fibers tend to absorb more quantity of water due to the presence of hydroxyl (-OH) groups in cellulose [48]. Therefore, the values reported, herein, agree with previous results and indicate that SP fiber absorbs less water amounts than the other natural fibers, including cotton. This is a positive factor since it favors its processability, cleaning, and mechanical performance stability. Besides, it is also important to note that the water content of SP fibers is higher than that of synthetic fibers based on polyesters, such as polyethylene terephthalate (PET), which shows a value of approximately 0.4% at 65% RH. However, excessively low moisture may cause static electricity due to friction [58] and it also produces dust and dirt attraction. For these reasons, synthetic fibers are habitually mixed with other hydrophilic material fibers [47].

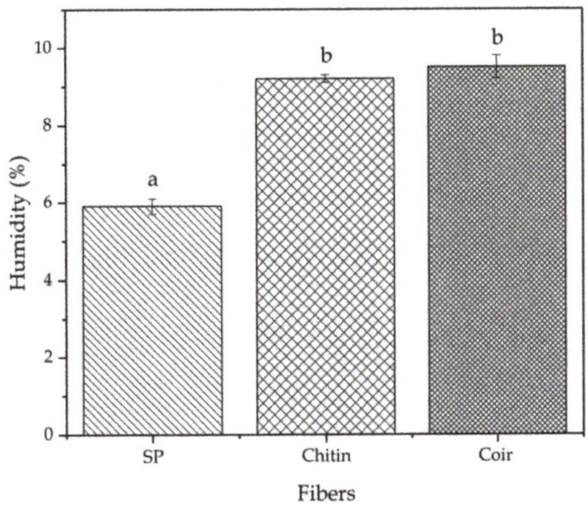

Figure 2. Percentage of moisture content (%) of soy protein (SP), chitin, and coir fibers. [a,b] Different letters indicate a significant difference among the samples ($p < 0.05$).

3.3. Antimicrobial Activity of the Natural Fibers

In the next step, the antimicrobial activity of the fibers was assessed against the *S. aureus* and *E. coli* bacteria and *T. mentagrophytes*, *E. stockdaleae*, and *C. albicans* fungi. The antimicrobial results of the natural fibers are gathered in Figure 3. One can observe in Figure 3a that chitin fibers showed a high antibacterial activity, especially against *E. coli*. In particular, after 18 h of contact, a reduction of 100% and 89.3% in the CFUs/mL of *S. aureus* and *E. coli* was respectively achieved. In this regard, Cheng et al. [28], reported that chitosan, a derivative from chitin, yielded a bacterial reduction for *S. aureus* of 98.57% and 98.59% with contact times of 30, and 50 min, respectively. In addition, for the same contact times, a reduction of 60.87% and 54.35% was observed for *E. coli*. Similarly, Qin et al. [65] developed chitosan fibers with silver sodium hydrogen zirconium phosphate by spinning solution, yielding a reduction of 98.60% in the growth of *S. aureus*. In another study, solution-spun blend fibers of chitin and cellulose were developed by Pang et al. [55]. In the fiber samples containing 6.46% of chitin, a reduction of 65.12% and 55.76% was obtained for *S. aureus* and *E. coli* growth. A high antibacterial activity was also observed by Cheah et al. [66], who developed a chitosan modified membrane with a quaternary amine (P-CS-GTMAC-A), having a reduction value of 76.7% for *E. coli*. The antibacterial activity of chitin and its derivatives involves electrostatic attraction. In the case of chitosan, for example, it has a positive charge that attracts bacterial cells charged negatively that produces the rupture of the bacterial cells and also causes the leak of the intracellular components [66,67]. In relation to the antifungal activity of chitin, shown in Figure 3b, one can observe that a reduction in the *T. mentagrophytes*, *E. stockdaleae*, and *C. albicans* strains of 88.8%, 79.1%, and 88.5% was respectively achieved after 18 h of contact. In the case of *C. albicans*, Qin et al. [65] observed a similar reduction trend by using chitosan fibers, reporting a reduction of 78.62% for this microorganism. Moreover, Surdu et al. [68] also analyzed the effect on the growth of *C. albicans* of cotton fabrics with different contents of chitosan, showing that the antimicrobial effect increased with the quantity of chitosan. In particular, for samples with 5 g/L of chitosan, the antimicrobial efficiency was nearly 83%, yielding an increase of 45% regarding the samples with the lowest chitosan value, that is, 1 g/L. Authors also analyzed the growth of other fungi of the Trichophyton family, that is, *T. interdigitale*, observing a reduction in the microbial growth of only 27% for cotton samples with 5 g/L of chitosan. This value is relatively lower compared with the one obtained in the present study, which can be related to the fact that chitosan was applied in the form of coating in the former study.

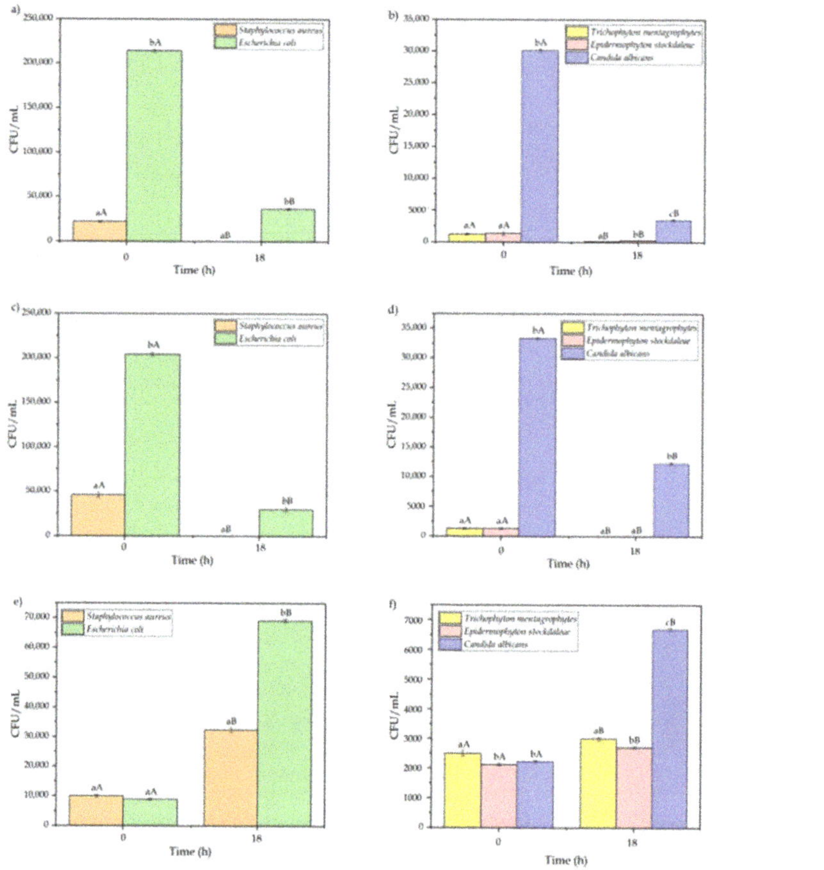

Figure 3. Antibacterial and antifungal activities in terms of colony-forming units (CFU/mL) of chitin fibers (**a**,**b**), soy protein (SP) fibers (**c**,**d**), and coir fibers (**e**,**f**). $^{a–c}$ Different letters in the same period for different samples indicate a significant difference ($p < 0.05$). ^{A,B} Different letter of the same sample in different periods indicate a significant difference ($p < 0.05$).

In relation to the SP fiber, its antibacterial activity against *S. aureus* and *E. coli* is shown in Figure 3c. Similar to the chitin fibers, at the beginning of the process, the bacterial growth was high, especially for *E. coli*. However, a reduction of 100% of the bacterial growth was observed after 18 h of contact. It is worth noting that SP alone has no antimicrobial effect and the present results differ from other studies regarding the use of SP fibers. For instance, soy protein isolated (SPI) film did not presented a decrease of the antimicrobial activity against Gram-positive *S. aureus* after 24 h of incubation [69]. In the same way, Xiao et al. [70] indicated that SPI did not presented antibacterial activity against both *S. aureus* and *E. coli*. On the other hand, in the study performed by Raeisi, et al. [71] it was observed for SPI/gelatin blends a slight decrease in the percentages of colonies (CFU) of 13% against *S. aureus*, while no colonies decrease was observed against *E. coli* after 24 h of incubation. Therefore, the antimicrobial performance of the SP fibers, reported herein, can be ascribed to the antimicrobial compounds added during the spinning process. These biocidal agents are mainly based on phenolics or acidic and bacteriocin since the functional groups present in the side chains of SP are known to alter their retention [72]. Some other examples of biocides added to the textiles include triclosan, silver, and quaternary ammonium compounds [73–76]. The mechanism used by antimicrobial textiles varies depending on their type, but in most cases, the main acting-mechanisms include preventing cell reproduction, damage on cell walls, protein denaturation, and enzyme blocking [77,78]. In relation to the antifungal activity of the SP fibers, which can be seen in Figure 3d, *T. mentagrophytes* and *E. stockdaleae* showed more than 1200 CFU/mL at the beginning of the test but both fungi were fully eliminated after 18 h of contact. In case of *C. albicans*, this reduction was of 63.3%, which still represents a good antifungal activity because it is above 50%. In comparison to other materials, Mishra et al. [79] reported the antibacterial activity of bamboo fibers, which presented lower antibacterial activities against *S. aureus* and *E. coli* with reduction values of 63.08%, and 68.44%, respectively.

As opposed to the chitin and SP fibers, in the case of coir, one can observe that the number of microorganisms increased after 18 h of contact with the fibers. In particular, in Figure 3e one can observe that the number of CFU/mL at the beginning of the test was under 10,000 CFU/mL for both bacteria. However, it increased by approximately 3 and 7 times for *S. aureus* and *E. coli*, respectively. This behavior agrees with the results of the study performed by Chandy et al. [80], in which coir fibers did not present any inhibition zone against *E. coli*. These results mean that the fibers do not show any antibacterial activity. Similar results can be observed in Figure 3f for fungi, showing no evidence of antifungal activity after 18 h of contact. On the contrary, an increase in the number of CFU/mL was observed for all the fungi tested. In particular, the growth of *T. mentagrophytes*, *E. stockdaleae*, and *C. albicans* increased by 16.6%, 21.35%, and 66.6%, respectively. In this regard, Sumi et al. [52] reported that coir fibers do not present good antimicrobial properties against the *Aspergillus niger* fungus.

Therefore, one can conclude that both chitin and SP fibers present a high antibacterial activity against these two types of microorganisms. Additionally, the high increase in the number of bacteria and fungi observed for the coir fibers can be related to its high moisture content, which represents a more favorable medium for the development of microorganisms. However, it is also worth mentioning that the test conditions applied for all of the three types of fibers were the same, taking into account that microorganism growth also depends on temperature, pH, nutrients, oxygen and moisture content, etc. [81].

3.4. Comfort of the Natural Fiber Fabrics

Thermal comfort responds to the thermal environment that, in turns, depends on the temperature of the air and the room, the speed of the air and its humidity as well as on the type of clothing and the activity that is carried out. Therefore, water vapor resistance in fabrics is an important parameter due to the necessity that water vapor on the skin can pass through their pores, while an uncomfortable sensation can be produced if water vapor remains on the skin. In terms of thermal resistance, this parameter will depend on the type of fabric, its thickness, and bulk density, among others, whereas heat transmission involves the three different ways of transfer, that is, convection, conduction, and radiation. Furthermore, air permeability is related to the initial warm or cool feeling when the user wears clothes. If the airflow value is high, it means that the feeling of warm or cool will be more intense [82].

Table 7 shows the results of the thermal comfort analysis obtained through the skin model performed on the chitin and SP fabrics. These values were compared with fabrics of commercial cotton and blends of cotton with polyester and acrylic polymer to ascertain their application in textiles. Cotton is also a natural fiber, but it currently faces some environmental issues derived from the use of pesticides and insecticides to guarantee the good growth of the plants [83]. Furthermore, the coir fabric was discarded because as shown above, it did not achieve the expected properties.

Table 7. The results obtained through skin model for fabrics of soy protein (SP), chitin, cotton, cotton-based fibers.

Fabrics	Water Vapor Resistance ($m^2 \cdot Pa/W$)	Thermal Resistance ($m^2 \cdot K/W$)	Air Permeability (mm/s)
Chitin	4.38 ± 0.11 [a]	0.0426 ± 0.0023 [a]	2716 ± 40 [a]
SP	5.01 ± 0.11 [b]	0.0463 ± 0.0017 [a]	1792 ± 70 [b]
Cotton	4.82 ± 0.07 [c]	0.0425 ± 0.0019 [a]	1638 ± 150 [c]
Cotton/Polyester (65/35)	4.85 ± 0.03 [c]	0.0417 ± 0.0010 [a]	1829 ± 90 [b]
Cotton/Acrylic (60/40)	4.44 ± 0.05 [a]	0.0379 ± 0.0008 [b]	1614 ± 120 [c]

[a–c] Different letters in the same column indicate a significant difference among the samples ($p < 0.05$).

One can observe that the water vapor resistance of the SP fabric was 5.01 $m^2 \cdot Pa/W$, which was slightly but still significantly higher than that of cotton, that is, 4.82 $m^2 \cdot Pa/W$, and also than the combinations of cotton with polyester and acrylic polymer, that is, 4.85, and 4.44 $m^2 \cdot Pa/W$, respectively. In the case of the chitin fabric, the water vapor resistance value was 4.38 $m^2 \cdot Pa/W$. These results are related to the moisture absorption capacity of each natural fiber, which agree with the moisture content shown in previous Figure 2, and also to its lower air permeability [82]. The SP fibers presented the lowest water absorption and showed the highest water vapor resistance. Moreover, according to the results obtained in this study, it can be concluded that the chitin fabric behaves similarly to that of cotton/acrylic, while the SP fabric would be more similar to that of neat cotton. These results indicate that the fabrics studied herein, being in a range of 4–5 $m^2 \cdot Pa/W$, would have good breathability to water vapor. In comparison with previous works, Mishra et al. [79] reported that the water vapor resistance for cotton fabrics was 3.82 m2·Pa/W, being this value lower than that of the chitin and SP fabrics obtained in this study. This result can be explained by the fact that the previous cotton fabrics showed a fineness of 1.23 dtex, which is also lower than the present natural fibers.

Regarding the thermal resistance of the fabrics, it can be observed that the SP one also presented the highest value, that is, 0.0463 $m^2 \cdot K/W$. In the case of the chitin fabric, it showed a value of 0.0426 $m^2 \cdot K/W$. However, it is important to mention that the thermal resistance of both fabrics was similar to those of cotton and cotton/polyester, which did not present significant differences, whereas the cotton-based fabric containing the acrylic polymer presented the lowest value. Thermal resistance is a very important parameter for fabrics intended for underwear applications, for example, in which cotton is mostly used. Therefore, in view of these results, it can be indicated that both the SP and chitin fabrics can be suitable for this type of application. For other types of textiles where the human body protection becomes more important due to climatic fluctuation, the SP fabric is the only one expected to provide the required properties.

Finally, the air permeability is the measure of air flow passed through a given area of a fabric, which is widely used in the textile industry for outdoor garment manufacturers to describe their functional performance. One can observe that the chitin fabric presented higher permeability than the pure cotton, showing a value of 2716 mm/s. In particular, this value was nearly 2 times higher than that of the cotton fabric, which was 1638 mm/s. Meanwhile, the SP fiber air permeability did not present a significant difference versus the cotton and cotton/polyester fabrics. The higher values attained herein can be related to the lower hairiness of these natural fibers due to they presented longer fiber lengths [82]. In the case of chitin, the high value of air permeability can be additionally related to high fineness of this natural fiber, which was 3.25 dtex, by which more air could penetrate through the fabric [41]. Therefore, considering the values obtained, herein, one can conclude that fibers of chitin and SP are more suitable to produce fabrics for summer clothing because in these fabrics the pass of the air is more favored.

Other important characteristic of fabrics is tactile sensation, which is crucial when determining the degree of comfort. The fabric hand properties allow knowing the performance of the material in contact with the skin, which depends on different factors such as fiber type, fabric structure, and, in some cases, the fabric finishing. For fabrics, there are two types of performance analysis. The first one is called utility performance and it is related to strength, color, and shrinkage, among others. The second one is quality performance, which is related to the appearance and comfort. It is important to consider that the second analysis is more difficult to conduct than the first one [84]. The assessment used to carry out the quality performance analysis is hand evaluation, being considered as a subjective method and known as Kawabata. Through this method the three PHVs, namely Koshi (stiffness), Numeri (smoothness), and Fukurami (fullness and softness), and also THVs were calculated in order to characterize the different fabrics. PHV is normally rated on a scale from 0 (weak) to 10 (strong), whereas THV is rated from 0 (not useful) to 5 (excellent) [85]. In Table 8 the results of the tactile comfort studied using Kawabata Evaluation System are gathered.

Table 8. Results obtained through Kawabata of the primary and total hand values (PHVs and THVs) of men's winter suit fabrics of soy protein (SP), chitin, cotton, and cotton-based fibers.

Fabric	PHVs						THVs	
	Koshi		Fukurami		Numeri			
	Value	Grade	Value	Grade	Value	Grade	Value	Grade
SP	7.71 ± 0.08 [a]	Strong	4.70 ± 0.21 [a]	Average	4.06 ± 0.05 [a]	Average	3.23 ± 0.06 [a]	Average
Chitin	8.40 ± 0.04 [b]	Strong	3.94 ± 0.15 [b]	Average	6.15 ± 0.04 [b]	Average	3.42 ± 0.07 [b,c]	Average
Cotton	7.90 ± 0.05 [b,c]	Strong	4.44 ± 0.11 [a,b]	Average	8.56 ± 0.07 [c]	Strong	3.72 ± 0.09 [c]	Good
Cotton/polyester (35/65)	7.44 ± 0.03 [c]	Strong	5.43 ± 0.09 [c]	Average	6.81 ± 0.12 [d]	Average	3.60 ± 0.07 [c]	Good
Cotton/acrylic (40/60)	8.16 ± 0.10 [b]	Strong	1.54 ± 0.19 [d]	Weak	3.77 ± 0.03 [e]	Weak	2.72 ± 0.13 [d]	Average

[a–e] Different letters in the same column indicate a significant difference among the samples ($p < 0.05$).

The fabric PHV concept covers the stiffness, fullness and softness of the fabric, and smoothness, which allows to quantify comfort based on the determination of the mechanical properties [86]. As seen in the Table 8, chitin presented values without significative differences, regarding cotton, and cotton/acrylic in Koshi parameter, which are related to the stiffness/elasticity when folded. In general, all samples values were ranged between 7.5 and 8.5, which indicates that they were high values. According to the reference used by Kawabata et al. [39], it can be considered that all of the samples presented a high-quality suiting for winter-autumn because their values are within the range of $4.5 \leq HV \leq 6.5$. In the case of Fukurami, which indicates the fullness and softness in terms of body feeling and richness, it can be observed that both the SP and chitin fabrics showed relatively similar values than pure cotton. Nevertheless, it is important to consider that, within the current samples, the only value that was in the range of $5 \leq HV \leq 8$ of high-quality suiting for winter-autumn was the cotton/polyester blend. Regarding Numeri, which determines the smoothness and it is ascribed to a smooth, flat, soft touch fabric, both chitin and SP fabrics showed lower values, that is, 4.06 and 6.15, respectively, than the pure cotton, which presented the highest value, that is, 8.56. However, the values attained for chitin fibers is in the range of $6 \leq HV \leq 8$, which is considered as high-quality suiting for winter-autumn. Furthermore, the THV was also determined since this parameter is used to indicate the quality of the fabric for a specified end use and it gives a general idea of a fabric hand [87,88]. Therefore, the higher this value, the better the fabric touch and the more comfortable is the textile. The neat cotton fabric showed the highest value, that is, a THV of 3.72. As indicated in the table, it is considered as good, being relative similar to that reported by Behera [82], that is, 3.01. While, chitin presented no significant difference compared to cotton.

The results also indicated that the cotton/polyester fabric presents a good value, whereas in the case of SP and chitin their values are in the range that corresponds to average. The fabric with the lowest THV was attained for the cotton/acrylic combination, with a value of 2.72, although it could be still considered as average for this type of application. However, it must be necessary a $THV \geq 4$ for winter-autumn and a $THV \geq 3.5$ for midsummer, according to the criteria for an ideal fabric [39]. Based on this evaluation, only pure cotton and the blend of cotton/polyester fulfill this requirement for midsummer with values of 3.72, and 3.60, respectively. As reported by Sun et al. [89], the PHVs for Koshi, Numeri, and Fukurami and THV were 4.78, 0.05, 3.04, and 2.03, respectively. Therefore, it can be said that the stiffness and softness values were within the good range, however, the smoothness value was low. Thus, THV was also low and the fabric can be considered weak for winter-autumn clothes. In the study carried out by Verdu et al. [90], the behavior of the cotton/polyester blend in proportions of 35/65 showed values of PHV for Koshi, Numeri, and Fukurami and THV of 5.8, 4.6, 4.4, and 3.1, respectively. All comfort values were within the range of average, which indicates that the fabric can be potentially applied for winter clothes, while the THV parameter indicated that the fabric has good quality. From the above, one can consider that the SP fabric and, more importantly, the chitin fabric presented Koshi (stiffness), Fukurami (fullness and softness), and total hand values (THV) close to those of cotton fabrics, which opens up their potential applications in sustainable textiles.

4. Conclusions

Over the last years, new chemical and synthetic fibers have been developed to fulfill different technical requirements of the textile industry. These fibers result from chemical processes applied to natural fibers or, more habitually, derive from synthetic polymers. However, synthetic fibers show several sustainability issues due to the polymers used for their manufacturing derives from petroleum, which is not a renewable source, and the resultant textiles are not biodegradable or compostable, that is, they do not disintegrate in natural environments or in controlled compost conditions, respectively. In this regard, the valorization of wastes or processing by-products of the agricultural and food industries can potentially increase the economic value of natural fibers and also provide enhanced environmental benefits that improve the overall sustainability of the textiles. The present study reported on the use of chitin and SP fibers as an environmentally friendly alternative to chemical and synthetic fibers, whereas fabrics of both fibers were also considered as potential candidates to replace cotton. In terms of mechanical properties, both natural fibers presented values similar to cotton for fineness and tenacity. Furthermore, the chitin and SP fibers were longer than common cotton fibers, which is something desirable in textiles since their yarns will be less prone to twist when knitted. Moreover, in the case of SP, these fibers showed the lowest water absorption value, which is a positive aspect to improve the mechanical stability and reduce microbial growth. The antimicrobial tests also demonstrated that the chitin and SP fibers are very promising to avoid the growth of bacteria and fungi. In the case of chitin, its antimicrobial properties are ascribed to the interaction between positively charged biopolymer molecules and negatively charged microbial cell membranes. For SP, these fibers were functionalized by the manufacturer with microbial resistant elements integrated in the fiber molecule chains during the spinning process. The thermal comfort analysis also revealed that thermal resistance of the chitin and SP fibers had similar values than cotton whereas, regarding air permeability, both fabrics showed higher values than the cotton fabric. The lowest value was observed for chitin due to its higher fineness, indicating that these novel fabrics allow the air pass more easily through the fabric and, thus, can be potentially considered for summer clothes. For water vapor, the SP fabric presented higher resistance than chitin, which is related to the low moisture content of these fibers. In terms of PHV, the stiffness for the SP and chitin fabrics turned out to be high, being more than the average, however softness and fullness had similar values to that of cotton. Alternatively, smoothness presented more variability among the tested fabrics, showing cotton the highest value. In addition, the THV parameter showed adequate values for SP and chitin, being in the range of average, while the cotton/polyester fabric showed the only value that corresponded to the good range to meet the requirements for winter-autumn cloths.

Author Contributions: Conceptualization, S.T.-G.; methodology, S.R.-L. and J.I.-M.; validation, J.I.-M. and S.T.-G.; formal analysis, S.R.-L., J.G.-C. and J.I.-M.; investigation, S.R.-L., M.F. and J.I.-M.; data curation, E.F., J.G.-C., J.I.-M. and S.R.-L.; writing—original draft preparation S.R.-L.; writing—review and editing, S.T.-G.; visualization, S.R.-L.; funding acquisition, S.T.-G. All authors have read and agreed to the published version of the manuscript.

Funding: This research work was funded by the Spanish Ministry of Science and Innovation (MICI) project number MAT2017-84909-C2-2-R.

Institutional Review Board Statement: Not applicable.

Informed Consent Statement: Not applicable.

Data Availability Statement: Data is contained within the article.

Acknowledgments: S.R-L is a recipient of a Santiago Grisolia grant from Generalitat Valenciana (GVA) (GRISOLIAP/2019/132). S.T.-G. acknowledges MICI for his Ramón y Cajal contract (RYC2019-027784-I).

Conflicts of Interest: The authors declare no conflict of interest.

References

1. Jiang, B.; Wang, L.; Wang, M.; Wu, S.; Wang, X.; Li, D.; Liu, C.; Feng, Z.; Chi, Y. Direct Separation and Purification of α-Lactalbumin from Cow Milk Whey by Aqueous Two-phase Flotation of Thermo-sensitive Polymer/Phosphate. *J. Sci. Food Agric.* **2021**. [CrossRef]
2. Jiang, B.; Na, J.; Wang, L.; Li, D.; Liu, C.; Feng, Z. Reutilization of food waste: One-step extration, purification and characterization of ovalbumin from salted egg white by aqueous two-phase flotation. *Foods* **2019**, *8*, 286. [CrossRef] [PubMed]
3. Liang, M.; Fu, C.; Xiao, B.; Luo, L.; Wang, Z. A fractal study for the effective electrolyte diffusion through charged porous media. *Int. J. Heat Mass Transf.* **2019**, *137*, 365–371. [CrossRef]
4. Xiao, B.; Huang, Q.; Wang, Y.; Chen, H.; Chen, X.; Long, G. A fractal model for capillary flow through a single tortuous capillary with roughened surfaces in fibrous porous media. *Fractals* **2020**. [CrossRef]
5. Rinaudo, M. Chitin and chitosan: Properties and applications. *Prog. Polym. Sci.* **2006**, *31*, 603–632. [CrossRef]
6. Lee, J.-W. Antimicrobial Agents and Applications on Polymeric Materials. *Text. Color. Finish.* **2008**, *20*, 39–56. [CrossRef]
7. Fan, L.; Du, Y.; Zhang, B.; Yang, J.; Cai, J.; Zhang, L.; Zhou, J. Preparation and properties of alginate/water-soluble chitin blend fibers. *J. Macromol. Sci. Part A Pure Appl. Chem.* **2005**, *42*, 723–732. [CrossRef]
8. Hsieh, S.H.; Huang, Z.; Huang, Z.; Tseng, Z. Antimicrobial and physical properties of woolen fabrics cured with citric acid and chitosan. *J. Appl. Polym. Sci.* **2004**, *94*, 1999–2007. [CrossRef]
9. Pan, Y.; Huang, X.; Shi, X.; Zhan, Y.; Fan, G.; Pan, S.; Tian, J.; Deng, H.; Du, Y. Antimicrobial application of nanofibrous mats self-assembled with quaternized chitosan and soy protein isolate. *Carbohydr. Polym.* **2015**, *133*, 229–235. [CrossRef] [PubMed]
10. Pillai, C.; Paul, W.; Sharma, C.P. Chitin and chitosan polymers: Chemistry, solubility and fiber formation. *Prog. Polym. Sci.* **2009**, *34*, 641–678. [CrossRef]
11. Ifuku, S.; Saimoto, H. Chitin nanofibers: Preparations, modifications, and applications. *Nanoscale* **2012**, *4*, 3308–3318. [CrossRef]
12. Dutta, P.K.; Ravikumar, M.; Dutta, J. Chitin and chitosan for versatile applications. *J. Macromol. Sci. Part C Polym. Rev.* **2002**, *42*, 307–354. [CrossRef]
13. Mohammad, F. Sustainable natural fibres from animals, plants and agroindustrial wastes—An overview. In *Sustainable Fibres for Fashion Industry*; Springer: Berlin/Heidelberg, Germany, 2016; pp. 31–44.
14. Yigit, S.; Dinjaski, N.; Kaplan, D.L. Fibrous proteins: At the crossroads of genetic engineering and biotechnological applications. *Biotechnol. Bioeng.* **2016**, *113*, 913–929. [CrossRef]
15. Burgos, N.; Valdés, A.; Jiménez, A. Valorization of agricultural wastes for the production of protein-based biopolymers. *J. Renew. Mater.* **2016**, *4*, 165–177. [CrossRef]
16. Leceta, I.; Etxabide, A.; Cabezudo, S.; de la Caba, K.; Guerrero, P. Bio-based films prepared with by-products and wastes: Environmental assessment. *J. Clean. Prod.* **2014**, *64*, 218–227. [CrossRef]
17. Nishinari, K.; Fang, Y.; Guo, S.; Phillips, G.O. Soy proteins: A review on composition, aggregation and emulsification. *Food Hydrocoll.* **2014**, *39*, 301–318. [CrossRef]
18. Herrero, A.M.; Jiménez-Colmenero, F.; Carmona, P. Elucidation of structural changes in soy protein isolate upon heating by Raman spectroscopy. *Int. J. Food Sci. Technol.* **2009**, *44*, 711–717. [CrossRef]
19. Tang, C.-H.; Wang, C.-S. Formation and Characterization of Amyloid-like Fibrils from Soy β-Conglycinin and Glycinin. *J. Agric. Food Chem.* **2010**, *58*, 11058–11066. [CrossRef] [PubMed]
20. Tang, C.-H.; Wang, S.-S.; Huang, Q. Improvement of heat-induced fibril assembly of soy β-conglycinin (7S Globulins) at pH 2.0 through electrostatic screening. *Food Res. Int.* **2012**, *46*, 229–236. [CrossRef]
21. Xia, W.; Zhang, H.; Chen, J.; Hu, H.; Rasulov, F.; Bi, D.; Huang, X.; Pan, S. Formation of amyloid fibrils from soy protein hydrolysate: Effects of selective proteolysis on β-conglycinin. *Food Res. Int.* **2017**, *100*, 268–276. [CrossRef] [PubMed]
22. Cho, S.Y.; Park, J.-W.; Batt, H.P.; Thomas, R.L. Edible films made from membrane processed soy protein concentrates. *LWT Food Sci. Technol.* **2007**, *40*, 418–423. [CrossRef]
23. Wang, Q.; Liu, W.; Tian, B.; Li, D.; Liu, C.; Jiang, B.; Feng, Z. Preparation and characterization of coating based on protein nanofibers and polyphenol and application for salted duck egg yolks. *Foods* **2020**, *9*, 449. [CrossRef]

24. Jiang, B.; Wang, X.; Wang, L.; Wu, S.; Li, D.; Liu, C.; Feng, Z. Fabrication and Characterization of a Microemulsion Stabilized by Integrated Phosvitin and Gallic Acid. *J. Agric. Food Chem.* **2020**, *68*, 5437–5447. [CrossRef] [PubMed]
25. Yi-You, L. The Soybean Protein Fibre-A Healthy & Comfortable Fibre for the 21st Century. *Fibres Text. East. Eur.* **2004**, *12*, 8–9.
26. Pekhtasheva, E.; Neverov, A.; Kubica, S.; Zaikov, G. Biodegradation and biodeterioration of some natural polymers. *Polym. Res. J.* **2012**, *5*, 77–108. [CrossRef]
27. Fernandes, J.C.; Tavaria, F.K.; Fonseca, S.C.; Ramos, Ó.S.; Pintado, M.E.; Malcata, F.X. In vitro screening for anti-microbial activity of chitosans and chitooligosaccharides, aiming at potential uses in functional textiles. *J. Microbiol. Biotechnol.* **2010**, *20*, 311–318. [CrossRef] [PubMed]
28. Cheng, X.; Ma, K.; Li, R.; Ren, X.; Huang, T. Antimicrobial coating of modified chitosan onto cotton fabrics. *Appl. Surf. Sci.* **2014**, *309*, 138–143. [CrossRef]
29. Szostak-Kotowa, J. Biodeterioration of textiles. *Int. Biodeterior. Biodegrad.* **2004**, *53*, 165–170. [CrossRef]
30. Zhao, Y.; Xu, Z.; Lin, T. Barrier textiles for protection against microbes. In *Antimicrobial Textiles*; Elsevier: Amsterdam, The Netherlands, 2016; pp. 225–245.
31. Han, S.; Yang, Y. Antimicrobial activity of wool fabric treated with curcumin. *Dyes Pigment.* **2005**, *64*, 157–161. [CrossRef]
32. Esquenazi, D.; Wigg, M.D.; Miranda, M.M.; Rodrigues, H.M.; Tostes, J.B.; Rozental, S.; Da Silva, A.J.; Alviano, C.S. Antimicrobial and antiviral activities of polyphenolics from Cocos nucifera Linn.(Palmae) husk fiber extract. *Res. Microbiol.* **2002**, *153*, 647–652. [CrossRef]
33. Tayel, A.A.; Moussa, S.; Wael, F.; Knittel, D.; Opwis, K.; Schollmeyer, E. Anticandidal action of fungal chitosan against Candida albicans. *Int. J. Biol. Macromol.* **2010**, *47*, 454–457. [CrossRef]
34. Ryan, K.J.; Ray, C.G. *Sherris Medical Microbiology*; Mcgraw-Hill: New York, NY, USA, 2003.
35. Ristić, T.; Zemljič, L.F.; Novak, M.; Kunčič, M.K.; Sonjak, S.; Cimerman, N.G.; Strnad, S. Antimicrobial efficiency of functionalized cellulose fibres as potential medical textiles. *Sci. Against Microb. Pathog. Commun. Curr. Res. Technol. Adv.* **2011**, *6*, 36–51.
36. Sekhri, S. *Textbook of Fabric Science: Fundamentals to Finishing*; PHI Learning Pvt. Ltd.: New Delhi, India, 2019.
37. Gandhi, K. *Woven Textiles: Principles, Technologies and Applications*; Woodhead Publishing: Sawston, UK, 2019.
38. Schindler, W.D.; Hauser, P.J. 3—Softening finishes. In *Chemical Finishing of Textiles*; Schindler, W.D., Hauser, P.J., Eds.; Woodhead Publishing: Sawston, UK, 2004; pp. 29–42. [CrossRef]
39. Kawabata, S.; Niwa, M.; Yamashita, Y. Recent developments in the evaluation technology of fiber and textiles: Toward the engineered design of textile performance. *J. Appl. Polym. Sci.* **2002**, *83*, 687–702. [CrossRef]
40. Behery, H. *Effect of Mechanical and Physical Properties on Fabric Hand*; Elsevier: Amsterdam, The Netherlands, 2005.
41. Raj, S.; Sreenivasan, S. Total wear comfort index as an objective parameter for characterization of overall wearability of cotton fabrics. *J. Eng. Fibers Fabr.* **2009**, *4*, 155892500900400406. [CrossRef]
42. Paul, R. *Functional Finishes for Textiles: Improving Comfort, Performance and Protection*; Elsevier: Amsterdam, The Netherlands, 2014.
43. Shishoo, R. *Textiles in Sport*; Elsevier: Amsterdam, The Netherlands, 2005.
44. Ramesh, M.; Palanikumar, K.; Reddy, K.H. Plant fibre based bio-composites: Sustainable and renewable green materials. *Renew. Sustain. Energy Rev.* **2017**, *79*, 558–584. [CrossRef]
45. Tomczak, F.; Sydenstricker, T.H.D.; Satyanarayana, K.G. Studies on lignocellulosic fibers of Brazil. Part II: Morphology and properties of Brazilian coconut fibers. *Compos. Part A Appl. Sci. Manuf.* **2007**, *38*, 1710–1721. [CrossRef]
46. Alonso Felipe, J.V. *Manual Control de Calidad en Productos Textiles y Afines*; Escuela Técnica Superior de Ingenieros Industriales UPM: Madrid, Spain, 2015.
47. Hari, P.K. Types and properties of fibres and yarns used in weaving. In *Woven Textiles*; Elsevier: Amsterdam, The Netherlands, 2020; pp. 3–34.
48. Yu, C. Natural textile fibres: Vegetable fibres. In *Textiles and Fashion*; Elsevier: Amsterdam, The Netherlands, 2015; pp. 29–56.
49. Netravali, A.N. *Biodegradable Natural Fiber Composites*; Woodhead Publishing: Cambridge, UK, 2005; pp. 271–309. [CrossRef]
50. McKenna, H.A.; Hearle, J.W.S.; O'Hear, N. Ropemaking materials. In *Handbook of Fibre Rope Technology*, 1st ed.; Woodhead Publishing: Sawston, UK, 2004; pp. 35–74.
51. Li, X.; Tabil, L.G.; Panigrahi, S. Chemical treatments of natural fiber for use in natural fiber-reinforced composites: A review. *J. Polym. Environ.* **2007**, *15*, 25–33. [CrossRef]
52. Sumi, S.; Unnikrishnan, N.; Mathew, L. Surface modification of coir fibers for extended hydrophobicity and antimicrobial property for possible geotextile application. *J. Nat. Fibers* **2017**, *14*, 335–345. [CrossRef]
53. Jackman, D.R.; Dixon, M.K. *The Guide to Textiles for Interiors*; Portage & Main Press: Winnipeg, MB, Canada, 2003.
54. Singh, J.P.; Verma, S. 3—Raw materials for terry fabrics. In *Woven Terry Fabrics*; Singh, J.P., Verma, S., Eds.; Woodhead Publishing: Sawston, UK, 2017; pp. 19–28. [CrossRef]
55. Pang, F.J.; He, C.J.; Wang, Q.R. Preparation and properties of cellulose/chitin blend fiber. *J. Appl. Polym. Sci.* **2003**, *90*, 3430–3436. [CrossRef]
56. Bunsell, A.R. *Handbook of Properties of Textile and Technical Fibres*; Woodhead Publishing: Sawston, UK, 2018.
57. Montalvo, J.; Von Hoven, T. Review of standard test methods for moisture in lint cotton. *J. Cotton Sci.* **2008**, *12*, 33.
58. ASTM International. *Standard Test Method for Moisture in Cotton by Oven-Drying1*; ASTM International: West Conshohocken, PA, USA, 2019.

59. Delhom, C.; Rodgers, J. Cotton Moisture—Its Importante, Measurements and Impacts. In Proceedings of the 33rd International Cotton Conference Bremen, New Orleans, LA, USA, 16–18 March 2016.
60. Hu, J.; Li, Y.; Yeung, K.-W.; Wong, A.S.; Xu, W. Moisture management tester: A method to characterize fabric liquid moisture management properties. *Text. Res. J.* **2005**, *75*, 57–62. [CrossRef]
61. Higgins, L.; Anand, S.; Hall, M.; Holmes, D. Effect of tumble-drying on selected properties of knitted and woven cotton fabrics: Part I: Experimental overview and the relationship between temperature setting, time in the dryer and moisture content. *J. Text. Inst.* **2003**, *94*, 119–128. [CrossRef]
62. General Administration of Quality Supervision, Inspection and Quarantine of the People's Republic of China. Conventional Moisture Regains of Textiles. GB 9994-2008. In *National Standard of the People's Republic of China*; General Administration of Quality Supervision, Inspection and Quarantine of the People's Republic of China: Beijing, China, 2008; pp. 1–8.
63. Mohanty, A.; Misra, M.; Drzal, L.T. Surface modifications of natural fibers and performance of the resulting biocomposites: An overview. *Compos. Interfaces* **2001**, *8*, 313–343. [CrossRef]
64. Cunha, A.G.; Gandini, A. Turning polysaccharides into hydrophobic materials: A critical review. Part 2. Hemicelluloses, chitin/chitosan, starch, pectin and alginates. *Cellulose* **2010**, *17*, 1045–1065. [CrossRef]
65. Qin, Y.; Zhu, C.; Chen, J.; Zhong, J. Preparation and characterization of silver containing chitosan fibers. *J. Appl. Polym. Sci.* **2007**, *104*, 3622–3627. [CrossRef]
66. Cheah, W.Y.; Show, P.-L.; Ng, I.-S.; Lin, G.-Y.; Chiu, C.-Y.; Chang, Y.-K. Antibacterial activity of quaternized chitosan modified nanofiber membrane. *Int. J. Biol. Macromol.* **2019**, *126*, 569–577. [CrossRef]
67. Jiao, Y.; Niu, L.N.; Ma, S.; Li, J.; Tay, F.R.; Chen, J.H. Quaternary ammonium-based biomedical materials: State-of-the-art, toxicological aspects and antimicrobial resistance. *Prog. Polym. Sci.* **2017**, *71*, 53–90. [CrossRef]
68. Surdu, L.; Stelescu, M.D.; Iordache, O.; Manaila, E.; Craciun, G.; Alexandrescu, L.; Christian Dinca, L. The improvement of the resistance to Candida Albicans and Trichophyton interdigitale of some cotton textile materials by treating with oxygen plasma and chitosan. *J. Text. Inst.* **2016**, *107*, 1426–1433. [CrossRef]
69. Li, S.; Donner, E.; Xiao, H.; Thompson, M.; Zhang, Y.; Rempel, C.; Liu, Q. Preparation and characterization of soy protein films with a durable water resistance-adjustable and antimicrobial surface. *Mater. Sci. Eng. C* **2016**, *69*, 947–955. [CrossRef]
70. Xiao, Y.; Liu, Y.; Kang, S.; Wang, K.; Xu, H. Development and evaluation of soy protein isolate-based antibacterial nanocomposite films containing cellulose nanocrystals and zinc oxide nanoparticles. *Food Hydrocoll.* **2020**, *106*, 105898. [CrossRef]
71. Raeisi, M.; Mohammadi, M.A.; Coban, O.E.; Ramezani, S.; Ghorbani, M.; Tabibiazar, M.; Noori, S.M.A. Physicochemical and antibacterial effect of Soy Protein Isolate/Gelatin electrospun nanofibres incorporated with Zataria multiflora and Cinnamon zeylanicum essential oils. *J. Food Meas. Charact.* **2020**, 1–11. [CrossRef]
72. Rani, S.; Kumar, R. A Review on Material and Antimicrobial Properties of Soy Protein Isolate Film. *J. Polym. Environ.* **2019**, *27*, 1613–1628. [CrossRef]
73. Cerkez, I.; Kocer, H.B.; Worley, S.; Broughton, R.; Huang, T. Multifunctional cotton fabric: Antimicrobial and durable press. *J. Appl. Polym. Sci.* **2012**, *124*, 4230–4238. [CrossRef]
74. Simoncic, B.; Tomsic, B. Structures of novel antimicrobial agents for textiles-a review. *Text. Res. J.* **2010**, *80*, 1721–1737. [CrossRef]
75. Gao, Y.; Cranston, R. Recent advances in antimicrobial treatments of textiles. *Text. Res. J.* **2008**, *78*, 60–72.
76. Le Ouay, B.; Stellacci, F. Antibacterial activity of silver nanoparticles: A surface science insight. *Nano Today* **2015**, *10*, 339–354. [CrossRef]
77. Vigo, T.L. *Antimicrobial Polymers and Fibers: Retrospective and Prospective*; ACS Publications: Washington, DC, USA, 2001.
78. Schindler, W.D.; Hauser, P.J. *Chemical Finishing of Textiles*; Elsevier: Amsterdam, The Netherlands, 2004.
79. Mishra, R.; Behera, B.; Pada Pal, B. Novelty of bamboo fabric. *J. Text. Inst.* **2012**, *103*, 320–329. [CrossRef]
80. Chandy, M.; Kumar, K.S. Characterisation of Silver Deposited Coir Fibers by Magnetron Sputtering. *CORD* **2015**, *31*, 7. [CrossRef]
81. Abatenh, E.; Gizaw, B.; Tsegaye, Z.; Wassie, M. The role of microorganisms in bioremediation—A review. *Open J. Environ. Biol.* **2017**, *2*, 038–046. [CrossRef]
82. Behera, B. Comfort and handle behaviour of linen-blended fabrics. *AUTEX Res. J.* **2007**, *7*, 33–47.
83. Shabbir, M. *Textiles and Clothing: Environmental Concerns and Solutions*; John Wiley & Sons: Hoboken, NJ, USA, 2019.
84. Raheel, M. *Modern Textile Characterization Methods*; Routledge: London, UK, 2017.
85. Varghese, N.; Thilagavathi, G. Handle and comfort characteristics of cotton core spun lycra and polyester/lycra fabrics for application as blouse. *J. Text. Appar. Technol. Manag.* **2014**, *8*, 1–13.
86. Sun, D.; Stylios, G.K. Cotton fabric mechanical properties affected by post-finishing processes. *Fibers Polym.* **2012**, *13*, 1050–1057. [CrossRef]
87. Jariyapunya, N.; Musilová, B.; Koldinská, M. Evaluating the Influence of Fiber Composition and Structure of Knitting Fabrics on Total Hand Value (THV). *Proc. Appl. Mech. Mater.* **2016**, 211–215. [CrossRef]
88. Ji, D.S.; Lee, J.J. Mechanical properties and hand evaluation of hemp woven fabrics treated with liquid ammonia. *Fibers Polym.* **2016**, *17*, 143–150. [CrossRef]
89. Sun, D.; Stylios, G. Investigating the plasma modification of natural fiber fabrics-the effect on fabric surface and mechanical properties. *Text. Res. J.* **2005**, *75*, 639–644. [CrossRef]
90. Verdu, P.; Rego, J.M.; Nieto, J.; Blanes, M. Comfort analysis of woven cotton/polyester fabrics modified with a new elastic fiber, part 1 preliminary analysis of comfort and mechanical properties. *Text. Res. J.* **2009**, *79*, 14–23. [CrossRef]

polymers

Article

Preparation and Characterization of a New Polymeric Multi-Layered Material Based K-Carrageenan and Alginate for Efficient Bio-Sorption of Methylene Blue Dye

Chiraz Ammar [1,2], Fahad M. Alminderej [3], Yassine EL-Ghoul [2,3,*], Mahjoub Jabli [4,5] and Md. Shafiquzzaman [6]

1 Department of Fashion Design, College of Design, Qassim University, Al Fayziyyah Buraidah 52383, Saudi Arabia; c.ammar@qu.edu.sa
2 Textile Engineering Laboratory, University of Monastir, Monastir 5019, Tunisia
3 Department of Chemistry, College of Science, Qassim University, Buraidah 51452, Saudi Arabia; f.alminderej@qu.edu.sa
4 Department of Chemistry, College of Science Al-Zulfi, Majmaah University, Zulfi 11932, Saudi Arabia; m.jabli@mu.edu.sa
5 Textile Materials and Processes Research Unit, Tunisia National Engineering School of Monastir, University of Monastir, Monastir 5019, Tunisia
6 Department of Civil Engineering, College of Engineering, Qassim University, Buraidah 51452, Saudi Arabia; shafiq@qec.edu.sa
* Correspondence: y.elghoul@qu.edu.sa; Tel.: +966-595519071

Citation: Ammar, C.; Alminderej, F.M.; EL-Ghoul, Y.; Jabli, M.; Shafiquzzaman, M. Preparation and Characterization of a New Polymeric Multi-Layered Material Based K-Carrageenan and Alginate for Efficient Bio-Sorption of Methylene Blue Dye. *Polymers* **2021**, *13*, 411. https://doi.org/10.3390/polym13030411

Academic Editors: Rafael Antonio Balart Gimeno, Daniel García García, Vicent Fombuena Borrás, Luís Jesús Quiles Carrillo and Marina Patricia Arrieta Dillon

Received: 22 December 2020
Accepted: 25 January 2021
Published: 28 January 2021

Abstract: The current study highlights a novel bio-sorbent design based on polyelectrolyte multilayers (PEM) biopolymeric material. First layer was composed of sodium alginate and the second was constituted of citric acid and k-carrageenan. The PEM system was crosslinked to non-woven cellulosic textile material. Resulting materials were characterized using FT-IR, SEM, and thermal analysis (TGA and DTA). FT-IR analysis confirmed chemical interconnection of PEM bio-sorbent system. SEM features indicated that the microspaces between fibers were filled with layers of functionalizing polymers. PEM exhibited higher surface roughness compared to virgin sample. This modification of the surface morphology confirmed the stability and the effectiveness of the grafting method. Virgin cellulosic sample decomposed at 370 °C. However, PEM samples decomposed at 250 °C and 370 °C, which were attributed to the thermal decomposition of crosslinked sodium alginate and k-carrageenan and cellulose, respectively. The bio-sorbent performances were evaluated under different experimental conditions including pH, time, temperature, and initial dye concentration. The maximum adsorbed amounts of methylene blue are 124.4 mg/g and 522.4 mg/g for the untreated and grafted materials, respectively. The improvement in dye sorption evidenced the grafting of carboxylate and sulfonate groups onto cellulose surface. Adsorption process complied well with pseudo-first-order and Langmuir equations.

Keywords: polyelectrolyte multi-layers; sodium alginate; k-carrageenan; cellulosic nonwoven textile; surface functionalization; characterization; bio-sorption; isotherms

1. Introduction

The discharge of dyes from industries and their elimination has received much attention as they can damage human health and animals [1–3]. Indeed, dye molecules are resistant to natural degradation, allergenic, carcinogenic, and stable in the presence of oxidizing agents [4–6]. This interesting topic requires the development of various technologies to treat colored waters. Biological treatment and coagulation/flocculation processes are viewed as ineffective to treat soluble dyes [7–9]. Adsorption appeared more effective as it is simple and economic and it is especially used to remove pollutants, which are not easily biodegradable. Thus, a specific attention is devoted to explore new adsorbents, which could be cheaper, proficient, and easy regenerated.

In this sense, several polymeric adsorbents were used for the removal of dyes from contaminated matters [10–12]. Concerning textile adsorbent materials, they are rarely investigated in the literature. Some sorbent material based synthetic functionalized textiles were studied for the removal of cationic dyes. Despite their pretreatment and/or functionalization, adsorption capacities were limited due to their hydrophobicity [13–15]. Cellulose, alginates, and k-carrageenan biopolymers were naturally abundant polysaccharides. Cellulose is recognized for its good hydrophilicity [16–21]. Alginate, a polysaccharide biopolymer, is an essential component of the cell wall of brown algae. It could be an excellent bio-sorbent of organic dyes due to its high contents of carboxylic and hydroxyl functional groups [22–25]. K-carrageenan is a natural sulfated polysaccharide extracted from red edible seaweeds. It is widely used in the food industry, owing to gelling, thickening, and stabilizing properties [26,27]. As examples of works conducted in previous literatures, Rahman et al., in a comprehensive review, discussed synthesis, characteristics, and methylene blue adsorption of various cellulose nanocrystal-based hydrogels [28]. They indicated that the addition of other polysaccharides including chitosan and alginate into cellulose displayed remarkably improved adsorption capacities for methylene blue. Guesmi et al. demonstrated that the addition of sodium alginate (5–20%) to hydroxyapatite improved well methylene blue adsorption [29] and the adsorption capacity increased from 77.51 to 142.85 mg/g, using hydroxyapatite and hydroxyapatite-alginate, respectively. In our previous work, we have demonstrated that the adsorption of methylene blue onto extracted cellulose, from Aegagropila L., reached 109 mg/g and it was only about 47mg/g for the raw marine macroalgae [30]. Yang et al. synthesized gel beads based on k-carrageenan and graphene oxide [31] and observed that the adsorption capacity of the gel beads for methylene blue attained 658.4 mg/g at 25 °C. As globally observed, the developed materials-based polysaccharides are outstanding adsorbents with excellent adsorption capacities of cationic dyes.

In line with this emergent topic, the current study proposes a novel adsorbent design based on polyelectrolyte multi-layered (PEM) biopolymeric material as a potent bio-sorbent. This is formed by an alternation of layers of two polyelectrolyte biopolymers. The first layer is composed of alginate polyanion and the second is obtained after the reticulation of the citric acid with k-carrageenan polyanion polymer. The PEM system is crosslinked to cellulosic natural material and applied as adsorbent of cationic dyes. Indeed, the proposed technique offers the adsorption characteristics of not only cellulose as the main support but also k-carrageenan and sodium alginate as immobilized biopolymers. The system provides massive hydroxyl, sulfonate, and carboxylate groups suggesting therefore the use of the resulting material as a bio-sorbent of cationic species. Therefore, the importance of our proposal lies in the preparation of a new PEM biomaterial based on alginate and k-carrageenan crosslinked to cellulosic material, which is very rich in anionic functional groups. In addition, the investigation of the layer-by-layer functionalization method in the adsorption of textile rejects is a novel study and was not reported in the literature. The prepared materials were characterized using FT-IR, SEM, TGA, and DTA. The bio-sorbent performances were evaluated under different experimental conditions including pH, time, temperature, and the initial dye concentration. The theoretical kinetic and isotherms equations were used to analyze the experimental data.

2. Experimental

2.1. Materials

A non-woven cellulosic textile material was used in this study. Its surface weight was 240 g/m^2 and the thickness had an average of 0.6 mm. The textile is a 3-dimensional calendared fiber network. Sodium alginate (AG) was supplied from Sigma-Aldrich with a medium molecular weight (30,000 g/mol) having a degree of deacetylation of 70%. Citric acid (CTR, 226.2 g/mol) and Kappa carrageenan (a fine white powder) were purchased from Sigma-Aldrich. All chemicals were used without further purification. Methylene blue (MB) was supplied by Central Drug House (India).

2.2. Preparation of PEM Bio-Sorbent

PEM bio-sorbent was prepared via an alternating grafting process of sodium alginate and k-carrageenan using citric acid as a crosslinking agent. The process of grafting of the cellulosic material is based on a pad-dry technique. PEM bio-sorbent was carried out by functionalizing the cellulosic material with alternating baths in sodium alginate/citric acid solution and then in a K-carrageenan solution. The PEM is deposited by alternating successive baths according to the "layer-by-layer deposition" method. Different pairs of layers (from 1 to 8) are prepared for bio-sorbent cellulosic material. Each impregnation was performed in a total volume of 50 mL. All polymer solutions are completely renewed after the deposition of 2 pairs of layers. The samples (4 cm × 4 cm) were treated in a solution of soluble sodium alginate polymer (0.4% *w/v*) and citric acid (0.4% *w/v*) under constant stirring at 170 rpm for 20 min at room temperature. After each impregnation, the samples were dried 20 min at 90 °C. After drying, the samples were impregnated in a solution of k-carrageenan (at 0.8% *w/v*) with stirring (170 rpm) for 20 min at room temperature. The cycle was repeated as many times as necessary. At each end of the cycle, a pair of layers was thus deposited and crosslinked on the cellulose material.

During the preparation process of PEM materials, the weight gain was calculated after depositing each pair of layers of sodium alginate and k-carrageenan. The following formula was applied illustrating the weight gain as a function of the number of pairs of grafted copolymer layers.

$$\% - Weight\ gain\ (n) = \frac{(m_n - m_i)}{m_i} \times 100 \quad (1)$$

where n is the number of pairs of layers and m_i and m_n are the sample weights before and after functionalization, respectively.

2.3. Characterizations

For the analysis of the chemical structure of prepared multilayered adsorbent, infrared spectroscopy analysis was conducted using a FT-IR spectrometer (Agilent Technologies, Cary 600 Series FTIR Spectrometer) via the attenuated total reflection mode (ATR). Spectra of cellulose material and multilayered bio-adsorbent polymeric system were recorded at a range of 4000 to 400 cm^{-1}, and with a resolution of 2 cm^{-1}.

Thermal stability and different thermal properties of natural cellulosic material and PEM bio-sorbent were determined by thermogravimetric measurements using TA Instruments apparatus. The fixed parameters for different analysis were a heating rate of 10 °C min^{-1} and a temperature from 25 to 600 °C.

Surface morphology was assessed using a scanning electron microscopy (FEI Quanta SEM). An accelerating voltage of 5 KV with various magnification essays was applied for the surface analyzing of different samples. The SEM analysis was preceded by a coating procedure of samples with a carbon layer to enhance their conductivity.

2.4. Adsorption Experiments

The adsorption tests were carried out in a batch reactor by stirring the colored synthetic solutions in the presence of each of the adsorbents at a constant agitation speed (150 rpm). We studied the effect of the main parameters influencing the adsorption capacity such as pH (ranged from 3 to 9), contact time (in a range of 0 to 120 min.), initial dye concentration (varied from 50 to 1000 mg/L), and temperature (22, 40, and 60 °C). The residual concentration of each of the dyes was determined using an UV/visible spectrophotometer (Shimadzu UV-2600).

The adsorbed methylene blue dye amount (q(mg/g)) onto the surface of PEM crosslinked to non-woven cellulosic textile material was calculated using the following formula:

$$q(mg/g) = \frac{(C_0 - C_e) \cdot V}{m} \quad (2)$$

where C_0 and C_e are the initial and residual concentration (mg/L), respectively, V is the volume of methylene blue dye (L) used for the adsorption, and m is the mass of the biosorbent (g) used for the adsorption.

3. Results and Discussion

3.1. Preparation of PEM Biopolymer Adsorbent

The design of PEM was carried out by grafting the cellulosic material with alternating layers of sodium alginate and K-carrageenan biopolymers crosslinked by citric acid (Figure 1). The layer-by-layer deposition technique was used to elaborate different multilayered bio-sorbent (from 1 to 8-layer pairs). The variation of the weight gain according to the number of grafted layers of sodium alginate and K-carrageenan was presented in Figure 2. We noticed the progressive increase of total weight gain with the number of polymeric layers grafted to the cellulosic material. Results revealed a more significant increase of the weight gain within 3-pairs-layers finishing the cellulosic sample. For the characterization and adsorption investigations, the samples treated with three polymeric layers of reticulated alginate and K-carrageenan will be investigated.

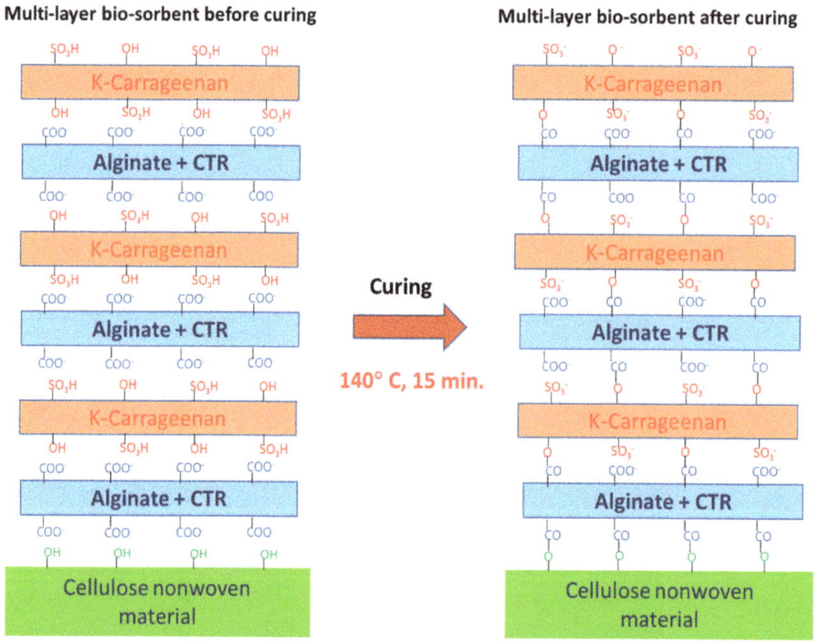

Figure 1. Illustration of the functionalization of cellulosic material upon polyesterification reactions via CTR (Citric acid) crosslinking, leading to the polyanionic PEM bio-sorbent system after curing at 140 °C for 15 min.

3.2. FT-IR Spectroscopy Analysis

FT-IR-ATR analysis was performed in order to characterize the chemical grafting and conception of the bio-sorbent material via the identification of the different functional groups appearing after functionalization. This characterization was applied to analyze both the untreated cellulosic sample and the PEM grafted adsorbent material. Spectra of untreated cellulosic sample and grafted PEM material (3-layers PEM) were showed in Figure 3. Different new peaks appeared in the PEM grafted cellulosic material proving the chemical grafting. A peak centered at 1705 cm^{-1} appeared referring to the ester groups confirming the polyesterification reaction established between carboxylic groups of citric acid and hydroxyl functions of both alginate and carrageenan biopolymers. This is in line with previous research studies, in which FT-IR were used to confirm the evidence of a polyesterification reaction between cellulosic material and different polysaccharides via polycarboxylic acids as crosslinking agents [32–34]. A peak closed to 1310 cm^{-1} appeared clearly, which corresponded to the symmetric stretching vibration of the carboxylic acid groups COO- of galacturonic acid of alginate grafted polymer [35]. In addition, a more significant wide peak around to 3290 cm^{-1} appeared with grafted PEM material was attributed to hydroxyl groups of cellulose, alginate, and carrageenan grafted polymers. Briefly, FT-IR analysis of the two spectra, permitted us to confirm the chemical interconnection of the PEM bio-sorbent system and indicated the efficiency of the applied grafting chemical process.

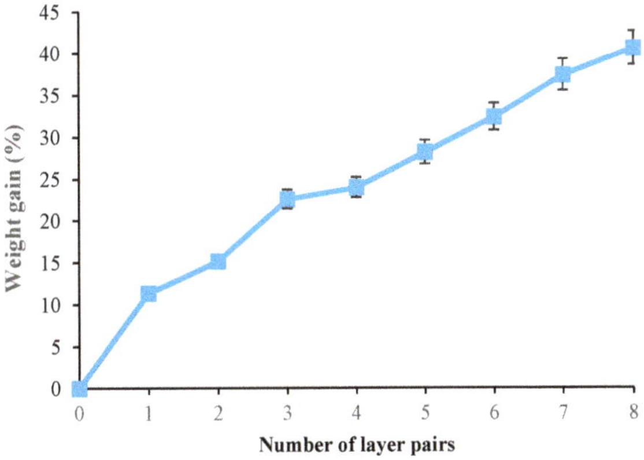

Figure 2. Weight gain variation as function of the number of layers grafted onto the cellulosic bio-sorbent material.

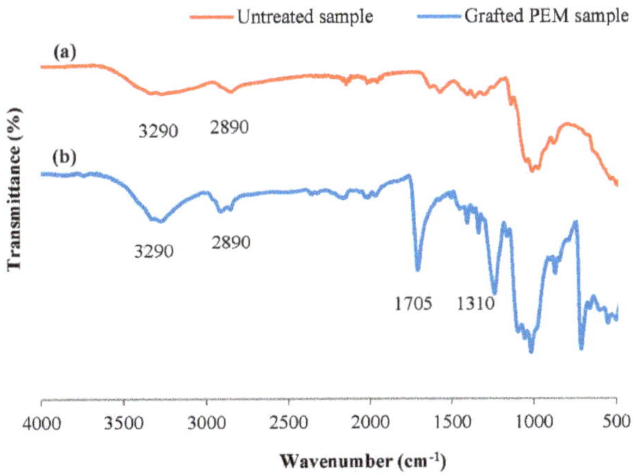

Figure 3. FT-IR-ATR spectra of virgin cellulosic sample (**a**) and PEM grafted bio-sorbent material (**b**).

3.3. TGA and DTA Analysis

Figure 4 showed thermograms of untreated and grafted PEM samples. The untreated cellulosic sample showed only one zone showing one temperature of degradation referring to the cellulosic material closed to 370 °C. However, two distinct parts appeared in the thermogram of the designed PEM sample. The first with a temperature of degradation around 250 °C was attributed to the crosslinked polymer of alginate and K-carrageenan. The second closed to 370 °C, with a high loss of weight referring to the degradation of the cellulosic sample. The observed weight loss for the two samples, around 100 °C, was due the evaporation of water absorbed inside their structures. At this temperature we noticed a higher weight loss in the case of the grafted PEM material. This was due to the increase of the hydrophilicity after biopolymers grafting onto the cellulosic samples. Furthermore, the prepared PEM samples presented a higher residual weight after degradation (20%) compared to the untreated one (0%). This proved well the improvement thermal stability of the PEM bio-sorbent material after grafting.

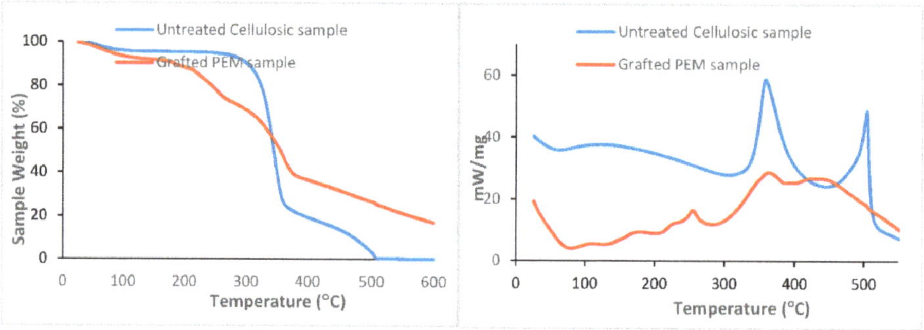

(a) (b)

Figure 4. TGA (**a**) and DTA (**b**) thermograms of virgin cellulosic sample and grafted PEM bio-sorbent material.

DTA results confirmed the TGA analysis and revealed one significant peak in the derivative signal ascribed to the cellulose degradation in the case of untreated sample. For the untreated PEM sample two principal peaks appeared, which were attributed to the degradation of the grafted polymers (at 250 °C) and the degradation of the cellulosic material (at 370 °C).

3.4. Morphological Analysis

Figure 5 showed the micrographs of untreated cellulosic sample and grafted PEM material with 3-layer pairs. We noticed a significant surface modification after grafting. The microspaces between fibers were filled with the layers of the two functionalizing polymers. The grafted PEM revealed a higher surface roughness compared to the virgin sample. This modification of the surface morphology confirmed the stability and the effectiveness of the applied method of grafting.

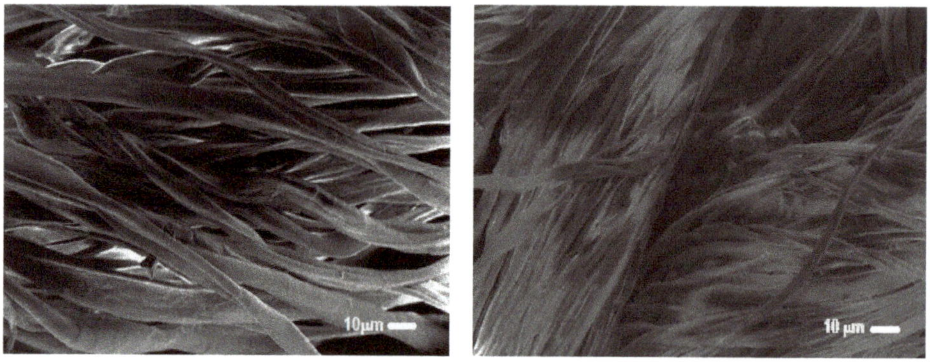

(a) (b)

Figure 5. SEM micrographs of (**a**) untreated cellulosic sample and (**b**) PEM grafted bio-sorbent material.

3.5. Application to the Adsorption of Methylene Blue

In this investigation, the prepared untreated and grafted materials were applied as adsorbents of methylene blue in synthetic medium by varying pH, time, initial dye concentration, and temperature. Figure 6a represents the evolution of the adsorbed methylene blue amount as a function of pH. The maximum adsorbed amount was reached at pH = 6. In fact, under high acidic conditions, the positively charged methylene blue ions opposed the positively charged adsorbent surface leading to low adsorbed rates. At higher pH values (pH < 6), the adsorbent surface became negative, favoring an electrostatic interaction with methylene blue. At basic conditions, the adsorbed dye amount decreases, which could be explained by the repulsion forces between dye molecules and biosorbent surface.

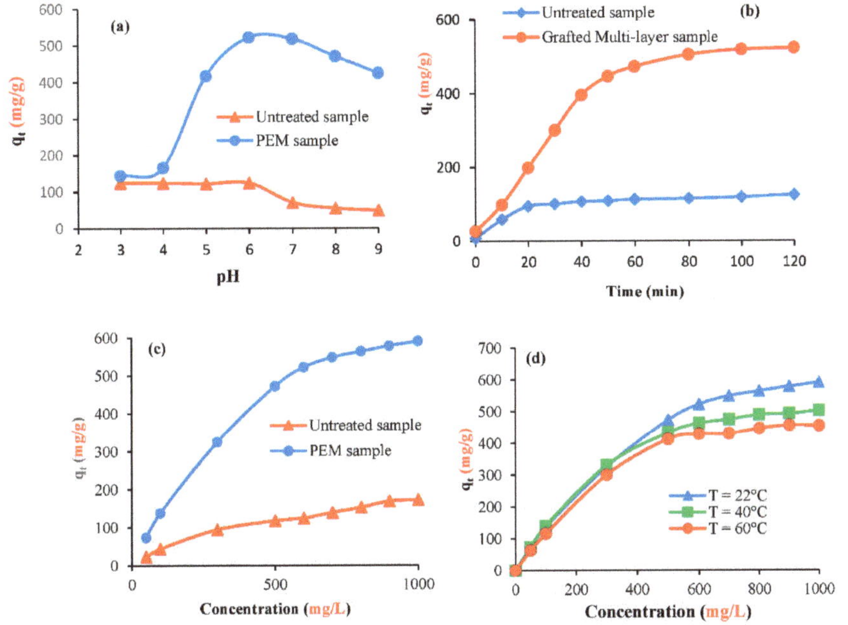

Figure 6. (**a**) Effect of pH on methylene blue adsorption (C_0 = 600 mg/L, t = 120 min), (**b**) effect of time, (**c**) effect of initial dye concentration, and (**d**) effect of temperature.

The time required to achieve equilibrium was observed at 120 min (Figure 6b). It is seen that the adsorption was fast during the first period of times (from 0 to 50 min) where more than 80% of target was accomplished. This trend could be explained by the fact that many adsorption sites are available during this first stage at the surface of the adsorbents. After this period of times, the adsorption attained a steady state, which could be interpreted by the saturation of the adsorption sites. Results showed also that the maximum adsorbed amount of methylene blue are 124.4 mg/g and 522.4 mg/g for the untreated materials and grafted ones, respectively. This difference in the sorption capacities could be explained by the addition of new functional groups (carboxylate and sulfonate groups) grafted on the surface of cellulose.

Scheme 1 provides a schematic representation of hydrogen bonding and ionic interaction between methylene blue molecule and PEM crosslinked to non-woven cellulosic textile material surface. Indeed, the free hydroxyl groups of cellulose non-woven could interact with nitrogen atom of methylene blue via hydrogen bonding. However, the sulfonate groups (SO_3^-) of K-carrageenan and/or carboxylate groups of both sodium alginate (COO^-) and CTR crosslinking agent could react with positive nitrogen atom (N^+) through ionic interaction.

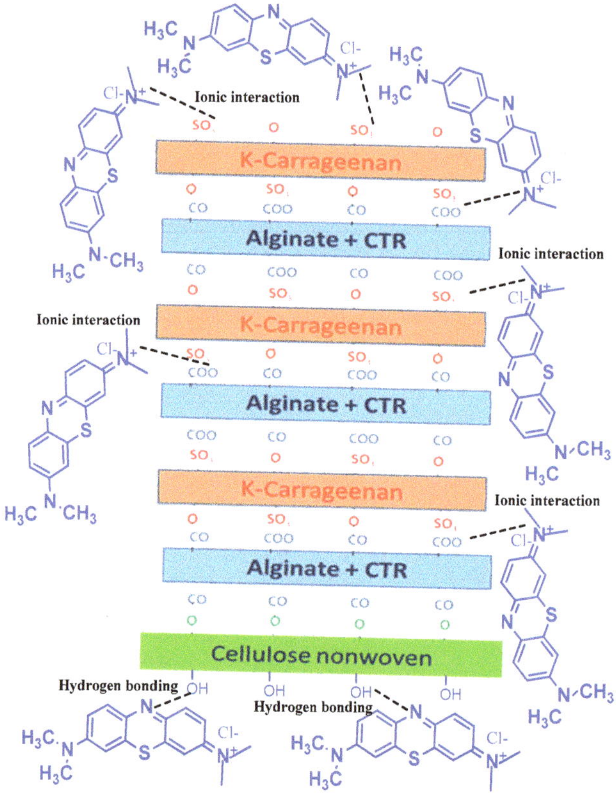

Scheme 1. A schematic representation of hydrogen bonding and ionic interaction between methylene blue molecule and the surface of PEM crosslinked to non-woven cellulosic textile material.

The adsorbed dye amount was found to greatly increase with the increase of initial methylene blue concentration. At equilibrium, it reached 171.4 mg/g for the untreated samples and 590.5 mg/g for the grafted samples (Figure 6c). This adsorbed amount depends on the temperature value and it decreased for example in the case of grafted samples from 590.5 mg/g at 22 °C to 453.5 mg/g at 60 °C (Figure 6d). This indicated that the adsorption of methylene blue was exothermic in this case. Indeed, at higher temperatures values, the dye could be desorbed from the samples.

3.6. Kinetic Modeling

Kinetics data were significant to recognize the attraction between adsorbates and adsorbents at equilibrium state. It could suggest if the studied mechanism is physical and/or chemical, and mass transport. Herein, modeling kinetic data was assessed using pseudo first order, pseudo second order, Elovich, and intradiffusion kinetic models. Results are depicted in Figure 7. The computed kinetic parameters for the different equations were summarized in Table 1. The obtained curves and computed kinetic parameters indicated that pseudo-first-order equation could describe well the kinetic data ($0.96 < R^2$). The correlation coefficients obtained within the pseudo-second-order were also high ($0.95 < R^2$). These results indicated that the adsorption process is so complex and could be considered as physical and chemical [36]. The divergence of the plots for the intraparticle diffusion model from the origin suggests that this model was not the sole rate-controlling step [37].

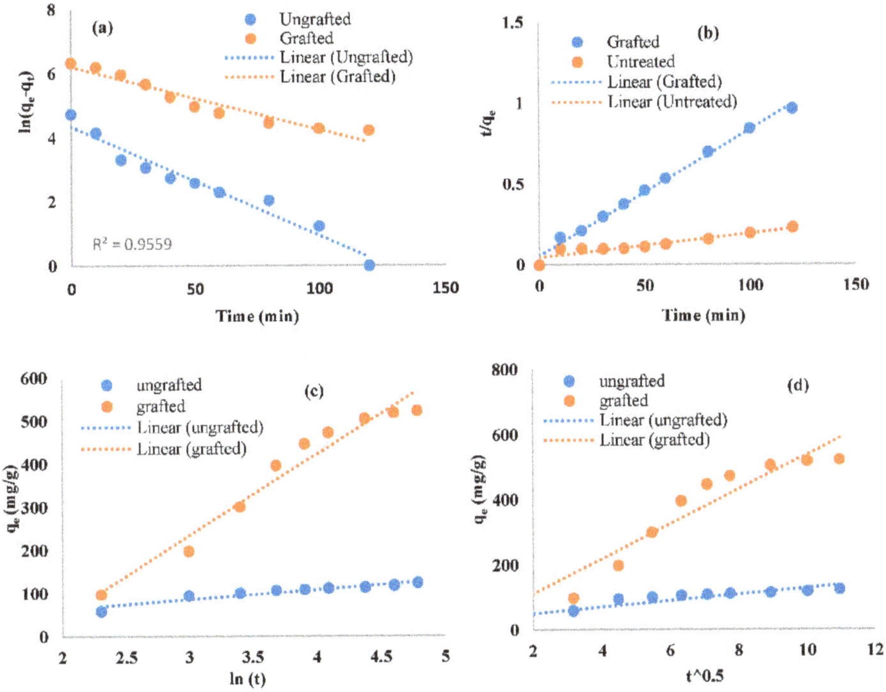

Figure 7. Kinetic data modeled using: (**a**) pseudo first order, (**b**) pseudo second order, (**c**) Elovich, and (**d**) intraparticular diffusion.

Table 1. Summarized kinetic parameters for the grafted and ungrafted samples.

Pseudo First Order			Pseudo Second Order			Elovich			Intraparticle Diffusion		
Parameters	Ungrafted	Grafted		Ungrafted	Grafted		Ungrafted	Grafted		Ungrafted	Grafted
q_e (mg/g)	121	530	q_e (mg/g)	127.982	954.645	β (mg/g/min)	0.043	0.005			
k_1 (1/min)	0.035	0.040	k_2	1.07×10^{-3}	1.5×10^{-5}	α (mg/g/min)	47.903	32.722	K	13.808	54.074
R^2	0.967	0.992	R^2	0.973	0.95	R^2	0.9	0.958	R^2	0.675	0.9217

Where: α (mg/g/ min) and β (mg/g/min) are Elovich constants; R^2 is the regression coefficient.

3.7. Isotherms Investigation

The relationship between the grafted materials and the studied adsorbate was evaluated using the equations of Langmuir, Freundlich, and Temkin (Figure 8). The calculated parameters were shown in Table 2. We observed that the Langmuir equation fitted well with the experimental data ($0.99 \leq R^2$) compared to the other studied equations. This trend suggested that the adsorption process is a monolayer, and the sorption sites are homogeneous having similar adsorption capacities [38]. The values of ($1/n$) revealed adsorption intensity or surface heterogeneity. Indeed, when the value of $1/n$ ranges from 0.1 to 1.0, it therefore reflects a good adsorption [39]. In our study $2.3 \leq n \leq 3.8$, which indicates that the adsorption of the molecules of methylene blue onto the surface of the grafted samples is so favorable. The decrease of the values of the adsorption energy constant (b_T), determined from the Temkin model, with the increase of temperature indicated the exothermic nature of the adsorption mechanism. This is consistent with the trends observed within the effect of temperature discussed above.

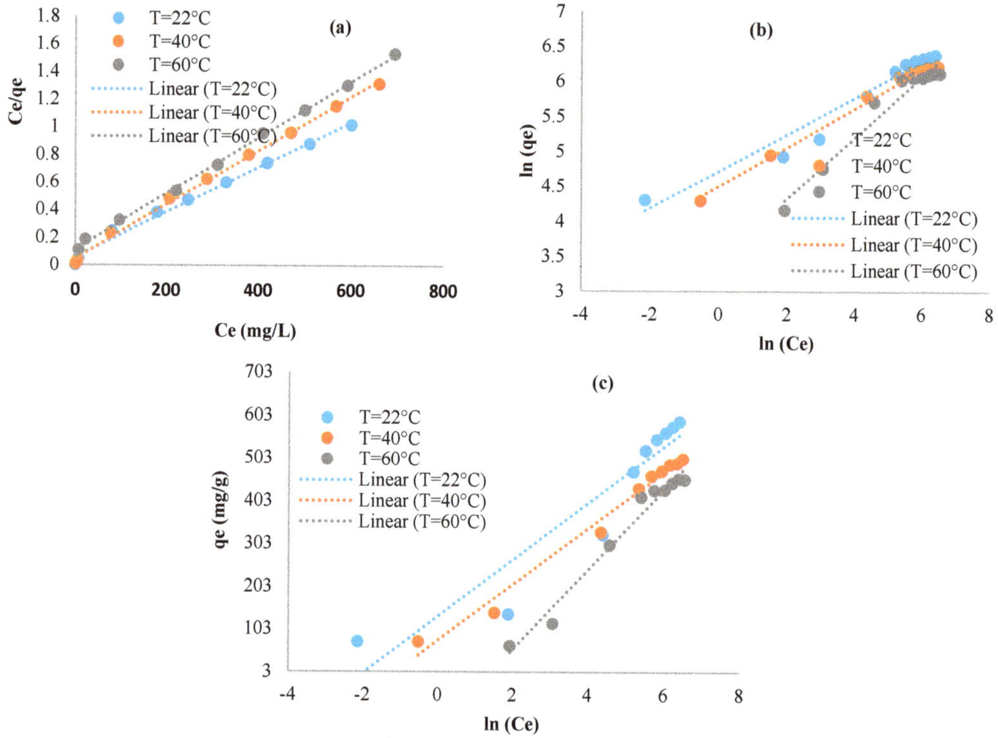

Figure 8. Isotherms data modeled using: (**a**) Langmuir, (**b**) Freundlich, and (**c**) Temkin.

Table 2. Summarized Langmuir, Freundlich, and Temkin parameters for the grafted samples.

Langmuir				Freundlich				Temkin			
	T = 22 °C	T = 40 °C	T = 60 °C		T = 22 °C	T = 40 °C	T = 60 °C		T = 22 °C	T = 40 °C	T = 60 °C
q_{max} (mg/g)	614.28	512.15	494.18	K_f (L/g)	112.1	90.3	31.8	b_T (J/mol)	36.98	39.69	29.66
K_L (L/g)	0.028	0.042	0.02	n	3.80	3.59	2.30	A_t (L/g)	7.55	3.24	0.25
R^2	0.99	0.99	0.99	R^2	0.9714	0.99	0.95	R^2	0.87	0.98	0.98

4. Conclusions

In this study, a new multilayered polymeric bio-sorbent was elaborated using the layer-by-layer grafting method. The prepared cellulosic material based k-carrageenan and alginate crosslinked biopolymers were then chemically and morphologically characterized. Chemical grafting via polyesterification reactions and the stability of the multilayer functionalization were confirmed by FT-IR, TGA-DTA, and SEM analysis. Methylene blue dye was used as an adsorbate for the bio-sorption evaluation on the PEM grafting material in batch experiments. The influence of the different process parameters was studied with respect to the sorption equilibrium. The designed bio-sorbent showed excellent sorption capacities of MB with a capacity above 522.4 mg/g. The improvement in dye sorption was evidently due to the presence of many carboxylate and sulfonate groups of the crosslinked k-carrageenan and alginates grafted on the cellulosic surface material. The correlation of the experimental data with the theoretical equations showed that the kinetic data could be described with both pseudo first order and pseudo second order equations suggesting that the adsorption process involved physical and chemical interactions. The adsorption phenomenon occurred in heterogeneous adsorption sites and it was exothermic and spontaneous. Langmuir isotherm fitted better to adsorption data, indicating that the adsorption process was localized on a monolayer, and all adsorption sites on the adsorbent were homogeneous and had the same adsorption capacity. In brief, we developed a simple and new polymeric material for removing one of the most dangerous contaminants from textile industrial aqueous discharges. Further investigation could be extended for the exploration of this material in the removal of other pollutants such as metals and pesticides.

Author Contributions: Data curation, C.A., Y.E.-G. and M.S.; Formal analysis, C.A., F.M.A., Y.E.-G. and M.J.; Investigation, C.A., F.M.A. and M.S.; Methodology, C.A., Y.E.-G. and M.S.; Project administration, F.M.A. and Y.E.-G.; Software, F.M.A. and M.S.; Supervision, F.M.A.; Validation, F.M.A., Y.E.-G. and M.J.; Writing—original draft, C.A., Y.E.-G. and M.J.; Writing—review and editing, Y.E.-G. and M.J. All authors have read and agreed to the published version of the manuscript.

Funding: The authors gratefully acknowledge Qassim University, represented by the Deanship of Scientific Research, on the financial support for this research under the number 5656-cos-2019-2-2-I, during the academic year 1440 AH/2019 AD.

Institutional Review Board Statement: Not applicable.

Informed Consent Statement: Not applicable.

Data Availability Statement: The data presented in this study are available on request from the corresponding author.

Conflicts of Interest: The authors declare no conflict of interest.

References

1. Venkat, S.M.; Vijay, B.P.V. Kinetic and equilibrium studies on the removal of Congo red from aqueous solution using Eucalyptus wood (Eucalyptus globulus) saw dust. *J. Taiwan Inst. Chem. Eng.* **2013**, *44*, 81–88.
2. Enamul, H.; Jong, W.J.; Sung, H.J. Adsorptive removal of methyl orange and methylene blue from aqueous solution with a metal-organic framework material, iron terephthalate (MOF-235). *J. Hazard. Mater.* **2011**, *185*, 507–511.
3. Karagozoglu, B.; Tasdemir, M.; Demirbas, E.; Kobya, M. The adsorption of basic dye (Astrazon Blue FGRL) from aqueous solutions onto sepiolite, fly ash and apricot shell activated carbon: Kinetic and equilibrium studies. *J. Hazard. Mater.* **2007**, *147*, 297–306. [CrossRef] [PubMed]
4. Mahmoud, D.K.; Salleh, M.A.M.; Karim, W.A.; Idris, A.; Abidin, Z.Z. Batch adsorption of basic dye using acid treated kenaf fibre char: Equilibrium, kinetic and thermodynamic studies. *Chem. Eng. J.* **2012**, *181*, 449–457. [CrossRef]
5. Fan, L.; Luo, C.; Li, X.; Lu, F.; Qiu, H.; Sun, M. Fabrication of novel magnetic chitosan grafted with graphene oxide to enhance adsorption properties for methyl blue. *J. Hazard. Mater.* **2012**, *215*, 272–279. [CrossRef] [PubMed]
6. Asgher, M.; Bhatti, H.N. Evaluation of thermodynamics and effect of chemical treatments on sorption potential of Citrus waste biomass for removal of anionic dyes from aqueous solutions. *Ecol. Eng.* **2012**, *38*, 79–85. [CrossRef]
7. Pagga, U.; Brown, D. The degradation of dyestuffs: Part II Behaviour of dyestuffs in aerobic biodegradation tests. *Chemosphere* **1986**, *15*, 479–491. [CrossRef]
8. Ishak, S.A.; Murshed, M.F.; Md Akil, H.; Ismail, N.; Md Rasib, S.Z.; Al-Gheethi, A.A.S. The Application of Modified Natural Polymers in Toxicant Dye Compounds Wastewater: A Review. *Water* **2020**, *12*, 2032. [CrossRef]
9. Bahrpaima, K.; Fatehi, P. Preparation and Coagulation Performance of Carboxypropylated and Carboxypentylated Lignosulfonates for Dye Removal. *Biomolecules* **2019**, *9*, 383. [CrossRef] [PubMed]
10. Jabli, M.; Gamha, E.; Sebeia, N.; Hamdaoui, M. Almond shell waste (Prunus dulcis): Functionalization with [dimethy-diallyl-ammonium-chloride-diallylamin-co-polymer] and chitosan polymer and its investigation in dye adsorption. *J. Mol. Liq.* **2017**, *240*, 35–44. [CrossRef]
11. Ali Khan, M.; Momina; Siddiqui, M.R.; Otero, M.; Alshareef, S.A.; Rafatullah, M. Removal of Rhodamine B from Water Using a Solvent Impregnated Polymeric Dowex 5WX8 Resin: Statistical Optimization and Batch Adsorption Studies. *Polymers* **2020**, *12*, 500. [CrossRef]
12. Sebeia, N.; Jabli, M.; Ghith, A.; Elghoul, Y.; Alminderej, F.M. Populus tremula, Nerium oleander and Pergularia tomentosa seed fibers as sources of cellulose and lignin for the bio-sorption of methylene blue. *Int. J. Biol. Macromol.* **2019**, *121*, 655–665. [CrossRef] [PubMed]
13. Sharma, M.; Singh, G.; Vaish, R. Diesel soot coated non-woven fabric for oil-water separation and adsorption applications. *Sci. Rep.* **2019**, *9*, 8503. [CrossRef] [PubMed]
14. Ren, Y.; Guo, J.; Lu, Q.; Xu, D.; Qin, J.; Yan, F. Polypropylene Nonwoven Fabric@Poly(ionic liquid)s for Switchable Oil/Water Separation, Dye Absorption, and Antibacterial Applications. *ChemSusChem* **2018**, *11*, 1092–1098. [CrossRef] [PubMed]

15. Haji, A.; Mousavi Shoushtari, A.; Abdouss, M. Plasma activation and acrylic acid grafting on polypropylene nonwoven surface for the removal of cationic dye from aqueous media. *Desalination Water Treat.* **2013**, *53*, 3632–3640. [CrossRef]
16. Salah, F.; El-Ghoul, Y.; Roudesli, S. Bacteriological effects of functionalized cotton dressings. *J. Text. Inst.* **2016**, *107*, 171–181. [CrossRef]
17. Salah, F.; El-Ghoul, Y.; Alminderej, F.M.; Golli-Bennour, E.; Ouanes, Z.; Maciejak, O.; Jarroux, N.; Majdoub, H.; Sakli, F. Development, characterization, and biological assessment of biocompatible cellulosic wound dressing grafted Aloe vera bioactive polysaccharide. *Cellulose* **2019**, *26*, 4957–4970. [CrossRef]
18. El-Ghoul, Y.; Alminderej, F.M. Bioactive and superabsorbent cellulosic dressing grafted alginate and Carthamus tinctorius polysaccharide extract for the treatment of chronic wounds. *Text. Res. J.* **2020**, *91*, 235–248. [CrossRef]
19. Salah, F.; El-Ghoul, Y.; Mahdhi, A.; Majdoub, H.; Jarroux, N.; Sakli, F. Effect of the deacetylation degree on the antibacterial and antibiofilm activity of acemannan from Aloe vera. *Ind. Crop. Prod.* **2017**, *103*, 13–18. [CrossRef]
20. El-Ghoul, Y.; Salah, F.; Majdoub, H.; Sakli, F. Synthesis and study of drug delivery system obtained via β-cyclodextrin functionalization of viscose/polyester dressings. *J. Ind. Text.* **2017**, *47*, 489–504. [CrossRef]
21. El-Ghoul, Y.; Ammar, C.; El-Achari, A. New polymer based modified cyclodextrins grafted to textile fibers; characterization and application to cotton wound dressings. *Int. J. Appl. Res. Text.* **2014**, *2*, 11–21.
22. Torres-Caban, R.; Vega-Olivencia, C.A.; Alamo-Nole, L.; Morales-Irizarry, D.; Roman-Velazquez, F.; Mina-Camilde, N. Removal of Copper from Water by Adsorption with Calcium-Alginate/Spent-Coffee-Grounds Composite Beads. *Materials* **2019**, *12*, 395. [CrossRef] [PubMed]
23. Siwek, H.; Bartkowiak, A.; Włodarczyk, M. Adsorption of Phosphates from Aqueous Solutions on Alginate/Goethite Hydrogel Composite. *Water* **2019**, *11*, 633. [CrossRef]
24. Zhang, J.; Deng, R.; Ren, B.; Yaseen, M.; Hursthouse, A. Enhancing the Removal of Sb (III) from Water: A Fe3O4@HCO Composite Adsorbent Caged in Sodium Alginate Microbeads. *Processes* **2020**, *8*, 44. [CrossRef]
25. Yang, C.-H.; Shih, M.-C.; Chiu, H.-C.; Huang, K.-S. Magnetic Pycnoporus sanguineus-Loaded Alginate Composite Beads for Removing Dye from Aqueous Solutions. *Molecules* **2014**, *19*, 8276–8288. [CrossRef]
26. Jancikova, S.; Dordevic, D.; Jamroz, E.; Behalova, H.; Tremlova, B. Chemical and Physical Characteristics of Edible Films, Based on κ- and ι-Carrageenans with the Addition of Lapacho Tea Extract. *Foods* **2020**, *9*, 357. [CrossRef]
27. Papoutsis, K.; Golding, J.B.; Vuong, Q.; Pristijono, P.; Stathopoulos, C.E.; Scarlett, C.J.; Bowyer, M. Encapsulation of Citrus By-Product Extracts by Spray-Drying and Freeze-Drying Using Combinations of Maltodextrin with Soybean Protein and ι-Carrageenan. *Foods* **2018**, *7*, 115. [CrossRef]
28. Rahman, M.M.; Rimu, S.H. Recent development in cellulose nanocrystal-based hydrogel for decolouration of methylene blue from aqueous solution: A review. *Int. J. Environ. Anal. Chem.* **2020**. [CrossRef]
29. Guesmi, Y.; Agougui, H.; Lafi, R.; Jabli, M.; Hafiane, A. Synthesis of hydroxyapatite-sodium alginate via a co-precipitation technique for efficient adsorption of Methylene Blue dye. *J. Mol. Liq.* **2018**, *249*, 912–920. [CrossRef]
30. Sebeia, N.; Jabli, M.; Ghith, A.; Elghoul, Y.; Alminderej, F.M. Production of cellulose from Aegagropila Linnaei macro-algae: Chemical modification, characterization and application for the bio-sorption of cationic and anionic dyes from water. *Int. J. Biol. Macromol.* **2019**, *135*, 152–162. [CrossRef]
31. Yang, M.; Liu, X.; Qi, Y.; Sun, W.; Men, Y. Preparation of κ-carrageenan/graphene oxide gel beads and their efficient adsorption for methylene blue. *J. Colloid Interface Sci.* **2017**, *506*, 669–677. [CrossRef] [PubMed]
32. Alminderej, M.F.; El-Ghoul, Y. Synthesis and study of a new biopolymer-based chitosan/hematoxylin grafted to cotton wound dressings. *J. Appl. Polym. Sci.* **2019**, *136*, 47625. [CrossRef]
33. Uranga, J.; Nguyen, B.T.; Si, T.T.; Guerrero, P.; de la Caba, K. The Effect of Cross-Linking with Citric Acid on the Properties of Agar/Fish Gelatin Films. *Polymers* **2020**, *12*, 291. [CrossRef] [PubMed]
34. El-Ghoul, Y. Biological and microbiological performance of new polymer-based chitosan and synthesized amino-cyclodextrin finished polypropylene abdominal wall prosthesis biomaterial. *Text. Res. J.* **2020**, *90*, 2690–2702. [CrossRef]
35. Zhao, M.; Yang, N.; Yang, B.; Jiang, Y.; Zhang, G. Structural characterization of water-soluble polysaccharides from Opuntia monacantha cladodes in relation to their anti-glycated activities. *Food Chem.* **2007**, *105*, 1480–1486. [CrossRef]
36. Gucek, A.; Sener, S.; Bilgen, S.; Mazmanci, A. Adsorption and kinetic studies of cationic and anionic dyes on pyrophyllite from aqueous solutions. *J. Coll. Interf. Sci.* **2005**, *286*, 53–60. [CrossRef]
37. Ho, Y.S.; McKay, G. The kinetics of sorption of basic dyes from aqueous solution by sphagnum moss peat. *Can. J. Chem. Eng.* **1998**, *76*, 822–827. [CrossRef]
38. Mall, I.D.; Srivastava, V.C.; Agarwal, N.K. Removal of Orange-G and Methyl Violet dyes by adsorption onto bagasse fly ash-kinetic study and equilibrium isotherm analyses. *Dye. Pigment.* **2006**, *69*, 210–223. [CrossRef]
39. Kuang, Y.; Zhang, X.; Zhou, S. Adsorption of Methylene Blue in Water onto Activated Carbon by Surfactant Modification. *Water* **2020**, *12*, 587. [CrossRef]

polymers

MDPI

Article

Production and Properties of Lignin Nanoparticles from Ethanol Organosolv Liquors—Influence of Origin and Pretreatment Conditions

Johannes Adamcyk [1,*], Stefan Beisl [1,2], Samaneh Amini [1], Thomas Jung [1], Florian Zikeli [1], Jalel Labidi [2] and Anton Friedl [1]

[1] Institute of Chemical, Environmental and Bioscience Engineering, TU Wien, 1060 Vienna, Austria; stefan.beisl@tuwien.ac.at (S.B.); e1229284@student.tuwien.ac.at (S.A.); thomas.jung@tuwien.ac.at (T.J.); florian.zikeli@tuwien.ac.at (F.Z.); anton.friedl@tuwien.ac.at (A.F.)

[2] Chemical and Environmental Engineering Department, University of the Basque Country UPV/EHU, 20018 Donostia-San Sebastián, Spain; jalel.labidi@ehu.eus

* Correspondence: johannes.adamcyk@tuwien.ac.at; Tel.: +43-1-58801-166254

Abstract: Despite major efforts in recent years, lignin as an abundant biopolymer is still underutilized in material applications. The production of lignin nanoparticles with improved properties through a high specific surface area enables easier applicability and higher value applications. Current precipitation processes often show poor yields, as a portion of the lignin stays in solution. In the present work, lignin was extracted from wheat straw, spruce, and beech using ethanol organosolv pretreatment at temperatures from 160–220 °C. The resulting extracts were standardized to the lowest lignin content and precipitated by solvent-shifting to produce lignin micro- and nanoparticles with mean hydrodynamic diameters from 67.8 to 1156.4 nm. Extracts, particles and supernatant were analyzed on molecular weight, revealing that large lignin molecules are precipitated while small lignin molecules stay in solution. The particles were purified by dialysis and characterized on their color and antioxidant activity, reaching ASC equivalents between 19.1 and 50.4 mg/mg. This work gives detailed insight into the precipitation process with respect to different raw materials and pretreatment severities, enabling better understanding and optimization of lignin nanoparticle precipitation.

Keywords: lignin; nanoparticles; biorefinery; organosolv pretreatment

Citation: Adamcyk, J.; Beisl, S.; Amini, S.; Jung, T.; Zikeli, F.; Labidi, J.; Friedl, A. Production and Properties of Lignin Nanoparticles from Ethanol Organosolv Liquors—Influence of Origin and Pretreatment Conditions. *Polymers* **2021**, *13*, 384. https://doi.org/10.3390/polym13030384

Academic Editor: Rafael Antonio Balart Gimeno

Received: 29 December 2020
Accepted: 22 January 2021
Published: 26 January 2021

1. Introduction

Lignin as the second most abundant biopolymer after cellulose has gained massive interest in recent years. It is estimated that approximately 2×10^{11} tons of lignocellulosic biomass residues are produced worldwide every year [1], making lignin a renewable phenolic compound available in large quantities. Since it is already produced in side-streams of pulp-productions and biorefineries, it has the potential to enable the transition from a fossil-based to a biobased economy by supplanting synthetic compounds currently produced from fossil resources. Since only around 40% of the lignin produced in pulping processes are needed to cover the internal energy demand of the processes [2,3], the possibility exists to vastly increase the amount of lignin used for material purposes providing efficient use of all biomass components and improving the sustainability of the process. Thus, the use of lignin as a material should be heavily expanded.

One challenge in the utilization of lignin is its diversity, as it is known to differ for different plant species, but also depending on the extraction process [4]. Lignin from the most common pulping processes, the Kraft and sulfite processes, contains a relatively-high amount of sulfur and may therefore not be suitable for many practical applications. In contrast, ethanol organosolv pretreatment yields lignin of high purity, with a structure closely resembling that of native lignin, and without sulfur contamination through the process [5,6], opening up possibilities for new applications [6]. A disadvantage of this process is its lower efficiency and high energy demand for solvent recovery [7]. Improvement of the process economy while still maintaining the high quality of the lignin produced will be necessary to compete with other pretreatment technologies on the one hand and fossil-based products that are aimed to be replaced by lignin on the other.

Although increased process severity tends to increase delignification and lignin concentration in the liquor [8], which would be desirable for increased process efficiency, simultaneously more sugar degradation products are formed [9], and the lignin changes structurally, which might be problematic for the final product. Thus, process efficiency needs to be balanced out with the final product quality.

Another possibility to improve the economy of a biorefinery based on organosolv pretreatment is finding high value applications for the produced lignin. Recent works have shown that colloidal lignin particles have many desirable and improved properties which lead to a wide range of possible applications [10], due to the high ratio of surface to volume. Especially the application of lignin for UV-protection in cosmetics [11–13] but also wood conservation [14] has been reported and commercial use in these areas seems within reach. Qian et al. [15] reported that the ability to block UV-radiation is improved for smaller particle sizes, which underlines the connection between improved properties and particle size. However, previous works have shown that a major portion of the lignin remained in solution after the precipitation of colloidal particles, resulting in a lower overall yield [16,17]. This raises the suspicion that lignin gets fractionated by molecular weight in the precipitation step, which was confirmed for commercial soda lignin by Sipponen et al. [18]. Further understanding of the precipitation of lignin should make it possible to improve the process towards higher yields while still maintaining small particle sizes.

Based on this, the present work investigates the extraction with ethanol organosolv pretreatment and the subsequent precipitation of lignin into colloidal particles. Three different raw materials (wheat straw, spruce wood, and beech wood), representing three combinations of guaiacyl (G), *p*-hydroxy phenol (H) and syringyl (S) lignins common in nature (GSH, G and GS lignins, respectively), and four temperatures (160–220 °C) were applied in the pretreatment process. Through this experimental plan, the influence of raw material and process severity on the lignin precipitation was studied. The extracts' compositions were characterized, and the prepared lignin particles were analyzed regarding their size and physico-chemical properties. The different process fractions were analyzed regarding their molecular weight to further investigate the precipitation process.

2. Materials and Methods

An overview of the experimental procedure is shown in Figure 1, indicating process steps, fractions, and analytics. Generally, lignin was extracted from the different raw materials and precipitated into colloidal particle suspensions by addition of an antisolvent. A part of these suspensions was centrifuged for analytics, the remainder was dialyzed to remove dissolved impurities. The last step was membrane filtration to produce a thin particle layer for color measurements.

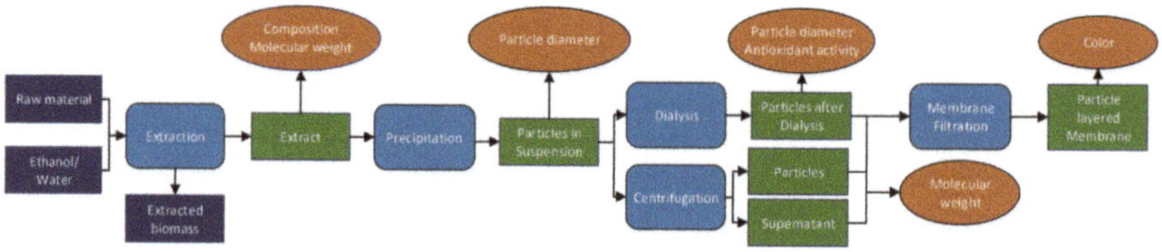

Figure 1. Schematic of the experimental plan conducted for each pretreatment condition and raw material. The analyzed process fractions are depicted in green.

The particle fraction separated via dialysis is termed "particles after dialysis" and the particle fraction separated via centrifugation is termed "particles" throughout the whole manuscript. The particles after the centrifugation are considered the same as the ones in the suspension directly after precipitation.

2.1. *Materials*

The wheat straw used was harvested in 2015 in Lower Austria, the spruce wood was supplied by Laakirchen Papier (Laakirchen, Austria), the beech wood by Lenzing AG (Lenzing, Austria). The particle size of all raw materials was reduced in a cutting mill with a 2 mm mesh, the resulting material was stored under dry conditions until the pretreatment. Ultra-pure water (18 MΩ/cm) and Ethanol (Merck, 96 vol %, undenaturated, Darmstadt, Germany) was used in the organosolv treatment.

2.2. *Pretreatment/Extraction*

The organosolv pretreatment was conducted in a 1 L stirred autoclave (Zirbus, HAD 9/16, Bad Grund, Germany) using a 60 wt% aqueous ethanol mixture as solvent under consideration of the water content in the raw material. The biomass content in the reactor based on dry matter was 8.3 wt%. The reactor was heated to the pretreatment temperature and then held at this temperature. After 60 min of total pretreatment time, the reactor was cooled to room temperature. The solid and liquid fractions were separated using a hydraulic press (Hapa, HPH 2.5, Achern, Germany) at 200 bar, the extract was then centrifuged (Thermo Scientific, Sorvall, RC 6+, Waltham, MA, USA) at 24,104× g for 20 min to remove residual solids. The particle-free extracts of the single batches were unified, the composition was analyzed, and it was stored at 5 °C until the precipitation experiments were performed.

2.3. *Precipitation*

To eliminate the influence of the lignin concentration on the precipitation, the concentration of all extracts was set to the lowest total lignin concentration (3.52 g/L) by dilution with 60 wt% aqueous ethanol. The precipitation experiments were conducted at 25 °C, the volume ratio of extract to antisolvent was kept constant at 1:5, where the antisolvent consisted of Ultra-pure water. Setup (b) as described in [16] was used for the precipitations. The resulting suspensions were stored at 5 °C.

2.4. *Downstream Processing*

Parts of the suspensions were centrifuged in a Thermo WX Ultra 80 ultracentrifuge (Thermo Scientific, Waltham, MA, USA) at 288,000× g for 60 min. The supernatants were decanted and stored at 5 °C, the liquid free particles were freeze dried and stored in a desiccator.

The rest of the suspensions were dialyzed for 7 days in excess of water (daily replaced) using Nadir® dialysis hoses with 10,000–20,000 Dalton cut-off. The particles after dialysis were stored at 5 °C.

2.5. *Analytics*

2.5.1. Extract Characterization

The organosolv extract was analyzed for carbohydrates, lignin, and degradation products. The carbohydrate content was determined by the sample preparation following the National Renewable Energy Laboratory (NREL) laboratory analytical procedure (LAP) "Determination of sugars, byproducts, and degradation products in liquid fraction process samples" [19] but with no neutralization of the samples after hydrolysis. A Thermo Scientific ICS-5000 HPAEC system equipped with a photo array detector (PAD) with deionized water as the eluent was used for the determination of arabinose, glucose, mannose, xylose, and galactose. The concentrations of the degradation products acetic acid, hydroxymethylfurfural (HMF) and furfural were determined with a Shimadzu LC-20A "prominence" HPLC system and a Shodex SH1011 (Showa Denko, Tokyo, Japan) column at 40 °C, with 0.005 M H_2SO_4 as eluent. The acid insoluble and acid soluble lignin contents were determined following the NREL LAP "Determination of Structural Carbohydrates and Lignin in Biomass" [20] using the dry matter of the extract obtained at 105 °C.

2.5.2. HP-SEC Analysis

Molar mass distributions of the standardized organosolv extracts, suspensions, centrifuged lignin particles, supernatants, and lignin particles after dialysis were determined by alkaline High Performance Size Exclusion Chromatography (HP-SEC) analysis (eluent: 10 mM NaOH) using three TSK-Gel columns in series at 40 °C (PW5000, PW4000, PW3000; TOSOH Bioscience, Darmstadt, Germany) with an Agilent 1200 HPLC system (flow rate: 1 mL/min, DAD detection at 280 nm, Santa Clara, CA, USA). Solid lignin fractions were dissolved in the eluent, and liquid lignin fractions were diluted with the eluent to adjust the pH for analysis. Calibration of the columns set was done using polystyrene sulfonate reference standards (PSS GmbH, Mainz, Germany). The molar masses at peak Maximum (M_p) were 78,400 Da, 33,500 Da, 15,800 Da, 6430 Da, 1670 Da, 891 Da, and 208 Da.

2.5.3. Hydrodynamic Diameter

Particle size measurements were conducted in a ZetaPALS (Brookhaven Instruments, Holtsville, NY, USA) using dynamic light scattering (DLS). The refractive index of the particles was set to 1.53 and the imaginary refractive index to 0.1. All reported mean values for hydrodynamic diameters are based on intensity based distributions.

2.5.4. Color

The colorimeter PCE-CMS 7, PCE Instruments (Albacete, Spain) was used to measure the Hunter color values (L, a, and b) of the lignin. The lignin particles after dialysis were filtered against a HFK™-131 ultrafiltration membrane (Koch Membrane Systems, Rimsting, Germany) resulting in layer densities of lignin ranging from around 1 g/m^2 to 14 g/m^2. All membranes were purged with 10 wt% aqueous ethanol and the color background (L = 99.08, a = 0.001, b = −4.911) was measured after purging. The color of each lignin sample was measured at three different lignin layer densities and was standardized to 3 g/m^2 via linear regression of the L, a, and b values.

2.5.5. Antioxidant Activity

All particle suspensions were filtrated using a PES-Filter (Merck, Darmstadt, Germany) with a pore size of 0.22 μm to remove agglomerates and yield comparable particle size distributions.

ABTS: The determination of 2,2′-azinobis-(3-ethylbenzothiazoline-6-sulfonic acid) (ABTS) radical inhibition was based on the method of Re et al. [21]. Twenty-five milliliters of ABTS stock solution was prepared in dark bottle by dissolving 69.7 mg of ABTS salt and 11.8 mg $K_2S_2O_8$ in water. The stock solution was left in a refrigerator and diluted with deionized water to obtain an initial absorbance of 0.7 at 734 nm in a Jasco V-630 spectrophotometer (JASCO, Tokyo, Japan) before use. To determine the radical scavenging ability, the sample was added to the diluted ABTS solution. The absorbance was measured after 6 min incubation at 734 nm. The percentage of inhibition was calculated from below equation:

$$AA[\%] = (Ac - At)/Ac \times 100\% \quad (1)$$

where: AA-antioxidant activity, Ac—absorbance of control sample, At—absorbance of tested sample.

FRAP: The ferric reducing antioxidant power (FRAP) test is based on the reduction of Fe(III) to Fe(II) by the antioxidant compound which forms a colored complex (593 nm) with 2,4,6-tripyridyl-s-triazine (Fe(II)-TPTZ) in acetate buffer at pH 3.6 [22]. Briefly, the reactive solution was freshly prepared with 25 mL of 300 mM acetate buffer (pH 3.6), 2.5 mL of 10 mM 2,4,6-tripyridyl-s-triazine in 40 mM HCl and 2.5 mL 20 mM $FeCl_3 \cdot 6H_2O$ in distilled water. Samples (0.1 mL) were mixed with 3 mL of the FRAP reactive solution. The reference standard used was ascorbic acid (ASC) and results are given as ASC equivalents.

3. Results

3.1. Extract Composition

The composition of the extracts is shown in Figure 2. The amount of each compound in the extract increases with temperature, the largest increase of acid insoluble lignin (AIL) and the degradation products happens from 200 to 220 °C for all raw materials, indicating a considerable increase in process severity. Lignin is the dominant compound in the dry matter of the extracts, making up between 50 wt% for extracts from straw and 70 wt% for extracts from the woods. From 160 to 200 °C, the total lignin content of the wheat straw extracts is higher than that of the woods due to the consistently higher amount of acid soluble lignin (ASL) in wheat straw extracts, while at 220 °C the lignin content in the extract from spruce is highest, followed by beech (Figure 2a). This could suggest that although the initial lignin content of wheat straw is lower than that of the woods, the extraction is more efficient already at lower temperatures, and a higher severity is required for efficient delignification of the woods when using ethanol organosolv pretreatment. Siika-aho et al. [23] applied an oxygen/carbonate process to three different raw materials at varying process severities and also found delignification of straw to be more efficient at low process severities compared to spruce and beech.

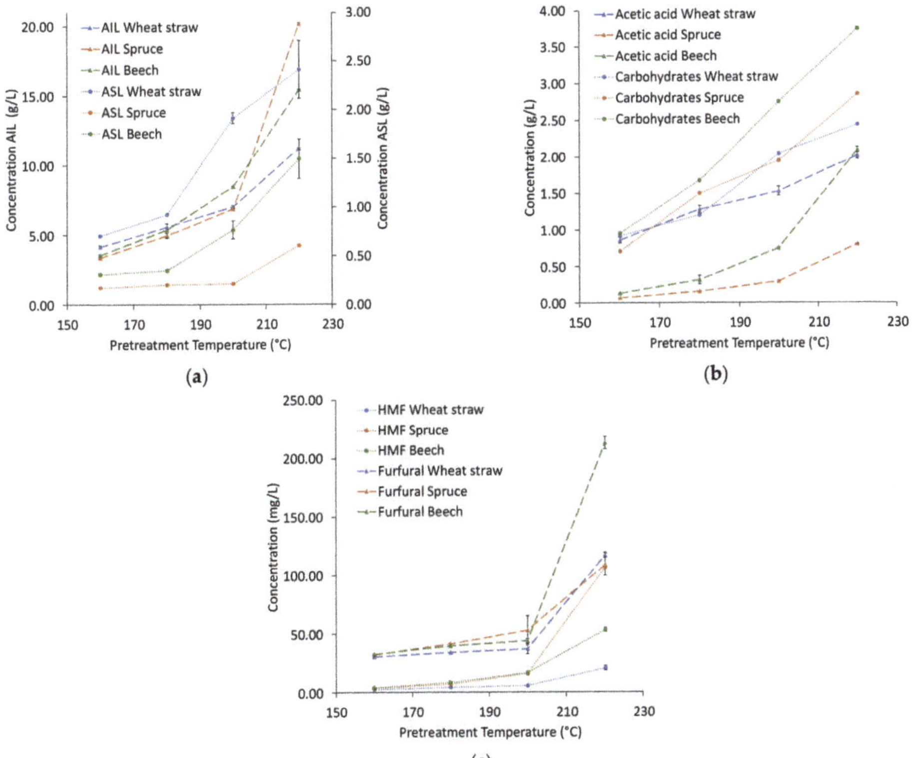

Figure 2. Concentration of lignin (**a**), acetic acid and carbohydrates (**b**), and degradation products (**c**) in organosolv extracts. Abbreviations: AIL—acid insoluble lignin, ASL—acid soluble lignin, HMF—hydroxymethylfurural.

The trends of the carbohydrates in the extracts (Figure 2b) are similar for all raw materials, though beech wood has consistently the highest concentrations. For acetic acid, extraction from wheat straw results in a higher concentration than from the woods in most cases, though there is a strong increase for beech from 200 °C to 220 °C. For the degradation products (Figure 2c), the amounts are of the same order of magnitude from 160 °C to 200 °C, at 220 °C hydroxymethylfurfural (HMF) of spruce and furfural of beech stand out.

To remove the influence of the lignin concentration in the precipitation step, the extracts were diluted to the lowest lignin concentration (3.52 g/L) with aqueous ethanol (60 wt %). As demonstrated in Figure 3a, the molecular weight distribution of the lignin in the extracts is influenced by the pretreatment temperature and the raw material. The changes between the raw materials become more visible in the molecular weight distributions (Figure 3b). There are four distinct peaks of different molecular weight, with maxima at approximately 900, 610, 410, and 255 Da. The two lower molecular weight fractions are more dominant in the wheat straw extracts than in those of the woods, which could be linked to the higher concentration of ASL found in wheat straw extracts. In earlier works the third peak was found to be strongly influenced by p-hydroxycinnamic acids like ferulic acid and p-coumaric acid, in monomeric form or connected to lignin fragments present in wheat straw [24]. Apart from the differences in the raw materials, the pretreatment temperature influences the molecular weight distribution of the dissolved lignin; large, dissolved lignin molecules are depolymerized at higher temperatures, and lignin fragments can again re-condensate at severe conditions [25]. The combination of raw material and temperature decides which effect is dominant. The M_W of all three raw materials show a trend to lower values with increasing temperature (Figure 3b), which is more pronounced for wheat straw than for the woods. In Figure 3c, the absorbance is multiplied with the molecular weight to highlight the strong influence of high molecular weight fractions on the M_W. These molecular weight distributions also indicate (stronger than the weight averaged values) that the temperature increase from 200 to 220 °C significantly influences the molecular weight.

3.2. Precipitation

The lignin dissolved in the extracts was precipitated into micro- and nanoparticles by addition of water as an antisolvent. The resulting suspensions were characterized on the particles' hydrodynamic diameter by dynamic light scattering. The particles and supernatant were separated by centrifugation for analysis; parts of the suspensions were dialyzed to remove impurities before further analytics. Visible agglomerates formed during dialysis were removed by filtration before determination of the antioxidant activity in order to analyze exclusively colloidal particles. Molecular weight of lignin particles after precipitation, dissolved lignin in the supernatants, and lignin particles after dialysis was measured. The yields of the precipitations could not be determined due to too low sample amounts.

The hydrodynamic diameter of the particles stays relatively similar for each raw material between 160 °C and 200 °C, with only a slight trend upwards and diameters between 67.8 nm and 105.9 nm. However, the extracts from 220 °C yielded significantly bigger particles with diameters ranging from 152.8 nm for spruce to 1156 nm for beech (Figure 4), suggesting that the precipitation is strongly influenced by changes in the lignin occurring above 200 °C of pretreatment temperature.

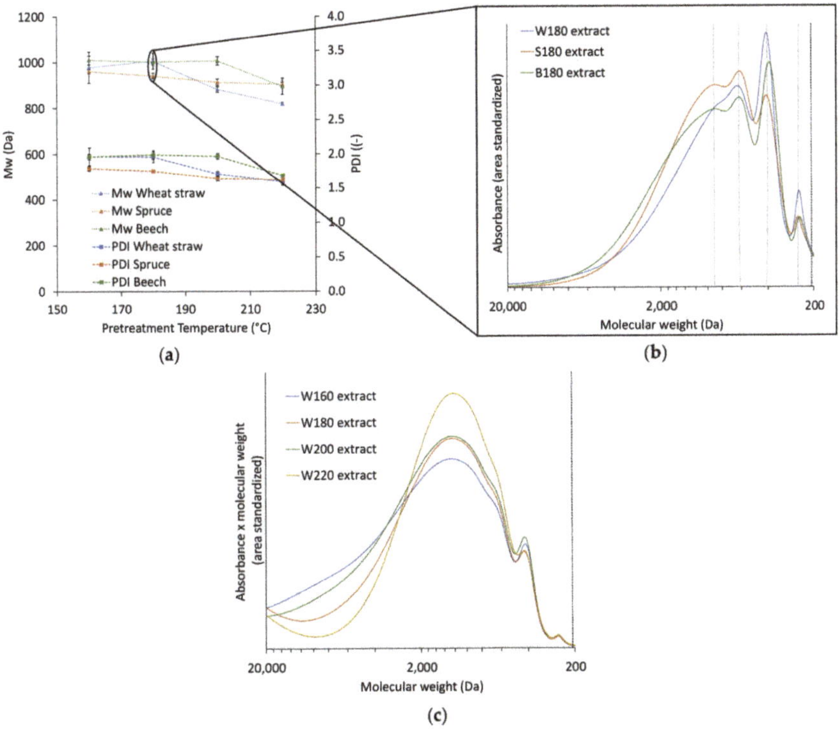

Figure 3. M_W and PDI of the extracts from the different raw materials (**a**), molecular weight distribution of extracts from different raw materials at 180 °C (**b**), and molecular weight distribution of wheat straw at different pretreatment temperatures (**c**). Abbreviations: W—wheat straw, S—spruce, B—beech; the number next to the letter is the temperature of the corresponding pretreatment temperature.

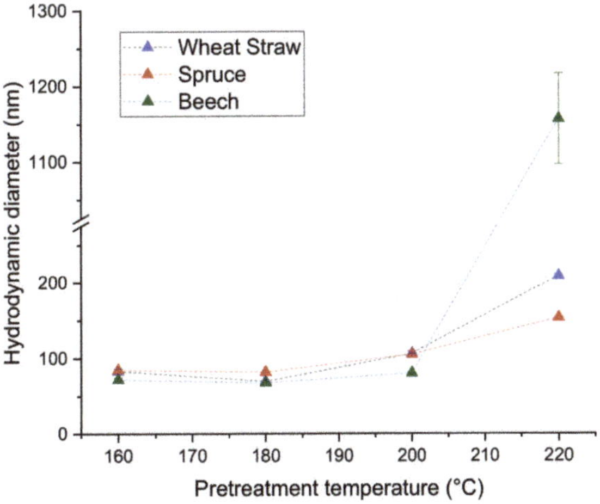

Figure 4. Hydrodynamic diameters of lignin particles.

This leads back to the differences in molecular weight distribution found in the extracts for different pretreatment temperatures and raw materials. In Figure 5, the M_Ws of all raw materials, pretreatment temperatures and fractions are presented. The results show that M_ws of the lignin in the extracts are lower than those of the particles as well as of the particles after dialysis, but higher than those of the supernatants for all experiments. This supports the assumption that during the solvent shifting precipitation mainly lignin of high molecular weight is precipitated into particles while lignin of lower molecular weight remains in solution. Figure 6 showcases this fractionated precipitation for beech wood pretreated at 180 °C. Sipponen et al. [18] also found this molar-mass-fractionation when precipitating wheat straw soda lignin from aqueous ethanol by water addition and attributed it to electronic interactions of the aromatic lignin structures, which lead to lower solubility of larger lignin molecules.

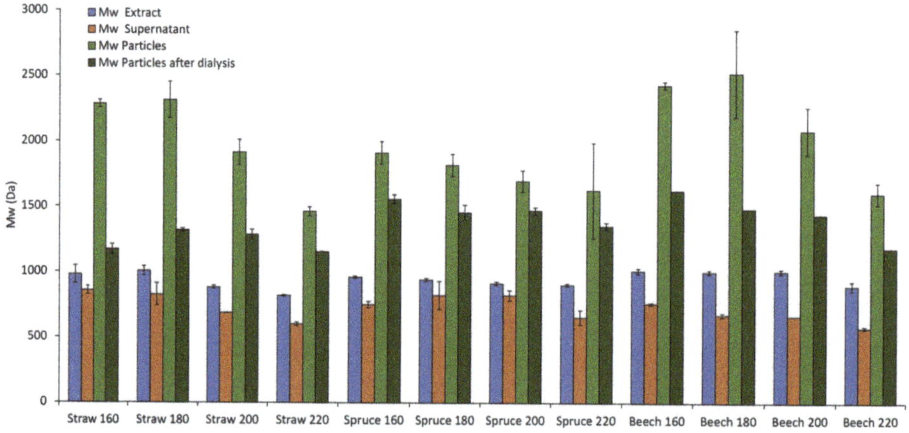

Figure 5. M_w in the different process fractions for wheat straw, spruce, and beech.

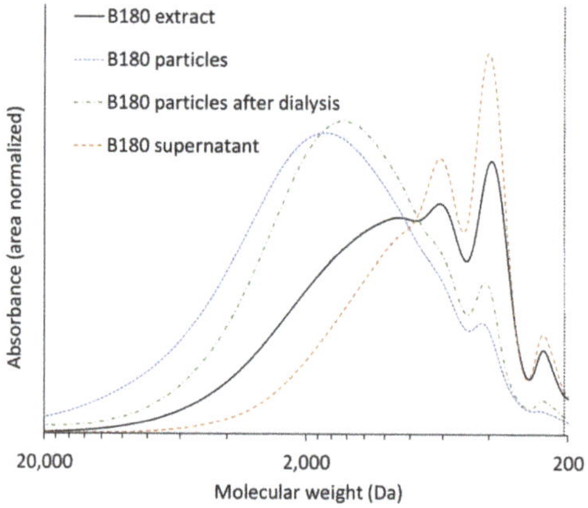

Figure 6. Molecular weight distributions of extract, particles, particles after dialysis, and supernatant from beech wood experiments at 180 °C pretreatment temperature.

Interestingly, the M_ws of the particles separated from the suspension directly after precipitation are consistently higher than those of the particles separated after dialysis (light and dark green columns in Figure 5). This indicates that some lignin with lower molecular weight is still in solution after precipitation but not removed by dialysis, despite a much higher membrane molecular weight cut-off. A possible explanation for this is that the gradual ethanol removal during dialysis lowers the lignin solubility. Smaller lignin molecules that are dissolved in the supernatant would precipitate onto the existing particles due to this, increasing their diameter and facilitating agglomeration. The results show that a portion of the lignin dissolved in the supernatant is removed by dialysis, since the M_w of the particles after dialysis is always higher than that of the extract.

The trend to lower M_ws at higher pretreatment temperatures found in the extracts (Figure 3a) is also present and more pronounced for the particles (Figure 5), which coincides with the increase in particle size at higher temperatures (Figure 4). A possible explanation for this is that at higher pretreatment temperatures more lignin is depolymerized into smaller molecules which have a higher solubility than the larger molecules in the same extract. Because of the higher solubility of smaller lignin molecules, the degree of local supersaturation after the mixing would be lower for extracts produced at 220 °C, causing slower precipitation and larger particles [26]. Figure 7c shows this decrease of particle size with increasing M_w. Zwilling et al. [27] also found this increase of particle size with decreasing molecular weight, and connected it with more hydrophilic character of low molecular weight lignin. However, in this work lignin structure was not investigated.

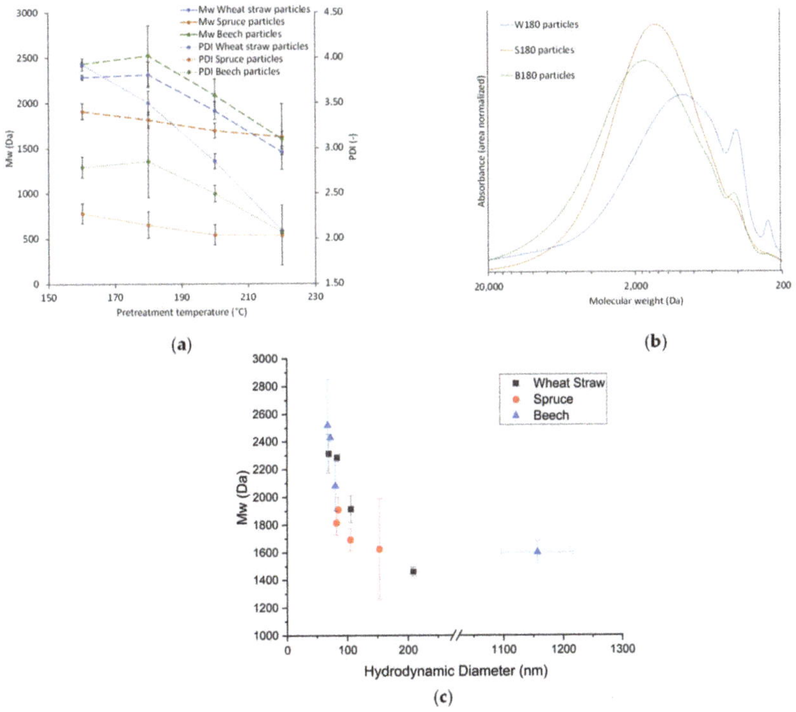

Figure 7. M_w and PDI of lignin particles (**a**), molecular weight distribution of particles from different raw materials at 180 °C pretreatment temperature (**b**), M_w of particles over diameter of particles (**c**).

Comparison of the particles' molecular weight distributions for the different raw materials (Figure 7) shows that spruce wood extracts yield particles with lower M_ws and PDIs than wheat straw or beech wood extracts. Figure 7b exemplifies this and additionally shows that the lignin particles from the woods have a similar distribution, while wheat straw results in a much broader distribution with more dominant peaks at 410 Da and 210 Da. At 220 °C, M_ws and PDIs of all raw materials are within the same range and generally lower, suggesting that the lignin of all raw materials is depolymerized to a similar degree (Figure 7a). Comparing Figures 3a and 7a shows that the differences between the M_ws or PDIs found for the particles are not present to the same extent in the respective extracts, which means that the precipitation amplifies differences between lignin from different raw materials. This might in turn have an influence on the particles' properties.

3.3. Lignin Properties

Since the particle size generally increased strongly for the precipitations from 220 °C extracts, and the precipitation did not give reasonable results for beech at that temperature, the color and antioxidant potential were only determined for lignin particles after dialysis from pretreatment temperatures 160 °C to 200 °C.

3.3.1. Color

For many of the possible commercial material applications of colloidal lignin particles, like in sunscreens or food packaging [10], it is important to consider the color of the produced lignin particles. For example, if lignin particles are being used in a product for their antioxidant activities to avoid color change in a product, a dark, brown particle color is obstructive to the application and a limiting factor to the maximum particle concentration applicable. Figure 8 shows the colors of equally thin layers of lignin particles on a membrane and the respective values in the CIELAB-color space. Higher pretreatment temperatures lead to darker lignin and stronger colors, demonstrated by the lowered brightness (L *) and increased reddening (+a *) and yellowing (+b *) values in the spider graph, which can be considered undesirable. The darkening of the lignin particles can be attributed to increase of condensation reactions at higher process severities. The color changes most for wheat straw and least for beech wood with increasing pretreatment temperature. The ratio of red to yellow color component is similar for all lignins, except for wheat straw at 160 °C, where the yellow component is more dominant. Generally, through variation of raw material and pretreatment temperature a wide array of colors in the brown spectrum can be achieved for lignin particles. The influence of the particle size on the color found by other groups [28] could not be investigated with the given experimental plan, since too many other contributing factors were varied, and the investigated particles where of similar diameters.

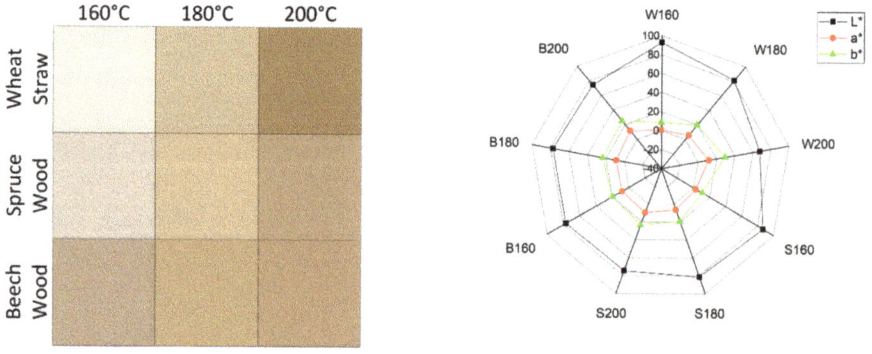

(a) (b)

Figure 8. Color of a 3 g/m^2 layer of lignin particles (**a**) and affiliated components in CIELAB-color space (**b**).

3.3.2. Antioxidant Activities

Before determining the antioxidant activities of the particles after dialysis, agglomerates and large particles were removed from the particle suspensions after dialysis by filtration, which resulted in suspensions with hydrodynamic diameters between 65.7 and 90.7 nm. Both methods used to determine the antioxidant activity (FRAP and ABTS) show similar trends (Figure 9): Higher pretreatment temperatures lead to lower antioxidant activity for beech, for spruce this effect is only witnessed when the temperature is increased from 180 to 200 °C. For straw, the antioxidant activity does not significantly change with the temperature for ABTS, for FRAP it reaches a maximum at 200 °C. The antioxidant potential of the beech lignin particles is highest at 160 and 180 °C according to both methods, at 200 °C wheat straw reaches similar values. The divergence in the results between ABTS and FRAPS makes it difficult to compare the antioxidant potential of particles from spruce and wheat straw at 160 and 180 °C.

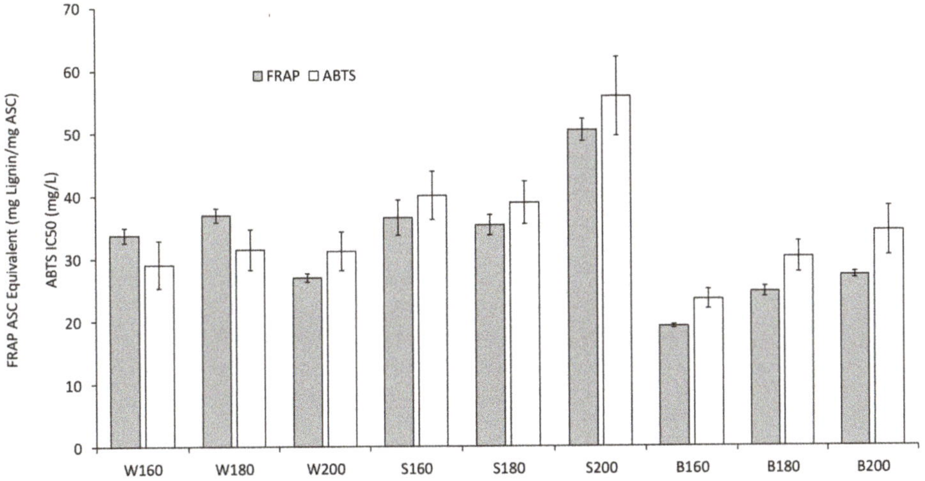

Figure 9. Antioxidant activities of particles after dialysis.

It is important to note that the antioxidant potential was measured for the particles (not solubilized lignin), and therefore is also affected by the surface area. Zhang et al. [29] investigated the antioxidant activities of lignin nanoparticles from corncob residue and found IC_{50} values between 101.2 and 296.6 mg/L. Due to the variation in raw materials and pretreatment temperatures combined with comparatively small differences in the particle diameters, the influence of the particle size on the antioxidant activity found in other works could not be investigated in this work.

4. Conclusions

The production and properties of colloidal lignin particles from organosolv extracts from different raw materials and the influence of the pretreatment temperature was investigated in the present work. All investigated raw materials (wheat straw, spruce wood, beech wood) were successfully applied for the production of colloidal particles with diameters below 110 nm whereas the origin of the lignin showed only minor influence in the size of the resulting particles.

Higher pretreatment temperatures increase the delignification of the raw materials but also favor depolymerization and structural alteration of the extracted lignin. This leads to increased particle sizes of over 110 nm and agglomeration at pretreatment temperatures of over 200 °C.

The investigation of the precipitation via solvent-shifting showed lignin fractionation by molecular weight, meaning that large lignin molecules precipitate while smaller lignin molecules stay in solution.

The resulting lignin nanoparticles were characterized on their application relevant properties of color and antioxidant potential. Higher pretreatment temperatures generally results in particles with darker colors and lower antioxidant activity especially for particles produced from beech and spruce. The antioxidant activity of the particles ranged from 19.1 and 50.4 mg Lignin/mg ASC equivalents.

From a process perspective, it is desirable to reach a high concentration of lignin in the extract, which can be achieved by higher pretreatment temperatures. However, this increase in process efficiency comes at the cost of lignin quality, which is demonstrated by the sharp increase in particle diameter, darker color of the resulting particles, and the decrease in antioxidant activity of the lignin particles. Therefore, optimal conditions and choice of raw material in a commercial lignin nanoparticle production process heavily depend on the final application.

Author Contributions: Conceived and designed the experiments, S.B. and J.A.; performed the experiments and analyzed the data, S.A., T.J., J.A., S.B., and F.Z.; wrote and edited the manuscript, J.A. with significant input and editing from A.F., J.L., S.B., and F.Z. All authors have read and agreed to the published version of the manuscript.

Funding: This research received no external funding.

Institutional Review Board Statement: Not applicable.

Informed Consent Statement: Not applicable.

Data Availability Statement: The data presented in this study are available on request from the corresponding author.

Acknowledgments: The authors acknowledge TU Wien Bibliothek for financial support through its Open Access Funding Program.

Conflicts of Interest: The authors declare no conflict of interest.

References

1. Zhang, M.-L.; Fan, Y.-T.; Xing, Y.; Pan, C.-M.; Zhang, G.-S.; Lay, J.-J. Enhanced biohydrogen production from cornstalk wastes with acidification pretreatment by mixed anaerobic cultures. *Biomass Bioenergy* **2007**, *31*, 250–254. [CrossRef]
2. Sassner, P.; Galbe, M.; Zacchi, G. Techno-economic evaluation of bioethanol production from three different lignocellulosic materials. *Biomass Bioenergy* **2008**, *32*, 422–430. [CrossRef]
3. Sannigrahi, P.; Ragauskas, A.J. Characterization of Fermentation Residues from the Production of Bio-Ethanol from Lignocellulosic Feedstocks. *J. Biobased Mater. Bioenergy* **2011**, *5*, 514–519. [CrossRef]
4. Calvo-Flores, F.G.; Dobado, J.A.; Isac-García, J.; Martín-MartíNez, F.J. *Lignin and Lignans as Renewable Raw Materials*; John Wiley & Sons, Ltd.: Chichester, UK, 2015.
5. Gordobil, O.; Herrera, R.; Yahyaoui, M.; İlk, S.; Kaya, M.; Labidi, J. Potential use of kraft and organosolv lignins as a natural additive for healthcare products. *RSC Adv.* **2018**, *8*, 24525–24533. [CrossRef]
6. Lora, J.H.; Glasser, W.G. Recent Industrial Applications of Lignin: A Sustainable Alternative to Nonrenewable Materials. *J. Polym. Environ.* **2002**, *10*, 39–48. [CrossRef]
7. Zhao, X.; Cheng, K.; Liu, D. Organosolv pretreatment of lignocellulosic biomass for enzymatic hydrolysis. *Appl. Microbiol. Biotechnol.* **2009**, *82*, 815–827. [CrossRef] [PubMed]
8. Pan, X.; Kadla, J.F.; Ehara, K.; Gilkes, N.; Saddler, J.N. Organosolv ethanol lignin from hybrid poplar as a radical scavenger: Relationship between lignin structure, extraction conditions, and antioxidant activity. *J. Agric. Food Chem.* **2006**, *54*, 5806–5813. [CrossRef] [PubMed]
9. Pan, X.; Gilkes, N.; Kadla, J.; Pye, K.; Saka, S.; Gregg, D.; Ehara, K.; Xie, D.; Lam, D.; Saddler, J. Bioconversion of hybrid poplar to ethanol and co-products using an organosolv fractionation process: Optimization of process yields. *Biotechnol. Bioeng.* **2006**, *94*, 851–861. [CrossRef] [PubMed]
10. Beisl, S.; Friedl, A.; Miltner, A. Lignin from Micro- to Nanosize: Applications. *Int. J. Mol. Sci.* **2017**, *18*, 2367. [CrossRef] [PubMed]
11. Gordobil, O.; Olaizola, P.; Banales, J.M.; Labidi, J. Lignins from Agroindustrial by-Products as Natural Ingredients for Cosmetics: Chemical Structure and In Vitro Sunscreen and Cytotoxic Activities. *Molecules* **2020**, *25*, 1131. [CrossRef] [PubMed]
12. Qian, Y.; Qiu, X.; Zhu, S. Lignin: A nature-inspired sun blocker for broad-spectrum sunscreens. *Green Chem.* **2015**, *17*, 320–324. [CrossRef]
13. Qian, Y.; Qiu, X.; Zhu, S. Sunscreen Performance of Lignin from Different Technical Resources and Their General Synergistic Effect with Synthetic Sunscreens. *ACS Sustain. Chem. Eng.* **2016**, *4*, 4029–4035. [CrossRef]
14. Zikeli, F.; Vinciguerra, V.; Taddei, A.R.; D'Annibale, A.; Romagnoli, M.; Scarascia Mugnozza, G. Isolation and characterization of lignin from beech wood and chestnut sawdust for the preparation of lignin nanoparticles (LNPs) from wood industry side-streams. *Holzforschung* **2018**, *72*, 961–972. [CrossRef]
15. Qian, Y.; Zhong, X.; Li, Y.; Qiu, X. Fabrication of uniform lignin colloidal spheres for developing natural broad-spectrum sunscreens with high sun protection factor. *Ind. Crops Prod.* **2017**, *101*, 54–60. [CrossRef]

16. Beisl, S.; Loidolt, P.; Miltner, A.; Harasek, M.; Friedl, A. Production of Micro- and Nanoscale Lignin from Wheat Straw Using Different Precipitation Setups. *Molecules* **2018**, *23*, 633. [CrossRef]
17. Beisl, S.; Adamcyk, J.; Friedl, A. Direct Precipitation of Lignin Nanoparticles from Wheat Straw Organosolv Liquors Using a Static Mixer. *Molecules* **2020**, *25*, 1388. [CrossRef]
18. Sipponen, M.H.; Lange, H.; Ago, M.; Crestini, C. Understanding Lignin Aggregation Processes. A Case Study: Budesonide Entrapment and Stimuli Controlled Release from Lignin Nanoparticles. *ACS Sustain. Chem. Eng.* **2018**, *6*, 9342–9351. [CrossRef]
19. Sluiter, A.; Hames, B.; Ruiz, R.; Scarlata, C.; Sluiter, J.; Templeton, D. Determination of Sugars, Byproducts, and Degradation Products in Liquid Fraction Process Samples: Laboratory Analytical Procedure (LAP). *Gold. Natl. Renew. Energy Lab.* **2008**, *11*, 65–71.
20. Sluiter, A.; Hames, B.; Ruiz, R.; Scarlata, C.; Sluiter, J.; Templeton, D.; Crocker, D. Determination of Structural Carbohydrates and Lignin in Biomass: Laboratory Analytical Procedure (LAP). *Lab. Anal. Proced.* **2008**, *1617*, 1–16.
21. Re, R.; Pellegrini, N.; Proteggente, A.; Pannala, A.; Yang, M.; Rice-Evans, C. Antioxidant activity applying an improved ABTS radical cation decolorization assay. *Free Radic. Biol. Med.* **1999**, *26*, 1231–1237. [CrossRef]
22. Benzie, I.F.F.; Strain, J.J. The Ferric Reducing Ability of Plasma (FRAP) as a Measure of Antioxidant Power: The FRAP Assay. *Anal. Biochem.* **1996**, *239*, 70–76. [CrossRef] [PubMed]
23. Siika-aho, M.; Varhimo, A.; Sirviö, J.; Kruus, K. Sugars from Biomass—High Cellulose Hydrolysability of Oxygen Alkali Treated Spruce, Beech and Wheat Straw. In Proceedings of the 6th Nordic Wood Biorefinery Conference 2015, Helsinki, Finland, 20–22 October 2015; pp. 71–77.
24. Zikeli, F.; Ters, T.; Fackler, K.; Srebotnik, E.; Li, J. Fractionation of wheat straw Dioxane lignin reveals molar mass dependent structural differences. *Ind. Crops Prod.* **2016**, *91*, 186–193. [CrossRef]
25. El Hage, R.; Brosse, N.; Sannigrahi, P.; Ragauskas, A. Effects of process severity on the chemical structure of Miscanthus ethanol organosolv lignin. *Polym. Degrad. Stab.* **2010**, *95*, 997–1003. [CrossRef]
26. Salaheldeen Elnashaie, S.; Danafar, F.; Hashemipour Rafsanjani, H. *Nanotechnology for Chemical Engineers*; Springer Singapore: Singapore, 2015; ISBN 9789812874955.
27. Zwilling, J.D.; Jiang, X.; Zambrano, F.; Venditti, R.A.; Jameel, H.; Velev, O.D.; Rojas, O.J.; Gonzalez, R. Understanding lignin micro- and nanoparticle nucleation and growth in aqueous suspensions by solvent fractionation. *Green Chem.* **2021**. [CrossRef]
28. Lee, S.C.; Yoo, E.; Lee, S.H.; Won, K. Preparation and Application of Light-Colored Lignin Nanoparticles for Broad-Spectrum Sunscreens. *Polymers* **2020**, *12*, 699. [CrossRef]
29. Zhang, X.; Yang, M.; Yuan, Q.; Cheng, G. Controlled Preparation of Corncob Lignin Nanoparticles and their Size-Dependent Antioxidant Properties: Toward High Value Utilization of Lignin. *ACS Sustain. Chem. Eng.* **2019**, *7*, 17166–17174. [CrossRef]

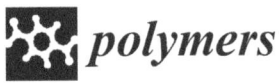

polymers

MDPI

Article

Carboxymethyl Bacterial Cellulose from Nata de Coco: Effects of NaOH

Pornchai Rachtanapun [1,2,3,*], Pensak Jantrawut [2,4], Warinporn Klunklin [1], Kittisak Jantanasakulwong [1,2,3], Yuthana Phimolsiripol [1,2,3], Noppol Leksawasdi [1,2,3], Phisit Seesuriyachan [1,2], Thanongsak Chaiyaso [1,2], Chayatip Insomphun [1,2], Suphat Phongthai [1,2], Sarana Rose Sommano [3,5], Winita Punyodom [3,6], Alissara Reungsang [7,8,9] and Thi Minh Phuong Ngo [10]

1 Faculty of Agro-Industry, School of Agro-Industry, Chiang Mai University, Chiang Mai 50100, Thailand; warinporn.k@cmu.ac.th (W.K.); jantanasakulwong.k@gmail.com (K.J.); yuthana.p@cmu.ac.th (Y.P.); noppol@hotmail.com (N.L.); phisit.s@cmu.ac.th (P.S.); thachaiyaso@hotmail.com (T.C.); chayatip@yahoo.com (C.I.); suphat.phongthai@cmu.ac.th (S.P.)
2 The Cluster of Agro Bio-Circular-Green Industry (Agro BCG), Chiang Mai University, Chiang Mai 50100, Thailand; pensak.amuamu@gmail.com
3 Center of Excellence in Materials Science and Technology, Chiang Mai University, Chiang Mai 50200, Thailand; sarana.s@cmu.ac.th (S.R.S.); winitacmu@gmail.com (W.P.)
4 Department of Pharmaceutical Sciences, Faculty of Pharmacy, Chiang Mai University, Chiang Mai 50200, Thailand
5 Plant Bioactive Compound Laboratory (BAC), Department of Plant and Soil Sciences, Faculty of Agriculture, Chiang Mai University, Chiang Mai 50200, Thailand
6 Department of Chemistry, Faculty of Science, Chiang Mai University, Chiang Mai 50200, Thailand
7 Department of Biotechnology, Faculty of Technology, Khon Kaen University, Khon Kaen 40002, Thailand; alissara@kku.ac.th
8 Research Group for Development of Microbial Hydrogen Production Process, Khon Kaen University, Khon Kaen 40002, Thailand
9 Academy of Science, Royal Society of Thailand, Bangkok 10300, Thailand
10 Department of Chemical Technology and Environment, The University of Danang—University of Technology and Education, Danang 550000, Vietnam; ntmphuong@ute.udn.vn
* Correspondence: pornchai.r@cmu.ac.th

Citation: Rachtanapun, P.; Jantrawut, P.; Klunklin, W.; Jantanasakulwong, K.; Phimolsiripol, Y.; Leksawasdi, N.; Seesuriyachan, P.; Chaiyaso, T.; Insomphun, C.; Phongthai, S.; et al. Carboxymethyl Bacterial Cellulose from Nata de Coco: Effects of NaOH. *Polymers* **2021**, *13*, 348. https://doi.org/10.3390/polym13030348

Academic Editor: Rafael Antonio Balart Gimeno

Received: 25 December 2020
Accepted: 15 January 2021
Published: 22 January 2021

Abstract: Bacterial cellulose from nata de coco was prepared from the fermentation of coconut juice with *Acetobacter xylinum* for 10 days at room temperature under sterile conditions. Carboxymethyl cellulose (CMC) was transformed from the bacterial cellulose from the nata de coco by carboxymethylation using different concentrations of sodium hydroxide (NaOH) and monochloroacetic acid (MCA) in an isopropyl (IPA) medium. The effects of various NaOH concentrations on the degree of substitution (DS), chemical structure, viscosity, color, crystallinity, morphology and the thermal properties of carboxymethyl bacterial cellulose powder from nata de coco (CMCn) were evaluated. In the carboxymethylation process, the optimal condition resulted from NaOH amount of 30 g/100 mL, as this provided the highest DS value (0.92). The crystallinity of CMCn declined after synthesis but seemed to be the same in each condition. The mechanical properties (tensile strength and percentage of elongation at break), water vapor permeability (WVP) and morphology of CMCn films obtained from CMCn synthesis using different NaOH concentrations were investigated. The tensile strength of CMCn film synthesized with a NaOH concentration of 30 g/100 mL increased, however it declined when the amount of NaOH concentration was too high. This result correlated with the DS value. The highest percent elongation at break was obtained from CMCn films synthesized with 50 g/100 mL NaOH, whereas the elongation at break decreased when NaOH concentration increased to 60 g/100 mL.

Keywords: bacterial cellulose; biopolymer; carboxymethyl cellulose; CMC; nata de coco; sodium hydroxide

1. Introduction

Primary cell walls of eukaryotic plants, algae and the oomycetes consist of cellulose as the major component. The cellulose consists of D-glucose units linked as a linear chain, ranging from several hundred to over ten thousand β (1→4) units [1]. Cellulose is insoluble in water. However, sodium carboxymethyl cellulose (CMC), which is one of cellulose's derivatives, can be dissolved in water [2]. CMC is an anionic, linear polymer, water-soluble cellulose reacted with monochloroacetic acid (MCA) or monochloroacetate acid (NaMCA). CMC is an important cellulose derivative applied in several industrial fields, such as the food industry, cosmetics, pharmaceuticals, detergents, textiles, [3] ceramics [4], etc. However, intensive utilization of wood has caused environmental problems and is expensive to manufacture. Therefore, there have been many studies about utilizing agricultural waste to be sources of CMC, such as cellulose from papaya peel [4], sugar beet pulp [5], sago waste [6], mulberry paper [7], *Mimosa pigra* peel [8] and durian husks [2,9], palm bunch and bagasse [10] and asparagus stalk ends [11]. The uses of CMC in food manufacturing require high purity of CMC grades ranging between 0.4 and 1.5 [12].

Cellulose has been obtained from bacterial cellulose (BC), including the genera *Agrobacterium*, *Rhizobium*, *Pseudomonas*, *Sarcina* and *Acetobacter* [13]. Acetobacter produced a pure bacterial cellulose aggregate containing impurities, such as hemicelluloses, pectin and lignin [14,15]. There are many desirable properties of bacterial cellulose, such as high purity, high degree of polymerization, high crystallinity, high wet tensile strength, high water-holding capacity and good biocompatibility [16,17]. *Acetobacter xylinum* is an acetic acid bacterium which can ferment and digest the carbon source in the coconut juice before converting it to the extracellular polysaccharide or cellulose [12]. The coconut juice is a waste product from the coconut milk industry known for its use in manufacturing cellulose [15,18]. Bacterial cellulose becomes a cellulosic white-to-creamy-yellow product called nata de coco. Therefore, the high intrinsic purity of bacterial cellulose from nata de coco together with the low environmental impact of the bacterium isolation means it is used in several applications, such as hydrogels [19], tissue engineering [20], cell culture [21], wound dressing and cancer therapy [22], CMC production [12], etc. However, the study of production and characterization of CMC films from nata de coco (CMCn) is limited.

This study aims to determine the effect of NaOH concentrations (20–60 g/100 mL) on the thermal properties, degree of substitution (DS), chemical structure, viscosity, crystallinity, and morphology of CMCn powder. With this aim, the effects of NaOH concentrations on solubility, mechanical properties (tensile strength, percentage elongation at break), percentage of transmittance, water vapor permeability (WVP) and the morphology of CMCn films were evaluated.

2. Materials and Methods

2.1. Materials

Coconut juice was collected from Muang District, Chiang Mai Province, Thailand. *Acetobacter xylinum* came from the Division of Biotechnology, Faculty of Agro-Industry, Chiang Mai University, Thailand. Di-ammonium hydrogen orthophosphate was obtained from Univar, Williamson Rd Ingleburn, Australia; magnesium sulphate from Prolabo, England; and commercial grade citric acid and sugar powder from the Thai Roong Ruang Sugar Group, Thailand. Glacial acetic acid and sodium hydroxide (artificial grade) were purchased from Lab-scan; hydrochloric acid and sodium chloride were purchased from Merck, Darmstadt, Germany; m-cresol purple, indicator grade, from Himedia, Marg, Mumbai, India; chloroacetic acid from Sigma-Aldrich, Burlington, MA, USA; isopropyl alcohol (IPA), ethanol and absolute methanol were purchased from Union Science, Muang District, Chiang Mai, Thailand.

2.2. Preparation of Bacterial Cellulose

Twenty liters of coconut juice were boiled in a pot and then 20 g of di-ammonium hydrogen orthophosphate, 10 g of magnesium sulphate, 30 g of citric acid and 100 g of sugar powder were added and mixed in. One and a half liters of the mixture were poured into each sterile plastic tray and then 55 mL of 95% *v*/*v* of ethanol was added. The mixture stood at room temperature for 30 min. One hundred milliliters of *Acetobacter xylinum* were then added to each tray with an aseptic technique. The tray was covered with fabric and paper to protect it from any contaminants. Nata de coco was obtained after the mixture had stood at room temperature for 10 days without being disturbed. The nata de coco was sliced, boiled 5 times in a pot with 15 L of water and dried in a hot air oven (Model: 1370FX-2E, Sheldon Manufacturing Inc., Cornelius, OR, USA) at 55 °C for 12 h. It was then ground using a grinder (Philips-HR1701, Simatupang, Jakarta, Indonesia), and screened through a 0.5 mm mesh sieve (35 mesh size). The nata de coco bacterial cellulose powders were contained in propylene bags until used.

2.3. Synthesis of Carboxymethyl Cellulose from Bacterial Cellulose

Fifteen grams of bacterial cellulose powder from nata de coco, NaOH solution (100 mL) in different concentrations (30, 40, 50 and 60 g/100 mL) and 450 mL of isopropanol (IPA) were mixed for 30 min. Eighteen grams of monochloroacetic acid (MCA) was added to initiate the carboxymethylation process and the mixture was stirred continuously at 55 °C for 30 min. The mixture beaker was enclosed by aluminum foil and stood in a hot air oven at 55 °C for 3.5 h. After the heating process, the mixture separated into solid and liquid phases. The solid phase was collected and suspended in 300 mL of absolute methanol, and then 80 mL/100 mL of acetic acid was added to neutralize it. The suspended mixture was filtered by a Buchner funnel. Undesirable by-products were removed from the final product by soaking 5 times in 300 mL of 70% *v*/*v* ethanol for 10 min. Then, the final product was washed again with 300 mL of absolute methanol. The carboxymethylated bacterial cellulose (CMCn) obtained from nata de coco was dried in an oven at 55 °C for around 12 h [8].

2.4. Degree of Substitution (DS)

The degree of substitution (*DS*) of CMCn is defined as the average number of hydroxyl groups in the cellulose structure which have been replaced with carboxymethyl and sodium carboxymethyl groups at C2, 3 and 6. The DS determination was described in a crosscarmellose sodium monograph in USP XXIII. The method included two steps: titration and residue on ignition [8]. The DS is calculated using the following Equation (1):

$$DS = A + S \tag{1}$$

where A is the degree of substitution with carboxymethyl acid and S is the degree of substitution with sodium carboxymethyl. A and S were calculated using information from the titration and ignition steps using the following Equations (2) and (3):

$$A = \frac{1150M}{(7120 - 412M - 80C)} \tag{2}$$

$$S = \frac{(162 + 58A)C}{(7102 - 80C)} \tag{3}$$

where M is the mEq of the base required for endpoint titration. C is the percentage of ash remaining after ignition. The reported DS values are means of three repetitions.

2.5. Titration

Precisely one gram of CMCn was weighed and added to a 500 mL Erlenmeyer flask. Three hundred milliliters of sodium chloride (NaCl) solution (10 g/100 mL) was added next. The Erlenmeyer flask was covered by a stopper and intermittently shaken for 5 min. The mixture was mixed with 5 drops of m-cresol purple, and then 15 mL of 0.1 N hydrochloric acid (HCl) was added. If the solution color did not change, 0.1 N HCl was gradually added until the solution changed to yellow. The solution was gradually titrated with 0.1 N NaOH. The solution changed from yellow to violet at the endpoint. The net amount of milliequivalent base required for neutralization of 1 g of CMCn (M) was calculated using Equation (4):

$$M(mEq) = mmole \times Valence \quad (4)$$

where m is 10^{-3}, mole is mass in grams per molecular weight of NaOH and the valence of NaOH is 1.

2.6. Residue on Ignition

A crucible was placed in an oven at 100 °C for 1 h and kept in a desiccator until the weight achieved a constant value (weighing apparatus, AR3130, Ohaus Corp. Pine, Brook, NJ, USA). Next, 1.000 g of CMCn was added to a crucible. To obtain black residue, a crucible containing CMCn was ignited in a 400 °C kiln (Carbolite, CWF1100, Scientific Promotion, Parsons Ln, Hope S33 6RB, England) for 1–1.5 h and then placed into a desiccator. The whole residue was wetted by adequate sulfuric acid and then heated until white fumes completely volatilized. White residue was obtained from the ignition crucible containing residue at 800 ± 25 °C and the residue was placed in desiccator to reach an accurate weight. All experiments were performed three times. The percentage of residue on ignition was calculated by following Equation (5).

$$C = \frac{Weight\ of\ residue}{Weight\ of\ CMC} \times 100 \quad (5)$$

2.7. Fourier Transform Infrared Spectroscopy (FTIR)

An infrared spectrophotometer (Bruker, Tensor 27, Fahrenheitstr 4, D-28359, Bremen, Germany) was used to evaluate the functional groups of bacterial cellulose from nata de coco and CMCn. The bacterial cellulose from nata de coco and CMCn samples (~2 mg) with KBr were used to make pellets. Transmission was measured at a wave number range of 4000–400 cm^{-1}.

2.8. X-Ray Diffraction (XRD)

X-ray diffraction patterns of bacterial cellulose from nata de coco and CMCn were recorded in a reflection mode on a JEOL JDX-80-30 X-ray diffractometer, Brucker, Milton, ON, Canada. The scattering angle (2Θ) was from 5° to 60° at a scan rate of 2°/min.

2.9. Viscosity

The bacterial cellulose from nata de coco and CMCn were analyzed by a Rapid Visco Analyzer (Model: RVA-4, Newport Scientific, Warriewood, Australia) to measure viscosity. Three grams of cellulose and CMCn was dissolved in 25 mL of distilled water, heated to 80 °C and continuously stirred for 10 min. The sample solutions were tested in 2 steps. In the first step, the samples were tested at 960 rpm for 10 s. Next, the temperature was set at 30, 40 and 50° C at 5 min-intervals with a speed of 160 rpm. All tests were repeated three times [8].

2.10. Scanning Electron Microscopy (SEM)

The morphology of bacterial cellulose from nata de coco, CMCn powder and CMCn films were analyzed using a scanning electron microscope (SEM) (Phillip XL 30 ESEM, FEI Company, Hillsboro, OR, USA) equipped with a large field detector. The acceleration voltage was 15 kV under low settings with 5000×.

2.11. Thermal Properties

The thermal properties of the melting point of bacterial cellulose from nata de coco (CMCn) were determined using a differential scanning calorimeter (DSC) (Perkin Elmer precisely, Inst. Model Pyris Diamond DSC, Hodogaya-Ku, Godocho, Kanagawa, Japan). Aluminum pans containing 10 mg of samples were heated from 40 to 240 °C. A heat rate was set as 10 °C min^{-1}. The tests were performed under N_2 gas with a flow rate of 50 mL min^{-1}. The data was represented in terms of a thermogram, which indicated the melting point. All samples were tested three times [8].

2.12. Film Preparation

To obtain a film-forming solution, 3 g (Weighing apparatus, AR3130, Ohaus Corp. Pine, Brook, NJ, USA) of CMCn was dissolved in 300 mL of 80 °C distilled water with constant stirring for 15 min. Glycerol (15 g/100 g) was added and the film-forming solution was casted onto plates (20 cm × 15 cm). The plates containing the film-forming solution was placed at room temperature for 48 h to obtain dry CMCn films. Then, the CMCn films were removed from the plates. The CMCn films were kept at 27 ± 2 °C with 65 ± 2% relative humidity (RH) for 24 h (Thai industrial standard, TIS 949-1990).

2.13. Film Solubility

The method to determine the percentage of solubility of CMCn films was modified from Phan et al. [23]. CMCn films were dried at 105 °C in a hot air oven for 24 h and kept in desiccators for 24 h. Then, about 0.2000 g of the CMCn films was weighed. This weight was recorded as an initial dry weight (W_i). Each weighed CMCn film was suspended in 50 mL of distilled water with shaking at 500 rpm for 15 min and poured onto filter paper (Whatman, No.93). The films then were dried in an oven at 105 °C for 24 h and weighed to obtain the final dry weight (W_f). The percentage of soluble matter (%SM) of the films was calculated by using the following Equation (6):

$$\%SM = \frac{\left(W_i - W_f\right)}{W_i} \quad (6)$$

2.14. Percent Transmittance

The percent transmittance of CMCn film was measured using a spectrophotometer (LaboMed, inc., Los Angeles, CA, USA) with a wavelength of 660 nm.

2.15. Mechanical Properties

The CMCn films were cut into rectangular strips (1.5 cm × 14 cm) for tensile strength and percentage elongation measurements. To precondition them, CMCn films were kept for 24 h at 27 ± 2 °C and relative humidity (RH) was controlled in the range of 65 ± 2% (Thai Industrial Standard; TIS 949-1990). Film thickness was measured at five different location in each sample by a micrometer, model GT-313-A (Gotech testing machine Inc., Taichung Industry Park, Taichung City, Taiwan). Tensile strength and water vapor permeability (WVP) was calculated using average film thickness. The tensile strength (TS) and percentage elongation at break (EB) were measured using a Universal Testing Machine Model 1000 (HIKS, Selfords, Redhill, England) as per the ASTM Method (ASTM, D882-80a, 1995a). The upper grips were separated. The machine was operated at 100 mm and 20 mm min^{-1} for the initial grip separation distance and crosshead speed, respectively. To calculate the TS, the maximum load at break was divided by the cross-sectional area of the sample. The EB was defined by the increasing of sample length at break point as the percentage of the initial length (100 mm). All mechanical tests were repeated 10 times.

2.16. Water Vapor Permeability (WVP)

The method for measuring the water vapor permeability (WVP) of the CMCn films was described in the ASTM method (ASTM, E96-93, 1993). Circular aluminum cups (8 cm diameter and 2 m depth) containing 10 g of silica gel were covered by the circle shape of CMCn films (7 cm diameter). Then, the aluminum cups were closed by paraffin wax, weighed and kept in a desiccator. NaCl saturated solution was used to maintain the conditions (25 °C, 75% RH) in the desiccator. The closed cups were weighed daily for 10 days. A slope was obtained from plotting a graph between weight gain (Y axis) and time (X axis). The water vapor transmission rate (WVTR) was calculated using the following Equation (7):

$$WVTR\left(\frac{g}{m^2}.day\right) = \frac{slope}{Film\ Area} \tag{7}$$

where the film area was 28.27 cm^2 and the WVP (g.m/m^2.mmHg.day) was calculated using the following Equation (8):

$$WVP\left(g.\frac{m}{m^2}.day.mmHg\right) = \frac{WVTR}{\Delta P} \times L \tag{8}$$

where L is the average film thickness (mm) and ΔP is the partial water vapor pressure difference (mmHg) across two sides of the film specimen (the vapor pressure of pure water at 23.6 °C = 21.845 mmHg). These samples were tested 3 times [8].

2.17. Statistical Analysis

All the data were presented as the mean ± SD. One-way ANOVA was used to evaluate the significance of differences at the significance level of p-value < 0.05. Statistical analysis was performed using SPSS software version 16.0 (SPSS Inc., Chicago, IL, USA).

3. Results and Discussion

3.1. Degree of Substitution (DS)

The effects of NaOH concentrations on the DS of CMCn is shown in Figure 1. As NaOH concentrations increased from 20 to 30 g/100 mL NaOH, the DS of CMCn increased as well, while the DS dropped at 40 to 60 g/100 mL NaOH concentration. The values of the obtained CMCn were from 0.30–0.92. This phenomenon can be explained by the carboxymethylation procedure, where two competitive reactions occurred concurrently. A cellulose hydroxyl reacts with sodium monochloroacetate (NaMCA) to obtain CMCn in the first reaction, as shown in Equations (9) and (10).

$$\text{CLL-OH} + \text{NaOH} \rightarrow \text{CLL-ONa} + H_2O \tag{9}$$

(Cellulose) (Reactive alkaline form)

$$\text{CLL-ONa} + \text{Cl-}CH_2\text{-COONa} \rightarrow \text{CLL-O-}CH_2\text{-COONa} + \text{NaCl} \tag{10}$$

(CMC)

NaMCA alternated to sodium glycolate as a byproduct by reacting with NaOH in the second reaction, as shown in Equation (11).

$$\text{NaOH} + \text{Cl-}CH_2\text{COONa} \rightarrow \text{HO-}CH_2\text{COONa} + \text{NaCl} \tag{11}$$

The second reaction overcame the first at stronger alkaline concentrations. In conditions of excessive alkalinity, DS was low because a side reaction dominates, causing the formation of sodium glycolate as a byproduct. This result agrees with Rachtanapun and Rattanapanone [8], who reported that degradation of CMC from *Mimosa pigra* occurred because of high concentrations of NaOH. Similar results were reported for the maximum DS value (0.87) of carboxymethyl cellulose from durian rind [2] and the maximum DS value (0.98) of carboxymethyl cellulose from asparagus stalk ends with a NaOH concentration of 30 g/100 mL [11].

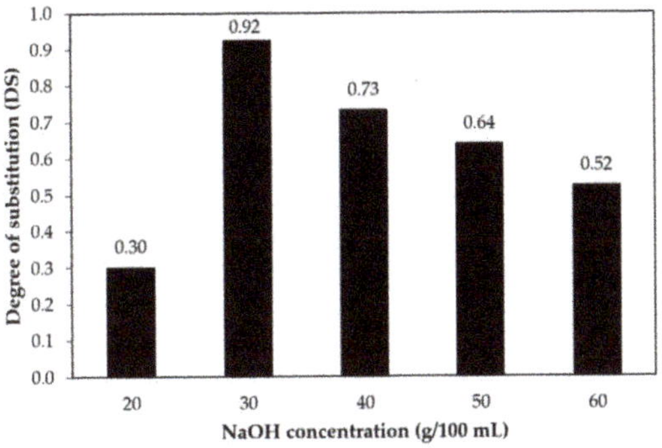

Figure 1. Effect of the amount of NaOH on the DS of CMCn. CMCn: carboxymethyl bacterial cellulose powder from nata de coco. DS: degree of substitution.

3.2. Percent Yield of Carboxymethyl Cellulose from Nata de Coco

The effect of NaOH concentrations on percent yield of carboxymethyl bacterial cellulose from nata de coco (CMCn) powder synthesized is shown in Figure 2. This study investigated the effect of NaOH in concentrations of 20–60 g/100 mL on the percent yield of bacterial cellulose powder synthesis. At 20–40 g/100 mL NaOH concentrations, the percentage yield increased. However, the percentage decreased at 50–60 g/100 mL NaOH concentrations. This resulted from the alkali-catalyzed degradation of bacterial cellulose. The trend of percentage yield of CMCn was similar to the DS results (Figure 2). This result was in agreement with carboxymethyl cellulose from durian husks [2] and asparagus stalk ends [11].

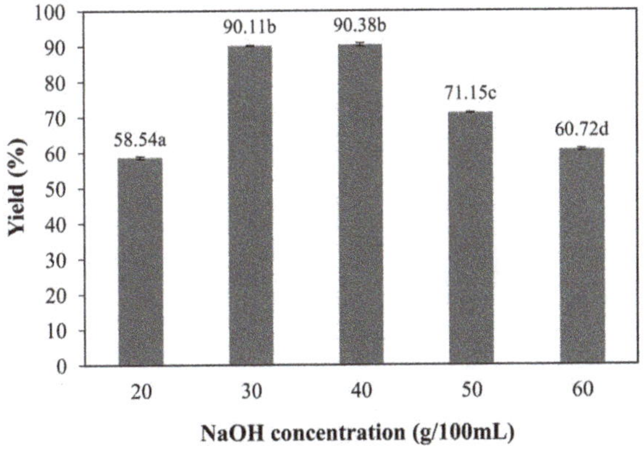

Figure 2. Percent yield of carboxymethyl cellulose from nata de coco synthesized with various NaOH concentrations (20, 30, 40, 50 and 60 g/100 mL). The different letter, e.g., 'a', 'b', 'c' or 'd' are statistically different ($p < 0.05$).

3.3. Fourier Transform Infrared Spectroscopy (FTIR)

The substitution reaction in carboxymethylation was identified using FTIR. The FTIR spectrums of bacterial cellulose from nata de coco and CMCn synthesized with a 30 g/100 mL NaOH concentration are shown in Figure 3. The same functional groups appeared in both cellulose and CMCn. A broad peak at 3441 cm^{-1} referred to the hydroxyl group (–OH stretching). A peak at 1420 cm^{-1} was related to $-CH_2$ scissoring. Strong peaks at 1604 cm^{-1} and 1060 cm^{-1} indicated carbonyl group (C=O stretching) and ether groups (–O– stretching), respectively [8]. In the CMCn sample, the obvious increase of the carbonyl group (C=O), methyl group ($-CH_2$) and ether group (–O–) peaks was observed, but the intensity of the hydroxyl group (–OH) was lower when compared to bacterial cellulose (Figure 3). This result indicated that cellulose molecules changed due to carboxymethylation occurrence [2,6,10,11,17].

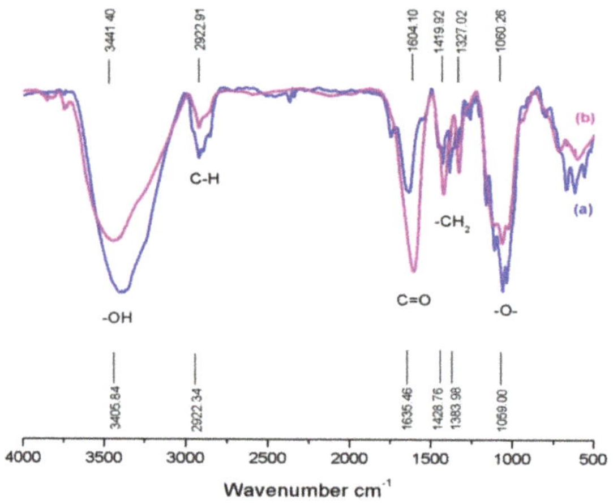

Figure 3. Fourier transform infrared spectroscopy of (**a**) bacterial cellulose from nata de coco and (**b**) CMCn synthesized with 30 g/100 mL NaOH.

3.4. *Effect of Various NaOH Concentrations on Viscosity of CMCn*

The effect of using various NaOH concentrations on the viscosity of CMC is shown in Figure 4. The peak viscosity of CMCn increased when the concentration of NaOH increased from 20 to 30 g/100 mL because hydroxyl groups on the cellulose molecule were substituted by more carboxymethyl groups, being hydrophilic groups. Cohesive forces reduced because CMCn temperature increased, whereas the rate of molecular interchange concurrently increased [8,24]. The viscosity of CMCn solution was in accordance with the DS value at the same temperature. A previous study found that NaOH concentrations influenced the viscosity of CMC [25]. They stated that more carboxymethyl groups (a higher DS value) increased CMC viscosity [8]. In addition, they reported that the degradation of CMC polymers causes decreasing viscosity with too much NaOH, leading to a lower DS [26].

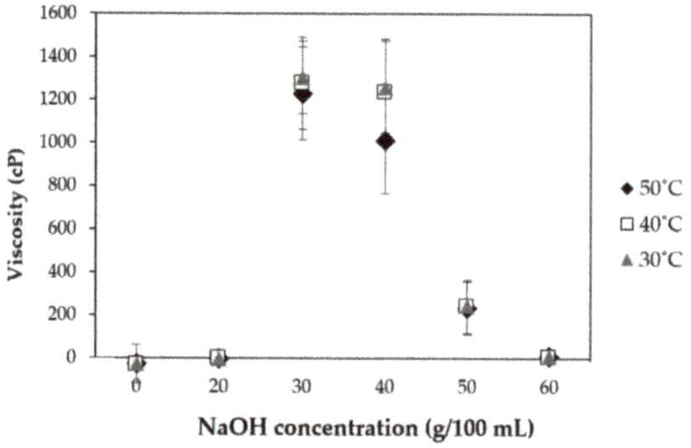

Figure 4. Effect of NaOH concentrations on the viscosity of bacterial cellulose and CMCn.

3.5. *Effects of Various NaOH Concentrations on Thermal Properties of CMCn Powder*

The effects of various NaOH concentrations on the thermal properties of bacterial cellulose from nata de coco and CMCn powder are shown in Figure 5. The melting temperature TM of bacterial cellulose was 167.4 °C, and those of CMCn synthesized with 20, 30, 40, 50 and 60 g/100 mL NaOH were 175.3, 188.2, 179.4, 164.7 and 151.0 °C, respectively.

The melting temperature of CMCn was increased at 20–30 g/100 mL because the substituent of carboxymethyl groups caused increases in the ionic character and intermolecular forces between the polymer chains. At 40–60 g/100 mL NaOH, the melting temperature decreased because of the side reaction predominating with the formation of sodium glycolate as a byproduct and chain breaking of the CMCn polymer. This result was in agreement with carboxymethyl cellulose from asparagus stalk ends [11].

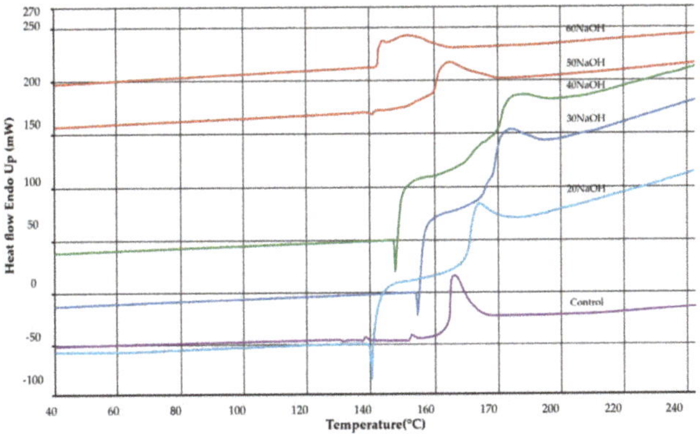

Figure 5. Differential scanning calorimetry of cellulose from nata de coco and CMCn synthesized with various amounts of NaOH.

3.6. X-Ray Diffraction (XRD)

The effect of NaOH concentrations on the amount of crystallinity in bacterial cellulose from nata de coco and CMCn powder is shown in Figure 6. The treatment of cellulose with NaOH caused a decrease in the amount of cellulose crystallinity because NaOH leaded to the dividing of hydrogen bond [18]. By comparison with cellulose without alkalization with NaOH, monochloroacetic acid molecules substituted into bacterial cellulose molecules more easily because the distance between each polymer molecule increased. Thus, before the carboxymethylation reaction, NaOH had an effect on cellulose structured to decrease the crystallinity of the CMCn. These results were also found in the reduction of crystallinity of CMC from asparagus stalk ends [11] and cavendish banana cellulose [26] due to alkalizing by 20 g/100 mL and 15 g/100 mL, respectively.

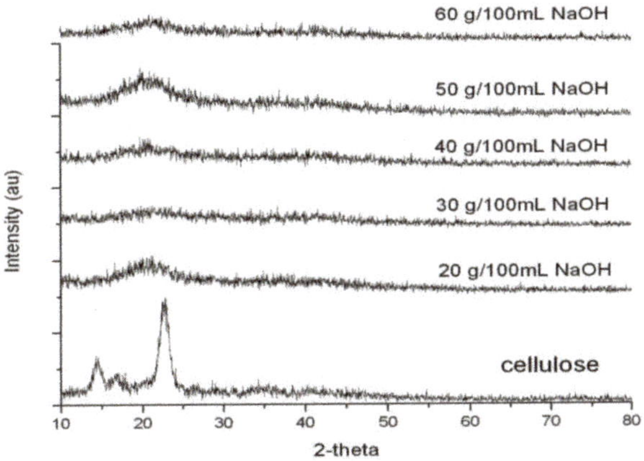

Figure 6. X-ray diffractograms of cellulose from nata de coco and CMCn synthesized with various amounts of NaOH.

3.7. Scanning Electron Microscopy of Cellulose from Nata de Coco and CMCn Powder

The scanning electron micrographs of bacterial cellulose from nata de coco and CMCn powder with 20, 30, 40, 50 and 60 g/100 mL NaOH are shown in Figure 7a–f. The micrograph of bacterial cellulose from nata de coco (Figure 7a) exhibited a compact appearance with a smooth surface without pores or cracks, and a lot of small fibers appeared. However, the microstructure of samples dramatically changed when the NaOH concentration increased. The morphology of CMCn powder with 20 g/100 mL NaOH (Figure 7b) showed small fibers with minimal damage. As shown in Figure 7c, the morphology of CMCn powder with 30 g/100 mL NaOH was compact and dense, with no signs of cracks and pits. The morphology of CMCn powder with 40 g/100 mL NaOH (Figure 7d) was deformed and had some cracks and pits. Moreover, the morphology of CMCn powder with 40 g/100 mL NaOH also correlated with the DS value, due to a chain degradation of the CMC polymer. This result is similar to those of other studies [2,8,11]. CMCn powder with 50 g/100 mL NaOH (Figure 7e) showed increased surface irregularities with a more indented and collapsed surface. Moreover, when increasing NaOH concentration to 60 g/100 mL (Figure 7d), the surface became rougher and totally deformed. Thus, it can be concluded that synthesis with increasing NaOH concentrations damages the surface area of bacterial cellulose powder. This result was similar to those of carboxymethyl rice starch [27] and carboxymethyl cassava starch [28] and carboxymethyl cellulose from asparagus stalk ends [11]. The principal parameter for weakening the structure of cellulose and causing a loss of crystallinity that allowed etherifying agents to reach the cellulose molecules for carboxymethylation processes was the alkalization [27]. Consequently, this result was consistent with the DS value. To indicate carboxymethyl reactions, the color formation was investigated by color measurement. The main effects on CMC color value were increasing a* (redness), b* (yellowness) and yellowness index (YI) values as NaOH concentrations increased (20–30 g/100 mL NaOH). The L* value of CMCn synthesized with various NaOH concentrations decreased as the concentrations increased from 20 to 30 g/100 mL NaOH. The a* and b* values decreased, but the L* value of CMCn increased as NaOH was increased to 40 g/100 mL NaOH. Moreover, the trend of YI of CMCn was decreased when increasing the NaOH concentration up to 40%. This phenomenon was probably due to the first step in the carboxymethylation of cellulose (Equation (9)), which delivered CMC or sodium glycolate [10]. At a high NaOH concentration (50 and 60 g/100 mL NaOH), all color values were decreased. A carboxymethylation reaction might have been a reason for the color change in this study [2,11,27]. The ΔE of cellulose and CMCn had same trends as the a* and b* values. The WI of cellulose and CMCn had the same trends as the L* value (Table 1).

Table 1. Color values of bacterial cellulose from nata de coco and CMCn synthesized with various amounts of NaOH.

NaOH (g/100 mL)	L*	a*	b*	ΔE	YI	WI	h_{ab}
0	75.62a	2.247a	16.62a	23.39a	31.42a	70.45a	82.34a
20	64.79b	8.23b	28.61b	40.29b	63.09b	54.70b	73.94b
30	65.54b	10.19c	28.96b	40.61b	63.26b	54.74b	70.60c
40	74.34a	7.23d	22.14c	29.08c	42.55c	65.99c	71.91d
50	70.01c	7.67e	21.64c	31.85d	44.22c	62.92d	70.49c
60	68.14d	7.47de	21.13c	32.89d	44.30c	61.67d	70.54c

Note: Obtained by Duncan's test ($p < 0.05$). L* = lightness, a* = redness, b* = yellowness, ΔE = total color difference, YI = yellowness index, WI = whiteness index, h_{ab} = hue. The different letter, e.g., 'a', 'b', 'c' or 'd' are statistically different ($p < 0.05$).

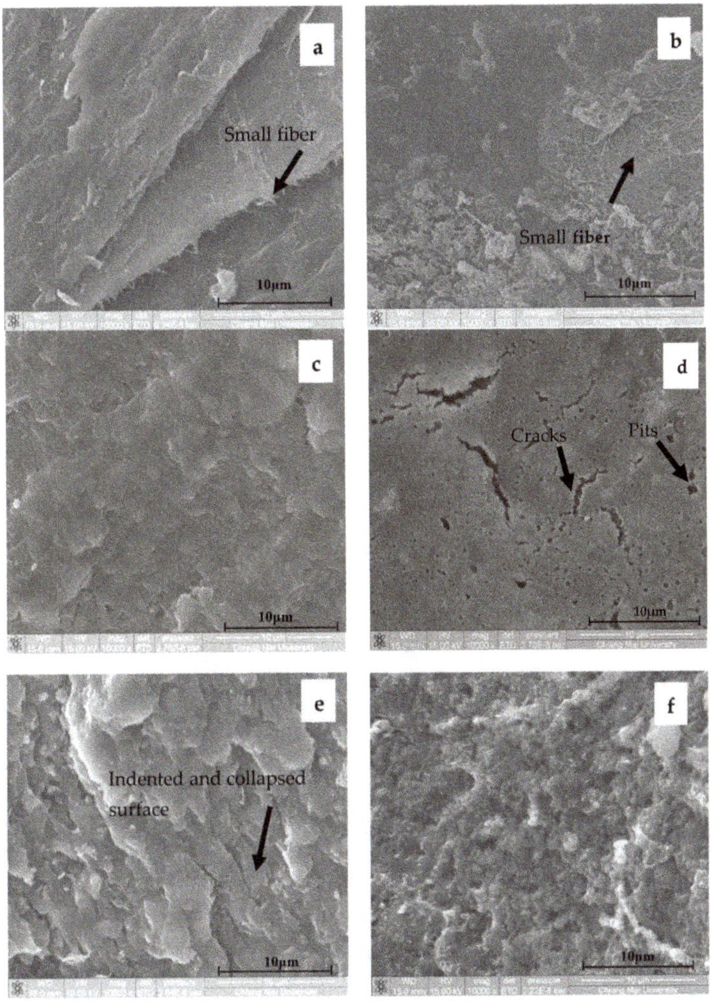

Figure 7. Scanning electron micrographs of (**a**) cellulose from nata de coco and CMCn powder: with (**b**) 20 g/100 mL NaOH, (**c**) 30 g/100 mL NaOH, (**d**) 40 g/100 mL NaOH, (**e**) 50 g/100 mL NaOH and (**f**) 60 g/100 mL NaOH.

3.8. Solubility and Transmittance

The effects of NaOH concentrations on the solubility of CMCn films are shown in Figure 8. At 20–30 g/100 mL NaOH, the solubility of CMCn film increased; however, at 40–60 g/100 mL NaOH, the film solubility decreased. The percentage of soluble matter could be indicated by the DS. The percentage of soluble matter rose as the DS value increased [6]. The effects of NaOH concentrations on the percent transmittance of the CMCn films are shown in Figure 9. The percent transmittance increased with NaOH concentrations of 20–30 g/100 mL and decreased at 40–60 g/100 mL. This result correlated with the percentage of solubility of the CMCn films.

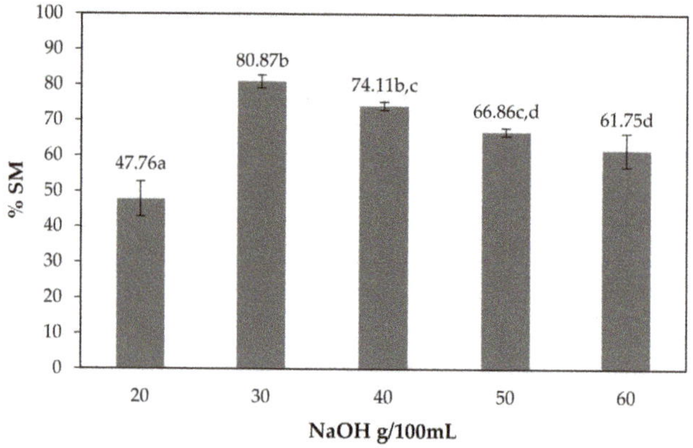

Figure 8. Effect of NaOH concentrations on the percentage of soluble matter of the CMCn films. The different letter, e.g., 'a', 'b', 'c' or 'd' are statistically different ($p < 0.05$).

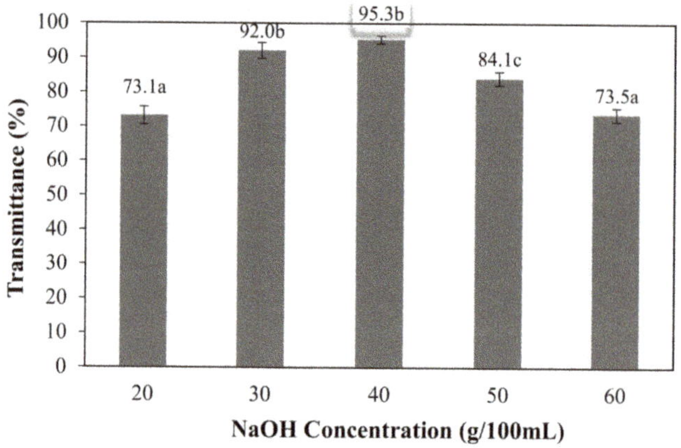

Figure 9. Effect of NaOH concentrations on the percentage of transmittance of the CMCn films. The different letter, e.g., 'a', 'b', 'c' or 'd' are statistically different ($p < 0.05$).

3.9. Scanning Electron Microscopy of Cellulose from Nata de Coco and CMCn Films

Cross-section views of scanning electron micrographs of CMCn films synthesized with various NaOH concentrations are shown in Figure 10a–e. As shown in Figure 10a, the morphology of CMCn films with 20 g/100 mL NaOH was a rather smooth surface without pits and cracks. CMCn films with 30 g/100 mL NaOH (Figure 10b) were compact and dense. The morphology of CMCn films synthesized with 40 g/100 mL NaOH (Figure 10c) was still compact but rougher and some scraps protruded on the surface. CMCn films with 50 g/100 mL NaOH (Figure 10d) showed increasing surface irregularities and defects such as cracks and pits. The morphology of CMCn films with 60 g/100 mL NaOH (Figure 10e) was deformed and revealed lots of cracks and some scraps.

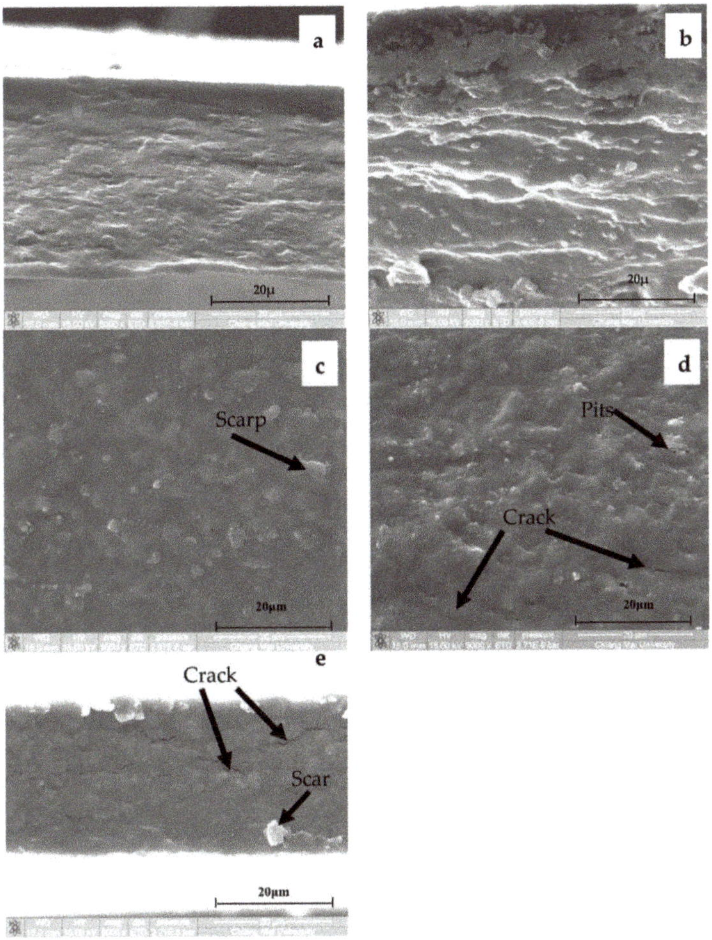

Figure 10. Scanning electron micrographs of CMCn films: with (**a**) 20 g/100 mL NaOH, (**b**) 30 g/100 mL NaOH, (**c**) 40 g/100 mL NaOH, (**d**) 50 g/100 mL NaOH and (**e**) 60 g/100 mL NaOH.

3.10. Water Vapor Permeability (WVP)

The effects of NaOH concentrations on the water vapor permeability (WVP) of the CMCn films are shown in Figure 11. When the concentration of NaOH increased, cellulose transformed to CMCn, causing a higher polarity. This result led to a rise of the WVP. Moreover, the morphology of CMCn films with 40–60 NaOH showed cracks and pits. Damage dramatically increased in the WVP of CMCn with 50–60 g/100 mL NaOH. This result can be explained through a SEM micrograph (Figure 10). SEM morphology demonstrated that CMCn films with 40–60 NaOH showed cracks and pits. The results were in agreement with studies of CMC from rice starch [24], which showed that the WVP of carboxymethyl rice starch films rose due to increases in NaOH concentrations. Furthermore, carboxylation influenced increases of polarity, reductions in crystallinity and changes in granule morphology [29].

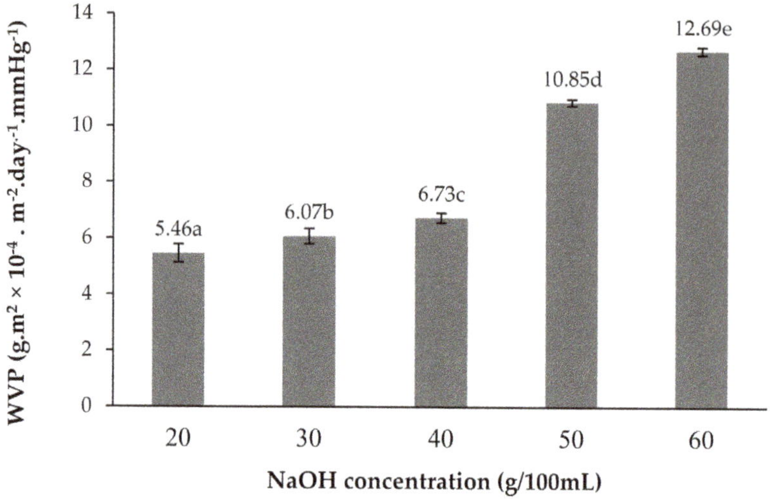

Figure 11. Effect of NaOH concentration (20, 30, 40, 50 and 60 g/100 mL) on the water vapor permeability (WVP) of CMCn films. The different letter, e.g., 'a', 'b', 'c' or 'd' are statistically different ($p < 0.05$).

3.11. Tensile Strength (TS) and Elongation at Break (%)

The tensile strength values of CMC films with various NaOH concentrations are shown in Figure 12a. When NaOH increased (20–30 g/100 mL NaOH), the tensile strength also rose. This is because TS values are correlated with increases in DS values because of the carboxymethyl group substitution, causing a rise in the ionic character and intermolecular forces between the polymer chains [23]. Moreover, this is related to why the morphology of CMCn film synthesized with 30 g/100 mL NaOH was compact and dense. However, at higher NaOH concentration, the TS dropped because of sodium glycolate, which is a secondary product from the CMC synthesis reaction and polymer degradation. These results are related to those of CMC films from *Mimosa pigra* [8], which shows that the TS of CMC films from *Mimosa pigra* rose when concentrations of NaOH increased (20–30 g/100 mL). Studies of CMC films from mulberry paper waste showed that the TS increased with increasing NaOH concentrations, but too-high NaOH concentrations (60 g/100 mL) could cause a hydrolysis reaction in the cellulose chain [7]. The percentage of elongation at break (EB) of CMCn films with various NaOH concentrations used in CMCn synthesis are shown in Figure 12b. CMCn films exhibited higher EB when NaOH concentrations increased (20–50 g/100 mL NaOH). However, when NaOH concentration was increased to 60 g/100 mL, the EB of CMCn films decreased. This phenomenon can be explained as a cellulose structure with decreased crystallinity caused the CMCn films to have increased elasticity due to higher concentrations of NaOH. Nevertheless, at 60 g/100 mL NaOH, the occurrence of cellulose hydrolysis resulted in lower flexibility of CMCn films [8].

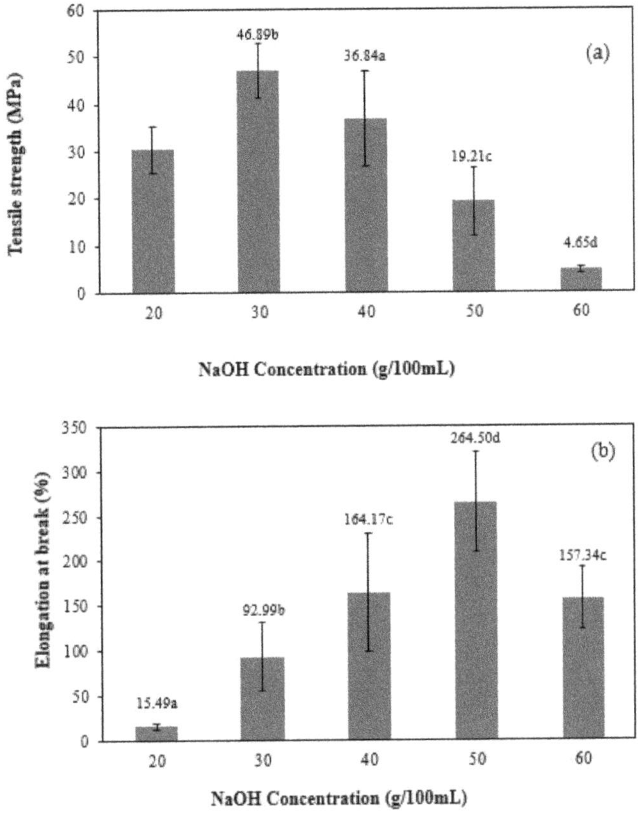

Figure 12. Effect of various NaOH concentrations (0, 20, 30, 40, 50 and 60 g/100 mL) on (**a**) tensile strength and (**b**) percent elongation at break of CMCn films. The different letter, e.g., 'a', 'b', 'c' or 'd' are statistically different ($p < 0.05$).

4. Conclusions

From the results, this study was successful in using nata de coco as a resource to obtain carboxymethyl bacterial cellulose (CMCn) using different NaOH concentrations. This demonstrates that the major parameter relating to the properties of CMCn is NaOH quantity. The DS of CMCn dramatically increased when NaOH concentrations increased from 20 to 30 g/100 mL in CMCn synthesis, and the DS value gradually dropped at NaOH concentrations of 40–60 g/100 mL. The mechanical properties and viscosity of CMCn films were correlated with the DS. CMCn consisting of high DS can have improved mechanical properties. Therefore, the CMCn films synthesized with 30 g/100 mL NaOH exhibited good mechanical properties, such as high TS and EB with no evidence of cracking and pitting when examined by SEM. CMC is widely applied in several fields, including in the food, nonfood and pharmaceutical industries, and thus the results from this study are useful. The investigation of methods to modify cellulose in this study, such as varying the NaOH content, supports new and potentially useful applications of cellulose from nata de coco.

Author Contributions: P.R. conceived and planned all experiments; conceptualization, investigation and methodology, P.R.; C.I., S.R.S. and T.M.P.N. performed the measurements; W.K., K.J., N.L., P.S., T.C., C.I., S.R.S., T.M.P.N. and P.R. wrote the manuscript with input from all authors; W.K., S.P., P.J., W.P. and A.R. analyzed the data; K.J. and N.L. interpreted the results; Y.P. and P.R. contributed reagents and materials; validation, P.R.; project supervision, P.R. All authors discussed the results, commented, and agreed to publish this manuscript. All authors have read and agreed to the published version of the manuscript.

Funding: We wish to thank the Center of Excellence in Materials Science and Technology, Chiang Mai University, for financial support under the administration of the Materials Science Research Center, Faculty of Science, Chiang Mai University. This research was supported by the Program Management Unit for Human Resources & Institutional Development, Research and Invitation, NXPO [Frontier Global Partnership for Strengthening Cutting-edge Technology and Innovations in Materials Science, and Faculty of Science (CoE64-P001) as well as the Faculty of Agro-Industry (CMU-8392(10)/COE64), CMU. The present study was partially supported by The Thailand Research Fund (TRF) Senior Research Scholar (Grant No. RTA6280001). This research work was also partially supported by Chiang Mai University.

Institutional Review Board Statement: Not applicable.

Informed Consent Statement: Not applicable.

Data Availability Statement: The data presented in this study are available on request from the corresponding author.

Acknowledgments: We wish to thank the Center of Excellence in Materials Science and Technology, Chiang Mai University, for financial support under the administration of the Materials Science Research Center, Faculty of Science, Chiang Mai University. This research was supported by the Program Management Unit for Human Resources & Institutional Development, Research and Invitation, NXPO [Frontier Global Partnership for Strengthening Cutting-edge Technology and Innovations in Materials Science, and Faculty of Science (CoE64-P001) as well as the Faculty of Agro-Industry (CMU-8392(10)/COE64), CMU. The present study was partially supported by The Thailand Research Fund (TRF) Senior Research Scholar (Grant No. RTA6280001). This research work was also partially supported by Chiang Mai University.

Conflicts of Interest: The authors declare no conflict of interest.

References

1. Crawford, D.L.; Barder, M.J.; Pometto, A.L.; Crawford, R.L. Chemistry of softwood lignin degradation by Strepto-myces viridosporus. *Arch. Microbiol.* **1982**, *131*, 140–145. [CrossRef]
2. Rachtanapun, P.; Luangkamin, S.; Tanprasert, K.; Suriyatem, R. Carboxymethyl cellulose film from durian rind. *LWT-Food Sci. Technol.* **2012**, *48*, 52–58. [CrossRef]
3. Olaru, N.; Stoleriu, A.; Timpu, D. Carboxymethylcellulose synthesis in organic media containing ethanol and/or acetone. *J. Appl. Polym. Sci.* **1998**, *67*, 481–486. [CrossRef]
4. Rachtanapun, P.; Eitssayeam, S.; Pengpat, K. Study of Carboxymethyl Cellulose from Papaya Peels as Binder in Ceramics. *Adv. Mater. Res.* **2010**, *93–94*, 17–21. [CrossRef]
5. Sun, R.; Hughes, S. Fractional extraction and physico-chemical characterization of hemicelluloses and cellulose from sugar beet pulp. *Carbohydr. Polym.* **1998**, *36*, 293–299. [CrossRef]
6. Pushpamalar, V.; Langford, S.J.; Ahmad, M.; Lim, Y.Y. Optimisation of reaction condition for preparing carboxymethyl cellu-lose from Sago waste. *Carbohydr. Polym.* **2006**, *64*, 312–318. [CrossRef]
7. Rachtanapun, P.; Kumthai, S.; Mulkarat, N.; Pintajam, N.; Suriyatem, R. Value added of mulberry paper waste by carboxymethy-lation for preparation a packaging film. *Mater. Sci. Eng.* **2015**, *87*, 012081. [CrossRef]
8. Rachtanapun, P.; Rattanapanone, N. Synthesis and characterization of carboxymethyl cellulose powder and films from Mimosa pigra. *J. Appl. Polym. Sci.* **2011**, *122*, 3218–3226. [CrossRef]
9. Suriyatem, R.; Auras, R.A.; Rachtanapun, P. Utilization of Carboxymethyl Cellulose from Durian Rind Agricultural Waste to Improve Physical Properties and Stability of Rice Starch-Based Film. *J. Polym. Environ.* **2019**, *27*, 286–298. [CrossRef]
10. Suriyatem, R.; Noikang, N.; Kankam, T.; Jantanasakulwong, K.; Leksawasdi, N.; Phimolsiripol, Y.; Insomphun, C.; Seesuriyachan, P.; Chaiyaso, T.; Jantrawut, P.; et al. Physical Properties of Carboxymethyl Cellulose from Palm Bunch and Bagasse Agricultural Wastes: Effect of Delignification with Hydrogen Peroxide. *Polymers* **2020**, *12*, 1505. [CrossRef]

11. Klunklin, W.; Jantanasakulwong, K.; Phimolsiripol, Y.; Leksawasdi, N.; Seesuriyachan, P.; Chaiyaso, T.; Insomphun, C.; Phongthai, S.; Jantrawut, P.; Sommano, S.R.; et al. Synthesis, Characterization, and Application of Carboxymethyl Cellulose from Asparagus Stalk End. *Polymers* **2021**, *13*, 81. [CrossRef] [PubMed]
12. Casaburi, A.; Rojo, Ú.M.; Cerrutti, P.; Vázquez, A.; Foresti, M.L. Carboxymethyl cellulose with tailored degree of substitution obtained from bacterial cellulose. *Food Hydrocoll.* **2018**, *75*, 147–156. [CrossRef]
13. Cannon, R.E.; Anderson, S.M. Biogenesis of Bacterial Cellulose. *Crit. Rev. Microbiol.* **1991**, *17*, 435–447. [CrossRef]
14. Moon, S.-H.; Park, J.-M.; Chun, H.-Y.; Kim, S.-J. Comparisons of physical properties of bacterial celluloses produced in different culture conditions using saccharified food wastes. *Biotechnol. Bioprocess Eng.* **2006**, *11*, 26–31. [CrossRef]
15. Kurosumi, A.; Sasaki, C.; Yamashita, Y.; Nakamura, Y. Utilization of various fruit juices as carbon source for production of bacterial cellulose by Acetobacter xylinum NBRC 13693. *Carbohydr. Polym.* **2009**, *76*, 333–335. [CrossRef]
16. Bielecki, S.; Krystynowicz, A.; Turkiewicz, M.; Kalinowska, H. Bacterial Cellulose. In *Biopolymers Online*; Wiley & Sons: Hoboken, NJ, USA, 2002.
17. Hong, F.F.; Guo, X.; Zhang, S.; Han, S.-F.; Yang, G.; Jönsson, L.J. Bacterial cellulose production from cotton-based waste textiles: Enzymatic saccharification enhanced by ionic liquid pretreatment. *Bioresour. Technol.* **2012**, *104*, 503–508. [CrossRef]
18. Adejoye, O.D.; Adebayo-Toyo, B.C.; Ogunjobi, A.A.; Olaoye, O.A.; Fadahunsi, F.I. Effect of carbon, nitrogen and mineral sources on growth of Pleurotus florida, a Nigeria edible mushroom. *Afr. J. Biotechnol.* **2006**, *5*, 1355–1359.
19. Halib, N.; Amin, M.C.I.; Ahmad, I. Unique Stimuli Responsive Characteristics of Electron Beam Synthesized Bacterial Cellulose/Acrylic Acid Composite. *J. Appl. Polym. Sci.* **2010**, *116*, 2920–2929.
20. Nimeskern, L.; Ávila, H.M.; Sundberg, J.; Gatenholm, P.; Müller, R.; Stok, K.S. Mechanical evaluation of bacterial nanocellulose as an implant material for ear cartilage replacement. *J. Mech. Behav. Biomed. Mater.* **2013**, *22*, 12–21. [CrossRef]
21. Luo, H.; Cui, T.; Gan, D.; Gama, M.; Zhang, Q.; Wan, Y. Fabrication of a novel hierarchical fibrous scaffold for breast cancer cell culture. *Polym. Test.* **2019**, *80*, 106107. [CrossRef]
22. Chu, M.; Gao, H.; Liu, S.; Wang, L.; Jia, Y.; Gao, M.; Wan, M.; Xu, C.; Ren, L. Functionalization of composite bacterial cellulose with C60nanoparticles for wound dressing and cancer therapy. *RSC Adv.* **2018**, *8*, 18197–18203. [CrossRef]
23. Phan, T.D.; Debeaufort, F.; Luu, D.; Voilley, A. Functional Properties of Edible Agar-Based and Starch-Based Films for Food Quality Preservation. *J. Agric. Food Chem.* **2005**, *53*, 973–981. [CrossRef] [PubMed]
24. Fried, J.R. *Polymer Science and Technology*; Prentice-Hall: New Jersey, NJ, USA, 1995; p. 389.
25. Heinze, T.; Pfeiffer, K. Synthesis and characterization of carboxymethylcellulose. *Angew. Makromol. Chem.* **1999**, *266*, 37–45. [CrossRef]
26. Adinugraha, M.P.; Marseno, D.W. Haryadi Synthesis and characterization of sodium carboxymethylcellulose from cavendish banana pseudo stem (Musa cavendishii LAMBERT). *Carbohydr. Polym.* **2005**, *62*, 164–169. [CrossRef]
27. Rachtanapun, P.; Chaiwan, W.; Watthanaworasakun, Y. Effect of sodium hydroxide concentration on synthesis and characterization of carboxymethyl rice starch (CMSr). *Int. Food Res. J.* **2012**, *19*, 923–931.
28. Sangseethong, K.; Ketsilp, S.; Sriroth, K. The Role of Reaction Parameters on the Preparation and Properties of Carboxymethyl Cassava Starch. *Starch-Stärke* **2005**, *57*, 84–93. [CrossRef]
29. Li, Y.; Shoemaker, C.F.; Ma, J.; Shen, X.; Zhong, F. Paste viscosity of rice starches of different amylose content and carboxymethylcellulose formed by dry heating and the physical properties of their films. *Food Chem.* **2008**, *109*, 616–623. [CrossRef]

polymers

MDPI

Article

Improving the Tensile and Tear Properties of Thermoplastic Starch/Dolomite Biocomposite Film through Sonication Process

Azlin Fazlina Osman [1,2,*], Lilian Siah [1], Awad A. Alrashdi [3], Anwar Ul-Hamid [4] and Ismail Ibrahim [1,2]

1 Faculty of Chemical Engineering Technology, Universiti Malaysia Perlis (UniMAP), Arau 02600, Malaysia; msliiian0820@gmail.com (L.S.); ismailibrahim@unimap.edu.my (I.I.)
2 Biomedical and Nanotechnology Research Group, Center of Excellence Geopolymer and Green Technology (CEGeoGTech), Universiti Malaysia Perlis (UniMAP), Arau 02600, Malaysia
3 Chemistry Department, Umm Al-Qura University, Al-Qunfudah University College, Al-Qunfudah Center for Scientific Research (QCSR), Al Qunfudah 21962, Saudi Arabia; aarashdi@uqu.edu.sa
4 Center for Engineering Research, Research Institute, King Fahd University of Petroleum & Minerals, Dhahran 31261, Saudi Arabia; anwar@kfupm.edu.sa
* Correspondence: azlin@unimap.edu.my

Abstract: In this work, dolomite filler was introduced into thermoplastic starch (TPS) matrix to form TPS-dolomite (TPS-DOL) biocomposites. TPS-DOL biocomposites were prepared at different dolomite loadings (1 wt%, 2 wt%, 3 wt%, 4 wt% and 5 wt%) and by using two different forms of dolomite (pristine (DOL(P) and sonicated dolomite (DOL(U)) via the solvent casting technique. The effects of dolomite loading and sonication process on the mechanical properties of the TPS-DOL biocomposites were analyzed using tensile and tear tests. The chemistry aspect of the TPS-DOL biocomposites was analyzed using Fourier transform infrared spectroscopy (FTIR) and X-Ray Diffraction (XRD) analysis. According to the mechanical data, biocomposites with a high loading of dolomite (4 and 5 wt%) possess greater tensile and tear properties as compared to the biocomposites with a low loading of dolomite (1 and 2 wt%). Furthermore, it is also proved that the TPS-DOL(U) biocomposites have better mechanical properties when compared to the TPS-DOL(P) biocomposites. Reduction in the dolomite particle size upon the sonication process assisted in its dispersion and distribution throughout the TPS matrix. Thus, this led to the improvement of the tensile and tear properties of the biocomposite. Based on the findings, it is proven that the sonication process is a simple yet beneficial technique in the production of the TPS-dolomite biocomposites with improved tensile and tear properties for use as packaging film.

Keywords: thermoplastic starch; dolomite; biocomposite; mechanical properties; sonication

Citation: Osman, A.F.; Siah, L.; Alrashdi, A.A.; Ul-Hamid, A.; Ibrahim, I. Improving the Tensile and Tear Properties of Thermoplastic Starch/Dolomite Biocomposite Film through Sonication Process. *Polymers* **2021**, *13*, 274. https://doi.org/10.3390/polym13020274

Received: 9 December 2020
Accepted: 8 January 2021
Published: 15 January 2021

1. Introduction

Nowadays, petroleum-based plastics are daily and widely used packaging materials. Unfortunately, this contributes negative effects to the environment as a result of the long degradation time of plastics leading to waste disposal problems [1]. Scientifically, the molecular structure and chemical bonding of the petroleum-based plastic caused them to be too durable and resistant to the natural process of degradation. For example, polyolefins such as polypropylene (PP) and polyethylene (PE) are very resistant to hydrolysis and totally non-biodegradable. Consequently, this causes a large accumulation of plastic wastes in the landfill, causing a serious waste management problem and soil contamination. Moreover, the toxic chemicals present in the plastic will also cause pollution to the environment. When plastics are burned, they emit dioxin, which is the most toxic substance that may harm the health, reproduction, development, immune systems, disrupt hormones, cause cancer, and create other issues [2]. To reduce those effects, recycling of the plastic wastes is performed. However, several problems may arise during the recycling process, especially during the collection, separation, and cleaning, due to possible contamination on the plastics and difficulty of finding an economically viable outlet, where incineration may give off some toxic gas [3].

Due to the above-mentioned problems, effort to replace the synthetic plastics with the natural-based plastics is gaining more traction from year to year. The development of new plastic materials from natural and renewable resources has been the object of intensive academic and industrial research [4–6]. Starch-based biodegradable plastics are an ideal replacement to petroleum-based plastics. Even though these biopolymers cannot replace synthetic polymers in every application, at least they can be used to produce specific products, normally for application in which the recovery of plastic is not viable, economically feasible, and controllable such as in one-time use plastic [4–7].

Thermoplastic starch (TPS) can be obtained through plasticization of the native starch granules. TPS can be used to form biodegradable biocomposites films for packaging application. However, TPS-based products usually have several disadvantages. They may exhibit poor mechanical performance due to severe issues of water sensitivity, environmental stress cracking, and high fragility over time. Their low mechanical properties are a major concern, especially when exposed to hydrolytic and oxidative conditions and elevated temperatures [4,8]. These restrict the use of the TPS packaging films for long-term application as opposed to the petroleum-based plastic. In order to improve the properties, the TPS needs to be reinforced with fillers. In this research, a mineral filler, which is dolomite (DOL), was used to reinforce the TPS film. In this context, dolomite particles dispersion in the TPS matrix is very important, since it can affect the mechanical properties of the resultant TPS-DOL biocomposite films. Previous research indicated that a good dispersion of filler leads to improvement in the mechanical properties of the TPS composites, such as tensile and tear strength [4,8]. Generally, the pristine dolomite particles exist in agglomerated form. This may hinder their homogenous dispersion throughout the TPS matrix. The poor dispersion of dolomite can inhibit their reinforcing capability; thus, biocomposites films with poor properties may be produced [9]. Due to these reasons, dolomite needs to undergo a pre-dispersion process to diminish particle agglomeration, thus ensuring a good dispersion of dolomite in the TPS matrix. According to previous research, sonication is a practical method for the pre-dispersion process of the particles and nanoparticles [10]. Thus, this method was employed in this study to obtain a good dispersion of dolomite in the TPS matrix.

To the best of our knowledge, no research related to TPS/dolomite biocomposite film has been conducted and reported. Our research group is the first to study and investigate the TPS/dolomite biocomposite system for potential use as packaging film. Both starch and dolomite are abundant sources of materials to be transformed into biocomposite or other products with several beneficial applications. Thus, research in this area is considered impactful toward sustainability. Of particular interest, an optimization study on the TPS/dolomite biocomposite system can be the first step to formulate the biocomposite with the best property profiles. Therefore, the objective of this study was to investigate the effects of dolomite loading (0, 1, 2, 3, 4, and 5 wt%) on the tensile and tear properties of the TPS/dolomite biocomposite film. The novelty of this work is the use of a simple technique for pre-dispersing the dolomite, which is tip sonication process to improve dispersion of dolomite particles in the TPS matrix. This technique involved no harmful chemicals, as the medium used is only water. Furthermore, the application of this pre-dispersion technique allows the enhancement in both tensile and tear properties of the TPS biocomposite incorporating dolomite. This article highlights all of these aspects.

2. Materials and Chemicals

Corn starch powder, dolomite filler, glycerol, and sodium bicarbonate have been used to produce the virgin TPS film and TPS-DOL biocomposites film. Corn starch was manufactured by Sigma-Aldrich (St. Louise, MO, USA) with code number S4126 in 2 kg poly bottle packing. It was supplied by Euroscience Sdn. Bhd (Kuala Lumpur, Malaysia) in white powder form. It consists of about 27% amylose and 73% amylopectin. Dolomite was kindly supplied by Perlis Dolomites Industries Sdn. Bhd (Padang Besar, Malaysia) and is known by the locals as 'batu reput'. It came in powder form with a particle size of 150 μ. Dolomite was used as filler in the TPS biocomposite film in pristine and sonicated form. Glycerol with a minimum concentration of 99.5% was manufactured by HmbG Chemicals (Hamburg, Germany) with code number C0347-91552409. This chemical was purchased through A.R. Alatan Sains Sdn Bhd (Alor Setar, Malaysia). Glycerol served as a plasticizer in the formulation of TPS film. Sodium bicarbonate was manufactured by HmbG Chemicals (Hamburg, Germany) with code number C0725-2134330 and supplied by A.R. Alatan Sains Sdn Bhd (Alor Setar, Malaysia). It acted as a thickener in the production of TPS film.

2.1. Preparation of Dolomite (DOL(P)) and Ultrasonicated Dolomite (DOL(U)

Dolomite powder having a particle size of 150 μ was ground into finer and smaller particle size by using a mortar and pestle. After that, it was sieved by using a 50 μ sieve with 100 mesh. The resultant powder is called pristine dolomite (DOL(P)).

The sonication of dolomite was done as a pre-dispersion process prior to biocomposite production to assist dolomite dispersion and distribution in the TPS matrix. The suspension of dolomite was prepared by mixing the powder form of dolomite with distilled water in a ratio of 1:100. The dolomite was sonicated by using a Branson Digital Ultrasonic Disrupter/Homogenizer, Model 450 D with 20% amplitude, 20 s pulse on, and 10 s pulse off for 120 min. The resultant sonicated dolomite is referred to as DOL(U).

2.2. Plasticization of Starch to Form Thermoplastic Starch (TPS)

Thermoplastic starch (TPS) was obtained through the combination of corn starch with plasticizers under a stirring and heating process. Firstly, distilled water, corn starch, and glycerol were combined in a mass ratio of 100:5:2 in one beaker. The amount of water, corn starch, and glycerol used to produce one film of TPS in a Teflon mold were 100 mL, 5 g, and 2 g, respectively. Next, the mixture was stirred continuously in a water bath under a temperature range of 75–85 °C by using a magnetic stirrer (500 rpm). This procedure took about 45 min or until the mixture turned into paste.

2.3. Preparation of TPS-DOL Biocomposites

The preparation of the TPS-DOL biocomposites films was done through the solution casting and drying process method. The suspension of dolomite (sonicated) was added into the TPS suspension and keep stirred in the range between 75 and 85 °C. Next, 0.5 g of sodium bicarbonate was dissolved in 30 mL of distilled water and then added into the TPS suspension, which was continuously stirred for another 5 min. After that, the biocomposite suspension was poured into Teflon-coated mold and put into the oven at 45 °C for 24 h for drying. The dried film of the TPS-DOL biocomposite is shown in Figure 1. It was removed from the Teflon-coated mold and stored in a desiccator to avoid moisture while conditioning it in standard environment prior to testing. The process was repeated for pristine dolomite (without sonication) to form control biocomposites films. All the film samples were prepared in the same volume and using the same-size molds. Therefore, all samples had almost similar thickness.

Figure 1. Dried film of the thermoplastic starch-dolomite (TPS-DOL) biocomposite in the Teflon-coated mould.

The formulations of TPS-DOL biocomposite films are shown in Table 1. In preliminary works, we have prepared the TPS biocomposite films with dolomite loading greater than 5 wt% (7 and 9 wt%). However, no good films were observed, which was perhaps due to agglomeration and the overcrowding of dolomite particles in the matrix. The films were brittle and cracked after being cut into dumbbell shape. Furthermore, the samples took a longer time for drying. Based on these observations, we did not proceed for testing.

Table 1. The formulation of thermoplastic starch-dolomite (TPS-DOL) biocomposite films with their respective acronyms.

TPS (wt%)	Dolomite Filler without Sonication Process (wt%)	Dolomite Filler Subjected to 120 min Sonication Process (wt%)	Acronym of Biocomposite Film
100	-	-	TPS
99	1	-	TPS-1wt%DOL(P)
99	-	1	TPS-1wt%DOL(U)
98	2	-	TPS-2wt%DOL(P)
98	-	2	TPS-2wt%DOL(U)
97	3	-	TPS-3wt%DOL(P)
97	-	3	TPS-3wt%DOL(U)
96	4	-	TPS-4wt%DOL(P)
96	-	4	TPS-4wt%DOL(U)
95	5	-	TPS-5wt%DOL(P)
95	-	5	TPS-5wt%DOL(U)

The overall experimental procedures to produce the TPS-DOL biocomposite films are summarized in Figure 2.

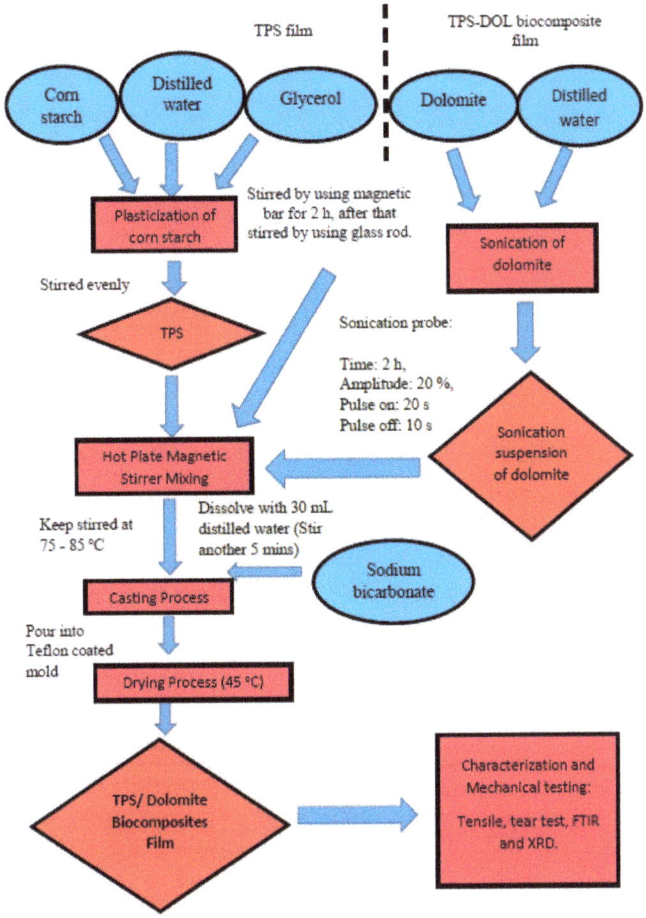

Figure 2. Flow chart of experimental procedures.

3. Characterization and Mechanical Testing

The tensile test, tear test, and FTIR analysis were conducted to test the properties and chemical structure of the TPS film and TPS-DOL biocomposite films.

3.1. Fourier Transform Infrared Spectroscopy (FTIR)

By using a Perkin Elmer RXI FTIR spectrophotometer (Waltham, MA, USA), the FTIR spectra of all the samples were analyzed. The chemical functional groups of TPS-DOL biocomposites (incorporating the DOL(P) and DOL(U)) were characterized by FTIR. The samples were recorded for 32 scans in the frequency range of 650–4000 cm^{-1} wavenumber with a resolution of 32 cm^{-1}.

3.2. X-ray Diffraction (XRD)

X-ray diffraction (XRD) was used to analyze the structure of dolomite before and after the sonication process. The analysis was also performed on the TPS film and TPS-DOL biocomposite films incorporating the DOL(P) and DOL(U) to determine their crystallinity. Each sample was placed in sample holder and scanned over a range of $2\theta = 10°–35°$ and with a scan speed of $2°/min$.

The degree crystallinity of corn starch, TPS, DOL(P), DOL(U), and TPS-DOL biocomposite films was evaluated by using the formula below [11].

$$\text{Degree of Crystallinity} = \frac{(\text{total area of cristalline peaks})}{(\text{total area of all peaks})} \times 100\%.$$

3.3. Tensile Test

The tensile properties such as the tensile strength, elongation at break, and Young's modulus of the TPS-DOL biocomposite films were determined through a tensile test using the Instron Machine 5569 (Norwood, MA, USA), according to angle test specimen (ASTM D 882) [12]. The test specimens were cut into dumbbell shape according to an ASTM D 638 Type V die cutter [13]. Details are shown in Figure 3 and Table 2.

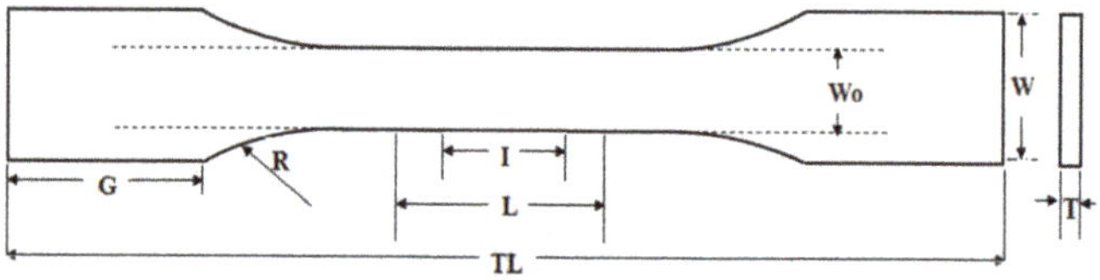

Figure 3. Schematic drawings of dumbell specimen according to ASTM D638 Type 5 [13].

Table 2. Dimension of a dumbbell specimen (ASTM D638 Type V).

Symbol	Description	Dimension (mm)
I	Gauge length	7.62
L	Length of narrow selection	9.53
Wo	Width of narrow selection	3.18
G	Grip length	19.05
W	Total width	9.53
TL	Total length	63.50
R	Radius of curvature	12.70

The samples were tested with a crosshead speed of 10 mm/min. Five replicates of each sample were tested, and the average values of tensile strength, elongation at break, and Young's modulus were recorded.

3.4. Tear Test

The tear strength of the TPS-DOL biocomposite films was determined by using an Instron Machine 5569. The samples were cut according to angle test specimen (ASTM D624 Type C) using a wood-base laser cutter [14]. Dimensions of the tear test specimen are shown in Figure 4 The test was run at a crosshead speed of 500 mm/min. Three replicates of each biocomposite film were tested and average values of tear strength were taken.

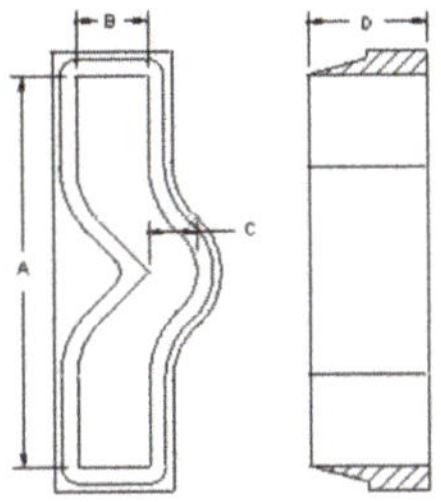

D-624-C-MET

#	Nominal Dimension
A	102.00mm
B	19.00mm
C	12.70mm
D	31.75mm

Figure 4. Schematic drawings of tear test specimen according to ASTM D624 Type C [14].

Setup for tensile and tear tests is shown in Figure 5. Both testings were carried out using the same Instron machine.

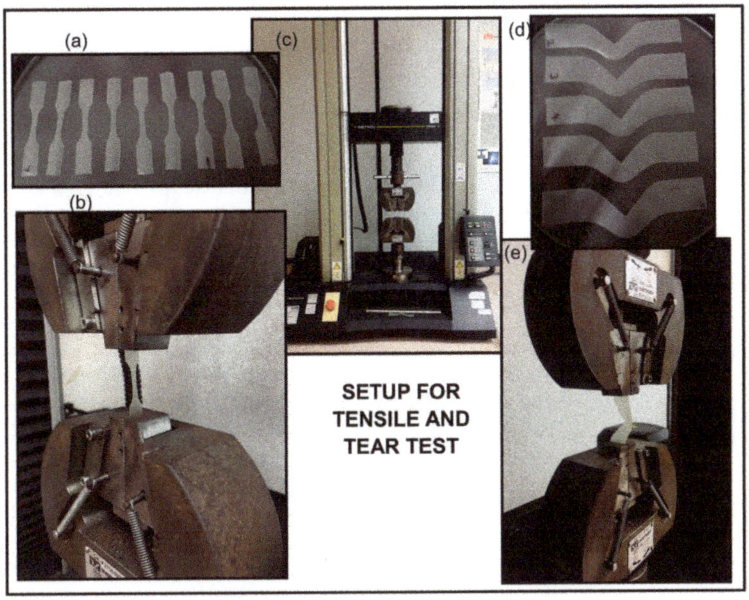

Figure 5. Setup for tensile and tear tests: (**a**) dumbbell-shaped samples for tensile test, (**b**) sample loading for tensile test, (**c**) Instron machine model 5569 to perform tensile and tear tests, (**d**) samples for tear test, (**e**) sample loading for tear test.

4. Results and Discussion

4.1. Functional Group Analysis of the Dolomite Filler (DOL(P) and DOL(U))

Fourier-transform infrared spectroscopy (FTIR) analysis was conducted for DOL(P), DOL(U), TPS, and TPS-DOL biocomposite films. Figure 6 illustrates the FTIR spectra of DOL(P) and DOL(U).

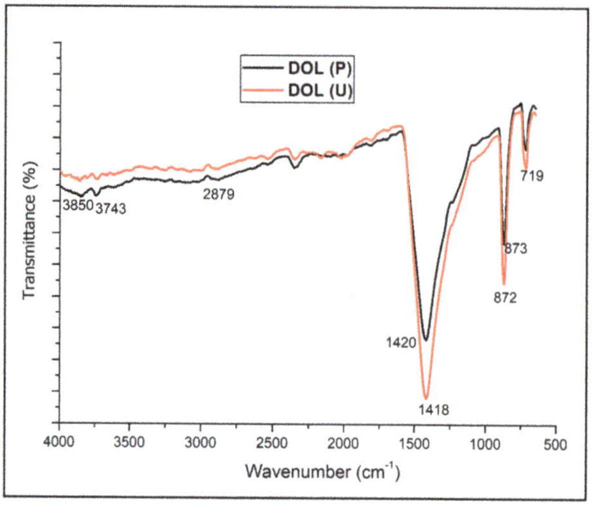

Figure 6. Fourier transform infrared spectroscopy (FTIR) spectra of pristine dolomite (DOL(P)) and sonicated dolomite (DOL(U)).

Generally, the FTIR spectra of DOL(P) indicate that the most prevalent component present in dolomite is carbonate [15]. The carbonate mineral group (magnesium and calcium carbonate) illustrates the strong peak at 1420 cm^{-1}, which is associated with lattice CO_3^{2-} [16]. On the other hand, the other main components of dolomite can be seen through the appearance of spectral peaks around 3000 and 719 cm^{-1} [15]. The FTIR absorption at around 3000 cm^{-1} relates to organic residue, while the band at around 2870 cm^{-1} represents the calcite combination band, and the peak at 719 cm^{-1} is attributed to magnesian calcite or calcite [16,17]. In agreement with our result, Chen et al. also stated that there are few FTIR absorptions at 1420, 873, and 719 cm^{-1}, which are verified as the presence of dolomite [17]. DOL(U) refers to the dolomite, which has underwent a 2 h sonication process. It is observable that the features of the FTIR spectra of DOL(P) and DOL(U) are quite similar. There are only slight changes in the peak position of the overall spectra, which suggest that there is no structural and chemical modification of the dolomite when subjected to the sonication process. However, there are more intense peaks around 1418, 872, and 719 cm^{-1} for DOL(U) when compared with DOL(P). This could be due to the smaller particle size but larger surface area of dolomite after it was subjected to the sonication process. It is widely understood that the infrared (IR) spectroscopy involves the analysis on how the infrared light interacts with matter. Therefore, the concentration of molecules in the tested sample affects the peak intensity in the IR spectra [18]. Particles with higher surface area may allow penetration of the IR light into a greater number of molecules. As a result, more intense peaks can be observed. Note that the aim of conducting the sonication process in this research was to break the dolomite particles into smaller and finer size so that the dolomite filler can be more efficiently dispersed throughout the TPS matrix. This can enhance the tensile and tear properties due to better interaction between dolomite filler and the TPS matrix.

4.2. Functional Groups Analysis on TPS Film and TPS-DOL Biocomposite Films

Figure 7 indicates the FTIR spectra of the TPS film, TPS-5wt%DOL (P) and TPS-5wt%DOL(U) biocomposite films. It can be observed that the FTIR signal of the TPS film indicates a strong spectral band in the range of 600 to 1500 cm^{-1}, which confirms the fingerprint of the starch material [5]. TPS film displays the characteristics FTIR absorptions at 3265 and 2925 cm^{-1}, which correspond to the O-H and C-H stretching, respectively. The appearance of the peak at 1645 cm^{-1} was due to the hydrophilic characteristics of the

TPS itself, as it can be tightly bound with water [19]. The spectral peak at 1140 cm^{-1} is contributed to the C-O-H bond. Within the region below 860 cm^{-1}, the spectrum of the TPS film reveals the complex vibrational mode, which was likely due to the skeletal vibrations of the pyranose ring in the glucose unit [19].

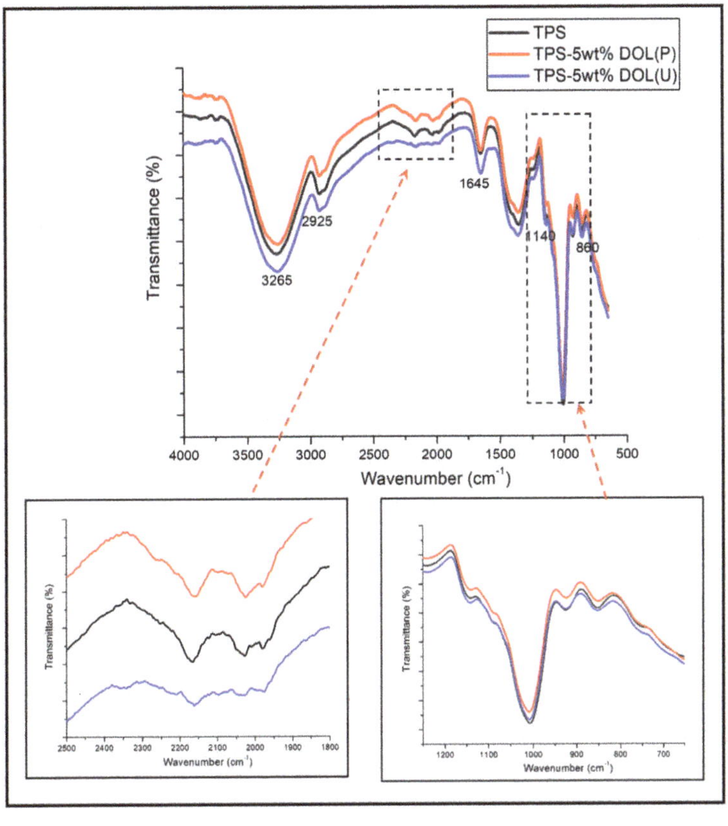

Figure 7. FTIR spectra of thermoplastic starch (TPS) film and TPS-5wt%DOL biocomposites films.

Overall, the FTIR spectra of the TPS film were almost similar to the TPS-5wt%DOL biocomposite films. Only slight changes can be observed in the range of 1800 to 2500 cm^{-1} and 750 to 1200 cm^{-1}. When benchmarked with the TPS and TPS-5wt%DOL(P) biocomposite film, we can see slight broadening of the peak that relates with molecules of starch in the TPS-5wt% DOL(U) film (refer to the bands around 750 to 1200 cm^{-1}). Interestingly, this was also accompanied by the broadening and weakening of peaks, which was related to dolomite (1800–2500 cm^{-1}). These could be a signal that some matrix–filler interactions have occurred (perhaps hydrogen bonding) between the starch and DOL(U) filler. As mentioned earlier, the size of dolomite became much smaller upon the sonication process. Consequently, these smaller particles could be better dispersed and diffused into the TPS chains, which interact well through the formation of hydrogen bonding. Better interaction between the DOL(U) filler and TPS molecules resulted in the TPS-5wt%DOL(U) biocomposite film that possesses better mechanical properties than the TPS film, as discussed in the following section.

4.3. Crystallinity Analysis of the Dolomite Filler (DOL(P) and DOL(U))

The structure of dolomite is ordered; thus, its crystalline structure can be analyzed through XRD. The structure of the DOL(P) and DOL(U) was characterized and compared by using XRD analysis. An XRD pattern was used to observe the changes in the dolomite structure before and after undergoing the sonication process. Figure 8 illustrates the XRD pattern of the DOL(P) and DOL(U).

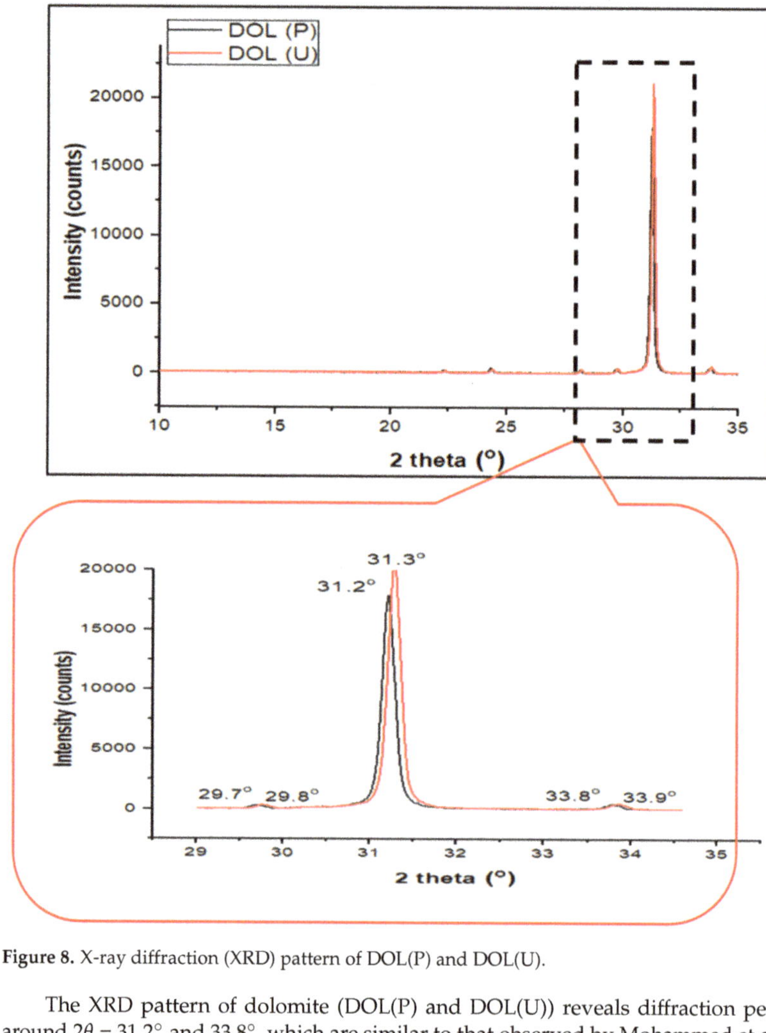

Figure 8. X-ray diffraction (XRD) pattern of DOL(P) and DOL(U).

The XRD pattern of dolomite (DOL(P) and DOL(U)) reveals diffraction peaks at around $2\theta = 31.2°$ and 33.8°, which are similar to that observed by Mohammed et al. [15]. However, when benchmarked with the DOL(P), the DOL(U) shows more intense diffraction peaks at both angles. This signifies that the crystallinity of dolomite was enhanced by the sonication process. To prove this, the degree of crystallinity of both types of dolomite was measured and tabulated in Table 3. The degree of crystallinity for DOL(P) and DOL(U) was found to be 76.57% and 77.10%, respectively. The higher degree of crystallinity of the DOL(U) could be as a result of the 2-h sonication process. The reduction in particle size of dolomite has increased its surface area. As a result, signals from X-ray diffraction that went through the crystalline lattices of dolomite particles became stronger [5].

Table 3. Crystallinity of the pristine dolomite (DOL(P)), sonicated dolomite (DOL(U)), corn starch, TPS film, TPS-5wt%DOL (P), and TPS-5wt%DOL (U) biocomposite films based on XRD analysis.

Samples	2θ (°)	Percent (%)	
		Crystallinity	Amorphous
DOL(P)	22.4, 24.4, 28.2, 29.7, 31.2, 33.8	76.57	23.43
DOL(U)	22.4, 24.3, 28.1, 29.8, 31.3, 33.9	77.10	22.90
Corn starch powder	11.4, 15.1, 17.1,18, 20, 22.9	58.03	41.97
TPS	17.2, 19.5, 22.6, 34.1	25.38	74.62
TPS-5wt%DOL (P)	20.4, 22.7, 31, 34.3	51.79	48.21
TPS-5wt%DOL (U)	17.4, 20.2, 22.8, 31.2, 31.6, 34.4	52.43	47.57

4.4. Crystallinity Analysis of the Corn Starch, TPS Film, and TPS-DOL Biocomposite Films

XRD can be used to determine the degree of crystallinity based on the ratio of crystalline to amorphous phase in the starch. Starch is a semi-crystalline material, which consists of two main components; amylose (forming crystalline region) and amylopectin (forming amorphous phase). The existence of these two phases can be proved through the XRD spectrum.

The degree of crystallinity values for the corn starch, TPS film, TPS-5wt%DOL(P), and TPS-5wt%DOL(U) films were also further studied by using XRD analysis. Table 2 indicates the percent crystallinity of all the samples, while their XRD pattern in the range of 10° to 35°(2θ) is illustrated in Figure 9.

XRD pattern of the corn starch powder indicates strong and sharp diffraction peaks at 2θ = 15.1°, 20°, and 22.9° and an unresolved doublet at 17.1° and 18°, which signify a typical crystalline structure of cereal starch (A-type crystallinity). Due to the double-helical arrangement of amylopectin chains, these A-type crystallites are usually denser and less hydrated. Similar results of the pure corn starch were proved by other researchers [5,20]. The peaks at 2θ = 15.1° and 18° that exist in the corn starch disappeared in the XRD signal of the cast TPS film, which means the original crystalline structure was destroyed during the solution mixing process. However, there is a minimum residue of A-type crystalline that could still be observed in the TPS film produced by the solvent casting technique [21]. In the preparation of the solvent cast TPS film, shear and thermal energies applied during the stirring process were adequate to melt granular the crystallites; therefore, most of the crystalline phase of the starch has been diminished.

The XRD pattern of the TPS film has shown the combination of crystallinities Type B and Type Vh as well as the low value of the degree of crystallinity, which is about 25.38%. There are diffraction peaks at 2θ of 17.2°, 19.5°, 22.6°, and 34.1°. Peaks at 17.2° are assigned to B-type crystals, while peaks at 19.5° and 22.6° belong to Vh-type crystals, which are related to amylose crystallization into a single helical structure [5,22].

Upon the addition of dolomite into the starch matrix, the peak at 2θ = 31° (for TPS-5wt% DOL(P)) and doublet peak at 2θ = 31.2° and 31.6° (for TPS-5wt%DOL(U)) appear. Those peaks are correlated with dolomite, further proving the existence of dolomite particles in the structure of the biocomposite films. As referred to Table 3, it can be seen that the addition of dolomite filler significantly increased the degree of crystallinity of the TPS film. Although the peak location of starch does not change obviously in the XRD pattern, the intensities increased significantly due the addition of dolomite into the TPS film (refer to the diffraction peaks 2θ at around 22° for TPS film and TPS-5wt%DOL(P)) film. The degree of crystallinity of the TPS film was increased from 25.38% to 51.79% when incorporated with DOL(P).

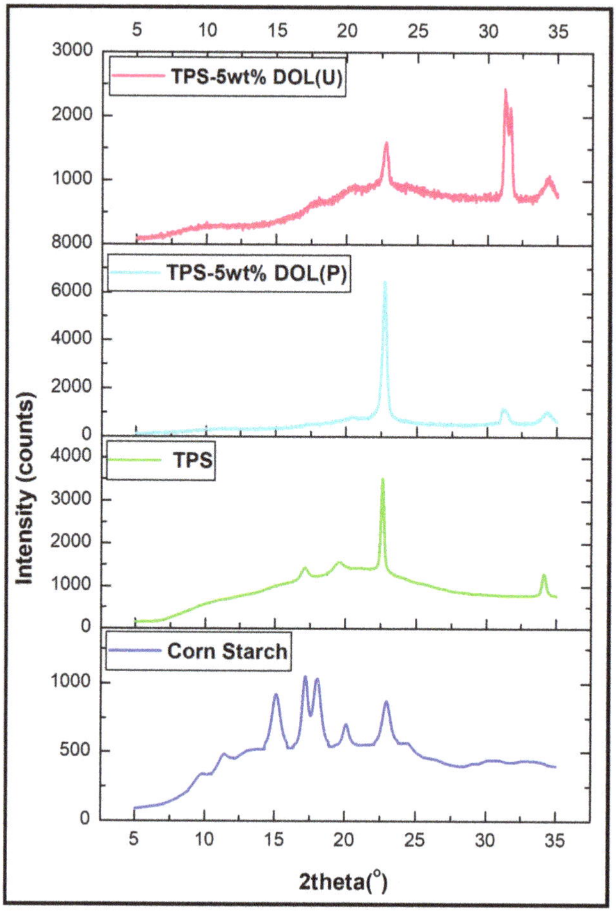

Figure 9. XRD pattern of corn starch, TPS film, and TPS-5wt%DOL (P) and TPS-5wt%DOL (U) biocomposite films.

According to Table 3, TPS-5wt%DOL(U) (52.43%) has a higher degree of crystallinity as compared with TPS-5wt%DOL(P) (51.79%). Based on Figure 9, the peak intensity for TPS-5wt%DOL(U) is higher than that of TPS-5wt%DOL(P), (refering to diffraction peaks at around $2\theta = 31°$). In addition, TPS-5wt%DOL(U) also showed a strong doublet peak at around 2θ at 31° due to the presence of well-dispersed dolomite particles. The de-agglomeration of dolomite upon the sonication process led to the increased surface area of the dolomite particles, thus producing a stronger diffraction peak when dispersed in the TPS matrix. Another reason could be related to the sonication process that facilitated the dispersion of dolomite particles throughout the TPS matrix and improved the matrix–filler interactions. Subsequently, this has encouraged the nucleation process for the TPS matrix crystallization and thus increased the degree of crystallinity of the biocomposites film [23]. The increase of degree of crystallinity is commonly associated with its improvement in tensile strength [5,21]. This is also the reason why the TPS-DOL(U) biocomposite film has better tensile strength when compared with TPS-DOL(P) biocomposite film, which will be shown and discussed in the next section.

4.5. Mechanical Properties Evaluation through Tensile Test

Figures 10–12 illustrate the effect of dolomite loading (0, 1, 2, 3, 4 and 5 wt%) and sonication process on the tensile properties of TPS biocomposite films. According to Figure 10, low loadings of dolomite cause a deterioration of the tensile strength of the TPS-DOL biocomposite film. The TPS film possesses the tensile strength of 2.64 MPa. Meanwhile, the tensile strength of the TPS-DOL (P) biocomposites with 1, 2 and 3 wt% filler was found to be 1.76 MPa, 1.73 MPa, and 1.98 MPa, respectively. On the other hand, TPS-DOL(P) biocomposite films with high loading of filler (4 and 5 wt%) show almost the same tensile strength value with the TPS film. This assessment was made based on the tensile strength and standard deviation values of the TPS film and TPS-DOL(P) biocomposite films containing 4 and 5 wt% DOL(P), which show statistically no significant difference. The results suggest that dolomite in pristine form was not a good reinforcing filler for the TPS film and thus could not improve the tensile strength of the host biopolymer when added in 1 to 5 wt%. Conversely, for TPS-DOL(U) biocomposite film cases, the tensile strength was enhanced when DOL(U) was added in 4 wt% and 5 wt%. Based on the results, the tensile strength of TPS-4wt%DOL(U) biocomposites was 3.06 MPa, which showed an improvement of about 15.9% when compared with the TPS film. However, when less DOL(U) was added into the TPS (1 wt%, and 2 wt%), the tensile strength was 1.88 MPa and 1.87 MPa, respectively, which is somewhat lower than the TPS film. The TPS-5wt% DOL(U) biocomposite has the highest tensile strength among all the samples (3.61 MPa) by demonstrating 36.74% higher in tensile strength when benchmarked with the TPS film.

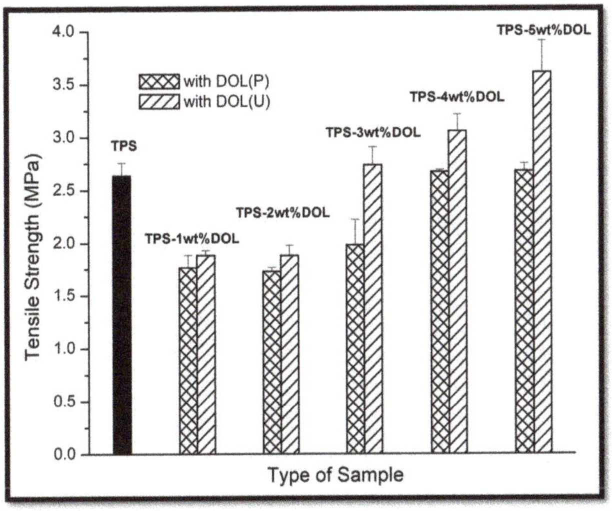

Figure 10. Tensile strength of TPS film and TPS-DOL biocomposite films.

Overall, the results indicate that the low loading of dolomites, either in pristine or sonicated form, could not improve the tensile strength of the TPS biocomposite film. This could be due to the insufficient content of dolomite particles to allow a reinforcing effect to the TPS chains. This happened due to the lack of dispersed particles to act as a stress-transferring medium when subjected to the tensile load. In fact, the lowering of tensile strength can be seen due to the presence of filler that induces inhomogeneity in the biocomposite system, rather than providing reinforcing effect. Syed Adam et al. have obtained and reported the same findings through their research on TPS biocomposite [7]. However, when a higher content of dolomite was added, there was a greater number of particles being diffused into the TPS chain and acting as reinforcing filler. Upon the sonication process, the dolomite filler turned into smaller size particles, which can be more easily dispersed throughout the TPS matrix. There were higher numbers of well-dispersed dolomite particles to interact with TPS molecular chains. As observed through the FTIR analysis, such interactions can be observed through the formation of hydrogen bonding between the dolomite filler and TPS molecular chains. As a result, a better stress-transforming mechanism between the matrix and filler occurred. Apparently, the tensile strength of the TPS-5wt%DOL(U) biocomposite film (3.61 MPa) was better than the tensile strength of the TPS-5wt%DOL(P) biocomposite film (2.68 MPa). This shows that the sonication process can assist the distribution and dispersion of dolomite filler throughout the TPS, which eventually improved the tensile strength of the biocomposite film. A better dispersion of dolomite could increase the efficiency of the filler to transfer the stress effectively within the composite structure. This is in agreement with the findings of Nik Adik et al. [24].

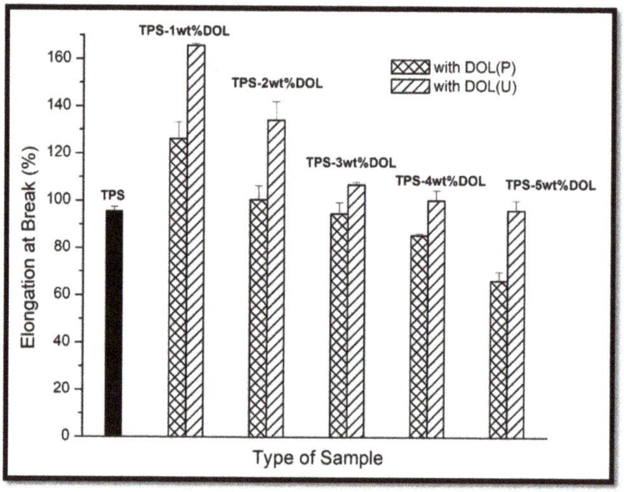

Figure 11. Elongation at break of TPS film and TPS-DOL biocomposite films.

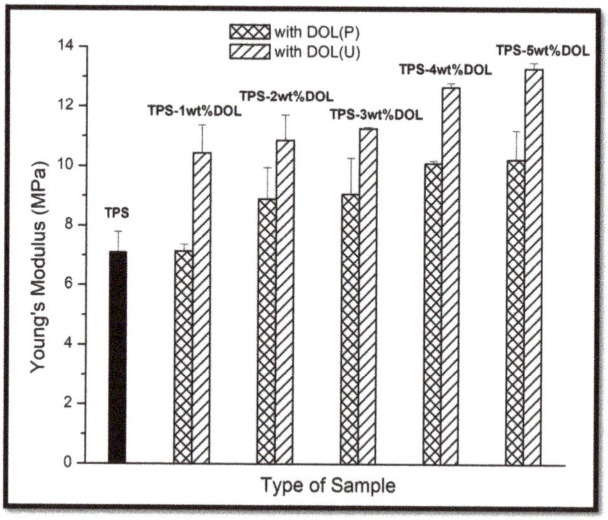

Figure 12. Young's modulus of TPS film and TPS-DOL biocomposite films.

Elongation at break values of the TPS film, TPS-DOL(P) biocomposite, and TPS-DOL(U) biocomposite films are compared in Figure 11. It can be seen that as the DOL(P) and DOL(U) loading increased in the TPS matrix, reduction in the elongation at break occurred. Elongation at break is a reflection of the ductility of the material: the direct opposite for brittleness. When dolomite filler is incorporated into a TPS matrix, it will increase its stiffness. The increase in stiffness resulted in a decrease in the ductility of the material. Thus, as the dolomite loading increases, the ductility decreases. Such a trend was also reported by other researchers [24,25]. The elongation at break of the TPS film was found to be 95.6%. With the addition of 1 wt% DOL(P) in TPS film, the elongation at break was 126.13%. The elongation at break was reduced significantly (25.43%) when the DOL(P) loading was increased to 2 wt%. When the loading of DOL (P) increased to 3 wt% and 4 wt%, the elongation at break value continued to decrease to 94.7% and 85.67%, respectively. TPS-5wt%DOL(P) has the lowest elongation at break among all the samples (66.37%), showing 30.58% reduction when benchmarked with the TPS film. Apparently, TPS filled with DOL(U) has a similar trend with TPS filled with DOL(P). The elongation at break of TPS-1wt%DOL(U) biocomposites was 165.77%, but it reduced significantly by 23.59% to 134.13% when the loading of DOL(U) was increased to 2 wt%. A further increase of DOL(U) loading to 3 wt% and 4 wt% resulted in the decrease of elongation at break to 106.9% and 100.37%, respectively. The TPS-DOL(U) biocomposite film achieved the lowest elongation at break (96.17%) when the filler loading was 5 wt%. As expected, the elongation at break value of the TPS-DOL(P) biocomposite films is lower than the TPS-DOL(U) biocomposite films when the same amount of filler was added. For instance, the elongation at break of TPS-1wt%DOL(P) (126.13%) was lower compared with TPS-1wt%DOL(U) (165.77%). This was probably due to the greater aggregation of dolomite filler when no sonication process was involved, thereby resulting in a greater stiffening effect to the TPS molecular chains. Subsequently, the flexibility of the biocomposite film reduced more significantly.

Young's modulus values of the TPS film, TPS-DOL(P), and TPS-DOL(U) biocomposite films at different dolomite content (1 wt%, 2 wt%, 3 wt%, 4 wt%, and 5 wt%) are demonstrated in Figure 12. It can be seen that the addition of DOL(P) and DOL(U) into the TPS matrix resulted in the increase of the Young's modulus of the biocomposites. In general, the Young's modulus of these samples was enhanced by the incorporation of dolomite filler in the TPS matrix. This was due to the stiffness of dolomite, which can restrict the chain mobility of the TPS matrix. The results are in tandem with findings reported by a few researchers [24–26]. It can be concluded that the TPS film has the lowest Young's modulus (7.1 MPa) among all the samples. The Young's modulus of TPS-5wt%DOL(P) biocomposite film was 10.23 MPa, which possesses a 44.08% increment when benchmarked with the TPS film. The TPS-5wt%DOL(U) biocomposite achieved the highest Young's modulus (which represents the stiffest samples) among all the samples. It possesses an 87.32% increment in Young's modulus when benchmarked with the TPS film. Moreover, the value of Young's modulus obtained for TPS-DOL(U) biocomposite films was higher than that of TPS-DOL(P) biocomposite films. The most significant difference can be observed when incorporating 5 wt% of dolomite into the TPS matrix. The TPS-5wt%DOL(U) (13.3 MPa) achieved a 30% increment in Young's modulus when compared with TPS-5wt%DOL(P) (10.23 MPa). As explained previously, dolomite that underwent a sonication process can be more homogeneously dispersed throughout the TPS matrix. Better filler–matrix interactions cause more restriction in TPS chain mobility and thus increase the Young's modulus of the biocomposite film [5,7].

4.6. Mechanical Properties Evaluation through Tear Test

Figure 13 compares the values of the tear strength of the TPS film and TPS-DOL biocomposite films. Generally, the tear strength of the TPS film (0.80 MPa) is lower than that of the TPS-DOL(P) and TPS-DOL(U) biocomposite films. This indicates that both pristine dolomite and sonicated dolomite have successfully improved the tear strength of the TPS. Both TPS-DOL(P) and TPS-DOL(U) biocomposite filmss achieved the highest tear strength when 5 wt% of filler was employed. TPS-5wt%DOL(U) biocomposite film shows the highest tear strength (2.30 MPa) among all the samples. It has achieved 187.50% higher tear strength when compared with the TPS film.

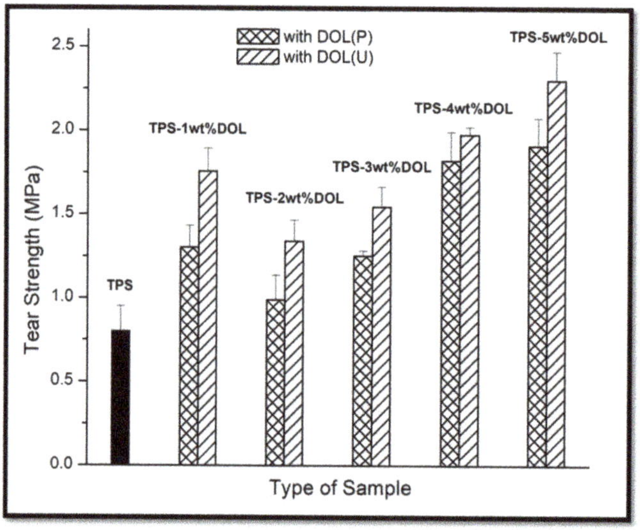

Figure 13. Tear strength of TPS film and TPS-DOL biocomposite films.

Obviously, when the same amount of filler was used, the TPS-DOL(U) biocomposite film possesses greater tear strength as compared to the TPS-DOL(P) biocomposite film. This happened due to the sonication process that helps to homogenously distribute dolomite particles into TPS molecule chains and improve the matrix–filler interactions [5]. The homogenous distribution of dolomite can increase the efficiency of filler to transfer the load within the TPS biocomposite structure. Obviously, the tear strength data of the biocomposite seem to follow the trend of the tensile strength data, in which the higher loading of dolomite filler causes a better reinforcing effect. The following statement explains this phenomenon: When a greater amount of dolomite particles penetrated through the polymer chain, they would develop more surface interactions with the TPS matrix; therefore, a better reinforcing effect can be seen. This is translated through the tensile and tear strength values. Osman, Edwards, and Martin have observed the same phenomenon in their composite system with mineral filler (clay) [27].

4.7. Morphology of Tensile Fractured Surface of the TPS Film and TPS-DOL Biocomposites

Figure 14 presents the SEM images of the tensile fractured surface of the TPS film and TPS-DOL biocomposite films. In its neat form (without filler), the TPS film shows a smooth and homogeneous surface. However, upon the addition of pristine dolomite (DOL(P)), the surface morphology of the matrix became rough and non-homogeneous with pores and pits. Apparently, at the same dolomite loading, the TPS-DOL biocomposites with sonicated filler (TPS-DOL(U)) exhibit smoother and more homogeneous surface morphology than the TPS-DOL biocomposites with pristine dolomite (TPS-DOL(P). These findings support the results of a test where the addition of DOL(U) in the TPS resulted in a more homogeneous biocomposite due to the well-dispersed filler distributed in the matrix. As a result, the tensile properties of the TPS film were enhanced. Li et al. have observed similar features in their poly (3-hydroxybutyrate-co-3-hydroxyvalerate) (PHBV)-based biocomposites with miscanthus biocarbon as the filler. A homogeneous dispersion of filler has been found to improve the mechanical and thermal stability of the host biopolymer. The biocomposite with a poor dispersion of filler indicated a rough fractured surface with the appearance of voids, suggesting the poor wetting of filler by the biopolymer matrix [28].

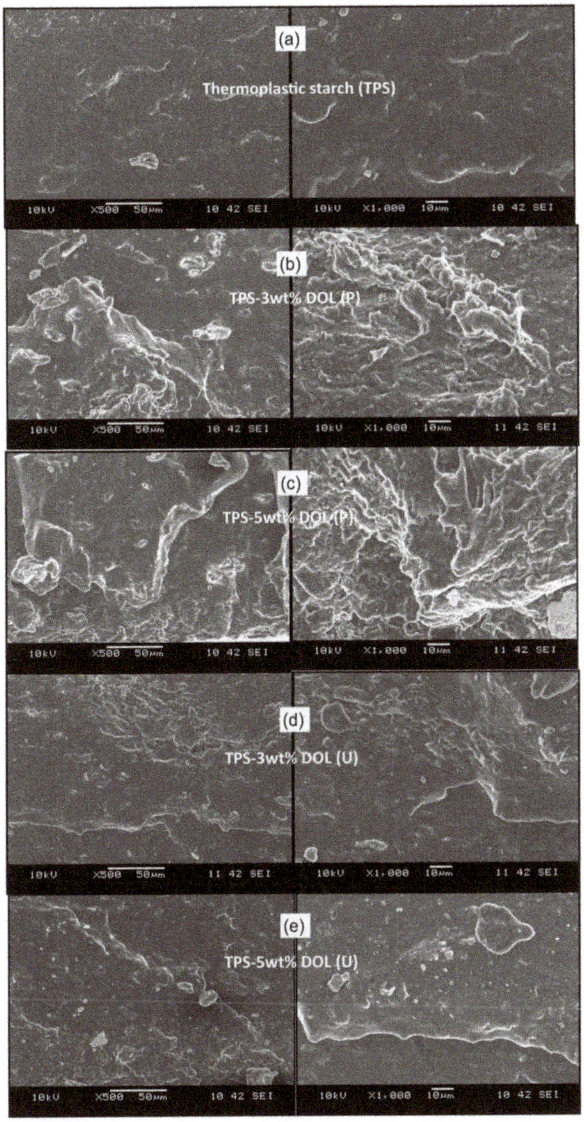

Figure 14. SEM images of tensile fractured surface of the (**a**) TPS film (**b**) TPS-3wt%DOL (P) (**c**) TPS-3wt%DOL (U) (**d**) TPS-5wt% DOL (P) and (**e**) TPS-5wt%DOL (U) biocomposite films at 500× magnification (left) and 1000× magnification (right).

5. Summary

The effects of dolomite loading and the sonication process on the mechanical properties of the TPS-DOL biocomposite films have been explored through this research. Indicatively, the use of high loading of dolomite (4 to 5 wt%) can improve both the tensile and tear properties of the biocomposite film, provided that the dolomite is being sonicated prior to being added into the TPS matrix. The use of dolomite in pristine form (DOL(P)) could not improve the tensile properties, even when added in 5 wt%. Conversely, upon the sonication process, the dolomite (DOL(U) brought an increment in both tensile and tear strength of the biocomposite film when added in 4 and 5 wt%. Based on the FTIR results, the spectral peaks of DOL(U) were more intense when compared with DOL(P). This means that when dolomite underwent the sonication process, a reduction in particle size occurred, causing a larger surface area of particles for the penetration of IR light into a greater number of molecules. In addition, the sonication process also reduced the particle size of the dolomite, assisting the dispersion and distribution of its particles throughout the TPS matrix. XRD analysis showed that the sonication process has increased the crystallinity of dolomite. This factor also contributed to the enhancement of the tensile and tear properties of the biocomposite film. The fractured surface morphology of the TPS-DOL (U) biocomposites is smoother and more homogeneous than the TPS-DOL (P) counterparts, further supporting the fact that the sonication process can improve the dispersion of the dolomite filler throughout the TPS matrix and produce more homogeneous biocomposites. In conclusion, sonicated dolomite has potential to be used as a reinforcing filler in the TPS film in order to improve its viability for packaging application. The low cost and abundancy of dolomite and the simple and environmental friendly method of sonication can be adopted to produce a sustainable biocomposite for future needs.

Author Contributions: Conceptualization, A.F.O.; data curation, L.S. and A.A.A.; formal analysis, L.S. and A.A.A.; investigation, A.F.O. and L.S.; methodology, A.F.O. and I.I.; project administration, A.F.O.; software, A.A.A.; writing—original draft, A.F.O.; writing—review & editing, A.U.-H. All authors have read and agreed to the published version of the manuscript.

Funding: This research was funded by the Fundamental Research Grant Scheme (FRGS) under a grant no: FRGS/1/2019/TK05/UNIMAP/02/6 from the Ministry of Education Malaysia.

Institutional Review Board Statement: Not applicable.

Informed Consent Statement: Not applicable for studies not involving humans.

Data Availability Statement: The data presented in this study are available on request from the corresponding author.

Conflicts of Interest: The authors declare no conflict of interest.

Abbreviations

ASTM D	American Society for Testing and Materials D Standard
DOL (P)	Pristine dolomite
DOL (U)	Sonicated dolomite
FTIR	Fourier Transform Infrared Spectroscopy
ISO	International Organization for Standardization
mm	Micronmeter
MPa	Mega Pascal
PE	Polyethylene
PP	Polypropylene
TPS	Thermoplastic starch
TPS-Xwt%DOL(P)	TPS biocomposite containing X wt% pristine dolomite
TPS-Xwt%DOL(U)	TPS biocomposite containing X wt% sonicated dolomite
XRD	X-ray Diffraction
μm	Micrometer

References

1. Haider, T.P.; Völker, C.; Kramm, J.; Landfester, K.; Wurm, F.R. Plastics of the Future? The Impact of Biodegradable Polymers on the Environment and on Society. *Angew. Chem. Int. Ed.* **2019**, *58*, 50–62. [CrossRef] [PubMed]
2. Jambeck, J.R.; Geyer, R.; Wilcox, C.; Siegler, T.R.; Perryman, M.; Andrady, A.; Narayan, R.; Law, K.L. Plastic waste inputs from land into the ocean. *Science* **2015**, *347*, 768–771. [CrossRef] [PubMed]

3. Hopewell, J.; Dvorak, R.; Kosior, E. Plastics recycling: Challenges and opportunities. *Philos. Trans. R. Soc. Lond. B Biol. Sci.* **2009**, *364*, 2115–2126. [CrossRef] [PubMed]
4. Balakrishnan, P.; Gopi, S.; Sreekala, M.S.; Thomas, S. UV resistant transparent bionanocomposite films based on potato starch/cellulose for sustainable packaging. *Starch* **2018**, *70*, 1700139. [CrossRef]
5. Osman, A.F.; Ashafee, A.M.T.; Adnan, S.A.; Alakrach, A. Influence of Hybrid Cellulose/Bentonite Fillers on Structure, Ambient and Low Temperature Tensile Properties of Thermoplastic Starch Composites. *Polym. Eng. Sci.* **2020**, *60*, 810–822. [CrossRef]
6. Husseinsyah, S.; Zailuddin, N.L.I.; Osman, A.F.; Lili, C.; Al-rashdi, A.A.; Alakrach, A. Methyl methacrylate (MMA) treatment of empty fruit bunch (EFB) to improve the properties of regenerated cellulose biocomposite films. *Polymers* **2020**, *12*, 2618. [CrossRef]
7. Syed Adam, S.N.S.; Osman, A.F.; Shamsudin, R. Tensile Properties, Biodegradability and Bioactivity of Thermoplastic Starch (TPS)/Bioglass Composites for Bone Tissue Engineering. *Sains Malays.* **2018**, *47*, 1303–1310. [CrossRef]
8. Ismail, I.; Osman, A.F.; Ping, T.L. Effects of ultrasonication process on crystallinity and tear strength of thermoplastic starch/cellulose biocomposites. *IOP Conf. Ser. Mater. Sci. Eng.* **2019**, *701*, 012045. [CrossRef]
9. Ni, L.; Mao, Y.; Liu, Y.; Cai, P.; Jiang, X.; Gao, X.; Cheng, X.; Chen, J. Synergistic reinforcement of waterborne polyurethane films using Palygorskite and dolomite as micro/nano-fillers. *J. Polym. Res.* **2020**, *27*, 23. [CrossRef]
10. Osman, A.F.; Kalo, H.; Hassan, M.S.; Hong, T.W.; Azmi, F. Pre-dispersing of montmorillonite nanofiller: Impact on morphology and performance of melt compounded ethyl vinyl acetate nanocomposites. *J. Appl. Polym. Sci.* **2016**, *133*, 1–15. [CrossRef]
11. Hulleman, S.H.D.; Kalisvaart, M.G.; Jansen, F.H.P.; Feil, H.; Vliegenthart, J.F.G. Origins of B-type crystallinity in glycerol-plasticized, compression moulded potato starches. *Carbohydr. Polym.* **1999**, *39*, 51–360. [CrossRef]
12. ASTM D882-18. *Standard Test Method for Tensile Properties of Thin Plastic Sheeting*; ASTM International: West Conshohocken, PA, USA, 2018.
13. ASTM D638-14. *Standard Test Method for Tensile Properties of Plastics*; ASTM International: West Conshohocken, PA, USA, 2014.
14. ASTM D624-00. *Standard Test Method for Tear Strength of Conventional Vulcanized Rubber and Thermoplastic*; ASTM International: West Conshohocken, PA, USA, 2020.
15. Mohammed, M.A.A.; Salmiaton, A.; Wan Azlina, W.A.K.G.; Mohamad Amran, M.S.; Taufiq-Yap, Y.H. Preparation and Characterization of Malaysian Dolomites as a Tar Cracking Catalyst in Biomass Gasification Process. *J. Energy* **2013**, 1–8. [CrossRef]
16. Wilson, K.; Hardacre, C.; Lee, A.F.; Montero, J.M.; Shellard, L. The application of calcined natural dolomitic rock as a solid base catalyst in triglyceride transesterification for biodiesel synthesis. *Green Chem.* **2008**, *10*, 654–659. [CrossRef]
17. Ji, J.; Ge, Y.; Balsam, W.; Damuth, J.E.; Chen, J. Rapid identification of dolomite using a Fourier Transform Infrared Spectrophotometer (FTIR): A fast method for identifying Heinrich events in IODP Site U1308. *Mar. Geol.* **2009**, *258*, 60–68. [CrossRef]
18. Workman, J., Jr.; Weyer, L. *Practical Guide to Interpretive Near-Infrared Spectroscopy*; CRC press: Boca Raton, FL, USA, 2007.
19. Flores-Morales, A.; Jiménez-Estrada, M.; Mora-Escobedo, R. Determination of the structural changes by FT-IR, Raman, and CP/MAS 13C NMR spectroscopy on retrograded starch of maize tortillas. *Carbohydr. Polym.* **2012**, *87*, 61–68. [CrossRef]
20. Hong, L.F.; Cheng, L.H.; Lee, C.Y.; Peh, K.K. Characterisation of physicochemical properties of propionylated corn starch and its application as stabilizer. *Food Technol. Biotechnol.* **2015**, *53*, 278–285. [CrossRef]
21. Muller, P.; Kapin, E.; Fekete, E. Effects of preparation methods on the structure and mechanical properties of wet conditioned starch/montmorillonite nanocomposite films. *Carbohydr. Polym.* **2014**, *113*, 569–576. [CrossRef]
22. Mendes, J.F.; Paschoalin, R.T.; Carmona, V.B.; Neto, A.R.S.; Marques, A.C.P.; Marconcini, J.M.; Oliveira, J.E. Biodegradable polymer blends based on corn starch and thermoplastic chitosan processed by extrusion. *Carbohydr. Polym.* **2016**, *137*, 452–458. [CrossRef]
23. Girones, J.; Lopez, J.P.; Mutje, P.; Carvalho, A.J.F.; Curvelo, A.A.S.; Vilaseca, F. Natural fiber-reinforced thermoplastic starch composites obtained by melt processing. *Compos. Sci. Techno.* **2012**, *72*, 858–863. [CrossRef]
24. Nik Adik, N.N.A.; Lin, O.H.; Akil, H.M.; Sandu, A.V.; Villagracia, A.R.; Santos, G.N. Effects of stearic acid on tensile, morphological and thermal analysis of polypropylene (PP)/dolomite (Dol) composites. *Mater. Plast* **2016**, *53*, 61–64.
25. Saidi, M.A.A.; Mazlan, F.S.; Hassan, A.; Rashid, R.A.; Rahmat, A.R. Flammability, thermal and mechanical properties of polybutylene terephthalate/dolomite composites. *J. Phys. Sci.* **2019**, *30*, 175–189. [CrossRef]
26. Adesakin, A.O.; Ajayi, O.O.; Imosili, P.E.; Attahdaniel, B.E.; Olusunle, S.O.O. Characterization and Evaluation of Mechanical Properties of Dolomite as Filler in Polyester. *Chem. Mater. Res.* **2013**, *3*, 36–40.
27. Osman, A.F.; Edwards, G.; Martin, D. Effects of Processing Method and Nanofiller Size on Mechanical Properties of Biomedical Thermoplastic Polyurethane (TPU) Nanocomposites. *Adv. Mat. Res.* **2014**, *911*, 115–119. [CrossRef]
28. Li, Z.; Reimer, C.; Wang, T.; Mohanty, A.K.; Misra, M. Thermal and Mechanical Properties of the Biocomposites of Miscanthus Biocarbon and Poly(3-Hydroxybutyrate-co-3-Hydroxyvalerate) (PHBV). *Polymers* **2020**, *12*, 1300. [CrossRef] [PubMed]

polymers

MDPI

Article

Synthesis, Characterization, and Application of Carboxymethyl Cellulose from Asparagus Stalk End

Warinporn Klunklin [1], Kittisak Jantanasakulwong [1,2,3], Yuthana Phimolsiripol [1,2,3], Noppol Leksawasdi [1,2,3], Phisit Seesuriyachan [1,2], Thanongsak Chaiyaso [1,2], Chayatip Insomphun [1,2], Suphat Phongthai [1,2], Pensak Jantrawut [3,4], Sarana Rose Sommano [3,5], Winita Punyodom [3,6], Alissara Reungsang [7,8,9], Thi Minh Phuong Ngo [10] and Pornchai Rachtanapun [1,2,3,*]

1 School of Agro-Industry, Faculty of Agro-Industry, Chiang Mai University, Chiang Mai 50100, Thailand; warinporn.k@cmu.ac.th (W.K.); jantanasakulwong.k@gmail.com (K.J.); yuthana.p@cmu.ac.th (Y.P.); noppol@hotmail.com (N.L.); phisit.s@cmu.ac.th (P.S.); thachaiyaso@hotmail.com (T.C.); chayatip@yahoo.com (C.I.); suphat.phongthai@cmu.ac.th (S.P.)
2 The Cluster of Agro Bio-Circular-Green Industry (Agro BCG), Chiang Mai University, Chiang Mai 50100, Thailand
3 Center of Excellence in Materials Science and Technology, Chiang Mai University, Chiang Mai 50200, Thailand; pensak.amuamu@gmail.com (P.J.); sarana.s@cmu.ac.th (S.R.S.); winitacmu@gmail.com (W.P.)
4 Department of Pharmaceutical Sciences, Faculty of Pharmacy, Chiang Mai University, Chiang Mai 50200, Thailand
5 Plant Bioactive Compound Laboratory (BAC), Department of Plant and Soil Sciences, Faculty of Agriculture, Chiang Mai University, Chiang Mai 50200, Thailand
6 Department of Chemistry, Faculty of Science, Chiang Mai University, Chiang Mai 50200, Thailand
7 Department of Biotechnology, Faculty of Technology, Khon Kaen University, Khon Kaen 40002, Thailand; alissara@kku.ac.th
8 Research Group for Development of Microbial Hydrogen Production Process, Khon Kaen University, Khon Kaen 40002, Thailand
9 Academy of Science, Royal Society of Thailand, Bangkok 10300, Thailand
10 Department of Chemical Technology and Environment, the University of Danang—University of Technology and Education, Danang 550000, Vietnam; ntmphuong@ute.udn.vn
* Correspondence: pornchai.r@cmu.ac.th

Citation: Klunklin, W.; Jantanasakulwong, K.; Phimolsiripol, Y.; Leksawasdi, N.; Seesuriyachan, P.; Chaiyaso, T.; Insomphun, C.; Phongthai, S.; Jantrawut, P.; Sommano, S.R.; et al. Synthesis, Characterization, and Application of Carboxymethyl Cellulose from Asparagus Stalk End. *Polymers* **2021**, *13*, 81. https://dx.doi.org/polym13010081

Received: 8 November 2020
Accepted: 21 December 2020
Published: 28 December 2020

Abstract: Cellulose from *Asparagus officinalis* stalk end was extracted and synthesized to carboxymethyl cellulose (CMC_{as}) using monochloroacetic acid (MCA) via carboxymethylation reaction with various sodium hydroxide (NaOH) concentrations starting from 20% to 60%. The cellulose and CMC_{as} were characterized by the physical properties, Fourier Transform Infrared spectroscopy (FTIR), Differential scanning calorimetry (DSC), Scanning electron microscopy (SEM) and X-ray diffraction (XRD). In addition, mechanical properties of CMC_{as} films were also investigated. The optimum condition for producing CMC_{as} was found to be 30% of NaOH concentration for the carboxymethylation reaction, which provided the highest percent yield of CMC_{as} at 44.04% with the highest degree of substitution (DS) at 0.98. The melting point of CMC_{as} decreased with increasing NaOH concentrations. Crystallinity of CMC_{as} was significantly deformed ($p < 0.05$) after synthesis at a high concentration. The L^* value of the CMC_{as} was significantly lower at a high NaOH concentration compared to the cellulose. The highest tensile strength (44.59 MPa) was found in CMC_{as} film synthesized with 40% of NaOH concentration and the highest percent elongation at break (24.99%) was obtained in CMC_{as} film treated with 30% of NaOH concentration. The applications of asparagus stalk end are as biomaterials in drug delivery system, tissue engineering, coating, and food packaging.

Keywords: agricultural waste; asparagus; biopolymer; carboxymethyl cellulose; CMC; degree of substitution; DS; cellulose extraction

1. Introduction

Asparagus (*Asparagus officinalis* L.) is a nutritious and perennial vegetable continually used as antifungal activities, anticancer and anti-inflammatory herbal medicine in Asia. In the processing of asparagus, the spears which are 2–3% of total weight of the asparagus are typically processed into three types of products: canned, fresh, and frozen [1]. The residues from the processing were used for animal feeding and produced low-value products due to the high content of cellulose [2]. Therefore, tons of asparagus stalk end are the agriculture by-product which can cause environmental pollution [2]. With abundant cellulose and photochemical properties of cellulose in asparagus by-products, it acts as a good source of new value-added products [1], including biological compounds and functions [3,4], a dietary fiber [5] and nanocellulose [6]. Agricultural by-products from asparagus are rich source of celluloses which can be isolated from asparagus stalk end. A little work has been carried out to extract the cellulose from the asparagus; however, the previous works have not comprehensively considered to synthesis and apply CMC from asparagus stalk end as a film or coating to a food industry.

Cellulose is the most abundant polymer present in the primary cell wall of plants, algae and the oomycetes [6,7]. The abundant cellulose consists of repeated links of β–D-glucopyranose which can convert into high value cellulose esters and ethers [7]. The cellulose is insoluble in water which limits many applications. To increase the utility of cellulose, the carboxymethylation process of cellulose is able to syntheses water-soluble cellulose derivatives called carboxymethylcellulose (CMC) [8]. The forecast for global carboxymethyl cellulose market is estimated to reach approximately USD 1.86 billion by 2025. Food market generated the highest revenue of the global CMC market in 2016 with a market share of 33.7% [9]. The productions of CMC synthesized from agricultural waste have been studied such as sago waste [10], *Mimosa pigra* peel [11], durian rind [8], cotton waste [12], corn husk [13], corncob [14], grapefruit peel [15], rice stubble [16], papaya peel [17], *Juncus* plant stems [18], and mulberry wastepaper [19].

Thus, CMC can be versatile applied in various field such as drug delivery carrier [20], wound healing application [21], active corrosion protection [22], binder in ceramics [23], composited in hydrogel film [24] and hydrogel food packaging [25], fresh fruit and vegetable coatings [26,27] and hydrocolloid in noodles, bakery products and other foods [28,29] and it can produce composite films. CMC films exhibit many worthwhile attributes than many other biopolymers such as biodegradability, ease of production, forming a stable cross-linked matrix for packaging, etc. However, no research has presented the production of CMC powder from asparagus stalk end (CMC_{as}) and its application such as packaging film so far.

Therefore, this study aimed to determine the effect of NaOH concentration at 20–60% on percent yield, the degree of substitution (DS), thermal properties, chemical structure, viscosity, crystallinity, morphology of CMC_{as}. The effect of NaOH concentration mechanical properties (tensile strength and percentage elongation at break) and morphology of CMC_{as} films was also carried out.

2. Materials and Methods

2.1. Materials

Asparagus stalk end was purchased from Nong Ngulueam Sub-District (Nakhon Prathom, Thailand). All chemicals used in the preparation and analysis of synthesized CMC were analytical reagent (AR) grade or the equivalent. Absolute methanol, ethanol, and sodium silicate from Union science Co., Ltd. (Chiang Mai, Thailand), hydrogen peroxide from QReC^{TM} (Auckland, New Zealand), sodium hydroxide and glacial acetic acid from Lab-scan, isopropyl alcohol (IPA), Monochloroacetic acid from Sigma-Aldrich (Steinheim, Germany).

2.2. Materials Preparation

Two grams of asparagus stalk end was cut into small pieces and dried in an oven 105 °C for 6 h. Then dried asparagus was weighed to calculate the percent dryness as described in Equation (1). Dried asparagus stalk end was grounded by using a hammer mill (Armfield, Ringwood, Hampshire, UK) to less than 1 mm. The dried powder was then stored in polypropylene (PE) bags at ambient temperature until used. The percent dryness was calculated by the following Equation (1)

$$\%Dryness = \frac{Weight\ of\ cellulose\ without\ moisture\ content}{Weight\ of\ cellulose\ content} \times 100 \quad (1)$$

2.3. Extraction of Cellulose from Asparagus Stalk End

Cellulose from dried asparagus stalk end powder was extracted according to the method of [11,16]. Briefly, the dried powder of asparagus stalk end was extracted with a ratio of cellulose to 10% (w/v) NaOH solution at 1:20 (w/v) treated at 100 °C for 3 h. The black slurry was filtered and rinsed with cold water until a neutral pH of rinsed water was obtained. To obtain the cellulose fiber, the washed fiber residue was dried in an oven at 55 °C for 24 h. Hydrogen peroxide and sodium silicate were used to bleach the cellulose fiber. The bleached cellulose was then grounded by using the hammer mill size 70 mash (Armfield, Ringwood, Hampshire, UK). The bleached cellulose powder was kept in PE bags at ambient temperature until used.

2.4. Carboxymethyl Cellulose (CMC) Synthesized from Asparagus Stalk End

50 mL of various concentrations of NaOH at 20, 30, 40, 50, and 60% (w/v) and 350 mL of isopropanol (IPA) were blended with 15 g of cellulose powder extracted from asparagus stalk end and this was mixed for 30 min. Subsequently, the carboxymethylation reaction was synthesized according to the method of [11]. The final CMC_{as} product was in powder form. The percent yield was calculated by the following Equation (2) [11]:

$$Yield\ of\ CMC\ (\%) = \frac{Weight\ of\ CMC\ (g)}{Weight\ of\ cellulose\ (g)} \times 100 \quad (2)$$

2.5. Determination of the Degree of Substitution (DS) of CMC_{as}

The degree of substitution (DS) of CMC_{as} presents the average number of hydroxyl group replaced by carboxymethyl and sodium carboxymethyl groups at C2, 3, and 6 in the cellulose structure. The DS of CMC_{as} was evaluated by the USP XXIII method for Crosscarmellose sodium which are titration and residue on combustion [11,23]. Calculation of the DS was showed in the following Equation (3);

$$DS = A + S \quad (3)$$

where A is the DS of carboxymethyl acid and S is the DS of sodium carboxymethyl which A and S were calculated using Equations (4) and (5):

$$A = \frac{1150M\ content}{(7120 - 412M - 80C)Content} \quad (4)$$

$$S = \frac{(162 + 58A)C\ content}{(7120 - 80C)content} \quad (5)$$

where M is consumption of the titration to end point (mEq) and C is the amount of ash remained after ignition (%).

2.6. The Determination of Percentage of Residue on Ignition

A crucible was dried in an oven at 100 °C for 1 h and transferred to a desiccator until the weight reached at an accurate value (Scale, AR3130, Ohaus Corp. Pine, NJ, USA) [23].

CMC_{as} was added into a crucible. The crucible containing CMC_{as} were ignited at 400 °C using a kiln (Carbolite, CWF1100, Scientific Promotion, Sheffield, UK) for approximately 1 to 1.5 h and placed into the desiccator to obtain black residue. Sufficient sulfuric acid was added to moisten the entire residues and heated up until the gas fumes (white smoke) was utterly disappeared. The crucible with the black residue was ignited at 800 ± 25 °C to produce white residue and placed in the desiccators to get a constant weight. The residue on ignition was presented in the percentage and calculated using Equation (6).

$$The\ percentage\ of\ residue = \frac{Weight\ of\ residue\ content}{Weight\ of\ CMC\ content} \times 100 \quad (6)$$

2.7. Fourier Transform Infrared Spectroscopy (FTIR)

The aldehyde groups of the cellulose from stalk asparagus and CMC_{as} were determined by using FTIR spectra (Bruker, Tensor 27, Billerica, MA, USA). 2 mg of dry sample was pressed into a pellet with KBr. Transmission level measured at the wavenumber range of 4000–400 cm^{-1} [28].

2.8. Viscosity of Cellulose and CMC_{as}

A Rapid Visco Analyzer (RVA-4, Newport Scientific, Warriewood, Australia) was used to determine the viscosity of samples. 1 g of cellulose or CMC_{as} sample was weight and dissolved in 25 mL of distilled water with stirring at 80 °C for 10 min—automatic stirring action set at 960 rpm for 10 s. The temperature of the sample was varied from 30, 40, 50 and 60 °C at 5 min intervals and held speed at 160 rpm until end of the test [11].

2.9. Thermal Conductivity Measurements

The thermal characterization of cellulose from asparagus stalk end and CMC_{as} were determined using Differential Scanning Calorimeter (DSC) (Perkin Elmer, Kitakyushu, Japan) according to the melting temperature determined from the thermogram. The determination was performed under a nitrogen atmosphere at a flow rate of 50 mL min^{-1}. Samples (10 mg) contained in aluminum pans were heated from 40 to 450 °C with a heating rate of 10 °C min^{-1} [11].

2.10. X-ray Diffraction (XRD)

The crystallinity of the samples was determined using X-ray diffraction patterns carried out by X-ray powder diffractometer (JEOL, JDX-80-30, Shimadzu, Kyoto, Japan) in the reflection mode. Prior to the test, the samples were dried in a hot air oven (Memmert, Büchenbach (Bavaria, Germany) at 105 °C for 3 h to produce a powder form. Scans were carried out in the range of scattering angle (2θ) from 10 to 60° with at a scan rate of 5°/min [13].

2.11. Scanning Electron Microscopy (SEM)

Scanning Electron Microscopy (SEM) (JEOL JSM-5910LV SEM; Tokyo, Japan) was used to analyze the surface morphology of cellulose from asparagus stalk end and CMC_{as}. The samples were analyzed through a large field detector. The acceleration voltage was used at 15 kV with 1500× original magnification [12].

2.12. Color Characteristics

The color characteristic of cellulose form asparagus stalk end and CMC_{as} was evaluated by using a Color Quest XE Spectrocolorimeter (Hunter Lab, Shen Zhen Wave Opto-electronics Technology Co., Ltd., Shenzhen, China) in order to express the CIELAB color as three values: L^* for the lightness from blackness (0) to whiteness (100), a^* from greenness (−) to redness (+), and b^* from blueness (−) to yellowness (+). The total color differences

(ΔE) take into account the comparisons between the L^*, a^* and b^* value of the sample and standard and calculated by the following Equation (7) [8]

$$\Delta E = \sqrt{(L^*_{Standard} - L^*_{Sample})^2 + (a^*_{Standard} - a^*_{Sample})^2 + (b^*_{Standard} - b^*_{Sample})^2} \quad (7)$$

The whiteness index (WI) was also calculated by the following Equation (8) to represent the degree of whiteness of samples [30].

$$WI = \sqrt{\left(100 - L^{*2}\right) + a^{*2} + b^{*2}} \quad (8)$$

2.13. Preparation of CMC_{as} Film

The film-forming solutions were prepared by dissolving 4 g of CMC_{as} in 180 mL of distilled water with a constantly stirred at 80 °C under magnetic stirring for 15 min. 15% (w/w) glycerol was added to the solution and stirred for additional 5 min. The solution was poured on acrylic plates (20 cm × 15 cm) and the dried at room temperature for 72 h. After that, the film was peeled from the plates and stored in polyethylene bags at ambient temperature [11].

2.14. Mechanical Properties of CMC_{as} Film

The thickness of CMC_{as} films was determined using a micrometer model GT-313-A (Gotech Testing Machine Inc., Taichung, Taiwan, China). The examined mechanical properties which were tensile strength (TS) and percent elongation at break (EB) of the CMC_{as} film were tested (10 measurement each) using a universal texturometer (Model 1000 HIK-S, UK) according to the standard procedure of ASTM D882-80a [31] with preconditioning for 24 h and determined at 27 ± 2 °C with a relative humidity (RH) of 65 ± 2% according to Thai industrial standard for oriented polypropylene film (TIS 949-2533). The rectangular CMC_{as} films were cut into 15 × 140 mm as test specimens. The specimens were carried out by using an initial grip separation distance of 100 mm and crosshead speed at 20 mm/min. The TS and EB were calculated by following Equations (9) and (10), respectively.

$$TS\ (MPa) = \frac{The\ maximum\ load\ (N)}{Weight\ of\ each\ film \times Thickness\ of\ each\ film} \quad (9)$$

$$EB\ (\%) = \frac{The\ length\ of\ the\ film\ rupture - The\ initial\ length\ of\ the\ film}{The\ initial\ length\ of\ the\ film} \quad (10)$$

2.15. Statistical Analysis

Statistic data were analyzed by a one-way analysis of variance (ANOVA) using SPSS software version 16.0 0 (SPSS, an IBM company, Chicago, IL, USA). Duncan's multiple range test was employed to evaluate significant differences among the treatments ($p < 0.05$). All measurements were analyzed in triplicated. The results were represented the mean values ± standard deviation. The figures present the standard deviation as the appropriate values. The error bars for some data points overlap the mean values.

3. Results and Discussion

3.1. Percent Yield of Carboxymethyl Cellulose from Asparagus Stalk End (CMC_{as})

The percent yield of CMC_{as} synthesized with various NaOH concentrations at 20, 30, 40, 50 and 60% (w/v) is shown in Figure 1. The percent yield of CMC_{as} increased when increasing of NaOH concentration from 20% to 30% (w/v) and then decreased with further increasing NaOH concentration at over 30% (w/v). The resulted phenomenon might be occurred due to a limitation of sodium monochloroacetate (NaMCA) as etherifying agents for substituting cellulose. Moreover, the NaOH concentration was not high enough to complete for converting the cellulose into alkali cellulose [32]. This result was similar to the work of synthesis carboxymethyl cellulose from durian rind [8].

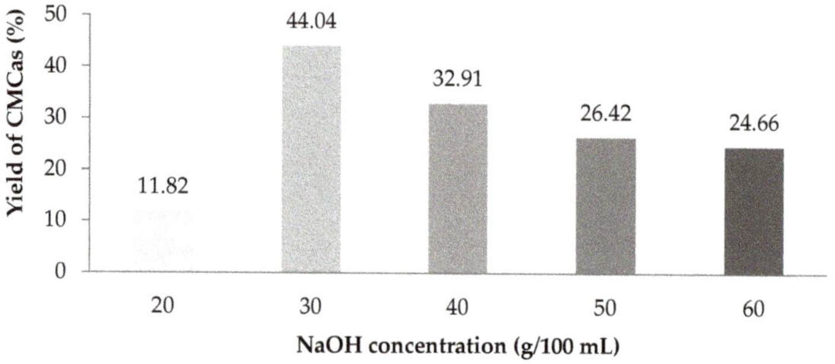

Figure 1. Percent yield of carboxymethyl cellulose from asparagus stalk end (CMC_{as}).

3.2. *Degree of Substitution (DS) of CMC_{as}*

The effect of various concentrations of NaOH on DS of CMC_{as} was carried out shown in Figure 2. The DS of CMC_{as} was in between 0.49 and 0.98 which reached the highest value at the concentration of NaOH at 30% (w/v). However, the DS of CMC_{as} was reduced at higher concentration of NaOH at 40–60% (w/v) ranged from 0.83–0.49. This phenomenon can be explained by investigating the carboxymethylation process, where two reaction occur concurrently. The first reaction involved a cellulose hydroxyl reacting with sodium monochloroacetate (NaMCA) in the presence of NaOH to obtain CMC_n shown in Equations (11) and (12).

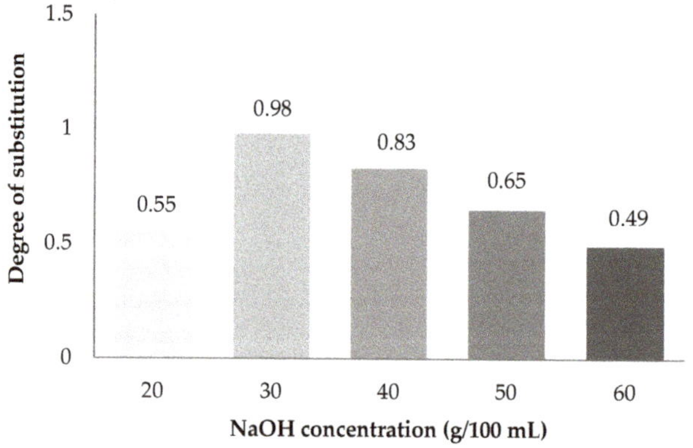

Figure 2. Effect of NaOH on degree of substitution of CMC_{as}.

$$\text{R-OH} + \text{NaOH} \rightarrow \text{R-ONa} + H_2O \quad (11)$$

$$\text{R-ONa} + \text{Cl-CH}_2\text{-COONa} \rightarrow \text{R-O-CH}_2\text{-COONa} + \text{NaCl} \quad (12)$$

The second reaction involves NaOH reacting with NaMCA to form sodium glycolate as by-product shown in Equation (13) [33].

$$\text{NaOH} + \text{Cl-CH}_2\text{COONa} \rightarrow \text{HO-CH}_2\text{COONa} + \text{NaCl} \quad (13)$$

The second reaction overwhelms the first reaction together with a strong alkaline concentration. If the alkaline level in the second reaction is too high, a side reaction will form a high level of sodium glycolate as a by-product, thus lowering the DS. This result was agreed with the result found by [11] who studied CMC from *Mimosa pigra* peel. The degradation was taken place due to a high concentration of NaOH [11]. Moreover, similar results have been reported in [8]. The maximum DS value (0.98) of CMC from durian rind was also got from the NaOH concentration at 30% and this was also related to the trend in percent yield presented in Figure 1.

3.3. Fourier Transform Infrared Spectroscopy (FTIR) of Cellulose from Asparagus Stalk End and CMC_{as}

FTIR was used to evaluate functional groups changes in the cellulose from asparagus stalk end and CMC_{as} structures. The substitution reaction of CMC_{as} during the carboxymethylation was confirmed by FTIR [28]. Cellulose and each CMC_{as} have similar functional groups with same absorption bands such as hydroxyl group (–OH stretching) at 3200–3600 cm^{-1}, C–H stretching vibration at 3000 cm^{-1}, carbonyl group (C=O stretching) at 1600 cm^{-1}, hydrocarbon groups (–CH_2 scissoring) at 1450 cm^{-1}, and ether groups (–O– stretching) at 1000–1200 cm^{-1} [34]. The FTIR spectra of the cellulose and CMC_{as} synthesized with various NaOH concentrations are shown in Figure 3. The carbonyl group (C=O), methyl group (–CH_2) and ether group (–O–) notably increased in the CMC_{as} sample; however, the absorption band of hydroxyl group (–OH) reduced when compared to those cellulose samples (Figure 3). This result confirmed that the carboxymethylation had been substituted on cellulose molecules [32,34].

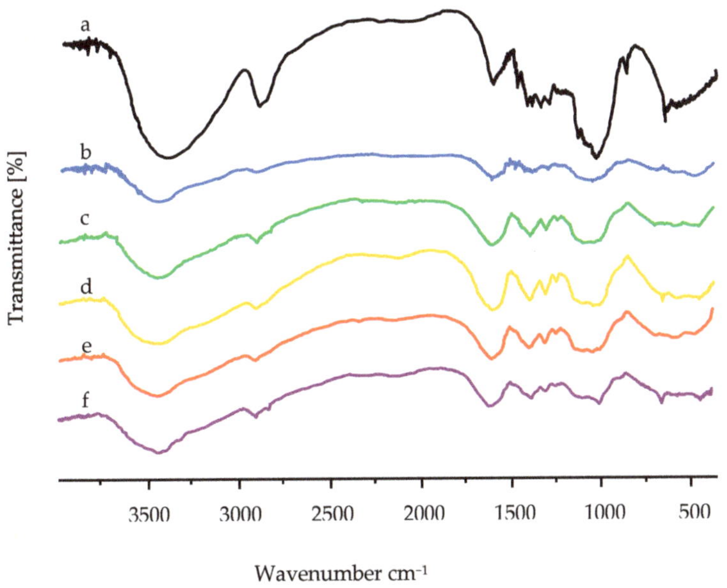

Figure 3. FTIR spectra of (**a**) cellulose, (**b**) CMC_{as} synthesized with 20% (*w*/*v*) NaOH concentration, (**c**) CMC_{as} synthesized with 30% (*w*/*v*) NaOH concentration, (**d**) CMC_{as} synthesized with 40% (*w*/*v*) NaOH concentration, (**e**) CMC_{as} synthesized with 50% (*w*/*v*) NaOH concentration and (**f**) CMC_{as} synthesized with 60% (*w*/*v*) NaOH concentration.

3.4. Effect of Various NaOH Concentrations on Viscosity of CMC_{as}

The relationship between different NaOH concentrations and viscosity of CMC_{as} solution is shown in Figure 4. The viscosity of the CMC_{as} solution (1 g/100 mL) reduced with increase in the temperature. Increasing temperature plays a role in reducing the cohesive forces while simultaneously increasing the rate of molecular interchange, causing the lower viscosity [11]. According to the literature, the viscosity of CMC_{as} solution affected by NaOH concentrations was changed in accordance with DS value at the same temperature [33,35]. The CMC_{as} synthesized with 20% (*w*/*v*) NaOH concentration was the highest in viscosity due to the swelling and gelatinization of cellulose comparable to previous studies [11]. As far as NaOH concentration rose above 20%, the viscosity fell down. The decreasing of viscosity at a higher NaOH concentration was because of the degradation of CMC polymer led to a lower DS value that provides a small number of hydrophilic groups reducing the ability of the polymer to bond among water molecules [32,36]. An increase of the DS value also increased the CMC's ability to immobilize water in a system [36]. Nevertheless, the viscosity of CMC depends on many influencing factors such as solution concentration [32], pH value, NaOH concentrations [36], and temperatures [11].

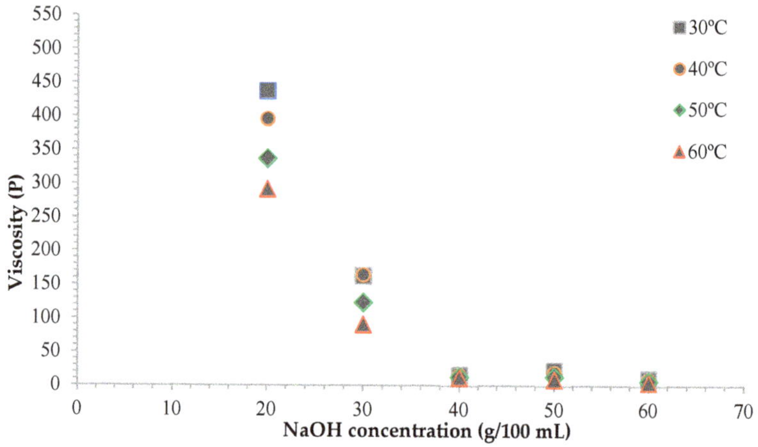

Figure 4. Effect of various NaOH concentrations on viscosity of CMC_{as}.

3.5. *Effect of Various NaOH Concentrations on Thermal Properties of CMC_{as}*

The effect of various NaOH concentrations on the thermal property of cellulose from asparagus stalk end and CMC_{as} are shown in Figure 5. The melting temperature (T_m) of cellulose and CMC_{as} synthesized with 20, 30, 40, 50, and 60 % (*w*/*v*) NaOH concentrations were 178.79, 206.61, 193.13, 182.34, 178.34 and 161.21 °C, respectively. The highest T_m of CMC_{as} increased when 20% (*w*/*v*) NaOH concentration was employed. The T_m at 20–30% (*w*/*v*) of NaOH concentrations were higher than T_m of cellulose due to the substituent of carboxymethyl groups influenced on increasing ionic character and intermolecular bonds between the polymer chains [37]. The T_m at 40–50% (*w*/*v*) of NaOH concentrations decreased due to the melting temperature of CMC_{as} slightly decreased as increasing the concentration of NaOH because of the increase in alkalization reaction together with a high substitution of carboxymethyl group. Moreover, the decrease in T_m was caused by an increase in number of the carbon atoms producing carbon skeleton, after a rapid quenching cooling with rate of 80 °C/min from the isotropic state at 200 °C [38]. As the T_m of 60% (*w*/*v*) of NaOH concentration decreased, side reaction was in the majority with sodium glycolate forming as a by-product and chain breaking of CMC_{as} polymer similar to the results presented in *Mimosa pigra* peel [11].

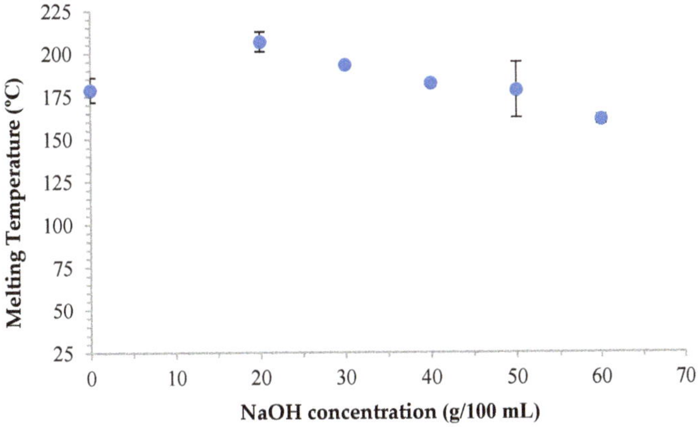

Figure 5. Thermal property of cellulose from asparagus stalk end and CMC_{as} synthesized with various NaOH concentrations.

3.6. X-ray Diffraction (XRD) of Cellulose from Asparagus Stalk End and CMC_{as}

The changes on the structure of cellulose from asparagus stalk end and CMC_{as} were evaluated by using XRD shown in Figure 6. The strength of hydrogen bonding and crystallinity contribute to the microstructure of CMC_{as} material. All CMC_{as} samples were less value in a peak of intensity (au) compared to cellulose from asparagus stalk end. Decrease in crystallinity index of cellulose may be due to the reorganization or cleavage of molecular according to alkalization of NaOH solution [32]. The increased aperture among cellulose polymer molecules was also caused by the substitution of monochloroacetic acid (MCA) molecules into the hydroxyl group of cellulose macromolecules easier than the cellulose without treating with an alkali solution [23].

These results are accord with results of various studies which were tested in cavendish banana cellulose [32] and durian rind [8]. The crystallinity index was also decreased when alkalizing by 15% (*w*/*v*) NaOH [32]. Thus, the crystallinity of CMC_{as} decreased by the effect of alkali solution prior to the carboxymethylation reaction on cellulose structure.

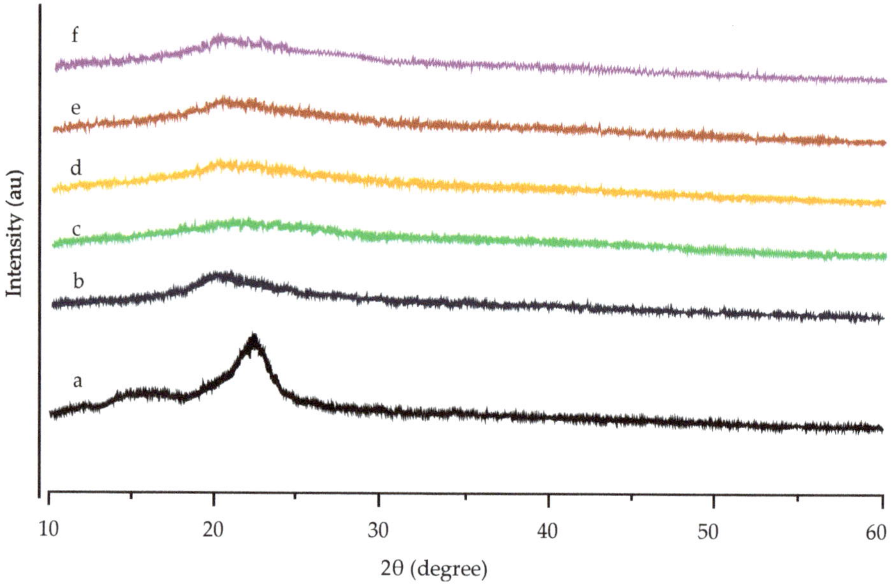

Figure 6. X-ray diffractograms of (**a**) cellulose from asparagus stalk end and CMC_{as} powder: with (**b**) 20% (*w*/*v*) NaOH, (**c**) 30% (*w*/*v*) NaOH, (**d**) 40% (*w*/*v*) NaOH, (**e**) 50% (*w*/*v*) NaOH and (**f**) 60% (*w*/*v*) NaOH.

3.7. Morphology of Cellulose from Asparagus Stalk End and CMC_{as}

Morphology of cellulose from asparagus stalk end and CMC_{as} powder with different NaOH concentrations were characterized by SEM with an acceleration voltage of 15 kV and 1500× original magnification (Figure 7a–f). The surface of cellulose from asparagus stalk end (Figure 7a) showed a smooth without pores or cracks and with a dense structure; however, the appearance of small fibers has been found in the cellulose. The increase of NaOH concentration affected changes in microstructure of the samples. The morphology of CMC_{as} powder treated with 20% of NaOH concentration emerged small fiber with minimal damage (Figure 7b). By increasing the content of NaOH from 30% to 60% (*w*/*v*), the surface uniformity decreased. The surface of CMC_{as} powder mixed with NaOH concentration at 30% started to peel without cracking as shown in Figure 7c. The surface of CMC_{as} powder has been begun to crack and deform when the powder treated with 40% of NaOH concentration (Figure 7d), due to the degradation of CMC polymer chain. Increasing the NaOH concentration at 50% (Figure 7e), the morphology of CMC_{as} powder increased irregularity surface with more collapsed and indented. The surface was more roughly and totally deformed when CMC_{as} treated with a higher NaOH concentration. Thus, this phenomenon could be inferred that CMC_{as} was synthesized with a high NaOH concentration causing a damage in surface area of cellulose powder. This result was comparable with a synthesized carboxymethyl from rice starch [23] and cassava starch [39]. Alkaline solution probably reduced the strength of structure, loss the crystallinity and changed the stability of the molecular organization of the cellulose causing the etherifying agents to have more access to the molecules for the carboxymethylation processes [36]. Consequently, this result was according to DS value.

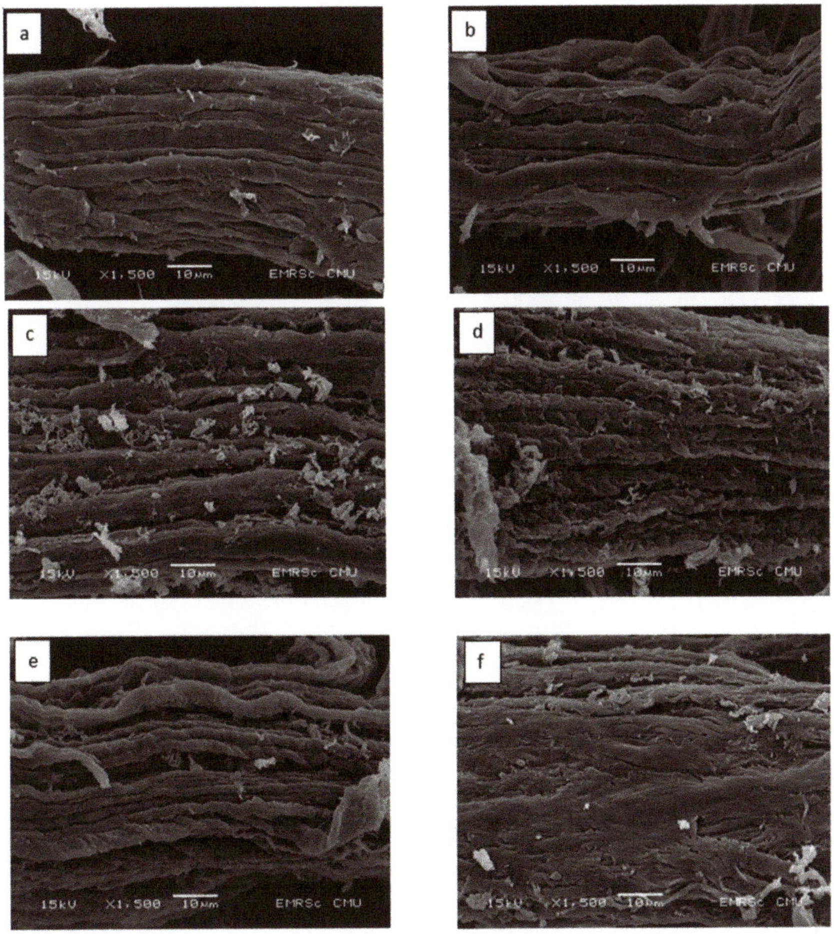

2θ (degree)

Figure 7. Scanning electron micrographs of (**a**) cellulose from asparagus stalk end and cmc_{as} powder: with (**b**) 20% (w/v) naoh, (**c**) 30% (w/v) naoh, (**d**) 40% (w/v) naoh, (**e**) 50% (w/v) naoh and (**f**) 60% (w/v) naoh. The acceleration voltage was 15 kv under low with 1500×.

3.8. Color of Cellulose from Asparagus Stalk End and CMC_{as}

Color measurement was carried out to evaluate the color formation caused by carboxymethylation reaction. The color values of CMC_{as} were decreasing in L^* value as NaOH concentrations increased from 20 to 60% (w/v) of NaOH concentration. The a* and b* values showed the highest increased in CMC_{as} treated with 40% (w/v) of NaOH concentration. Moreover, an increase in NaOH concentration up to 40% decreased the yellowness of CMC_{as} probably due to the first reaction of carboxymethylation (Equation (12)) to produce CMC or sodium glycolate [36]. At higher NaOH concentration (50 and 60%, w/v), the decreasing of all color values was occurred. Therefore, the carboxymethylation reaction might be the reason the color values of cellulose and CMC_{as} has been changed [35]. The ΔE of the cellulose and CMC_{as} had similar trend compared to a* and b* values. These results are akin to the results from the previous studies [23]. The WI of the cellulose and CMC_{as} had same trend with L^* value (Table 1).

Table 1. Color values of cellulose and CMC_{as} synthesized with various NaOH concentration.

Type of Sample	*L**	*a**	*b**	*WI*	*YI*	*ΔE*	H_{ab}
Cellulose	71.51 ± 0.37^{b}	3.01 ± 0.05^{c}	21.31 ± 0.26^{b}	61.83 ± 0.33^{b}	31.21 ± 1.89^{c}	29.62 ± 0.24^{ns}	5.87 ± 0.17^{a}
20 g/100 mL NaOH-CMC_{as}	73.88 ± 0.45^{a}	2.69 ± 0.28^{d}	20.77 ± 0.50^{c}	64.54 ± 0.55^{a}	45.02 ± 1.05^{a}	37.46 ± 21.89	6.33 ± 0.90^{a}
30 g/100 mL NaOH-CMC_{as}	71.78 ± 0.74^{b}	$2.87 \pm 0.08^{c,d}$	21.70 ± 0.38^{b}	61.99 ± 0.62^{b}	45.02 ± 1.05^{a}	29.67 ± 0.35	6.31 ± 0.10^{a}
40 g/100 mL NaOH-CMC_{as}	$68.51 \pm 0.44^{c,d}$	3.69 ± 0.25^{b}	22.78 ± 0.17^{a}	58.52 ± 0.42^{d}	45.20 ± 1.01^{a}	32.90 ± 0.32	$4.94 \pm 0.14^{c,d}$
50 g/100 mL NaOH-CMC_{as}	69.30 ± 0.47^{c}	3.54 ± 0.14^{b}	21.56 ± 0.29^{b}	59.79 ± 0.35^{c}	41.52 ± 1.66^{b}	31.50 ± 0.44	5.06 ± 0.17^{b}
60 g/100 mL NaOH-CMC_{as}	67.97 ± 1.33^{d}	4.00 ± 0.16^{a}	21.28 ± 0.31^{b}	58.34 ± 1.25^{d}	40.17 ± 1.59^{b}	32.43 ± 0.94	4.42 ± 0.22^{d}

Mean ± Standard deviation values within a column in the same group followed by the different letters (a–d) are significantly different ($p < 0.05$). ns = not significant difference.

3.9. Mechanical Properties of CMC_{as} Films

The tensile strength (TS) and percentage of elongation at break (EB) of CMC_{as} films synthesized with various NaOH concentrations were determined by using the ASTM Standard Method D882-80a (Table 2). The TS value increased gradually with increasing NaOH concentration up to 40% (*w*/*v*) of NaOH. The TS value was correlated with an increasing DS value due to the substituted carboxymethyl groups in anhydroglucopyranose unit affecting an increase in intermolecular force among the polymer chains and the ionic character [32]. Nevertheless, the TS value of CMC_{as} films started to decline at a high NaOH concentration due to the polymer degradation and formation of by-product generated from sodium glycolate increased the reduction of CMC content and thus decreasing the intermolecular forces.

The EB of CMC_{as} films also increased with increasing NaOH concentrations up to 30% (*w*/*v*) of NaOH and drop down after that. At high NaOH concentrations, the crystallinity of the cellulose structure decreased with increasing a flexibility. In addition, CMC_{as} films has been obtained the lower flexibility at 60% (*w*/*v*) of NaOH concentrations because of an occurrence of hydrolysis reaction in cellulose chain [36].

Table 2. Mechanical properties for CMCas films synthesized with various NaOH concentrations.

Type of Film	Tensile Strength (MPa)	Elongation at Break (%)
20 g/100 mL NaOH-CMC_{as}	36.30 ± 1.32^{c}	13.00 ± 7.22^{c}
30 g/100 mL NaOH-CMC_{as}	41.78 ± 3.28^{b}	24.99 ± 3.79^{a}
40 g/100 mL NaOH-CMC_{as}	44.59 ± 1.73^{a}	17.32 ± 4.21^{b}
50 g/100 mL NaOH-CMC_{as}	28.97 ± 2.06^{d}	9.52 ± 2.57^{c}
60 g/100 mL NaOH-CMC_{as}	27.55 ± 1.72^{d}	2.38 ± 0.88^{d}

Mean ± Standard deviation values within a column in the same group, followed by the different letters (a–d) are significantly different ($p < 0.05$).

4. Conclusions

This study revealed that *Asparagus officinalis* stalk end could be favorably used as a raw material to produce carboxymethyl cellulose. The alkali solution is the main parameter that affected all characteristics of CMC_{as} synthesized with various NaOH concentrations. The percent yield of CMC_{as} initially increased with NaOH concentrations increase up to 30% (*w*/*v*) and then gradually decreased with further NaOH concentrations increase related with the DS value of CMC_{as}. The lightness (L^*) of CMC_{as} has been changed after treated with increasing NaOH concentration. The crystallinity of CMC_{as} was found to decrease after synthesized with alkali solution. The highest tensile strength and elongation at break values of CMC_{as} films were found in 40% (*w*/*v*) NaOH-synthesized CMC_{as} films and 30% (*w*/*v*) NaOH-synthesized CMC_{as} films, respectively.

Author Contributions: Conceived and planned the whole experiments, P.R.; Conceptualization, Investigation and Methodology, P.R.; wrote the manuscript with input from all authors, W.K., K.J., N.L., P.S., T.C., C.I., S.R.S., T.M.P.N. and P.R.; analyzed the data, W.K., S.P., P.J., W.P. and A.R.; contributed reagents and materials, Y.P. and P.R.; Validation, P.R.; supervised the project, P.R. All authors have read and agreed to the published version of the manuscript.

Funding: The present study was partially supported by TRF Senior Research Scholar (Grant No. RTA6280001). This research work was partially supported by Chiang Mai University.

Acknowledgments: We wish to thank the Center of Excellence in Materials Science and Technology, Chiang Mai University for financial support under the administration of the Materials Science Research Center, Faculty of Science, and Chiang Mai University. We also acknowledge partial financial supports and/or in-kind assistance from Program Management Unit—Brain Power (PMUB), the Office of National Higher Education Science Research and Innovation Policy Council (NXPO) in Global Partnership Project, Basic research fund, Thailand Science Research and Innovation (TSRI), Chiang Mai University (CMU), Faculty of Science (CoE64-P001) as well as Faculty of Agro-Industry (CMU-8392(10)/COE64), CMU.

Conflicts of Interest: The authors declare that there are no conflicts of interest regarding the publication of this paper.

References

1. Nindo, C.; Sun, T.; Wang, S.; Tang, J.; Powers, J. Evaluation of drying technologies for retention of physical quality and antioxidants in asparagus (*Asparagus officinalis*, L.). *LWT Food Sci. Technol.* **2003**, *36*, 507–516. [CrossRef]
2. Huang, Y.; Kennedy, J.; Peng., Y.; Wang, X.; Yuan, X.; Zhao, B.; Zhao, Q. Optimization of ultrasonic circulating extraction of polysaccharides from *Asparagus officinalis* using response surface methodology. *Int. J. Biol. Macromol.* **2011**, *49*, 181–187.
3. Guo, Q.; Wang, N.; Liu, H.; Li, Z.; Lu, L.; Wang, C. The bioactive compounds and biological functions of Asparagus officinalis L.—A review. *J. Funct. Foods* **2020**, *65*, 103727. [CrossRef]
4. Noperi-Mosqueda, L.C.; López-Moreno, F.J.; Navarro-León, E.; Sánchez, E.; Blasco, B.; Moreno, D.A.; Soriano, T.; Ruiz, J.M. Effects of asparagus decline on nutrients and phenolic compounds, spear quality, and allelopathy. *Sci. Hortic.* **2020**, *261*, 109029. [CrossRef]
5. Fuentes-Alventosa, J.M.; Jaramillo-Carmona, S.; Rodríguez-Gutiérrez, G.; Rodríguez-Arcos, R.; Fernández-Bolaños, J.; Guillén-Bejarano, R.; Espejo-Calvo, J.A.; Jiménez-Araujo, A. Effect of the extraction method on phytochemical composition and antioxidant activity of high dietary fibre powders obtained from asparagus by-products. *Food Chem.* **2009**, *116*, 484–490. [CrossRef]
6. Wang, Q.; Yao, Q.; Liu, J.; Sun, J.; Zhu, Q.; Chen, H. Processing nanocellulose to bulk materials: A review. *Cellulose* **2019**, *26*, 7585–7617. [CrossRef]
7. Updegraff, D.M. Semimicro determination of cellulose in biological materials. *Anal. Biochem.* **1969**, *32*, 420–424. [CrossRef]
8. Rachtanapun, P.; Luangkamin, S.; Tanprasert, K.; Suriyatem, R. Carboxymethyl Cellulose Film from Durian Rind. *LWT Food Sci. Technol.* **2012**, *48*, 52–58. [CrossRef]
9. Grand View Research, Inc. *Carboxymethyl Cellulose Market Analysis By Application (Cosmetics & Pharmaceuticals, Food & Beverages, Oil & Gas, Paper & Board, Detergents), By Region, And Segment Forecasts, 2018–2025 (Report No GVR-1-68038-463-5)*; Grand View Research, Inc.: San Francisco, CA, USA, 2017. Available online: https://www.grandviewresearch.com/industry-analysis/carboxymethyl-cellulose-cmc- (accessed on 8 November 2020).
10. Pushpamalar, V.; Langford, S.J.; Ahmad, M.; Lim, Y.Y. Optimisation of Reaction Condition for Preparing Carboxymethyl Cellulose from Sago Waste. *Carbohydr. Polym.* **2006**, *64*, 312–318. [CrossRef]
11. Rachtanapun, P.; Rattanapanone, N. Synthesis and Characterization of Carboxymethyl Cellulose Powder and Films from *Mimosa Pigra* Peel. *J. Appl. Polym. Sci.* **2011**, *122*, 3218–3226. [CrossRef]
12. Haleem, N.; Arshad, M.; Shahid, M.; Tahir, M.A. Synthesis of carboxymethyl cellulose from waste of cotton ginning industry. *Carbohydr. Polym.* **2014**, *113*, 249–255. [CrossRef] [PubMed]
13. Mondal, I.H.; Yeasmin, M.S.; Rahman, S. Preparation of food grade carboxymethyl cellulose from corn husk agrowaste. *Int. J. Biol. Macromol.* **2015**, *79*, 144–150. [CrossRef] [PubMed]
14. Jia, F.; Liu, H.-J.; Zhang, G.-G. Preparation of Carboxymethyl Cellulose from Corncob. *Procedia Environ. Sci.* **2016**, *31*, 98–102. [CrossRef]
15. Karataş, M.; Arslan, N. Flow behaviours of cellulose and carboxymethyl cellulose from grapefruit peel. *Food Hydrocoll.* **2016**, *58*, 235–245. [CrossRef]
16. Rodsamran, P.; Sothornvit, R. Rice stubble as a new biopolymer source to produce carboxymethyl cellulose-blended films. *Carbohydr. Polym.* **2017**, *171*, 94–101. [CrossRef]
17. Rachtanapun, P. Blended Films of Carboxymethyl Cellulose from Papaya Peel/Corn Starch Film Blends. *Kasetsart J.* **2009**, *43*, 259–266.
18. Kassab, Z.; Syafri, E.; Tamraoui, Y.; Hannache, H.; Qaiss, A.E.K.; Achaby, M.E. Characteristics of sulfated and carboxylated cellulose nanocrystals extracted from Juncus plant stems. *Int. J. Biol. Macromol.* **2020**, *154*, 1419–1425. [CrossRef]
19. Rachtanapun, P.; Kumthai, S.; Mulkarat, N.; Pintajam, N.; Suriyatem, R. Value added of mulberry paper waste by carboxymethylation for preparation a packaging film. *IOP Conf. Ser. Mater. Sci. Eng.* **2015**, *87*, 012081. [CrossRef]
20. Inphonlek, S.; Sunintaboon, P.; Leonard, M.; Durand, A. Chitosan/carboxymethylcellulose-stabilized poly(lactide-co-glycolide) particles as bio-based drug delivery carriers. *Carbohydr. Polym.* **2020**, *242*, 116417. [CrossRef]
21. Koneru, A.; Dharmalingam, K.; Anandalaksh, R. Cellulose based nanocomposite hydrogel films consisting of sodium carboxymethylcellulose–grapefruit seed extract nanoparticles for potential wound healing applications. *Int. J. Biol. Macromol.* **2020**, *148*, 833–842. [CrossRef]
22. Calegari, F.; Da Silva, B.C.; Tedim, J.; Ferreira, M.; Berton, M.A.; Marino, C.E. Benzotriazole encapsulation in spray-dried carboxymethylcellulose microspheres for active corrosion protection of carbon steel. *Prog. Org. Coatings* **2020**, *138*, 105329. [CrossRef]
23. Rachtanapun, P.; Eitssayeam, S.; Pengpat, K. Study of Carboxymethyl Cellulose from Papaya Peels Binder in Ceramics. *Adv. Mater. Res.* **2010**, *93–94*, 17–21. [CrossRef]
24. Ghorpade, V.S.; Yadav, A.V.; Dias, R.J. Citric acid cross linked β-cyclodextrin/carboxymethyl cellulose hydrogel films for controlled delivery of poorly soluble drugs. *Carbohydr. Polym.* **2017**, *164*, 339–348. [CrossRef]
25. Gregorova, A.; Saha, N.; Kitano, T.; Saha, P. Hydrothermal effect and mechanical stress properties of carboxymethylcellulose based hydrogel food packaging. *Carbohydr. Polym.* **2015**, *117*, 559–568. [CrossRef] [PubMed]

26. Kowalczyk, D.; Kordowska-Wiater, M.; Złotek, U.; Skrzypek, T. Antifungal resistance and physicochemical attributes of apricots coated with potassium sorbate added carboxymethyl cellulose-based emulsion. *Int. J. Food Sci. Technol.* **2018**, *53*, 728–734. [CrossRef]
27. Kowalczyk, K.; Kordowska-Wiater, M.; Kałwa, K.; Skrzypek, T.; Sikora, M.; Łupina, K. Physiological, qualitative, and microbiological changes of minimally processed Brussels sprouts in response to coating with carboxymethyl cellulose/candelilla wax emulsion. *J. Food Process. Preserv.* **2019**, *43*, e14004. [CrossRef]
28. Tang, Y.; Yang, Y.; Wang, Q.; Tang, Y.; Li, F.; Zhao, J.; Zhang, Y.; Ming, J. Combined effect of carboxymethylcellulose and salt on structural properties of wheat gluten proteins. *Food Hydrocoll.* **2019**, *97*, 105189. [CrossRef]
29. Yuliarti, O.; Mei, K.H.; Ting, Z.K.X.; Yi, K.Y. Influence of combination carboxymethylcellulose and pectin on the stability of acidified milk drinks. *Food Hydrocoll.* **2019**, *89*, 216–223. [CrossRef]
30. Rhim, J.; Wu, Y.; Weller, C.; Schnepf, M. Physical character-istics of a composite film of soy protein isolate and propylene-glycol alginate. *J. Food Sci.* **1999**, *64*, 149–152. [CrossRef]
31. ASTM (D 882-80a). Standards test methods for tensile properties of thin plastic sheeting. In *Annual Book of ASTM Standard*; ASTM: West Conshohochen, PA, USA, 1995; pp. 182–190.
32. Adinugraha, M.P.; Marseno, D.W.; Hayadi. Synthesis and characterization of sodium carboxymethyl cellulose from cavendish banana pseudo stem (*Musa cavendishii* LAMBERT). *Carbohydr. Polym.* **2005**, *62*, 164–169. [CrossRef]
33. Kirk, R.E.; Othmer, D.F. *Cellulose in Encyclopedia of Chemical Technology*; Wiley: Hoboken, NJ, USA, 1967; Volume 4, pp. 593–683.
34. Célino, A.; Gonçalves, O.; Jacquemin, F.; Fréour, S. Qualitative and quantitative assessment of water sorption in natural fibres using ATR-FTIR spectroscopy. *Carbohydr. Polym.* **2014**, *101*, 163–170. [CrossRef] [PubMed]
35. Suriyatem, R.; Noikang, N.; Kankam, T.; Jantanasakulwong, K.; Leksawasdi, N.; Phimolsiripol, Y.; Insomphun, C.; Seesuriyachan, P.; Chaiyaso, T.; Jantrawut, P.; et al. Physical Properties of Carboxymethyl Cellulose from Palm Bunch and Bagasse Agricultural Wastes: Effect of Delignification with Hydrogen Peroxide. *Polymers* **2020**, *12*, 1505. [CrossRef] [PubMed]
36. Asl, S.A.; Mousavi, M.; Labbaf, M. Synthesis and characterization of carboxymethyl cellulose from sugarcane bagasse. *J. Food Process. Technol.* **2017**, *8*, 687.
37. El-sayed, S.; Mahmoud, K.H.; Fatah, A.A.; Hassen, A. DSC, TGA and dielectric properties of carboxymethyl cellulose/polyvinyl alcohol blends. *Phys. B Condens. Matter* **2011**, *406*, 4068–4076. [CrossRef]
38. Teramoto, Y. Functional Thermoplastic Materials from Derivatives of Cellulose and Related Structural Polysaccharides. *Molecules* **2015**, *20*, 5487–5527. [CrossRef]
39. Sangseethong, K.; Chatakanonda, P.; Wansuksri, R.; Sriroth, K. Influence of reaction parameters on carboxymethylation of rice starches with varying amylose contents. *Carbohydr. Polym.* **2015**, *115*, 186–192. [CrossRef]

polymers

MDPI

Article

The Effect of Halloysite Nanotubes on the Fire Retardancy Properties of Partially Biobased Polyamide 610

David Marset [1], Celia Dolza [1], Eduardo Fages [1], Eloi Gonga [1], Oscar Gutiérrez [1], Jaume Gomez-Caturla [2], Juan Ivorra-Martinez [2,*], Lourdes Sanchez-Nacher [2] and Luis Quiles-Carrillo [2,*]

1 Textile Industry Research Association (AITEX), Plaza Emilio Sala 1, 03801 Alcoy, Spain; dmarset@aitex.es (D.M.); cdolza@aitex.es (C.D.); EFages@aitex.es (E.F.); egonga@aitex.es (E.G.); ogutierrez@aitex.es (O.G.)

2 Technological Institute of Materials (ITM), Universitat Politècnica de València (UPV), Plaza Ferrándiz y Carbonell 1, 03801 Alcoy, Spain; jaugoca@epsa.upv.es (J.G.-C.); lsanchez@mcm.upv.es (L.S.-N.)

* Correspondence: juaivmar@doctor.upv.es (J.I.-M.); luiquic1@epsa.upv.es (L.Q.-C.); Tel.: +34-966-528-433 (L.Q.-C.)

Received: 15 December 2020; Accepted: 17 December 2020; Published: 19 December 2020

Abstract: The main objective of the work reported here was the analysis and evaluation of halloysite nanotubes (HNTs) as natural flame retardancy filler in partially biobased polyamide 610 (PA610), with 63% of carbon from natural sources. HNTs are naturally occurring clays with a nanotube-like shape. PA610 compounds containing 10%, 20%, and 30% HNT were obtained in a twin-screw co-rotating extruder. The resulting blends were injection molded to create standard samples for fire testing. The incorporation of the HNTs in the PA610 matrix leads to a reduction both in the optical density and a significant reduction in the number of toxic gases emitted during combustion. This improvement in fire properties is relevant in applications where fire safety is required. With regard to calorimetric cone results, the incorporation of 30% HNTs achieved a significant reduction in terms of the peak values obtained of the heat released rate (HRR), changing from 743 kW/m^2 to about 580 kW/m^2 and directly modifying the shape of the characteristic curve. This improvement in the heat released has produced a delay in the mass transfer of the volatile decomposition products, which are entrapped inside the HNTs' lumen, making it difficult for the sample to burn. However, in relation to the ignition time of the samples (TTI), the incorporation of HNTs reduces the ignition start time about 20 s. The results indicate that it is possible to obtain polymer formulations with a high renewable content such as PA610, and a natural occurring inorganic filler in the form of a nanotube, i.e., HNTs, with good flame retardancy properties in terms of toxicity, optical density and UL94 test.

Keywords: PA610; halloysite nanotubes (HNTs); nanocomposites; flame retardant; cone calorimeter

1. Introduction

The current social awareness about the environmental problems derived from the use of non-renewable polymers and additives is generating a great change in the industry. Moreover, governments are beginning to become concerned about this problem and are starting to develop legislation that favors environmental protection and the use of materials that reduce the harmful impact on nature [1,2]. In particular, a great effort has been made in recent decades to develop and use new materials and additives that are biodegradable and possess sustainable properties as well as a reduced carbon footprint by reducing greenhouse gases during their production [3–6].

In this context, until relatively recently, high performance polymers such as polyamides (PAs) were materials derived entirely from oil; however, new technologies and research have succeeded in obtaining them from monomers, both fully and partially renewable [7,8]. Monomers of biological origin include, for example, brazilic acid, sebacic acid, 1,4-diaminobutane (putrescine), and 1,5-diaminopentane (cadaverine) [9–12]. From these types of monomers, different kinds of polyamides can be obtained. It is always desirable to generate polymers with properties similar to those of petrochemical counterparts. This fact is vital for industry, because obtaining biobased polyamides (bio-PA) that behave similarly to PA6 and PA66 regarding stiffness, and similarly to PA12 regarding flexibility, is increasingly important for economic and ecological issues [13,14].

Currently, more than 6 million tons of polyamides are required annually, with growing demand [15]. In particular, polyamide 6 (PA6) and polyamide 66 (PA66) make up approximately 90% of the total polyamide use in the plastic industry [16]. However, as noted, the development of a "green" route for the production of biobased polyamides (bio-PA) has generated increasing interest due to the inevitable stoichiometric waste associated with the classic petrochemical production routes, which are commonly thought to cause global warming and other environmental problems [8,17]. For example, in some applications, PA6 can be replaced with a biobased variant called PA610, which possesses very similar properties, but it is more ductile as well as it exhibits high renewable content. Because of the fact that the dicarboxylic acid can readily be condensed with the petroleum-based 1.6-hexamethylenediamine (HMDA) obtained from butadiene, the material PA610, containing 60–63 natural content, may be obtained [18].

Within the polymer industry, polyamide 6 (PA6) is a highly relevant engineering polymer, which finds applicability in certain areas where high flame retardant and fire retardant properties are required [19]. The development of flame retardant additives that are attractive from an ecological point of view has become the focus of much effort. In particular, halogen-free flame retardants based on renewable sources have attracted great interest [20]. Many of these are phosphorus-based additives [21,22], such as aluminum hypophosphite (AlHP), with very good results for blends considering sustainable polymers such as polylactic acid [23] or polyvinyl alcohol [24], and others such as PA6 [25]. Other elements that are having large application as halogen-free retardants are ammonium polyphosphate (APP) or antimony trioxide, with great application at present in polyolefins [26,27]. Other materials like polyurethane foams, ammonium polyphosphate [28], or triphenyl phosphate (TPhP) [29,30] are often used as intumescent due to their low toxicity, those being halogen-free materials, and highly efficient. In other cases, flame-retardant additives containing elements other than phosphorus, such as expanded graphite [31], nanoclays, and nanosilicates [32,33] or graphite oxides [34], have been utilized. In this context, the search for this type of more natural additives has found elements such as halloysite nanotubes (HNTs) [35].

Blends containing halloysite nanotubes (HNTs) present great potential in the generation of natural fire retardancy polymeric materials distinguished by their high sustainability and low emission of toxic gases and fumes during combustion [36,37]. Normally, one of the main disadvantages of nanofillers is the low dispersion it presents in a polymer matrix, directly affecting both mechanical and flame-retardant properties [38,39]. However, because of their polar structure, HNTs can be efficiently dispersed in different polyamides [40,41]. Over the last two decades, nanocomposites based on inorganic clay minerals have attracted a lot of attention. In addition, due to their nanoscale structure, nanocomposites may exhibit significant improvements in aspects such as mechanical properties, reduced gas permeability, increased thermal stability, and improved flame retardancy compared to the properties of the polymer without these nanocomposites [23–26]. For this reason, the use of halloysite nanotubes is becoming more attractive. In this context, some authors have incorporated HNTs in polyamides of petrochemical origin such as PA 6 or PA66, providing promising results for blends between 5% and 40% of HNTs regarding mechanical properties and fire protection [42,43]. On the other hand, trying to get away from oil products, authors such as Sahnaoune et al. [44] are introducing these types of natural fillers into biobased polyamides such as Polyamide 11, taking into account as a disadvantage the fact that this type of polyamide is more expensive and has less applicability in today's industry. For this reason, the aim of this project is to find a balance between sustainability and direct application in industry for Polyamide 610.

In previous works, the impact of the incorporation of HNTs into biobased PA610 at different levels on mechanical, thermal, and morphological properties has been deeply studied [45]. However, the main objective of the work reported here was the use of different levels of halloysite nanotubes (HNTs) to improve the flame-retardant properties of PA610. The objective was to determine how the incorporation of nanotubes into PA610 affects its fire retardancy and flame-retardant properties in high-performance injected parts. In order to determine the properties of the blends, several samples have been characterized using cone calorimetry, limiting oxygen index (LOI), and a calorimetric pump.

2. Materials and Methods

2.1. Materials

Partially BioBased Polyamide 610 (PA610) was supplied by NaturePlast (Ifs, France), in the form of pellets. According to the manufacturer, this is a biobased medium-viscosity injection-grade homopolyamide with a density of 1.06 g/cm^3 and a viscosity number (VN) of 160 cm^3/g. This polyamide has 63% of biological content. As flame retardant, the halloysite nanotubes (HNTs) were supplied by Sigma Aldrich (Madrid, Spain) with CAS number 1332-58-7. This material had an average tube diameter of 50 nm and inner lumen diameter of 15 nm. Typical specific surface area of this halloysite was 65 m^2/g.

2.2. Sample Preparation

Prior to processing, the biobased PA610 and Halloysite nanotubes were dried at 60 °C for 48 h in the dehumidifying dryer MDEO. Both components were mechanically pre-homogenized in a zipper bag. The materials were then fed into the main hopper of a co-rotating twin-screw extruder (Construcciones Mecánicas Dupra, S.L., Alicante, Spain). The screws featured 25 mm diameter with a length-to-diameter ratio (L/D) of 24. The extrusion process was carried out at 20 rpm, setting the temperature profile, from the hopper to the die, as follows: 215–225–235–245 °C. The different PA610/HNTs composites were extruded through a round die to produce strands and, subsequently, pelletized using an air-knife unit. In all cases, residence time was approximately 1 min. The four prepared compositions are shown in Table 1.

Table 1. Summary of compositions according to the weight content (wt.%) of polyamide 610 (PA610) and halloysite nanotubes (HNTs).

Code	PA610 (wt.%)	HNTs (wt.%)
PA610	100	0
PA610/10HNTs	90	10
PA610/20HNTs	80	20
PA610/30HNTs	70	30

In the final step, the compounded pellets were shaped into square plates of 150 × 150 × 5 mm^3 by injection molding in a Meteor 270/75 from Mateu and Solé (Barcelona, Spain). The temperature profile in the injection molding unit was 220 °C (hopper), 225 °C, 230 °C, and 235 °C (injection nozzle). A clamping force of 75 tons was applied while the cavity filling and cooling times were set to 2 and 20 s, respectively.

2.3. Material Characterization

2.3.1. Cone Calorimeter Test (CCT)

The cone calorimeter model was 82121 (FIRE Ltd., Surrey, UK) and the tests were performed according to ISO 5660 standard procedures. The dimensions of the samples were 100 × 100 × 5 mm^3. Each sample was wrapped in aluminum foil (0.0025 to 0.04 mm thick) and horizontally exposed to an external heat flux of 50 kW/m^2, 25 mm conical distance, and 20 min test time.

2.3.2. Limiting Oxygen Index (LOI) and UL94

LOI was carried out in an 82121 model (FIRE Ltd., UK) according to the standard oxygen index test stated in the UNE-EN ISO 4589-2 norm. Type I test pieces and the ignition procedure (A) related only to the upper surface were used. Prior to the test, the specimens were conditioned at 23 °C and 50% relative humidity for 24 h. The size of the samples used was 150 × 10 × 4 mm^3. Three samples were studied using the LOI test.

The UL-94 horizontal burn tests were carried out following the testing procedure UL 94:2006; EN 60695-11-10:1999/A1:2003 with a test specimen bar that was 150 mm long, 10 mm wide, and about 5 mm thick.

2.3.3. Toxicity and Opacity Test

The smoke density chamber tests the opacity of the emitted fumes according to the EN ISO 5659-2 norm. This test allows to obtain the optical density using a simple chamber; simultaneously, the toxicity of the fumes was determined according to the UNE-EN 17084 norm. The test was carried out in a chamber model NBS Smoke Chamber (Concept Equipment Ltd., Arundel, UK), and a FTIR model MG2030 MKS Instruments, Inc. (San Diego, USA) to study the toxicity.

The dimensions of the samples were 75 × 75 × 5 mm^3. The samples must be conditioned to a constant mass at a temperature of 23 °C and a relative humidity of 50% in accordance with ISO 291. Each sample was wrapped in aluminum foil (0.04 mm thick) and exposed to a radiation of 50 kW/m^2, 25 mm conical distance, and 600 s test time. Regarding toxicity, a flow rate of 4 L/min was extracted for 30 s at minutes 4 and 8 of the test. Three replicates of each material were performed.

Regarding the calculation of the specific optical density (Ds), Equation (1) was used:

$$D_s(t) = \frac{V}{AL}\log_{10}\frac{100}{T(t)}[Adimensional] \quad (1)$$

where *Ds*(*t*) is the specific optical density; $\frac{V}{AL}$ is the ratio between the volume of the camera (*V*), the exposed area of the specimen (*A*), and the length of the light path (*L*). This ratio is equivalent to 132 and, finally, *T*(*t*) is the value of transmittance measured in %.

In relation to the toxicity, the concentration of toxic gas (CO_2, CO, HF, HCl, HCN, NO_2, SO_2, HBr) and optical density of smoke were recorded following the previous norm (UNE-EN 17084). During the smoke toxicity test, the concentrations of eight toxic gases were used as quantifying terms for the conventional index of toxicity (CIT). CIT was calculated using Equation (2):

$$CIT_G = 0.0805\sum_{i=1}^{i=8}\frac{c_i}{C_i} \quad (2)$$

where, CIT_G was the conventional toxicity index for general products, c_i was the concentration of the gas in the chamber, and C_i was the reference concentration of the gas.

2.3.4. Calorific Value

The equipment used for this test was a PARR 6200 calorimeter (Parr Instrument Company, Moline, IL, USA). The samples in pellet form were turned into fine powder by a grinder and liquid nitrogen was used in order to avoid thermal decomposition in the samples. The temperature of the distilled water was set at 26 °C and the closing pressure was set between 3.0 and 3.5 MPa with no air in the inside. The reagents used were distilled water, pressurized oxygen with purity greater than 99.5%, a standardized benzoic acid tablet, and a pure iron wire of 0.1 mm.

3. Results

3.1. Cone Calorimeter Test (CCT)

The calorimetric cone test (CCT) provides a great deal of information on the fire behaviour of the material under study when exposed to a variable radiation source [46]. The CCT is based on the principle of oxygen consumption, simulating the combustion of polymers in real fire situations, demonstrating great utility in research studies, and allowing the development of new materials with excellent fire-retardant properties [47,48]. Table 2 summarizes the main results obtained from this test in relation to thermal parameters and Table 3 summarizes the results of smoke parameters.

Table 2. Summary of thermal parameters obtained with the calorimetric cone test (CCT) on the PA610 and HNTs samples.

Code	TTI(s)	$t_{sos.\ inflammability}$ (s)	pHRR (kW/m²)	tpHRR (s)	EHC (MJ/kg)	THR (MJ/m²)	FRI
PA610	73.5 ± 0.5	621 ± 3	743 ± 4	272 ± 3	31.7 ± 1.9	128.1 ± 10.2	1 ± 0.1
PA610/10HNTs	45.5 ± 0.3	853 ± 4	800 ± 10	290 ± 4	52.2 ± 2.4	160.8 ± 8.3	0.46 ± 0.2
PA610/20HNTs	47.0 ± 0.4	694 ± 4	738 ± 10	300 ± 2	32.3 ± 1.6	164.3 ± 7.4	0.50 ± 0.3
PA610/30HNTs	45.0 ± 0.2	695 ± 3	581 ± 8	268 ± 2	39.3 ± 1.5	147.0 ± 9.9	0.68 ± 0.2

Table 3. Summary of smoke parameters obtained with the CCT on the PA610 and HNT samples.

Code	SEA (m²/kg)	CO_2 Yield$_{max}$ (kg/kg)	CO Yield$_{max}$ (kg/kg)	Total Smoke (m²/m²)
PA610	360 ± 15	2.2 ± 0.5	0.037 ± 0.003	915.5 ± 16.5
PA610/10HNTs	1344 ± 32	3.6 ± 0.3	0.035 ± 0.002	1993.1 ± 25.1
PA610/20HNTs	403 ± 12	2.2 ± 0.4	0.067 ± 0.004	1245.4 ± 23.6
PA610/30HNTs	517 ± 26	2.7 ± 0.2	0.056 ± 0.002	1190.1 ± 19.8

In relation to the ignition time of the samples (TTI), the incorporation of halloysite reduces the ignition start time by about 25 s in all the cases studied. Furthermore, the total duration of the inflammation does not present great variations, except for PA610/HTN10, which reaches values of 850s. Authors like Marney et al. [49] showed similar behaviour in PA6 mixtures with HNTs, where a reduction in the ignition time of the halloysite mixtures was observed. This factor may be due to the early release of the internal elements of the HNTs, which leads to the formation of small combustible molecules and results in an accelerated decomposition.

The flame retardant effect of aluminum phosphinate is in combination with zinc borate, borophosphate, and nanoclay in polyamide-6Mehmet. It is sometimes difficult to evaluate the performance of incorporating a flame-retardant additive because the results obtained are expressed in terms of time or released energy, thus generating problems of comparison. To overcome this issue, Vahabi et al. [50] have defined a dimensionless concept called the "Flame Retardancy Index" (*FRI*), which allows a very simple comparison between the pure polymer and its fire retardancy composite. This dimensionless concept is born from the following equation:

$$FRI = \frac{\left[THR \cdot \frac{pHRR}{TTI}\right]_{Neat\ Polymer}}{\left[THR \cdot \frac{pHRR}{TTI}\right]_{Composite}} \tag{3}$$

Thanks to the use of the *FRI*, a better comparison can be made between the different values obtained from the calorimetric cone test, obtaining a direct comparison between the pure polymer and its compounds.

If the FRI values in Table 2 are analyzed, it can be seen directly whether the HNTs improve the fire-retardant characteristics. Initially, it is expected that by introducing the flame-retardant additive and dividing the term calculated for the neat polymer by that of its composite with HNTs, a dimensionless quantity greater than 1 is obtained. However, it can be seen how the incorporation of 10% of HNTs into the mixture reduces the value to 0.46, obtaining a very poor value. The incorporation of a higher amount of HNTs progressively improved the previous results. In particular, 30% of HNTs improve by more than 47% of the previous value, but both are below the unit. Following the guidelines of Vahabi's work, any compound that has an FRI value below 1 is taken as the "Poor" level of performance in terms of fire retardancy. Following these general premises, HNTs seem to not provide a fire retardancy improvement when compounded with PA610. This factor may be related to the large fire-retardant capacity of PA, an engineering plastic with intrinsic fire retardancy properties, due to its base structure.

Some characteristic results of the calorimetric cone test are presented above.

3.1.1. Heat Release Rate (HRR)

The HRR measured by cone calorimeter is a very important parameter as it expresses the intensity of fire [51]. In this section, Table 2 and Figure 1 show the results related to the heat released by the samples through the CCT. Referring to the maximum peak of heat released (pHRR), most of the samples exhibit a decrease in the heat released thanks to the presence of halloysite, with the exception of PA610/HTN10, which presents a higher peak than the base compound without additive, possibly due to a low concentration of the additive. This reduction of the peak indicates that little energy has been released to the system, verifying the possible application of HNTs as fire retardancy additives in certain applications. In relation to the time when the maximum pHRR value is produced, there seems to be no apparent relationship between this value and the amount of halloysite.

On the other hand, Figure 1 shows the evolution of the heat release rate as a function of time. A heat reduction can be observed for the first 10 min of the test, reaching maximum values around 300 s. The results show a slight increase in heat release in the PA610/HTN10 sample and a very slight reduction in the heat released in the PA610/HTN20 sample, both compared to the polymer without load. However, in the case of 30% halloysite, a significant reduction is achieved in terms of the maximum values obtained, going from 743 to about 580 kW/m^2 and directly modifying the shape of the curve. This improvement in the heat released is produced by the carbonization of the halloysite, forming a layer that acts as a barrier that slows down the combustion of the polyamide, making it difficult for the sample to burn. In this context, authors like Marney et al. [42] showed similar results with the incorporation of HNTs in petrochemical polyamides. This happens due to the formation of a thin layer of carbon (or skin) on the surface, which breaks during the first stages of combustion, as shown by the small plateau at about 100, verifying how the modification in the shape of the curve is directly related to the formation of carbon in the external layers [52]. In addition, the HRR increases until it reaches a peak because of the increasing amount of combustible volatile compounds caused by the rise in total temperature of the substrate. The apparent temperature increases because the material properties are changing so that the unexposed surface reaches temperatures close to that of the exposed surface at the time of ignition. When the release of combustible volatiles has been exhausted, the combustion reaction ceases, and the peak is usually followed by a sharp linear decrease to zero.

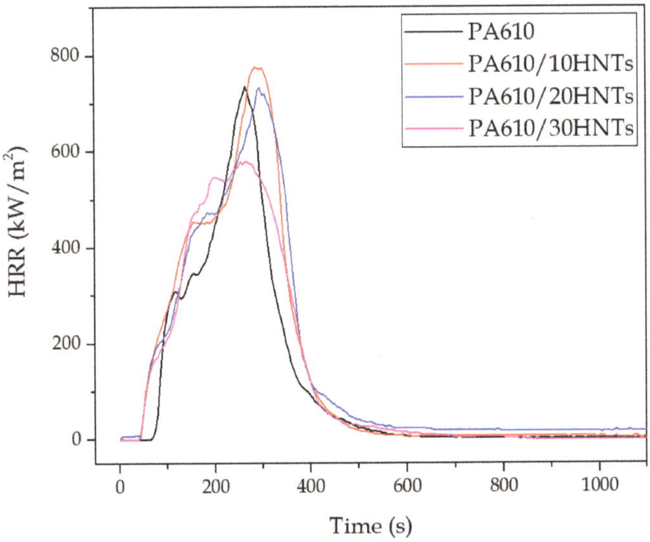

Figure 1. Heat release rate as a function of time.

The integration of the HRR curves allows to obtain the THR (total heat realised) values. The PA610 produced 128.1 MJ/m^2; the introduction of HNTs slightly increased the heat to 164.3 for the PA610/20HNTs. The range of values obtained matches with the proposed by Doğan et al. [53] for a polyamide-6 and different flame-retardant additives.

3.1.2. Effective Heat Combustion (EHC)

In the CCT, the effective heat of combustion (EHC) represents the heat released during combustion per unit mass. Figure 2 shows the graph of the results obtained as a function of time. A slight increase in the effective heat of combustion in the samples with HNTs can be seen, which is determined by the values of HRR and mass loss of the different materials during the test. Particularly, it can be seen how the PA610/10HNTs sample has provoked an increase of 20 MJ/kg compared to the sample of PA without any load. Qin et al. [54] showed similar results with the incorporation of montmorillonite to polyamide 66, where the average EHC value of PA66 increased after the addition of nano-loads, which were obtained from decomposition processes. This increase in EHC may be related to a low load concentration in the mixture, which generates an increase in the amount of energy released. This can be seen directly in samples where the amount of HNTs is higher, as it can be observed in 20% and 30% HNT samples, where the EHC is close to the PA610.

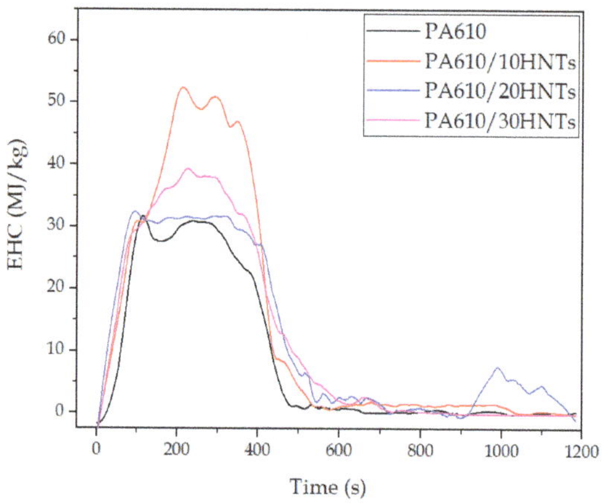

Figure 2. Effective heat of combustion as a function of time.

3.1.3. CO and CO_2 Production

Another important aspect to consider during a combustion produced by fire is the amount of CO and CO_2 that is released from the burned products, mainly because a great release could cause anoxia conditions, making it difficult to evacuate the site. Table 3 shows the maximum production values for these gases. The analyzer of the equipment used is able to determine the concentration of CO and CO_2 during the course of the test, making the analysis of the emission of these gases deeper.

Figure 3 shows the values of CO_2 produced during the test in relation to the quantity released in kg per kg of material analysed and according to the release rate. Regarding the emitted quantity, a slight increase can be observed in the 20% and 30% HNT samples compared to the base material, while in the case of the 10% HNT sample, the emission of CO_2 doubles the quantity emitted by the base polyamide. On the other hand, referring to the release rate, although it seems that the amount of total CO_2 emitted is greater as the additive load is increased, as observed with the 10% and 20% HNT samples, a significant reduction in the maximum peak for the CO_2 rate in the case of the 30% HNT sample can be observed. This fact could mean that the multi-layered porous nature of the HNT structure may prevent those evolved gases from entering the combustion zone of the burned sample. In addition, the heat feedback from the flame zone to ensure a faster polymer decomposition may also be restricted by the insulating nature of the carbon structure in the HNTs [55].

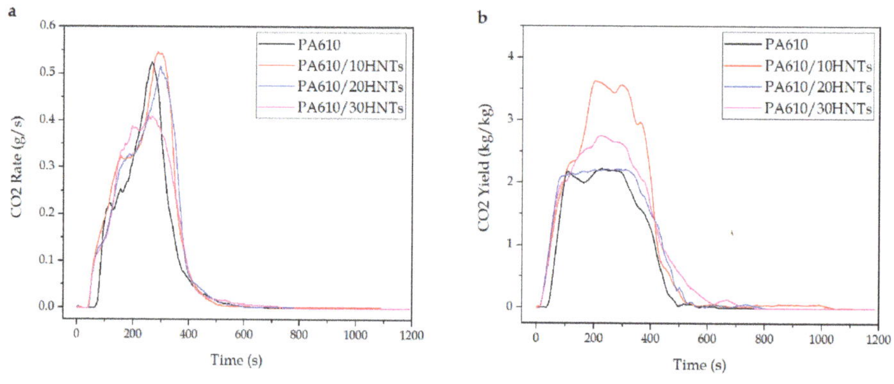

Figure 3. CO_2 production during the CCT test: (**a**) CO_2 rate and (**b**) CO_2 yield.

In relation to the values obtained for CO, Figure 4 shows how the values obtained both in total production and ratio are relatively low for all samples. However, from a global point of view, it can be seen how the incorporation of HNTs into the PA610 matrix means a slight decrease in the amount of CO generated during combustion. However, these values are not as representative as those obtained for the CO_2 produced. This factor indicates that the compounds are burned reasonably efficiently (since carbon monoxide can be measured due to an incomplete combustion) [42]. In this sense, Gilman et al. [56] studied silicate nanocomposites in PA6 and suggested that if EHC, smoke extinguishing area (SEA), and CO yields did not change, flame inhibition of the condensed phase was involved, and this was accompanied by a decrease in PHRR and mass loss ratio (MLR) as well as a change in carbon yield.

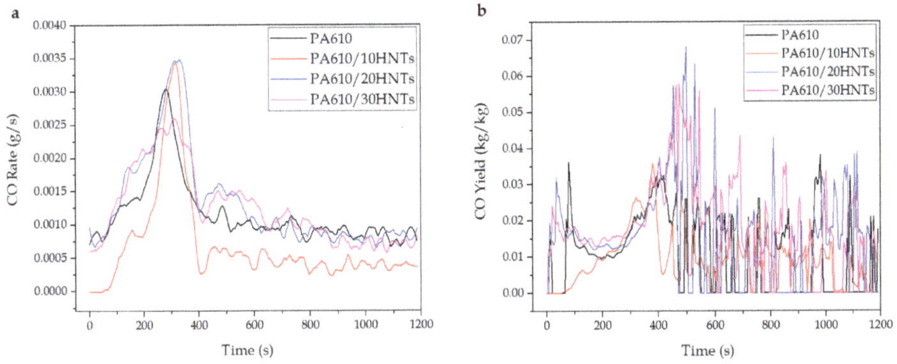

Figure 4. CO production during the CCT test: (**a**) CO rate and (**b**) CO yield.

3.1.4. Rate of Smoke Production (RSP)

The smoke performance of a fire retardancy material is a vital parameter in terms of fire safety. In the case studied, the significant increase in the amount of smoke produced is related to HNTs incorporation into the blends. This increase can be linked to the formation of a carbonized layer of material during combustion produced by halloysite itself. It is this outer layer on the surface of the test piece that could cause the combustion to release a greater amount of smoke. It should be noted that an increase in the amount of carbonized residue is observed at the end of the test in materials with halloysite. The appearance of carbonized combustion products generates an emission of darker fumes, which causes an increase in the RSP as shown in Figure 5. Similar results were reported by Levchik et al. [57] where they showed that the amount of smoke produced by the incorporation of HNTs was almost double than that of the original polymer because HNTs appear to aid the process of carbon formation in the composite material by acting as a "glue" between the HNTs, thus ensuring the formation of a consistent and strong porous carbon layer over the polymer.

The increase in the RSP value during the test is strongly related to the smoke formation as determined by the smoke extinguishing area (SEA) (Figure 6) during the combustion of HNT-containing compounds; however, it does not seem to depend on the amount of HNT. The presence of HNTs seems to accelerate the rate of PA610 smoke production during the first stages of the combustion process, being that this effect is especially marked in the 10% HNT sample. Similar results were reported by Marney et al. [42].

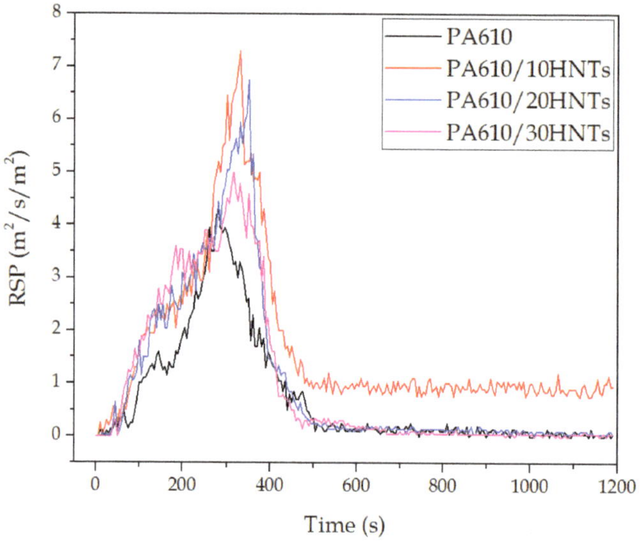

Figure 5. Smoke production rate of samples with HNTs.

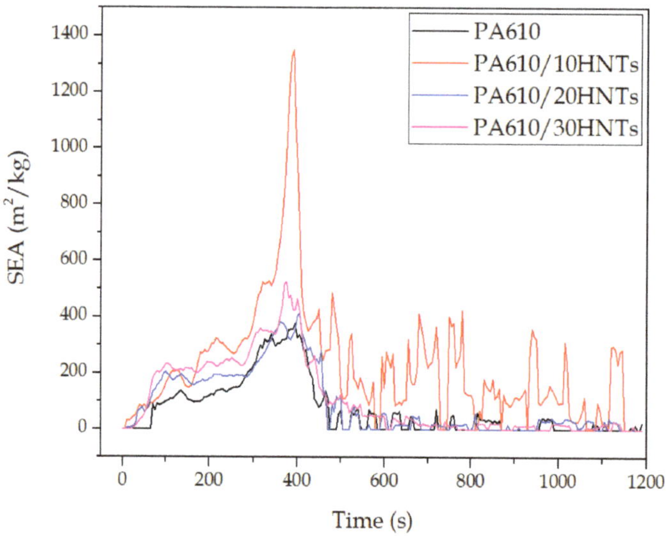

Figure 6. Specific extinction area for PA610 samples loaded with HNTs.

3.1.5. MASS

Regarding the amount of final mass obtained after the test, Figure 7 shows the results obtained in terms of mass loss ratio and percentage of residual mass. In particular, Figure 7a shows an increase in residual mass with the incorporation of HNTs. This increase is closely related to the amount of carbonized mass remaining as waste at the end of the test, which is largely made up of carbonized halloysite. The creation of this layer of carbonized material has been reported by various authors even with the use of other flame retardants such as APP or TiO_2 [58]. An intumescent carbon layer may appear on the surface of materials during combustion, creating a physical protective barrier able to contain heat and mass transfers. The carbon layer limits the diffusion of oxygen to the underlying part of the material or insulates it from heat and combustible gases; it also further delays the pyrolysis of the material. The mass loss decrease was attributed to the formation of carbon and the morphological structure at the surface of the materials [59,60]. Similarly, the result of each test specimen before and after exposure to cone radiation can be seen in Figure 8. In the case of PA610, total disappearance of the material can be seen, while in the rest of the tests, a layer of carbonized material appears, formed mainly by halloysite remains. These remains may be evidence of the increase in smoke generation previously observed.

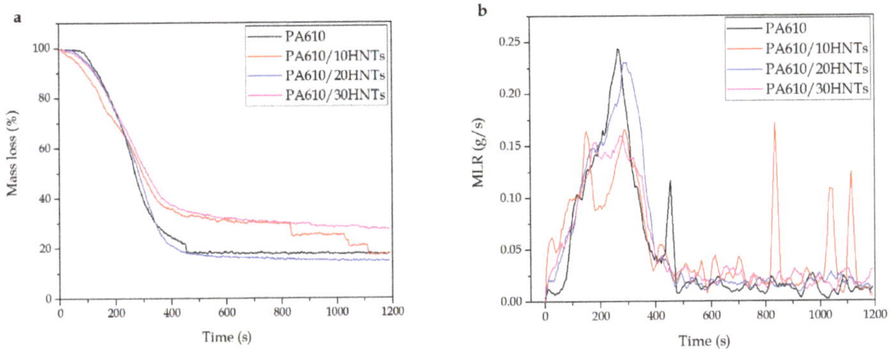

Figure 7. (**a**) Percentage of residual mass during the testing of PA610 samples with HNTs; (**b**) Loss of mass ratio of PA610 samples with HNTs.

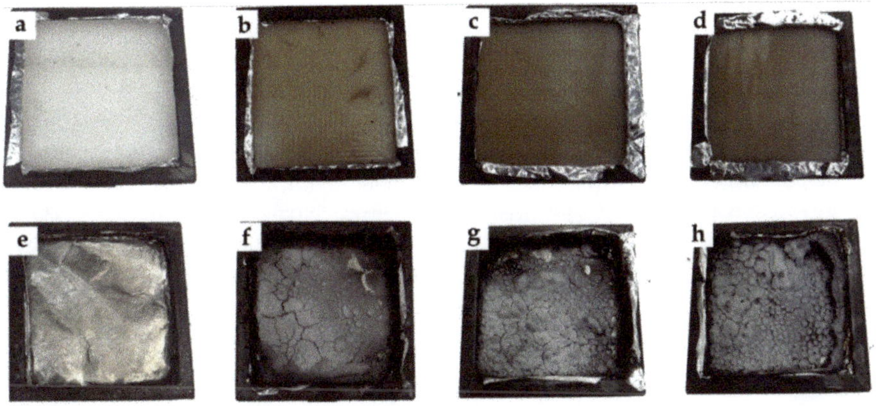

Figure 8. Visual difference between samples before and after the CCT test: (**a**) PA1010 before, (**b**) PA1010 /10HNTs before, (**c**) PA1010/20HNTs before, (**d**) PA1010/30HNTs before and (**e**) PA1010 after, (**f**) PA1010/10HNTs after, (**g**) PA1010/20HNTs after, (**h**) PA1010/30HNTs after.

In this sense, the incorporation of HNTs in relation to the residual mass is closely related to the values obtained in TGA degradation behavior in the previous work [45], focused on mechanical, thermal, morphological, and thermomechanical properties. The previous TGA results on the PA610/HNT system indicated a slight increase in the onset degradation temperature from 417.4 °C up to 419.6 °C, as well as the maximum degradation rate temperature (from 461.5 °C up to 466.6 °C). The residual weight is, as expected, close to 10, 20, and 30 wt.%, as HNTs do not decompose below 700 °C. In the previous work, scanning electron microscopy images (FESEM) also revealed good particle dispersion (especially for 10 wt.% and 20 wt.% HNTs). It can be seen how the residual mass of the compositions are almost coincident with the nominal HNTs content.

On the other hand, Figure 7b shows the mass loss ratio (MLR), which is a relevant factor in terms of selecting a fire retardancy additive, as it provides valuable information about the behavior against fire of the analyzed materials. The results obtained in the MLR are closely related to those observed in the smoke extinction area (SEA) in Figure 6, where the samples with halloysite present lower mass loss than the PA610 samples. Some authors have shown how the HNTs reduce the fire hazard of the butadiene-acrylonitrile rubber (NBR) vulcanizates. They clearly extend the time to ignition (TTI), and positively influence the average mass loss rate (MLR) parameter. Because of the synergetic relation between flame retardants and halloysite nanotubes, the parameters connected with the amount of heat released during NBR composites' combustion are reduced [36].

3.1.6. The Maximum Average Heat Rate of Emission (MARHE)

The maximum average heat emission index (MARHE) is a parameter used in the EN 45545-2:2013+A1:2015 standard to classify the materials to be tested for railway applications. This value is obtained by dividing the maximum heat emission value recorded (in kW) by the area of the test specimen (0.01 m^2). The maximum heat emission value per unit area (kW/m^2) alongside other results such as opacity and smoke toxicity are used to classify the tested material according to the applicable risk stated in the standard norm stated above. In this context, Table 4 shows MARHE's results obtained in the tests carried out.

Table 4. Summary of results of maximum average heat emission index (MARHE).

Code	MARHE (kW/m^2)
PA610	337.8 ± 5.2
PA610/10HNTs	409.4 ± 8.1
PA610/20HNTs	396.9 ± 7.5
PA610/30HNTs	363.4 ± 9.8

The values obtained in all cases were very high. The PA610 shows values above 335 kW/m^2 and the incorporation of HNTs increases this value up to 409 kW/m^2 for a concentration of 10% HNTs. In the European context, the MARHE value must be inferior to 90 kW/m^2 for these materials to be used in railway applications; the green composites studied here exhibit higher MARHE values (330–400). Additional flame-retardant additives or protective coatings for these biocomposites should be able to reduce their MARHE value, but for the time being, these green composites are not suitable for European rail applications due to the limitation exposed [61].

3.2. *Limiting Oxygen Index (LOI) and UL94 Results*

This test is defined as the minimum percentage of oxygen needed in a mixture in order to maintain the combustion of the sample after ignition. LOI tests are widely used to evaluate fire-retardant properties of materials, especially for the screening of fire-retardant polymer formulations. In this context, Figure 9 shows the LOI values obtained for PA610 with different concentrations of HNTs in their structure.

The mean of the results obtained for the PA610 without HNTs stands at 27.2%. This value is relatively high for this type of polymer. Other authors have reported values close to 20–25% for PA6 [42,53], verifying the good application that this type of biopolyamide can have from the point of view of flammability. The inclusion of HNTs means a slight decrease in the LOI values, standing at 26% for a 30% load of HNTs. Authors like Li et al. [62] showed how the incorporation of 2% HNTs into PA6 slightly improved the LOI values of this polymer, but always remained in values below 25%. In general, from the experimental results obtained for LOI, a decrease in the oxygen limit value could be observed. This means that the increase of halloysite in the polyamide makes the combustion of the material easier in low oxygen concentration conditions. In spite of this fact, variations in numerical results do not suppose a considerable change leaving aside the facility for ignition of the compound (differences smaller than a 2% of concentration in volume of O_2). Similar results have been reported by Sol et al. [63], where the increase of HNTs in the mixtures supposed a slight decrease in the LOI values, but always stayed at values superior to 24% of LOI.

The incorporation of HNTs in large quantities does not improve flammability for PA610 due to its good initial results. Authors like Vahabi et al. [64] have reported that the best results of HNTs as flame retardants are found in percentages close to 10%, verifying the results obtained in this experiment relating to LOI values.

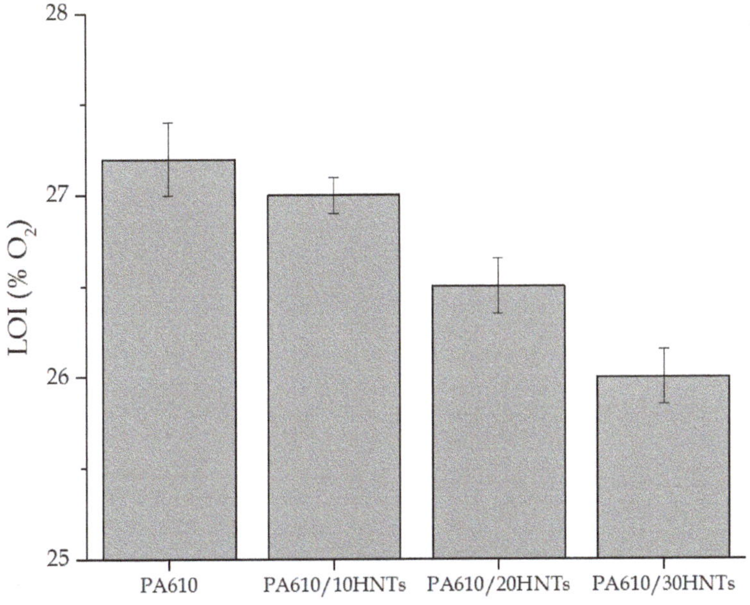

Figure 9. Graphic representation of the limiting oxygen index (LOI) values of each sample.

The UL-94 test shows that all the samples comply remarkably with the test, providing in all cases a V-0 classification. This factor verifies that the incorporation of HNT does not impair the original properties of PA610, showing promising results.

3.3. Smoke Density and Toxicity Analysis

Among the parameters that can be obtained by means of the CCT, the specific optical density (Ds) is a measurement of the degree of opacity of smoke, taken as the negative decimal logarithm of the relative light transmission. Figure 10 shows the optical density evolution along time. Except for the 20% HNTs mixture, the incorporation of nanotubes generates a slight decrease in the maximum values obtained for optical density as the halloysite load in the polyamide matrix increases. In addition, an increase in the time needed to generate smoke can be seen in the samples with HNTs. This type of behavior is very important in applications where properties against fire are needed, verifying the applicability of these types of natural loads in different fields.

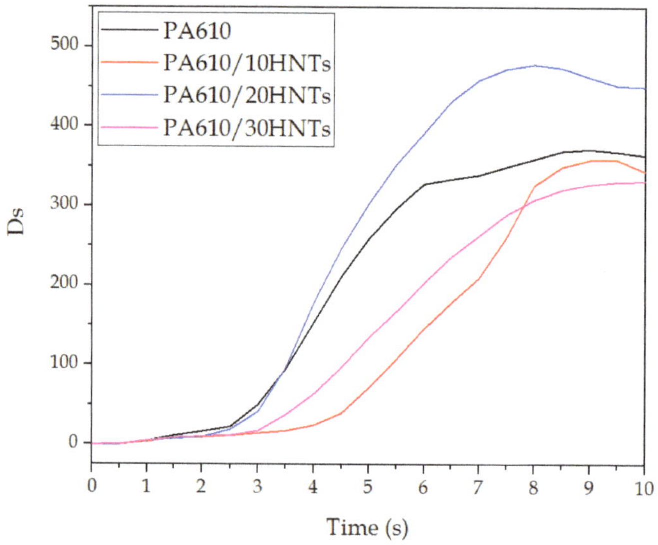

Figure 10. Evolution of optical density as a function of time.

In addition, the representation of density evolution as a function of time is shown according to the UNE-EN ISO 5659 standard. It is also important to point out the specific optical density values after 10 min of testing, as well as the maximum value (Ds_{max}). In this context, Table 5 shows values obtained by Equation (1). These values are closely related to those that can be seen in Figure 10.

Table 5. Summary of results of the maximum specific optical density values (Ds_{max}) and at 10 min of testing (Ds_{10}).

Code	Ds_{10}	Ds_{max}
PA610	364.3 ± 4.2	372.2 ± 5.9
PA610/10HNTs	344.2 ± 5.5	358.8 ± 4.8
PA610/20HNTs	450.5 ± 6.1	478.5 ± 5.4
PA610/30HNTs	332.1 ± 4.0	332.1 ± 4.5

The results obtained here are very similar to those obtained by Zhang et al. [65]. These authors reported that the addition of 5% in weight of HNT in a thermosetting resin managed to reduce the density of the smoke emitted by the mixtures. This result means that the incorporation of this additive generates a synergetic effect on the suppression of smoke from the resin, PA610 being the case studied. The reduction in optical density may be related to good dispersion of HNTs in the PA610 matrix. As already verified by mechanical properties, reducing agglomerates makes the HNTs act more adequately.

The smoke toxicity test procedure for railway industry applications follows EN 45545-2/ISO 5659-2 standards [66] performed at 50 kW/m^2. Related to toxicity, one of the main causes of death in cases where fire is involved is toxic gases generation [48]. Table 6 shows the results obtained in terms of toxicity of the analysed fumes, such as CO_2, CO, HCl, HF, HCN, NO_2, SO_2, and NO. It can be seen how the incorporation of HNTs produces, in most cases, a clear reduction in the emission of certain gases, such as CO_2 and NO_2. In particular, it can be seen how the 30% HNTs mixture reduces this value from 437.6 kg/kg to 151.2 kg/kg in the case of CO_2. Those are quite promising values referring to this type of material. Attia et al. [67] showed similar gas reduction results with the incorporation of different HNT loads in an ABS matrix. In addition, certain charges present in silicates, such as iron oxides, can participate in the flame-retardant mechanism by trapping radicals during polymer degradation, thus improving thermal stability and flammability properties for these nanocomposites [68,69].

Table 6. Volumetric fraction of the combustion gases.

	Volumetric Fraction (µL/L)							
	PA		PA/10HNTs		PA/20HNTs		PA/30HNTs	
	4 min	8 min	4 min	8 min	4 min	8 min	4 min	8 min
CO_2	169.51	437.59	253.99	479.68	114.16	102.94	112.31	151.19
CO	0.18	0.10	2.83	2.97	0.17	0.01	0.36	0.23
HCl	0.11	0.15	0.12	0.33	0.02	0.02	0.27	0.20
HF	0.05	0.04	0.16	0.17	0.05	0.07	0.03	0.06
HCN	0.16	0.03	0.81	0.95	0.38	0.46	0.12	0.06
NO_2	0.42	0.58	0.06	0.81	0.31	0.16	0.29	0.29
SO_2	33.15	29.38	58.73	57.71	22.85	22.51	25.63	26.03
NO	0.36	0.65	0.23	1.22	0.15	0.13	0.18	0.25

In order to make a deeper analysis of the final toxicity of the samples, the CIT_G was obtained. This value is a dimensionless index that provides information about the overall toxicity of all the combustion gases analyzed.

Table 7 shows the results obtained from the CIT_G index. It can be seen that most of the samples with halloysite present a notable reduction in emitted gases concentration compared to pure PA610. However, the 10% HNTs sample shows a slight increase in the final concentration of emitted gases, which results in a higher index in comparison to the base material (PA).

Table 7. Summary of the results of the conventional index of toxicity (CIT_G) index.

Code	CIT_G	
	4 min	8 min
PA610	0.125 ± 0.005	0.111 ± 0.005
PA610/10HNTs	0.221 ± 0.005	0.220 ± 0.005
PA610/20HNTs	0.087 ± 0.005	0.085 ± 0.005
PA610/30HNTs	0.096 ± 0.005	0.098 ± 0.005

3.4. *Calorific Value*

The calorific value test is obtained by means of a calorimetric pump. It is used to establish the combustion power in MJ/kg of the material to be characterized, and to analyze the existing difference made by the incorporation of fire retardancy additives. To carry out this test, the material is first introduced in a crucible together with an ignition wire, inside an airtight container with oxygen under pressure. It is then filled with pressurized oxygen gas. Next, the determination of combustion energy is carried out, so it is introduced inside a container with water and two electrodes are connected to each side of the previously introduced ignition wire. Finally, by means of agitation made by the equipment, the temperature of water is maintained homogeneous so that the temperature probe determines the increase in centigrade degrees of water produced by the material combustion.

Table 8 shows combustion energy and temperature difference values for the characterized samples. It can be seen how the incorporation of HNTs supposes a clear reduction of the combustion heat, verifying the application of these types of natural loads as fire retardancy additives. The combustion energy value for PA610 is 33.2 MJ/kg, while the samples with HNTs as additives, particularly the mixture with 30% HNTs, reduce the combustion value to 23.9 MJ/kg. This reduction is a clear advantage of this type of nanocomposites, since they have direct impact on flammability. During combustion, a halloysite-rich barrier is formed, delaying mass loss/transfer (i.e., less fuel available in the flame zone). Additionally, the refractory nature of the HNT limits the heat conduction and results in the reduction of available energy, as proposed by Marney or Smith [42,70]. There have been proposed several fire retardancy mechanisms that HNTs can exert. One of the proposed mechanisms is the typical formation of a char layer that prevents direct contact of the polymer with oxygen and slows down the escape of volatile products. Another mechanism considers that iron oxides contained in HNTs can trap free radicals during the decomposition, thus allowing to delay the burning process as observed in other iron-containing nanoparticle systems for improved flammability [71]. Nevertheless, the amount of iron oxide in HNTs is relatively low (0.29 wt.%), and there would be not enough iron oxide to play main role in fire retardancy with HNTs [72]. Du et al. [68] reported that the particular nanotube structure of HNTs allows entrapment of the decomposition products in the lumen, with a positive effect on delaying the mass transport and thus, leading to increased thermal stability. This phenomenon that allows to improve the fire retardancy properties has been represented in Figure 11. Authors like Hajibeygi et al. [73] obtained similar energy values for polyamide, besides corroborating the reduction of heat release thanks to the incorporation of additives such as zinc oxide (ZnO) nanoparticles. On the other hand, regarding the incorporation of natural nanocomposites as fire retardancy additives, Majka et al. [74] showed very similar heat release results for polyamide 6 with montmorillonite (MMT), justifying the incorporation of natural additives as fire retardancy elements.

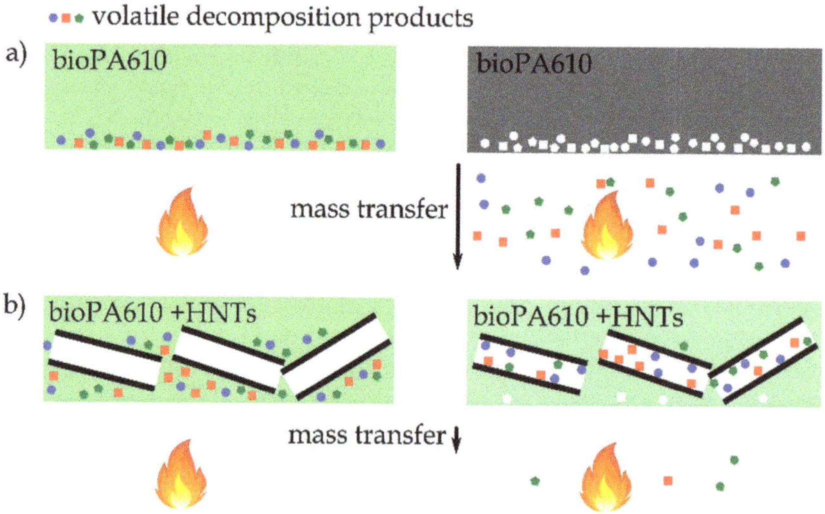

Figure 11. Scheme of the fire retardancy enhancement by HNTs by the entrapment of volatile decomposition products and delaying the mass transfer in (**a**) bioPA610 and (**b**) bioPA610+HNTs.

This improvement in heat release may be closely related to the good dispersion obtained by HNTs in PA610. This factor can be seen directly in the mechanical properties section, and in particular, in the morphology of the samples from the previous work [45].

Table 8. Summary of calorific values results obtained in the PA610/HNT samples.

Code	Heat Release (MJ/kg)	ΔT (°C)
PA610	33.2 ± 0.2	1.7 ± 0.1
PA610/10HNTs	30.2 ± 0.1	1.5 ± 0.1
PA610/20HNTs	27.3 ± 0.1	1.4 ± 0.1
PA610/30HNTs	23.9 ± 0.2	1.2 ± 0.1

4. Conclusions

This work shows that the incorporation of nano-loads with fire-retardant properties, such as HNTs, can be effectively used as new reinforcement elements in partially biobased PA610 parts prepared by conventional industrial processes for thermoplastic materials, such as injection molding.

In relation to cone calorimetric values, the incorporation of 30% HNTs achieved a significant reduction in terms of the maximum values obtained of HRR, changing from 743 kW/m^2 to about 580 kW/m^2 and directly modifying the shape of the curve. This improvement in the heat released is produced by the entrapment of the volatile decomposition products into the HNTs' lumen, which slows down the combustion of the polyamide, making it difficult for the sample to burn. However, in relation to ignition time of the samples (TTI), the incorporation of HNTs reduces the ignition start time about 20 s. This factor causes the FIR value of the compounds with HNTs to be below 1, showing poor behavior in terms of fire-retardant characteristics. In addition, the incorporation of these natural nano-loads favors the delay of the time needed to provoke an increase in generated smoke. From the point of view of gas emission and average toxicity, the CITg index, the incorporation of 20 and 30% HNTs, generates a clear reduction in emitted gases concentration, close to 50%, compared to neat PA610.

On the other hand, considering released energy during combustion, a clear reduction can be seen in all the cases where HNTs have been added as nano-loads. The incorporation of 30% of HNTs generates a reduction of 9 MJ/kg of energy released in relation to pure PA610. The good dispersion analysed in the previous work between the HNTs and the PA610 matrix could be the cause of the good results obtained in energy release. Because it prevents HNTs agglomerates from being generated and avoids the loss of fire properties.

The presence of HNTs in the PA610 matrix means a notable reduction in the number of toxic gases emitted. Especially in the case of the 30% HNTs mixture, which reduces the quantity of CO_2 emitted from 437.6 kg/kg to 151.2 kg/kg. The results obtained indicate that it is possible to obtain compounds with high renewable content such as PA610, and natural inorganic filler with nanotube structure, that is, HNTs, which exhibit acceptable fire-retardant properties, at the same time that it is possible to obtain a highly balanced material from the mechanical, thermal, and flame-retardant point of view. This improvement in properties against fire is very relevant in many applications where fire safety is crucial.

Author Contributions: Conceptualization, L.S.-N., L.Q.-C., and E.F.; methodology, L.Q.-C., C.D., and D.M.; validation, O.G., E.F., and E.G.; formal analysis, L.Q.-C., J.G.-C., and D.M.; investigation, E.M., L.S.-N., and D.M.; data curation, J.I.-M., D.M., and C.D.; writing—original draft preparation, L.Q.-C. and D.M.; writing—review and editing, J.G.-C. and L.Q.-C.; supervision, E.G., L.Q.-C., and L.S.-N.; project administration, E.F., L.S.-N., and L.Q.-C. All authors have read and agreed to the published version of the manuscript.

Funding: This research was funded by the Ministry of Science, Innovation, and Universities (MICIU) project numbers MAT2017-84909-C2-2-R and AGL2015-63855-C2-1-R.

Acknowledgments: J.I.-M. wants to thank Universitat Politècnica de València for his FPI grant from (SP20190011) and Spanish Ministry of Science, Innovation and Universities for his FPU grant (FPU19/01759). AITEX wants to thank the European Regional Development Fund (ERDF) from the European Union, for co-funding the project "NABITEX—Innovative technical textiles based on SUDOE natural fibers to be applied in Habitat Sector" through the Interreg SUDOE Program. This publication (communication) is the sole responsibility of its author. The Management Authority is not responsible for the use that may be made of the information herein disclosed

Conflicts of Interest: The authors declare no conflict of interest.

References

1. Ivorra-Martinez, J.; Manuel-Mañogil, J.; Boronat, T.; Sanchez-Nacher, L.; Balart, R.; Quiles-Carrillo, L. Development and Characterization of Sustainable Composites from Bacterial Polyester Poly (3-Hydroxybutyrate-co-3-hydroxyhexanoate) and Almond Shell Flour by Reactive Extrusion with Oligomers of Lactic Acid. *Polymers* **2020**, *12*, 1097. [CrossRef] [PubMed]
2. Agüero, A.; Morcillo, M.d.C.; Quiles-Carrillo, L.; Balart, R.; Boronat, T.; Lascano, D.; Torres-Giner, S.; Fenollar, O. Study of the influence of the reprocessing cycles on the final properties of polylactide pieces obtained by injection molding. *Polymers* **2019**, *11*, 1908. [CrossRef]
3. España, J.; Samper, M.; Fages, E.; Sánchez-Nácher, L.; Balart, R. Investigation of the effect of different silane coupling agents on mechanical performance of basalt fiber composite laminates with biobased epoxy matrices. *Polym. Compos.* **2013**, *34*, 376–381. [CrossRef]
4. Garcia, D.; Balart, R.; Sanchez, L.; Lopez, J. Compatibility of recycled PVC/ABS blends. Effect of previous degradation. *Polym. Eng. Sci.* **2007**, *47*, 789–796. [CrossRef]
5. Quiles-Carrillo, L.; Duart, S.; Montanes, N.; Torres-Giner, S.; Balart, R. Enhancement of the mechanical and thermal properties of injection-molded polylactide parts by the addition of acrylated epoxidized soybean oil. *Mater. Des.* **2018**, *140*, 54–63. [CrossRef]
6. Juárez, D.; Ferrand, S.; Fenollar, O.; Fombuena, V.; Balart, R. Improvement of thermal inertia of styrene–ethylene/butylene–styrene (SEBS) polymers by addition of microencapsulated phase change materials (PCMs). *Eur. Polym. J.* **2011**, *47*, 153–161. [CrossRef]
7. Carole, T.M.; Pellegrino, J.; Paster, M.D. Opportunities in the Industrial Biobased Products Industry. In Proceedings of the Twenty-Fifth Symposium on Biotechnology for Fuels and Chemicals, Breckenridge, CO, USA, 4–7 May 2003; pp. 871–885.
8. Quiles-Carrillo, L.; Boronat, T.; Montanes, N.; Balart, R.; Torres-Giner, S. Injection-molded parts of fully bio-based polyamide 1010 strengthened with waste derived slate fibers pretreated with glycidyl-and amino-silane coupling agents. *Polym. Test.* **2019**, *77*, 105875. [CrossRef]
9. Jasinska, L.; Villani, M.; Wu, J.; van Es, D.; Klop, E.; Rastogi, S.; Koning, C.E. Novel, fully biobased semicrystalline polyamides. *Macromolecules* **2011**, *44*, 3458–3466. [CrossRef]
10. Wu, J.; Jasinska-Walc, L.; Dudenko, D.; Rozanski, A.; Hansen, M.R.; Van Es, D.; Koning, C.E. An investigation of polyamides based on isoidide-2, 5-dimethyleneamine as a green rigid building block with enhanced reactivity. *Macromolecules* **2012**, *45*, 9333–9346.
11. Kind, S.; Neubauer, S.; Becker, J.; Yamamoto, M.; Völkert, M.; von Abendroth, G.; Zelder, O.; Wittmann, C. From zero to hero–production of bio-based nylon from renewable resources using engineered Corynebacterium glutamicum. *Metab. Eng.* **2014**, *25*, 113–123. [CrossRef]
12. Nguyen, A.Q.; Schneider, J.; Reddy, G.K.; Wendisch, V.F. Fermentative production of the diamine putrescine: System metabolic engineering of Corynebacterium glutamicum. *Metabolites* **2015**, *5*, 211–231. [CrossRef] [PubMed]
13. Winnacker, M.; Rieger, B. Biobased Polyamides: Recent Advances in Basic and Applied Research. *Macromol. Rapid Commun.* **2016**, *37*, 1391–1413. [CrossRef] [PubMed]
14. Quiles-Carrillo, L.; Montanes, N.; Boronat, T.; Balart, R.; Torres-Giner, S. Evaluation of the engineering performance of different bio-based aliphatic homopolyamide tubes prepared by profile extrusion. *Polym. Test.* **2017**, *61*, 421–429. [CrossRef]
15. Fischer, M. Polyamides (PA). *Kunstst. Plast Eur.* **2004**, *94*, 90–95.
16. Marchildon, K. Polyamides–Still Strong After Seventy Years. *Macromol. React. Eng.* **2011**, *5*, 22–54. [CrossRef]
17. Nakayama, A. Development of biobased polyamides. *Sen'i Gakkaishi* **2010**, *66*, 368–372. [CrossRef]
18. Ogunsona, E.O.; Misra, M.; Mohanty, A.K. Sustainable biocomposites from biobased polyamide 6, 10 and biocarbon from pyrolyzed miscanthus fibers. *J. Appl. Polym. Sci.* **2017**, *134*, 44221. [CrossRef]
19. Savas, L.A.; Dogan, M. Flame retardant effect of zinc borate in polyamide 6 containing aluminum hypophosphite. *Polym. Degrad. Stab.* **2019**, *165*, 101–109. [CrossRef]
20. Wang, Y.; Liu, C.; Lai, J.; Lu, C.; Wu, X.; Cai, Y.; Gu, L.; Yang, L.; Zhang, G.; Shi, G. Soy protein and halloysite nanotubes-assisted preparation of environmentally friendly intumescent flame retardant for poly (butylene succinate). *Polym. Test.* **2020**, *81*, 106174.

21. Schartel, B. Phosphorus-based flame retardancy mechanisms—Old hat or a starting point for future development? *Materials* **2010**, *3*, 4710–4745. [CrossRef]
22. Levchik, S.V.; Weil, E.D. A review of recent progress in phosphorus-based flame retardants. *J. Fire Sci.* **2006**, *24*, 345–364. [CrossRef]
23. Tang, G.; Wang, X.; Xing, W.; Zhang, P.; Wang, B.; Hong, N.; Yang, W.; Hu, Y.; Song, L. Thermal degradation and flame retardance of biobased polylactide composites based on aluminum hypophosphite. *Ind. Eng. Chem. Res.* **2012**, *51*, 12009–12016. [CrossRef]
24. Yuan, B.; Bao, C.; Guo, Y.; Song, L.; Liew, K.M.; Hu, Y. Preparation and characterization of flame-retardant aluminum hypophosphite/poly (vinyl alcohol) composite. *Ind. Eng. Chem. Res.* **2012**, *51*, 14065–14075. [CrossRef]
25. Xu, M.J.; Liu, C.; Ma, K.; Leng, Y.; Li, B. Effect of surface chemical modification for aluminum hypophosphite with hexa-(4-aldehyde-phenoxy)-cyclotriphosphazene on the fire retardancy, water resistance, and thermal properties for polyamide 6. *Polym. Adv. Technol.* **2017**, *28*, 1382–1395. [CrossRef]
26. Duquesne, S.; Futterer, T. *Intumescent Systems*; Wiley Online Library: New York, NY, USA, 2014; pp. 293–346.
27. Tirri, T.; Aubert, M.; Aziz, H.; Brusentsev, Y.; Pawelec, W.; Wilén, C.-E. Sulfenamides in synergistic combination with halogen free flame retardants in polypropylene. *Polym. Degrad. Stab.* **2019**, *164*, 75–89. [CrossRef]
28. Chen, Y.; Li, L.; Xu, L.; Qian, L. Phosphorus-containing silica gel-coated ammonium polyphosphate: Preparation, characterization, and its effect on the flame retardancy of rigid polyurethane foam. *J. Appl. Polym. Sci.* **2018**, *135*, 46334. [CrossRef]
29. Shi, Y.-Q.; Fu, T.; Xu, Y.-J.; Li, D.-F.; Wang, X.-L.; Wang, Y.-Z. Novel phosphorus-containing halogen-free ionic liquid toward fire safety epoxy resin with well-balanced comprehensive performance. *Chem. Eng. J.* **2018**, *354*, 208–219. [CrossRef]
30. Thirumal, M.; Singha, N.K.; Khastgir, D.; Manjunath, B.; Naik, Y. Halogen-free flame-retardant rigid polyurethane foams: Effect of alumina trihydrate and triphenylphosphate on the properties of polyurethane foams. *J. Appl. Polym. Sci.* **2010**, *116*, 2260–2268. [CrossRef]
31. Zhu, Z.-M.; Rao, W.-H.; Kang, A.-H.; Liao, W.; Wang, Y.-Z. Highly effective flame retarded polystyrene by synergistic effects between expandable graphite and aluminum hypophosphite. *Polym. Degrad. Stab.* **2018**, *154*, 1–9. [CrossRef]
32. Horrocks, R.; Sitpalan, A.; Zhou, C.; Kandola, B.K. Flame retardant polyamide fibres: The challenge of minimising flame retardant additive contents with added nanoclays. *Polymers* **2016**, *8*, 288. [CrossRef]
33. Kausar, A. Flame retardant potential of clay nanoparticles. In *Clay Nanoparticles*; Elsevier: Amsterdam, The Netherlands, 2020; pp. 169–184.
34. Qi, Y.; Wu, W.; Liu, X.; Qu, H.; Xu, J. Preparation and characterization of aluminum hypophosphite/reduced graphene oxide hybrid material as a flame retardant additive for PBT. *Fire Mater.* **2017**, *41*, 195–208. [CrossRef]
35. Jasinski, E.; Bounor-Legaré, V.; Taguet, A.; Beyou, E. Influence of halloysite nanotubes onto the fire properties of polymer based composites: A review. *Polym. Degrad. Stab.* **2020**, 109407. [CrossRef]
36. Rybiński, P.; Janowska, G. Influence synergetic effect of halloysite nanotubes and halogen-free flame-retardants on properties nitrile rubber composites. *Thermochim. Acta* **2013**, *557*, 24–30. [CrossRef]
37. Chrissafis, K.; Bikiaris, D. Can nanoparticles really enhance thermal stability of polymers? Part I: An overview on thermal decomposition of addition polymers. *Thermochim. Acta* **2011**, *523*, 1–24. [CrossRef]
38. Laoutid, F.; Bonnaud, L.; Alexandre, M.; Lopez-Cuesta, J.-M.; Dubois, P. New prospects in flame retardant polymer materials: From fundamentals to nanocomposites. *Mater. Sci. Eng. R Rep.* **2009**, *63*, 100–125. [CrossRef]
39. Pascual, J.; Fages, E.; Fenollar, O.; García, D.; Balart, R. Influence of the compatibilizer/nanoclay ratio on final properties of polypropylene matrix modified with montmorillonite-based organoclay. *Polym. Bull.* **2009**, *62*, 367–380. [CrossRef]
40. Prashantha, K.; Schmitt, H.; Lacrampe, M.-F.; Krawczak, P. Mechanical behaviour and essential work of fracture of halloysite nanotubes filled polyamide 6 nanocomposites. *Compos. Sci. Technol.* **2011**, *71*, 1859–1866. [CrossRef]
41. Francisco, D.L.; de Paiva, L.B.; Aldeia, W.; Lugão, A.B.; Moura, E.A. Investigation on mechanical behaviors of polyamide 11 reinforced with halloysite nanotubes. In *Characterization of Minerals, Metals, and Materials 2019*; Springer: Berlin, Germany, 2019; pp. 693–701.

42. Marney, D.; Russell, L.; Wu, D.; Nguyen, T.; Cramm, D.; Rigopoulos, N.; Wright, N.; Greaves, M. The suitability of halloysite nanotubes as a fire retardant for nylon 6. *Polym. Degrad. Stab.* **2008**, *93*, 1971–1978. [CrossRef]
43. Karahan Toprakci, H.A.; Turgut, A.; Toprakci, O. Nailed-bat like halloysite nanotube filled polyamide 6, 6 nanofibers by electrospinning. *Polym. Plast. Technol. Mater.* **2020**, *132*, 1–14. [CrossRef]
44. Sahnoune, M.; Taguet, A.; Otazaghine, B.; Kaci, M.; Lopez-Cuesta, J.M. Fire retardancy effect of phosphorus-modified halloysite on polyamide-11 nanocomposites. *Polym. Eng. Sci.* **2019**, *59*, 526–534. [CrossRef]
45. Marset, D.; Dolza, C.; Boronat, T.; Montanes, N.; Balart, R.; Sanchez-Nacher, L.; Quiles-Carrillo, L. Injection-Molded Parts of Partially Biobased Polyamide 610 and Biobased Halloysite Nanotubes. *Polymers* **2020**, *12*, 1503. [CrossRef] [PubMed]
46. Price, D.; Pyrah, K.; Hull, T.R.; Milnes, G.J.; Ebdon, J.R.; Hunt, B.J.; Joseph, P. Flame retardance of poly (methyl methacrylate) modified with phosphorus-containing compounds. *Polym. Degrad. Stab.* **2002**, *77*, 227–233. [CrossRef]
47. Park, W.-H.; Yoon, K.-B. Optimization of pyrolysis properties using TGA and cone calorimeter test. *J. Therm. Sci.* **2013**, *22*, 168–173. [CrossRef]
48. Zheng, T.; Ni, X. Loading an organophosphorous flame retardant into halloysite nanotubes for modifying UV-curable epoxy resin. *RSC Adv.* **2016**, *6*, 57122–57130.
49. Marney, D.; Yang, W.; Russell, L.; Shen, S.; Nguyen, T.; Yuan, Q.; Varley, R.; Li, S. Phosphorus intercalation of halloysite nanotubes for enhanced fire properties of polyamide 6. *Polym. Adv. Technol.* **2012**, *23*, 1564–1571. [CrossRef]
50. Vahabi, H.; Kandola, B.K.; Saeb, M.R. Flame retardancy index for thermoplastic composites. *Polymers* **2019**, *11*, 407. [CrossRef] [PubMed]
51. Tewarson, A. Heat release rate in fires. *Fire Mater.* **1980**, *4*, 185–191.
52. Porter, D.; Metcalfe, E.; Thomas, M. Nanocomposite fire retardants-a review. *Fire Mater.* **2000**, *24*, 45–52. [CrossRef]
53. Doğan, M.; Bayramlı, E. The flame retardant effect of aluminum phosphinate in combination with zinc borate, borophosphate, and nanoclay in polyamide-6. *Fire Mater.* **2014**, *38*, 92–99. [CrossRef]
54. Qin, H.; Su, Q.; Zhang, S.; Zhao, B.; Yang, M. Thermal stability and flammability of polyamide 66/montmorillonite nanocomposites. *Polymer* **2003**, *44*, 7533–7538. [CrossRef]
55. Straka, P.; Nahunkova, J.; Brožová, Z. Kinetics of copyrolysis of coal with polyamide 6. *J. Anal. Appl. Pyrolysis* **2004**, *71*, 213–221. [CrossRef]
56. Gilman, J.W.; Jackson, C.L.; Morgan, A.B.; Harris, R.; Manias, E.; Giannelis, E.P.; Wuthenow, M.; Hilton, D.; Phillips, S.H. Flammability properties of polymer−layered-silicate nanocomposites. Polypropylene and polystyrene nanocomposites. *Chem. Mater.* **2000**, *12*, 1866–1873. [CrossRef]
57. Levchik, S.V.; Weil, E.D. Combustion and fire retardancy of aliphatic nylons. *Polym. Int.* **2000**, *49*, 1033–1073.
58. Chen, X.; Wang, Y.; Jiao, C. Influence of TiO_2 particles and APP on combustion behavior and mechanical properties of flame-retardant thermoplastic polyurethane. *J. Therm. Anal. Calorim.* **2018**, *132*, 251–261. [CrossRef]
59. Schartel, B.; Hull, T.R. Development of fire-retarded materials—Interpretation of cone calorimeter data. *Fire Mater. Int. J.* **2007**, *31*, 327–354. [CrossRef]
60. Lin, M.; Li, B.; Li, Q.; Li, S.; Zhang, S. Synergistic effect of metal oxides on the flame retardancy and thermal degradation of novel intumescent flame-retardant thermoplastic polyurethanes. *J. Appl. Polym. Sci.* **2011**, *121*, 1951–1960. [CrossRef]
61. Nakamura, R.; Netravali, A.; Morgan, A.; Nyden, M.; Gilman, J. Effect of halloysite nanotubes on mechanical properties and flammability of soy protein based green composites. *Fire Mater.* **2013**, *37*, 75–90. [CrossRef]
62. Li, L.; Wu, Z.; Jiang, S.; Zhang, S.; Lu, S.; Chen, W.; Sun, B.; Zhu, M. Effect of halloysite nanotubes on thermal and flame retardant properties of polyamide 6/melamine cyanurate composites. *Polym. Compos.* **2015**, *36*, 892–896. [CrossRef]
63. Sun, J.; Gu, X.; Coquelle, M.; Bourbigot, S.; Duquesne, S.; Casetta, M.; Zhang, S. Effects of melamine polyphosphate and halloysite nanotubes on the flammability and thermal behavior of polyamide 6. *Polym. Adv. Technol.* **2014**, *25*, 1552–1559. [CrossRef]

64. Vahabi, H.; Saeb, M.R.; Formela, K.; Cuesta, J.-M.L. Flame retardant epoxy/halloysite nanotubes nanocomposite coatings: Exploring low-concentration threshold for flammability compared to expandable graphite as superior fire retardant. *Prog. Org. Coat.* **2018**, *119*, 8–14. [CrossRef]
65. Zhang, Z.; Xu, W.; Yuan, L.; Guan, Q.; Liang, G.; Gu, A. Flame-retardant cyanate ester resin with suppressed toxic volatiles based on environmentally friendly halloysite nanotube/graphene oxide hybrid. *J. Appl. Polym. Sci.* **2018**, *135*, 46587. [CrossRef]
66. Standard, I. *5659-2, Plastics–Smoke Generation-Part 2: Determination of Optical Density by a Single-Chamber Test*; The International Organization for Standardization: Geneva, Switzerland, 2006.
67. Attia, N.F.; Hassan, M.A.; Nour, M.A.; Geckeler, K.E. Flame-retardant materials: Synergistic effect of halloysite nanotubes on the flammability properties of acrylonitrile–butadiene–styrene composites. *Polym. Int.* **2014**, *63*, 1168–1173. [CrossRef]
68. Du, M.; Guo, B.; Jia, D. Thermal stability and flame retardant effects of halloysite nanotubes on poly (propylene). *Eur. Polym. J.* **2006**, *42*, 1362–1369. [CrossRef]
69. Kashiwagi, T.; Grulke, E.; Hilding, J.; Harris, R.; Awad, W.; Douglas, J. Thermal degradation and flammability properties of poly (propylene)/carbon nanotube composites. *Macromol. Rapid Commun.* **2002**, *23*, 761–765.
70. Smith, R.J.; Holder, K.M.; Ruiz, S.; Hahn, W.; Song, Y.; Lvov, Y.M.; Grunlan, J.C. Environmentally benign halloysite nanotube multilayer assembly significantly reduces polyurethane flammability. *Adv. Funct. Mater.* **2018**, *28*, 1703289. [CrossRef]
71. Zhu, J.; Uhl, F.M.; Morgan, A.B.; Wilkie, C.A. Studies on the mechanism by which the formation of nanocomposites enhances thermal stability. *Chem. Mater.* **2001**, *13*, 4649–4654. [CrossRef]
72. Goda, E.S.; Yoon, K.R.; El-sayed, S.H.; Hong, S.E. Halloysite nanotubes as smart flame retardant and economic reinforcing materials: A review. *Thermochim. Acta* **2018**, *669*, 173–184. [CrossRef]
73. Hajibeygi, M.; Shabanian, M.; Omidi-Ghallemohamadi, M.; Khonakdar, H.A. Optical, thermal and combustion properties of self-colored polyamide nanocomposites reinforced with azo dye surface modified ZnO nanoparticles. *Appl. Surf. Sci.* **2017**, *416*, 628–638. [CrossRef]
74. Majka, T.M.; Witek, M.; Radzik, P.; Komisarz, K.; Mitoraj, A.; Pielichowski, K. Layer-by-Layer Deposition of Copper and Phosphorus Compounds to Develop Flame-Retardant Polyamide 6/Montmorillonite Hybrid Composites. *Appl. Sci.* **2020**, *10*, 5007. [CrossRef]

polymers

MDPI

Article

Hydroxypropyl Methylcellulose E15: A Hydrophilic Polymer for Fabrication of Orodispersible Film Using Syringe Extrusion 3D Printer

Pattaraporn Panraksa [1], Suruk Udomsom [2], Pornchai Rachtanapun [3], Chuda Chittasupho [1,4], Warintorn Ruksiriwanich [1,4] and Pensak Jantrawut [1,4,*]

1 Department of Pharmaceutical Sciences, Faculty of Pharmacy, Chiang Mai University, Chiang Mai 50200, Thailand; pattaraporn.prs@gmail.com (P.P.); chuda.c@cmu.ac.th (C.C.); Yammy109@gmail.com (W.R.)
2 Biomedical Engineering Institute, Chiang Mai University, Chiang Mai 50200, Thailand; suruk_u@cmu.ac.th
3 Division of Packaging Technology, School of Agro-Industry, Faculty of Agro-Industry, Chiang Mai University, Chiang Mai 50100, Thailand; pornchai.r@cmu.ac.th
4 Cluster of Research and Development of Pharmaceutical and Natural Products Innovation for Human or Animal, Chiang Mai University, Chiang Mai 50200, Thailand
* Correspondence: pensak.amuamu@gmail.com or pensak.j@cmu.ac.th; Tel.: +66-53944309

Received: 29 October 2020; Accepted: 11 November 2020; Published: 12 November 2020

Abstract: Extrusion-based 3D printing technology is a relatively new technique that has a potential for fabricating pharmaceutical products in various dosage forms. It offers many advantages over conventional manufacturing methods, including more accurate drug dosing, which is especially important for the drugs that require exact tailoring (e.g., narrow therapeutic index drugs). In this work, we have successfully fabricated phenytoin-loaded orodispersible films (ODFs) through a syringe extrusion 3D printing technique. Two different grades of hydroxypropyl methylcellulose (HPMC E5 and HPMC E15) were used as the film-forming polymers, and glycerin and propylene glycol were used as plasticizers. The 3D-printed ODFs were physicochemically characterized and evaluated for their mechanical properties and in vitro disintegration time. Then, the optimum printed ODFs showing good mechanical properties and the fastest disintegration time were selected to evaluate their drug content and dissolution profiles. The results showed that phenytoin-loaded E15 ODFs demonstrated superior properties when compared to E5 films. It demonstrated a fast disintegration time in less than 5 s and rapidly dissolved and reached up to 80% of drug release within 10 min. In addition, it also exhibited drug content uniformity within United States Pharmacopeia (USP) acceptable range and exhibited good mechanical properties and flexibility with low puncture strength, low Young's modulus and high elongation, which allows ease of handling and application. Furthermore, the HPMC E15 printing dispersions with suitable concentrations at 10% *w/v* exhibited a non-Newtonian (shear-thinning) pseudoplastic behavior along with good extrudability characteristics through the extrusion nozzle. Thus, HPMC E15 can be applied as a 3D printing polymer for a syringe extrusion 3D printer.

Keywords: 3D printing; syringe extrusion 3D printing; hydroxypropyl methylcellulose; orodispersible film; phenytoin

1. Introduction

Over the last few decades, there has been a growing interest in the use of three-dimensional (3D) printing technology within the medical and pharmaceutical fields to fabricate the customizable solid dosage forms that suit different needs, preferences and individual characteristics of each patient [1]. Three-dimensional printing is a manufacturing method that can fabricate 3D-printed products of any shape and size on-demand from digital design software through depositing materials layer-by-layer [2]. This technology involves three commonly used techniques: printing-based inkjet (IJ) systems, nozzle-based deposition systems or extrusion (solid or semi-solid)-based printing technique and laser-based writing systems. Among these, the extrusion-based printing technique has been recognized as the most popular technique for the fabrication of solid oral dosage forms owing to their excellent capability to print with a wider selection of polymers and drugs at room temperature and the capability to incorporate high amounts of drugs with low-cost [1,3]. Numerous studies have been performed regarding the benefits of extrusion-based 3D printing to design various novel dosage forms such as polypills, gastro-floating tablets and orodispersible films (ODFs) [4]. ODFs are the relatively novel dosage form prepared by using hydrophilic polymers and designed to rapidly disintegrate within a minute in the buccal cavity, without requiring water [5]. This dosage form exhibits several advantages over other oral dosage forms, including ease of administration to pediatric and geriatric patients experiencing dysphagia (swallowing difficulty), dose flexibility and improving the bioavailability of drugs due to high vascularity and high permeability in the buccal cavity [6]. The major advantages of preparing an ODF by 3D printing over the standard film solvent casting are the ability to print objects with different filling (hollow, matrix or full) and the dose of drugs can be controlled by calculating the material consumption during the resizing of the printed object at the design stage, which is proper for personalized therapy. Moreover, 3D-printed films can be formulated with less amount of time.

A variety of hydrophilic polymers such as polyvinyl alcohol (PVA), polyvinylpyrrolidone (PVP), polyethylene glycol (PEG), hydroxypropyl cellulose (HPC) and hydroxypropyl methylcellulose (HPMC) are used as film-forming polymers for the preparation of ODFs, and most of them can also be used as printing materials for extrusion-based 3D printers [7,8]. Hydroxypropyl methylcellulose (HPMC), also known as hypromellose, is widely implemented in pharmaceutical manufacturing as a binder, thickening agent, hydrophilic matrix material and film-forming material. It is classified into several grades based on viscosity, degree of hydroxypropyl substitution and degree of methoxy substitution. The low viscosity HPMC grades (e.g., HPMC E3, HPMC E5 and HPMC E15) are often used for ODFs preparation and suitable for extrusion-based 3D printing of oral dosage forms [8–10]. Moreover, the use of HPMC, which is a hydrophilic polymer, can further be advantageous in terms of enhancing the solubility and dissolution of poorly water-soluble drugs in the manufacture of solid dispersion. However, there are still limited studies available on the preparation of 3D-printed ODFs while using low viscosity HPMC as a film-forming polymer. In a previous study, levocetirizine dihydrochloride ODFs consisting of HPMC E15 and pregelatinized starch were prepared using semi-solid extrusion (SSE) 3D printer. The 3D-printed ODFs exhibited good flexibility and rapid drug release in vitro by dissolving completely in two minutes [11]. In addition, previous work on the extrusion-based 3D-printing of HPMC in the pharmaceutical field can be found [12]. The extrusion-based (fused-deposition modeling) 3D printer was used to fabricate 3D-printed tablets by using their developed HPMC filament.

Phenytoin, which was selected as a model drug in this study, is an antiepileptic drug widely used in the treatment of partial seizures, generalized seizures and status epilepticus [13]. It belongs to the Biopharmaceutical Classification System (BCS) class II drug in which its bioavailability is limited due to poor water solubility (32 µg/mL). Various approaches were employed to overcome the solubility problem. Solid dispersion of the drug in a hydrophilic polymer is also one of the promising techniques for enhancing its solubility. There are many available dosage forms of phenytoin in the market, such as oral suspension, chewable tablets, capsules, and intravenous injections. However, the commercial production of orodispersible phenytoin dosage form has not yet been available. The development of phenytoin ODF has numerous advantages over conventional dosage forms such as convenience for patient administration, accurate drug dosing, rapid onset of action with increased bioavailability due to bypassing hepatic first-pass effect and noninvasiveness. Moreover, phenytoin ODF can be used for dysphasic and schizophrenic patients and can be taken without water due to their ability to disintegrate within a few minutes to release medication in the mouth.

Consequently, the present study aimed to assess the possibility of phenytoin ODFs fabrication by syringe extrusion 3D printer using two different low viscosity grades of HPMC (HPMC E5 and HPMC E15) as film-forming polymers. The extrudability and printability of different grades of HPMC were investigated and discussed. Then, the developed 3D-printed ODFs were evaluated for their physicochemical properties, mechanical properties, in vitro disintegration time and in vitro release profiles. The most important impact of our work is the use of the syringe extrusion 3D printer developed by Biomedical Engineering Institute, Chiang Mai University, Thailand and the use of a polymer (HPMC E15) with optimum viscosity as a single printing material to fabricate phenytoin-loaded ODFs for leading to personalized medicine. Until now, there has been no previous study/experiment that has used this syringe extrusion 3D printer to fabricate 3D-printed products. Our syringe extrusion 3D printer was developed to print varieties of fluid gels like materials, such as hydrogels and pastes. Moreover, it can control the printing material temperature by using a temperature control system on the syringe socket. This temperature control system will help control the viscosity of printing material and keep the printing material in a semi-solid state, which makes the material printable through the 3D printer.

2. Materials and Methods

2.1. Materials

The model drug, 5,5-diphenylhydantoin (phenytoin: PT), with purity >98%, was purchased from Merck (Darmstadt, Germany). Hydroxypropyl methylcellulose E5 (HPMC E5, AnyCoat®-C AN5, substitution type 2910, viscosity 5 mPa·s) and Hydroxypropyl methylcellulose E15 (HPMC E15, AnyCoat®-C AN15, substitution type 2910, viscosity 15 mPa·s) were purchased from Lotte Fine Chemical Co., Ltd. (Seoul, South Korea). Refined glycerin was purchased from Srichand United Dispensary Co., Ltd. (Bangkok, Thailand). Propylene glycol (PG) was purchased from Dow Chemical Company (Midland, MI, USA). Ethanol ≥95% (Bangkok Alcohol Industrial Co., Ltd., Bangkok, Thailand) and distilled water were used as the solvents for preparing the printing dispersions. All of the other reagents were analytical grade.

2.2. Determination of Extrudability and Printability

2.2.1. Rheological Characterization

To investigate the printability of the different concentrations of two different grades of HPMC that were proposed for use in syringe extrusion 3D printing, the rheological and dimensional accuracy tests of 10, 12.5, 15 and 20 % *w/v* of HPMC E5 and HPMC E15 were conducted. The samples were prepared by dispersing HPMC E5 and HPMC E15 in ethanol–water mixtures (9:1) and stirred for 4 h with a magnetic stirrer.

The rheological behaviors of HPMC E5 and HPMC E15 were investigated using a plate-and-plate Brookfield Rheometer (Brookfield Rheometer R/S, P25 DIN plate, Brookfield engineering laboratories, Middleboro, MA, USA). The measurements were carried out by measuring the shear stress and viscosity with varying the shear rate ranging from 0 to 100 s^{-1} at 25 °C. The gap between plate and base was set at 1 mm. All the experiments were carried out in triplicate. The flow behavior index (n) and consistency coefficient (K) were calculated using the power-law model equation:

$$\tau = K\dot{\gamma}^n$$

where τ is the shear stress (Pa), $\dot{\gamma}$ is the shear rate (s^{-1}), and K is a consistency coefficient (Pa·s^n)

2.2.2. 3D Printer, Design and Printing Parameters

In this study, the syringe extrusion 3D printer (Figure 1) developed by the Biomedical Engineering Institute of Chiang Mai University was used to fabricate the 3D-printed ODFs. This customized syringe extrusion 3D printer is based on a core-XY 3D printer in which an extrusion nozzle can move in an X and Y axis and a separate building plate on a Z-axis while printing layer-by-layer for generating 3D structure to prevent vibration on the sample. A custom syringe-based extruder was designed and built in-house for precision deposition. In Figure 1, the syringe-based extruder used a stepper motor to move a plunger of a 10 mL syringe via a direct lead screw drive. The syringe extrusion 3D printer was controlled by a computer and a user interface on the printer. Before 3D printing, an object was designed using an open-source program and was divided into numerous two-dimensional (2D) layers with a defined thickness, infill and speed of printing. These 2D layers can be piled up by selectively adding the desired materials in a highly reproductive layer-by-layer manner under the instruction of computer-aided design (CAD) models.

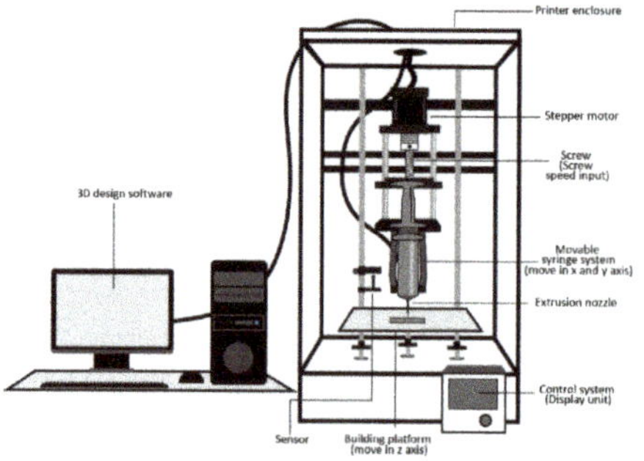

Figure 1. Schematic diagram of syringe extrusion 3D printer.

The 3D printing process was performed using a syringe extrusion 3D printer. Initially, as shown in Figure 2, the 3D designs and models of the printed ODF with the dimension of 32.5 mm width × 32.5 mm length × 0.19 mm height were obtained using Tinkercad® software (2020, Autodesk Inc., San Rafael, CA, USA). The 3D models were constructed by a simplified constructive solid geometry method and exported as a 3D printer readable stereolithography (.stl) file format to Repetier-Host software version 2.1.6 (Hot-World GmbH & Co. KG., Willich, Germany). Then, the .stl file was sliced and converted to a 3D printable code (G-code) by the open-source Slic3r software version 1.3.1 (GNU Affero general public license, version 3). Thereafter, the samples were transferred into a 10 mL disposable syringe (14.5 mm diameter), and the 3D models were printed with a syringe extrusion 3D printer equipped with a single head printing extruder nozzle with a diameter of 0.51 mm (21 G). The printing process was conducted at 25 °C, 10 mm/s printing speed and 120 mm/s nozzle traveling speed, and the printing parameters were preset as the following: the layer height was 0.19 mm, the fill angle was 45°, the perimeters were 2, and the infill was defined as rectilinear with 100% ratio (For the measurement of the diameter of printed filaments, the infill was set as 0% of the volume, and the perimeter was 1).

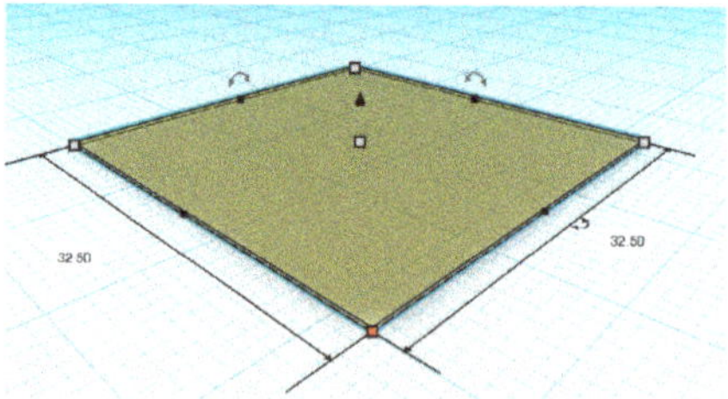

Figure 2. Computer-aided design (CAD) of the 3D-printed orodispersible films (ODF) (length 32.50 mm, width 32.50 mm and height 0.19 mm).

2.2.3. Dimensional Accuracy Tests

The diameters of printed filaments (0% infill) and area of printed ODFs (100% infill) were determined using photos taken by a digital camera (Canon EOS 750D with an 18–55 mm lens, Canon, Inc., Tokyo, Japan) and measured with ImageJ software version 1.52 (National Institutes of Health, Bethesda, MD, USA; available at https://imagej.nih.gov/ij/index.html). Then, the shape fidelity factor (SFF), which is the ratio between the 3D-printed area and CAD model area, was calculated using the following equation:

$$Shape\ fidelity\ factor = \frac{Printed\ area\ \left(cm^2\right)}{CAD\ model\ area\ (cm^2)}$$

2.3. Fabrication of 3D-Printed Orodispersible Film

In order to prepare the printing dispersions for fabrication, the following steps were taken. First, HPMC and phenytoin were dispersed in specific proportions as listed in Table 1 in the mixed solvent of ethanol and distilled water (9:1, *v/v*) and stirred for at least 3 h with a magnetic stirrer at room temperature (25 ± 2 °C). Subsequently, the two different plasticizers (glycerin and propylene glycol) were added into the dispersions at 10% *w/w* of the polymer. The dispersions were gently stirred for 10 min and then left until all of the air bubbles disappeared. Ten milliliters of the dispersion were loaded into a 10 mL disposable syringe attached with 21 G needle tips and then printed with the parameters described in Section 2.2.2. The total printing time was approximately 3.5 min for each film. After printing, the printed ODFs were placed at room temperature for 15–30 min to complete the drying process.

Table 1. Composition of different formulations of 3D-printed ODFs.

Formulation Code	Polymer (% *w/v*)		Phenytoin (% *w/v*)	Plasticizer (% *w/w* of Polymer)	
	HPMC E5	HPMC E15		Glycerin	Propylene Glycol
E5	20	-	15	-	-
E5-PT	20	-	15	-	-
E5-PT-PG	20	-	15	-	10
E5-PT-G	20	-	15	10	-
E15	-	10	15	-	-
E15-PT	-	10	15	-	-
E15-PT-PG	-	10	15	-	10
E15-PT-G	-	10	15	10	-

2.4. Characterization of 3D-Printed ODFs

2.4.1. Morphological Characterization

The morphology of the printed ODFs was investigated by scanning electron microscopy (SEM) with a JEOL scanning electron microscope (JSM-5410LV, JEOL, Ltd., Peabody, MA, USA) at 10 kV under low vacuum mode. The film characterizations were performed without any coating solution at magnifications of ×500. The thickness and surface of the films were evaluated.

2.4.2. Film Thickness and Weight Variation

The thickness and weight variation were determined by selecting ten printed ODFs randomly. The thickness of each 3.25 cm × 3.25 cm sized film was measured at three points by an outside micrometer (3203-25A, Insize Co, Ltd., Suzhou New District, Jiangsu, China). Weight variation was evaluated by an analytical balance (PA214, Ohaus Corporation, Parsippany, NJ, USA). The average thickness (in mm) and weight (in g) of printed ODFs with the standard deviation were calculated. The measurements were conducted in triplicate.

2.4.3. Mechanical Properties of 3D-Printed ODFs

Mechanical strength tests were carried out on square ODFs pieces (3.25 cm × 3.25 cm) with a texture analyzer TX.TA plus (Stable Micro Systems, Surrey, UK) using a cylindrical stainless probe (2 mm diameter) with a plane flat-faced surface (probe contact area =3.14 mm^2). The film was fixed on the plate with a cylindrical hole of 9.0 mm-diameter (area of the sample holder hole =63.56 mm^2), and a cylindrical probe was moved down at a velocity of 1.0 mm/s (Figure 3). Measurement started when the probe had contacted the sample surface (triggering force). The probe moved on at constant speed until the film was torn. The maximum force (N), distance (mm) and slope of the linear region of the force–time curve (N/s) were recorded. All of the experiments were conducted in triplicate for each sample at room condition (25 °C, 70% relative humidity). The mechanical strength of the film was characterized by puncture strength (MPa), elongation to break (%) and Young's modulus (MPa), which were calculated using the following equations [14,15]:

$$Puncture\ strength = \frac{F_{max}}{Ar_p}$$

where F_{max} is the maximum force required to tear the film (N) and Ar_p is the probe contact area (mm^2).

$$Elongation\ to\ break = \left(\frac{\sqrt{a'^2 + b^2} + r}{a} - 1\right) \times 100$$

where a is the radius of the film in the radius of sample holder cylindrical hole, a' is the length of the film sample is not torn by the probe $(a - r)$, b is the probe displacement, and r is the probe radius.

$$Young's\ modulus = \frac{Slope\ of\ force - time\ curve\ (N/s)}{Film\ thickness\ \times Probe\ speed}$$

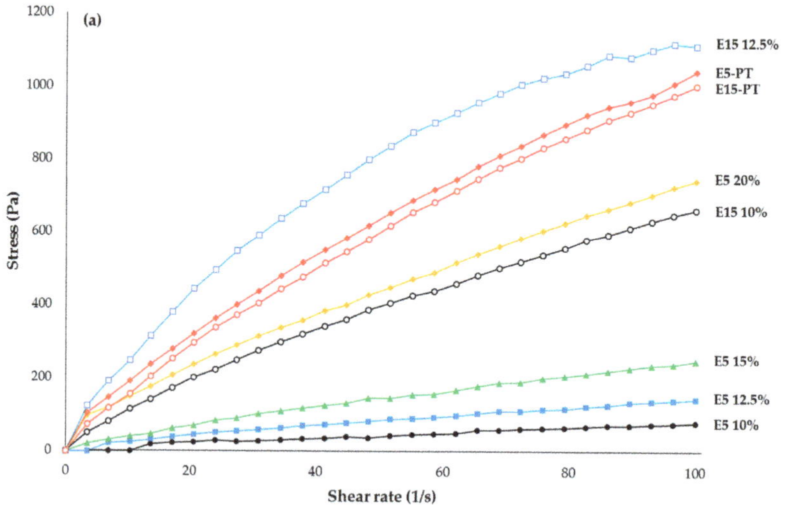

Figure 3. *Cont.*

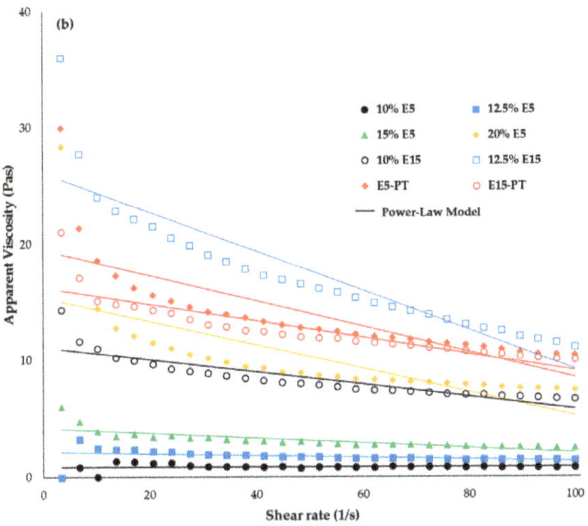

Figure 3. Rheological behaviors of hydroxypropyl methylcellulose (HPMC) E5 and HPMC E15 at different concentrations: (**a**) stress versus shear rate; (**b**) viscosity versus shear rate and fitting power-law model.

2.4.4. In Vitro Disintegration Time Study

The disintegration test method used in this study was modified from Preis et al. (2014) [16]. The printed ODF was clamped between the sample holder and the magnetic clip (attached to the bottom side of the film). The magnetic clip had a weight of 3 g (0.03 N) that represented the approximate minimal force applied by the human tongue. Then, the attached film was half-immersed (50%) in 70 mL of simulated salivary fluid pH 6.8, which was prepared according to Marques et al. (2011) [17], at 37 ± 0.5 °C. The time required for the film to break and the magnetic clip to drop down was recorded visually and noted as in vitro disintegration time. All studies were performed in triplicate for each formulation. The optimum formulations in terms of mechanical properties and disintegration time were selected for further experiments.

2.5. Phenytoin Content

Three printed ODFs of size 3.25 × 3.25 cm dimension of each selected formulation (E5-PT-G and E15-PT-G) were taken in the vials containing 5 mL of distilled water and stirred at 600 rpm until the film was completely dissolved, after which 15 mL of ethanol was added and then set aside to ensure complete solubility of phenytoin. The solution was properly diluted with Tris buffer pH 6.8 with 1% *w/v* sodium lauryl sulfate (SLS) [18] and filtered through a 0.45 μm membrane filter. The phenytoin content was then determined by using HPLC. An HPLC system (HP 1100 Series HPLC, Agilent Technologies, Inc., Santa Clara, CA, USA) equipped with a Capcell Pak AQ 250 mm × 4.6 mm C18 column having a particle size of 5 μm (Shiseido Co. Ltd., Tokyo, Japan) was utilized. The mobile phase consisted of 23% *v/v* acetonitrile, 27% *v/v* methanol and 50% *v/v* of pH 3.0 phosphate buffer solution. The pump flow rate was maintained at 1.0 mL/min, and the injection volume was 10 μL. The UV detection wavelength was set as 240 nm. The phenytoin contents were calculated from the standard calibration curve of phenytoin in Tris buffer pH 6.8 with 1% *w/v* SLS, which demonstrated linearity with a high correlation coefficient (r^2 = 0.9999). The following regression equation was obtained: $y = 5.7615x - 1.8235$, where y was the peak area and x was the concentration of phenytoin (μg/mL). All the measurements were performed in triplicate, and the average percentages of phenytoin content with the standard deviation were calculated.

2.6. *In Vitro Phenytoin Release Study*

The in vitro release test of printed ODFs was carried out in 300 mL of Tris with 1% *w/v* SLS buffer solution (pH 6.8) maintained at 37 ± 0.5 °C at a stirring rate of 50 rpm. At predetermined time intervals (1, 3, 5, 10, 15, 30 and 60 min), the samples were withdrawn, and then an equivalent volume of fresh dissolution medium was replaced in order to maintain sink conditions throughout the experiment. Withdrawn samples were filtered and analyzed by HPLC, as described in the drug content studies. The cumulative percentage of drug release was calculated using the standard calibration curve of phenytoin in Tris buffer pH 6.8 with 1% *w/v* SLS. All experiments were performed in triplicate.

2.7. *Statistical Analysis*

All the data were presented as mean ± SD. One-way ANOVA was used to evaluate the significance of differences at the significance level of p-value < 0.05. Statistical analysis was performed using SPSS software version 16.0 (SPSS, Inc., Chicago, IL, USA).

3. Results and Discussion

3.1. *Rheological Characterization and Printability Study*

The flowability and extrudability of printing dispersions are the important factors to ensure that the ODFs containing the required amount of phenytoin are printed precisely. The rheological behaviors of 10, 12.5, 15 and 20 % *w/v* of HPMC E5 and 10%, 12.5% HPMC E15 are shown in Figure 3. As observed in the figure, all the HPMC dispersions exhibited a non-Newtonian (shear-thinning) pseudoplastic behavior since their viscosity values decreased as well as shear stress increased when the shear rate increased from 0 to 100 s^{-1}. Furthermore, the power-law model was fitted to the obtained results (shear stress–shear rate flow curves), and the power-law model parameters (flow behavior index (n) and consistency coefficient) are shown in Table 2. The results showed that n values were less than 1.0 (0.66 to 0.78) for all HPMC E5 and HPMC E15 dispersions indicating the shear-thinning pseudo-plasticity nature, and the n values decreased with increasing of polymer concentration, indicating more intense shear-thinning. For the consistency coefficient, the results exhibited that the consistency coefficient values of HPMC E5 were lower than those of HPMC E15 with the same concentration, indicating that HPMC E5 was less pseudoplastic and less viscous than HPMC E15. The viscosities of 10, 12.5, 15 and 20 % *w/v* of HPMC E5 were 0.80 ± 0.04, 1.70 ± 0.03, 3.02 ± 0.18 and 9.84 ± 0.08 Pas and the viscosities of 10 and 12.5 % *w/v* of HPMC E15 were 8.10 ± 0.51 and 16.85 ± 0.76 Pas, respectively. As expected, the viscosities were found to be increased with increasing concentration of each HPMC. Moreover, a higher viscosity value was also observed when phenytoin was incorporated into the formulations. In the case of E5 20% and E15 10% formulations, after phenytoin was incorporated, the viscosities were found to be 13.39 ± 0.74 and 12.17 ± 0.47 Pas, respectively. Additionally, the results indicated that, for lower molecular weight polymer (HPMC E5), the higher polymer mass was required to achieve the same viscosity as that of the high molecular weight polymer (HPMC E15).

Table 2. Power-law model parameters and shape fidelity factor of different concentrations of HPMC E5 and HPMC E15.

Formulation	Concentration (% *w/v*)	Flow Behavior Index	Consistency Coefficient (Pa·s^n)	Shape Fidelity Factor	Diameter of Printed Filament (mm)
E5	10	0.75	2.37	1.07 ± 0.01	1.15 ± 0.17
	12.5	0.72	5.09	1.06 ± 0.02	0.97 ± 0.02
	15	0.76	7.31	1.07 ± 0.03	0.81 ± 0.02
	20	0.66	34.07	1.03 ± 0.02	0.55 ± 0.03
E15	10	0.78	18.69	1.05 ± 0.01	0.59 ± 0.07
	12.5	0.67	55.91	1.03 ± 0.01	0.56 ± 0.03
	15	NA	NA	NA	NA
	20	NA	NA	NA	NA

Note: NA (not applicable) means the printing dispersions cannot be extruded through the nozzle.

According to Table 2, it was found that the diameters of the printed filaments were decreased when the concentration of HPMC E5 and HPMC E15 increased. The increase in the viscosity values could lead to the resolution improvement of printed ODFs in which the ODFs could be printed more accurately and more precisely. Shape fidelity factors of all printed formulations were close to one, suggesting that a stable shape of gel filament was deposited in the desired place during extrusion. Nonetheless, the results found that the addition of the drug could affect the viscosities and printability of the printing dispersion. The E15(12.5%)-PT formulation could not be extruded through the extruder nozzle smoothly due to its high viscosity and fast drying of the printing dispersion at the tip of the nozzle, ultimately causing a blockage of the extruder nozzle, whereas the E5(20%)-PT and E15(10%)-PT formulations were well extruded through the nozzle. Their diameters of printed filaments were found to be mostly close to the actual extruder nozzle diameters (0.51 ± 0.02 and 0.52 ± 0.01, respectively), thereby making them more suitable for this customized syringe extrusion 3D printer.

Thus, the results indicated that 20 and 10 % *w/v* were the suitable final polymer concentrations in printing dispersions for HPMC E5 and HPMC E15, respectively. Then, the E5(20%)-PT and E15(10%)-PT formulations, which exhibited similar rheological behavior and viscosity (approximately 12–13 Pas) and enabled extrusion through nozzles, were subsequently selected for plasticizers-loading and evaluated for their mechanical properties and disintegration time.

3.2. Film Preparation and Characterization

The control E5 and E15 ODFs were slightly yellowish to colorless and transparent, whereas the phenytoin-loaded ODFs (E5-PT, E5-PT-PG, E5-PT-G, E15-PT, E15-PT-PG and E15-PT-G ODFs) were white in color and opaque due to phenytoin content. It was also observed that E5-PT-G, E15-PT, E15-PT-PG and E15-PT-G ODFs were smooth, uniform and flexible while the E5-PT and E5-PT-PG were more brittle and easily cracked during the drying process due to lack of plasticizers or an inadequate amount of plasticizer. Moreover, it was observed that the printed surface of E5-PT-G, E15-PT, E15-PT-PG and E15-PT-G ODFs showed the rectilinear pattern printed diagonally without the interruption and did not exhibit any noticeable difference upon visual inspection.

Additionally, the thickness and weight variation of all printed films were carried out in order to ensure the consistency of the printing process and printed ODFs. The average thickness of all printed ODFs varied in the range of 0.013–0.061 mm while the average thickness of phenytoin-loaded ODFs varied in the range of 0.056–0.061 mm, which falls within a typical range for fast dissolving oral films (0.05–0.15 mm) and would be adequate for handling to avoid any discomfort when applied on the buccal mucosa [19]. Therefore, as can be observed in Table 3, the average thickness and weight of printed ODFs was increased when the drug and/or plasticizers were incorporated into the film formulations; however, there were no significant differences ($p > 0.05$) observed between the thickness and weight of phenytoin-loaded E5 ODFs and phenytoin-loaded E15 ODFs.

Table 3. Weight and thickness of 3D-printed ODFs.

Film	Weight (g ± SD)	Thickness (mm ± SD)
E5	0.0544 ± 0.0005 ^{a}	0.030 ± 0.003 ^{a}
E5-PT	0.0717 ± 0.0055 ^{b}	0.059 ± 0.003 ^{b}
E5-PT-PG	0.0730 ± 0.0032 ^{b}	0.060 ± 0.003 ^{b}
E5-PT-G	0.0736 ± 0.0052 ^{b}	0.061 ± 0.001 ^{b}
E15	0.0263 ± 0.0017 ^{c}	0.013 ± 0.002 ^{c}
E15-PT	0.0516 ± 0.0006 ^{a}	0.056 ± 0.002 ^{b}
E15-PT-PG	0.0554 ± 0.0003 ^{a}	0.061 ± 0.001 ^{b}
E15-PT-G	0.0549 ± 0.0002 ^{a}	0.061 ± 0.001 ^{b}

Note: 20% *w/v* of HPMC E5 and 10% *w/v* of HPMC E15 were used. For each test, means with the same letter are not significantly different. Thus, means with different letters, e.g., 'a' or 'b' are statistically different ($p < 0.05$).

3.3. *Mechanical Properties of 3D-Printed ODFs*

The mechanical properties of all printed ODFs are displayed in Table 4, whereas the mechanical properties of E5-PT and E5-PT-G ODFs were not performed since they cracked during the drying process. The mechanical properties of the ODFs were taken to ensure that the ODFs possess suitable mechanical properties for easy handling and ease of removal from the packaging during use. In this study, the puncture strength, elongation to break and Young's modulus were used as parameters to characterize the mechanical properties of printed ODFs. The puncture strength is a measure of the toughness of film or the required force to apply on the film surface and puncture the film until breaking point [20]. The elongation to break is a parameter that reflects the ductility and stretchability of film (the ability of a film to be stretched without being torn). The Young's modulus is a parameter that reflects the elasticity and flexibility of film [21].

Table 4. Mechanical parameters and disintegration time of 3D-printed ODFs.

Film	Puncture Strength (MPa ± SD)	Elongation (% ± SD)	Young's Modulus (MPa ± SD)	Disintegration Time (s ± SD)
E5	2.78 ± 0.37 ^{a}	6.36 ± 0.78 ^{a}	216.65 ± 22.57 ^{a}	ND
E5-PT	ND	ND	ND	ND
E5-PT-PG	ND	ND	ND	ND
E5-PT-G	0.32 ± 0.04 ^{b}	0.44 ± 0.04 ^{b}	46.97 ± 2.93 ^{b}	5.51 ± 1.10 ^{a}
E15	2.38 ± 0.35 ^{c}	9.37 ± 1.47 ^{c}	316.14 ± 25.93 ^{c}	ND
E15-PT	0.38 ± 0.01 ^{b}	0.64 ± 0.02 ^{b}	44.16 ± 1.89 ^{b}	3.39 ± 0.67 ^{b}
E15-PT-PG	0.29 ± 0.04 ^{b}	0.81 ± 0.16 ^{b}	27.78 ± 3.55 ^{b}	3.81 ± 1.33 ^{b}
E15-PT-G	0.33 ± 0.02 ^{b}	1.11 ± 0.34 ^{b}	27.16 ± 3.54 ^{b}	3.56 ± 1.20 ^{b}

Note: ND (not determined) means films cannot be performed for E5-PT and E5-PT-PG ODFs, and the disintegration time of control E5 andE15 ODFs were not evaluated. For each test, means with the same letter are not significantly different. Thus, means with different letters, e.g., 'a' or 'b' are statistically different ($p < 0.05$).

Mechanically, the films in which phenytoin was incorporated (E5-PT-G, E15-PT, E15-PT-PG and E15-PT-G ODFs) were found to have significantly lower puncture strength, elongation to break and Young's modulus than the control E5 and E15 ODFs ($p < 0.05$), indicating that the E5-PT-G, E15-PT, E15-PT-PG and E15-PT-G ODFs were becoming less ductile and more flexible. This may be due to the disruption of the polymeric matrix continuity when drug particles were incorporated. This result was consistent with the previous studies showing that the addition of the drug to the formulations significantly influenced the mechanical properties of the film by decreasing the tensile strength and increasing the film's brittleness [22–24].

Furthermore, this study also showed that, without any plasticizers, the phenytoin-loaded E5 films (E5-PT) could not be formed, whereas the E5 films with a plasticizer (E5-PT-G) showed no significant difference in the puncture strength, elongations to break and Young's modulus value ($p > 0.05$) when compared with the phenytoin-loaded E15 films (E15-PT, E15-PT-G and E15-PT-PG). Additionally, it was observed that the E15-PT-PG and E15-PT-G ODFs showed slightly lower puncture strength, lower Young's modulus and higher elongation to break than the films without any plasticizers (E15-PT ODFs), but the difference was not significant ($p > 0.05$). Thus, the results indicated that the addition of glycerin at a concentration of 10% (*w/w* of polymer) as a plasticizer in the ODFs formulation had the effect on improving the mechanical properties of both E5 and E15 formulations, whereas it may require a higher concentration of propylene glycol for E5 formulation in order to avoid cracking of the films, to induce sufficient flexibility and to obtain the similar mechanical properties as E5-PT-G ODFs. Due to the hydrogen-bonding capabilities of plasticizers (e.g., glycerin and propylene glycol), the addition of suitable amounts of plasticizers can improve the flexibility of films by allowing the interaction of hydrogen bonding between polymer chain functional groups and the plasticizing agent, resulting in the reduction of polymer–polymer interaction [25].

3.4. Disintegration Time

The results of in vitro disintegration time of E5-PT-G, E15-PT, E15-PT-PG and E15-PT-G ODFs are presented in Table 4. For the film disintegration tests, there is still no official guidance available for determining the disintegration time of ODFs in the pharmacopeia. Nevertheless, according to the United States Pharmacopeia (USP) and CDER guidance regarding the orally disintegrating tablets (ODTs), disintegration time (which may be applied to ODFs), a time limit of 30 s or less for disintegration is specified [26]. In this study, the results showed that E5-PT-G, E15-PT, E15-PT-PG and E15-PT-G ODFs disintegrated within 6 s, which passes the time limit for disintegration. Moreover, all of the phenytoin-loaded E15 ODFs disintegrated significantly faster than the E5-PT-G ODFs ($p < 0.05$). This finding was consistent with the results of previous studies reporting that the formulation with the higher concentration of HPMC had a significantly longer disintegration time than that of formulations containing less concentration of HPMC [27].

According to the mechanical properties and disintegration time results, the optimal formulations, E15-PT, E15-PT-PG and E15-PT-G, which exhibited good mechanical properties and faster disintegration time, were selected for further experiments.

3.5. Morphological Characteristics of 3D-Printed ODFs

Scanning electron microscopy images of E15, E15-PT, E15-PT-PG and E15-PT-G ODFs are shown in Figure 4. The E15-PT, E15-PT-PG and E15-PT-G ODFs demonstrated uniform and well-distributed extruded dispersions with a highly porous structure in the films. Generally, 3D printing of polymer inks with in situ evaporation of solvents has allowed fabrication of 3D porous structures with stringent requirements of rheological properties of the printing ink, e.g., high viscosity and high vapor pressure [28]. In our study, spontaneous solidification via in situ evaporation of mixed solvent of ethanol and distilled water (9:1, *v/v*) generated porosity at micro-to-nano scales. The addition of different plasticizers (PG and glycerin) did not affect the morphology of printed ODFs. In addition, there was no significant difference in the thickness observed from SEM images compared to the thickness measured by using a micrometer.

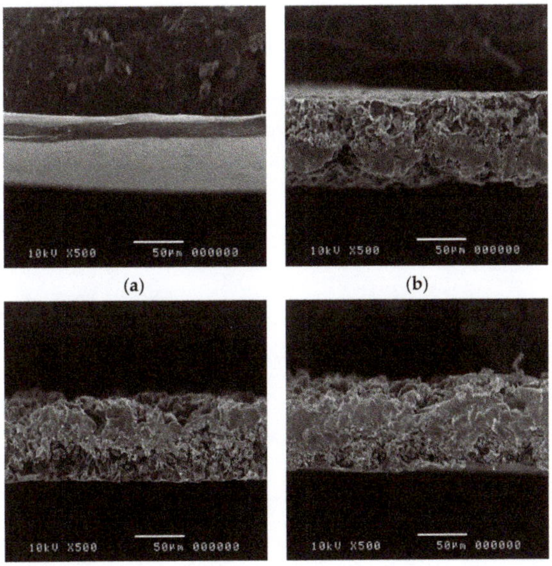

(c) (d)

Figure 4. Scanning electron microscopy images of E15 (**a**), E15-PT (**b**), E15-PT-PG (**c**) and E15-PT-G (**d**).

3.6. Phenytoin-Loading Content

In order to assess the uniformity of the printed ODFs and the robustness and precision of the film manufacturing process (3D printing process), phenytoin-loading content was evaluated. In this study, the theoretical phenytoin content in the printed ODFs size of 3.25 cm × 3.25 cm was 30 mg/film. Considering the theoretical content as 100%, the phenytoin-loading content was found to be 99.4 ± 5.2, 101.3 ± 3.8 and 99.7 ± 2.4% for E15-PT, E15-PT-PG and E15-PT-G ODFs, respectively. The results indicated that the phenytoin-loading content values of all phenytoin-loaded E15 ODFs were in agreement with the theoretical drug-loading (30 mg) and met acceptable criteria (95.0–105.0%) endorsed by the USP [18], confirming the uniformity of printed ODFs and the precision of 3D printing process of our customized syringe extrusion 3D printer.

3.7. In Vitro Phenytoin Release Studies

In vitro phenytoin release experiments were carried out in Tris with 1% *w/v* SLS buffer solution pH 6.8 at 37 ± 0.5 °C. The dissolution profiles of selected 3D-printed ODFs (E15-PT, E15-PT-PG and E15-PT-G ODFs) are shown in Figure 5. All of the selected ODFs demonstrated a similar dissolution profile pattern showing the rapid phenytoin release up to 80% within 10 min, followed by a slow constant release rate to complete drug release (100% level) in 60 min. However, the dissolution rate of all selected films was not significantly different at all time points ($p > 0.05$). The rapid release behavior of phenytoin from all selected printed ODFs could be attributed to the highly porous structure of printed ODFs, fast disintegration time within 5 s, which creates more pores to allow the drug to diffuse out of the films and the hygroscopic nature of HPMC, which increases the water uptake capability of the films. According to the obtained results, this study indicated that the improvement of the dissolution rate of poorly water-soluble drugs, such as phenytoin, in the printed ODFs can be achieved. At present, there have been not many pharmaceutical products adopted 3D printing in manufacturing. In this study, we have successfully used 3D printing technology for fabricating phenytoin-loaded ODF. Although 3D printing has been used to prepare the ODFs to enhance the solubility of poorly water-soluble drugs, there has not been any research works reporting the preparation of ODFs containing phenytoin. According to the literature, Najafi et al. formulated phenytoin sodium mucoadhesive film using Carbopol 934, sodium carboxymethyl cellulose and HPMC as film formers by solvent casting method. They reported that the best formulation released 80% of the drug within 120 min [29]. However, phenytoin-loaded ODF developed by using the 3D printing technique in this study significantly enhanced drug disintegration and dissolution to less than 5 s and up to 80% in 10 min drug release, respectively, which was shown to be a novel pharmaceutical preparation possessing optimal mechanical strength and drug release profile compared with the previous reports.

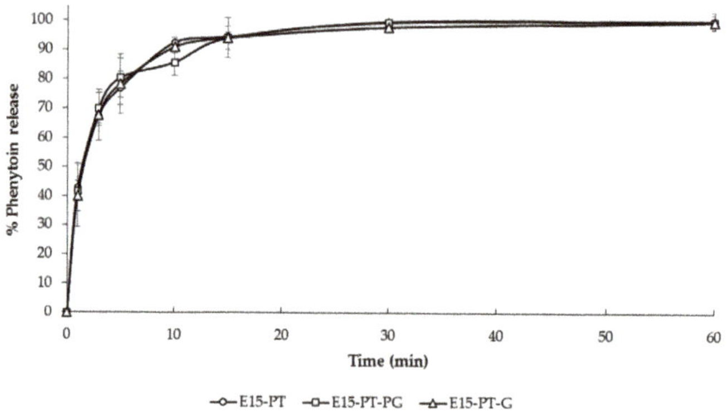

Figure 5. In vitro phenytoin release from E15-PT, E15-PT-G and E15-PT-PG 3D-printed ODFs (n = 3) in Tris with 1% *w/v* SLS buffer solution (pH 6.8).

4. Conclusions

In this study, the orodispersible films containing the required amount of phenytoin (30 mg) were successfully fabricated by 3D printing technology. The rheological properties and dimensional accuracy tests for different concentrations of HPMC E5 and HPMC E15 dispersions were performed to determine the optimal concentration and viscosity that were appropriate for the customized syringe extrusion 3D printer. The results showed that 20% *w/v* of HPMC E5 and 10% *w/v* of HPMC E15 were the most suitable polymers for incorporating drug and printing. Among all developed 3D-printed ODFs, the phenytoin-loaded E15 ODFs exhibited good physical appearance, good mechanical strength, rapid disintegration time (within 5 s) and a rapid release (up to 80% within 10 min), showing the improvement of solubility and dissolution rate of a poorly water-soluble drug, phenytoin. Overall, this study indicated that HPMC E15 is feasible to be applied as one of the 3D printing polymers for extrusion-based 3D printer and for the fabrication of orodispersible films. Additionally, this approach could be explored further for the stability of the film and the individualization of the ODFs for each patient by adjusting the volume of printing formulations.

Author Contributions: Conceptualization, P.P. and P.J.; methodology, P.P. and P.J.; software, S.U.; supervision, P.J.; validation, P.P. and P.J.; writing—original draft, P.P., S.U., P.R., C.C., W.R. and P.J.; writing—review and editing, P.P. and P.J. All authors have read and agreed to the published version of the manuscript.

Funding: This study received financial support from the National Research Council of Thailand (NRCT) and partial funding from Chiang Mai University.

Acknowledgments: This research project is supported by the National Research Council of Thailand (NRCT): NRCT5-RGJ63007-079.

Conflicts of Interest: The authors declare no conflict of interest.

References

1. Redekop, W.K.; Mladsi, D. The Faces of Personalized Medicine: A Framework for Understanding Its Meaning and Scope. *Value Health* **2013**, *16*, S4–S9. [CrossRef]
2. Goole, J.; Amighi, K. 3D printing in pharmaceutics: A new tool for designing customized drug delivery systems. *Int. J. Pharm.* **2016**, *499*, 376–394. [CrossRef]
3. Cerda, J.R.; Arifi, T.; Ayyoubi, S.; Knief, P.; Ballesteros, M.P.; Keeble, W.; Barbu, E.; Healy, A.M.; Lalatsa, A.; Serrano, D.R. Personalised 3D Printed Medicines: Optimising Material Properties for Successful Passive Diffusion Loading of Filaments for Fused Deposition Modelling of Solid Dosage Forms. *Pharmaceutics* **2020**, *12*, 345. [CrossRef]
4. Jamróz, W.; Szafraniec, J.; Kurek, M.; Jachowicz, R. 3D Printing in Pharmaceutical and Medical Applications—Recent Achievements and Challenges. *Pharm. Res.* **2018**, *35*, 176. [CrossRef]
5. Hoffmann, E.M.; Breitenbach, A.; Breitkreutz, J. Advances in orodispersible films for drug delivery. *Expert Opin. Drug Deliv.* **2011**, *8*, 299–316. [CrossRef]
6. Lopez, F.L.; Ernest, T.B.; Tuleu, C.; Gul, M.O. Formulation approaches to pediatric oral drug delivery: Benefits and limitations of current platforms. *Expert Opin. Drug Deliv.* **2015**, *12*, 1727–1740. [CrossRef]
7. Dixit, R.; Puthli, S. Oral strip technology: Overview and future potential. *J. Control. Release* **2009**, *139*, 94–107. [CrossRef]
8. Azad, M.A.; Olawuni, D.; Kimbell, G.; Badruddoza, A.Z.M.; Hossain, S.; Sultana, T. Polymers for Extrusion-Based 3D Printing of Pharmaceuticals: A Holistic Materials–Process Perspective. *Pharmaceutics* **2020**, *12*, 124. [CrossRef]
9. Li, C.L.; Martini, L.G.; Ford, J.L.; Roberts, M. The use of hypromellose in oral drug delivery. *J. Pharm. Pharmacol.* **2005**, *57*, 533–546. [CrossRef]
10. Dahiya, M.; Saha, S.; Shahiwala, A. A review on mouth dissolving films. *Curr. Drug Deliv.* **2009**, *6*, 469–476. [CrossRef]
11. Yan, T.-T.; Lv, Z.-F.; Tian, P.; Lin, M.-M.; Lin, W.; Huang, S.-Y.; Chen, Y. Semi-solid extrusion 3D printing ODFs: An individual drug delivery system for small scale pharmacy. *Drug Dev. Ind. Pharm.* **2020**, *46*, 531–538. [CrossRef] [PubMed]

12. Kadry, H.; Al-Hilal, T.A.; Keshavarz, A.; Alam, F.; Xu, C.; Joy, A.; Ahsan, F. Multi-purposable filaments of HPMC for 3D printing of medications with tailored drug release and timed-absorption. *Int. J. Pharm.* **2018**, *544*, 285–296. [CrossRef] [PubMed]
13. Golan, D.E.; Tashjian, A.H.; Armstrong, E.J. *Principles of Pharmacology: The Pathophysiologic Basis of Drug Therapy*; Lippincott Williams & Wilkins: Philadelphia, PA, USA, 2011; p. 231.
14. Preis, M.; Knop, K.; Breitkreutz, J. Mechanical strength test for orodispersible and buccal films. *Int. J. Pharm.* **2014**, *461*, 22–29. [CrossRef] [PubMed]
15. Morales, J.O.; McConville, J.T. Manufacture and characterization of mucoadhesive buccal films. *Eur. J. Pharm. Biopharm.* **2011**, *77*, 187–199. [CrossRef]
16. Preis, M.; Gronkowsky, D.; Grytzan, D.; Breitkreutz, J. Comparative study on novel test systems to determine disintegration time of orodispersible films. *J. Pharm. Pharmacol.* **2014**, *66*, 1102–1111. [CrossRef]
17. Marques, M.R.C.; Loebenberg, R.; Almukainzi, M. Simulated Biological Fluids with Possible Application in Dissolution Testing. *Dissolution Technol.* **2011**, *18*, 15–28. [CrossRef]
18. USP. *The United States Pharmacopeia*, 41st ed.; United States Pharmacopeial Convention: Rockville, MD, USA, 2018; p. 3286.
19. Patil, S.L.; Mahaparale, P.R.; Shivnikar, M.A.; Tiwari, S.S.; Pawar, K.V.; Sane, P.N. Fast dissolving oral films: An innovative drug delivery system. *Int. J. Res. Rev. Pharm. Appl. Sci.* **2012**, *2*, 482–496.
20. Abedin, N.H.Z.; Roos, Y.; Kerry, J.P. Use of beef, pork and fish gelatin sources in the manufacture of films and assessment of their composition and mechanical properties. *Food Hydrocoll.* **2012**, *29*, 144–151. [CrossRef]
21. Karki, S.; Kim, H.; Na, S.-J.; Shin, D.; Jo, K.; Lee, J. Thin films as an emerging platform for drug delivery. *Asian J. Pharm. Sci.* **2016**, *11*, 559–574. [CrossRef]
22. Cilurzo, F.; Cupone, I.E.; Minghetti, P.; Selmin, F.; Montanari, L. Fast dissolving films made of maltodextrins. *Eur. J. Pharm. Biopharm.* **2008**, *70*, 895–900. [CrossRef]
23. Centkowska, K.; Ławrecka, E.; Sznitowska, M. Technology of Orodispersible Polymer Films with Micronized Loratadine—Influence of Different Drug Loadings on Film Properties. *Pharmaceutics* **2020**, *12*, 250. [CrossRef] [PubMed]
24. Jantrawut, P.; Chaiwarit, T.; Jantanasakulwong, K.; Brachais, C.H.; Chambin, O. Effect of Plasticizer Type on Tensile Property and In Vitro Indomethacin Release of Thin Films Based on Low-Methoxyl Pectin. *Polymers* **2017**, *9*, 289. [CrossRef] [PubMed]
25. Akil, A.; Agashe, H.; Dezzutti, C.S.; Moncla, B.J.; Hillier, S.L.; Devlin, B.; Shi, Y.; Uranker, K.; Rohan, L.C. Formulation and characterization of polymeric films containing combinations of antiretrovirals (ARVs) for HIV prevention. *Pharm. Res.* **2015**, *32*, 458–468. [CrossRef] [PubMed]
26. Banarjee, T.; Ansari, V.A.; Singh, S.; Mahmood, T.; Akhtar, J. A review on fast dissolving films for buccal delivery of low dose drugs. *Int. J. Life Sci. Rev.* **2015**, *1*, 117–123.
27. Bin Liew, K.; Tan, Y.T.F.; Peh, K.-K. Effect of polymer, plasticizer and filler on orally disintegrating film. *Drug Dev. Ind. Pharm.* **2014**, *40*, 110–119. [CrossRef]
28. Karyappa, R.; Ohno, A.; Hashimoto, M. Immersion precipitation 3D printing (ip3DP). *Mater. Horiz.* **2019**, *6*, 1834–1844. [CrossRef]
29. Najafi, R.B.; Rezaei, Z.; Najm, O. Formulation and evaluation of phenytoin sodium buccoadhesive polymeric film for oral wounds. *Iran. J. Pharm. Sci.* **2011**, *7*, 69–77.

polymers

MDPI

Article

Methyl Methacrylate (MMA) Treatment of Empty Fruit Bunch (EFB) to Improve the Properties of Regenerated Cellulose Biocomposite Films

Salmah Husseinsyah [1], Nur Liyana Izyan Zailuddin [1], Azlin Fazlina Osman [1,2,*], Chew Li Li [1], Awad A. Alrashdi [3] and Abdulkader Alakrach [4]

1 Faculty of Chemical Engineering Technology, Universiti Malaysia Perlis (UniMAP), Arau 02600, Perlis, Malaysia; irsalmah@unimap.edu.my (S.H.); liyana.babyblue@yahoo.com (N.L.I.Z.); lilichew1012@gmail.com (C.L.L.)

2 Biomedical and Nanotechnology Research Group, Center of Excellence Geopolymer and Green Technology (CEGeoGTech), Universiti Malaysia Perlis (UniMAP), Arau 02600, Perlis, Malaysia

3 Chemistry Department, Umm Al-Qura University, Al-qunfudah University College, Al-qunfudah Center for Scientific Research (QCSR), Al Qunfudah 21962, Saudi Arabia; aarashdi@uqu.edu.sa

4 Chemistry Department, Faculty of Pharmacy, Qasyoun Private University, Damascus 20872, Syria; abdoakrash@hotmail.com

* Correspondence: azlin@unimap.edu.my

Received: 23 October 2020; Accepted: 4 November 2020; Published: 6 November 2020

Abstract: The empty fruit bunch (EFB) regenerated cellulose (RC) biocomposite films for packaging application were prepared using ionic liquid. The effects of EFB content and methyl methacrylate (MMA) treatment of the EFB on the mechanical and thermal properties of the RC biocomposite were studied. The tensile strength and modulus of elasticity of the MMA treated RC biocomposite film achieved a maximum value when 2 wt% EFB was used for the regeneration process. The treated EFB RC biocomposite films also possess higher crystallinity index. The morphology analysis indicated that the RC biocomposite film containing MMA treated EFB exhibits a smoother and more homogeneous surface compared to the one containing the untreated EFB. The substitution of the –OH group of the EFB cellulose with the ester group of the MMA resulted in greater dissolution of the EFB in the ionic liquid solvent, thus improving the interphase bonding between the filler and matrix phase of the EF RC biocomposite. Due to this factor, thermal stability of the EFB RC biocomposite also successfully improved.

Keywords: empty fruit bunch; regenerated cellulose; ionic liquid; methyl methacrylate

1. Introduction

Biopolymers that derive from renewable resources have sparked great interest to the world community due to their benefits to the environment and sustainability. Their properties and characteristics demonstrated capability as substitutes for the common traditional petrochemical plastics for various applications [1–4]. Biopolymers are plenty and can be derived from various sources. Generally, biopolymers can be divided into poly (amino acids) and proteins, poly-, di- and monosaccharides such as chitin [5], starch, glucose, fructose and cellulose [6,7]. Cellulose which have the formula of $(C_6H_{10}O_5)_n$ are recognized as renewable materials that can be found naturally abundant on Earth. These natural resources exist vastly in plants, animals and some microorganisms and can cover a range or variety of properties for instance from biodegradable and biocompatibility to being economical and low cost [5–8]. They are now considered as one of the most promising polymeric materials with their derived products applicable to numerous parts and sectors, principally in fibers, polymers, paints, papers and films industry [5,6]. Despite such outstanding characteristics shown, there are also some limitations or disadvantages that are raised specifically in terms of cellulose processing. In order to produce cellulosic based film, the cellulose needs to be dissolved prior to the casting process. However, the dissolution process of cellulose is challenging since cellulose are known to be non-soluble in common solvents (water and conventional organic solvents). The insolubility of cellulose is due to the presence or appearance of intra- and intermolecular hydrogen bonding and Van der Waals interaction which closely pack the chains together, and also the partially crystalline structure of cellulose [8,9]. These restraints have motivated further discoveries and investigations for suitable cellulose dissolution methods to facilitate or aid in the use of cellulose, for example, in applications of regenerated cellulose [8,10,11]. Regenerated cellulose (RC) can be referred to as the chemical dissolution of insoluble natural cellulose followed by the recovery of the material from the solution [10,11]. Regenerated celluloses (RCs) are produced or formed commercially using various methods, for example, cellulose carbamate process, lyocell process and viscose process. These different routes can result in fibers with a range of mechanical properties like fiber with low elongation but high tenacity and modulus, and fiber with great elongation but low or small modulus and tenacity [10,11]. Ionic liquids (ILs) are recognized as purely ionic, salt-like materials that are liquid at unusually low temperatures or in simple context they are ionic compounds which are liquid below 100 °C [12,13]. ILs demonstrate exceptional performance or behavior for a variety of applications in the chemical industry, for instance catalysts, chemical synthesis, separation and preparation of materials [14]. Xia et el. have recently reviewed recent advances in processing and volarization of lignocellulosic materials in ILs, including the use of ILs as pre-treatment solvents for lignocellulose to improve and enhance accessibility for lignocellulose-based biocomposite production. They summarized that current research trends have shown that the ILs have a huge potential in multifarious, efficient and environmentally friendly utilization of lignin, lignocellulose and cellulose in green technology advancement. By utilizing ILs as the solvent, additive and dispersant, the production of high performance materials generated from natural plants or biomass materials can be increased and applied in various industries and fields [15]. Other findings also agreed that ILs are considered as an adequate solvent to dissolve cellulose for the cellulose regeneration process and this approach be employed for advanced applications [10,12]. The recovery of ILs after cellulose regeneration could be achieved via methods like evaporation, reverse osmosis, ion exchange and salting out for reuse and recycling purposes [16]. Other solvents that can be used for dissolving cellulose to generate regenerated cellulose are N-methylmorpholine-N-oxide (NMMO) [17], N-dimethylacetamide (DMAC) / lithium chloride (LiCl) [10], 1-butyl-3-methylimidazolium chloride (BMIMCl) [18], 1-allyl-3-methylimidazolium chloride (AMIMCl) [19], 1-ethyl-3-methylimidazolium acetate (EMIMAc) [20] and 1-ethyl-3-methylimidazolium chloride (EMIMCl) [21]. The solvent 1-butyl-3-methylimidazolium chloride (BMIMCl) was first determined or established by Swatloski and co-workers to have good dissolving power for cellulose [22]. This finding generates a way towards a new class of cellulose solvent system, and accelerates or prompts the investigation into the potential of other ILs in dissolving cellulose.

Oil palm empty fruit bunches (OPEFBs) are known as one of the most abundant biomasses from palm oil plantations. It is considered as a lignocellulosic residue of palm oil plants, which is regarded as an important agricultural source in Southeast Asian countries. For instance, Malaysia has produced about 15.8 million tonnes of OPEFB annually due to the vigorous activity of oil palm cultivation [10,23]. Upon oil extraction from fresh fruit bunches (FFB), empty fruit bunch (EFB) is attained as biomass with advantageous properties such as being inexhaustible, renewable, biodegradable and environmental friendly [10]. EFB consists of cellulose content ranging from 42.7–65.0%, where as much as 60.6% of this cellulose can be turned into value added products [24]. Cellulose which is derived from EFB can be employed for several applications such as making bioplastic film, paper and polymer composite products. Chemical pre-treatments are usually implemented for cellulose extraction and also to enhance or improve the cellulose dissolution [10,25]. Cellulose pre-treatments are usually carried out on natural fibers to remove the lignin and partly eliminate the hemicelluloses. It also disrupts or breaks the crystalline structure of cellulose, allowing the cellulose to swell and dissolve in suitable solvent [10,25]. Some examples of pre-treatment methods include alkaline treatment [10,25], acid hydrolysis [26], hot compressed water [27] and enzymes [28]. Since cellulosic fibers are innately hydrophilic, they can easily degrade and absorb moisture from the environment. Therefore, chemical modification is applied and utilized to increase hydrophobicity and moisture resistance. Examples of chemical modification methods are silane functionalization [29], acetylation [30] and polymer grafting [31]. However, in the production of regenerated cellulose biocomposite, chemical treatment can also facilitate the regeneration process by reducing the hydrogen bonds of the cellulose and improving its solubility in the ionic liquid [25].

To-date, the mechanical and thermal analyses data of the RC film derived from the EFB based cellulose and produced through the use of 1-butyl-3-methylimidazolium chloride (BMIMCl) as ionic liquid and methyl methacrylate (MMA) as chemical treatment not available in the literature, but is truly important to contribute to the progress of the RC biocomposite films intended for use in biodegradable packaging applications. This is because the combination of both chemicals is practical and efficient to produce RC biocomposite with improved performance. BMIMCl has excellent cellulose-dissolving ability while MMA is an industrially relevant chemical with good capability to treat cellulose and improve its dissolution process during the regeneration process. The MMA was previously used to treat chitosan and *Nypa frutican* based-cellulose for the production of biocomposite [32,33]. However, this is the first attempt to utilize MMA in the chemical treatment of the EFB. In order to produce the EFB RC biocomposite film with improved mechanical and thermal properties, the role of MMA as the EFB's modifier must be clarified while the best EFB loading needs to be determined. This research was aimed at conducting an investigation into all of these aspects, thus revealing the never reported findings for future development of biodegradable plastics.

2. Experimental

2.1. Materials

Empty fruit bunch (EFB) fibers produced as biomass from palm oil trees were collected and received from the Malaysian Palm Oil Board (MPOB) in Bangi, Selangor, Malaysia. The EFB fibers were ground using a ring mill into powder form. The average particle size of EFB was 45 microns which was determined using Malvern (Malvern, United Kingdom) particle size analyzer. Microcrystalline cellulose (MCC) was purchased from Aldrich (St. Louise, MO, USA). Sulfuric acid, ethanol and sodium hydroxide (NaOH) were acquired from HmbG Chemicals (Hamburg, Germany). Acetic acid was purchased from BASF Chemical Company (Ludwigshafen, Germany) while sodium chlorite ($NaClO_2$) was supplied by Sigma Aldrich (St. Louise, MO, USA). Furthermore, 1-butyl-3-methyl-imidazolium chloride and methyl methacrylate (MMA) were purchased from Merck (Darmstadt, Germany).

2.2. Pre-Treatment of EFB

Firstly, EFB powder was treated using 4 wt% of NaOH solution for 1 h at 70 °C. The treatment was performed under mechanical stirring at the speed of 750 rpm. This procedure was repeated 3 times, and the EFB was washed with distilled water and filtered several times after each treatment to remove the soluble alkaline. Then, the alkaline treated EFB was subjected to a bleaching process for further delignification. The bleaching process was carried out in solution containing acetate buffer (solution of 2.7 g NaOH and 7.5 mL glacial acid in 100 mL distilled water), 1.7% w/v aqueous sodium chlorite and distilled water. The EFB was mixed together with the bleaching solution which was stirred for 60 min at the temperature of 70 °C. The bleached fibers were then washed with distilled water and filtered. Acid hydrolysis was carried out using 65 wt% of H_2SO_4 at the temperature of 50 °C under mechanical stirring for 45 min. The EFB attained was washed with cold distilled water to stop the acid hydrolysis reaction. The EFB was washed and filtered repeatedly until pH 7 was reached. The filtered EFB was dried in an oven for a day at 70 °C.

2.3. EFB Modification

Chemical treatment was carried out on pre-treated EFB using methyl methacrylate (MMA). Firstly, the MMA solution was prepared by dissolving 3% of MMA in ethanol (v/v). Then, the EFB was slowly added into the MMA solution and the mixture was stirred continuously for 1 h. The solution was left overnight. The next day, the EFB was filtered and dried in an oven at 70 °C for 24 h.

2.4. Preparation of EFB Regenerated Cellulose (RC) Biocomposite Films

The concept of this RC biocomposite is the same as self-reinforced composite (SRC). Both EFB cellulose and MCC were used as the reinforcement and matrix when they were partially dissolved in the 1-butyl-3-methylimidazolium chloride (BMIMCl) solvent system. The dissolved cellulose was transformed into the matrix phase surrounding the remaining non-dissolved cellulose. EFB RC biocomposite films with 1, 2, 3 and 4 wt% EFB (untreated and MMA treated), plus a constant amount of microcrystalline cellulose (MCC) (3 wt%) were prepared for this study. BMIMCl was used as solvent in the regeneration process of the EFB cellulose. Firstly, the BMIMCl was heated in order to dissolve it into liquid form. Furthermore, 3 wt% of MCC and 1 wt% of EFB was dispersed in BMIMCl. The mixture was heated at the temperature of 100 °C and stirred for 30 min until a homogeneous cellulose solution was achieved. The cellulose solution was cast onto a glass plate mold. As the film sample developed, it was soaked in a water bath to allow the diffusion of ionic liquid out of the sample. Water was changed a few times to ensure complete removal of ionic liquid from the film sample. The film attained was dried in an oven for 24 h at 50 °C. Similar steps were repeated to produce the EFB RC biocomposite films with different contents of untreated and treated EFB. Note that the purpose of adding MCC into the mixture was to provide a strengthening effect to the RC biocomposite films. Preliminary work was done in which the RC films were produced without the addition of MCC. Unfortunately, the films produced were too brittle. In addition, we have tried to prepare the film with 0% EFB; however, no good film has been successfully produced. These factors were the reason for preparing the RC biocomposite films with both MCC and EFB. As MCC was added in constant amounts (3 wt%) in all the RC biocomposite films, the changes in the morphology and properties of the films reported in the discussion part were assumed due to the EFB contents.

The overall process to produce the EFB RC biocomposite films is illustrated in the schematic diagram in Figure 1.

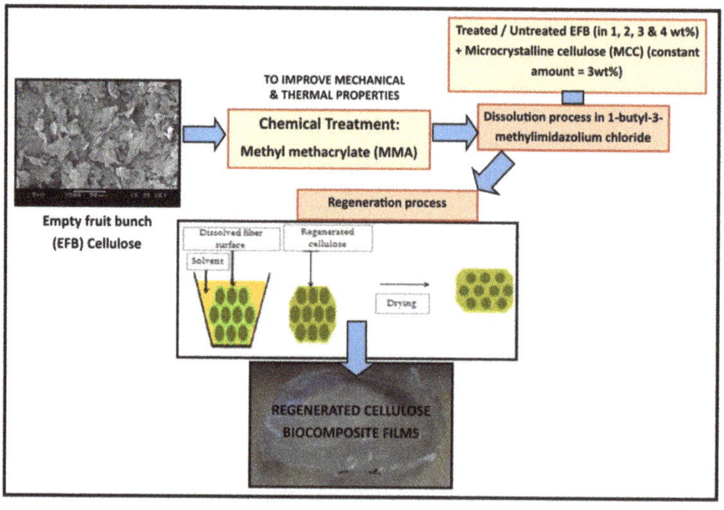

Figure 1. Procedures to prepare EFB RC biocomposite films.

3. Texting and Characterization

3.1. Tensile Testing

Instron Universal Testing Machine (Norwood, MA, USA), model 5590 was used to carry out the tensile test which was based on ASTM D 882. The tensile test was operated at the speed of 10 mm/min. The EFB RC biocomposite films were cut into rectangular shapes with dimensions of 15 mm × 100 mm. The thickness of the sample was measured using a digital vernier caliper along its gauge length area three times (at center, upper and bottom sides). The mean value was taken and recorded. Five replicates of each material were used for the tensile test, and the average values of tensile strength, modulus of elasticity and elongation at break were attained and recorded. Statistical analysis was performed using two-tailed Student's t-test for unpaired data to compare among materials. A significance level of 0.05 or less was accepted as statistically significant.

3.2. Scanning Electron Microscopy (SEM)

Scanning electron microscope (SEM) model JEOL JSM-6460LA (Akishima, Japan) was utilized to analyze and study the morphology of EFB RC biocomposite films. The fractured surface of the biocomposite film was subjected to morphological analysis to study the deformation behavior of the film samples. The films were coated with a thin layer of palladium for conductive purpose. Scanning was done at a voltage of 10 kV.

3.3. X-Ray Diffraction (XRD)

XRD analysis was conducted using a Bruker DS X-ray diffractometer (Billerica, MA, USA) utilizing X-ray sources of Cu-Kα radiation (λ = 1.5418 Å). The analysis was carried out under normal atmospheric conditions and at ambient temperature. Samples used for XRD testing were strips of films with dimensions of 10 mm × 15 mm [32]. The crystallinity index (CrI) for both untreated and treated EFB RC biocomposite films were calculated based on [10,34]:

$$\text{Crystallinity Index (CrI)} = (I/I') \,/\, (I) \times 100 \quad (1)$$

where I = the height of peak assigned to (200) planes, measured in the range 2θ = 20–23°, and I′ = the height of peak assigned to (100) planes, located at 2θ = 12–16°.

3.4. Fourier Transform Infrared (FTIR)

A Perkin Elmer L1280044 FTIR spectrometer (Waltham, MA, USA) was used to study the functional groups of untreated and treated EFB RC biocomposite films. Four scans were conducted on the films in the wavenumber range from 4000–650 cm^{-1} with resolution of 4 cm^{-1} using attenuated total reflectance (ATR) method.

3.5. Thermogravimetric Analysis (TGA)

TGA Pyris Diamond Perkin Elmer (Waltham, MA, USA) was used to analyze the thermal behavior of both untreated and treated EFB RC biocomposite films. The film samples used were weighed to about 5 ± 2 mg. The films were placed in aluminum pans and heated from 25 to 700 °C with a heating rate of 10 °C/min. The testing was performed in nitrogen atmosphere with nitrogen flow rate of 50 mL/min.

4. Results and Discussion

4.1. Tensile Properties

Figure 2 shows the effect of EFB content on the tensile strength of the untreated and MMA treated EFB-based RC biocomposite films (EFB RC biocomposite). The tensile strength for both untreated and treated EFB RC biocomposite films demonstrated similar trends in which the value increased initially when the EFB content increased from 1 to 2 wt%. Further increase in the EFB content resulted in the decrease of the tensile strength of the EFB RC. Large accumulations of EFB particles led to non-homogeneous dispersion of the EFB during the formation of the EFB RC biocomposite film. This would develop more stress concentrated areas that hasten the failure of the biocomposite's tensile deformation. The highest tensile strength value (34 MPa) was obtained when 2 wt% MMA treated EFB was employed to form the RC biocomposite. Obviously, at similar EFB content, the treated EFB RC biocomposite films obtained higher tensile strength compared to the untreated biocomposite films. For instance, when 2 wt% MMA treated EFB was used to form the RC biocomposite; a 26% improvement in tensile strength was obtained vs. the untreated RC biocomposite film. It is proven that the treatment of EFB with MMA has enhanced and improved the interfacial adhesion between the non-dissolved (filler phase) and dissolved (matrix phase) cellulose, allowing greater stress transferring mechanisms throughout the biocomposite structure.

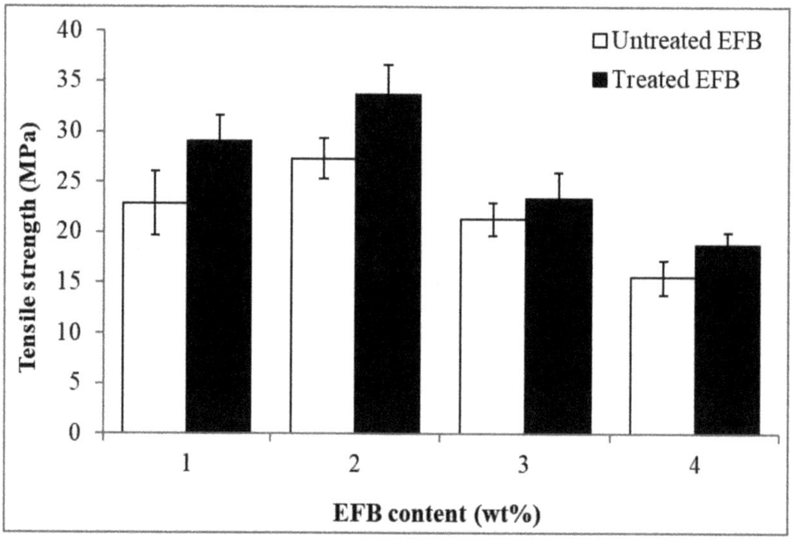

Figure 2. Effect of EFB content on tensile strength of untreated and MMA-treated EFB RC biocomposite films.

Figure 3 displays the effect of EFB content on the elongation at break of the untreated and MMA treated EFB RC biocomposite films. Generally, the elongation at break values of the biocomposites does not show any clear trend. The fluctuating of the elongation at break value could be related to environmental influences such as moisture, absorption rate of the film and different levels of void content due to random dispersion of the EFB particles in the biocomposite structure. These factors affect the flexibility and conformation of the biopolymer chains when subjected to tensile deformation. In addition to that, the use of single material (EFB cellulose which functioned as both matrix and reinforcement) to form the biocomposite can cause an uncontrollable portion of the dissolved cellulose (matrix) and undissolved cellulose (filler) in the RC biocomposite structure. In this case, the "real composition" of filler phase may not be proportional with the EFB content, thus the portion that contributes to elasticity may also vary. These reasons might explain the unexpected elongation at break results of these biocomposite films. However, regardless of the EFB content, the MMA treated films exhibit lower elongation at break compared to the untreated biocomposite films. Enhancement in the interfacial bonding between matrix and filler obtained through the chemical treatment has restrained the flexibility and mobility of the polymer chain.

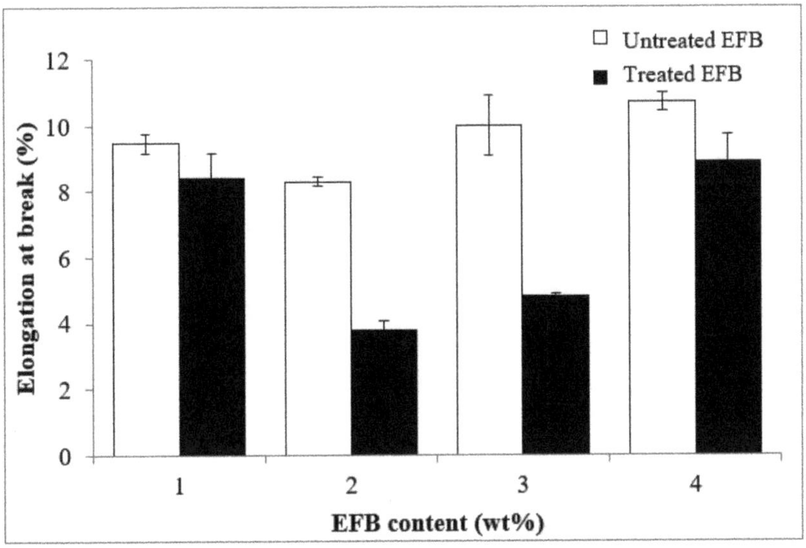

Figure 3. Effect of EFB content on elongation at break of untreated and MMA-treated EFB RC biocomposite films.

The effect of EFB content on the modulus of elasticity of the untreated and MMA treated EFB RC biocomposite films is illustrated in Figure 4. The modulus of elasticity for the untreated and treated EFB RC biocomposite films increased as the EFB content increased from 1 to 2 wt%. Addition of more EFB content (3 and 4 wt%) resulted in the declination of modulus of elasticity for both types of film. At 1, 2 and 3 wt% EFB content, the treated RC films demonstrated higher modulus of elasticity than the untreated biocomposite films. As mentioned earlier, MMA treatment of EFB has improved the matrix-filler interactions, thereby increasing the stiffness of the biocomposite films. The higher stiffness required higher stress to prompt movement, thereby resulting in the increase in the modulus of elasticity of the biocomposite films. However, when 4 wt% EFB was employed, the modulus of elasticity value of the RC biocomposite film with the untreated EFB statistically shows no significant difference to the RC biocomposite film with the MMA-treated EFB. Even though the EFB has been treated with the MMA, the high content and overcrowding of the cellulose might hinder the aligning of the cellulosic molecular chains in an orderly manner during the regeneration process, hence more flexible chains are formed. This results in no significant increase in the modulus of elasticity of the treated EFB RC biocomposite when compared with the untreated one.

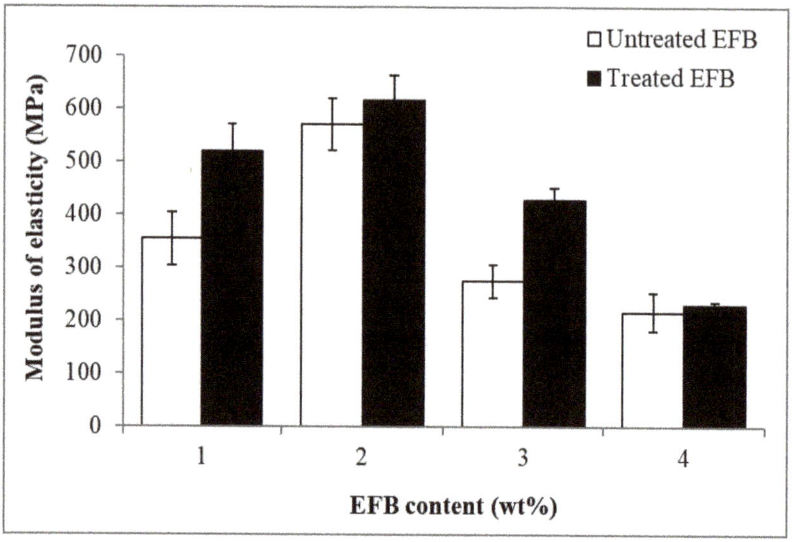

Figure 4. Effect of EFB content on modulus of elasticity of untreated and MMA-treated EFB RC biocomposite films.

4.2. X-Ray Diffraction (XRD)

Figure 5 shows the XRD pattern of the untreated and treated RC biocomposite films with MMA at 2 and 4 wt% of EFB content, while Table 1 summarizes their crystallinity index (CrI). The diffraction peaks of all the samples were observed in the 2θ range of 11.0–13.0° and 19.0–21.0°. These peaks suggest the cellulose II structure, as stated by Zhang et al. [35]. The intensity of the peaks of the treated films is higher compared to the untreated biocomposite films, thus the CrI is also higher. This was due to the treatment of EFB with MMA which has improved and enhanced the dissolution and regeneration of cellulose, thus a higher degree of cellulose II was formed.

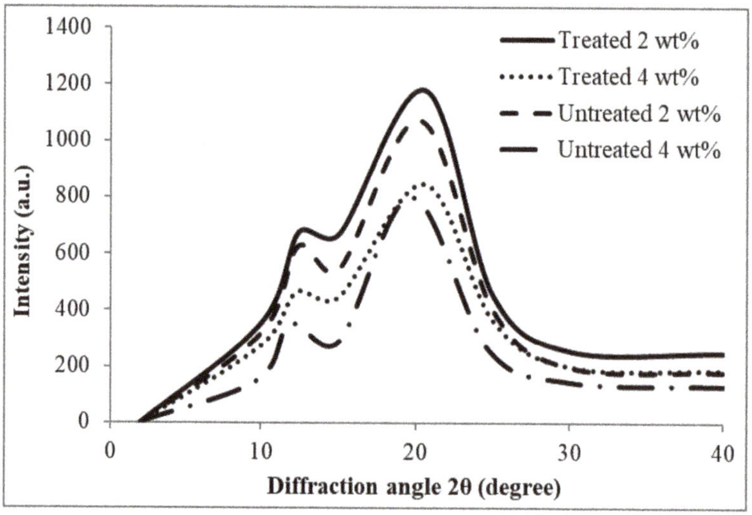

Figure 5. XRD pattern of untreated and MMA-treated RC biocomposite films containing 2 and 4 wt% EFB.

Table 1. The crystallinity index (CrI) of untreated and MMA-treated EFB RC biocomposite films containing 2 and 4 wt% EFB.

Biocomposite Films	Untreated	Treated
	CrI (%)	CrI (%)
2 wt% EFB RC	41.9	45.0
4 wt% EFB RC	38.5	40.7

4.3. Morphology Study

Figure 6a, and Figure 6b–e represent the SEM images of the EFB powder and fractured surface of the RC biocomposite films at 2 and 4 wt% of EFB content, respectively. The SEM image of the EFB powder indicates that it contains particles with irregular shape. Upon regeneration process, the particulate form of the EFB disappeared as observed through the images of the biocomposite films (Figure 6b–e). A smoother and more homogeneous surface can be observed in the EFB RC biocomposite films that contain 2 wt% EFB as compared to the ones containing 4 wt% EFB (untreated and MMA treated). This shows that, lower content of EFB (2 wt%) resulted in better dissolution and dispersion of the EFB in the structure of the RC biocomposite. The presence of higher EFB content (4 wt%) resulted in poorer dispersion of the non-dissolved EFB (filler phase) and lower rate of dissolution of the regenerated cellulose (matrix phase). Similar results were attained by Liu et al. [12], which concluded that the homogeneous surface was obtained in films regenerated from cellulose solution containing low cellulose content. Furthermore, it can also be witnessed that the chemical treatment of the EFB using MMA resulted in a more homogeneous and smoother surface morphology of the RC biocomposite. The micrographs also indicate that fewer aggregates were observed. This implied that the reinforcement was well-surrounded by the regenerated cellulose matrix, due to improved interactions between both filler and matrix phases. The untreated RC biocomposite film with 4 wt% of EFB content (Figure 6c) illustrates a rough and heterogeneous morphology, which could be the cause of its tensile property reduction as mentioned previously.

Figure 6. *Cont.*

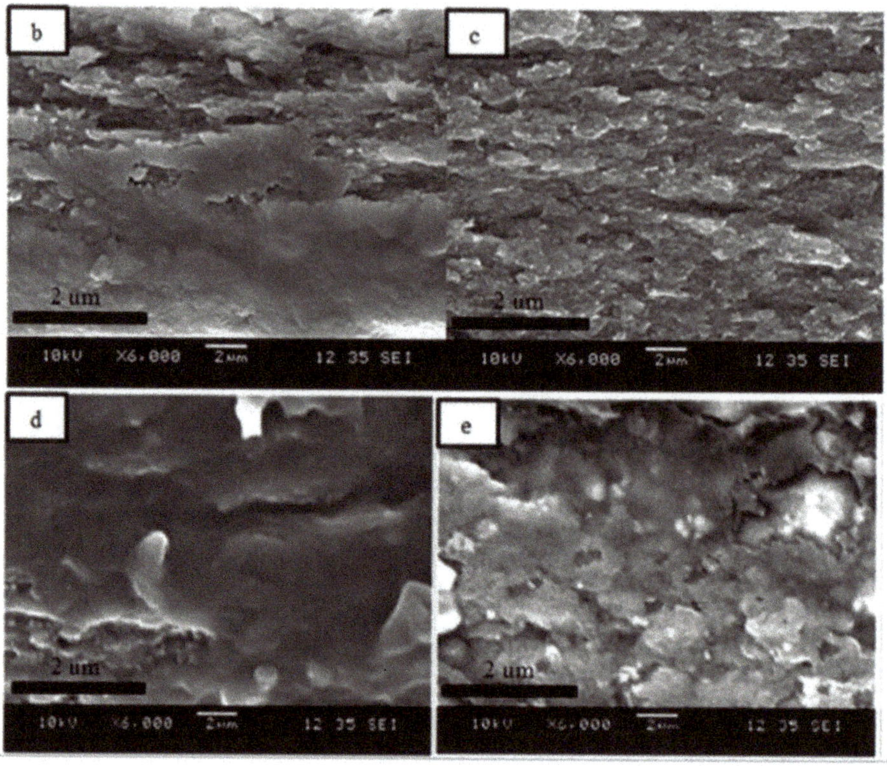

Figure 6. (**a**) SEM of EFB powder. Tensile fractured surface of (**b**) untreated RC biocomposite film containing 2 wt% EFB, (**c**) untreated RC biocomposite film containing 4 wt% EFB, (**d**) treated RC biocomposite film containing 2 wt% EFB and (**e**) treated RC biocomposite film containing 4 wt% EFB as observed through SEM.

4.4. Thermal Gravimetric Analysis

Thermal gravimetric analysis (TGA) curves of the untreated and MMA-treated EFB RC biocomposite film containing 2 wt% of EFB were illustrated in Figure 7, while Table 2 tabulated the TGA data for all those materials. At the early stages of the thermal decomposition process (which is below 100 °C), the weight loss was due to the volatilization and vaporization of water [34,35]. At temperature range of 95–191 °C, the weight loss for both types of biocomposite film remained almost unchanged. This could be due to the removal of the remaining hemicellulose and extractives upon the alkaline treatment [34]. At temperatures higher than 190 °C, the decomposition of the RC biocomposite films generally occurred in two stages and this trend was also observed by Yeng et al. [34] and Reddy et al. [35]. At the first stage, a sharp weight loss can be observed in the temperature range of 191–282 °C for the treated RC film and 182–285 °C for the untreated film. This is associated with the degradation process of the crystalline cellulose. The second degradation stage occurred at 289–427 °C for treated film and 288–425 °C for untreated film, indicating that the cellulose almost completely burnt due to the continuing thermal oxidative degradation [35,36]. Apparently, the treated biocomposite films exhibit higher maximum degradation temperature (T_{dmax}) as opposed to the untreated films. This is ascribed to the better compatibility between the matrix and filler phase. Most obvious at temperatures between 300 and 600 °C, the treated films demonstrate lower weight loss. This suggests that the MMA treatment of the EFB can improve the thermal stability of the EFB RC biocomposite film. Other researchers also realized the benefit of chemical treatment of fiber in improving its thermal stability. For instance, Chen et al. [37] have analyzed and reported the thermal properties of the cellulose fibers from rice straw, and concluded that the treated cellulose fibers possess higher degradation temperatures and higher thermal stability than the untreated ones.

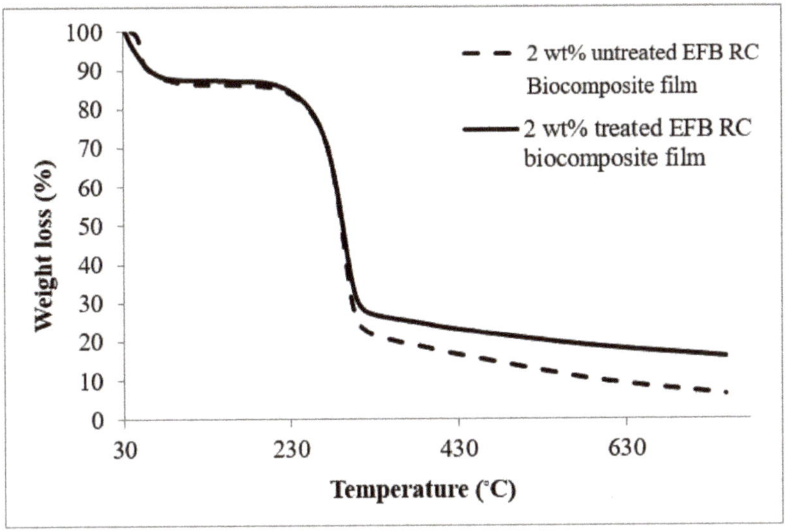

Figure 7. Thermal gravimetric analysis (TGA) curves of untreated and MMA-treated EFB RC biocomposite films containing 2 wt% EFB.

Table 2. TGA data of EFB RC biocomposite films containing 2 wt% EFB.

Biocomposite Films	T_{dmax} (°C)	Weight Loss (%)	
		300 °C	600 °C
2 wt% untreated EFB RC	293	65.1	81.6
2 wt% treated EFB RC	295	59.3	78.6

4.5. Fourier Transform Infrared Spectroscopy Analysis (FTIR)

Figure 8a,b illustrates the FTIR spectra of the untreated and MMA-treated EFB RC biocomposite films. Both untreated and treated EFB RC biocomposite films demonstrate similar peaks, suggesting that both films have comparable chemical compositions. The broad peak at 3373 cm^{-1} relates to the appearance of –OH stretching vibration. The band which can be seen at 2961 and 2930 cm^{-1}, is attributed to the C-H group's stretching vibration [10,38]. Vibration of CH_2 groups is implied by the peak at 2857 cm^{-1}. The peak at 1725 cm^{-1} corresponds to the carbonyl groups (C=O) appears due to the chemical treatment of the EFB using MMA. The reason for the appearance of this peak in the spectra of both untreated and treated EFB RC biocomposite films is that there might be some residual of the lignin which also exhibits these functional groups. The absorption band observed at 1463 cm^{-1} relates to the CH_3 and CH_2 deformation vibration, while the peak at 1270 cm^{-1} displays the presence of C-O bond stretching vibration [10,38]. The absorption band detected at 1188 cm^{-1} for the treated biocomposite films indicated the anti-symmetric C-O-C stretch from the ester [10]. The band which is seen at 1118 cm^{-1} suggests the C-O-C pyranose ring skeletal vibration. The appearance of the band at 1020 cm^{-1} in both untreated and treated films indicates the presence of C-O bond stretching vibration of C-O-C linkage in the anhydroglucose unit (AGU). The band at ~877 cm^{-1} is associated with β-glycosidic linkages between glucose in the EFB cellulose as a result of ring vibration and –OH bending of the glycosidic C-H group. This type of response can only be seen in the regenerated cellulose. Figure 9 demonstrates the schematic reaction between EFB and MMA. Upon chemical treatment, the hydrogen bonds of the EFB cellulose reduced due to replacement of its –OH group with the ester group of the MMA. As a result, the EFB cellulose can be better dissolved in the solvent, leading to an improved regeneration process and greater interphase bonding between the filler phase and matrix phase of the EFB RC biocomposite. As observed through the SEM analysis, more homogeneous RC biocomposite was obtained that also led to the improved mechanical and thermal properties of the EFB RC biocomposite film.

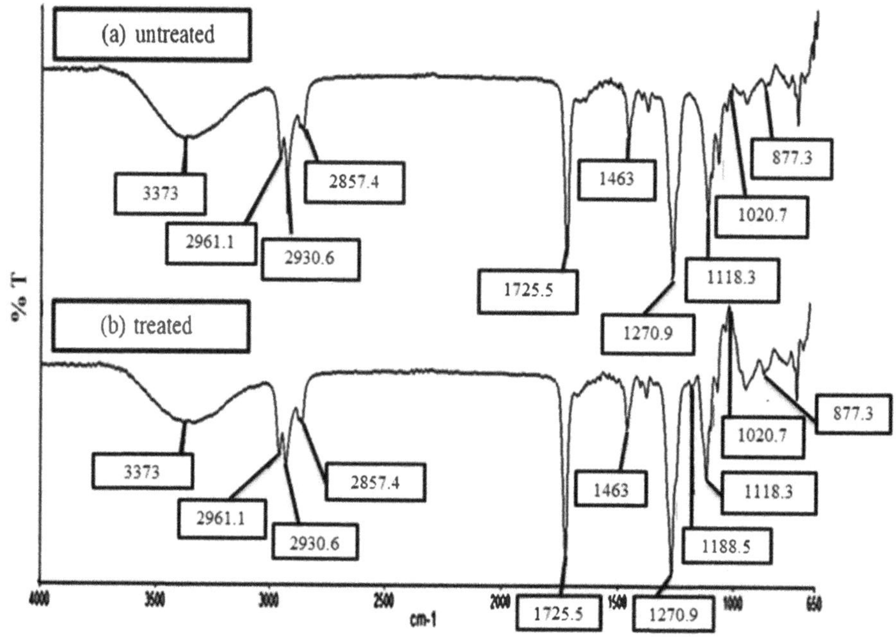

Figure 8. FTIR spectra of (**a**) untreated and (**b**) treated RC EFB biocomposite films with MMA.

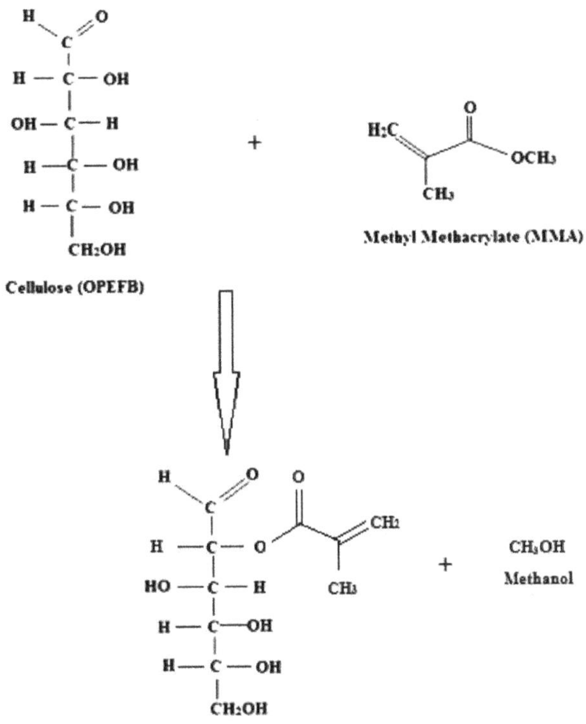

Figure 9. The schematic reaction between EFB and MMA.

5. Conclusions

The effects of EFB content and chemical treatment of EFB using methyl methacrylate (MMA) on the mechanical and thermal properties of the EFB RC biocomposite films were studied. Results demonstrated that the treated EFB films exhibit higher and greater tensile strength and modulus of elasticity, but lower elongation at break compared to the untreated films. The treated EFB RC films also attained higher crystallinity index. Surface morphology of the treated biocomposite films indicates smoother surface due to the MMA treatment. Thermal analysis demonstrates that the treated biocomposite film possesses higher T_{dmax} and lower weight loss at 600 °C. This suggests that the treated biocomposite film has greater thermal stability. The presence of ester in the treated EFB RC biocomposite films with MMA was proven and affirmed by FTIR analysis. The substitution of the –OH unit of the EFB cellulose with the ester unit of the MMA has facilitated the dissolution process of the EFB during the regeneration process due to weakening of the hydrogen bonding of the cellulose. A more homogeneous biocomposite structure developed which led to the improvement of both tensile properties and thermal stability of the EFC RC biocomposite film.

Author Contributions: Conceptualization, S.H. and A.F.O.; methodology, C.L.L.; software, A.A.; validation, N.L.I.Z., A.F.O. and S.H.; formal analysis, A.F.O.; investigation, A.A.; resources, A.A.A.; data curation, A.A.A.; writing—original draft preparation, N.L.I.Z.; writing—review and editing, A.F.O.; visualization, C.L.L.; supervision, S.H.; project administration, A.F.O.; funding acquisition, A.A.A. All authors have read and agreed to the published version of the manuscript.

Funding: This research received no external funding.

Acknowledgments: The authors would like to thank the School of Materials Engineering for the facilities and Madam Marliza Mosthapa for her kind help in experimental works.

Conflicts of Interest: This manuscript, or its contents in some other form, has not been published previously by any of the authors and/or is not under consideration for publication in another journal at the time of submission. All authors declare that there is no conflict of interest regarding the publication of this paper.

References

1. Haider, T.P.; Völker, C.; Kramm, J.; Landfester, K.; Wurm, F.R. Plastics of the Future? The Impact of Biodegradable Polymers on the Environment and on Society. *Angew. Chem. Int. Ed.* **2019**, *58*, 50–62. [CrossRef] [PubMed]
2. Balakrishnan, P.; Gopi, S.; Sreekala, M.S.; Thomas, S. UV resistant transparent bionanocomposite films based on potato starch/cellulose for sustainable packaging. *Starch-Starke* **2018**, *70*, 1700139. [CrossRef]
3. Osman, A.F.; Ashafee, A.M.T.; Adnan, S.A.; Alakrach, A. Influence of Hybrid Cellulose/Bentonite Fillers on Structure, Ambient and Low Temperature Tensile Properties of Thermoplastic Starch Composites. *Polym. Eng. Sci.* **2020**, *60*, 810–822. [CrossRef]
4. Syed Adam, S.N.S.; Osman, A.F.; Shamsudin, R. Tensile Properties, Biodegradability and Bioactivity of Thermoplastic Starch (TPS) / Bioglass Composites for Bone Tissue Engineering. *Sains Malay.* **2018**, *47*, 1303–1310. [CrossRef]
5. Yang, Y.; Shen, H.; Wang, X.; Qiu, J. Preparation of Nanolignocellulose/Chitin Composites with Superior Mechanical Property and Thermal Stability. *J. Bioresour. Bioprod.* **2019**, *4*, 251–259.
6. Hassan, M.; Bai, J.; Dou, D. Biopolymers; Definition, Classification and Applications. *Egypt. J. Chem.* **2019**, *62*, 1725–1737. [CrossRef]
7. Miao, X.; Lin, J.; Bian, F. Utilization of discarded crop straw to produce cellulose nanofibrils and their assemblies. *J. Bioresour. Bioprod.* **2020**, *5*, 26–36. [CrossRef]
8. Sanjay, M.R.; Arpith, G.R.; Laxmana Naik, L.; Gopalakrishna, K.; Yogesha, B. Applications of Natural Fibers and Its Composites: An Overview. *Nat. Resour.* **2016**, *7*, 108–114. [CrossRef]
9. Ismail, I.; Osman, A.F.; Ping, T.L. Effects of ultrasonication process on crystallinity and tear strength of thermoplastic starch/cellulose biocomposites. *IOP Conf. Ser. Mat. Sci. Eng.* **2019**, *701*, 012045.
10. Zailuddin, N.L.I.; Osman, A.F.; Rahman, R. Morphology, Mechanical Properties and Biodegradability of All-Cellulose Composite Films from Oil Palm Empty Fruit Bunch. *SPF Polymers* **2020**, *1*, 4–14. [CrossRef]
11. Huber, T.; Mussig, J.; Curnow, O.; Pang, S.; Bickerton, S.; Staiger, M.P. A Critical review of all-cellulose composites. *J. Mater. Sci.* **2012**, *47*, 1171–1186. [CrossRef]
12. Liu, Z.; Wang, H.; Li, Z.; Lu, X.; Zhang, X.; Zhang, S.; Zhou, K. Characterization of the regenerated cellulose films in ionic liquids and rheological properties of the solutions. *Mater. Chem. Phys.* **2011**, *128*, 220–227. [CrossRef]
13. Jiang, G.; Huang, W.; Li, L.; Wang, X.; Pang, F.; Zhang, Y.; Wang, H. Structure and properties of regenerated cellulose fibers from different technology processes. *Carbohydr. Polym.* **2012**, *87*, 2012–2018. [CrossRef]
14. Earle, M.J.; Seddon, K.R. Ionic liquids. Green solvents for the future. *Pure Appl. Chem.* **2000**, *72*, 1391–1398. [CrossRef]
15. Xia, Z.; Li, J.; Zhang, J.; Zhang, X.; Zheng, X.; Zhang, J. Processing and Valorization of Cellulose, Lignin and Lignocellulose Using Ionic Liquids. *J. Bioresour. Bioprod.* **2020**, *5*, 79–95. [CrossRef]
16. Zhu, S.; Wu, Y.; Chen, Q.; Yu, Z.; Wang, C.; Jin, S.; Ding, Y.; Wu, G. Dissolution of cellulose with ionic liquids and its application: A mini-review. *Green Chem.* **2006**, *8*, 325–327. [CrossRef]
17. Biganska, O.; Navard, P. Kinetics of precipitation of cellulose from cellulose-NMMO-water solutions. *Biomacromolecules* **2005**, *6*, 1948–1953. [CrossRef] [PubMed]
18. Soheilmoghaddam, M.; Wahit, M.U. Development of regenerated cellulose/halloysite nanotube bionanocomposite films with ionic liquid. *Int. J. Biol. Macromol.* **2013**, *58*, 133–139. [CrossRef]
19. Pang, J.; Liu, X.; Zhang, X.; Wu, Y.; Sun, R. Fabrication of cellulose film with enhanced mechanical properties in ionic liquid 1-allyl-3-methylimidaxolium chloride (AmimCl). *Materials* **2013**, *6*, 1270–1284. [CrossRef]
20. Liu, X.; Pang, J.; Zhang, X.; Wu, Y.; Sun, R. Regenerated cellulose film with enhanced tensile strength prepared with ionic liquid 1-ethyl-3-methylimidazolium acetate (EMIMAc). *Cellulose* **2013**, *20*, 1391–1399. [CrossRef]
21. Wang, L.; Gao, L.; Cheng, B.; Ji, X.; Song, J.; Lu, F. Rheological behaviors of cellulose in 1-ethyl-3-methylimidazolium chloride/ dimethylsulfoxide. *Carbohydr. Polym.* **2014**, *110*, 292–297. [CrossRef]

22. Swatloski, R.P.; Spear, S.K.; Holbrey, J.D.; Rogers, R.D. Dissolution of cellose with ionic liquids. *J. Am. Chem. Soc.* **2002**, *124*, 4974–4975. [CrossRef]
23. Ying, T.Y.; Teong, L.K.; Abdullah, W.N.W.; Peng, L.C. The effect of various pretreatment methods on oil palm empty fruit bunch (EFB) and kenaf core fibers for sugar production. *Procedia Environ. Sci.* **2014**, *20*, 328–335. [CrossRef]
24. Nazir, M.S.; Wahjoedi, B.A.; Yussof, A.W.; Abdullah, M.A. Eco-friendly extraction and characterization of cellulose from oil palm empty fruit bunches. *Bioresources* **2013**, *8*, 2161–2172. [CrossRef]
25. Zailuddin, N.L.I.; Osman, A.F.; Rozyanty Rahman, R. Effects of Formic Acid Treatment on Properties of Oil Palm Empty Fruit Bunch (OPEFB)-Based All Cellulose Composite (ACC) Films. *J. Eng Sci.* **2020**, *16*, 75–95.
26. Kobayashi, H.; Ito, Y.; Komanoya, T.; Hosaka, Y.; Dhepe, P.L.; Kasai, K.; Hara, K.; Fukuoka, A. Synthesis of sugar alcohols by hydrolytic hydrogenation of cellulose over supported metal catalysts. *Green Chem.* **2011**, *13*, 326–333. [CrossRef]
27. Tolonen, L.K.; Zuckerstätter, G.; Penttilä, P.A.; Milacher, W.; Habicht, W.; Serimaa, R.; Kruse, A.; Sixta, H. Structural changes in microcrystalline cellulose in subcritical water treatment. *Biomacromolecules* **2011**, *12*, 2544–2551. [CrossRef]
28. Kim, J.W.; Kim, K.S.; Lee, J.S.; Park, S.M.; Cho, H.Y.; Park, J.C.; Kim, J.S. Two-stage pretreatment of rice straw using aqueous ammonia and dilute acid. *Bioresource Technol.* **2011**, *102*, 8992–8999. [CrossRef]
29. Thakur, M.K.; Gupta, R.K.; Thakur, V.K. Surface modification of cellulose using silane coupling agent. *Carbohydr. Polym.* **2014**, *111*, 849–855. [CrossRef]
30. Wang, B.; Li, J.; Zhang, J.; Li, H.; Chen, P.; Gu, Q.; Wang, Z. Thermo-mechanical properties of the composite made of poly (3-hydroxybutyrate-co-3-hydroxyvalerate) and acetylated chitin nanocrystals. *Carbohydr. Polym.* **2013**, *95*, 100–106. [CrossRef]
31. Yu, H.Y.; Qin, Z.Y. Surface grafting of cellulose nanocrystals with poly (3-hydroxybutyrate-co-3-hydroxyvalerate). *Carbohydr. Polym.* **2014**, *101*, 471–478. [CrossRef] [PubMed]
32. Tanjung, F.A.; Husseinsyah, S.; Hussin, K.; Tahir, I. Chemically chitosan modified with methyl methacrylate and its effect on mechanical and thermal properties of polypropylene composites. *Indones. J. Chem.* **2013**, *13*, 114–121. [CrossRef]
33. Rasidi, M.S.M.; Husseinsyah, S.; Leng, T.P. Chemical modification of Nypa fruticans filled polylactic acid/recycled low-density polyethylene biocomposites. *Bioresources* **2014**, *9*, 2033–2050.
34. Segal, L.G.J.M.A.; Creely, J.J.; Martin, A.E.; Conrad, C.M. An empirical method for estimating the degree of crystallinity of native cellulose using the X-ray diffractometer. *Text. Res. J.* **1959**, *29*, 786–794. [CrossRef]
35. Zhang, H.; Wu, J.; Zhang, J.; He, J. 1-Allyl-3-methylimidazolium chloride room temperature ionic liquid: A new and powerful nonderivatizing solvent for cellulose. *Macromolecules* **2005**, *38*, 8272–8277. [CrossRef]
36. Reddy, K.O.; Zhang, J.; Zhang, J.; Rajulu, A.V. Preparation and properties of self-reinforced cellulose composite films from Agave microfibrils using an ionic liquid. *Carbohydr. Polym.* **2014**, *114*, 537–545.
37. Chen, X.; Yu, J.; Zhang, Z.; Lu, C. Study on structure and thermal stability properties of cellulose fibers from rice straw. *Carbohydr. Polym.* **2011**, *85*, 245–250. [CrossRef]
38. Chen, Y.; Pang, L.; Li, Y.; Luo, H.; Duan, G.; Mei, C.; Xu, W.; Zhou, W.; Liu, K.; Jiang, S. Ultra-thin and highly flexible cellulose nanofiber/silver nanowire conductive paper for effective electromagnetic interference shielding. *Compos. Part A-Appl. Sci.Manuf.* **2020**, *135*, 105960. [CrossRef]

polymers

MDPI

Article

Biodegradable Starch/Chitosan Foam via Microwave Assisted Preparation: Morphology and Performance Properties

Xian Zhang, Zhuangzhuang Teng and Runzhou Huang *

Co-Innovation Center of Efficient Processing and Utilization of Forest Products, College of Materials Science and Engineering, Nanjing Forestry University, Nanjing 210037, China; shinezhang17@163.com (X.Z.); RowanZJ@163.com (Z.T.)
* Correspondence: runzhouhuang@njfu.edu.cn

Received: 19 October 2020; Accepted: 2 November 2020; Published: 6 November 2020

Abstract: The effects of chitosan (CTS) as the reinforcing phase on the properties of potato starch (PS)-based foams were studied in this work. The formic acid solutions of CTS and PS were uniformly mixed in a particular ratio by blending and then placed in a mold made of polytetrafluoroethylene for microwave treatment to form starch foam. The results showed that the molecular weight and concentration of CTS could effectively improve the density and compressive properties of starch-based foams. Furthermore, orthogonal experiments were designed, and the results showed that when the molecular weight of CTS in foams is 4.4×10^5, the mass fraction is 4 wt%, and the mass ratio of CTS–PS is 3/4.2; the compressive strength of foams is the highest at approximately 1.077 mPa. Furthermore, Fourier transform infrared spectroscopy analysis demonstrated the interaction between starch and CTS, which confirmed that the compatibility between CTS and PS is excellent.

Keywords: chitosan; potato starch; microwave; foam; orthogonal experiments

1. Introduction

Global pollution is increasingly becoming a serious concern, causing extreme climate changes and natural disasters. The huge pollution of traditional plastic products in the environment has extensively raised people's concern. Traditional plastics such as polystyrene (PS), soft (hard) polyurethane, polyvinyl chloride, and polyolefin are widely used as the matrix of foaming materials, but due to the poor operability and degradability of these plastics, most of the foaming materials are wasted, leading to increasingly serious white pollution. With people's increased awareness of environmental protection, considerable attention has been paid to the development of environmentally friendly materials. Therefore, bio-based and biodegradable polymers as replacement materials for traditional nonbiodegradable plastics have been extensively studied [1–4]. Biodegradable plastics in the natural environment are decomposed into carbon dioxide (CO_2) and oxygen (O_2) by the action of water (H_2O), light, and microorganisms, and they hardly pollute the environment. Conventional plastics are nondegradable, which degrade slowly and require a long time, even hundreds of years, to degrade completely. They also impact the soil negatively during this time. Therefore, the exploration of degradable plastics that can replace traditional plastics holds great practical significance.

Starch, a rich polysaccharide substance found on earth, is an important raw material for the synthesis of biodegradable plastics. Starch has excellent biodegradability, and it can be decomposed by microorganisms in the natural environment and finally be metabolized into H_2O and CO_2. Starch-based plastics are beneficial in being nontoxic, harmless, widely available, and cheap and also possess relatively superior properties. Use of these plastics effectively can alleviate the problem of white pollution and the crisis of biochemical energy shortage [5,6]. Biodegradable porous materials prepared from starch are being widely used in the fields of food packaging and medical tissue engineering [7]. To ensure complete degradability of the system, various biodegradable natural polymers such as cellulose, lignin, and chitosan (CTS) and polyesters such as polylactic acid (PLA), polycaprolactone, polybutylene succinate, and polyvinyl alcohol have been blended with starch in an attempt to synthesize biodegradable plastics [8–13]. Loercks et al. [14] successfully developed a degradable thermoplastic starch-based product by compounding starch with a hydrophobic biodegradable polymer, which greatly enhanced its aerobic degradation rate. Ferri et al. [15] reported that the addition of glycerin, polyethylene glycol, and other plasticizers to the blend of thermoplastic starch (TPS) and PLA could significantly reduce the cost, enhance mechanical performance, and broaden the application ranges of PLA-TPS blends. Bénézet et al. [16] prepared fiber-reinforced foams by the extrusion method and reported that the addition of fibers increases the expansion index and significantly reduces the water absorption of starch-based foams; the foams with 10% hemp fibers presented excellent mechanical properties.

CTS consists of β-(1,4)-linked 2-amino-deoxy-d-glucopyranose, and it can be synthesized by deacetylation of chitin [17,18]. It is soluble in acidic solutions because of the protonation of its $-NH_2$ group at the C–2 position of the glucosamine unit. Owing to its biodegradability and unique physicochemical properties, it is widely used in the preparation of hydrogels, films, fibers, or sponges and in the field of biomedicine [19]. Although CTS is a novel substance that possesses excellent film-forming potential, permeability, and antibacterial properties, it has many deficiencies. Dang et al. [20] prepared TPS or CTS films by using different CTS concentrations and confirmed that CTS is distributed on the surface of films. Due to the high crystallinity and hydrophobicity of CTS and its ability to interact with starch-forming intermolecular hydrogen bonds, CTS on the membrane surface improves the water vapor and oxygen barrier properties of materials and decreases the hydrophilicity of the membrane surface.

Starch and chitosan are natural high molecular polymers with excellent biocompatibility and biodegradability, which can be completely degraded without causing harm to the environment [21]. Currently, most research has focused on the synthesis of the starch–CTS membrane [22–25], and the application of CTS-starch foams in the field has been rare. The foam prepared by starch and chitosan is a kind of biological material, which is easy to be processed and molded. The chitosan in foam can promote the formation of blood coagulation and thrombosis, inhibit the growth of various bacteria and fungi, and promote the repair of damaged tissues [26]. Therefore, the foam is expected to be a good hemostatic material and packaging material for food and medicine. In this paper, we aimed to manufacture a new foam material by using CTS and potato starch (PS) in a polytetrafluoroethylene mold for microwave treatment, analyze the effects of molecular weight and concentration of CTS on the morphological characteristics and functional properties, such as physicochemical, mechanical, and thermal properties, of starch-based foam, and determine the optimum proportion of CTS and PS in foams to achieve excellent performance. We hope to enable further development of starch-based foams for use as active materials in biomedical applications as well as food and drug packaging.

2. Materials and Methods

2.1. Materials

PS (with approximately 75% amylopectin and 25% amylose) and CTS (low viscosity ≤200 mPa·s; medium viscosity: 200–400 mPa·s; high viscosity ≥400 mPa·s) were purchased from Shanghai Meryer Company (Shanghai, China). Formic acid (96%) was purchased from Shanghai Macklin Company (Shanghai, China).

2.2. Sample Preparation

(1) Determination of the molar mass of CTS

Acid hydrolysis was used to determine the molecular weight of three CTS samples used in this study. The dried CTS powder was dissolved in 0.2 M acetic acid (HAc)/0.1 M sodium acetate (NaAc) solvent (pH = 4) and different concentrations of the solution were prepared; CTS powder was dissolved completely in the solvent by moderately heating the solution at 40 °C. The 10 mL solvent and cooled CTS solution were injected into the Ubbelohde viscometer (capillary diameter = 0.8 mm) (Shanghai Liangjing Glass Instrument Factory, Shanghai, China), and the vertical state Ubbelohde viscometer was placed in a constant temperature water bath at 30 °C for 15 min. The outflow time of solvent and CTS solution (t, t_0) was measured with a stopwatch, and the mean value of the three measurements was recorded (the difference was not more than 0.2 s).

(2) Preparation of CTS-starch based foams

The starch-based foaming materials were formulated using CTS and PS by heating in a microwave oven. A WD900ASL23-2 Galanz microwave oven (Galanz Shanghai Company Limited, Shanghai, China) with microwave emission frequency of 2450 MHz was used to heat CTS-starch compounds under microwaves at 100% power of 900 watts. The container carrying the CTS-starch mixtures was located at the center of the microwave chamber and irradiated by microwaves for the same time (40 s). Different CTS-starch mixtures were packed in special containers and the specific experimental process is shown in Figure 1. After microwave treatment, the samples were conditioned overnight at 25 °C before further characterization.

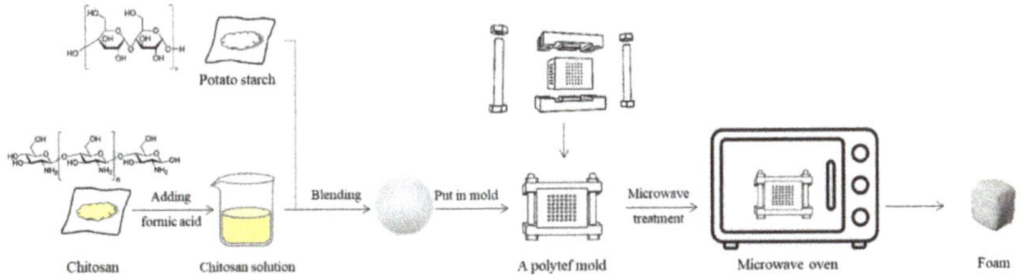

Figure 1. Schematic diagram of preparation of CTS-starch foams.

(3) Experiment design

The comprehensive experiment was designed with three factors at three levels respectively ($3^3 = 27$ combinations, without considering the reproducibility of experiments), namely the viscosity of CTS (A), the mass fraction of CTS (B), and the weight ratio of CTS solution/starch (C). Table 1 shows the experimental design under different factors and levels.

Table 1. Formula of CTS-starch foams.

Factor / Level	Viscosity of CTS (A) (mPa.s)	Mass Fraction of CTS (B) (wt%)	Weight Ratio of CTS/Starch (C) (g)
1	≤200	3.8	3/3.8
2	200–400	4	3/4
3	≥400	4.2	3/4.2

On the basis of the comprehensive experiment, orthogonal test analysis and multi-factor analysis of variance were carried out to explore the interaction factors' influence on the mechanical properties, which would further help the experiment to determine the best process parameter of each level and to determine the relative importance of individual parameters and the combination of process parameters with high performance.

2.3. Characterization

(1) Calculation of CTS molecular weight

When the viscosity of polymer in a wide range of molecular weight is measured with the same viscometer, the relative value of solution viscosity to solvent viscosity or the relative viscosity, η_r, is expressed as t/t_0 (t_0 and t are the outflow time of dilute solution and pure solvent, respectively, under the same temperature and measured using the same viscometer,). The specific viscosity, η_{sp}, which reflects the internal friction effect between polymer and polymer and between pure solvent and polymer, is expressed as follows:

$$\eta_{sp} = (\eta - \eta_0)/\eta_0 = (\eta_0/\eta_0) - 1 = \eta_r - 1 \quad (1)$$

The intrinsic viscosity $[\eta]$ is defined as the reduced specific viscosity (η_{sp}/c) or the relative viscosity of the inherent relative viscosity ($ln\ \eta_r/c$) of a solution in the case of infinite dilution of solution mass concentration (c), which is expressed as follows:

$$[\eta] = \lim_{c=0} \eta_{sp}/c = \lim_{c=0} ln\eta_r/c \quad (2)$$

The intrinsic viscosity $[\eta]$ of polymer solution exhibits a specific relation with the molecular weight of polymer at a specific temperature. According to the Mark–Houwink–Sakurada (MHS) equation, the following empirical formula is used for calculating $[\eta]$ and the molecular weight of polymer solution [27–29]:

$$[\eta] = KM^a, \quad (3)$$

where M is the molecular weight, K and α are constants for given solute–solvent system and temperature, respectively. In this study, K and α were 6.59×10^{-3} and 0.88, respectively.

The relationship between $ln\ \eta_r/c$ (or η_{sp}/c) and mass concentration c is linear. Thus, a straight line can be drawn by plotting $ln\ \eta_r/c$ (or η_{sp}/c). When c is extrapolated from Equation (2), the line should cross a point on the y-axis; the intercept is the intrinsic viscosity [η]; the molecular mass of polymer can be calculated using Equation (3).

(2) Morphology of starch-based foams

The starch-based composite structure was investigated using scanning electron microscopy (SEM) (FEI 200 Quanta FEG, pressure 0.9 Torr) (FEI Company, Hillsboro, OR, USA). Foam samples were sputter-coated with platinum for 20 min after cutting with a razorblade into 3-mm thick slices, and the transversal cross-section area was examined using SEM.

(3) Pore size, porosity, and density of CTS–PS foams

Image analysis (Image-J) software (National Institutes of Health, Bethesda, MD, USA) was used to measure the average diameters of nearly 100 pores. Plot of pore size distribution enabled the comparison of cellular structure for different formulations of starch foams. The finished foam was cut into five samples of $24 \times 24 \times 32$ mm^3 size, and then the exact size of the sample was measured to ensure that the error was less than 0.02 mm. The sample mass was weighed with an electronic balance to make the result accurate to 0.001 g, and finally the density of the material was calculated using the following formula:

$$\rho = \frac{M}{V} \times 10^3 \quad (4)$$

where ρ = density (g/cm^3), M = foam weight (g), V = foam volume (cm^3)

(4) Mechanical property

According to the standard GB/T 8813-2008, the compressive strength of foam materials was tested using a CMT 4000 universal testing machine (Mechanical Testing & Simulation Company, Eden Prairie, MN, USA). The vertical direction of foam growth was consistent with the direction of compression. The compressive strength of the foam material is the ratio of the maximum compressive stress to the original cross-sectional area of the sample, when the relative deformation of the material is less than 10%, which is expressed in kPa.

The specific formula is as follows:

$$P = F/S = F \times 10^4/(l \times w) \quad (5)$$

where P = compressive strength (kPa), F = compressive stress (kN), S = cross-sectional area (cm^3), l = length (mm), w = width (mm).

(5) Fourier transform infrared analysis

The dried samples and potassium bromide were ground into a fine powder in an agate mortar and pressed into tablets. The Fourier transform infrared (FTIR) spectrum of material was measured using the VERTEX 80 V type FTIR spectrometer (German Bruker company, Karlsruhe, Germany) and the spectrum at a range of 4000–400 cm^{-1} was acquired for every sample. The processing software OMNIC 8.0 (Thermo Nicolet Corporation, Fitchburg, WI, USA) was used to perform automatic baseline correction of the collected infrared spectrum to ensure its baseline level.

(6) Thermal characterization

The total mass loss on a TG209F3 analyzer (Netzsch Corporation, Bavarian, Germany) was assessed using the thermogravimetric (TG) method. For the 6 mg samples dried for 24 h at 70 °C, the experimental temperature was set as 30 ± 3 °C to 800 °C in nitrogen atmosphere. The heating rate was 10 °C/min, and the experimental flow rate was 30 mL/min.

(7) Differential scanning calorimetry (DSC) test

The glass transition temperature of sample was evaluated by a 200 F3 Differential Scanning Calorimeter (Netzsch Corporation, Bavarian, Germany). The dried sample of around 3 mg was placed in an aluminum crucible and sealed. An empty aluminum box was used as a control and nitrogen was used as purge gas. The temperature ranged from 20 to 180 °C and the heating rate was 10 °C/min.

(8) Solubility of foams in water

The samples were dried in a drying oven at 60 °C for 24 h and then soaked in deionized water at room temperature. The shape of the sample in water was recorded until it dissolved.

3. Results and Discussion

3.1. Molecular Weight of CTS

The viscosity of CTS solution at a particular concentration directly reflects its molecular weight. Assuming other factors to be fixed, the higher the relative molecular weight of CTS, the higher will be the viscosity of the CTS solution, and the smaller the relative molecular weight is, the smaller will be the viscosity of the CTS solution. Figure 2 illustrates the trend line of the CTS sample, with a low viscosity extrapolated to the intersection point of the straight line and y-axis to be 502.39, implying that the characteristic viscosity of the CTS, $[\eta]$ was 502.39, and the molecular weight of the CTS sample calculated according to the MHS equation was 3.5×10^5. Likewise, the molecular weights of other two types of CTS were also estimated through the same calculation; Table 2 summarizes the values of the molecular weight. The correlation coefficient (R^2) of CTS with different viscosities was greater than 0.95, which indicates the reliability of the calculated results. The molecular weight of CTS with viscosity between 200 and 400 mPa·s was 4.4×10^5 and that of CTS with a viscosity of 400 mPa·s was 5.2×10^5 (Table 2), and the difference was significant; the trend was completely consistent with the above-mentioned.

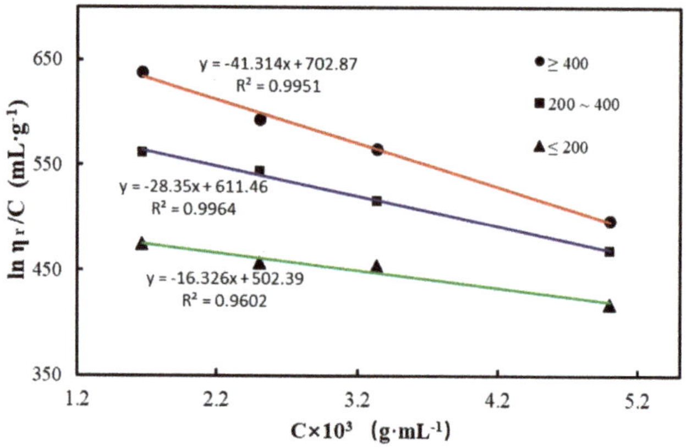

Figure 2. Intrinsic viscosity of CTS in 0.2 M HAc/0.1 M NaAc at 30 °C.

Table 2. Molecular weight of chitosan with different viscosities.

Viscosity (mPa.s)	$[\eta]$ (mL·g^{-1})	R^2	Molecular Weight (Da)
≤200	502.39	0.995	3.5×10^5
200–400	611.46	0.996	4.4×10^5
≥400	702.87	0.960	5.2×10^5

3.2. Morphology of Starch-Based Foams

Figure 3 presents SEM observations of the starch-based foam prepared using the molecular weight of CTS as a variable. Figure 3a,b indicate that the size and distribution of pores on the foam surface produced by microwave radiation are not as apparent as those observed in the inner layer structure. The pores on the surface layer were small and unevenly distributed, whereas the bubbles in the inner structure were significantly larger and more uniform. Starch-based foams made of CTS with different molecular weights also demonstrated distinct characteristics. Comparison of the three types of foam surface structure (Figure 3a,c) revealed the largest number of pores in the foam with CTS of molecular weight 4.4×10^5 (CTS-4.4/PS) (Figure 3e); the quantity of pores in foam with the CTS of molecular weight 3.5×10^5 (CTS-3.5/PS) was relatively smaller (Figure 3a) than that in the CTS of molecular weight 5.2×10^5 (Figure 3c). In short, the increasing order of the number of pores according to the SEM observation is as follows: Figure 3e > Figure 3a > Figure 3c. The trend observed in the pore quantity on the inner structure was exactly the opposite of that observed in the surface morphology. From the SEM images in Figure 3b,d,f, we can clearly observe that the starch-based foam made from medium-viscosity CTS (Figure 3d) has the largest number of pores, relatively smaller pore size (a rough comparison by the yellow circles in Figure 3b,d,f), and relatively regular bubble morphology compared with other foam types.

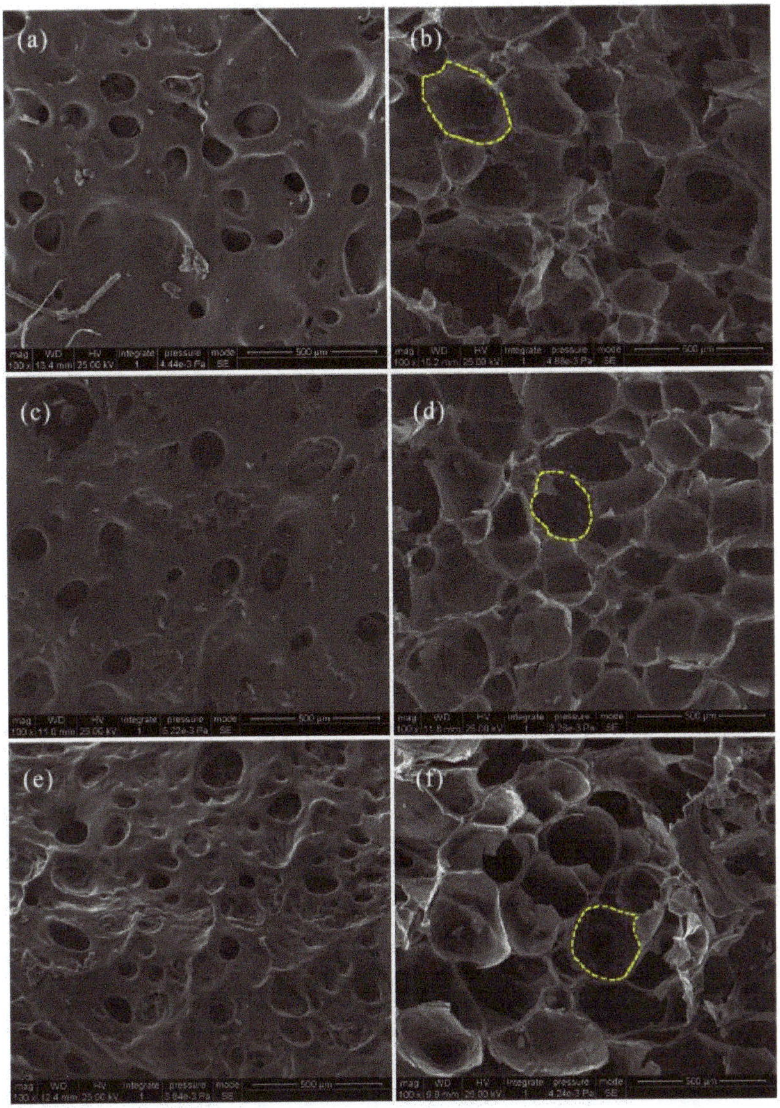

Figure 3. The surface morphologies of foams made from low (**a**), medium (**c**), and high (**e**) viscosity CTS and the corresponding inner morphologies (**b**,**d**,**f**). Scale bar = 500 µm

3.3. Pore Size, Porosity, and Density of CTS–PS Foams

The pore size and porosity were measured using Image-J software, and Table 3 presents the specific values. We observed that at a fixed amount of CTS solution and with an increase in the starch mass, the bubble porosity and porosity decrease in all the CTS–PS composites. The decrease in foam hole and porosity in the same volume mold with increase in starch content is quite reasonable. However, the pore size and porosity of foams with different levels of starch quality also changed with changes in the molecular weight of CTS. When the mass ratio of CTS solution and starch was 3:4, the bubble size and porosity of CTS-4.4/PS foam composites were the smallest (281.7 µm and 55.39%, respectively), whereas the difference in bubble size between CTS-3.5/PS and CTS-5.2/PS foams was not significant, which indicates that the molecular weight of CTS affects the starch-based foams to a certain extent; however, the effect is limited.

Table 3. The pore size and porosity of foams with CTS solution for 4 wt%.

CTS/PS	CTS-3.5		CTS-4.4		CTS-5.2	
	Pore Size (μm)	Porosity (%)	Pore Size (μm)	Porosity (%)	Pore Size (μm)	Porosity (%)
3/3.8	413.4 (14.0)	69.12	298.8 (24.2)	53.24	407.8 (38.5)	72.29
3/4	392.5 (23.9)	66.98	281.7 (27.0)	55.39	394.2 (54.2)	71.72
3/4.2	391.7 (30.2)	66.12	275.8 (36.3)	51.62	389.8 (28.4)	68.28

According to the density of the starch-based foam shown in Figure 2, when the mass fraction of CTS solution was 4%, the density of foams prepared from CTS solution with the same molecular weight gradually increased with the increase in starch content. However, when the mass ratio of CTS solution and starch was constant, the change was different from that observed earlier, and the density of CTS-4.4/PS foams was apparently higher (Figure 4). Figure 4 shows the influence of mass fraction of different CTS solutions on starch-based foams, when the mass ratio of CTS/PS was 3/4.2. Notably, the mass fraction of CTS solution had only a minor effect on the density of the entire foam material, whereas the molecular weight of CTS exerted a greater effect on the same. Table 4 summarizes the specific values of all composites. Combined with the aforementioned SEM observations on the pore size and porosity of foam materials, we can assume that by using a particular mass fraction of the CTS solution, the pore structure and density of materials can be improved by appropriately increasing the starch content. This is because when the starch content increases, the formic acid in the CTS solution and PS generates more esterification reactions in the microwave, and since the volume of the mold is constant, it appears as a prepared foam in a macroscopic view. Thus, the pore size and porosity gradually decrease, and correspondingly, the density of foams gradually increases.

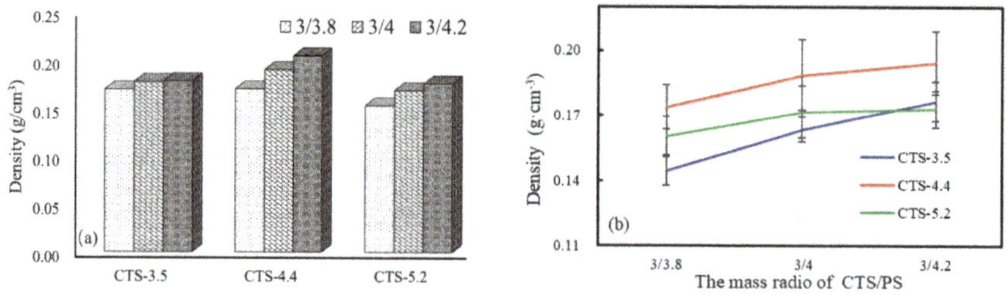

Figure 4. The density of foams with CTS solution for 4.2 wt%.

Table 4. The density values of CTS–PS foams.

Mass Fraction of CTS (wt%)	CTS/PS (g)	CTS-3.5	CTS-4.4	CTS-5.2
3.8	3/3.8	0.170 (0.018)	0.170 (0.008)	0.153 (0.013)
	3/4	0.177 (0.013)	0.190 (0.013)	0.169 (0.008)
	3/4.2	0.178 (0.017)	0.205 (0.037)	0.176 (0.008)
4	3/3.8	0.154 (0.018)	0.187 (0.011)	0.151 (0.009)
	3/4	0.174 (0.010)	0.186 (0.017)	0.169 (0.009)
	3/4.2	0.175 (0.015)	0.197 (0.011)	0.173 (0.008)
4.2	3/3.8	0.144 (0.007)	0.174 (0.010)	0.161 (0.009)
	3/4	0.164 (0.006)	0.189 (0.016)	0.172 (0.012)
	3/4.2	0.177 (0.009)	0.194 (0.015)	0.173 (0.008)

3.4. The Compressive Property of Foams

Biodegradable foam must possess definite mechanical strength to maintain its integrity during transportation, and the density and compressive strength of foams are somewhat correlated. Figure 5 illustrates the curves for the compressive strength of PS-based foams with change in the molecular weight of CTS and mass ratio of CTS/PS to 3/4.2 at 10% compressive ratio. The variation trend of these curves was basically consistent, as shown in the figure. The CTS-4.4/PS foams exhibited greater density and superior compressive performance compared with CTS-3.5PS and CTS-5.2PS foams. The increase in the density of the foam increases the bearing capacity of a single bubble hole and the effective bearing area within the unit section, which is generally reflected in the increased overall compressive strength of the foams.

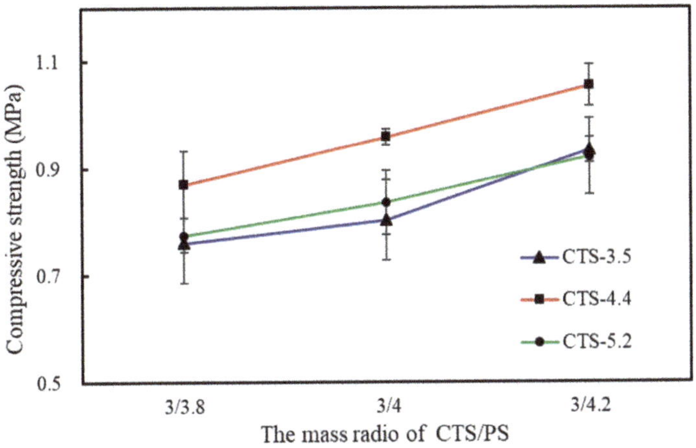

Figure 5. Compressive strength of foams with chitosan solution for 4.2 wt%.

The foams composed of CTS and starch exhibit such strength mainly because of the formation of a high intermolecular hydrogen bond between NH_3^+ in the main chain of CTS and OH^+ in the starch molecule. The amino group of CTS is easily protonated to form NH_3^+ in formic acid solution, and the ordered crystal structure in the starch molecule is destroyed during the microwave irradiation treatment, which leads to the formation of a hydrogen bond between the hydroxyl group of starch and amino group of CTS. In addition, electrostatic interactions between NH_3^+ in CTS and phosphate groups in PS may occur. Phosphorus, the most important element in PS, exists in the form of a covalent bond in PS. Glucose-6-phosphate ($pKa_1 = 0.94$, $pKa_2 = 6.11$) is the main structure of esterified phosphate in PS, and its acidity is stronger than that of orthophosphate [30,31]. Therefore, phosphate groups in PS possess a negative charge in aqueous solutions and do not combine with other negatively charged substances. The amino groups in the CTS solution possess a positive charge after being protonated and interact with the negatively charged phosphate groups in PS, which further improves the compressive performance of the entire CTS-PS foams. Under the condition of 4.2 wt% CTS, the compressive strength of CTS-PS foam increased with the increase in starch quality and the maximum compressive value was obtained at a mass ratio of 3/4.2. The molecular weight of CTS demonstrated a certain influence on the compressive strength of the starch-based foams. Among the foam composites prepared from CTS with three different molecular weights, the CTS-4.4/PS foams had the highest compressive strength value and those of CTS-3.5/PS and CTS-5.2/PS foams were extremely close (Table 5). The interaction between starch and CTS has been reported to enhance the functional properties [32]; however, when the number of NH_3^+ species increases beyond a critical value, synthesis of uniform CTS/PS foams becomes challenging and results in weak interaction at the boundary and poor compressive performance.

Table 5. Compressive strength of starch-based foams.

Samples	CTS-3.8 wt%			CTS-4 wt%			CTS-4.2 wt%		
	3/3.8	3/4	3/4.2	3/3.8	3/4	3/4.2	3/3.8	3/4	3/4.2
CTS-3.5/PS	0.768	0.811	0.983	0.860	0.916	0.958	0.761	0.803	0.934
CTS-4.4/PS	0.847	0.953	1.062	0.938	0.969	1.077	0.870	0.958	1.053
CTS-5.2/PS	0.745	0.792	0.939	0.822	0.892	0.918	0.775	0.836	0.920

To determine the effect of each factor on the foam compression performance, we further designed an orthogonal experiment and variance analysis. Table 6 presents the experiment results. According to range (R) analysis, R (C) > R (A) > R (B), and combined with the variance analysis in Table 7, we found that factor A (CTS molecular weight) and factor C (the mass ratio of CTS to starch) were all significantly affected, whereas factor B (CTS mass fraction) displayed no significant difference in the experimental results ($P > 0.05$), which indicates that the mass fraction of CTS fluctuating slightly above 4 wt% does not considerably affect the compressive performance of the foams. According to the results of intuitive and variance analyses, the optimal combination of foam preparation was $A_2B_2C_3$, implying that the CTS molecular weight of 4.4×10^5, CTS mass fraction of 4%, and the CTS/PS mass ratio of 3/4.2 are optimal for foam synthesis. We verified the compression performance of the CTS starch-based foam under these conditions, which was higher (1.077 mPa) than those observed in other conditions.

Table 6. Design and results of orthogonal test.

	Molecular Weight of CTS (A)	Mass Fraction of CTS (B)	Weight Ratio of CTS/PS (C)	Compressive Strength/mPa
1	1	1	1	0.768
2	1	2	2	0.916
3	1	3	3	0.934
4	2	1	2	0.953
5	2	2	3	1.077
6	2	3	1	0.870
7	3	1	3	0.939
8	3	2	1	0.822
9	3	3	2	0.836
K_1	2.618	2.660	2.460	
K_2	2.900	2.815	2.705	
K_3	2.597	2.640	2.950	
k_1	0.873	0.887	0.820	
k_2	0.967	0.938	0.902	
k_3	0.866	0.880	0.983	
R	0.101	0.058	0.163	
Optimal level	A_2	B_2	C_3	
Optimal combination		$A_2B_2C_3$		

Table 7. Variance analysis of orthogonal test.

Factor	Square Sum (SS)	Degree of Freedom (DF)	Mean Square (MS)	F-Value	Significance Probability (P)
A	0.030	2	0.015	40.411	0.024
B	0.005	2	0.002	6.112	0.141
C	0.043	2	0.021	56.613	0.017
Error	0.001	2			

3.5. FTIR Analysis

To study the interaction between CTS and PS, an infrared spectrum of each material was obtained. Figure 6 illustrates an extremely wide stretching vibration band at 3413 cm^{-1} for the dried PS corresponding to the characteristic absorption peak of the hydroxyl group in the starch. A moderate intensity peak at 2930 cm^{-1} was associated with $-CH_2$ antisymmetric stretching vibration, and an absorption band at 1647 cm^{-1} was attributed to water absorption in the amorphous region of PS; the absorption peak near 1380 cm^{-1} was ascribed to –CH_3 bending vibration and –CH_2 twisted vibration. C–C, C–O, and C–H stretching vibrations and bending vibration of C–OH formed obvious absorption peaks in the 1300–800 cm^{-1} region of the spectrum. CTS molecules also displayed a strong and wide absorption peak at 3428 cm^{-1}, corresponding to the stretching vibration of O–H and N–H; the absorption peaks at 1660, 1500, and 1425 cm^{-1} corresponded to amide-I belt stretching vibration, –NH_2 deformation vibration, and –CH_2 and –CH_3 bending vibrations, respectively. The characteristic absorption peak at 1157 cm^{-1} has been ascribed to the CTS C–O–C asymmetric vibration [33,34].

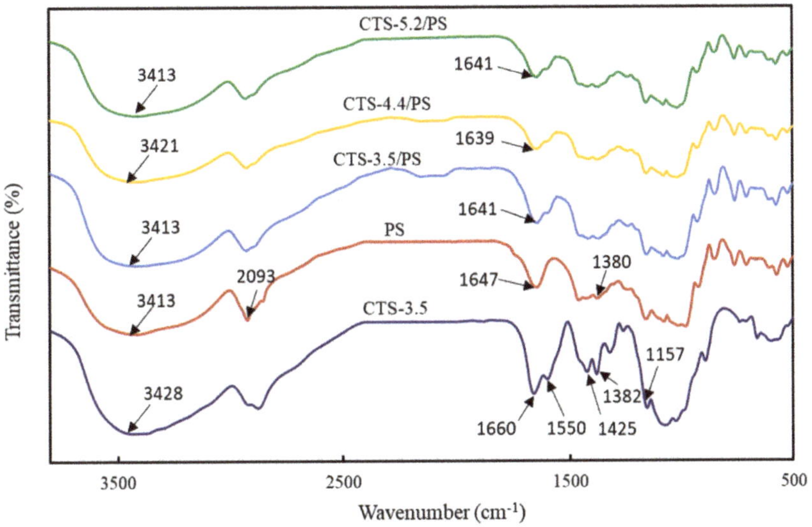

Figure 6. Infrared spectrogram of materials.

In a study, the interaction between CTS and starch in starch-based biological foam was shown to cause clear changes in the surroundings of polymer groups, which led to variations in the frequency and intensity of relevant absorption bands, and a unique response was obtained with changes in the characteristic spectra peak wave numbers [35]. Relative to the infrared spectra of pure CTS and starch, the amide peaks of CTS at 1660 cm^{-1} and the characteristic peaks of starch at 1646 cm^{-1} moved to a low-frequency region, and the intensity of the absorption band increased. Correspondingly, the absorption band strength of foam composites at 3413 cm^{-1} increased, which indicates that the hydrogen bond density in the CTS-starch system increased and inter- and intra- molecular hydrogen bonds formed between amino and hydroxyl groups in molecular chains of CTS and starch, thus improving the compatibility between CTS and starch [36,37].

3.6. *Thermogravimetric Analysis (TGA)*

The thermogravimetric curves and corresponding thermogravimetric rate curves of raw materials and starch-based foams are shown in Figure 7. From Figure 7a,b, it was clear that all materials exhibited three stages of heat loss during the whole pyrolysis process. The weight loss in the first stage below 100 °C could be attributed to the evaporation of water. Polysaccharides usually have a strong affinity for water, become hydrated easily, and form large molecules with a messy structure. Hydration properties of these polysaccharides depend on their primary and supramolecular structures [38]. The peak temperatures (Tp, shown in Table 8) for the water evaporation rates of chitosan with different molecular weights were 72.11, 74.70, and 72.17 °C, respectively. Moreover, the foams prepared by chitosan with different molecular weights reached their peaks at these temperatures. The water evaporation rate of chitosan increased with an increase in the molecular weight even though the increment was small. At the same time, the characteristic temperature points of the chitosan–starch foams moved to a higher temperature at this stage. This was due to the cross-linking between chitosan and starch to form new hydrogen bonds that inhibited the increase in water diffusion and thus amplified the water penetration resistance.

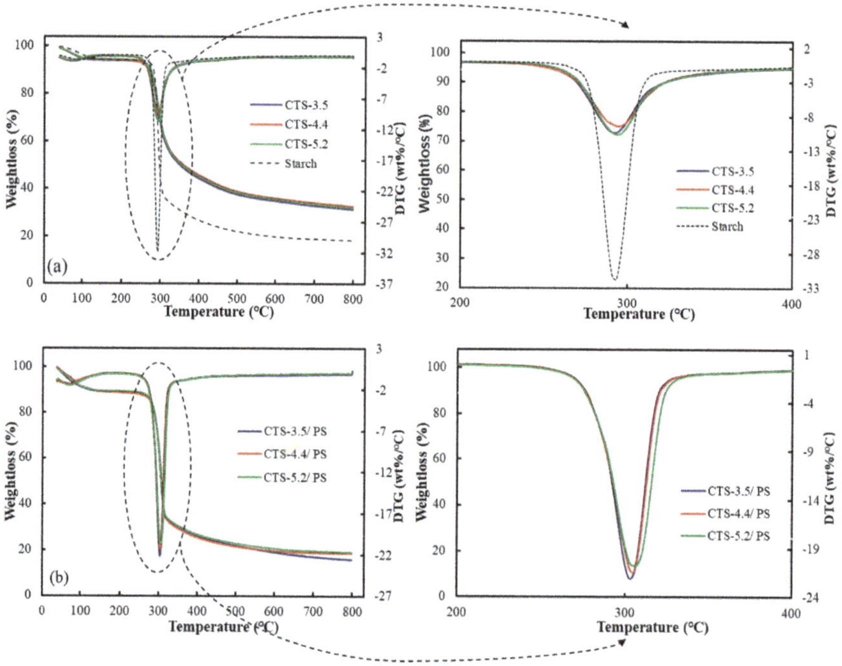

Figure 7. Thermal decomposition process of materials (**a**): raw materials, (**b**): CTS–PS foams.

Table 8. Thermal decomposition parameters of materials.

Samples	T_1 [a] (°C)	T_P (°C)	Residue (%)
CTS-3.5	72.11	292.98	31.44
CTS-4.4	74.70	295.40	32.80
CTS-5.2	72.17	292.94	32.02
PS	92.77	293.64	18.41
CTS-3.5-PS	72.77	302.88	15.97
CTS-4.4-PS	74.22	305.31	18.83
CTS-5.2-PS	73.60	305.22	19.18

[a] T_1 = Temperature at the peak below 100 °C on derivative thermogravimetry (DTG) curve, Tp = Temperature at the peak of the whole DTG curve, Residue = Mean value of residual mass.

The second thermogravimetric stage (between 200 to 400 °C) involved the depolymerization of chitosan molecular chains and the cleavage of starch molecular chains; this included the dehydration of sugar rings and the depolymerization and decomposition of glucose units [39]. Due to the similar depolymerization temperatures of starch and chitosan molecular chains, chitosan–starch foams presented the peak of the mixed weight loss rate of starch and chitosan molecules. The corresponding peak temperatures for the three kinds of foams were 302.88, 305.31, and 305.22 °C. The pyrolysis temperatures of the foams were higher than the temperatures observed for pure starch and chitosan. This indicated that the molecular interaction between the two biomolecules increased the molecular depolymerization temperature. Finally, the high-temperature "tail" between 340 and 500 °C was the third stage of the pyrolysis of chitosan and the vaporization and elimination of starch volatile products [40]. TGA curves showed the high-temperature characteristics of the biological starch-based foams. They depicted that the hydrogen bonding between chitosan and starch resulted in better thermal stability than that observed for neat starch and chitosan. Moreover, the maximum weight loss rate was associated with the molecular weight of chitosan. The value Tp of CTS-4.4/PS was higher than that of the others. This result could be attributed to increased hydrogen bonding and charge interactions.

3.7. DSC Analysis

The DSC diagrams of chitosan and starch and CTS/PS foams are shown in Figure 8. The convex peak represents the heat absorption of materials during the glass transition, the intersection point between the peak baseline and the tangent of the rising stage is the initial temperature of glass transition (To), the apex of the peak is the peak temperature (Tp), the peak stage of tangent and baseline by the intersection of terminated for gelation temperature (Tc) and the area of the peak for absorption of heat enthalpy (ΔH); these values are listed in Table 9. As can be seen from Table 9, To, Tp, Tc for chitosan were 34.65, 69.57, 104.72 °C, respectively, and ΔH was 133.53 J/g. The To, Tp, ΔH of porous foams after microwave treatment showed a backward trend with the increase in molecular weight of chitosan. Tester et al. [41] believed that the vitrification transition temperatures (To, Tp, and Tc) were associated with the perfection and stability of crystal. The advancement of glass transition temperature indicated that the temperature required to dissolve the crystal was reduced, indicating the worse stability of the crystal, and the backward shift was the opposite. Therefore, it can be concluded that the molecular weight of chitosan affected the stability of crystals in starch-based foam: the higher the molecular weight of chitosan, the greater the stability of the crystal in foams. The content of crystals could be determined by comparison of enthalpy and the enthalpy of samples increased with the increase in the crystal content. In addition, the improvement of crystal integrity, the enhancement of interaction between amylose or between amylose and amylopectin, the formation of chitosan and starch complex, and the formation of orderly crystalline regions may also increase the enthalpy value [42]. As shown in Table 9, the enthalpy of pure starch is 131.33 J/g, which is higher than that of some chitosan–starch composites; this is due to the easy formation of a double helix structure between short-chain molecules in starch, making the internal structure of starch stronger and more close. CTS/PS foams increased with the increase in the molecular weight of chitosan, which may be caused by the formation of partially ordered crystal regions between starch molecules and chitosan complexes and the longer the molecular chain of chitosan is, the slower its crystal structure is destroyed in the glass transition.

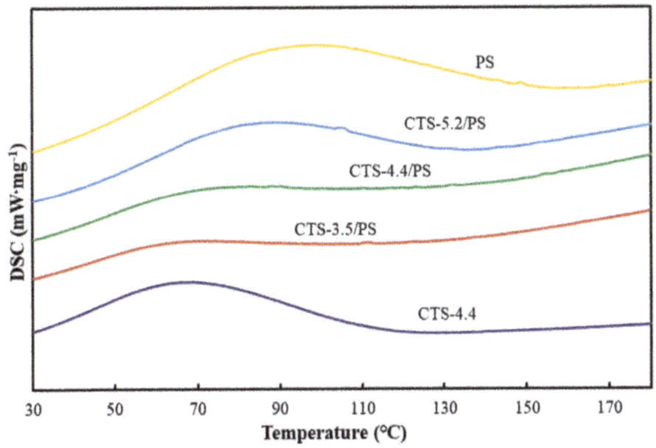

Figure 8. DSC diagrams of raw materials and CTS/PS foams.

Table 9. The glass transition temperature and enthalpy of samples.

Samples	To (°C)	Tp (°C)	Tc (°C)	ΔH (J/g)
CTS-4.4	34.65	69.57	104.72	133.53
PS	43.67	94.48	153.09	131.33
CTS-3.5/PS	27.93	62.38	104.67	86.32
CTS-4.4/PS	39.02	72.08	126.83	123.91
CTS-5.2/PS	50.96	86.32	116.99	167.86

To = initial temperature of glass transition, Tp = the apex of the peak is the peak temperature, Tc = the peak stage of tangent and baseline by the intersection of terminated for glass transition temperature; ΔH = the area of the peak for absorption of heat enthalpy.

3.8. Water Solubility Analysis

Figure 9 shows the solubility of CTS-starch-based foams in deionized water. The overall structure of starch-based foams was uniform and the edges were clearly visible (Figure 8a) when foams were submerged in water. After 3 days, the volume of foams increased and the edges appeared to soften. Furthermore, the turbidity of deionized water increased relative to that in the beginning of the experiment. After 10 days, the exterior foam color turned milky white, but the structure of foams remained intact. The foams began to dissolve in water after 18 days, the volume of foams gradually decreased, and water gradually mixed until foams were completely dissolved after 30 days. The whole dissolution process of CTS-starch-based foams demonstrated that the interaction between CTS and starch molecules can keep the morphological structure of foams intact for 10 days in water, but this interaction is effective for a limited time period.

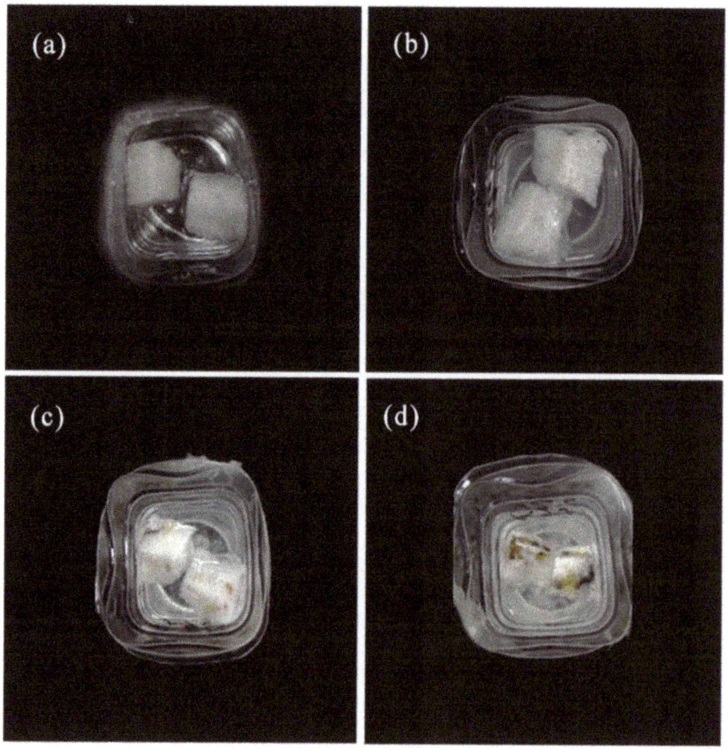

Figure 9. *Cont.*

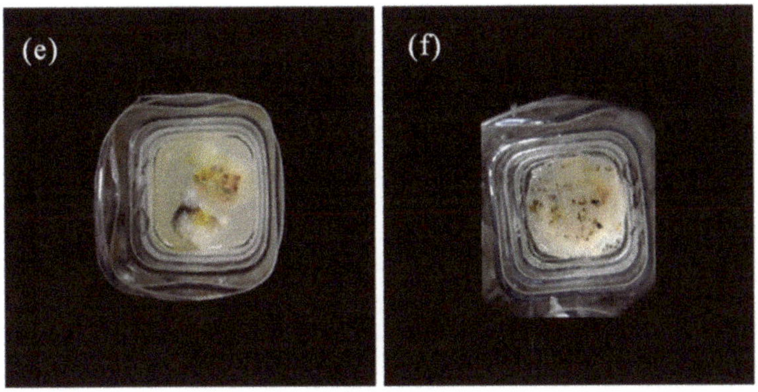

Figure 9. The morphology of CTS-starch-based foams in deionized water (**a**): start time, (**b**): 3 days later, (**c**): 10 days later, (**d**): 18 days later, (**e**): 24 days later, (**f**): 30 days later.

4. Conclusions

Biological foams containing CTS and PS were prepared successfully in a new polytef mold in a microwave oven. The effects of different viscosities and concentrations of CTS and different mass ratios of CTS-starch on the morphological, physicochemical, mechanical, and thermal properties of CTS–PS foams were comprehensively studied. The research conclusions can be summarized as follows:

(1) The viscosity of CTS solution at a particular concentration directly reflects the molecular weight of CTS. The molecular weights of CTS with different viscosities were 3.5×10^5, 4.4×10^5, and 5.2×10^5. The results confirmed that high viscosity corresponds to high relative molecular weight of CTS.

(2) Analysis of morphology, pore size, and density of starch-based biological foams indicated that CTS foams comprising CTS of molecular weight 4.4×10^5 exhibit small pores and low density. Moreover, the foams with a larger proportion of starch demonstrated small pore size and low density. Interestingly, the reverse was found to be true in the case of compressive properties; foams comprising CTS of molecular weight 4.4×10^5 exhibited the highest compressive strength, and foams with a larger proportion of starch exhibited higher compressive strength. In addition, orthogonal experiments were set up to confirm that the compressive performance of foams is highest when the molecular weight of CTS is 4.4×10^5, the mass fraction is 4 wt%, and the mass ratio of CTS-PS is 3/4.2.

(3) The change in the characteristic peaks of the foams composed of CTS and PS in the FTIR diagram indicated that the amino group in CTS interacted with the hydroxyl group in the starch. In TGA and DSC analysis, it could be seen that the interaction between starch and chitosan improved the thermal stability of the foam, and in the water solubility test, the interaction maintained the integrity of the foam's morphological structure in water within a certain period of time, and the foams completely degraded after 30 days.

Author Contributions: R.H. designed the study and conducted parts of experiments and data analysis. R.H. and X.Z. wrote the initial manuscript. X.Z. and Z.T. performed the composite property testing/data analysis. All authors have read and agreed to the published version of the manuscript.

Funding: This collaborative study was supported by Major projects of Natural Science Foundation of Jiangsu (18KJA220002); China Postdoctoral Science Foundation: Special Program (2017T100313); China Postdoctoral Science Foundation: General Program (2016M601821). Natural Science Foundation of China (Grant No. 51606103). And the Postgraduate Research and Practice Innovation Program of Jiangsu Province (2019NFUSPITP0004).

Conflicts of Interest: There is no conflict of interest among authors.

References

1. Nofar, M.; Park, C.B. Poly (lactic acid) foaming. *Prog. Polym. Sci.* **2014**, *39*, 1721–1741. [CrossRef]
2. Ventura, H.; Laguna-Gutiérrez, E.; Rodriguez-Perez, M.A.; Ardanuy, M. Effect of chain extender and water-quenching on the properties of poly(3-hydroxybutyrate-co-4-hydroxybutyrate) foams for its production by extrusion foaming. *Eur. Polym. J.* **2016**, *85*, 14–25. [CrossRef]
3. Feng, Z.; Yu, L.; Hong, Y.; Wu, J.; Jian, Z.; Li, H.; Qi, R.; Jiang, P. Preparation of enhanced poly (butylene succinate) foams. *Polym. Eng. Sci.* **2016**, *56*, 1275–1282. [CrossRef]
4. Gao, Z.; Zhan, W.; Wang, Y.; Yun, G.; Li, W.; Guo, Y.; Lu, G. Aldehyde-functionalized mesostructured cellular foams prepared by copolymerization method for immobilization of penicillin G acylase. *Microporous Mesoporous Mater.* **2015**, *202*, 90–96. [CrossRef]
5. Jordi, G.; Espinach, F.X.; Neus, P.; Josep, T.; Manel, A.; Pere, M. High-Performance-tensile-strength alpha-grass reinforced starch-based fully biodegradable composites. *Bioresources* **2013**, *8*, 6121–6135.
6. Guimares, J.L.; Wypych, F.; Saul, C.K.; Ramos, L.P.; Satyanarayana, K.G. Studies of the processing and characterization of corn starch and its composites with banana and sugarcane fibers from Brazil. *Carbohydr. Polym.* **2010**, *80*, 130–138. [CrossRef]
7. Moraes, A.R.F.E.; Pola, C.C.; Bilck, A.P.; Yamashita, F.; Tronto, J.; Medeiros, E.A.A.; Soares, N.D.F.F. Starch, cellulose acetate and polyester biodegradable sheets: Effect of composition and processing conditions. *Mater. Sci. Eng. C* **2017**, *78*, 932–941. [CrossRef]
8. Shariatinia, Z.; Fazli, M. Mechanical properties and antibacterial activities of novel nanobiocomposite films of chitosan and starch. *Food Hydrocoll.* **2015**, *46*, 112–124. [CrossRef]
9. Akrami, M.; Ghasemi, I.; Azizi, H.; Karrabi, M.; Seyedabadi, M. A new approach in compatibilization of the poly(lactic acid)/thermoplastic starch (PLA/TPS) blends. *Carbohydr. Polym.* **2016**, *144*, 254–262. [CrossRef] [PubMed]
10. Mahieu, A.; Terrié, C.; Agoulon, A.; Leblanc, N.; Youssef, B. Thermoplastic starch and poly(ε-caprolactone) blends: Morphology and mechanical properties as a function of relative humidity. *J. Polym. Res.* **2013**, *20*, 229. [CrossRef]
11. Yun, I.S.; Hwang, S.W.; Shim, J.K.; Seo, K.H. A study on the thermal and mechanical properties of poly (butylene succinate)/ thermoplastic starch binary blends. *Int. J. Precis. Eng. Manuf. Green Technol.* **2016**, *3*, 289–296. [CrossRef]
12. Aydın, A.A.; Ilberg, V. Effect of different polyol-based plasticizers on thermal properties of polyvinyl alcohol (PVA): Starch blends films. *Carbohydr. Polym.* **2015**, *136*, 441–448. [CrossRef] [PubMed]
13. Li, F.; Guan, K.; Liu, P.; Li, G.; Li, J. Ingredient of biomass packaging material and compare study on cushion properties. *Int. J. Polym. Sci.* **2014**, *2014*, 146509. [CrossRef]
14. Loercks, J.; Pommeranz, W.; Schmidt, H.; Timmermann, R.; Grigat, E.; Schulz-Schlitte, W. Thermoplastic Starch Utilizing a Biodegradable Polymer as Melting Aid. U.S. Patent 6472497, 29 October 2002.
15. Ferri, J.M.; Garcia-Garcia, D.; Sánchez-Nacher, L.; Fenollar, O.; Balart, R. The effect of maleinized linseed oil (MLO) on mechanical performance of poly(lactic acid)-thermoplastic starch (PLA-TPS) blends. *Carbohydr. Polym.* **2016**, *147*, 60–68. [CrossRef] [PubMed]
16. Bénézet, J.-C.; Stanojlovic-Davidovic, A.; Bergeret, A.; Ferry, L.; Crespy, A. Mechanical and physical properties of expanded starch, reinforced by natural fibres. *Ind. Crop. Prod.* **2012**, *37*, 435–440. [CrossRef]
17. Garcia, M.A.; Pinotti, A.; Zaritzky, N.E. Physicochemical, water vapor barrier and mechanical properties of corn starch and chitosan composite films. *Starch Stärke* **2006**, *58*, 453–463. [CrossRef]
18. Vicentini, D.S.; Smania, A.; Laranjeira, M.C.M. Chitosan/poly (vinyl alcohol) films containing ZnO nanoparticles and plasticizers. *Mater. Sci. Eng. C Mater.* **2010**, *30*, 503–508. [CrossRef]
19. Rinaudo, M. Chitin and chitosan: Properties and applications. *Cheminform* **2007**, *31*, 603–632. [CrossRef]
20. Dang, K.M.; Yoksan, R. Morphological characteristics and barrier properties of thermoplastic starch/chitosan blown film. *Carbohydr. Polym.* **2016**, *150*, 40–47. [CrossRef]
21. Zare, E.N.; Makvandi, P.; Tay, F.R. Recent progress in the industrial and biomedical applications of tragacanth gum: A review. *Carbohydr. Polym.* **2019**, *212*, 450–467. [CrossRef]
22. Mei, J.; Yuan, Y.; Wu, Y.; Li, Y. Characterization of edible starch–chitosan film and its application in the storage of Mongolian cheese. *Int. J. Biol. Macromol.* **2013**, *57*, 17–21. [CrossRef]

23. Bonilla, J.; Talón Argente, E.; Atarés, L.; Vargas, M.; Chiralt, A. Effect of the incorporation of antioxidants on physicochemical and antioxidant properties of wheat starch–chitosan films. *J. Food Eng.* **2013**, *118*, 271–278. [CrossRef]
24. Mathew, S.; Abraham, T.E. Characterisation of ferulic acid incorporated starch-chitosan blend films. *Food Hydrocoll.* **2008**, *22*, 826–835. [CrossRef]
25. Zhong, Y.; Li, Y.; Zhao, Y. Physicochemical, microstructural, and antibacterial properties of β-chitosan and kudzu starch composite films. *J. Food Sci.* **2012**, *77*, E280–E286. [CrossRef] [PubMed]
26. Makvandi, P.; Ghomi, M.; Padil, V.V.T.; Shalchy, F.; Ashrafizadeh, M.; Askarinejad, S.; Pourreza, N.; Zarrabi, A.; Zare, E.N.; Kooti, M.; et al. Biofabricated Nanostructures and Their Composites in Regenerative Medicine. *ACS Appl. Nano Mater.* **2020**, *3*, 6210–6238. [CrossRef]
27. Flory, P.J. *In Principles of Polymer Chemistry*; Cornell University Press: Ithaca, NY, USA, 1953.
28. Sabu, T. *Physical Chemistry of Macromolecules*; Wiley: Ithaca, NY, USA, 1961.
29. Kasaai, M.R.; Arul, J.; Charlet, G. Intrinsic viscosity–molecular weight relationship for chitosan. *J. Polym. Ence Part B Polym. Phys.* **2015**, *38*, 2591–2598. [CrossRef]
30. Swinkels, I.J.J.M. Composition and properties of commercial native starches. *Starch Stärke* **1985**, *37*, 1–5. [CrossRef]
31. Charles, W.P. *Biochemistry for the Pharmaceutical Sciences*; Jones & Bartlett Learning: Burlington, MA, USA, 2019.
32. Zamudio-Flores, P.B.; Torres, A.V.; Salgado-Delgado, R.; Bello-Pérez, L.A. Influence of the oxidation and acetylation of banana starch on the mechanical and water barrier properties of modified starch and modified starch/chitosan blend films. *J. Appl. Polym. Sci.* **2010**, *115*, 991–998. [CrossRef]
33. Park, P.J.; Je, J.Y.; Kim, S.K. Free radical scavenging activities of differently deacetylated chitosans using an ESR spectrometer. *Carbohydr. Polym.* **2004**, *55*, 17–22. [CrossRef]
34. Shukur, M.F.; Yusof, Y.M.; Zawawi, S.M.M.; Illias, H.A.; Kadir, M.F.Z. Conductivity and transport studies of plasticized chitosan-based proton conducting biopolymer electrolytes. *Phys. Scr.* **2013**, *2013*, 014050. [CrossRef]
35. Xu, Y.X.; Kim, K.M.; Hanna, M.A.; Nag, D. Chitosan-starch composite film: Preparation and characterization. *Ind. Crop. Prod.* **2005**, *21*, 185–192. [CrossRef]
36. Mathew, S.; Brahmakumar, M.; Abraham, T.E. Microstructural imaging and characterization of the mechanical, chemical, thermal, and swelling properties of starch-chitosan blend films. *Biopolymers* **2010**, *82*, 176–187. [CrossRef]
37. Yusof, Y.M.; Shukur, M.F.; Illias, H.A.; Kadir, M.F.Z. Conductivity and electrical properties of corn starch–chitosan blend biopolymer electrolyte incorporated with ammonium iodide. *Phys. Scr.* **2014**, *89*, 35701. [CrossRef]
38. Kittur, F.S.; Harish Prashanth, K.V.; Udaya Sankar, K.; Tharanathan, R.N. Characterization of chitin, chitosan and their carboxymethyl derivatives by differential scanning calorimetry. *Carbohydr. Polym.* **2002**, *49*, 185–193. [CrossRef]
39. Perez, J.J.; Francois, N.J. Chitosan-starch beads prepared by ionotropic gelation as potential matrices for controlled release of fertilizers. *Carbohydr. Polym.* **2016**, *148*, 134–142. [CrossRef]
40. Zawadzki, J.; Kaczmarek, H. Thermal treatment of chitosan in various conditions. *Carbohydr. Polym.* **2010**, *80*, 394–400. [CrossRef]
41. Tester, R.F.; Morrison, W.R. Swelling and gelatinization of cereal starches. II, Waxy rice starches. *Cereal Chem.* **1990**, *67*, 558–563.
42. Liu, H.; Lv, M.; Wang, L.; Li, Y.; Fan, H.; Wang, M. Comparative study: How annealing and heat-moisture treatment affect the digestibility, textural, and physicochemical properties of maize starch. *Starch Stärke* **2016**, *68*, 268. [CrossRef]

polymers

MDPI

Article

Natural Inspired Carboxymethyl Cellulose (CMC) Doped with Ammonium Carbonate (AC) as Biopolymer Electrolyte

Mohd Ibnu Haikal Ahmad Sohaimy [1] and Mohd Ikmar Nizam Mohamad Isa [1,2,*]

1 Energy Storage Research, Frontier Research Materials Group, Advanced Materials Team, Ionic & Kinetic Materials Research Laboratory (IKMaR), Faculty of Science & Technology, Universiti Sains Islam Malaysia, Nilai 71800, Malaysia; ibnuhyqal@gmail.com

2 Advanced Nano Materials, Advanced Materials Team, Ionic State Analysis (ISA) Laboratory, Faculty of Science & Marine Environment, Universiti Malaysia Terengganu, Kuala Nerus 21030, Malaysia

* Correspondence: ikmar_isa@usim.edu.my

Received: 30 July 2020; Accepted: 21 September 2020; Published: 26 October 2020

Abstract: Green and safer materials in energy storage technology are important right now due to increased consumption. In this study, a biopolymer electrolyte inspired from natural materials was developed by using carboxymethyl cellulose (CMC) as the core material and doped with varied ammonium carbonate (AC) composition. X-ray diffraction (XRD) shows the prepared CMC-AC electrolyte films exhibited low crystallinity content, X_c (~30%) for sample AC7. A specific wavenumber range between 900–1200 cm^{-1} and 1500–1800 cm^{-1} was emphasized in Fourier transform infrared (FTIR) testing, as this is the most probable interaction to occur. The highest ionic conductivity, σ of the electrolyte system achieved was 7.71×10^{-6} Scm^{-1} and appeared greatly dependent on ionic mobility, μ and diffusion coefficient, D. The number of mobile ions, η, increased up to the highest conducting sample (AC7) but it became less prominent at higher AC composition. The transference measurement, t_{ion} showed that the electrolyte system was predominantly ionic with sample AC7 having the highest value (t_{ion} = 0.98). Further assessment also proved that the H^+ ion was the main conducting species in the CMC-AC electrolyte system, which presumably was due to protonation of ammonium salt onto the complexes site and contributed to the overall ionic conductivity enhancement.

Keywords: biopolymer; carboxymethyl cellulose; solid polymer electrolyte; ionic transport

1. Introduction

Apart from increased energy and power density of energy storage, one of the remaining challenge in advancement of energy storage technologies such as portable electronic devices, smart grids and electric vehicles is to lower the production cost as well as to reduce the environmental effect [1–3]. Thus, one of the steps in order to achieve this is to transform the electrolyte used in energy storage. The electrolyte is a medium that facilitates the movement of ions across the electrode in order to complete the circuit and usually the electrolyte is in liquid form since ionic transport is easier in liquid phase. However, the development of solid polymer electrolyte (SPE) film has garnered much attention as it offers outstanding advantages compared to other form of electrolytes [3–5]. Several types of polymer were used to investigate the possibilities for energy storage material such as polyethylene oxide (PEO), polyvinyl chloride (PVC), poly(methyl methacrylate) (PMMA), polyvinylidene fluoride (PVDF) and polyacrylonitrile (PAN) [4,6–8]. Alternatively, polymers originated from natural sources (biopolymer) such as starch, chitosan, gelatin and agar [9–12] are also becoming an attractive substitute for synthetic polymers partly due to superior mechanical, electrical properties and for being cheaper [9]. Equally, cellulose in particular is another good choice since it is literally found everywhere but cellulose in its original form is hard to utilize due to its inability to dissolved and shaped. However, derivatives of the base polymer can overcome this problem.

Currently several cellulose derivatives biopolymer can be found commercially such as methyl cellulose (MC), hydroxyethyl cellulose (HEC) and carboxymethyl cellulose (CMC) [13–15]. CMC in particular is a good SPE biopolymer candidate. This is attributable to its good film-forming abilities due to the presence of a hydrophilic carboxyl group (–CH_2COONa) which allows the biopolymer to dissolve in cheap solvent (water) [16]. On top of that, CMC has a naturally high degree of amorphous phase [17]. These allow easier transport of conducting ions like lithium (Li^+) and proton (H^+). The major flaw with SPE is the low ionic conductivity due to inherent solid properties of the polymer itself, which impede ionic mobility especially at room temperature [18]. For a biopolymer electrolyte to be utilized in commercialized energy storage technology, the ionic conductivity needs to be enhanced to at least a minimum value of ~10^{-4} Scm^{-1}, and one of the most prominent methods for ionic conductivity enhancement is by introducing the ionic dopant into the polymer electrolyte. Ammonium salts are the dopant materials focused upon in this research. Ammonium nitrate (A–N) and ammonium fluoride (A–F) are some examples of ammonium salts [17,19] and are the most common proton donors. Deprotonated of H^+ from the ammonium ion group ($NH_4{}^+$) of ammonium salts helps to improve ionic conductivity of the electrolyte system. Thus in theory, ammonium carbonate (AC) which have two group of ammonium ions could inject more H^+ into the system.

The ionic conductivity is the most vital parameters in order to determine the viability of the electrolyte system. Nevertheless, the value calculated only gives a general overview of the electrolyte performance. Fundamentally, the ionic conductivity of the electrolyte is the product of ionic density, η, ionic mobility, μ and the elementary charge, *e*. Thus, it can be beneficial to acquire a detailed transport number for the electrolyte system as it can help to evaluate and later suggest an appropriate technique to improve the ionic conductivity of the electrolyte system. Several methods have been employed previously by researchers to calculate the transport number [20,21]. The commonly used one is by using the Fourier transform infrared (FTIR) deconvolution approach, which is also the method chosen for this study [22–24]. Apart from that, a high cation transference number is also crucial in electrolyte system as it can reduce charge concentration gradient and subsequently produce higher power density when assembled in a battery [25].

This paper aims to develop SPE from a biopolymer, which is CMC doped with ammonium carbonate (AC) as the ionic source, then to find the ionic conductivity and the transport parameters contributing towards the ionic conductivity behavior in CMC-AC biopolymer films. Ionic conductivity was measured using electrical impedance spectroscopy (EIS). The X-ray diffraction (XRD) and Fourier transform infrared (FTIR) spectrum of the electrolyte was de-convoluted to quantitatively determine the crystallinity content and the transport number (number of mobile ions, η mobility of mobile ions, μ and diffusion coefficient, *D*), respectively. A direct current (dc) polarization technique was utilized to confirm the ionic transference number of CMC-AC biopolymer films. From these analyses, the correlation between the results towards the ionic conductivity behavior can be ascertained.

2. Materials and Methods

2.1. Preparation of Carboxymethyl Cellulose-Ammonium Carbonate (CMC-AC) Electrolytes Films

Carboxymethyl cellulose (CMC) (Across, NJ, USA) were dissolved in 100 mL of distilled water until homogenous before adding (1–11 wt.%) of ammonium carbonate salts (AC) (Merck, Darmstadt, Germany) which is calculated using Equation (1), where the x represents the weight of added salt and y the weight of CMC. The compositions of the CMC-AC biopolymer films are shown in Table 1. The CMC-AC biopolymer electrolyte solution was casted into petri dishes and left for a period of time (1 month) at room temperature to dry into thin film. Three (3) sample replicates were produced for each composition. The CMC-AC biopolymer films obtained are free standing with no phase separation observed (Figure S1).

$$\text{composition} = \frac{x}{x+y} \times 100 \quad (1)$$

Table 1. The carboxymethyl cellulose-ammonium carbonate (CMC-AC) electrolyte composition.

Sample	Weight (g) (± 0.01)		
	CMC	AC	wt.%
AC0	2.00	0.00	0
AC1		0.02	1
AC3		0.06	3
AC5		0.11	5
AC7		0.15	7
AC9		0.20	9
AC11		0.25	11

2.2. Electrical Impedance Spectroscopy (EIS)

A HIOKI 3532-50 LCR Hi-Tester (HIOKI, Nagano-ken, Japan) used to investigate the conductivity of CMC-AC biopolymer film. Each samples tested in the frequency range of 50 Hz to 1 MHz by utilizing stainless steel as blocking electrodes at a temperature range of 303–363 K.

2.3. X-ray Diffractometer (XRD)

The degree of crystallinity of CMC-AC biopolymer films obtained using Rigaku Miniflex II X-ray diffractometer (Rigaku, Tokyo, Japan) with $CuK\alpha$ radiation sources. The radiation sources directed the X-ray beam onto the films and scanned at angle, 2θ between 5° and 80°.

2.4. Fourier-Transform Infrared (FTIR)

Complexation or interaction between CMC and AC salts were determined by using a Thermo Nicolet 380 FTIR spectroscopy (Thermo Fischer, Madison, WI, USA) equipped with attenuated total reflection (ATR). The spectrum recorded in a wavenumber ranged from 800–4000 cm^{-1} and with resolution of 4 cm^{-1}.

2.5. Transference Number Measurement (TNM)

The dc polarization techniques with 1.5 V fixed voltage were applied across the samples held by a stainless steel holder. A digital multimeter and computer were also connected to the circuit to collect the current (I) value against time (t). The transference number, t_{ion} of CMC-AC biopolymer films was calculated from the testing.

3. Results

3.1. XRD Analysis

Figure 1a shows the X-ray diffraction (XRD) result of pure AC and CMC-AC biopolymer films. From the figure, pure AC salt displays several sharp peaks which represent the crystalline nature of AC salt with three prominent peaks can be observed at angle $2\theta = 22.54°$, 30.2° and 34.54°. The XRD pattern for CMC-AC biopolymer films show a similar pattern after the addition of AC salts where the amorphous hump is centered at $2\theta = 21.1°$ (Figure 1a). It can be noticed that none of crystalline peak observed from AC salt appeared for all films, which indicates the AC salts dissolved within CMC amorphous phase to form an electrolyte system with improved electrolyte properties [26]. The amorphous phase will lead to higher ionic conductivity due to greater ionic diffusion, since the amorphous polymers have a flexible polymeric backbone [27]. Several reports have proven that the high amorphous phase has led to ionic conductivity improvement [28–30].

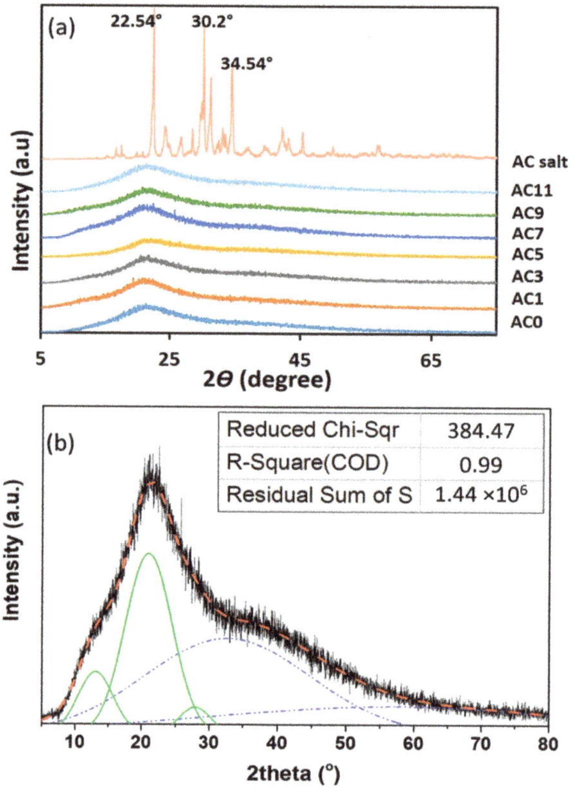

Figure 1. (**a**) The X-ray diffraction (XRD) diffractograms of all CMC-AC biopolymer films. (**b**) The de-convoluted plot of XRD for sample AC7 where green line represents the crystalline phase and blue line represents the amorphous phase.

De-convolution techniques on XRD spectra can reveal the explicit details of the amorphous or crystallinity phase of electrolyte films [31]. The de-convoluted XRD diffractograms show several peaks where a narrow and sharper peak (solid green line) represents the crystalline phase and a broader peak (dotted blue line) represents the amorphous phase (Figure 1b). From the de-convoluted peak, the degrees of crystallinity (X_c) of each CMC-AC biopolymer films were calculated using Equation (2) and the results are given in Table 2.

$$\text{degree of crystallinity, } X_c(\%) = \frac{A_c}{A_c + A_a} \times 100\% \quad (2)$$

Table 2. The respective area of crystalline (A_c) and amorphous (A_a) phase and degree of crystallinity (X_c) of each CMC-AC biopolymer films.

Sample	Coefficient of Determination, COD (R^2)	Residual Sum of Square ($\times10^6$) (a.u.)	Peak Area (a.u.)		X_c (%)
			A_c	A_a	
AC0	0.99	1.08	5547.07	7300.18	43.18
AC1	0.99	1.19	5056.57	6127.60	45.21
AC3	0.99	1.06	4333.50	8210.93	34.55
AC5	0.99	0.87	2764.22	4944.47	35.86
AC7	0.99	1.44	5744.59	12,860.22	30.88
AC9	0.99	1.07	5273.33	5759.21	47.80
AC11	0.99	1.08	5738.11	6768.42	45.88

The overlapping peaks and area forms under the diffractograms were determined for either amorphous (A_a) or crystalline (A_c) phases before calculating the X_c, and the results are tabulated in Table 2. The degree of crystallinity, X_c for the CMC-AC biopolymer films is in between 46% to 30% as shown in the table. The irregularity (random arrangement) of the polymer structure due to the increase in amorphous phase can led to lower energy barrier between hopping sites. Thus, the increasing amorphous phase allows for easier ion conduction which in this case is the H^+ ion through the polymer matrix and subsequently increases the ionic conductivity value. The degree of crystallinity, X_c trend of current work is similar compared to other works which the highest conducting sample has the lowest X_c [32,33].

3.2. FTIR Analysis

Figure 2 shows the FTIR spectrum of the CMC-AC biopolymer films in the range of (a) 1500–1800 cm^{-1} and (b) 900–1200 cm^{-1}. Overall the FTIR spectrum can be seen in Figure S2 in the Supplementary Materials. These ranges are responsible for the vibration of carboxyl group (COO) of CMC, where it is the probable site for interaction between CMC and AC [17,19,34]. The FTIR peak centered at 1056 cm^{-1} and 1591 cm^{-1} correspond to the vibration peak of C–O and C=O of the COO, respectively. As seen in Figure 2, there is no shifting of the peak indicating that the AC did not directly interact with that particular site. However, the C=O peak shows an observable reduction in intensity as the AC composition increases which implies that the AC salt weakly interacts with the COO. The presence of free lone pair electron at C=O attract free ions to interact. The small bump appeared around ~1650 cm^{-1} is believed due to the AC salt. The AC salt believed to dissociate to NH_4^+ and CO_3^{2-} ions. This allows a protonation process between the loosely bonded H^+ from NH_4^+ structure to the complex site [19]. The protonated H^+ then able to transport to the conduction site (C=O) and subsequently improve the ionic conductivity of the electrolyte films. This can be further clarify from the impedance and transport analysis.

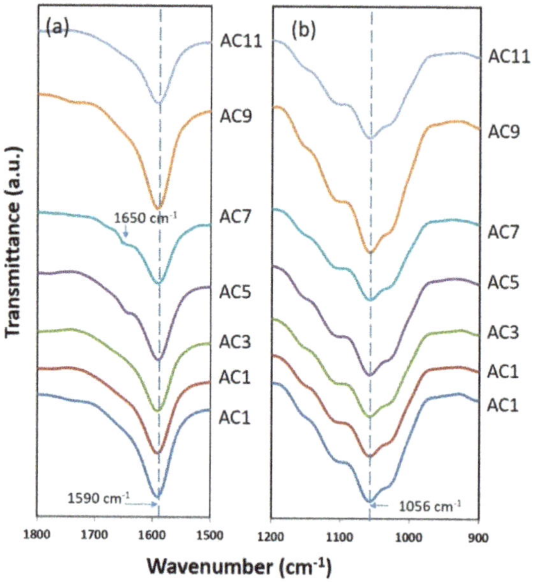

Figure 2. Fourier transform infrared (FTIR) spectra of CMC-AC biopolymer films range from (**a**) 1500–1800 cm^{-1} and (**b**) 900–1200 cm^{-1}.

3.3. Ionic Conductivity and Transport Properties

The real (Z_r) and imaginary (Z_i) impedance data obtained from EIS testing presented in the form of Cole-Cole plot (Figure 3a,b). This plot reveals semi-circle at high frequency (left side of the plot) and slanted line at low frequency (right side of the plot). This shape can be explained in terms of the electronics component where the semi-circle represents parallel combination of capacitor and resistor [35]. The resistor represents the ionic migration through the polymer matrix while the capacitor represents the polarized polymer chains [35]. The slanted line is due to the electrode polarization effect where electrical double layer capacitance occur due to the blocking stainless steel electrode used in this testing [36]. The bulk resistance, R_b value for each CMC-AC biopolymer film was obtained from the intersection of the semi-circle and slanted impedance line value at the *x*-axis [15]. The value of ionic conductivity is calculated using Equation (3). The *t* and *A* in the equation correspond to film thickness and the contact area, respectively. The ionic conductivity value at room temperature (303 K) against AC composition (wt.%) is plotted in Figure 3c. As seen in the figure, the ionic conductivity increases for sample AC1 (1 wt.%) to AC7 (7 wt.%) with the value of 7.71×10^{-6} Scm^{-1} and later dropped for sample AC9 (9 wt.%) and sample AC11 (11 wt.%). This was attributed to the diminishing effect of bulk resistance, which was apparent with the size changes to the impedance semi-circle as seen in Figure 3a,b, that indicates lower resistance for the mobile ions to move [17]. Table 3 shows the comparison of ionic conductivity of current work with other polymer-doped ammonium salt electrolyte systems. This work exhibits higher ionic conductivity when compared to other single ammonium and di-ammonium salt system. Figure 3d shows the ionic conductivity at elevated temperature for selected biopolymer films. In the investigated temperature range, the ionic conductivity increases as temperature increases. This is due to two reasons, (1) the AC absorbs the thermal energy and allows more ions dissociation, (2) the CMC backbone vibrate and create space for ionic conduction thus increase ionic conductivity at high temperature. This indicates the CMC-AC biopolymer films is thermally assisted which is proven from good fitting (R^2 = 0.93–0.98) of the ionic conductivity to the Arrhenius relationship (Equation (4)). Transport properties analysis can further reveal the details behind the ionic conductivity improvement of the CMC-AC biopolymer system.

$$\text{Ionic conductivity, } \sigma = \frac{t}{R_b \times A} \tag{3}$$

$$\sigma = \sigma_o exp^{\left(\frac{-E_a}{kT}\right)} \tag{4}$$

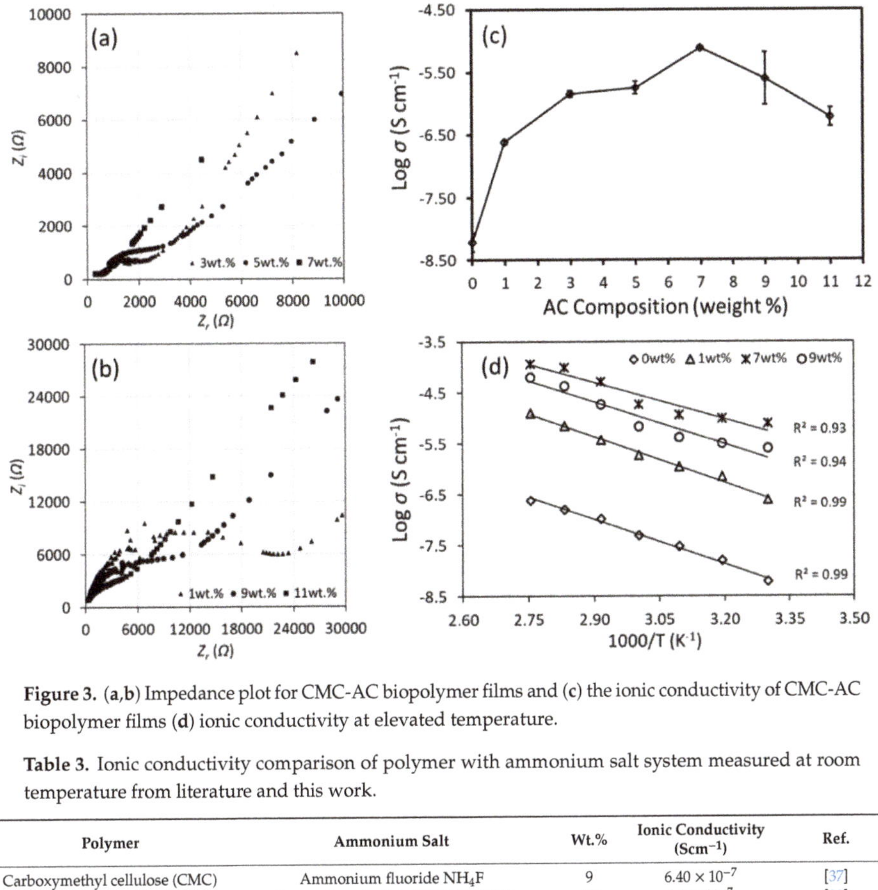

Figure 3. (**a**,**b**) Impedance plot for CMC-AC biopolymer films and (**c**) the ionic conductivity of CMC-AC biopolymer films (**d**) ionic conductivity at elevated temperature.

Table 3. Ionic conductivity comparison of polymer with ammonium salt system measured at room temperature from literature and this work.

Polymer	Ammonium Salt	Wt.%	Ionic Conductivity (Scm^{-1})	Ref.
Carboxymethyl cellulose (CMC)	Ammonium fluoride NH_4F	9	6.40×10^{-7}	[37]
Polyvinyl Alcohol (PVA)	Ammonium Acetate NH_4CH_3COO	20	1.3×10^{-7}	[38]
Chitosan (CS)	Ammonium Iodide NH_4NO_3	45	3.73×10^{-7}	[39]
Polyvinyl Alcohol (PVA)	Ammonium Sulfate $(NH_4)_2SO_4$	5	1.92×10^{-9}	[40]
Polyethylene Oxide (PEO)	Ammonium Adipate $(NH_4)_2C_4H_8(COO)_2$	1	1.87×10^{-7}	[41]
Carboxymethyl cellulose (CMC)	Ammonium Carbonate $(NH_4)_2CO_3$	7	7.71×10^{-6}	Current work

Arof et al. successfully calculated the transport number of their electrolyte system through FTIR peak (de-convoluted technique) [42]. Thus, the same technique was applied to the current work. The de-convoluted FTIR spectra were assigned into free ions and contact/aggregates ions to determine the area of de-convoluted peak before calculating the transport number of the electrolyte system. The FTIR spectra range of 1500–1750 cm^{-1} was selected for de-convoluted process to find the peak originated from ammonium (NH_4^+) ion. Figure 4a shows the de-convoluted peak of each CMC-AC biopolymer film. The peak centered at ~1589 cm^{-1} is the peak of the carboxylic group (COO–) of

CMC [17]. The free ions peak centered from 1594–1643 cm^{-1} while the contact ion centered from 1650–1684 cm^{-1}. The percentage of free ions were first calculated using Equation (5), where A_f is the area of free ions peak and A_c is the area of contact ions peak. Afterward, the percentage of free ions obtained were used to determine the transport number (number of mobile ions, η ionic mobility, μ and diffusion coefficient, D) using Equations (6)–(8) [43].

$$\text{Percentage of free ions } (\%) = \frac{A_f}{A_f + A_c} \times 100\% \quad (5)$$

$$\eta = \frac{M \times N_A}{V_{Total}} \times \% \text{ of free ions} \quad (6)$$

$$\mu = \frac{\sigma}{\eta e} \quad (7)$$

$$D = \frac{\mu k_B T}{e} \quad (8)$$

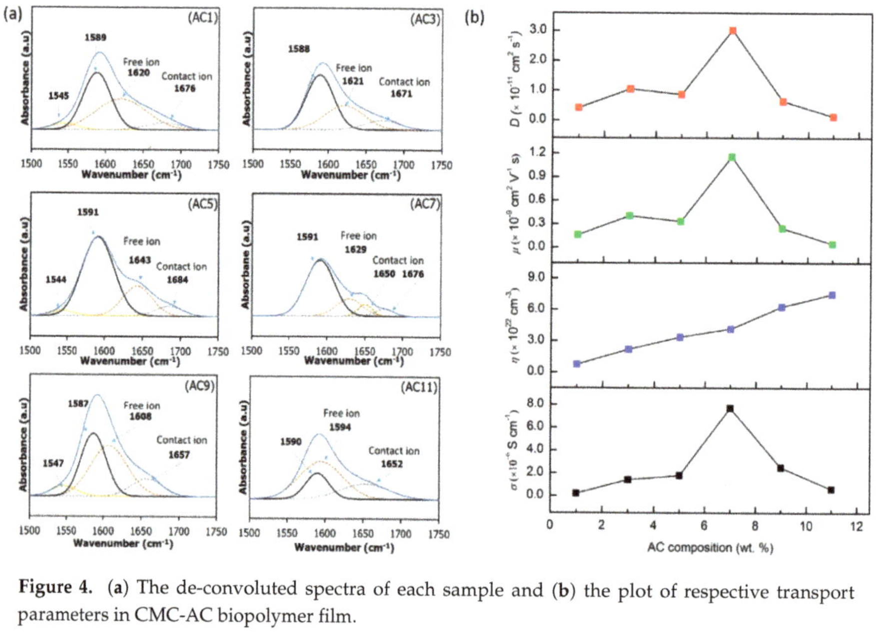

Figure 4. (**a**) The de-convoluted spectra of each sample and (**b**) the plot of respective transport parameters in CMC-AC biopolymer film.

In Equation (6), M, N_A, and V_{Total} stands for the number of moles of salts used, Avogadro's number and the total volume of electrolytes films respectively, where V_{Total} is equal to mass, m divided by density, ρ of each material used. In Equation (7), σ is the ionic conductivity of the electrolyte films and e is the electric charge. In Equation (8), k_B is the Boltzmann constant and T is the absolute temperature in kelvin.

The peak for free ions originated from NH_4^+ ions, which have deprotonated into NH_3^+ ions. The peak position found from the de-convoluted plot is in good agreement with other reports [44,45]. The percentage of free ions peak is calculated and tabulated in Table 4. From the table, the CMC-AC biopolymer films has a percentage of free ions is in the range between 59% and 85%. For sample AC1–AC7, the percentage of free ions decreases from 85.48% for sample AC1 to 59.53% for sample AC7 which corresponds to the highest conducting sample, before it starts to increase for sample AC9 and AC11. This is in inverse manner compared to other reports where the reports show that the percentage of free ions is usually with the highest ionic conductivity sample [46–48]. However, the percentage value only gives a general view of the transport number. Further analysis of the transport number (Equations (6)–(8)) will further help to explain the transport behavior.

Table 4. The respective area and percentage of free ions and contact ions of each sample.

Sample	Coefficient of Determination, COD (R^2)	Residual Sum of Square ($\times 10^{-2}$) (a.u.)	Free Ion		Contact Ion	
			Area (a.u.)	(%)	Area (a.u.)	(%)
AC1	0.99	1.16	126.07	85.48	21.42	14.52
AC3	0.99	1.74	79.73	73.52	28.71	26.48
AC5	0.99	6.14	69.66	75.37	22.77	24.63
AC7	0.99	2.67	36.97	59.53	25.12	40.47
AC9	0.99	1.70	152.35	77.81	43.46	22.19
AC11	0.99	0.46	128.31	68.51	58.97	31.49

Figure 4b presents the calculated transport number. From the figure, the value of η increase with each AC composition added for sample AC1 to AC7 (low AC composition film). At low composition, AC salt is able to dissociate into its respective ions (NH_4^+ and CO_3^-) and has a small probability to form an ion cluster (contact ions), thus allowing one of the protons (H^+) which is loosely bound to the NH_4^+ structure to be protonated and transported across the medium. As a result, the μ and D value is increased. At higher AC composition films (AC9-AC11), the value of η continues to increase in contrast to the ionic conductivity, which dropped along with μ and D value. The ions packed closely in the electrolyte when a huge amount ionic dopant supplied into the system. Consequently, this makes the de-protonation and protonation process at the ammonium (NH_4^+) ions occurs almost spontaneously and reduces the effective mobility of mobile ions, and consequently the diffusion rate thus prove μ and D dependency [47]. The dependency of μ and D towards ionic conductivity value is similar to the factors affecting ionic conductivity in a machine-learning framework proposed by Shi et al. [1], which mentioned diffusion coefficient and ionic radius as some of the factors. In this work, the diffusion coefficient improvement is believed due to small H^+ ionic radii. Scheme 1 illustrates the behavior of ionic transport in the CMC-AC biopolymer at (a) low composition and (b) at high composition.

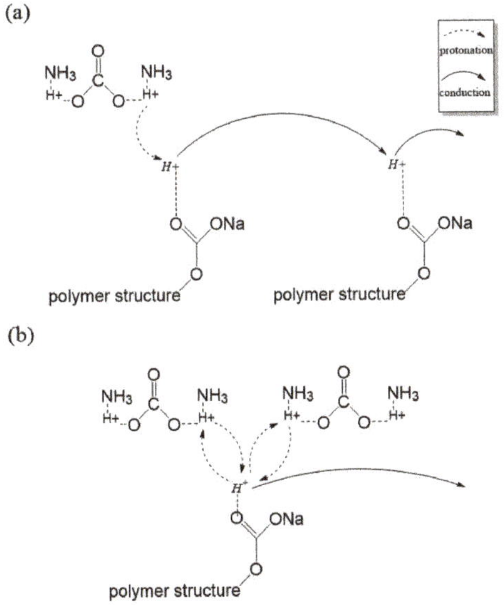

Scheme 1. The best ionic transport illustration for (**a**) at low AC composition and (**b**) high AC composition.

3.4. Transference Number Analysis

In this analysis, the biopolymer films were subjected to polarization currents (dc), which the current will become saturated indicating ionic transference, t_{ion} value [48]. See Figure S3 for testing setup. Figure 5 shows the plot of normalized current against time for sample AC7. From the figure, the initial total current decreases against time and eventually reached a steady state current. The value of t_{ion} for each sample was normalized before calculated using Equation (9) and the corresponding value is tabulated in Table 4.

$$t_{ion} = \frac{\left(I_{initial} - I_{steady-state}\right)}{I_{initial}} \tag{9}$$

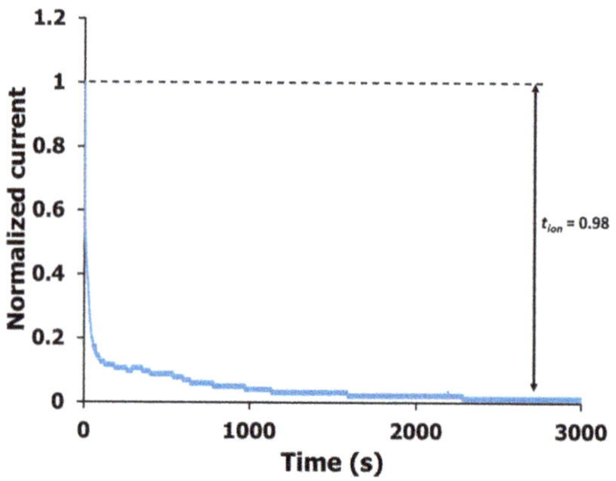

Figure 5. The plot of polarized current against time for AC7 electrolyte film.

The current become saturated due to ion accumulation at the blocking electrodes (increase resistance) while the remaining current flow is due to electron conduction. From the table, the sample with the highest ionic conductivity value (AC7) has the highest t_{ion} (0.98). This verified that the optimal CMC-AC biopolymer film system (AC7) is primarily ionic which is desirable for an intercalation process at the electrode [1]. However, the ionic conduction in the electrolyte system can be either due to cations or anions. The number of mobile ions, η calculated previously is actually the overall number for both ionic species and ionic mobility, μ and diffusion coefficient, D is the mean value for cations and anions [42]. Thus it can be differentiated to cationic and anionic mobility (μ_+ μ_-) and cationic and anionic diffusion coefficient (D_+, D_-) and calculated using Equations (11) and (13), respectively [42,43]. Table 5 shows the results of the respective calculated cationic and anionic value obtained for CMC-AC biopolymer system where the value of μ_+ and D_+ is higher than μ_- and D_-. This indicates that in the current work the primary conducting species is cations, which further proved that the CMC-AC biopolymer film is also a proton (H^+) conductor. The results obtained in this study is similar to the previous electrolyte system reported in the literature using another type of ammonium salt [49,50].

$$D = D_+ + D_- = \frac{k_B T \sigma}{\eta e^2} \tag{10}$$

$$t_{ion} = \frac{D_+}{D_+ + D_-} \tag{11}$$

$$\mu = \mu_+ + \mu_- = \frac{\sigma}{\eta e} \tag{12}$$

$$t_{ion} = \frac{\mu_+}{\mu_+ + \mu_-} \tag{13}$$

Table 5. The transference value and respective percentage of cation/anion of CMC-AC biopolymer films.

Sample	Transference, t_{ion} (± 0.01)	Diffusion Coefficient, D ($\times 10^{-13}$ cm^2 s^{-1})		Ionic Mobility, μ ($\times 10^{-11}$ cm^{-2} v^{-1} s^{-1})	
		Cation	Anion	Cation	Anion
AC1	0.78	32.4	9.13	12.4	3.49
AC3	0.82	86.0	18.9	32.9	7.22
AC5	0.92	79.6	6.92	30.5	2.65
AC7	0.98	297.0	6.06	114.0	2.32
AC9	0.90	58.1	6.46	22.2	2.47
AC11	0.92	12.1	1.05	4.6	0.40

4. Conclusions

To conclude, the CMC-AC biopolymer film has been successfully prepared. Both CMC and AC salts dissolved completely and had low degree of crystallinity, X_c for the highest conducting sample (AC7) as proven by the XRD de-convolution result, while the AC salt appeared to have weak interaction with carboxyl group of CMC. The highest ionic conductivity value for the system was 7.71×10^{-6} Scm^{-1} and thermally assisted (Arrhenius relations). The ionic conductivity of the current work was higher than other diammonium salts (ammonium adipate and ammonium sulfate) as presented in this work. It was also noted that the best composition (wt.%) of diammonium salt in a polymer electrolyte was lower compared to a single ammonium salt which was due to abundance of ammonium ions in a mol of ammonium salts. The transport number analysis showed that the CMC-AC biopolymer films were influenced by the ionic mobility, μ, and the diffusion coefficient, D, since both values achieved the maximum value for sample AC7, while the increased value of η for higher salt composition films led to lower μ and D values. The transference number also verified that the conducting species was mainly cation, which in this system was the proton (H^+). The H^+ was able to transport across the polymer matrix through the high amorphous phase of the electrolyte films and subsequently increased the ionic conductivity. This shows a promising prospect for CMC-AC biopolymer film in electrochemical applications. However, further enhancement is still needed to increase the ionic conductivity for suitable electrochemical applications (~10^{-4} Scm^{-1}).

Supplementary Materials: The following are available online at http://www.mdpi.com/2073-4360/12/11/2487/s1, Figure S1: The physical appearance of CMC-AC biopolymer films, Figure S2: The overall FTIR spectrum of CMC-AC electrolyte, Figure S3: Transference measurement testing setup.

Author Contributions: Conceptualization, M.I.N.M.I.; methodology, M.I.N.M.I. and M.I.H.A.S.; validation, M.I.N.M.I. and M.I.H.A.S.; formal analysis, M.I.N.M.I. and M.I.H.A.S.; investigation, M.I.H.A.S.; resources, M.I.H.A.S.; data curation, M.I.H.A.S.; writing—original draft preparation, M.I.N.M.I. and M.I.H.A.S.; writing—review and editing, M.I.N.M.I. and M.I.H.A.S.; visualization, M.I.H.A.S.; supervision, M.I.N.M.I.; project administration, M.I.N.M.I.; funding acquisition, M.I.N.M.I. All authors have read and agreed to the published version of the manuscript.

Funding: Fundamental Research Grant Scheme (FRGS) from Ministry of Education, VOT 59452 and Universiti Sains Islam Malaysia, Grant Scheme 166919 funded this research.

Acknowledgments: The authors would like to thank Universiti Sains Islam Malaysia for internal grant scheme (166919), Ministry of Higher Education for FRGS (59452) grants scheme. Special thanks to Advanced Materials team (AMt) members for consultations and assistances provided. The authors hope that everyone to stay strong, be safe and remain physical distance during this hard time dealing with COVID-19 pandemic. We will get through this very soon, Insha-Allah.

Conflicts of Interest: The authors declare no conflict of interest.

References

1. Shi, S.; Gao, J.; Liu, Y.; Zhao, Y.; Wu, Q.; Ju, W.; Ouyang, C.; Xiao, R. Multi-scale computation methods: Their applications in lithium-ion battery research and development. *Chin. Phys. B* **2016**, *25*, 018212. [CrossRef]

2. Goodenough, J.B.; Kim, Y.S. Challenges for rechargeable batteries. *J. Power Sources* **2011**, *196*, 6688–6694. [CrossRef]
3. Kerdphol, T.; Qudaih, Y.; Mitani, Y. Optimum battery energy storage system using PSO considering dynamic demand response for microgrids. *Int. J. Electr. Power Energy Syst.* **2016**, *83*, 58–66. [CrossRef]
4. Yue, L.P.; Ma, J.; Zhang, J.J.; Zhao, J.W.; Dong, S.M.; Liu, Z.H.; Cui, G.L.; Chen, L.Q. All solid-state polymer electrolytes for high-performance lithium ion batteries. *Energy Storage Mater.* **2016**, *5*, 139–164. [CrossRef]
5. Chai, M.N.; Isa, M.I.N. Electrical Characterization and Ionic Transport Properties of Carboxyl Methylcellulose-Oleic Acid Solid Polymer Electrolytes. *Int. J. Polym. Anal. Charact.* **2013**, *18*, 280–286. [CrossRef]
6. Zhang, Y.X.; Zuo, M.; Song, Y.H.; Yan, X.P.; Zheng, Q. Dynamic rheology and dielectric relaxation of poly(vinylidene fluoride)/poly(methyl methacrylate) blends. *Compo. Sci. Technol.* **2015**, *106*, 39–46. [CrossRef]
7. Khan, M.M.A.; Rafiudin; Inamudin. PVC based polyvinyl alcohol zinc oxide composite membrane: Synthesis and electrochemical characterization for heavy metal ions. *J. Ind. Eng. Chem.* **2013**, *19*, 1365–1370. [CrossRef]
8. Jacques, E.; Kjell, M.H.; Zenkert, D.; Lindbergh, G.; Behm, M.; Willgert, M. Impact of electrochemical cycling on the tensile properties of carbon fibres for structural lithium-ion composite batteries. *Compos. Sci. Technol.* **2012**, *72*, 792–798. [CrossRef]
9. Chatterjee, B.; Kulshrestha, N.; Gupta, P.N. Nano composite solid polymer electrolytes based on biodegradable polymers starch and poly vinyl alcohol. *Measurement* **2016**, *82*, 490–499. [CrossRef]
10. Alves, R.; Donoso, J.P.; Magon, C.J.; Silva, I.D.A.; Pawlicka, A.; Silva, M.M. Solid polymer electrolytes based on chitosan and europium triflate. *J. Non-Cryst. Solids.* **2016**, *432*, 307–312. [CrossRef]
11. Basu, T.; Tarafdar, S. Influence of gamma irradiation on the electrical properties of LiClO4-gelatin solid polymer electrolytes: Modelling anomalous diffusion through generalized calculus. *Radiat. Phys. Chem.* **2016**, *125*, 180–198. [CrossRef]
12. Raphael, E.; Avellaneda, C.O.; Manzolli, B.; Pawlicka, A. Agar-based films for application as polymer electrolytes. *Electrochim. Acta* **2010**, *55*, 1455–1459. [CrossRef]
13. Liao, H.Y.; Zhang, H.Y.; Hong, H.G.; Li, Z.H.; Qin, G.; Zhu, H.P.; Lin, Y.X. Novel cellulose aerogel coated on polypropylene separators as gel polymer electrolyte with high ionic conductivity for lithium-ion batteries. *J. Membrane Sci.* **2016**, *514*, 332–339. [CrossRef]
14. Colò, F.; Bella, F.; Nair, J.R.; Destro, M.; Gerbaldi, C. Cellulose-based novel hybrid polymer electrolytes for green and efficient Na-ion batteries. *Electrochim. Acta* **2015**, *174*, 185–190. [CrossRef]
15. Hafiza, M.N.; Isa, M.I.N. Conduction mechanism via correlated barrier hopping in EC-plasticized 2-hydroxyehyl cellulose-ammonium nitrate solid polymer electrolyte. *IOP Conf. Ser. Mater. Sci. Eng.* **2018**, *440*, 012039. [CrossRef]
16. Sasso, C.; Beneventia, D.; Zeno, E.; Conil, M.P.; Chaussya, D.; Belgacema, M.N. Carboxymethyl cellulose: A conductivity enhancer and film-forming agent for processable polypyrrole from aqueous medium. *Synth. Met.* **2011**, *161*, 397–403. [CrossRef]
17. Kamarudin, K.H.; Isa, M.I.N. Structural and DC Ionic conductivity studies of carboxymethyl cellulose doped with ammonium nitrate as solid polymer electrolytes. *Int. J. Phys. Sci.* **2013**, *8*, 1581–1587. [CrossRef]
18. Sohaimy, M.I.H.; Isa, M.I.N. Conductivity and dielectric analysis of cellulose based solid polymer electrolytes doped with ammonium carbonate (NH4CO3). *Appl. Mech.* **2015**, *719–720*, 67–72. [CrossRef]
19. Ramlli, M.A.; Isa, M.I.N. Structural and Ionic Transport Properties of Protonic Conducting Solid Biopolymer Electrolytes Based on Carboxymethyl Cellulose Doped with Ammonium Fluoride. *J. Phys. Chem. B* **2016**, *120*, 11567–11573. [CrossRef]
20. Rice, M.J.; Roth, W.L. Ionic transport in super ionic conductors: A theoretical model. *J. Solid State Chem.* **1972**, *4*, 294–310. [CrossRef]
21. Bandara, T.M.W.J.; Mellander, B.E. *Ionic Liquids: Theory, Properties, New Approaches*; InTech: London, UK, 2011; pp. 383–406. [CrossRef]
22. Petrowsky, M.; Frech, R. Concentration dependence of ionic transport in dilute organic electrolyte solutions. *J. Phys. Chem. B* **2008**, *112*, 8285–8290. [CrossRef] [PubMed]
23. Fuzlin, A.F.; Bakri, N.A.; Sahraoui, B.; Samsudin, A.S. Study on the effect of lithium nitrate in ionic conduction properties based alginate biopolymer electrolytes. *Mater. Res. Express* **2020**, *7*, 015902. [CrossRef]

24. Othman, L.; Md Isa, K.B.; Osman, Z.; Yahya, R. Ionic Transport Studies of Gel Polymer Electrolytes Containing Sodium Salt. *Mater. Today Proc.* **2017**, *4*, 5112–5129. [CrossRef]
25. Ye, L.; Feng, Z. *Polymer Electrolytes Fundamentals and Applications*; Woodhead Publishing: Cambridge, UK, 2010; pp. 550–582. [CrossRef]
26. Pradeepa, P.; Raj, S.E.; Sowmya, G.; Mary, J.K.; Prabhu, M.R. Optimization of hybrid polymer electrolytes with the effect of lithium salt concentration in PEO/PVdF-HFP blends. *Mater. Sci. Eng. B* **2016**, *205*, 6–17. [CrossRef]
27. Mohamad, A.A.; Mohamad, N.S.; Yahya, M.Z.A.; Othman, R.; Ramesh, S.; Alias, Y.; Arof, A.K. Ionic conductivity studies of poly(vinyl alcohol) alkaline solid polymer electrolyte and its use in nickel-zinc cells. *Solid State Ion.* **2003**, *156*, 171–177. [CrossRef]
28. Mejenom, A.A.; Hafiza, M.N.; Isa, M.I.N. X-Ray diffraction and infrared spectroscopic analysis of solid biopolymer electrolytes based on dual blend carboxymethyl cellulose-chitosan doped with ammonium bromide. *ASM Sci. J.* **2018**, *11*, 37–46.
29. Kumar, R.; Sharma, S.; Pathak, D.; Dhiman, N.; Arora, N. Ionic conductivity, FTIR and thermal studies of nano-composite plasticized proton conducting polymer electrolytes. *Solid State Ion.* **2017**, *305*, 57–62. [CrossRef]
30. Dhatarwal, P.; Sengwa, R.J. Dielectric relaxation, Li-ion transport, electrochemical, and structural behaviour of PEO/PVDF/LiClO4/TiO2/PC-based plasticized nanocomposite solid polymer electrolyte films. *Compos Commun.* **2020**, *17*, 182–191. [CrossRef]
31. Hafiza, M.N.; Isa, M.I.N. Solid Polymer electrolyte production from 2-hydroxyethyl cellulose: Effect of ammonium nitrate composition on its structural properties. *Carbohydr. Polym.* **2017**, *165*, 123–131. [CrossRef]
32. Shukur, M.F.; Ithnin, R.; Kadir, M.F.Z. Electrical characterization of corn starch-LiOAc electrolytes and application in electrochemical double layer capacitor. *Electrochim. Acta* **2014**, *136*, 204–216. [CrossRef]
33. Ramesh, S.; Lu, S.C. Effect of lithium salt concentration on crystallinity of poly(vinylidene fluoride-co-hexafluoropropylene)-based solid polymer electrolytes. *J. Mol. Struct.* **2011**, *994*, 403–409. [CrossRef]
34. Chai, M.N.; Isa, M.I.N. The Oleic Acid Composition Effect on the Carboxymethyl Cellulose Based Biopolymer Electrolyte. *J. Cryst. Process Technol.* **2013**, *3*, 1–4. [CrossRef]
35. Hema, M.; Selvasekerapandian, S.; Sakunthala, A.; Arunkumar, D.; Nithya, H. Structural, vibrational and electrical characterization of PVA–NH_4Br polymer electrolyte system. *Phys. B Condens. Matter* **2008**, *403*, 2740–2747. [CrossRef]
36. Samsudin, A.S.; Isa, M.I.N. Characterization of carboxy methylcellulose doped with DTAB as new types of biopolymer electrolytes. *Bull. Mater. Sci.* **2012**, *35*, 1123–1131. [CrossRef]
37. Ramlli, M.A.; Isa, M.I.N. Solid Biopolymer Electrolytes Based Carboxymethyl Cellulose Doped With Ammonium Fluoride: Ionic Transport and Conduction Mechanism. *Polym. Renew. Resour.* **2015**, *6*, 55–63. [CrossRef]
38. Khandale, A.P.; Bhoga, S.S.; Gedam, S.K. Study on ammonium acetate salt-added polyvinyl alcohol-based solid proton-conducting polymer electrolytes. *Ionics* **2013**, *19*, 1619–1626. [CrossRef]
39. Buraidah, M.H.; Arof, A.K. Characterization of chitosan/PVA blended electrolyte doped with NH_4I. *J. Non-Cryst. Solids.* **2011**, *357*, 3261–3266. [CrossRef]
40. Hassan, M.A.; Gouda, M.E.; Sheha, E. Investigations on the Electrical and Structural Properties of PVA Doped with $(NH_4)_2SO_4$. *J. Appl. Polym.* **2010**, *116*, 1213–1217. [CrossRef]
41. Kamlesh, P.; Nidhi, A.; Mrigank, M.D.; Chaturvedi, S.K. Effect of Plasticizers on Structural and Dielectric Behaviour of [PEO + $(NH_4)_2C_4H_8(COO)_2$] Polymer Electrolyte. *J. Polym.* **2013**, *2013*, 752596. [CrossRef]
42. Arof, A.K.; Amirudin, S.; Yusof, S.Z.; Noor, I.M. A method based on impedance spectroscopy to determine transport properties of polymer electrolytes. *Phys. Chem. Chem. Phys.* **2014**, *16*, 1856–1867. [CrossRef]
43. Hafiza, M.N.; Isa, M.I.N. Correlation between structural, ion transport and ionic conductivity of plasticized 2-hydroxyethyl cellulose based solid biopolymer electrolyte. *J. Membr. Sci.* **2020**, *597*, 117176. [CrossRef]
44. Ritthidej, G.C.; Phaechamud, T.; Koizumi, T. Moist heat treatment on physicochemical change of chitosan salt films. *Int. J. Pharm.* **2002**, *232*, 11–22. [CrossRef]
45. Fadzallah, I.A.; Majid, S.R.; Careem, M.A.; Arof, A.K. A study on ionic interaction in chitosan-oxalic acid polymer electrolyte membranes. *J. Membr. Sci.* **2014**, *463*, 65–72. [CrossRef]

46. Noor, I.S.; Majid, S.R.; Arof, A.K. Poly(vinyl alcohol)-LiBOB complexes for lithium-air cells. *Electrochim. Acta* **2013**, *102*, 149–160. [CrossRef]
47. Md Isa, K.B.; Osman, Z.; Arof, A.K.; Othman, L.; Zainol, N.H.; Samin, S.M.; Chong, W.G.; Kamarulzaman, N. Lithium ion conduction and ion-polymer interaction in PVdF-HFP based gel polymer electrolytes. *Solid State Ion.* **2014**, *268*, 288–293. [CrossRef]
48. Chai, M.N.; Isa, M.I.N. Novel Proton Conducting Solid Bio-polymer Electrolytes Based on Carboxymethyl Cellulose Doped with Oleic Acid and Plasticized with Glycerol. *Sci. Rep.* **2016**, *6*, 27328. [CrossRef] [PubMed]
49. Vahini, M.; Muthuvinayagam, M. AC impedance studies on proton conducting biopolymer electrolytes based on pectin. *Mater. Lett.* **2018**, *218*, 197–200. [CrossRef]
50. Rani, M.S.A.; Ahmad, A.; Mohamed, N.S. A comprehensive investigation on electrical characterization and ionic transport properties of cellulose derivative from kenaf fibre-based biopolymer electrolytes. *Poly. Bull.* **2018**, *75*, 5061–5074. [CrossRef]

polymers

MDPI

Article

Biocomposites Based on Plasticized Wheat Flours: Effect of Bran Content on Thermomechanical Behavior

Franco Dominici [1], Francesca Luzi [1], Paolo Benincasa [2], Luigi Torre [1] and Debora Puglia [1,]*

[1] Civil and Environmental Engineering Department, University of Perugia, Strada di Pentima 4, 05100 Terni, Italy; franco.dominici@unipg.it (F.D.); francesca.luzi@unipg.it (F.L.); luigi.torre@unipg.it (L.T.)

[2] Department of Agricultural, Food and Environmental Sciences, University of Perugia, Borgo XX Giugno 74, 06121 Perugia, Italy; paolo.benincasa@unipg.it

* Correspondence: debora.puglia@unipg.it; Tel.: +39-0744-492916

Received: 31 August 2020; Accepted: 26 September 2020; Published: 29 September 2020

Abstract: In the present work, the effect of different bran content on the overall thermomechanical behavior of plasticized wheat flours (thermoplastic wheat flour; TPWF) was investigated. Refined flour (F0) with negligible bran fiber content, F1 flour (whole grain flour, 20% wt. bran), F3 (50% wt. bran) and F2 (F1:F3, 50:50) film samples were realized by extrusion process. The effect of TPWF blending with two different biopolymers (polycaprolactone and poly butyrate adipate terephthalate), combined with the presence of citric acid as compatibilizer was also considered. Results from FESEM analysis and tensile characterization demonstrated that PCL was able to reach improved compatibility with the plasticized flour fraction at intermediate bran content (F2 based formulation) when 25% wt. of biopolymeric phase was added. Additionally, it was proved that improvements can be achieved in both thermal and mechanical performance when higher shear rate (120 rpm) and low temperature profiles ($T_{set}2$ = 130–135–140 °C) are selected. Disintegrability of the TPWF basic formulations in compositing conditions within 21 days was also confirmed; at the same time, an absence of any phytotoxic event of compost itself was registered. The obtained results confirmed the suitability of these materials, realized by adding different bran contents, to mechanically compete with bioplastics obtained by using purified starches.

Keywords: bran content; plasticized wheat flour; citric acid; biobased blends

1. Introduction

The increasing cost of petrol-based plastics and the public concern about their contribution to environmental pollution have raised the interest towards biobased and biodegradable materials, which help to dispose of by-products from agricultural production and food industries. In the case of bioplastics, purified starch from many agricultural sources (e.g., cereals, tubers, etc.) is often used as a basic ingredient. However, there is literature on the use of wheat flours to obtain bioplastics as an energetically and economically cheap alternative to purified starch [1,2]. Previous research from our group demonstrated that the tensile properties of thermoplastic films depended on wheat grain hardness and baking properties of refined flours [3,4]. However, the use of wholegrain flours has received limited attention, even if this approach could be of relevance for the reinforcement effect, due to bran, on the overall performance of plasticized starches [5]. Bran represents the outer portion of the grain, including the pericarp and seed teguments, containing a relevant amount of lignin and cellulose, the so-called fiber. It accounts for around 15–25% wt. of the total grain weight and generally comes out as a by-product of grain milling. It is normally used as animal feed [6]; however, the progressive decrease of the whole national livestock that occurred in the last decade has led to an increase of bran stocks to get rid of, with possible valorization as biological chemicals and energy source [7]. The effect of bran particle size on functionality of the gluten network was already explored in wholegrain flour and its baking properties [8]: It was found that a deleterious effect on time of dough development, gluten strength, starch gelatinization and the retrogradation was intensified by the presence of all constituents of the grain in the wheat mass formulation when compared to refined flour. In particular, it was evidenced that the quality of the protein and the differences between the particle sizes with respect to the stability and development time are broadly correlated with the quality of the gluten network. Additionally, it was proven that the presence of fibers limited the availability of water to the starch in the wholegrain samples, and that this effect was especially strong for flour with finer particle size, which also had the highest rate of absorption.

In another paper, Liu et al. [9] studied the adverse effects of wheat bran on gluten network formation, which may lead to the reduction in gluten viscoelasticity and quality deterioration of fiber enriched flour products. With the properties of starch, such as degree of gelatinization, gel stability, and retrogradation, being strongly influenced by the availability of water in the formed mass system [10], it is considered extremely important to investigate the role of bran content and its grinding level even on the thermomechanical behavior of plasticized flours.

Since it is expected that the bran level could, in general, increase the strength of the matrix to the expense of the deformability of the plasticized flour [11] by creating macroscopic defects in the material, the possibility of using biobased polymers in combination with thermoplastic flour to recover the plasticity has been also considered. The literature reports results on the effect of fiber reinforcement on mechanical behavior of thermoplastic starch blended with different polyesters [12], but no examples are available on how the presence of bran additive could tune the mechanical behavior; according to this, we here attempted to verify, for the first time, how different contents of grinded bran could affect the deformability of the flour; furthermore, blending with low melting polymeric fractions (polycaprolactone and polybutylene adipate terephthalate) was considered to increase the limited deformability of the polymeric matrix.

2. Materials and Methods

2.1. Materials

Soft wheat produced in the Umbria region (Italy) was chosen as the reference wheat variety to obtain, by grinding and selective extraction, flours with different bran contents. The milling products were kindly supplied by Molini Spigadoro (Bastia Umbra, Italy). The chemicals, glycerol, magnesium stearate, D-sorbitol, water, polyvinyl alcohol (≥99% hydrolyzed) (PVA) and citric acid (CA) were supplied by Sigma Aldrich. Polybutylene adipate terephthalate (PBAT) Ecoflex F Blend C1200 was supplied by BASF. Polycaprolactone (PCL) Capa 6500 was kindly provided by Perstorp.

2.2. Preparation of Milled Products, Their Plasticization and Blending

Four wheat flour milling products, namely F0, F1, F2, F3, were considered. The detailed compositions are reported in Table 1.

Table 1. Content of plasticizable fraction (PF) and fiber in% wt. of the samples.

Sample	Flour *	Bran **	Plasticizable Fraction (PF)	Unplasticizable Fraction (UF)
F0	100	0	100.0	0.0
F1	80	20	86	14
F2	65	35	75.5	24.5
F3	50	50	65.0	35.0

* FLOUR (PF/UF = 100/0); ** BRAN (PF/UF = 30/70).

F0 was a refined flour with negligible percent bran content; F1 was a wholegrain flour as it would come out from milling the whole grain (including the pericarp and seed coats); F3, with a bran content of 50% (wt.), represented the outcome generally obtained as the grinding tail of milling; F2 was obtained by mixing F1 and F3 in equal parts (50:50 wt.).

Plasticization of the samples was carried out with an Xplore Microcompounder 5 & 15 cc extruder (DSM, Sittard, The Netherlands) by considering suitable contents of glycerol, water and other process facilitating additives, as reported in our previous work [3]: Flour (68%, *w/w*), glycerol (23%, *w/w*), magnesium stearate (1.8%, *w/w*), sorbitol (5.2%, *w/w*), PVA in aqueous solution PVA/water 1:20 (2%, *w/w*). The initial process parameters were set as reported in our previous paper on refined flours: Temperature profile ($T_{set}2$) in the three heating zones of the extruder at 130–135–140 °C and mixing at 30 rounds per minute (rpm) for 6 min [4].

The doses of the reagents for the plasticization were adapted by calculating the plasticizable fraction of material (PF) (starch, proteins and other components), excluding the fiber and the other not plasticizable constituents (UF). Indeed, fiber content, which represents the fibrous portion of flour, does not participate to the plasticization process, while the bran content, which has plasticizable fraction, should be taken into account. The fiber, not participating directly in the plasticization reaction, was considered in the formulation only for the evaluation of absorbed water, estimated to be at 15% by weight of fiber [13]. In order to give an explanation of the adopted methodology for plasticization, a detailed recipe for the F1 sample is given as an example. F1 flour, as typical wholegrain flour, consists of 80% flour and of 20% bran. The plasticizable fraction (PF) of F1 is 86% wt. (given by $0.8 \times 1.0 + 0.2 \times 0.3$, where 1.0 is the PF of the flour and 0.3 is the PF of the bran). In a similar way, PF values have been calculated for all the samples and summarized in Table 1.

In the case of blends based on TPWF, two types of biodegradable polymers, PBAT and PCL, were initially considered at a weight amount of 20% wt. In the case of polycaprolactone, the research was extended to blends containing 25, 30 and 40% wt. of the polymeric component. Further attempts to optimize the formulations were made by changing the quantities and types of plasticizers. The amount of glycerol was reduced from 23 to 17% wt. and, at the same time, an additional water fraction of 17% wt. was added to provide hydroxyl groups functional to plasticization, less available due to glycerol reduction. Moreover, the plasticizing and compatibilizing effect of citric acid added at 0.8% wt. was also evaluated. Then, an optimization of the parameters was tried by varying the temperature and mixing speed. The effects of increasing the temperature profile were tested by setting $T_{set}3$ to 135–140–145 °C. An attempt was also made increasing the rotation speed of the screws from 30 to 120 rpm. All the samples were used to produce specimens of film with a thickness of about 300 μm with the aid of a Film Device Machine (DSM, Sittard, The Netherlands) coupled to the extruder. The list of samples is summarized in Table 2.

2.3. *Characterization of Flours and TPWF-Based Composites*

2.3.1. Alveographic Properties

F0 flour was used for the measurement of alveographic parameters as it is a good approximation to the plasticizable fraction of the processed samples. The tests were carried out by using a Chopin alveograph (Alveolink NG, Villeneuve-la-Garenne, France) in constant hydration (HC) mode, following the recommendations of the ISO 27,971 standard. Average values of the main alveographic parameters, Tenacity (P), Extensibility (L), Baking strength (W), and Configuration ratio (P/L), were determined with five replicates.

2.3.2. Thermogravimetric Analysis

Thermal degradation of the milling products F0, F1 and F3, having different content of bran, was evaluated carrying out thermal dynamic tests, from 30 °C to 600 °C at 10 °C min^{-1} by thermogravimetric analysis (TGA, Seiko Exstar 6300, Tokyo, Japan). About 5 mg of each sample was used, and dynamic tests were performed under nitrogen flow (200 mL min^{-1}). Mass loss (TG) and derivative mass loss (DTG) curves for each tested material were evaluated.

Table 2. Samples produced: Main constituents and processing parameters.

Sample	Flour	Glycerol (% wt.)	Biopolymer (% wt.)	Citric Acid (% wt.)	T_{set} *	Screw Speed (Rpm)
TPF0	F0	23	0	0	2	30
TPF1	F1	23	0	0	2	30
TPF2	F2	23	0	0	2	30
TPF3	F3	23	0	0	2	30
TPF2_CA	F2	23	0	0.8	2	30
TPF2_20BAT	F2	23	20 PBAT	0	2	30
TPF2_20CL	F2	23	20 PCL	0	2	30
TPF2_CA20BAT	F2	23	20 PBAT	0.8	2	30
TPF2_CA20CL	F2	23	20 PCL	0.8	2	30
F0_CA2	F0	17	0	0.8	2	30
F0_CA3	F0	17	0	0.8	3	30
F0_CA20CL2	F0	17	20 PCL	0.8	2	30
F0_CA20CL3	F0	17	20 PCL	0.8	3	30
F2_CA20CL2	F2	17	20 PCL	0.8	2	30
F2_CA20CL3	F2	17	20 PCL	0.8	3	30
F2_20CL	F2	17	20 PCL	0.8	2	30
F2_20CL120R	F2	17	20 PCL	0.8	2	120
F2_25CL	F2	17	25 PCL	0.8	2	30
F2_25CL120R	F2	17	25 PCL	0.8	2	120
F2_30CL	F2	17	30 PCL	0.8	2	30
F2_30CL120R	F2	17	30 PCL	0.8	2	120
F2_40CL	F2	17	40 PCL	0.8	2	30
F2_40CL120R	F2	17	40 PCL	0.8	2	120
F0_25CL120R	F0	17	25 PCL	0.8	2	120
F1_25CL120R	F1	17	25 PCL	0.8	2	120
F3_25CL120R	F3	17	25 PCL	0.8	2	120

* T_{set} 2 = 130–135–140 °C; 3 = 135–140–145 °C.

2.3.3. Tensile Tests

A universal electronic dynamometer LR30K Plus (LLOYD Instruments, Bognor Regis, UK) was used to carry out a mechanical characterization of the materials. Tensile tests were performed by setting a crosshead speed of 5 mm min^{-1} on 20 × 150 mm rectangular specimens about 300 mm thick, in accordance with ISO 527 standards. Ultimate tensile strength (σ) and strain at break (ε_b) were calculated from the resulting stress–strain curves with the support of a software specific to the test machine: NEXYGEN Plus Materials Testing. The measurements were done, after conditioning the samples at room temperature for 24 h at 50% relative humidity (RH), testing at least five specimens for each formulation.

2.3.4. Morphological Evaluation

A first visual analysis was performed on TPWF/bran-based film samples. Moreover, a morphological characterization of composites was carried out using a field emission scanning electron microscope (FESEM) Supra 25 by Zeiss (Oberkochen, Germany). Micrographs of fractured surfaces obtained by cry fracturing the samples in liquid nitrogen were taken with an accelerating voltage of 5 kV at different magnifications. Previously, the samples were gold sputtered to provide electric conductivity.

2.3.5. Disintegration in Compost

The compost mineralization of the films was evaluated on the basis of the ISO 20,200 standard. A certain amount of compost inoculum, supplied by Gesenu Spa, was mixed together with synthetic organic waste, prepared with an appropriate amount of sawdust, rabbit feed, starch, sugar, oil and urea to thus constitute the soil for composting. The soil moisture content was maintained at values of 50% RH by adding water and mixing at regular intervals of time, as indicated by the legislation, while aerobic and thermal conditions were guaranteed during the test. Based on ISO 20200, a sample can be considered disintegrated when it reaches 90% mass disintegration in at least 90 days in contact with the composting soil in the ripening phase. The disintegration percentage after a time t in compost is calculated as reported in Equation (1):

$$D_t = \frac{m_i - m_r}{m_i} \times 100 \quad (1)$$

where m_i is the initial mass of the sample and m_r is the mass of the extracted sample, after drying, at a given time t.

2.3.6. Evaluation of Phytotoxicity

The phytotoxicity of the compost obtained from the disintegration test of the films was assessed at 40 days from the start of composting and, following the obtained results, the evaluation was repeated at 60 days. A germination test was carried out on cress seeds (*Lepidum sativum* L.), a test plant normally used for this purpose, as required by the IPLA, DIVAPRA, ARPA methods "Compost Analysis Methods", 1998. This method involves the evaluation of the effect of an aqueous extract of compost, picked up from disintegration tests, on seed germination. It was decided to evaluate three composts, those obtained with plastic films derived from flours F0, F1 and F3, assuming that F2 would give an intermediate result between F1 and F3. For each compost, the following standard procedure was used. Each sample to be tested (200 g) was brought to a humidity of 85% and left for two hours in contact with the added water. It was then centrifuged at 6000 rpm for 15 min and the supernatant was filtered under pressure at 3.5 atm with a sterilizing membrane. The aqueous extract was diluted up to concentrations of 50% and 75%. Five aliquots, each of 1 mL, of each of the two dilutions of the obtained samples (plus the same number of controls with water) were placed in 9 cm diameter Petri dishes containing bibulous paper. 10 seeds of *Lepidium Sativum* were added to each capsule, soaked for one hour in distilled water. The capsules were placed to incubate at 27 °C for 24 h. After this period, the germinated seeds were counted, and the root length of the buds was measured. The germination index (I_g) was calculated as indicated in Equation (2):

$$I_g(\%) = \frac{(G_c \times L_c)}{(G_t \times L_t)} \times 100 \quad (2)$$

where:

G_c = average number of germinated seeds in the sample
G_t = average number of seeds germinated in the control
L_c = average root length in the sample
L_t = average root length in the control

The values of the germination indices for 50% and 75% dilutions after 40 and 60 days of maturation of the compost extracted from the disintegration soils of the samples TPF0, TPF1 and TPF3 were analyzed.

2.3.7. Statistical Analysis

Data were analysed by analysis of variance (ANOVA), using the Statgraphics Plus 5.1. Program (Manugistics Corp. Rockville, MD, USA). To differentiate samples, Fisher's least significant difference (LSD) was used at the 95% confidence level.

3. Results and Discussion

Wheat Flour Characterization

The refined flour F0, having a moisture content of 14.5% wt. and a protein amount of 11.8%, was tested with the Chopin's alveograph at constant hydration (HC), showing the following alveographic parameters: Tenacity (P = 64 mm H_2O), extensibility (L = 99 mm), baking strength (W = 182 × 10^{-4} J), configuration ratio (P/L = 0.65), elasticity index (I_e = 47.4%). These alveographic parameters are characteristic of a standard flour with moderate strength and standard quality for basic baking uses.

The results of the thermogravimetric analysis (Figure 1) show that, with the exception of weight loss due to water evaporation at around 80 °C, there were no significant weight losses due to thermal degradation within the temperature range for plasticization up to 150–160 °C. As reported in [14], the shape of this low temperature peak can be varied due to bran addition: It was shown that flour rich mixtures exhibit distinct features with an initial peak attributed to starch and a secondary shoulder attributed to gluten. In general, they observed that a gradual shift occurred in the gluten shoulder, in conjunction with the addition of bran to the mixture. In our case, we observed that in bran rich mixtures, i.e., F3 formulation, the peak was not symmetric and a shift to a lower temperature range was observed due to a modified moisture release from the flour during heating.

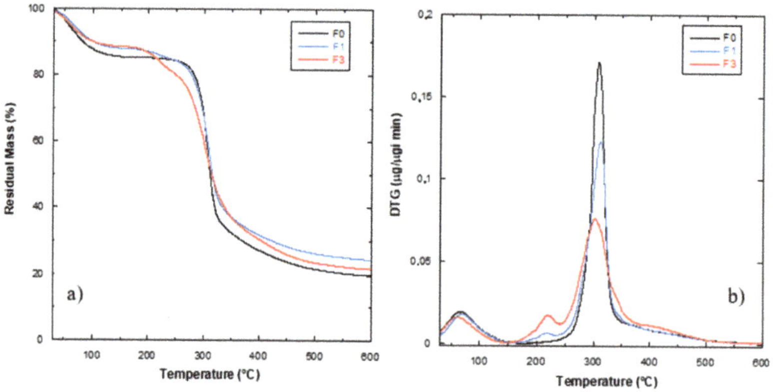

Figure 1. Mass loss (TG) (**a**) and derivative mass loss (DTG) (**b**) curves of F0, F1 and F3 samples.

The thermogravimetric analysis evidenced also the typical degradative pattern for cereal flours. The three wheat flours showed similar TG curves, with small differences in terms of residual weight at the end of the test, in line with the typical mass loss values (17% *w*/*w* for wheat flour) and the additional residue due to the presence of the bran component (Figure 1a): After water evaporation, the second main step centered at 300 °C corresponds to the decomposition of starch, while the third step (T > 400 °C) corresponds to the formation of inert carbonaceous residues [1] (Figure 1b). In the case of increasing bran amount, the F1 and F3 samples showed the presence of a further peak, in the range 200–250 °C, identified as the starting point for decomposition of lignocellulosic components (mainly cellulose and hemicellulose) in the bran fraction [15].

Having established that the flours with variable bran content could be processed and plasticized in the selected temperature range without losing thermal stability, films based on F0, F1, F2 and F3 flours were realized by extrusion, as detailed in Section 2.2. The visual analysis of the films in the top row (Figure 2a), produced without the addition of biopolymers, showed a progressive browning as the fraction of bran in the flour increased. However, films with low fiber content showed good transparency, satisfying a fundamental characteristic for some packaging applications. The yellowing/browning of the TPWF films, which it is normally caused by the non-enzymatic reactions that occur during plasticization, is emphasized by the presence of the bran, which adds opacity and darkening [16–18].

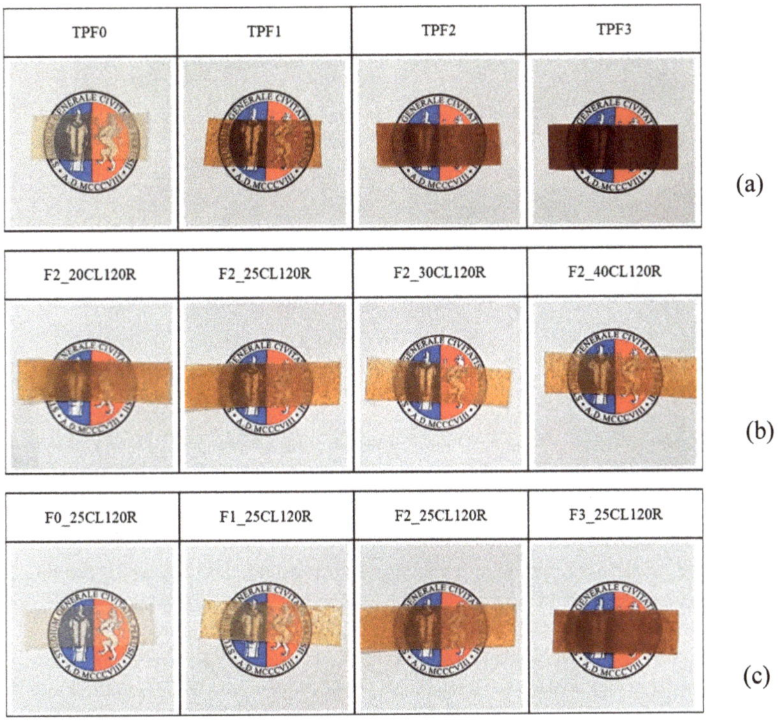

Figure 2. Visual observation of (**a**) thermoplastic wheat flour (TPWF) based films made with flours with increasing bran content (from F0 to F3), (**b**) F2 based films blended with 20, 25, 30 and 40% wt. of PC and (**c**) films based on plasticized F0, F1, F2 and F3 blended with 25% wt. of polycaprolactone (PCL).

In Figure 2b, the effect of the addition of PCL in the formulation based on F2 flour produces an improvement in transparency, which was enhanced as the proportion of biopolymer in the blend increased. In the last line of the picture (Figure 2c), images of the films obtained after the optimization of compositions (25% wt. of PCL fraction) and processing parameters (120 rpm) are included. The film based on refined flour F0 shows good transparency, which remained acceptable even in the F1-based film, despite the yellowing due to the presence of bran fibers. The level of transparency of F2_25CL120R is also acceptable, although the darkening caused by the abundant bran fraction produced a color change on browning tones and a sensible reduction in transparency [19]. In the case of F3-based film, the transparency was compromised to an extent that makes the film unsuitable for applications requiring visibility of the underlying objects, while its use for opaque packaging or other applications, such as mulching or shading sheets, where opacity is functional, can be envisaged for the F3_25CL120R composition.

The morphologies of the fractured surfaces for the TPWF films were observed by FESEM (Figure 3) and differences were found for the four milling products with different bran contents. In detail, the F0 flour (Figure 3a) appeared well plasticized with a uniform surface, no separate starch granules were noted and the absence of bran particles was evident. Plasticization of the F1 flour (Figure 3b) was also well achieved, since a smooth and homogeneous plastic phase was found in the analyzed surface. Bran fibers with a particle/lamellar appearance were uniformly distributed and well bonded to the plasticized starch, suggesting the realization of a composite material with good characteristics. Similarly to the previous ones, the plasticization of the F2 flour was also performed with good results.

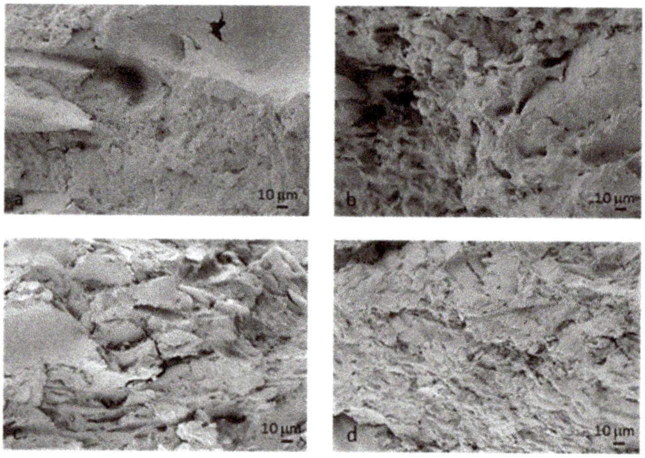

Figure 3. FESEM images of fractured cross-sections of (**a**) TPF0, (**b**) TPF1, (**c**) TPF2 and (**d**) TPF3 films, acquired at 1000× magnification.

In this case (Figure 3c), we noted the prevalent presence of lamellar particles of bran that resulted oriented, as alternating layers with the plastic phase, due to the orienting effect of the production process. F3, with the highest fiber fraction among the selected flours, highlighted the prevalent presence of bran particles (Figure 3d), with the plasticized starch having reduced adherence to the bran particles. Fibrous agglomerates and not well plasticized starch particles were noted: Due to the hindering effect of the large amount of bran fiber, wheat flour granules were less capable of forming hydrogen bonds with plasticizers through their hydroxyl groups, leaving some domains unreacted with unplasticized starch particles [3,20]. In general, while observed morphologies in TPF1 and TPF2 can confirm the ductility of the films, to the expense of low strength (that actually increased in TPF2 due to increased filler content), TPF3 appears saturated with the reinforcement phase and a matrix phase close to the wettability limit of the fibers. In this case, behavior that maximizes strength and stiffness but limiting the elastic-plastic characteristics can be expected.

The observations made by analyzing the sample morphologies were confirmed by checking the results of tensile characterization made on the same series of materials In Table 3, the results of the tensile tests carried out on plasticized flour samples and their bioblends are included.

Table 3. Results of tensile tests made on plasticized flours and their blends with PCL and polybutylene adipate terephthalate (PBAT), by varying the biopolymer amounts, processing temperatures and mixing rate.

Sample	Flour	Glycerol (% wt.)	Biopolymer (% wt.)	Citric Acid (% wt.)	Tset *	Screw Speed (rpm)	Ultimate Tensile Strength (MPa)	Strain at Break (%)
TPF0	F0	23	0	0	2	30	1.23 ± 0.05	54.13 ± 4.39
TPF1	F1	23	0	0	2	30	2.63 ± 0.10	32.23 ± 0.64
TPF2	F2	23	0	0	2	30	2.62 ± 0.40	23.84 ± 2.45
TPF3	F3	23	0	0	2	30	3.83 ± 0.20	19.64 ± 1.60
TPF2_CA	F2	23	0	0.8	2	30	1.61 ± 0,08	34.89 ± 2.44
TPF2_20BAT	F2	23	20 PBAT	0	2	30	2.21 ± 0,17	6.93 ± 0.47
TPF2_20CL	F2	23	20 PCL	0	2	30	1.96 ± 0.06	16.80 ± 1.98
TPF2_CA20BAT	F2	23	20 PBAT	0.8	2	30	1.50 ± 0.05	10.44 ± 0.17
TPF2_CA20CL	F2	23	20 PCL	0.8	2	30	1.66 ± 0.34	60.39 ± 13.00
F0_CA2	F0	17	0	0.8	2	30	1.54 ± 0.02	75.47 ± 11.66
F0_CA3	F0	17	0	0.8	3	30	1.05 ± 0.14	31.08 ± 6.07
F0_CA20CL2	F0	17	20 PCL	0.8	2	30	2.89 ± 0.23	61.80 ± 11,87
F0_CA20CL3	F0	17	20 PCL	0.8	3	30	2.69 ± 0.22	47.04 ± 18.15
F2_CA20CL2	F2	17	20 PCL	0.8	2	30	3,10 ± 0,24	12,42 ± 1,43
F2_CA20CL3	F2	17	20 PCL	0.8	3	30	2,72 ± 0,23	13,83 ± 1,12
F2_20CL	F2	17	20 PCL	0.8	2	30	3.10 ± 0.24	12.42 ± 1.43
F2_20CL120R	F2	17	20 PCL	0.8	2	120	3.17 ± 0.20	14.15 ± 2.46
F2_25CL	F2	17	25 PCL	0.8	2	30	3.05 ± 0.03	16.27 ± 1.47
F2_25CL120R	F2	17	25 PCL	0.8	2	120	3.70 ± 0.16	18.87 ± 2.77
F2_30CL	F2	17	30 PCL	0.8	2	30	3.69 ± 0.43	12.49 ± 1.03
F2_30CL120R	F2	17	30 PCL	0.8	2	120	3.43 ± 0.36	11.73 ± 0.99
F2_40CL	F2	17	40 PCL	0.8	2	30	4,57 ± 0.36	30.75 ± 1.97
F2_40CL120R	F2	17	40 PCL	0.8	2	120	4,63 ± 0.13	52.64 ± 4.02
F0_25CL120R	F0	17	25 PCL	0.8	2	120	2.68 ± 0.11	57.11 ± 6.29
F1_25CL120R	F1	17	25 PCL	0.8	2	120	3.58 ± 0.14	37.39 ± 4.07
F3_25CL120R	F3	17	25 PCL	0.8	2	120	3.91 ± 0.31	10.04 ± 3.92

* T_{set} 2 = 130–135–140 °C; 3 = 135–140–145 °C.

The refined F0 flour, following the plasticization process, shows mechanical properties in line with other flours, with comparable alveographic characteristics, tested in previous works [3]. As evidenced in Figure 4a, good elongation values (54%) correspond to a moderate tensile strength (1.23 MPa), which is the main drawback of TPWFs. The selection of flours containing different bran fractions offers the advantage of having a fibrous filler, which can be effective as a reinforcement phase and, at the same time, has a plasticizable fraction, able to guarantee good compatibility and bonding at the interface with the starch matrix upon plasticization. The bran plays the role of reinforcement by preventing creep and deformation of the thermoplastic phase. As the bran fraction and consequently the fiber content increased, the samples showed decreasing strain values. TPF1 showed an ε_b of 32.2%, which decreased to 23.8% with TPF2, further dropping to 19.6% in the case of the TPF3 sample. On the other hand, when the percentage of bran increased, the tensile strength increased as well, reaching an σ value more than tripled in the case of the F3-based sample (3.83 MPa) when compared to refined F0 flour.

The tensile strength of TPF1 was more than doubled (2.63 MPa) compared to the sample without fiber TPF0; TPF2 shows the same tensile strength value (2.62 MPa) as TPF1, albeit with a higher bran content. This result suggested the possibility of an improvement of the mechanical properties for reference TPF2, which could be achieved by improving the dispersibility of the bran fiber in the plasticized matrix. To pursue this goal, the characteristics of the matrix must be enhanced by improving the compatibility with the reinforcement phase. Comparing the values obtained for all formulations, it can be commented that values for the maximum tensile strength and elongation at break changed significantly ($p < 0.05$).

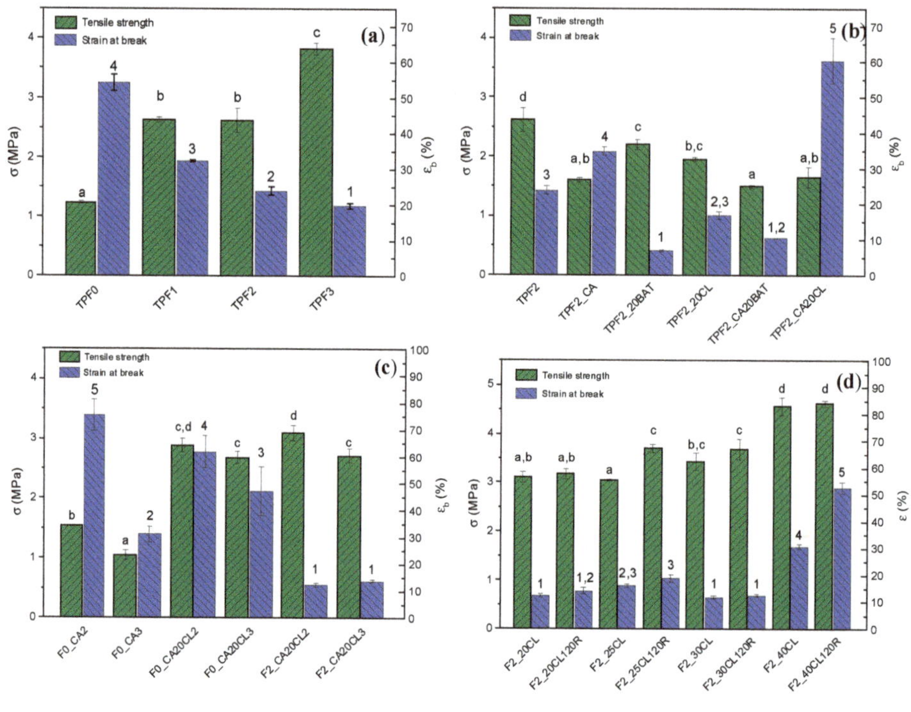

Figure 4. Results of tensile strength and strain at break for (**a**) TPWF/bran film samples, (**b**) TPWF/bran and biopolymeric blends, (**c**) samples processed with two different plasticization temperature profiles (T2, T3), (**d**) samples processed with two different mixing rates (30 rpm, 120 rpm). (a–d, stress values) (1–5, strain values) Different superscripts within the same column group (stress or strain values) indicate significant differences among formulations ($p < 0.05$).

In order to further improve the characteristics of the F2 based TPWF, two actions were considered: The improvement of the intrinsic characteristics of the composite by using plasticizers/compatibilizers and the addition of another matrix in blend that could enhance the characteristics of the entire composite system. The selection of the matrices was carried out taking into account some preliminary criteria, such as physical–chemical compatibility, conservation of the biodegradability of materials and process compatibility, possible plasticization and blending in one step to minimize energy waste and environmental impact, according to an eco-sustainable development perspective. Polybutylene adipate terephthalate (PBAT) and polycaprolactone (PCL) were selected for this specific purpose and initially used at a nominal percentage of 20% wt. Citric acid (CA) was indeed considered as a suitable compatibilizer for TPWF: CA, other than having a plasticizing effect, can be also effective in the compatibilization between plasticized starch, bran fiber and biopolymers [21,22]. A nominal percentage of 0.8% wt. was chosen, on the basis on the results of previous literature works [23]. The effects on the composites were firstly evaluated by adding individually CA, PBAT and PCL, and then the formulations with the concurrent use of biopolymer and compatibilizer were studied.

In Figure 4b, it is demonstrated that the citric acid, added alone to the TPF2 formulation, has a plasticizing effect, increasing the deformation to 34.9% but reducing the strength to 1.61 MPa. The TPF2_20BAT blend showed an increase in tensile strength, beside a reduction in deformability, which highlighted the poor compatibility between TPWF and PBAT. A decrease of both values of strength and deformation at break was also found in the TPF2_20CL formulation, with only PCL in blend. The use of citric acid was found to have no positive effects when added in the presence of PBAT, while its role was effective when combined with PCL. TPF2_20CL showed an increase in strain at break up to 60.4% compared to the value of 16.8% for the same sample without CA. The addition of citric acid induced reactions able to favor the adhesion between PCL and TPS (trans-esterification), improving the wettability (hydrolysis) and inhibiting the formation of cross-linking (sulfhydryl (SH)-SS exchange) during flour plasticization [21,22,24–26].

Considering that glycerol has a much higher plasticizing effect, due to the presence of three hydroxyl groups, compared to water, and even a "lubricating" effect that lowers the stiffness of the plasticized system, it was planned to replace 6% wt. of glycerol with 17% wt. of water [27]. Furthermore, an attempt was made to improve the mechanical performance of the produced films by varying the plasticization temperature from T2 to T3 [4,28]. The increase of the temperature had the purpose of improving the tensile stress resistance of the materials, by intensifying the formation of bonds and crosslinking, typical of the plasticizing process, conferring strength and rigidity to the TPWF. To better understand the effect of these variations on mechanical properties of TPWF-based samples, both the samples of refined flour F0 and those of F2 flour were tested. The three pairs of samples processed at T2 and T3 (Figure 4c) showed that the increase in temperature, in the presence of water and CA, generally causes a worsening of the mechanical properties, by lowering the stress and strain values. At higher temperatures, the hydrolytic phenomena induced by CA prevailed over the effects of transesterification and cross-linking, supported by the kinetics of the plasticization reaction of the flours at T3. It should be noted that the new dosage with the partial replacement of glycerol with water produced a notable increase in strength (+74%), that moved from 1.66 MPa of TPF2_CA20CL to 2.89 MPa of F2_CA20CL as expected.

A further attempt to optimize the formulations was made by increasing the fraction of PCL in the blend to evaluate the ideal TPWF/PCL ratio (Figure 4d). Furthermore, the effect of shear stresses during plasticization was evaluated by processing samples at 30 and 120 rpm. It is known that the shear stresses applied during the plasticization phase can produce effects on the destruction of starch granules and, consequently, on the mechanical characteristics of the materials. In order to take into account the effect of the rheological characteristics of the system, the tests were repeated for materials with different PCL fractions [29,30]. A higher speed of rotation of the screws during plasticization in the extruder produced a general improvement of the mechanical properties, both in terms of strength and, albeit to a lesser extent, of strain. In particular, the increase of shear stresses raised the strength (+23%) from 3.0 to 3.7 MPa in the sample F2_25CL120R. In Figure 5, the SEM micrographs of the samples processed at different screw speeds show a completely different morphology; higher shear stresses, produced with 120 rpm of screw speed, improved plasticization (Figure 5b); resulting images showed smoother and more uniform fracture surfaces, free of granules and fibrous conglomerates, with uniform distribution of the separate phases of PCL and TPWF.

Figure 5. FESEM morphologies of F2_25CL30R (**a**) and F2_25CL120R (**b**) samples.

Finally, samples of all flours were produced using the optimized formulation and process parameters. In Figure 6, the progressive improvement in tensile strength is closely related to the increase in the fraction of bran fiber. The lowest σ value (2.68 MPa) was obtained for F0_25CL120R, which was free of fiber, and rose to 3.91 MPa with F3_25CL120R. The intermediate fiber contents also corresponded to intermediate strength values for F1_25CL120R and F2_25CL120R, equal to 3.58 and 3.70 MPa, respectively. On the other hand, the increase in the bran fraction produced a progressive decrease in the strain at break, which reached 57.1% with F0 flour and dropped to 10.0% for the F3 flour-based material. Samples on F1 and F2, flour-based, showed intermediate ε_b values of 37.4% and 18.9%, respectively. A clear effect of the bran fiber fraction on the mechanical properties of the TPWF-based composites was noted: The addition of fiber produced a 45% increase in strength but caused a drop to 18% of maximum elongation reached without fiber. This result suggests that the composite F3_25CL120R has reached the fiber saturation (35% wt.) and a further increase of the reinforcement phase would lead to brittle behavior of the material. Materials made with quantities of fibers between the tested extremes show intermediate values of σ and ε_b indicating the possibility of designing the mechanical properties of the material to be produced according to the formulation.

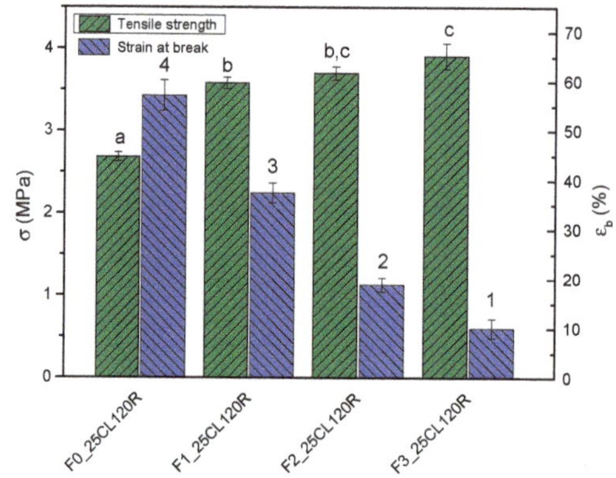

Figure 6. Results of tensile strength and strain at break for TPWF/PCL blends based on F0, F1 and F3 flours, containing 25% wt. of PCL and processed at 120 rpm. (a–c, stress values) (1–4, strain values) Different superscripts within the same column group (stress or strain values) indicate significant differences among formulations ($p < 0.05$).

The thermal stability of the optimized formulations was determined by thermogravimetric analysis. Figure 7 presents the thermal degradation profile (TG/DTG curves) of the TPWF/PCL blends based on F0, F1 and F3 flours, containing 25% wt. of PCL and processed at 120 rpm. Thermal degradation of blends presented four mass loss stages (Figure 7). Up to approximately 130 °C, there is a mass loss due to the presence of water, while following weight loss, observed between 130 and 230 °C, can be related to the evaporation of glycerol and other volatile compounds present in TPWF [31]. Then, the starch chains began to degrade at about 230 °C [32]; after that, the fourth stage of thermal degradation of the blends occurred from 350 to 430 °C, due to the degradation of PCL chains. The main differences in these profiles was found for the signal of the plasticized TPWF; while the maximum degradation rate of the polymeric PCL phase was almost constant in intensity for all the three blends, the second main weight loss accounted for the reduced amount of plasticized fraction. It essentially followed the trend that TPWF with more bran content (F3 based blend) showed reduced degradation rates. The increased amount of bran was also responsible for increased value of remaining mass at the end of the test, as observed in Figure 7b, due to the charred fraction of fiber.

Figure 8a shows the visual images of the samples during the progress of the disintegration process under composting conditions, while Figure 8b shows the trend of the mass disintegration rate in compost for the three tested formulations. All materials reach 90% disintegration after 15 days under composting conditions. The TPF3 film showed different disintegration kinetics, presenting lower disintegration values, in comparison with TPF0 and TPF1 films, between the 2nd and 4th day under composting conditions. This behavior can be justified considering that higher fiber content was present in F3 film, which slowed down the decomposition process of the plasticized fraction. Starting from the 10th day of the test, both degradation kinetics and final disintegration degree of the three systems were aligned and samples completely disintegrated within 21 days, confirming the compostability, at lab conditions, of the studied materials.

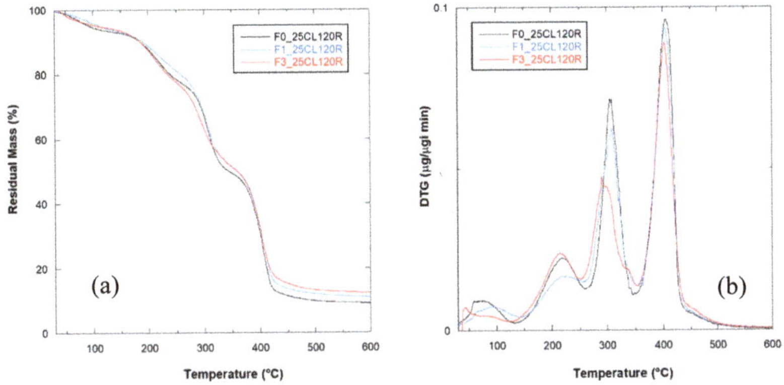

Figure 7. TG (**a**) and DTG (**b**) curves for TPWF/PCL blends based on F0, F1 and F3 flours, containing 25% wt. of PCL and processed at 120 rpm.

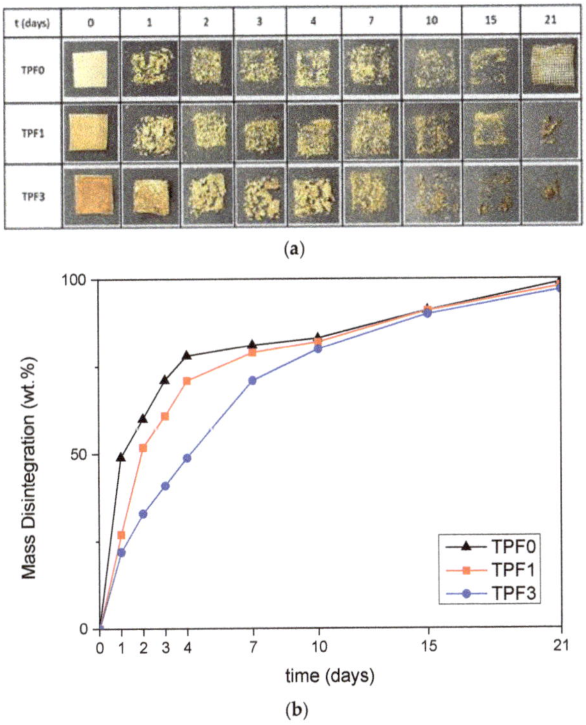

Figure 8. Visual images of the TPWF samples based on F0, F1 and F3 flours (**a**) and evolution of disintegration degree (**b**) at different compositing times.

The results obtained 40 days after the start of composting (Table 4) indicate an effect of both the type of compost and the concentration of the extract. All the composts tested were found to have a depressing effect on the germination and growth of watercress sprouts, as the germination index, I_g (%), which was always less than 70%, considered the minimum acceptable value. For all compost, a lower germination index corresponded to a higher concentration of the extract. Among the compost, the sample derived from refined flour (F0) is the one that gave the greatest phytotoxic effect, while the compost obtained from both plasticized flours containing bran contents (TPF1 and TPF3) were found to have a less depressing effect on germination performance, but still more toxic than desired. According to this, it was assumed that the revealed phytotoxicity was due to the incomplete maturation of the compost. For this reason, the germination test was repeated with compost extract taken 60 days after composting. In these conditions, it can be seen that, at 60 days, all compost allowed an acceptable germination index (i.e., >70%). In particular, no compost inhibited germination, which was always close to 100% even at the highest extract concentration, while root growth on compost extract obtained from refined flour (F0) was slightly reduced, but always within acceptable limits. On the other hand, there was a kind of hormetic or stimulating effect of the compost extract obtained from F1 flour when used at the lowest concentration (50% dilution); in this case the Ig (%) was 108%. This result is not surprising because it is known in the literature that various substances, both synthetic and natural (e.g., NaCl and other salts, herbicides and allelopathic substances), can have a depressive effect at high concentrations or a stimulating effect at low concentrations.

Table 4. Germination test results on compost extract taken 40 and 60 days after the incubation.

	40 Days		**60 Days**	
Samples	**G×L**	**Ig (%)**	**G×L**	**Ig (%)**
C	37.0		233	
TPF3_50	20.0	54	229	98
TPF1_50	21.0	57	251	108
TPF0_50	9.5	26	183	79
TPF3_75	13.3	36	197	85
TPF1_75	12.5	34	213	91
TPF0_75	1.9	5	171	73

4. Conclusions

The objective of this work was the study of thermomechanical behavior of eco-sustainable and biodegradable materials obtained by plasticizing wheat-milling products containing fractions of bran fiber as filler/reinforcement. Four flours, with different contents of bran fraction, were obtained by sampling along the wheat milling line. The standard alveographic characteristics of reference refined flour allowed the production of film samples, plasticized in the extruder, both with the refined flour F0 and with the milling products F1, F2 and F3 with fiber content of about 15, 25 and 35% wt. The TPWF/bran fiber composites proved to have acceptable mechanical characteristics, which can be improved by the use of suitable quantities of PCL in blend, with citric acid as compatibilizer and with the partial replacement of glycerol with water. Process parameter optimization tests have shown that the lowest plasticization temperature profile (T2) and the highest mixing rate (R120) produced materials with better mechanical properties. In light of the obtained results, we concluded that it is possible to design formulations and manage the process parameters to obtain eco-sustainable and compostable materials from plasticization of raw wheat flour/bran fiber reinforced, at affordable costs, with characteristics designed for different application sectors requiring different mechanical performance.

Author Contributions: Conceptualization, P.B., F.D., D.P.; investigation, F.D., F.L., D.P.; writing—original draft preparation, P.B., F.D., D.P.; writing—review and editing, P.B., D.P., F.D., L.T.; supervision, L.T.; funding acquisition, D.P. All authors have read and agreed to the published version of the manuscript.

Funding: This research received no external funding.

Conflicts of Interest: The authors declare no conflict of interest.

References

1. Leblanc, N.; Saiah, R.; Beucher, E.; Gattin, R.; Castandet, M.; Saiter, J.-M. Structural investigation and thermal stability of new extruded wheat flour based polymeric materials. *Carbohyd. Polym.* **2008**, *73*, 548–557. [CrossRef] [PubMed]
2. Soccio, M.; Dominici, F.; Quattrosoldi, S.; Luzi, F.; Munari, A.; Torre, L.; Puglia, D. PBS-based green copolymer as efficient compatibilizer in Thermoplastic inedible Wheat Flour/Poly (Butylene Succinate) Blends. *Biomacromolecules* **2020**, *21*, 3254–3269. [CrossRef] [PubMed]
3. Benincasa, P.; Dominici, F.; Bocci, L.; Governatori, C.; Panfili, I.; Tosti, G.; Torre, L.; Puglia, D. Relationships between wheat flour baking properties and tensile characteristics of derived thermoplastic films. *Ind. Crop. Prod.* **2017**, *100*, 138–145. [CrossRef]
4. Puglia, D.; Dominici, F.; Kenny, J.M.; Santulli, C.; Governatori, C.; Tosti, G.; Benincasa, P. Tensile behavior of thermoplastic films from wheat flours as function of raw material baking properties. *J. Polym. Environ.* **2016**, *24*, 37–47. [CrossRef]
5. Cano, A.; Jiménez, A.; Cháfer, M.; Gónzalez, C.; Chiralt, A. Effect of amylose:amylopectin ratio and rice bran addition on starch films properties. *Carbohyd. Polym.* **2014**, *111*, 543–555. [CrossRef]
6. Prückler, M.; Siebenhandl-Ehn, S.; Apprich, S.; Höltinger, S.; Haas, C.; Schmid, E.; Kneifel, W. Wheat bran-based biorefinery 1: Composition of wheat bran and strategies of functionalization. *LWT-Food Sci. Technol.* **2014**, *56*, 211–221. [CrossRef]
7. ElMekawy, A.; Diels, L.; De Wever, H.; Pant, D. Valorization of Cereal Based Biorefinery Byproducts: Reality and Expectations. *Environ. Sci. Technol.* **2013**, *47*, 9014–9027. [CrossRef]
8. Bressiani, J.; Oro, T.; Da Silva, P.; Montenegro, F.; Bertolin, T.; Gutkoski, L.; Gularte, M. Influence of milling whole wheat grains and particle size on thermo-mechanical properties of flour using Mixolab. *Czech. J. Food Sci.* **2019**, *37*, 276–284. [CrossRef]
9. Liu, N.; Ma, S.; Li, L.; Wang, X. Study on the effect of wheat bran dietary fiber on the rheological properties of dough. *Grain Oil Sci. Technol.* **2019**, *2*, 1–5. [CrossRef]
10. De Bondt, Y.; Liberloo, I.; Roye, C.; Goos, P.; Courtin, C.M. The impact of wheat (Triticum aestivum L.) bran on wheat starch gelatinization: A differential scanning calorimetry study. *Carbohyd. Polym.* **2020**, *241*, 116262. [CrossRef]
11. Dobircau, L.; Sreekumar, P.A.; Saiah, R.; Leblanc, N.; Terrié, C.; Gattin, R.; Saiter, J.M. Wheat flour thermoplastic matrix reinforced by waste cotton fibre: Agro-green-composites. *Compos. Part A-Appl. S.* **2009**, *40*, 329–334.
12. Pérez-Pacheco, E.; Canto-Pinto, J.C.; Moo-Huchin, V.M.; Estrada-Mota, I.A.; Estrada-León, R.J.; Chel-Guerrero, L. Thermoplastic Starch (TPS)-Cellulosic Fibers Composites: Mechanical Properties and Water Vapor Barrier: A Review. *Compos. Renew. Sustain. Mater.* **2016**, 85–105.
13. Hanis-Syazwani, M.; Bolarinwa, I.F.; Lasekan, O.; Muhammad, K. Influence of starter culture on the physicochemical properties of rice bran sourdough and physical quality of sourdough bread. *Food Res.* **2018**, *2*, 340–349.
14. Roozendaal, H.; Madian, A.; Frazier, R.A. Thermogravimetric analysis of water release from wheat flour and wheat bran suspensions. *J. Food Eng.* **2012**, *111*, 606–611. [CrossRef]
15. El-Sayed, S. Thermal decomposition, kinetics and combustion parameters determination for two different sizes of rice husk using TGA. *Eng. Agric. Environ. Food* **2019**, *12*, 460–469.
16. Yu, A.-N.; Li, Y.; Yang, Y.; Yu, K. The browning kinetics of the non-enzymatic browning reaction in L-ascorbic acid/basic amino acid systems. *Food Sci. Technol.* **2018**, *38*, 537–542.
17. Edoardo Capuano, E.; Ferrigno, A.; Acampa, I.; Ait-Ameur, L.; Fogliano, V. Characterization of the Maillard reaction in bread crisps. *Eur. Food Res. Technol.* **2008**, *228*, 311–319.
18. Rosell, C.M. The Science of Doughs and Bread Quality. In *Flour and Breads and Their Fortification in Health and Disease Prevention*; Academic Press: Cambridge, MA, USA, 2011; pp. 3–14.
19. Majewsky, L.; Cunha, A.G. Evaluation of Suitability of Wheat Bran as a Natural Filler in Polymer Processing. *Bioresources* **2018**, *13*, 7037–7052.

20. Follain, N.; Joly, C.; Dole, P.; Roge, B.; Mathlouthi. M. Quaternary starch based blends: Influence of a fourth component addition to the starch/water/glycerol system. *Carbohydr. Polym.* **2006**, *63*, 400–407. [CrossRef]
21. Dominici, F.; Gigli, M.; Armentano, I.; Genovese, L.; Luzi, F.; Torre, L.; Munari, A.; Lotti, N. Improving the flexibility and compostability of starch/poly(butylene cyclohexanedicarboxylate)-based blends. *Carbohydr. Polym.* **2020**, *246*, 116631. [CrossRef]
22. Carvalho, A.J.F.; Zambon, M.D.; da Silva Curvelo, A.A.; Gandini, A. Thermoplastic starch modification during melt processing: Hydrolysis catalyzed by carboxylic acids. *Carbohydr. Polym.* **2005**, *62*, 387–390. [CrossRef]
23. Genovese, L.; Dominici, F.; Gigli, M.; Armentano, I.; Lotti, N.; Torre, L.; Munari, A. Processing, thermo-mechanical characterization and gas permeability of thermoplastic starch/poly(butylene trans-1,4-cyclohexanedicarboxylate) blends. *Polym. Degr. Stab.* **2018**, *157*, 100–107. [CrossRef]
24. Olivato, J.B.; Grossmann, M.V.E.; Yamashita, F.; Eiras, D.; Pessan, L.A. Citric acid and maleic anhydride as compatibilizers in starch/poly(butylene adipate-co-terephthalate) blends by one-step reactive extrusion. *Carbohydr. Polym.* **2012**, *87*, 2614–2618. [CrossRef]
25. Jiugao, Y.; Ning, W.; Xiaofei, M. The Effects of Citric Acid on the Properties of Thermoplastic Starch Plasticized by Glycerol. *Starch/Stärke* **2005**, *57*, 494–504. [CrossRef]
26. Lagrain, B.; Thewissen, B.G.; Brijs, K.; Delcour, J.A. Mechanism of gliadin–glutenin cross-linking during hydrothermal treatment. *Food Chem.* **2008**, *107*, 753–760. [CrossRef]
27. Basiak, E.; Lenart, A.; Debeaufort, F. How Glycerol and Water Contents Affect the Structural and Functional Properties of Starch-Based Edible Films. *Polymers* **2018**, *10*, 412. [CrossRef]
28. Khamthong, P.; Lumdubwong, N. Effects of heat-moisture treatment on normal and waxy rice flours and production of thermoplastic flour materials. *Carbohydr. Polym.* **2012**, *90*, 340–347. [CrossRef]
29. Jbilou, F.; Ayadi, F.; Galland, S.; Joly, C.; Dole, P.; Belard, L.; Degraeve, P. Effect of Shear Stress Extrusion Intensity on Plasticized Corn Flour Structure: Proteins Role and Distribution. *J. Appl. Polym. Sci.* **2012**, *123*, 2177–2186. [CrossRef]
30. Sasimowski, E.; Majewski, L.; Grochowicz, M. Influence of the Design Solutions of Extruder Screw Mixing Tip on Selected Properties of Wheat Bran-Polyethylene Biocomposite. *Polymers* **2019**, *11*, 2120. [CrossRef]
31. Carmona, V.B.; Corrêa, A.C.; Marconcini, J.M.; Mattoso, L.H.C. Properties of a Biodegradable Ternary Blend of Thermoplastic Starch (TPS), Poly(ε-Caprolactone) (PCL) and Poly(Lactic Acid) (PLA). *J. Polym. Environ.* **2014**, *23*, 83–89. [CrossRef]
32. Sin, L.T.; Rahman, W.A.W.A.; Rahmat, A.R.; Mokhtar, M. Determination of thermal stability and activation energy of polyvinyl alcohol-cassava starch blends. *Carbohydr. Polym.* **2011**, *83*, 303–305. [CrossRef]

polymers

MDPI

Article

Encapsulation Effect on the In Vitro Bioaccessibility of Sacha Inchi Oil (*Plukenetia volubilis* L.) by Soft Capsules Composed of Gelatin and Cactus Mucilage Biopolymers

María Carolina Otálora [1,*], Robinson Camelo [2], Andrea Wilches-Torres [1], Agobardo Cárdenas-Chaparro [2] and Jovanny A. Gómez Castaño [2,*]

1 Grupo de Investigación en Ciencias Básicas (NÚCLEO), Facultad de Ciencias e Ingeniería, Universidad de Boyacá, 150001 Tunja, Boyacá, Colombia; andreawilches@uniboyaca.edu.co

2 Grupo Química-Física Molecular y Modelamiento Computacional (QUIMOL), Facultad de Ciencias, Universidad Pedagógica y Tecnológica de Colombia (UPTC), 150001 Tunja, Boyacá, Colombia; luis.camelo@uptc.edu.co (R.C.); agobardo.cardenas01@uptc.edu.co (A.C.-C.)

* Correspondence: marotalora@uniboyaca.edu.co (M.C.O.); jovanny.gomez@uptc.edu.co (J.A.G.C.)

Received: 30 July 2020; Accepted: 23 August 2020; Published: 2 September 2020

Abstract: Sacha inchi (*Plukenetia volubilis* L.) seed oil is a rich source of polyunsaturated fatty acids (PUFAs) that are beneficial for human health, whose nutritional efficacy is limited because of its low water solubility and labile bioaccessibility (compositional integrity). In this work, the encapsulation effect, using blended softgels of gelatin (G) and cactus mucilage (CM) biopolymers, on the PUFAs' bioaccessibility of *P. volubilis* seed oil was evaluated during in vitro simulated digestive processes (mouth, gastric, and intestinal). Gas chromatography–mass spectrometry (GC–MS) and gas chromatography with a flame ionization detector (GC–FID) were used for determining the chemical composition of *P. volubilis* seed oil both before and after in vitro digestion. The most abundant compounds in the undigested samples were α-linolenic, linoleic, and oleic acids with 59.23, 33.46, and 0.57 (g/100 g), respectively. The bioaccessibility of α-linolenic, linoleic, and oleic acid was found to be 1.70%, 1.46%, and 35.8%, respectively, along with the presence of some oxidation products. G/CM soft capsules are capable of limiting the in vitro bioaccessibility of PUFAs because of the low mucilage ratio in their matrix, which influences the enzymatic hydrolysis of gelatin, thus increasing the release of the polyunsaturated content during the simulated digestion.

Keywords: softgels; mucilage; biopolymers; in vitro digestion; bioaccessibility

1. Introduction

Sacha inchi (*Plukenetia volubilis* L.) is a plant belonging to the Euphorbiaceae family that grows in the Amazon rainforest in Northeastern Peru and Northwestern Brazil. It has aroused interest in the food industry because of the high quantity and quality of the edible oils contained in its seeds. Its fruit consists of a star-shaped capsule that is approximately 3–5 cm in size, and normally contains between four and six dark brown edible oval seeds that are 1.5–2 cm in size [1]. The seeds of this plant are an excellent source of edible oil (41–54%). This oil contains lipids (35–60%), free fatty acids (1.2%), and phospholipids (0.8%) [2]. The high nutritional value of sacha inchi oil is due to its high polyunsaturated fatty acid (PUFA) and monounsaturated fatty acid (MUFA) content, which varies between 77.5% and 84.4%, and 8.4% and 13.2%, respectively [2–4]. α-Linolenic acid (ALA; C18:3, ω-3) is the major fatty acid (47–51%), followed by linoleic acid (LA; C18:2, ω-6, 34–37%) and oleic acid (ω-9, 9–10%) [2]. These fatty acids are considered beneficial because of their antioxidant, antithrombotic, antidyslipidemic, and anticancer effects [5–7]. Although PUFAs are beneficial, they are very sensitive to oxidative damage when exposed to oxygen, and are affected by heat, light, and humidity [8]. Therefore, it is essential to establish adequate systems for the transport and encapsulation of fatty acids, while maintaining their nutritional properties until they are released within the body.

Encapsulation with hydrocolloid biopolymers is an effective and widely used technique in the food industry in order to protect dietary supplements against oxidation and loss of nutritional value. Its effectiveness is as a result of the hermeticity of its walls and the safe supply of bioactive compounds, due to its rapid disintegration in biological fluids at body temperature [9,10]. This characteristic makes soft capsules (or softgels) one of the most widely used encapsulation methods because of their safe and nutritionally-accepted means of delivering aqueous liquid or semi-solid dosage formulations. Although there are several studies on the encapsulation of omega-3 fatty acids using different techniques [11–14], the use of soft capsules for the encapsulation of polyunsaturated fatty acids in scientific studies is scarce [15,16].

Softgels are generally manufactured with animal-derived gelatin (G), water, non-volatile plasticizers, and minor additives such as opacifiers and dyes [17,18]. The choice of G as a traditional material in the walls of softgels is related to its biodegradable nature and its ability to form thermo-responsive hydrogels [19,20]. An emerging trend in the food industry has recently sought to partially or completely replace G with other non-animal natural hydrocolloids. This has allowed plant mucilage to emerge as an attractive structural biopolymer alternative to soft capsules [21]. The mucilage extracted from the cladodes of *Opuntia ficus-indica* is a heteropolysaccharide matrix that has proven to be a suitable natural structuring material in soft capsules because of its high fiber content and desirable functional properties [22–24].

Bioaccessibility is defined as the amount of a bioactive compound that can cross the intestinal barrier as a result of its release from the matrix by the action of digestive enzymes [25]. Therefore, the stability of the encapsulated polyunsaturated fatty acids that are released into the gastrointestinal tract and are available for intestinal absorption is directly related to the biocompatibility and biodegradability rates of the capsule matrix [26]. As a result, the behavior of the capsule structure in relation to changes in pH and the presence of digestive enzymes and bile salts is an important factor that should be properly evaluated. This makes in vitro methods that simulate gastrointestinal media particularly advantageous for the quantification of the biocompatibility and biodegradability of soft capsules containing such nutrients. The advantages of in vitro methods include speed, cost efficiency, and a lack of ethical restrictions when compared to in vivo methods [27]. The use of in vitro conditions provides the opportunity to evaluate the suitability of a matrix's ability to carry functional compounds of interest for the food and pharmaceutical industries. The bioaccessibility of polyunsaturated fatty acids has been evaluated using in vitro models in different studies [28,29].

The primary interest of this study was to determine the effect of gelatin/cactus mucilage as a softgel wall material in the in vitro digestion of encapsulated sacha inchi oil (SIO). The bioaccessibility analysis was performed using a comparative study that involved the quantification and identification of bioactive compounds using gas chromatography with a mass spectrometry (MS) detector and a flame ionization (FID) detector. The quantification was performed both before and after different laboratory-scale digestion processes. To the best of our knowledge, this study is the first evaluation of soft capsule wall matrices in the bioaccessibility of edible SIOs, which may be a topic of relevant interest to the food supplement industry.

2. Materials and Methods

2.1. Materials and Reagent

Sacha inchi fruits were supplied by local farmers in Miraflores (Boyacá, Colombia). The seeds were selected manually by discarding those that presented physical damage. They were packaged in polyethylene bags and stored at 18 °C until use. Cactus mucilage (CM) was extracted from the cladodes of *Opuntia ficus-indica* provided by local farmers in Duitama (Boyacá, Colombia), following the methodology reported by Quinzio et al. [30], and were used without further purification. Food grade gelatin (Type B, bloom strength 285, pI 4.2–6.5, Mw 40,000–50,000 Da, and ~99% purity) was provided by Gelco (Medellín, Antioquia, Colombia). Glycerol was provided by Merck (Darmstadt, Germany) for use as a plasticizer. Commercial samples of digestive enzymes and bile salts (α-amylase, pepsin from porcine gastric mucosa, and pancreatin from porcine pancreas) were obtained from Sigma–Aldrich (Auckland, New Zealand).

2.2. *Sacha Inchi Oil (SIO) Extraction*

The SIO extraction was performed using the Soxhlet methodology with chloroform as a solvent. Soxhlet extraction was selected as the most conventional, economical, and easiest process to implement. To maximize the SIO extraction, approximately 15 g of seed material was ground in an analytical mill (IKA A11 basic S1). Then, 1 g of seed sample was placed in a cellulose cartridge (33 × 80 mm) for extraction using a Soxhlet extractor SER 148/3 Velp Scientifica (Usmate Velate, Italy) for 1.5 h with 60 mL of solvent.

2.3. *Characterization Sacha Inchi Oil*

2.3.1. Fatty Acid Identification with Gas Chromatography Coupled to Mass Spectrometry (GC–MS)

The chemical composition of the SIO sample was determined by gas chromatography coupled to mass spectrometry (GC/MS) using a 6890 N GC–MS instrument (Agilent Technologies Inc., Palo Alto, CA, USA) coupled to an Agilent 5973 N inert mass selective (IMS) detector. A capillary column DB-1MS (30 m × 0.25 mm ID with 0.25 μm film thickness) was employed for the analysis. The elution program started at a temperature of 70 °C, held for 2 min, and then increased to 320 °C at a speed of 8 °C/min and was held for 29 min. Both the IMS detector and the injector port temperatures were 320 and 250 °C, respectively. The injection volume used was 3.0 μL, and helium was used as the carrier gas at a flow rate of 1 mL/min in a splitless mode. The components were identified using a commercial library higher than 85% (WileyW9N08, Mass Spectral Database of the National Institute of Standards and Technology (NIST)).

2.3.2. Fatty Acid Profile with Gas Chromatography with a Flame Ionization Detector (GC–FID)

The SIO sample was also analyzed using a 6890 N GC–FID instrument (Agilent Technologies Inc.). A capillary column DB-225 (60 m × 0.25 mm ID with 0.25 μm film thickness) was employed for the analysis. The elution program started at a temperature of 75 °C and increased to 220 °C at a speed of 5 °C/min for 50 min. Both the detector and injector port temperatures were 220 and 250 °C, respectively. The injection volume used was 0.2 μL, and helium was used as the carrier gas at a flow rate of 1 mL/min in a 100:1 split mode.

2.4. *Soft Capsule Preparation with SIO Inclusion*

The soft capsule design was developed according to the method reported by Camelo et al. [21]. CM (1.0 g) and food grade G (21.0 g) were separately dissolved in 100 mL of distilled water at 18 °C and 40 °C for 2 h and 30 min, respectively. Both solutions were constantly stirred at 300 rpm using a magnetic stirrer (C-MAG HS 7S000, IKA, Staufen im Breisgau, Germany) in order to ensure complete solubilization. Afterwards, the gelatin solution was mixed separately with glycerol (Gly) at a concentration of 15% (*w*/*v*) at 60 °C for 2 h to remove residual air bubbles and to obtain a homogeneous solution. The ratio of G/CM (3:1 *w*/*w*) was homogenized at room temperature for 1 h under constant magnetic stirring. The biopolymer solution was poured into an elliptically-shaped mold (22 mm length and 11 mm diameter) and dried in a Memmert UM 400 drying oven (Schwabach, Germany) at 25 °C with a relative humidity of 40% for 1 h. Subsequently, 2 mL of sacha inchi oil was injected into the formed soft capsule. The syringe hole in the capsule was then sealed by carefully applying heat using a small hot spatula.

2.5. Fatty Acids' Polyunsaturated Bioaccessibility

The encapsulated oil was submitted to an in vitro digestion process using mouth, gastric, and intestinal simulation. This method was implemented as described by Pacheco et al. [31]. To simulate mouth digestion, 5 g of soft capsules were weighed and added to 9 mL of simulated saliva (1.59 mM $CaCl_2$, 21.1 mM KCl, 14.4 mM $NaHCO_3$, 0.2 mM $MgCl_2$, pH adjusted to 7.0 using 1.0 N HCl, and the α-amylase enzyme). This mixture was incubated in a Schutzart DIN 60529-IP 20 shaking water bath (Memmert, Germany) for 5 min at 37 °C, and agitated at 185 rpm. Gastric digestion was then initiated by adding 36 mL of pepsin solution (25 mg/mL in 0.02 N HCl) to the samples. The mixture's pH was adjusted to 2.0 using 1.0 N HCl, and was incubated while continuously shaking at 130 rpm at 37 °C for 60 min. To simulate intestinal digestion, the gastric-digested mixture's pH was adjusted to 6.0 using 1 M $NaHCO_3$. Afterwards, 0.25 mL of pancreatin solution (2 g/L) and biliary salts (12 g/L) dissolved in aqueous 0.1 M $NaHCO_3$ were added and then the mixture was incubated at 37 °C for 120 min while constantly stirring at 45 rpm.

After intestinal digestion, the samples were immediately centrifuged at 5000 rpm for 10 min. The centrifuged samples were separated into two layers: an opaque sediment layer at the bottom, and a thin oily layer at the top. The oily layer was centrifuged again at 4000 rpm for 3 min and filtered using a Millipore membrane (0.45 μm), and then analyzed for its polyunsaturated fatty acid content and composition using GC–FID and GC–MS. The percentage of bioaccessibility was calculated using this equation:

$$\text{Bioaccessibility}(\%) = \frac{(\text{Content of fatty acids polyunsaturated present in the digestion product})}{(\text{Content of fatty acids polyunsaturated present in the encapsulted matrix})} \times 100 \quad (1)$$

3. Results

3.1. Characterization of Sacha Inchi Oil

3.1.1. Fatty Acid Identification with Gas Chromatography Coupled to Mass Spectrometry (GC–MS)

As shown in the GC–MS chromatograms in Figure 1, the fatty acid profiles revealed 13 main chemical constituents in the SIO sample prior to encapsulation. The peaks numbered 1, 2, and 3 in the oil, eluted at retention times of 31.34, 33.15, and 33.19 min, were identified as hexadecanoic acid methyl ester, 9,12–octadecadienoic acid methyl ester, and (Z,Z,Z)-9,12,15-octadecatrienoic acid methyl ester, respectively. This demonstrates that the oil is a rich source of polyunsaturated fatty acid. Peak number 4 in the chromatograms, eluted at a retention time of 33.52 min, was identified as (Z)-11-octadecenoic acid methyl ester and octadecenoic acid methyl ester. The peaks numbered 5, 6, 7, 8, and 9 were observed in the oil eluted at times of 33.52, 36.91, 37.68, 38.41, and 39.11 min, respectively, and were identified as eicosane. In the sample, the compounds hexadecane (peak number 10), 1,4-phthalazinedione 2,3 dihydro-6-nitro (peak number 11), cyclotrisiloxane hexamethyl (peak number 12), and 5-methyl-2-phenylindolizine (peak number 13) were eluted at retention times of 39.79, 40.45, 41.08, and 41.74 min, respectively.

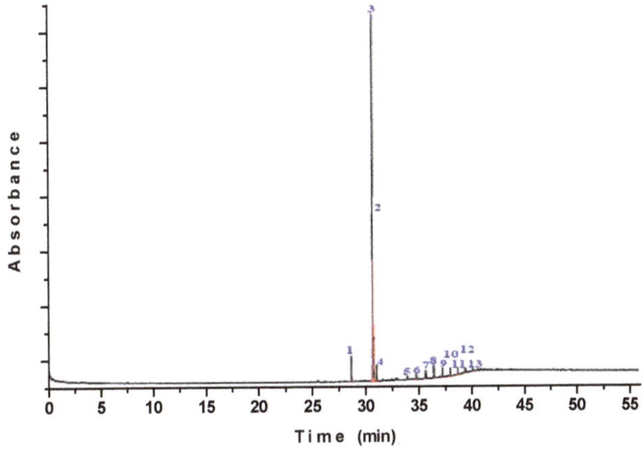

Figure 1. GC–MS chromatograms of sacha inchi oil before the simulated gastrointestinal digestion.

3.1.2. Fatty Acid Profile with Gas Chromatography with a Flame Ionization Detector (GC–FID)

The fatty acid contents analyzed by GC–FID in the unencapsuslated SIO sample (i.e., prior to encapsulation) are shown in Table 1. The presence of α-linolenic (C18:3 ω-3), linoleic (C18:2 ω-6), palmitic (C16:0), stearic (C18:0), and oleic (C18:1 ω-9) acids, in decreasing order of abundance, indicated a rich source of polyunsaturated fatty acids. These values agreed with the results of Chirinos et al. [3] and Gutiérrez et al. [32]. The content of ω-3 (59.23 g/100 g) is important and desirable because of its contribution to preventing several diseases such as obesity, diabetes, allergies, Alzheimer's, and coronary and neurodegenerative diseases [33]. The ω-6/ω-3 ratio of 0.56 has many health and nutritional benefits, such as the reduction of chronic diseases, and in cardiovascular and hypertension disease prevention [34,35]. The low ratio of linoleic/α-linolenic is similar to the results obtained in both irradiated and non-irradiated sacha inchi oils [32].

Table 1. Fatty acid composition (g/100 g) of sacha inchi oil before the simulated gastrointestinal digestion.

Fatty Acid	g/100 g-Sample
α-Linolenic (C18:3 ω-3)	59.23
Linoleic (C18:2 ω-6)	33.46
Oleic (C18:1 ω-9)	0.57
Stearic (C18:0)	2.66
Palmitic (C16:0)	4.07

3.2. *Chemical Content of Encapsulated SIO after Simulated In-Vitro Digestion*

In order to study the impact of soft capsules on the bioaccessibility of the fatty acids, the encapsulated sacha inchi oil was subjected to in vitro digestion (i.e., mouth–gastric–intestinal media simulation), and its chemical content was analyzed using GC–MS and GC–FID spectrometry.

3.2.1. Products Derived from the Degradation of Polyunsaturated Fatty Acids after In Vitro Digestion

The chromatogram profile of the main SIO compounds and their degradation by-products using GC–MS showed a maximum of 17 chemical constituents, as displayed in Figure 2. No peak attributable to 9,12–octadecadienoic acid methyl ester was detected in the chromatograms of the encapsulated oil after digestion, and therefore its bioaccessibility was considered negligible. Conversely, the presence of (Z,Z,Z)-9,12,15-octadecatrienoic acid methyl ester was detected before and after in vitro digestion in peaks number 3 and 13, which were eluted at a retention time of 33.20 and 32.06 min, respectively. The presence of this fatty acid was detected in the chromatograms of the encapsulated oil after digestion in a 43.96% peak area, making its bioaccessibility considerably significant. The presence of peaks with different retention times was also evidenced. For instance, 10,13-octadecadienoic acid methyl ester was identified as a degradation by-product of 9,12–octadecadienoic acid methyl ester [36], along with derivatives such as propanal, alcoholic amines, and aromatic compounds. All were related as by-products of simulated gastrointestinal digestion, demonstrating a low bioaccessibility of the encapsulated oil.

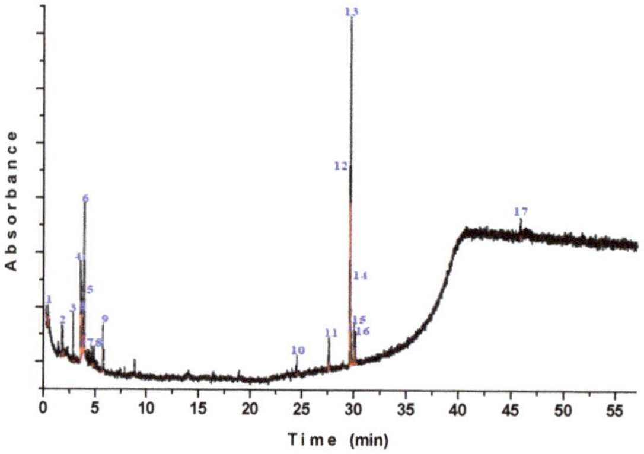

Figure 2. GC–MS chromatograms of sacha inchi oil after the simulated gastrointestinal digestion. Abbreviations: (**1**) Ethanol 2-(2-methoxyethoxy); (**2**) 3-[2-Diethylaminoethyl]-2,4-pentanedione; (**3**) Diethyl carbamoyl *t*-butoxy sulfide; (**4**) Ethane 1,2-bis(methylthio); (**5**) Ethane 1,2-bis(methylthio); (**6**) Cyclotetrasiloxane octamethyl; (**7**) Cycloserine; (**8**) 1-Propanol 2-amino; (**9**) 2-Pentanamine *N*-(1-methylbutyl); (**10**) *n*-Hexylmethylamine; (**11**) Methylpent-4-enylamine; (**12**) 10,13-Octadecadienoic acid methyl ester; (**13**) (Z,Z,Z)-9,12,15-Octadecatrienoic acid, methyl ester; (**14**) Epinephrine; (**15**) 1-Octanamine *N*-methyl; (**16**) 2-Amino-1-(*o*-hydroxyphenyl)propane; (**17**) Tetrasiloxane, decamethyl.

3.2.2. Fatty Acid Content under the Simulated In Vitro Digestion

The relative concentration of the encapsulated fatty acids before and after the simulated gastrointestinal digestion, determined using GC–FID spectrometry, is shown in Figure 3. It was found that the bioaccessibility of the α-linolenic, linoleic, and oleic polyunsaturated fatty acids was 1.70%, 1.46%, and 35.8% respectively, while the saturated stearic and palmitic acids presented bioaccessibility values of 2.26% and 1.72%, respectively.

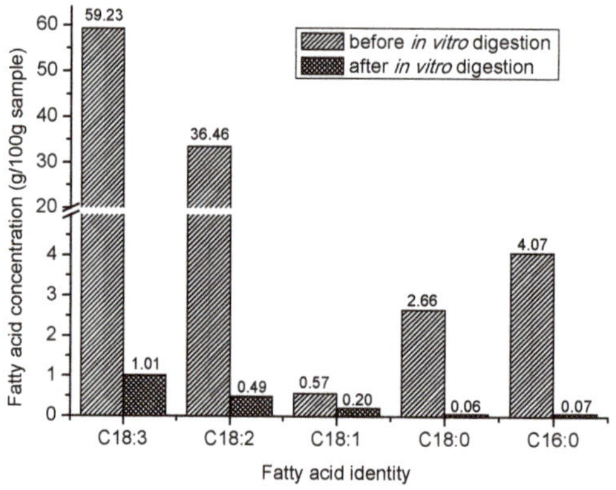

Figure 3. Relative concentration of fatty acids in sacha inchi oil (SIO) samples before and after the simulated gastrointestinal digestion.

4. Discussion

As seen in Figure 3, during the digestion of the SIO encapsulated in G/CM softgels, there was a significant decrease in the content of the two most abundant PUFAs (i.e., α-linolenic and linoleic acid content), which entailed a reduction in the nutritional and functional value of this natural oil. As can be inferred by comparing the GC–MS chromatograms of the SIO before and after in vitro digestion (Figures 2 and 3), the amounts of unsaturated fatty acids originally present in the SIO stimulated a higher generation of oxidation by-products during digestion [28]. Similar results were reported by Nieva-Echevarría, Goicoechea, and Guillén [37] during the in vitro gastrointestinal digestion of flaxseed oil. The amount of non-encapsulated oil (i.e., the oil that remained on the surface of the capsule) may also have affected the stability of the bioactive compound. In other words, the oil that was present at the surface of the capsule might have undergone oxidation and thereby affected the oxidative stability of the sample [38]. Alpizar-Reyes et al. [39] observed a similar situation when sesame seed oil was microencapsulated with tamarind seed mucilage.

Compared with the free sacha inchi oil [40], during in vitro digestion, the soft capsules showed a low protection of the encapsulated PUFAs against gastric conditions because of the nature of the wall materials and the G/CM ratio in the matrix. This behavior was associated with the intermolecular interactions between the functional groups of CM combined with G, which gradually disappeared as a result of the repulsion forces of the biopolymers [21,41,42], as well as the rapid degradation of the gelatin by the proteolytic enzymes present in the stomach [20], which reduced the barrier properties of the matrix. This was insufficient at protecting the oil against oxidation, and in turn affected the porosity of the biopolymer matrix and the oil release rate during in vitro digestion. Furthermore, the low amount of CM hydrocolloid in the soft capsule allowed for the degradation and facilitation of the enzymatic hydrolysis of G under gastric conditions [40].

These results seem to contradict the results of other published studies. Cortés-Camargo et al. [43] found that there was a delayed release of lemon essential oil microencapsulated using mesquite gum–chia mucilage mixtures. Papillo et al. [44] reported a high in vitro bioaccessibility of curcuminoids that were microencapsulated using gum arabic and maltodextrins as encapsulating agents. Da Silva Stefani et al. [45] reported a good bioavailability of nanoencapsulated linseed oil using chia seed mucilage as a structuring material. Jannasari et al. [42] studied the microencapsulation of vitamin D using gelatin and cress seed mucilage, and found release rates in the gastric and intestinal media of 28% and 70%, respectively. In addition, Barrow et al. [46] reported a high bioavailability of omega-3 fish oil microencapsulated using the technique of complex coacervation (140–180 mg EPA/DHA/g powder) in contrast with softgel capsules (data not shown). These results suggest that the G/CM matrix is not an encapsulating biopolymer that is sufficiently resistant to gastric conditions, thus reducing the bioaccessibility of the bioactive compounds carried by G/CM softgels.

5. Conclusions

In this work, oil samples extracted from sacha inchi seeds were encapsulated in softgels composed of gelatin (G) and cactus mucilage (CM) biopolymers, and then exposed to simulated gastric conditions. The nutritional composition of the oil samples was evaluated before and after in vitro digestion by means of GC–MS and GC–FID spectrometry. In this way, the protective capacity of the contents of sacha inchi oil offered by the G/CM biopolymeric wall of the softgel against digestive processes was evaluated.

α-Linolenic (C18:3 ω-3), linoleic (C18:2 ω-6), oleic (C18:1 ω-9), stearic (C18:0), and palmitic (C16:0) acids were the main fatty acids present in the non-encapsulated sacha inchi oil. It was found that the content of polyunsaturated fatty acids (PUFAs), especially α-linolenic (C18: 3 ω-3) and linoleic (C18: 2 ω-6), carried by the G/CM softgels, decreased significantly during in vitro digestion (bioaccessibility equal to 1.70% or 1.46%, respectively), which revealed a reduction in the nutritional value of the encapsulated oil after undergoing gastric processes. The low protective capacity of the

G/CM wall material was attributed to the low concentration of the CM hydrocolloid, which left the gelatin biopolymer exposed to enzymatic hydrolysis.

Although the bioaccessibility of the PUFAs obtained was relatively low, we believe that the use of a mixture of proteins (gelatin) and heteropolysaccharides (cactus mucilage) for the manufacture of microcapsules can act as a suitable delivery system for the incorporation of other bioactive compounds within acidic food matrices before being subjected to digestive processes, for example, by encapsulating functional agents and subsequent release (by shacking) in media such as fruit juices or dairy drinks. This result will undoubtedly be interesting for certain applications in the food and pharmaceutical industries.

Author Contributions: Data curation, R.C. and A.C.-C.; formal analysis, J.A.G.C. and M.C.O.; investigation, A.W.-T. and M.C.O.; project administration, M.C.O.; writing (original draft preparation), M.C.O.; writing (review and editing), J.A.G.C. and M.C.O. All authors have read and agreed to the published version of the manuscript.

Funding: This work was funded by the Universidad de Boyacá and the Universidad Pedagógica y Tecnológica de Colombia through the interinstitutional Project SGI 2384 of the Vicerrectoría de Investigaciones of the Universidad Pedagógica y Tecnológica de Colombia.

Conflicts of Interest: The authors declare that there is no conflict of interest.

References

1. Fu, Q.; Niu, L.; Zhang, Q.; Pan, B.Z.; He, H.; Xu, Z.F. Benzyladenine treatment promotes floral feminization and fruiting in a promising oilseed crop Plukenetia volubilis. *Ind Crop. Prod.* **2014**, *59*, 295–298. [CrossRef]
2. Gutiérrez, L.F.; Rosada, L.M.; Jiménez, A. Chemical composition of Sacha Inchi (*Plukenietia volúbilis* L.) seeds and characterisation of their lipid fraction. *Grasas y Aceites* **2011**, *62*, 76–83. [CrossRef]
3. Chirinos, R.; Zuloeta, G.; Pedreschi, R.; Mignolet, E.; Larondelle, Y.; Campos, D. Sacha inchi (*Plukenetia volubilis*): A seed source of polyunsaturated fatty acids, tocopherols, phytosterols, phenolic compounds and antioxidant capacity. *Food Chem.* **2013**, *141*, 1732–1739. [CrossRef] [PubMed]
4. Maurer, N.E.; Hatta-Sakoda, B.; Pascual-Chagman, G.; Rodriguez-Saona, L.E. Characterisation and authentication of a novel vegetable source of omega-3 fatty acids, Sacha Inchi (*Plukenetia volubilis* L.) oil. *Food Chem.* **2012**, *134*, 1173–1180. [CrossRef]
5. Kumar, B.; Smita, K.; Sánchez, E.; Stael, C.; Cumbal, L. Andean Sacha Inchi (*Plukenetia volubilis* L.) shell biomass as new biosorbents for Pb^{2+} and Cu^{2+} ions. *Ecol. Eng.* **2016**, *93*, 152–158. [CrossRef]
6. Garmendia, F.; Pando, R.; Ronceros, G. Effect of Sacha Inchi oil (*Plukenetia volúbilis* L.) on the lipid profile of patients with hyperlipoproteinemia. *Rev. Peru. Med. Exp. Salud Publica* **2011**, *28*, 628–632. [CrossRef]
7. Gonzalez-Aspajo, G.; Belkhelfa, H.; Haddioui-Hbabi, L.; Bourdy, G.; Deharo, E. Sacha Inchi oil (*Plukenetia volubilis* L.) effect on adherence of Staphylococus aureus to human skin explant and keratinocytes in vitro. *J. Ethnopharmacol.* **2015**, *171*, 330–334. [CrossRef] [PubMed]
8. Timilsena, Y.P.; Vongsvivut, J.; Tobin, M.J.; Adhikari, R.; Barrow, C.; Adhikari, B. Investigation of oil distribution in spray-dried chia seed oil microcapsules using synchrotron-FTIR microspectroscopy. *Food Chem.* **2019**, *275*, 457–466. [CrossRef] [PubMed]
9. Russo, P.; Zacco, R.; Rekkas, D.M.; Politis, S.; Garofalo, E.; Gaudio, P.; Aquino, R.P. Application of experimental design for the development of soft-capsules through a prilling, inverse gelation process. *J. Drug Deliv. Sci. Technol.* **2019**, *49*, 577–585. [CrossRef]
10. Donato, E.M.; Martins, L.A.; Fröehlich, P.E.; Bergold, A.M. Development and validation of dissolution test for lopinavir, a poorly water-soluble drug, in soft gel capsules, based on in vivo data. *J. Pharm. Biomed.* **2008**, *47*, 547–552. [CrossRef] [PubMed]
11. Silva, K.F.; Carvalho, A.G.; Rabelo, R.; Hubinger, M. Sacha inchi oil encapsulation: Emulsion and alginate beads characterization. *Food Bioprod. Process.* **2019**, *116*, 118–129. [CrossRef]
12. Chen, F.; Fan, G.; Zhang, Z.; Zhang, R.; Deng, Z.; McClements, D.J. Encapsulation of omega-3 fatty acids in nanoemulsions and microgels: Impact of delivery system type and protein addition on gastrointestinal fate. *Int. Food Res. J.* **2017**, *100*, 387–395. [CrossRef] [PubMed]
13. Rasti, B.; Erfanian, A.; Selamat, J. Novel nanoliposomal encapsulated omega-3 fatty acids and their applications in food. *Food Chem.* **2017**, *230*, 690–696. [CrossRef] [PubMed]

14. Eratte, D.; McKnight, S.; Gengenbach, T.R.; Dowling, K.; Barrow, C.J.; Adhikari, B.P. Co-encapsulation and characterisation of omega-3 fatty acids and probiotic bacteria in whey protein isolate–gum Arabic complex coacervates. *J. Funct. Foods* **2015**, *19*, 882–892. [CrossRef]
15. Chen, R.; Guo, X.; Liu, X.; Cui, H.; Wang, R.; Han, J. Formulation and statistical optimization of gastric floating alginate/oil/chitosan capsules loading procyanidins: In vitro and in vivo evaluations. *Int. J. Biol. Macromol.* **2018**, *108*, 1082–1091. [CrossRef]
16. Hu, J.; Liu, S.; Deng, W. Dual responsive linalool capsules with high loading ratio for excellent antioxidant and antibacterial efficiency. *Colloids Surf. B* **2020**, *190*, 110978. [CrossRef]
17. Gullapalli, R.P.; Mazzitelli, C.L. Gelatin and Non-Gelatin Capsule Dosage Forms. *J. Pharm. Sci.* **2017**, *106*, 1453–1465. [CrossRef]
18. Gullapalli, R.P. Soft gelatin capsules (softgels). *J. Pharm. Sci.* **2010**, *99*, 4107–4148. [CrossRef]
19. Gómez-Mascaraque, L.G.; Soler, C.; López-Rubio, A. Stability and bioaccessibility of EGCG within edible micro-hydrogels. Chitosan vs. gelatin, a comparative study. *Food Hydrocoll.* **2016**, *61*, 128–138. [CrossRef]
20. Nawong, S.; Oonsivilai, R.; Boonkerd, N.; Truelstrup Hansen, L. Entrapment in food-grade transglutaminase cross-linked gelatin–maltodextrin microspheres protects Lactobacillus spp. during exposure to simulated gastro-intestinal juices. *Int. Food Res. J.* **2016**, *85*, 191–199. [CrossRef]
21. Camelo Caballero, L.R.; Wilches-Torres, A.; Cárdenas-Chaparro, A.; Gómez Castaño, J.A.; Otálora, M.C. Preparation and physicochemical characterization of softgels cross-linked with cactus mucilage extracted from cladodes of *Opuntia Ficus-Indica*. *Molecules* **2019**, *24*, 2531. [CrossRef] [PubMed]
22. Otálora, M.C.; Carriazo, J.G.; Iturriaga, L.; Nazareno, M.A.; Osorio, C. Microencapsulation of betalains obtained from cactus fruit (Opuntia ficus-indica) by spray drying using cactus cladode mucilage and maltodextrin as encapsulating agents. *Food Chem.* **2015**, *187*, 174–181. [CrossRef] [PubMed]
23. Otálora, M.C.; Carriazo, J.G.; Osorio, C.; Nazareno, M.A. Encapsulation of cactus (Opuntia megacantha) betaxanthins by ionic gelation and spray drying: A comparative study. *Int. Food Res. J.* **2018**, *111*, 423–430. [CrossRef] [PubMed]
24. Otálora, M.C.; Gómez Castaño, J.A.; Wilches-Torres, A. Preparation, study and characterization of complex coacervates formed between gelatin and cactus mucilage extracted from cladodes of Opuntia ficus-indica. *LWT-Food Sci. Technol.* **2019**, *112*, 108234. [CrossRef]
25. Shim, S.M.; Ferruzzi, M.G.; Kim, Y.-C.; Janle, E.M.; Santerre, C.R. Impact of phytochemical-rich foods on bioaccessibility of mercury from fish. *Food Chem.* **2009**, *112*, 46–50. [CrossRef]
26. Frutos, G.; Prior-Cabanillas, A.; París, R.; Quijada-Garrido, I. A novel controlled drug delivery system based on pH-responsive hydrogels included in soft gelatin capsules. *Acta Biomater.* **2010**, *6*, 4650–4656. [CrossRef]
27. Bernardes, A.L.; Moreira, J.A.; Tostes, M.D.G.V.; Costa, N.M.B.; Silva, P.I.; Costa, A.G.V. In vitro bioaccessibility of microencapsulated phenolic compounds of jussara (Euterpe edulis Martius) fruit and application in gelatine model-system. *LWT Food Sci. Technol.* **2019**, *102*, 173–180. [CrossRef]
28. Nieva-Echevarría, B.; Goicoechea, E.; Manzanos, M.J.; Guillén, M.D. 1H NMR and SPME-GC/MS study of hydrolysis, oxidation and other reactions occurring during in vitro digestion of non-oxidized and oxidized sunflower oil. Formation of hydroxy-octadecadienoates. *Int. Food Res. J.* **2017**, *91*, 171–182. [CrossRef]
29. Nieva-Echevarría, B.; Goicoechea, E.; Guillén, M.D. Behaviour of non-oxidized and oxidized flaxseed oils, as models of omega-3 rich lipids, during in vitro digestion. Occurrence of epoxidation reactions. *Int. Food Res. J.* **2017**, *97*, 104–115. [CrossRef]
30. Quinzio, C.; Corvalán, M.; López, B.; Iturriaga, L. Studying stability against coalescence in tuna mucilage emulsions. *Acta Hortic.* **2009**, *811*, 427–431. [CrossRef]
31. Pacheco, C.; González, E.; Robert, P.; Parada, J. Retention and pre-colon bioaccessibility of oleuropein in starchy food matrices, and the effect of microencapsulation by using inulin. *J. Funct. Foods* **2018**, *41*, 112–117. [CrossRef]
32. Gutiérrez, L.F.; Quiñones-Segura, Y.; Sanchez-Reinoso, Z.; Díaz, D.L.; Abril, J.I. Physicochemical properties of oils extracted from γ-irradiated Sacha Inchi (*Plukenetia volubilis* L.) seeds. *Food Chem.* **2017**, *237*, 581–587. [CrossRef] [PubMed]
33. Molendi-Coste, O.; Legry, V.; Leclercq, I.A. Why and How Meet n-3 PUFA Dietary Recommendations? *Gastroent. Res. Pract.* **2011**, 1–11. [CrossRef] [PubMed]

34. Guillén, M.D.; Ruiz, A.; Cabo, N.; Chirinos, R.; Pascual, G. Characterization of sacha inchi (*Plukenetia volubilis* L.) oil by FTIR spectroscopy and H-1 NMR. Comparison with linseed oil. *J. Am. Oil Chem. Soc.* **2003**, *80*, 755–762. [CrossRef]
35. Simopoulos, A.P. The importance of the ratio of omega-6/omega-3 essential fatty acids. *Biomed. Pharmacother.* **2002**, *56*, 365–379. [CrossRef]
36. Yen, T.Y.; Stephen Inbaraj, B.; Chien, J.T.; Chen, B.H. Gas chromatography–mass spectrometry determination of conjugated linoleic acids and cholesterol oxides and their stability in a model system. *Anal. Biochem.* **2010**, *400*, 130–138. [CrossRef]
37. Nieva-Echevarría, B.; Goicoechea, E.; Guillén, M.D. Effect of adding alpha-tocopherol on the oxidation advance during in vitro gastrointestinal digestion of sunflower and flaxseed oils. *Int. Food Res. J.* **2019**, *125*, 108558. [CrossRef]
38. Tonon, R.V.; Pedro, R.B.; Grosso, C.R.; Hubinger, M.D. Microencapsulation of flaxseed oil by spray drying: Effect of oil load and type of wall material. *Dry. Technol.* **2012**, *30*, 1491–1501. [CrossRef]
39. Alpizar-Reyes, E.; Varela-Guerrero, V.; Cruz-Olivares, J.; Carrillo-Navas, H.; Alvarez-Ramirez, J.; Pérez-Alonso, C. Microencapsulation of sesame seed oil by tamarind seed mucilage. *Int. J. Biol. Macromol.* **2020**, *145*, 207–215. [CrossRef]
40. Silva Soares, B.; Pinto Siqueira, R.; de Carvalho, M.G.; Vicente, J.; Garcia-Rojas, E.E. Microencapsulation of sacha inchi oil (*Plukenetia volubilis* L.) using complex coacervation: Formation and structural characterization. *Food Chem.* **2019**, *298*, 125045. [CrossRef] [PubMed]
41. Maderuelo, C.; Zarzuelo, A.; Lanao, J.M. Critical factors in the release of drugs from sustained release hydrophilic matrices. *J. Control. Release* **2011**, *154*, 2–19. [CrossRef] [PubMed]
42. Jannasari, N.; Fathi, M.; Moshtaghian, S.J.; Abbaspourrad, A. Microencapsulation of vitamin D using gelatin and cress seed mucilage: Production, characterization and in vivo study. *Int. J. Biol. Macromol.* **2019**, *129*, 972–979. [CrossRef] [PubMed]
43. Cortés-Camargo, S.; Acuña-Avila, P.E.; Rodríguez-Huezo, M.E.; Román-Guerrero, A.; Varela-Guerrero, V.; Pérez-Alonso, C. Effect of chia mucilage addition on oxidation and release kinetics of lemon essential oil microencapsulated using mesquite gum—Chia mucilage mixtures. *Int. Food Res. J.* **2019**, *116*, 1010–1019. [CrossRef]
44. Papillo, V.A.; Arlorio, M.; Locatelli, M.; Fuso, L.; Pellegrini, N.; Fogliano, V. In vitro evaluation of gastro-intestinal digestion and colonic biotransformation of curcuminoids considering different formulations and food matrices. *J. Funct. Foods* **2019**, *59*, 156–163. [CrossRef]
45. Da Silva Stefani, F.; de Campo, C.; Paese, K.; Stanisçuaski Guterres, S.; Haas Costad, T.M.; Hickmann Flôres, S. Nanoencapsulation of linseed oil with chia mucilage as structuring material: Characterization, stability and enrichment of orange juice. *Int. Food Res. J.* **2019**, *120*, 872–879. [CrossRef]
46. Barrow, C.J.; Nolan, C.; Jin, Y. Stabilization of highly unsaturated fatty acids and delivery into foods. *Lipid Technol.* **2008**, *19*, 108–111. [CrossRef]

polymers

MDPI

Article

Cutin from *Solanum Myriacanthum* Dunal and *Solanum Aculeatissimum* Jacq. as a Potential Raw Material for Biopolymers

Mayra Beatriz Gómez-Patiño [1], Rosa Estrada-Reyes [2], María Elena Vargas-Diaz [3] and Daniel Arrieta-Baez [1,*]

1 Instituto Politécnico Nacional-CNMN, Unidad Profesional Adolfo López Mateos, Col. Zacatenco, México City CDMX CP 07738, Mexico; bethzem86@gmail.com

2 Laboratorio de Fitofarmacología, Dirección de Investigaciones en Neurociencias, Instituto Nacional de Psiquiatría Ramón de la Fuente Muñiz, Calzada México-Xochimilco 101, San Lorenzo Huipulco, Tlalpan, Ciudad de México 14370, Mexico; restrada@imp.edu.mx

3 Instituto Politécnico Nacional-Departamento de Química Orgánica, Escuela Nacional de Ciencias Biológicas, Prolongación de Carpio y Plan de Ayala S/N, Colonia Santo Tomás D.F. 11340, Mexico; evargasvd@yahoo.com.mx

* Correspondence: darrieta@ipn.mx; Tel.: +52-1-55-5729-6000 (ext. 57507)

Received: 4 August 2020; Accepted: 26 August 2020; Published: 28 August 2020

Abstract: Plant cuticles have attracted attention because they can be used to produce hydrophobic films as models for novel biopolymers. Usually, cuticles are obtained from agroresidual waste. To find new renewable natural sources to design green and commercially available bioplastics, fruits of *S. aculeatissimum* and *S. myriacanthum* were analyzed. These fruits are not used for human or animal consumption, mainly because the fruit is composed of seeds. Fruit peels were object of enzymatic and chemical methods to get thick cutins in good yields (approximately 77% from dry weight), and they were studied by solid-state resonance techniques (CPMAS ^{13}C NMR), attenuated total reflection-Fourier transform infrared spectroscopy (ATR-FTIR), atomic force microscopy (AFM) and direct injection electrospray ionization mass spectrometry (DIESI-MS) analytical methods. The main component of *S. aculeatissimum* cutin is 10,16-dihydroxypalmitic acid (10,16-DHPA, 69.84%), while *S. myriacanthum* cutin besides of 10,16-DHPA (44.02%); another two C18 monomers: 9,10,18-trihydroxy-octadecanoic acid (24.03%) and 18-hydroxy-9S,10R-epoxy-octadecanoic acid (9.36%) are present. The hydrolyzed cutins were used to produce films demonstrating that both cutins could be a potential raw material for different biopolymers.

Keywords: cutin; cuticles; bioplastics; biopolymers; solanum: CPMAS ^{13}C NMR

1. Introduction

The cuticle is the outer membrane that covers the aerial parts of plants, such as the stem, leaves, flowers, and fruit. In the evolution of plants, the cuticle plays a critical role against the loss of water from internal tissues [1–3]. In the same way, this biopolymer plays an essential physiological role as it is considered a first barrier that prevents the entry of pathogens and pesticides [4,5]. The cuticle consists primarily of cutin (a C16 and C18 long-chain hydroxy acid polyester), cell wall polysaccharides (cellulose, hemicellulose, and pectin), as well as epicuticular fatty acids [6–8]. Cutin, and other important natural biopolymers such as lignin, cellulose, and chitin, has been shown to have important bioplastic properties [9–13]. For this reason, they have been considered as models for plastic materials with biodegradable characteristics that eventually could replace conventional plastics derived from petroleum in specific industrial uses [14].

Raw renewable materials have a high impact on the cost of bio-based plastic production and, in this regard, different efforts have been directed to use biomass to get or produce biopolymers. Biopolymers such as starch, cellulose, lignocellulosic materials, and proteins; bio-derived monomers like polylactic acid (PLA), polyglycolic acid (PGA), and biodegradable polymers from petrochemicals (aliphatic polyesters, aromatic co-polyesters and polyvinyl alcohols) have been investigated as sources for bioplastics [15,16].

In this sense, the cutins of some fruits for human consumption, such as tomatoes, citrus have shown good physicochemical characteristics as biopolymers [9,17–20]. However, the ethical problem that could be generated has led researchers to use industrial waste products. In fact, the vegetable food processing industry generates a significant amount of waste worldwide [21]. Thus, from agro-industrial residues, biomaterials have been generated, and some of them have been used in the food packaging industry, and other bioplastic applications [22,23].

In the present work, we have searched for fruits that are not for human consumption, which present the same chemical characteristics of the cuticles used for biopolymers applications in order to be considered promising candidates as a raw material to produce bioplastics. In this sense, the cuticles of the fruits of two species of the *Solaneum* genus were studied. *S. aculeatissimum* and *S. myriacanthum* are shrubs that grow in the wild, covered with thin spines up to 18 mm long, that produce small fruits of approximately 2–3 cm, which are mainly filled with seeds. *S. aculeatissimum* is native to Brazil, but it could be found in tropical Africa and Asia. In Mexico, it is distributed in the states of Jalisco, Oaxaca, Chiapas, Veracruz, and Puebla [24]. It is considered a toxic plant due to the alkaloids present in the seeds and leaves [25,26]. Its fruit is green when it is immature and red when it is ripe. *S. myriacanthum* is native to central and south America, although it is also distributed in Asia, mainly in India. In Mexico, it is distributed in the states of Chiapas, Veracruz, Oaxaca, and Puebla [24]. Anthelmintic properties are attributed to the extracts of the fruit [27]. Its fruit is green when immature and yellow when ripe.

S. aculeatissimum and *S. myriacanthum* cutins were extracted and analyzed by means of CPMAS ^{13}C NMR, ATR-FTIR, AFM and DIESI-MS, and their components and physicochemical characteristics were determined and compared with other cutin components. From their hydrolyzed components, films were obtained and characterized and, from these results, both Solanum species could be considered as a raw material for biopolymers used in different fields of the plastic industry.

2. Materials and Methods

2.1. Chemicals

Trifluoroacetic acid (TFA), KOH, and other reagents were purchased from Sigma-Aldrich (St. Louis, MO, USA). The enzymes *Aspergillus niger* pectinase (EC 3.2.1.15) (specific activity $\geq$ 5 unit/mg protein), *A. niger* cellulase (EC 3.2.1.4) (specific activity $\geq$ 0.3 units/mg protein) and *A. niger* hemicellulose (EC 3.2.1.4) (specific activity 2.3 units/mg protein) were purchased from Sigma Chemicals (St Louis, MO, USA).

2.2. Isolation of Cutin

S. aculeatissimum and *S. myriacanthum* fruits were collected in Cuetzalan, Puebla (20°01′48.7″ N 97°29′04.3″ W) in September 2018. Fruits were washed with tap water, cut, and the seeds were removed to obtain the cuticle. The cutin was obtained using a previously reported protocol [9]. Briefly, the cuticle was treated with *A. niger* pectinase (EC 3.2.1.15, St Louis, MO, USA) (10 mg/mL) for 1 week. After this, cell wall polysaccharides were removed with an enzymatic digestion using *A. niger* cellulose (EC 3.2.1.4, St Louis, MO, USA) (80 mg/mL) for 1 week and *A. niger* hemicellulose (EC 3.2.1.4, St Louis, MO, USA) (80 mg/mL) for 1 week. To complete the extraction, a Soxhlet procedure was done with methylene chloride:methanol (1:1 *v*/*v*, 48 h, St Louis, MO, USA) to remove residual compounds of cutin such as monosaccharides and waxes. Five hundred grams of *S. aculeatissimum* dried peel yielded 386 g (77.2%) of cutin and 500 g of *S. myriacanthum* dried peel yielded 394 g (78.8%) of cutin. The resulting cutins were analyzed by Cross Polarization Magic-Angle Spinning (CPMAS ^{13}C NMR), attenuated total reflection-Fourier transform infrared spectroscopy (ATR-FTIR) and atomic force microscopy (AFM).

2.3. *Treatment of Cutin with Trifluoroacetic Acid (TFA)*

One hundred and fifty milligrams of the obtained cutin from the *S. aculeatissimum* or *S. myriacanthum* fruits was added to an aqueous TFA solution 2.0 mol·L^{-1} and stirred at 115 ± 5 °C for 2 h in separated experiments. Each reaction was filtered, and the insoluble material was washed using chloroform-methanol (1:1, *v*/*v*, St Louis, MO, USA) for 1 h to obtain the TFA-hydrolyzed cutin (TFA-HC). The TFA-HC was separated by filtration, dried, and analyzed by CPMAS ^{13}C NMR. The TFA solution was co-evaporated with methanol and the resulting solids were redissolved in methanol to give a clear brown solution, which was later analyzed by DIESI-MS (Bruker Daltonics, Biellerica, MA, USA) and solution-state NMR (Billerica, MA, USA) [28].

2.4. *Alkaline Hydrolysis of the Cutin with KOH/MeOH*

Fifty milligrams of *S. aculeatissimum* or *S. myriacanthum* cutin was added to 50 mL of 1.5 mol·L^{-1} methanolic KOH solution, and the mixture was stirred at room temperature for 24 h. After this time, the reaction was filtered, neutralized, and monomers were extracted with $CHCl_3$-MeOH. The dried extract was weighed, dissolved in $CHCl_3$-MeOH and analyzed by DIESI-MS (Bruker Daltonics, Biellerica, MA, USA).

2.5. *Preparation of Cutin Films*

Twenty-five milligrams of hydrolyzed *S. aculeatissimum* or *S. myriacanthum* cutin were added to 5 mL of ultrapure methanol:chloroform (1:1, *v*/*v*, St Louis, MO, USA) solution. The solution was sonicated for 30 s and it was deposited in plastic Petri dishes for making films using the casting method. On the other hand, 5 µL of the solution were deposited on a watch glass to study the structures of self-assembled layers. After this, films were kept in a chemical hood to remove residual solvents from the films [29].

2.6. *NMR Spectroscopy*

S. aculeatissimum and *S. myriacanthum* cutin and TFA-HC were analyzed using standard CPMAS ^{13}C NMR experiments carried out on a Varian Instruments Unityplus 300 widebore spectrometer (Palo Alto, CA, USA) equipped for solid-state NMR. The resonance frequency was 74.443 MHz, with a customary acquisition time of 30 ms, a delay time of 2 s between successive acquisitions and a CP contact time of 1.5 ms. Typically, each 30 mg sample was packed into a 5 mm rotor and supersonic MAS probe from Doty Scientific (Columbia, SC, USA), then spun at 6.00 (±0.1 kHz) at room temperature for approximately 10 h. No spinning sidebands were observed upon downfield from the major carbonyl, aromatic, or aliphatic carbon peaks, presumably due to motional averaging and/or excessive broadening of such features.

Soluble products derived from the TFA hydrolysis were examined using ^{1}H NMR. Experiments were conducted on a Bruker Instruments ASCEND 750 spectrometer (Billerica, MA, USA). The resonance frequency was 750.12 MHz, with a typical acquisition time of 2.1845 s and a delay time of 1.0 s between successive acquisitions. The ^{1}H and ^{13}C chemical shifts are given in units of δ (ppm), using tetramethylsilane (TMS) as internal standard.

2.7. *ATR-FTIR Spectroscopy*

Attenuated Total Reflectance Fourier transform infrared spectroscopy (ATR-FTIR) spectra were recorded with a BOMEM 157 FTIR spectrometer (Bomem Inc., Quebec, Canada) equipped with a deuterated triglycinesulfate (DTGS) detector. The instrument was under a continuous dry air purge to eliminate atmospheric water vapor. The spectra were recorded in the region of 4000 to 400 cm^{-1}.

2.8. *Atomic Force Microscopy*

The samples for the Atomic Force Microscopy analysis (AFM) were prepared by fixing to a metallic disk with double-sided tape. The images themselves were taken using a MultiMode AFMV (Bruker, SantaBarbara, CA, USA) in air with an RTESP cantilever, and operating the AFM in tapping mode. The size of each image was 5 × 5 μm^2. The roughness parameters R_q and R_a were determined using the expressions $R_q = \sqrt{\Sigma Z21} = N$ and $R_a = 1 = N\Sigma Nj = 1\ jZjj$, where R_q is the root mean square average of the height deviations, Ra is the arithmetic average of the absolute values of the surface height deviations, Z is the height value, and N is the number of data points. These parameters were obtain using the NanoScope Analysis image software.

2.9. *Mass Spectrometry*

Direct Ionization analysis (DIESI-MS) was done on a Bruker MicrOTOF-QII system, using an electrospray ionization (ESI) interface (Bruker Daltonics, Biellerica, MA, USA) operated in the negative ion mode. A solution of 10 μL of the sample resuspended in 1 mL of methanol was filtered with a 0.25 μm polytetrafluoroethylene (PTFE) filter and diluted 1:100 with methanol. Diluted samples were directly infused into the ESI source and analyzed in negative mode. Nitrogen was used with a flow rate of 4 L/min (0.4 Bar) as a drying and nebulizer gas, with a gas temperature of 180 °C and a capillary voltage set to 4500 V. The spectrometer was calibrated with an ESI-TOF tuning mix calibrant (Sigma-Aldrich, Toluca, Estado de México, México).

MS/MS analysis was performed using negative electrospray ionization (ESI$^-$), and the obtained fragments were analyzed by a Bruker Compass Data Analysis 4.0 (Bruker Daltonics, Technical Note 008, 2004, Bruker Daltonics, Biellerica, MA, USA). An accuracy threshold of 5 ppm was established to confirm the elemental compositions.

3. Results

S. aculeatissimum and *S. myriacanthum* fruits were collected in the Cuetzalan, Puebla (México) region (Figure 1). Usually, the mature fruit is a globose berry 2–3 cm in diameter, and it is composed of the peel (≈0.2 mm thick) that represent a 25–30% percent of the fruit, 10% of a polysaccharides layer, and 50–60% of seeds. Once the peels were washed and dried, cutins were obtained by previously published methods. Cutin obtained from the dry peels were in very high yield (≈77%) in relation to other fruit cutins, such as tomato or citrus fruits (≈0.5%) [9,30]. Most of the compounds hydrolyzed with the enzymatic treatment were identified as monosaccharides (Glu and Fru, data not shown), and cutins were characterized by solid-state resonance techniques (CPMAS ^{13}C NMR), ATR-FTIR, and AFM.

Figure 1. Photographs of the plants and fruits of (**A**) *S. aculeatissimum* Jacq, and (**B**) *S. myriacanthum* Dunal.

3.1. CPMAS ^{13}C NMR Analysis

Cutins obtained from *S. aculeatissimum* and *S. myriacanthum* were analyzed by CPMAS ^{13}C NMR, and their spectra are shown in Figure 2. According to our previous studies in fruit cuticles, typical resonances of aliphatic-aromatic polyesters are exhibited: bulk methylenes (20–35 ppm), oxygenated aliphatic carbons (55–85 ppm), aromatics and olefins (105–155 ppm), and carbonyl groups (172 ppm) signals.

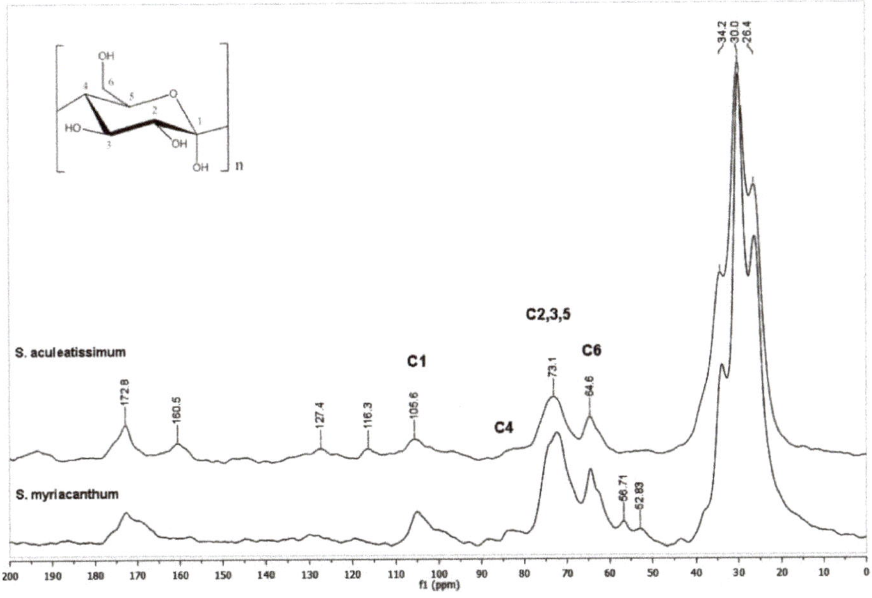

Figure 2. CPMAS ^{13}C NMR spectra of cutins from *S. aculeatissimum* and *S. myriacanthum*.

Some of these signals belong to the carbohydrate moieties (C6 at 60 ppm, C2,3,5 at 70–75 ppm, C4 at 83 ppm, and C1 at 101–105 ppm), some of these peaks could overlap with oxygenated aliphatic signals. However, every cutin showed unique NMR characteristics: more aromatic peaks are evident in *S. aculeatissimum* (Figure 2, upper spectrum), while in *S. myriacanthum* (Figure 2, lower spectrum) peaks at 52 and 56 ppm are present.

To garner more information about these materials, cutins from both fruits were the object of a TFA hydrolysis. It has been demonstrated that TFA hydrolysis could be used to remove non-cellulosic polysaccharides with the advantage that TFA is easy to remove by evaporation rather than a loss-prone neutralization step [28].

S. aculeatissimum cutin was found to be resistant to TFA hydrolysis. There was a minimal weight loss, and as seen in Figure 3, and most of the peaks remain as in the spectra without TFA treatment. However, *S. myriacanthum* shows ≈ 8% of weight loss, and this can be attributed to the disappearance of compounds with peaks at 50 ppm. The soluble part obtained from the TFA hydrolysis was studied, and it was found that these peaks belong to an epoxidated C18 long-chain aliphatic acid (see Supplementary Material).

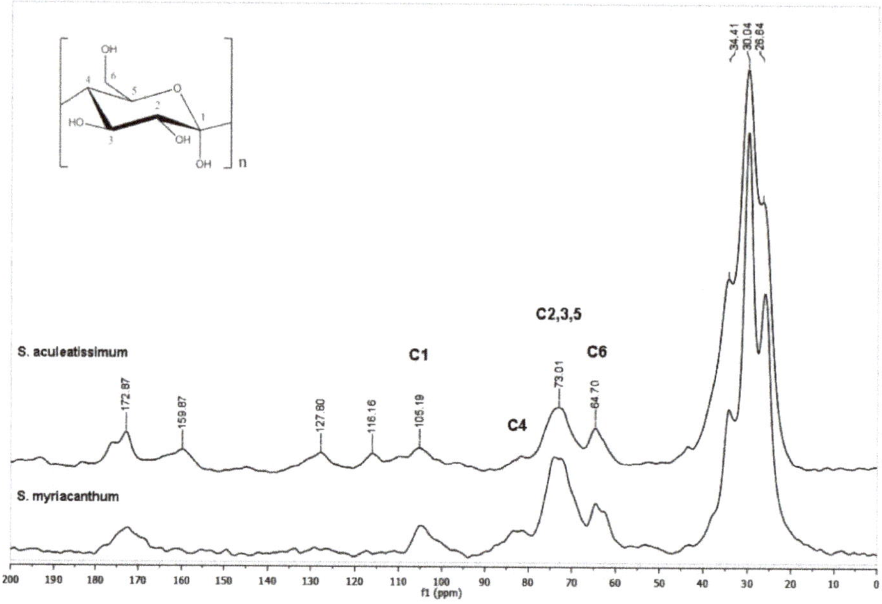

Figure 3. CPMAS ^{13}C NMR spectra of cutins from *S. aculeatissimum* and *S. myriacanthum* after trifluoracetic acid (TFA) hydrolysis (TFA-HC).

3.2. Infrared Spectroscopy Analysis of the S. Aculeatissimum and S. Myriacanthum Cutins

Isolated cutins have been characterized in situ at their functional chemical groups as well as their interactions at the cuticular levels with exogenous chemicals [8,31]. The ATR FT-IR analysis of *S. aculeatissimum* and *S. myriacanthum* cutins (Figure 4) were characterized as follows: hydroxyl groups of the polysaccharide domain and residual carboxylic acids showed its absorption maxima as broadband at 3860 cm^{-1}, characteristic intense bands corresponding to the asymmetrical and symmetrical stretching vibrations of the methylene CH_2 region at 2905 and 2850 cm^{-1}, with the bending vibrations at 1462 and 1350 cm^{-1}, which came from the aliphatic components present in the cutin. Another group of signals associated with the cutin matrix is that from 1600 to 1750 cm^{-1} attributed to the carbonyl C=O stretching band in ester groups, and their asymmetric stretching vibrations of C–CO–O at 1100 cm^{-1}. These assignations agree with those reported and used in the study of non-isolated plant cutins [8].

Figure 4 shows that IR spectra for both cutins are very similar, and the most intense bands correspond to the main domains of this polyester: aliphatic and polysaccharides groups. However, two groups of signals are making the difference between them. For *S. aculeatissimum* cutin, a group of bands at 1500 to 1650 cm^{-1} related to aromatic and C=C functional groups are less intense in *S. myriacanthum* cutin, due to the low presence of aromatics. On the other hand, the group of signals at 1100 cm^{-1} is broader and more intense in *S. myriacanthum*, possibly because of a poliesterification with at least two different long-chain acids. These observations agreed with the CPMAS ^{13}C NMR analysis.

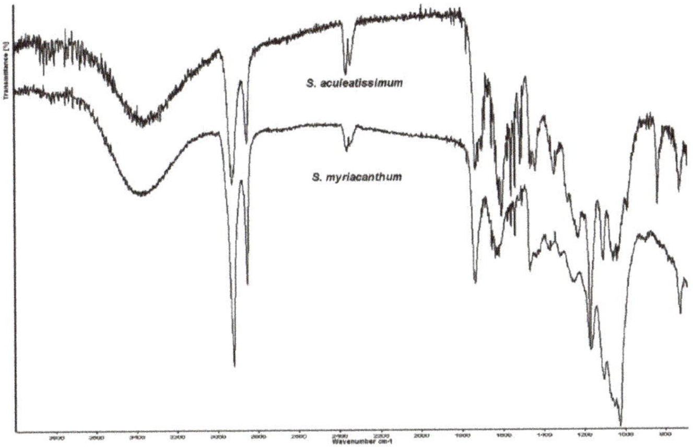

Figure 4. Attenuated Total Reflectance Fourier transform infrared spectroscopy (ATR-FTIR) spectra of *S. aculeatissimum* and *S. myriacanthum*.

3.3. Atomic Force Microscopy Analysis

The cuticles obtained from the fruits of *S. aculeatissimum* and *S. myriacanthum* were analyzed through AFM. Figure 5 shows that these cutins are thicker than other fruit cutins, such as tomatoes, lemon, orange. The AFM amplitude error images showed that cutin surfaces are composed mainly of fibers that give the characteristic roughness (Figure 5C,D). The fibers are more homogeneous in the *S. aculeatissimum* cutin with an average thickness of 34 nm, while in the *S. myriacanthum* cutin they are irregularly present, with fibers ranged from 125 to 23 nm that were observed. The roughness study showed that *S. aculeatissimum* has a R_q of 1.8 nm and a R_a of 1.3 nm, while the cutin of *S. myriacanthum* showed a lower roughness with a R_q of 3.4 nm and a R_a of 2.6 nm.

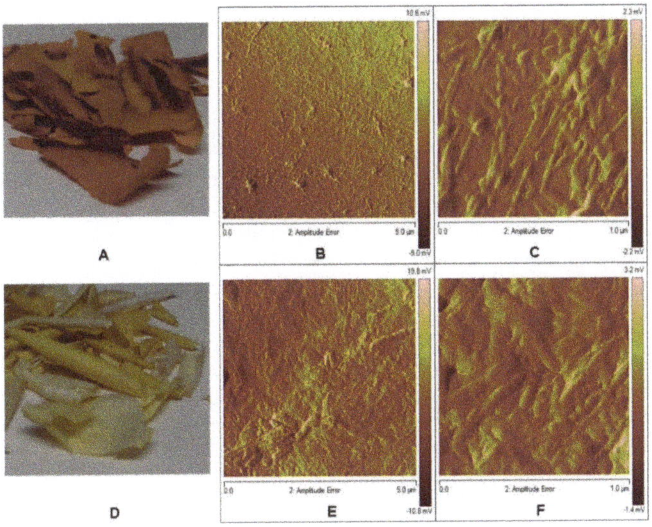

Figure 5. Atomic force microscopy tapping mode topographical images from (**A**) *S. aculeatissimum* (**B**,**C**): 5.0 and 1.0 μm, respectively) and (**D**) *S. myriacanthum* (**E**,**F**): 5.0 and 1.0 μm, respectively).

3.4. Alkaline Hydrolysis (KOH/MeOH)

To study the main aliphatic components present in the *S. aculeatissimum* and *S. myriacanthum* cutins, a complementary analysis was done with alkaline hydrolysis. In both cases, around ≈93% of the cuticular material was hydrolyzed. Soluble products from the alkaline hydrolysis were analyzed by means of direct-injection electrospray ionization mass spectrometry in negative mode (DIESI-MS, see Supplementary Material) and the compounds identified by the *ms/ms* analysis, are reported in Table 1.

Table 1. Identification of the Main Compounds in the Soluble Fraction of the Alkaline Hydrolysis.

Name	$[M-H]^{-}_{obs}$	$[M-H]^{-}_{exact}$	Formula	Error	%RA	
					SA	SM
Coumaric acid	163.0386	163.0389	$C_9H_8O_3$	2.1	0.18	—
Capric acid	171.1352	171.1379	$C_{10}H_{20}O_2$	3.5	0.36	0.25
Coniferaldehyde	177.0557	177.0546	$C_{10}H_{10}O_3$	3.6	0.54	—
n-Nonanedioic acid	187.1001	187.0964	$C_9H_{16}O_4$	4.1	0.18	0.25
Ferulic acid	193.0473	193.0495	$C_{10}H_{10}O_4$	2.5	0.18	—
Lauric acid	199.1689	199.1692	$C_{12}H_{24}O_2$	2.5	0.18	0.25
Myristic acid	227.2003	227.2005	$C_{14}H_{28}O_2$	1.9	0.72	1.02
n-Pentadecanoic acid	241.2144	241.2162	$C_{15}H_{30}O_2$	3.9	0.72	1.02
Palmitic acid	255.2322	255.2318	$C_{16}H_{32}O_2$	2.0	2.88	3.29
Hexyl 2-(4-hydroxy-3-methoxy-phenyl) acetate	265.1481	265.1434	$C_{15}H_{22}O_4$	4.2	—	1.77
16-hydroxypalmitic acid	271.2257	271.2267	$C_{16}H_{32}O_3$	2.6	6.12	1.52
Linoleic acid	279.2330	279.2318	$C_{18}H_{32}O_2$	3.1		2.02
10,16-DHPA	287.2209	287.2216	$C_{16}H_{32}O_4$	2.6	69.84	44.02
Heptadecanedioic acid	299.2228	299.2216	$C_{17}H_{32}O_4$	3.4	2.88	1.26
8-hydroxyhexadecane dioic acid	301.2017	301.2009	$C_{16}H_{30}O_5$	4.2	5.76	1.52
18-hydroxy-9S,10R-epoxy-octadecanoic acid	313.2387	313.2373	$C_{18}H_{34}O_4$	3.9	—	9.36
9,10,18-trihydroxy-octadecanoic acid	331.2487	331.2479	$C_{18}H_{36}O_5$	2.4	2.88	24.03
2,3-Divanillyl-1,4-butanediol	361.1563	361.1645	$C_{20}H_{26}O_6$	4.5	2.16	4.05

$[M-H]^{-}_{exact}$: Molecular Weight exact, $[M-H]^{-}_{obs}$: Molecular Weight observed, % RA: % Relative Area. Error [ppm]: Absolute value of the deviation between measured mass and theoretical mass of the selected peak in [ppm].

Even when most of the compounds are present in both cutins, some differences can be observed. The main constituent identified in *S. aculeatissimum* cutin was 10,16-dihydroxyhexadecanoic acid (10,16-DHPA), an important monomer present in different cutins such as tomato, citrus cuticles and green pepper [32], in a 69.84% of the relative abundance. Aromatic and some derivatives compounds were detected in agreement with the CPMAS ^{13}C NMR data. 10,16-DHPA was found in *S. myriacanthum* cutin. However, two other significant monomers are present: 9,10,18-trihydroxy-octadecanoic acid and 18-hydroxy-9S,10R-epoxy-octadecanoic acid in 24.03 and 9.36%, respectively. According to the TFA-hydrolysis analysis, the epoxilated long-chain aliphatic acid was hydrolyzed and obtained almost pure, according to the NMR analysis (see Supplementary Material). This observation could suggest that it is present in a different domain from the other components. The predominance of C16 long-chain acids in cutins is very common and corroborates previous cutin reports. However, it is important to highlight that most of the 25% of the main monomers in *S. myriacanthum* cutin are C18 acids. The presence of these C16 and C18 monomers could be the reason for the broadband esterification detected at 100 cm^{-1} in the ATR-FTIR spectrum. Aromatic compounds are not present as in *S. aculeatissimum* cutin, which agrees with the NMR analysis.

3.5. Analysis of the Films Prepared from Hydrolyzed Cutins

To demonstrate that *S. aculeatissimum* and *S. myriacanthum* cutins could be a good material for biopolymer, films were prepared by simple blending in solvents. Representative photographs of the films prepared from the hydrolyzed cutins are shown in Figure 6. Both samples have a waxy consistency, but their surface was quite homogeneous. Films were characterized through ATR-FTIR and AFM.

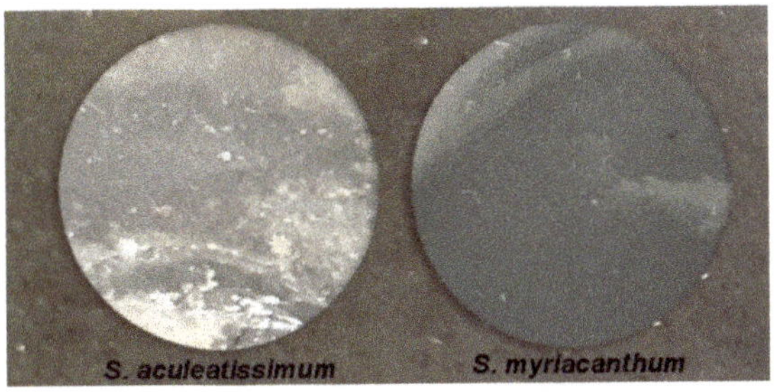

Figure 6. Photographs of the films prepared from hydrolyzed cutins.

3.6. ATR-FTIR Analysis of the Films

Analysis using Fourier transform infrared (FTIR) spectroscopy indicated that films from hydrolyzed cutins keep the spectral features of the original cutins. However, most of the signals demonstrate that bands associated with ester groups disappeared, especially the absorption at 1630 cm^{-1} that belongs to the stretching of C=O of ester groups. However, the presence of the band at 1127 cm^{-1}, ascribed to the asymmetric stretching vibrations of C–CO–O, demonstrates that part of this polyester network remains, or monomers were partially polymerized (Figure 7).

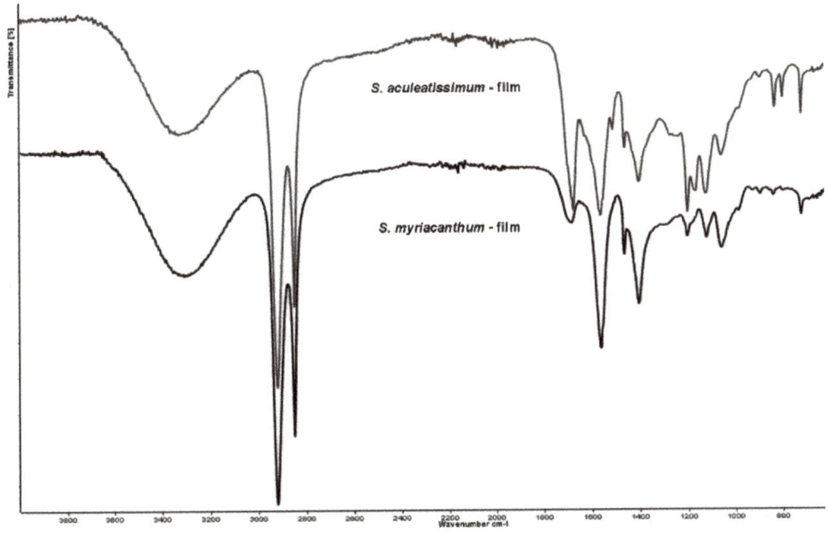

Figure 7. ATR-FTIR spectra of the films from hydrolyzed cutins of *S. aculeatissimum* and *S. myriacanthum*.

3.7. AFM Analysis of the Films

Atomic Force Microscopy (AFM) analysis shows a different topography from that observed in the original cutins. There is no occurrence of fibers that could be attributed to the cellulose or pectin presence [33]. According to DIESI-MS analysis, there is not a presence of sugars or some oligo- or polysaccharides in the soluble hydrolyzed cutins. The roughness study showed that film from *S. aculeatissimum* cutin has a value of a R_q 0.527 nm and a R_a 0.406 nm, while that obtained from *S. myriacanthum* cutin showed values of R_q 0.973 nm and Ra 0.584 nm (Figure 8).

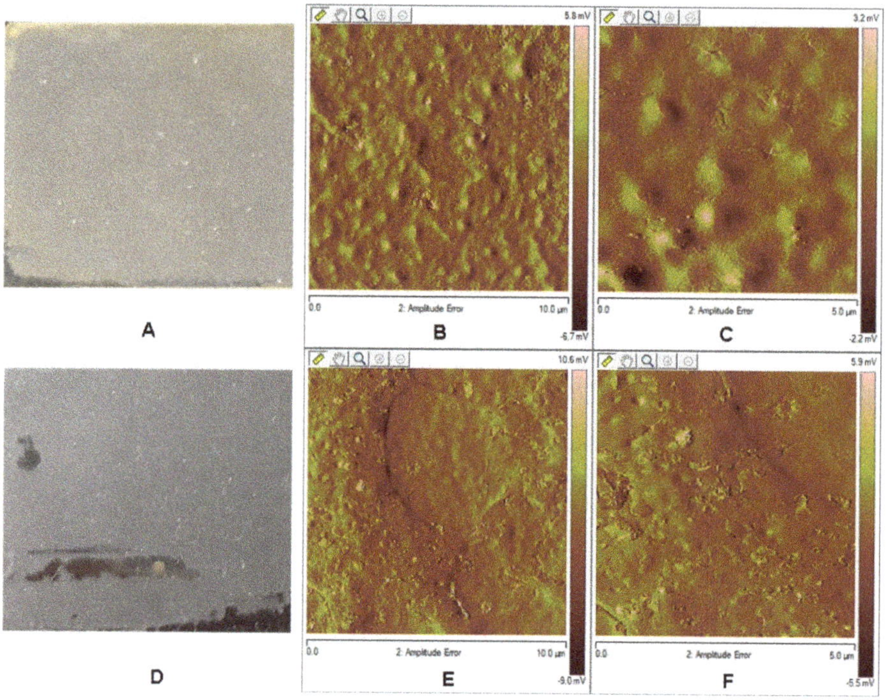

Figure 8. Atomic force microscopy tapping mode topographical images from (**A**) *S. aculeatissimum* (**B**,**C**): 5.0 and 1.0 μm, respectively) and (**D**) *S. myriacanthum* (**E**,**F**): 5.0 and 1.0 μm, respectively).

The homogeneity could be attributed to a good organization and a high degree of order of the monomers, oligomers or polymers present in the hydrolyzed cutin. This characteristic is highly important to get films with small porous or cavities distributed along the surface.

4. Conclusions

In this work, we have demonstrated that *S. aculeatissimum* and *S. myriacanthum* cuticles have a good percentage of cutin—around 70%, from the dry weight, more than other studied fruits such as tomato or citrus fruits (≈0.5%). These cutins have the same monomers composition reported in other fruits cuticles such as tomato, citric fruits, or pepper, were the main component was 10,16-DHPA. Films obtained from the hydrolyzed cutins showed a good homogeneity. Furthermore, the fact that these fruits are not for human or animal consumption makes it feasible for them to be considered as a potential raw material to produce sustainable composite materials as an alternative to traditional plastics.

Supplementary Materials: The Supplementary Materials are available online at http://www.mdpi.com/2073-4360/12/9/1945/s1.

Author Contributions: M.B.G.-P. and D.A.-B. conceived and designed the main ideas of this paper, carried out the NMR and DIESI-MS experiments, analyzed the experimental results, and wrote the paper. R.E.-R., and M.E.V.-D. carried out the cuticle and compounds extraction experiments and help to discuss the results. The authors read and approved the final manuscript. Investigation, D.A.B, R.E.-R., M.E.V.-D. and M.B.G.-P.; Project administration, D.A.B and M.B.G.-P.; Supervision, D.A.B and M.B.G.-P. All authors have read and agreed to the published version of the manuscript.

Funding: This research was funded by Consejo Nacional de Ciencia y Tecnologia (CONACyT) for funding Project No. 253570 and SIP-IPN grants No. 20201066 and 20201968.

Conflicts of Interest: The authors declare no conflict of interest. The funders had no role in the design of the study; in the collection, analyses, or interpretation of data; in the writing of the manuscript, or in the decision to publish the results.

References

1. Bargel, H.; Koch, K.; Cerman, Z.; Neinhuis, C. Structure-Function Relationships of the Plant Cuticle and Cuticular Waxes—A Smart Material? *Funct. Plant Biol.* **2006**, *33*, 893–910. [CrossRef] [PubMed]
2. Dominguez, E.; Heredia-Guerrero, J.A.; Heredia, A. The Biophysical Design of Plant Cuticles: An Overview. *N. Phytol.* **2011**, *189*, 938–949. [CrossRef]
3. Fich, E.A.; Segerson, N.A.; Rose, J.K. The Plant Polyester Cutin: Biosynthesis, Structure, and Biological Roles. *Annu. Rev. Plant Biol.* **2016**, *67*, 207–233. [CrossRef] [PubMed]
4. Kolattukudy, P.E. Polyesters in Higher Plants. *Adv. Biochem. Eng. Biotechnol.* **2001**, *71*, 1–49. [PubMed]
5. Burghardt, M.; Riederer, M. Cuticular transpiration. In *Biology of the Plant Cuticle*; Riederer, M., Müller, C., Eds.; Blackwell Publishing: Oxford, UK, 2006; pp. 292–311.
6. Walton, T.J.; Kolattukudy, P.E. Determination of the structures of cutin monomers by a novel depolymerization procedure and combined gas chromatography and mass spectrometry. *Biochemistry* **1972**, *11*, 1885–1897. [CrossRef]
7. Waltson, T. Waxes, cutin and suberin. In *Lipids, Membranes and Aspects of Photobiology*; Harwood, E.J., Boyer, J., Eds.; Academic Press: London, UK, 1990; pp. 105–158.
8. Heredia-Guerrero, J.A.; Benitez, J.J.; Dominguez, E.; Bayer, I.S.; Cingolani, R.; Athanassiou, A.; Heredia, A. Infrared and Raman spectroscopic features of plant cuticles: A review. *Front. Plant Sci.* **2014**, *5*, 305. [CrossRef]
9. Arrieta-Baez, D.; Cruz-Carrillo, M.; Gómez-Patiño, M.B.; Zepeda-Vallejo, L.G. Derivatives of 10,16-dihydroxyhexadecanoicacid isolated from tomato (*Solanum lycopersicum*) as potential material for aliphatic polyesters. *Molecules* **2011**, *16*, 4923–4936. [CrossRef]
10. Tedeschi, G.; Guzman-Puyol, S.; Ceseracciu, L.; Paul, U.C.; Picone, P.; Di, C.M.; Athanassiou, A.; Heredia-Guerrero, J.A. Multifunctional Bioplastics Inspired by Wood Composition: Effect of Hydrolyzed Lignin Addition to Xylan-Cellulose Matrices. *Biomacromolecules* **2020**, *21*, 910–920. [CrossRef]
11. Yang, J.; Ching, Y.C.; Chuah, C.H. Applications of Lignocellulosic Fibers and Lignin in Bioplastics: A Review. *Polymers* **2019**, *11*, 751. [CrossRef]
12. Reichert, C.L.; Bugnicourt, E.; Coltelli, M.B.; Cinelli, P.; Lazzeri, A.; Canesi, I.; Braca, F.; Martinez, B.M.; Alonso, R.; Agostinis, L.; et al. Bio-Based Packaging: Materials, Modifications, Industrial Applications and Sustainability. *Polymers* **2020**, *12*, 1558. [CrossRef]
13. Liu, D.; Yan, X.; Si, M.; Deng, X.; Min, X.; Shi, Y.; Chai, L. Bioconversion of Lignin into Bioplastics by Pandoraea Sp. B-6: Molecular Mechanism. *Environ. Sci. Pollut. Res. Int.* **2019**, *26*, 2761–2770. [CrossRef] [PubMed]
14. Kershaw, P. *Exploring the Potential for Adopting Alternative Materials to Reduce Marine Plastic Litter*; United Nations Environment Programme: Nairobi, Kenya, 2018; pp. 1–124.
15. Coles, R.; Kay, M.; Song, J. Bioplastics. In *Food and Beverage Packaging Technology*, 2nd ed.; Coles, R., Kirwan, C.M., Eds.; Blackwell Publishing Ltd.: Hoboken, NJ, USA, 2011; pp. 295–319.
16. Mostafa, N.A.; Farag, A.A.; Abo-dief, H.L.; Tayeb, A.M. Production of biodegradable plastic from agricultural wastes. *Arab. J. Chem.* **2018**, *11*, 546–553. [CrossRef]
17. Giosafatto, C.V.; Di, P.P.; Gunning, P.; Mackie, A.; Porta, R.; Mariniello, L. Characterization of Citrus Pectin Edible Films Containing Transglutaminase-Modified Phaseolin. *Carbohydr. Polym.* **2014**, *106*, 200–208. [CrossRef] [PubMed]
18. Baron, R.D.; Perez, L.L.; Salcedo, J.M.; Cordoba, L.P.; Sobral, P.J. Production and Characterization of Films Based on Blends of Chitosan from Blue Crab (Callinectes Sapidus) Waste and Pectin from Orange (Citrus Sinensis Osbeck) Peel. *Int. J. Biol. Macromol.* **2017**, *98*, 676–683. [CrossRef] [PubMed]
19. Naz, S.; Ahmad, N.; Akhtar, J.; Ahmad, N.M.; Ali, A.; Zia, M. Management of Citrus Waste by Switching in the Production of Nanocellulose. *IET Nanobiotechnol.* **2016**, *10*, 395–399. [CrossRef] [PubMed]

20. Arrieta-Baez, D.; Hernandez Ortiz, J.V.; Teran, J.C.; Torres, E.; Gomez-Patino, M.B. Aliphatic Diacidic Long-Chain C16 Polyesters From 10,16-Dihydroxyhexadecanoic Acid Obtained from Tomato Residual Wastes. *Molecules* **2019**, *24*, 1524. [CrossRef]
21. Gómez-Patiño, M.B.; López-Simeón, R.; Espinosa-Domínguez, S.; Hernández-Guerrero, M.; Arrieta-Baez, D.; Beltrán-Conde, H.; Campos-Terán, J.; Reyes-Duarte, D. Aprovechamiento de residuos agroindustriales: Composición, modificación enzimática y evaluación de sus potenciales aplicaciones in Obtención Enzimática de Ingredientes Funcionales, Compuestos Bioactivos y Nutraceúticos a Partir de Recursos Naturales Iberoamericanos. In *Biblioteca de las Ciencias*, 40th ed.; Fabián, G.C.S., Gasca, F.J.P., Eds.; Consejo Superior de Investigaciones Científicas: Madrid, Spain, 2012.
22. Benitez, J.J.; Osbild, S.; Guzman-Puyol, S.; Heredia, A.; Heredia-Guerrero, J.A. Bio-Based Coatings for Food Metal Packaging Inspired in Biopolyester Plant Cutin. *Polymers* **2020**, *12*, 942. [CrossRef]
23. Heredia-Guerrero, J.A.; Heredia, A.; Dominguez, E.; Cingolani, R.; Bayer, I.S.; Athanassiou, A.; Benitez, J.J. Cutin From Agro-Waste As a Raw Material for the Production of Bioplastics. *J. Exp. Bot.* **2017**, *68*, 5401–5410. [CrossRef]
24. Cuevas-Reyes, L. Taxonomía de la Familia *Solanaceae* en el Municipio de Coacoatzintla. Bachelor's Thesis, Facultad de Biología, Universidad Veracruzana, Veracruz, Mexico, 2018.
25. Kohara, A.; Nakajima, C.; Hashimoto, K.; Ikenaga, T.; Tanaka, H.; Shoyama, Y.; Yoshida, S.; Muranaka, T. A Novel Glucosyltransferase Involved in Steroid Saponin Biosynthesis in Solanum Aculeatissimum. *Plant Mol. Biol.* **2005**, *57*, 225–239. [CrossRef]
26. Ikenaga, T.; Kikuta, S.; Itimura, K.; Nakashima, K.; Matsubara, T. Growth and Production of Steroid Saponin in Solanum Aculeatissimum during One Vegetation Period. *Planta Med.* **1988**, *54*, 140–142. [CrossRef]
27. Yadav, A.K.; Tangpu, V. Anthelmintic Activity of Ripe Fruit Extract of Solanum Myriacanthum Dunal (Solanaceae) Against Experimentally Induced Hymenolepis Diminuta (Cestoda) Infections in Rats. *Parasitol. Res.* **2012**, *110*, 1047–1053. [CrossRef] [PubMed]
28. Arrieta-Baez, D.; Stark, R.E. Using Trifluoroacetic Acid to Augment Studies of Potato Suberin Molecular Structure. *J. Agric. Food Chem.* **2006**, *54*, 9636–9641. [CrossRef] [PubMed]
29. Gomez-Patiño, M.B.; Cassani, J.; Jaramillo-Flores, M.E.; Zepeda-Vallejo, L.G.; Sandoval, G.; Jimenez-Estrada, M.; Arrieta-Baez, D. Oligomerization of 10,16-Dihydroxyhexadecanoic Acid and Methyl 10,16-Dihydroxyhexadecanoate Catalyzed by Lipases. *Molecules* **2013**, *18*, 9317–9333. [CrossRef] [PubMed]
30. Hernandez, V.B.L.; Arrieta-Baez, D.; Cortez, S.P.I.; Méndez-Méndez, J.V.; Berdeja, M.B.M.; Gómez-Patiño, M.B. Comparative studies of cutins from lime (*Citrus aurantifolia*) and grapefruit (*Citrus paradisi*) after TFA hydrolysis. *Phytochemistry* **2017**, *144*, 78–86. [CrossRef]
31. Luque, P.; Ramírez, F.J.; Heredia, A.; Bukovac, M.J. Fouriertrans-form IR studies on the interaction of selected chemicals with isolates cuticles. In *Air Pollutants and the Leaf Cuticle, NATO ASI Series, G36*; Percy, K.E., Cape, J.N., Jagels, R., Simpson, C.J., Eds.; Springer: Berlin, Germany, 1994; pp. 217–223.
32. Gerard, H.C.; Osman, S.F.; Fett, W.F.; Moreau, R.A. Separation, identification and quantification of monomers from cutin polymers by high performance liquid chromatography and evaporative light scattering detection. *Phytochem. Anal.* **1992**, *3*, 139–144. [CrossRef]
33. Posé, S.; Paniagua, C.; Matas, A.J.; Gunning, A.P.; Morris, V.M.; Quesada, M.A.; Mercado, J.A. A nanostructural view of the cell wall disassembly process during fruit ripening and postharvest storage by atomic force microscopy. *Trends Food Sci. Technol.* **2019**, *87*, 47–58. [CrossRef]

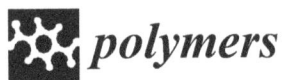

polymers

MDPI

Article

Controlled Release, Disintegration, Antioxidant, and Antimicrobial Properties of Poly (Lactic Acid) /Thymol/Nanoclay Composites

Marina Ramos [1,*], Elena Fortunati [2], Ana Beltrán [1], Mercedes Peltzer [3,4], Francesco Cristofaro [5], Livia Visai [5,6], Artur J.M. Valente [7], Alfonso Jiménez [1], José María Kenny [2] and María Carmen Garrigós [1]

1 Nutrition & Food Sciences, Department of Analytical Chemistry, University of Alicante, 03080 Alicante, Spain; ana.beltran@ua.es (A.B.); alfjimenez@ua.es (A.J.); mc.garrigos@ua.es (M.C.G.)
2 Civil Environmental Engineering Department, University of Perugia, UdR INSTM, Strada di Pentima 4, 05100 Terni, Italy; elenafortunati@gmail.com (E.F.); jose.kenny@unipg.it (J.M.K.)
3 Departamento de Ciencia y Tecnología, Universidad Nacional de Quilmes, Bernal, Buenos Aires B1876BXD, Argentina; mercedes.peltzer@unq.edu.ar
4 Consejo Nacional de Investigaciones Científicas y Técnicas (CONICET), Ciudad Autónoma de Buenos Aires (CABA), Buenos Aires C1425FQB, Argentina
5 Department of Molecular Medicine, Center for Health Technologies (C.H.T.), UdR INSTM, University of Pavia, 27100 Pavia, Italy; francesco.cristofaro01@universitadipavia.it (F.C.); livia.visai@unipv.it (L.V.)
6 Department of Occupational Medicine, Toxicology and Environmental Risks, Istituti Clinici Scientifici (ICS) Maugeri, Società Benefit S.p.A IRCCS, 27100 Pavia, Italy
7 Department of Chemistry, University of Coimbra, CQC, 3004-535 Coimbra, Portugal; avalente@ci.uc.pt
* Correspondence: marina.ramos@ua.es

Received: 13 July 2020; Accepted: 17 August 2020; Published: 20 August 2020

Abstract: Nano-biocomposite films based on poly (lactic acid) (PLA) were prepared by adding thymol (8 wt.%) and a commercial montmorillonite (D43B) at different concentrations (2.5 and 5 wt.%). The antioxidant, antimicrobial, and disintegration properties of all films were determined. A kinetic study was carried out to evaluate the thymol release from the polymer matrix into ethanol 10% (*v*/*v*) as food simulant. The nanostructured networks formed in binary and ternary systems were of interest in controlling the release of thymol into the food simulant. The results indicated that the diffusion of thymol through the PLA matrix was influenced by the presence of the nanoclay. Disintegration tests demonstrated that the incorporation of both additives promoted the breakdown of the polymer matrix due to the presence of the reactive hydroxyl group in the thymol structure and ammonium groups in D43B. Active films containing thymol and D43B efficiently enhanced the antioxidant activity (inhibition values higher than 77%) of the nano-biocomposites. Finally, the addition of 8 wt.% thymol and 2.5 wt.% D43B significantly increased the antibacterial activity against *Escherichia coli* and *Staphylococcus aureus 8325-4*, resulting in a clear advantage to improve the shelf-life of perishable packaged food.

Keywords: PLA; nanocomposites; functional properties; thymol; migration; films

1. Introduction

The use of biopolymers in packaging applications is considered a suitable alternative to the petrochemical counterparts contributing to the limitation of the environmental problems caused by the accumulation of plastic waste [1]. Among the different bio-based polymers that can be used for such purpose, poly (lactic acid) or PLA is a biodegradable polyester obtained from 100% renewable resources. It is also a highly versatile material with many commercial applications in the textile, medical, automotive and packaging sectors. PLA shows some desirable properties such as its inherent biodegradability, biocompatibility, stiffness, high strength, thermo-plasticity, and high transparency [2]. However, PLA-based films show some significant drawbacks which limit their performance in processing and the final use, such as poor barrier properties to gases, low thermal and mechanical resistance, slow crystallization and brittleness [3].

These shortcomings can be overcome by the development of nano-biocomposites based on PLA reinforced with nanoparticles, which has been introduced as a promising option [4]. Souza et al. [5] assumed that the incorporation of nanoclays into PLA-based formulations is a promising alternative to improve their barrier and mechanical properties without altering the transparency and compostability of the final material. Among the different nanoparticles, an interesting approach to improve PLA properties is the reinforcement with nanoclays such as montmorillonites, resulting in the development of PLA nanocomposites [6]. The use of this type of nanomaterials in food packaging applications can also modify the internal atmosphere due to the modification of barrier properties, thereby delaying ripening and extending the shelf-life in packaged foods [7].

The development of new nanocomposites with unique characteristics is desirable, especially for the food industry due to their full range of applications [8]. The satisfaction of consumers' requirements for healthier and highly nutritional foods, as well as the increase in the need for long shelf-life for fresh food, have resulted in a high interest in the use of functional packaging systems with a controlled release of active substances embedded in eco-friendly plastic materials [9]. Active packaging systems incorporate agents with specific functionalities into the polymer matrix [10,11], reacting with or releasing specific components to increase food quality, taking advantage of these interactions [2]. Low density polyethylene (LDPE) was combined with thymol, eugenol, and carvacrol embedded in montmorillonite or halloysite to produce active nanocomposite films [12]. These formulations improved the main properties of LDPE as well as food quality by the increase in antimicrobial and antioxidant performance, and enhanced barrier properties. Montmorillonite was also used in alginate-based films with lemon essential oil to obtain antibacterial and antifungal films [13]. Consequently, the combination of natural compounds extracted from plant species with antimicrobial and antioxidant properties with bio-based polymers to enhance their functional properties and to extend food shelf-life is a promising strategy to be applied in the food sector [14–20]. In particular, thymol which is the main phenolic compound present in thyme and oregano essential oils has been used due to its antimicrobial and antioxidant properties that make it useful to be incorporated as an active compound in packaging formulations [21–24]. To the best of our knowledge, the combination of thymol and montmorillonite in PLA-based films to obtain active nanocomposites with both antimicrobial and antioxidant properties has not been extensively studied. This combination can be considered an interesting approach in food science to reduce the oxidative and microbial deterioration of food products to increase food quality, safety and shelf-life while maintaining the material properties. Several works on PLA nanocomposites have been already published [25–29], but the present work could contribute to fill a gap in the development of new sustainable nanocomposite films with combined antioxidant and antibacterial performance by controlling the release of the active agent. In addition, it is important to validate the potential use of these materials in active packaging applications.

In a previous work, these novel nano-biocomposite films based on PLA with D43B and thymol for active packaging were successfully characterized in their physicochemical properties [30]. However, the evaluation of the controlled release of the active agent into a specific food-grade simulant (ethanol 10% *v*/*v*), disintegration, antioxidant and antimicrobial properties of the obtained nano-biocomposite films has not been published yet. Therefore, the aim of this study is the evaluation of the functional and compostability properties of these films for their potential use as active food packaging materials, demonstrating also the effective release of thymol from the active nano-biocomposite PLA films.

2. Materials and Methods

2.1. Materials

Commercial PLA-4060D (Tg = 58 °C, 11–13 wt.% D-isomer) was supplied in pellets by Natureworks Co., (Minnetonka, MN, USA). Thymol (99.5%), 2,2-Diphenyl-1-picrylhydrazyl (DPPH, 95%), methanol and ethanol (High-performance liquid chromatography, HPLC grade) were supplied by Sigma-Aldrich (Madrid, Spain). The commercial nanoclay used was Dellite®43B (D43B) (Laviosa Chimica Mineraria S.p.A. Livorno, Italy), a dimethyl-benzyldihydrogenated tallow ammonium modified montmorillonite with a cation exchange capacity (CEC) of 95 meq/100 g clay, a bulk density of 0.40 g cm^{-3} and a particle size distribution between 7–9 μm.

2.2. Nano-Biocomposite Films Preparation

PLA-based nano-biocomposite films incorporated with D43B (as nanofiller) and thymol (as active agent) were developed by melt blending in a Haake Polylab QC mixer (ThermoFischer Scientific, Walham, MA, USA) at 160 °C as already described in a previous work [30]. Five formulations were obtained by combining thymol (8 wt.%) and D43B at two different loadings (2.5 and 5 wt.%) in PLA matrices. Neat PLA was used as control. Homogenous and transparent films were obtained by compression-molding as described in [29]. The final mean film thickness was 190 ± 5 μm measured with a Digimatic Micrometer Series 293 MDC-Lite (Mitutoyo, Japan) at five random positions.

In our previous work [29], it was concluded that the addition of thymol did not significantly affect the thermal stability of PLA, but some decrease (around 15%) in elastic modulus was observed due to the slight plasticizing effect induced by the active additive. The incorporation of D43B and thymol to PLA did not result in an apparent enhancement in oxygen barrier properties, but the tensile behavior was improved due to the intercalation and partial exfoliation of nanoparticles through the polymer matrix, as observed by X-Ray diffraction (XRD). Moreover, the intrinsic transparency of PLA was not affected by the addition of both components and most of the thymol initially added to PLA (around 70–75%) remained in the nanocomposites after processing, ensuring the potential applicability of these films as active systems.

2.3. Migration and Mathematical Diffusion Analysis

Migration tests for thymol released from the different nano-biocomposite films were performed, in triplicate, into ethanol 10% (*v*/*v*) as food simulant following the legislation for food contact materials EU/10/2011 [31] and the European Standard EN 13130-2005 [32]. Double-sided, total immersion migration tests were performed with films (12 cm^2). 20 mL of food simulant were put in contact with the nano-biocomposite films (area-to-volume ratio around 6 dm^2 L^{-1}) at 40 °C in an oven (J.P. Selecta, Barcelona, Spain) for 10 days. A blank sample (pure simulant) was also studied.

On the other hand, the release mechanism of thymol from PLA nano-biocomposite films was also studied over time by modelling the obtained results for 15 days in ethanol 10% (*v*/*v*). Different approaches have been applied to assess the migration of additives and contaminants from food packaging films: (i) a quantitative assessment of $M_{F,\infty}$ to analyze the diffusion process by using Equation (1) to fit the experimental data,

$$\frac{M_{F,t}}{M_{P,o}} = \left(\frac{M_{F,\infty}}{M_{P,0}}\right)\cdot\left(1 - e^{-k' t}\right) \tag{1}$$

where $M_{P,0}$ is the initial amount of thymol inside the polymeric matrix, determined by using the HPLC-UV method optimized in our previous study [30]; k' is a constant; and $M_{F,t}$ is the mean of thymol released mass to the food simulant at a defined time; (ii) a Fick's approach (diffusion-controlled systems) to assess the migration of thymol from films by using mathematical modelling (Equation (2)).

This differential equation provides a general description of the migration of an additive or contaminant from an amorphous polymer packaging film:

$$\frac{\partial C}{\partial t} = \frac{\partial}{\partial x}\left(D\frac{\partial C}{\partial x}\right) \quad (2)$$

where D is the diffusion coefficient; c is the concentration of the released species; t is time (s); and x is the space coordinate.

Apparent partition coefficients (α) can be calculated by Equation (3) from the values obtained for $M_{F,\infty}/M_{P,0}$ by fitting Equation (1),

$$\frac{M_{F,\infty}}{M_{P,0}} = \frac{1}{(1+\alpha)} \quad (3)$$

where α is defined as:

$$\alpha = \frac{V_F}{K_{P,F}V_P} \quad (4)$$

where V_P and V_F are the volumes of the polymer sample (P) and food simulant (F), respectively; and $K_{P,F}$ is the partition coefficient of thymol in the system between the samples and the solution, which can be assumed as constant at low concentrations and also for its relative solubility at the equilibrium between PLA and the food simulant [33].

Crank solved Equation (2) and formulated initial and boundary conditions, from which Equation (5) was obtained as a solution. This equation was applied for a plane sheet of thickness l, and the initial condition for testing $-l/2 < x < l/2$, considering constant the thymol's concentration released, a boundary condition of a partition coefficient between both phases and for one-dimensional diffusion of thymol in a limited volume solution [34],

$$\frac{M_{F,t}}{M_{F,\infty}} = 1 - \sum_{n=1}^{\infty} \frac{2\alpha(1+\alpha)}{1+\alpha+\alpha^2 q_n^2} \exp\left[\frac{-Dq_n^2 t}{l^2}\right] \quad (5)$$

where q_n is the non-zero positive roots of $tanq_n = -\alpha\, q_n$ and l is the polymer matrix half-thickness.

The root mean square error (RMSE) was used to estimate the quality of the model fitting and it was calculated by following Equation (6),

$$\text{RMSE} = \left[\sum_{i=1}^{n} \frac{(y_i-\hat{y}_i)^2}{n}\right]^{\frac{1}{2}} \quad (6)$$

where y_i and $\hat{y}_i$ are, respectively, the experimental and predicted residual value; and n is the number of experimental points per migration curve.

High performance liquid chromatography coupled to an ultraviolet spectrophotometry detector (HPLC-UV) was used to determine the amount of thymol released from films at different migration times. An Agilent 1260 Infinity-HPLC Diode Array Detector (DAD) (Agilent, Santa Clara, CA, USA) and an Agilent Eclipse Plus C18 (100 mm × 4.6 mm × 3.5 µm) column were used. The mobile phase was composed of acetonitrile/water (40:60) at 1 mL min^{-1} flow rate. 20 µL of the extracts were injected and thymol was detected at λ = 274 nm. Analyses were performed in triplicate. Calibration standards were run at different concentrations between 12.5 and 780 mg kg^{-1} directly obtained from a stock solution (1000 mg kg^{-1}) using appropriately diluted standards of thymol in ethanol 10% (*v*/*v*). The method was validated by the calculation of the main analytical parameters affecting the determination of thymol in the studied food simulant. Limit of detection (LOD) and limit of quantification (LOQ) values were determined by using regression parameters from the calibration curve (3 $S_{y/x/a}$ and 10 $S_{y/x/a}$, respectively; where $S_{y/x}$ is the standard deviation of the residues and a is the slope of the calibration curve) obtaining values of 0.29 mg_{Thymol} kg^{-1} and 0.96 mg_{Thymol} kg^{-1}, respectively. An excellent linearity was obtained with a determination coefficient R^2 of 0.9994.

2.4. Determination of Antioxidant Activity

The antioxidant activity of the obtained migration extracts was evaluated to study the effect of thymol released from the nano-biocomposite films into ethanol (10%, *v/v*) after 10 days by using the spectrophotometric method based on the formation of the stable radical 2,2-diphenyl-1-picrylhydrazyl (DPPH), as described elsewhere [30].

2.5. Bacterial Strains, Culture Conditions and Antibacterial Activity

Escherichia coli (*E. coli*) was provided by the Zooprofilattico Institute of Pavia (Italy) as an isolate strain whereas *Staphylococcus aureus 8325-4* (*S. aureus 8325-4*) was obtained from Timothy J. Foster (Department of Microbiology, Dublin, Ireland). *E. coli* and *S. aureus 8325-4* were grown overnight under aerobic conditions at 37 °C in Luria Bertani Broth (LB) and Brian Heart Infusion (BHI) (Difco Laboratories Inc., Detroit, MI, USA), respectively. The final density of these cultures was established at 1×10^{10} cells mL^{-1}, determined by comparison of the optical density at 600 nm.

The evaluation of the antibacterial activity of the films was performed in 100 µL of a diluted cell suspension (1×10^3 cells mL^{-1}) of *E. coli* and *S. aureus 8325-4* maintained overnight. Bacterial strains were added to 3×3 mm^2 samples, seeded at the bottom of a 96-well tissue culture plate and incubated at three different temperatures: 4 °C, 24 °C and 37 °C for 3 h and 24 h, respectively. These temperatures were selected to simulate refrigeration conditions (4 °C), ambient temperature (24 °C) and the usual incubation temperature of microbiological tests (37 °C). Regarding time, 3 h simulated a fast contact between the polymer sample and the bacteria whereas 24 h is the usual time used to evaluate the microbiological growth in antimicrobial tests.

Furthermore, 96-well flat-bottom sterile polystyrene culture plates were used as control under the same experimental conditions. At the end of each incubation period, bacterial suspensions were serially diluted and, after the incubation for 24/48 h at 37 °C, cell survival was expressed as the percentage of colony-forming unit, CFU, of bacterial growth on active nano-biocomposite films compared to those obtained for the neat PLA film.

2.6. Study of Disintegrability Under Composting Conditions

Disintegration tests were performed by following the ISO 20,200 Standard [35]. Film samples for disintegration tests were cut in pieces (20×20 mm^2) and they were buried at 5 cm depth in perforated boxes with a solid synthetic bio-waste and incubated at 58 °C for 35 days. Different time intervals were selected to recover samples from their burial, and they were further tested after 0, 2, 4, 7, 10, 14, 21, 28 and 35 days. The degree of disintegration (%) was calculated by normalizing the sample weight at different stages of incubation to the initial weight, following Equation (7):

$$\text{Disintegrability } (\%) = \frac{W_i - W_t}{W_i} \cdot 100 \quad (7)$$

where W_i is the initial dry plastic weight, and W_t is the dry plastic weight at the end of the test.

The degradation of the chemical structure of the nano-biocomposite films during the disintegration tests was estimated by comparing Fourier transform infrared spectroscopy (FTIR) spectra and differential scanning calorimetry (DSC) thermograms at different test stages. Furthermore, the changes in the visual appearance of the samples were also evaluated.

DSC analysis of samples at different disintegration times was carried out from −25 to 180 °C at a heating rate of 10 °C min^{-1} by using a DSC Mettler Toledo 822/e equipment (Schwarzenbach, Switzerland) working under nitrogen atmosphere (50 mL min^{-1}). FTIR spectra of degraded samples were recorded by using a Jasco FT-IR 615 spectrometer (Jasco Inc., Easton, MD, USA) in attenuated total reflection (ATR) mode at 400–4000 cm^{-1} range.

2.7. Statistical Analysis

Statistical analysis of results was performed by using the SPSS commercial software (Version 15.0, Chicago, IL, USA). A one-way analysis of variance (ANOVA) was carried out. Differences between means were assessed based on confidence intervals using the Tukey test at a $p < 0.05$ significance level.

3. Results

3.1. Migration Test and Antioxidant Activity of the Active PLA Nano-Biocomposite Films

The use of nanofillers in active packaging systems has revealed their ability in controlling the release of active additives from polymer matrices, at suitable rates, improving their action [7]. In this work, the effect of the nanoclay in controlling the release kinetics of thymol from the PLA nano-biocomposite films to ethanol 10% (*v*/*v*) at 40 °C was evaluated. The amounts of thymol migrated in the simulant after 10 days were 285.0 ± 3.3, 275.5 ± 13.8 and 235.3 ± 19.4 $mg_{thymol}\ kg^{-1}{}_{food\ simulant}$ for PLA/T, PLA/T/D43B2.5 and PLA/T/D43B5, respectively. These results indicate that the formulation with the highest amount of D43B (PLA/T/D43B5) showed the lowest migration rate, retaining a higher amount of thymol in the polymer structure after 10 days. This behavior is in agreement with previously reported studies [36] indicating that the tortuosity effect imposed by the presence of D43B in the diffusion of the active compound through the matrix plays an important role in the release of thymol from PLA-based nano-biocomposite films. Campos-Requena et al. [37] observed that the intercalated morphology of organic modified montmorillonite/LDPE nanocomposite films had some influence on the release rate of thymol resulting in a decrease by approximately 15%, providing a controlled release material.

The antioxidant capacity of thymol released from the developed nano-biocomposite films was measured by analyzing the extracts obtained after 10 days by using the DPPH method. The inhibition values obtained were 77.8 ± 0.1, 77.0 ± 0.4, and 77.8 ± 0.8% for PLA/T, PLA/T/D43B2.5, and PLA/T/D43B5, respectively, showing the efficient antioxidant performance of thymol added in PLA-based nano-biocomposite films. This high antioxidant activity can be attributed to the capability of the thymol phenolic hydroxyl groups to convert the phenolic oxygen anion in an alkaline environment [38]. Moreover, thymol has been reported to be a better antioxidant in lipids than its isomer carvacrol due to the more significant steric hindrance of the phenolic group [39].

Mathematical Modelling of Thymol Released from PLA Nano-Biocomposite Films

Release studies are necessary and highly relevant when active substances with antimicrobial or antioxidant properties are incorporated into packaging materials to enhance the safety and quality of food during long storage. The kinetics of these processes is indicative of the transport phenomena inside the matrix and gives an accurate estimation of the effective release rate of the active compounds. In this process, the migrant starts to diffuse through the amorphous portion of the polymer matrix toward the interface with the driving force of the concentration gradient. The migrant's concentration is partitioned between the two media until its potential chemical values, in both the polymer and the food, reach equilibrium. Thus, the release is the result of diffusion, dissolution and equilibrium processes, which are often described by the Fick's second law [40].

Consequently, the release mechanism of thymol was evaluated by modelling the results obtained at different times (up to 15 days) in ethanol 10% (*v*/*v*). Figure 1a–c depicts the normalized plots obtained for the average amount of thymol released to the food simulant at a defined time, $M_{F,t}$, by the amount of thymol released at the equilibrium at time $t \to \infty$, $M_{F,\infty}$, vs. time t (hours).

The quantitative assessment of $M_{F,\infty}$ allowed the analysis of the diffusion process. For such purpose, Equation (1) was used to fit the experimental data. The corresponding data for α and $K_{P,F}$ (Table 1) were computed considering that V_F was 20 cm^3 of ethanol 10% (*v*/*v*) and the area of the PLA-based films used in these tests was 12 cm^2.

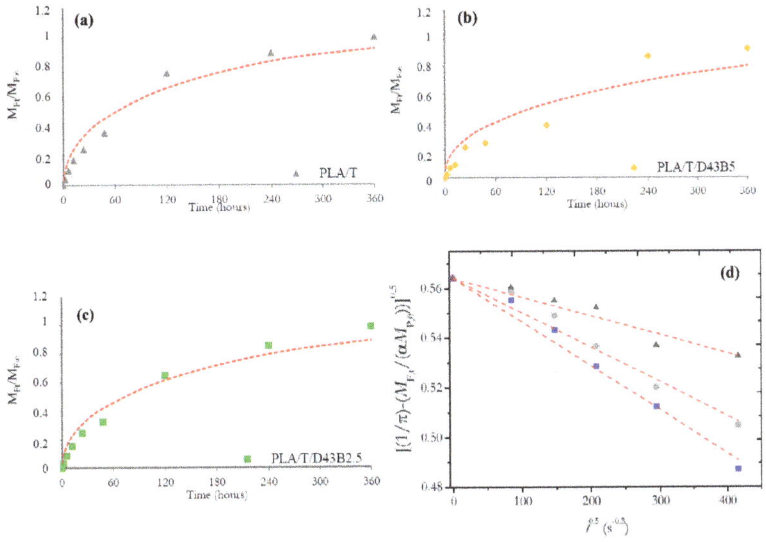

Figure 1. Normalized release of thymol from different polymer matrices: (**a**) PLA/T, (**b**) PLA/T/D43B2.5, and (**c**) PLA/T/D43B5; (**d**) plots of $\left[\frac{1}{\pi} - \frac{1}{\alpha} \times \frac{M_{F,t}}{M_{F,0}}\right]^{0.5}$ versus $t^{0.5}$ for the migration of thymol from: PLA/T (■), PLA/T/D43B2.5 (•), and PLA/T/D43B5 (▲), into ethanol 10% (*v*/*v*).

Table 1. Parameters obtained for the release of thymol from PLA-based films into ethanol 10% (*v*/*v*).

	PLA/T	PLA/T/D43B2.5	PLA/T/D43B5
$M_{P,0}$ (mg)	16.60 ± 0.20	17.82 ± 0.09	17.21 ± 0.05
$M_{F,\infty}$ (mg)	6.25 ± 0.22	6.31 ± 0.27	5.29 ± 0.29
l/cm	0.0167	0.0215	0.0180
A	1.65	1.82	2.25
$K_{P,F}$	60.3	42.4	41.1
	Equations (5) and (6):		
D ($cm^2\ s^{-1}$)	3.36×10^{-11}	4.86×10^{-11}	2.25×10^{-11}
RMSE	0.0773	0.0698	0.114
	Equations (6) and (8):		
D′ ($cm^2\ s^{-1}$)	5.95×10^{-12}	7.45×10^{-12}	5.82×10^{-12}
RMSE	0.00362	0.00306	0.00370

According to the results shown in Table 1, two main conclusions can be obtained: (i) the cumulative amount of thymol released into ethanol 10% (*v*/*v*) decreased from 38% (without nanoclay) to 35 and 31% for 2.5 and 5 wt.% of D43B, respectively; (ii) the analysis of the partition coefficients (α and $K_{P,F}$) showed their influence in the thymol diffusion mechanism. Therefore, it can be assumed that the release of thymol is governed by the Fick's second law (Equation (2)) and the diffusion coefficients (D, $cm^2\ s^{-1}$) were calculated (Table 1) from the least-square fit of Equation (5) to the experimental data (solid lines in Figure 1a–c). The RMSE (Equation (6)) of the experimental and estimated values between the calculated (y_i) and observed ($\hat{y}_i$) results for $M_{F,t}/M_{F,\infty}$ was minimized, providing a reliable indication of their fit, and it could be considered promising if the experimental error is taking into account the $M_{F,t}/M_{F,\infty}$ ratio. A good fit between calculated and experimental data was obtained for PLA/T and PLA/T/D34B2.5, with RMSE values of 0.0773 and 0.0698, respectively. However, a higher value was obtained for PLA/T/D43B5 (RMSE: 0.114), particularly for long-range times.

A more in-depth analysis of the fitting between experimental and calculated values showed that these results allow estimating the kinetics of thymol release at short times. Positive deviations of the fitting line for short times (i.e., $M_{F,t}/M_{F,\infty} < 0.60$) and negative deviations for $M_{F,t}/M_{F,\infty} > 0.60$ were observed in all cases. For such purpose, Equation (8) can be used for a linear regression analysis and as a simplified release model derived from Equation (5) [41]:

$$\left[\frac{1}{\pi} - \frac{1}{\alpha}\cdot\frac{M_{F,t}}{M_{P,0}}\right]^{0.5} = -\frac{D'^{0.5}}{\alpha\cdot l}\cdot t^{0.5} + \frac{1}{\pi^{0.5}} \tag{8}$$

Diffusion coefficients for short times, (D', cm^2 s^{-1}), were computed by using the linear fitting of Equation (8) to the experimental data (Figure 1d). Results obtained showed a very good fit between computed and experimental values for the first term in Equation (8) as a function of $t^{0.5}$, with coefficients of determination (R^2) higher than 0.999, suggesting that the experimental release data are well described by the proposed diffusion model for short-range times.

However, it was observed that the total data range cannot be fully characterized by a Fickian diffusion process, probably due to the lack of fitting caused by the last points in the plot resulting in poor results in the fitting to the first data, called short-range time. However, it was also observed that better fitting values were obtained after application of Equation (8). Therefore, the discrepancy in D values obtained (D and D') from Equation (5) and Equation (8) was a good indication that a non-Fickian release model was observed in this system. The failure of the Fickian solution process to predict the release kinetics may be due to the polymer matrix's structural modifications caused by the progressive deterioration of the PLA matrix due to the direct contact with ethanol, which can act as a plasticizer in this system. This could result in the erosion of the polymer by opening the PLA's internal structure, creating interstitial spaces that could favor the release of thymol over time due to the concentration gradient [42,43]. Moreover, as the diffusion rate increased ($D > D'$), the intermolecular interactions between ethanol molecules and PLA chains were enhanced at long times [44]. No reports on the diffusion coefficient for thymol in nanocomposite films based on PLA have been published, but higher values than those obtained in this study were reported in polypropylene (PP) films using ethanol 10% (*v*/*v*) as food simulant (1.75×10^{-10} cm^2 s^{-1}) [45]. Torres et al. [46] evaluated the thymol release from LDPE films in ethanol 10% (*v*/*v*). They indicated that the diffusion coefficient values ranged from 7.5×10^{-8} to 1.8×10^{-8} cm^2 s^{-1}, which were higher than those obtained in this work. These differences could be mainly due to the lower density and linear structure of LDPE compared to PLA, resulting in higher mass transport properties.

D values calculated in this work are consistent with the thymol release profiles shown in Figure 1a–c. The presence of D43B led to a decrease in the thymol release during the study at 40 °C, which was consistent with our previous study showing a shift in the XRD peak of D43B, suggesting the intercalated morphology of the PLA-D43B nanocomposites. In this case, a shift of the clay diffraction peak to lower angles of $2\theta = 4.6°$ was observed, corresponding to an interlayer distance of 35.6 Å [30]. Moreover, Campos-Requena et al. [37] reported that the influence of the chemical modification of clays on the active compound profile release could be a factor in the modelling of an active packaging system with controlled release of volatile compounds.

3.2. Antibacterial Activity

The antibacterial activity of the developed active nano-biocomposite films was evaluated against Gram-positive (*S. aureus* 8325-4) and Gram-negative bacteria (*E. coli*) by using the direct contact method. Table 2 shows the cell viability related to neat PLA, expressed as the percentage of microorganisms proliferated onto the PLA-based films after 3 and 24 h incubated at 4, 24 (room temperature), and 37 °C, respectively. Significant differences in cell viability for both bacterial strains were observed for almost all formulations compared to neat PLA at the tested experimental conditions. Exciting features could be observed when comparing the results obtained for active formulations containing thymol (PLA/T, PLA/T/D43B2.5, and PLA/T/D43B5) and their non-active counterparts (PLA/D43B2.5 and PLA/D43B5). These results are in accordance with those reported by Liu et al. [47] who found that the antibacterial activity of phenolic monoterpenes, including thymol, is related to their ability to be released through the polymer matrix over time allowing their continuous availability and diffusion through the bacterial cell membrane. Thymol can attach to the cell surface, and thereafter, penetrate the phospholipid bilayer of the cell membrane. The relative position of the hydroxyl group is crucial for the bioactivity of thymol, which explains its superior antimicrobial action as compared to other plant phenolics, such as 2-amino-p-cymene, which has a similar structure than thymol except for the hydroxyl group [48].

Table 2. Antibacterial activity of PLA-based nano-biocomposite films obtained at 2 incubation times (3 h and 24 h) and 3 different temperatures (4, 24 and 37 °C) against *E. coli* and *S. aureus* 8325-4, expressed as cell viability (%).

Formulation	*Cell Viability (%)*			
	***S. Aureus* 8325-4**		***E. Coli***	
	3 h	**24 h**	**3 h**	**24 h**
		At 4 °C		
PLA	100.0 ± 1.7 ^{a}	100.0 ± 1.8 ^{a}	100.0 ± 6.7 ^{a}	100.0 ± 2.5 ^{a}
PLA/T	83.5 ± 3.3 ^{b}	85.5 ± 1.1 ^{c}	65.4 ± 5.4 ^{b}	60.6 ± 1.0 ^{c}
PLA/D43B2.5	98.0 ± 1.1 ^{a}	88.6 ± 0.9 ^{b,c}	83.5 ± 2.9 ^{a}	74.0 ± 4.2 ^{b}
PLA/D43B5	91.1 ± 1.3 ^{c}	94.4 ± 1.7 ^{a,b}	88.0 ± 3.2 ^{a}	77.8 ± 3.5 ^{b}
PLA/T/D43B2.5	68.0 ± 0.3 ^{d}	60.9 ± 2.0 ^{d}	53.7 ± 3.3 ^{b}	52.0 ± 0.5 ^{c,d}
PLA//T/D43B5	70.0 ± 0.5 ^{d}	61.6 ± 1.0 ^{d}	54.4 ± 2.9 ^{b}	50.0 ± 0.7 ^{d}
		At 24 °C		
PLA	100.0 ± 1.0 ^{a}	100.0 ± 0.5 ^{a}	100.0 ± 3.8 ^{a}	100.0 ± 2.2 ^{a}
PLA/T	72.8 ± 2.2 ^{b}	62.4 ± 1.5 ^{b}	69.5 ± 1.6 ^{b}	66.2 ± 1.4 ^{c}
PLA/D43B2.5	83.4 ± 1.1 ^{a}	83.4 ± 1.4 ^{c}	75.3 ± 2.8 ^{b}	82.6 ± 1.9 ^{b}
PLA/D43B5	75.4 ± 1.4 ^{c}	81.2 ± 1.7 ^{c}	70.3 ± 1.3 ^{b}	80.0 ± 3.3 ^{b}
PLA/T/D43B2.5	56.1 ± 1.2 ^{d}	59.4 ± 1.4 ^{b}	53.1 ± 0.2 ^{c}	57.9 ± 0.7 ^{c,d}
PLA//T/D43B5	56.6 ± 2.0 ^{d}	58.7 ± 0.8 ^{b}	59.6 ± 2.0 ^{c}	60.2 ± 1.5 ^{d}
		At 37 °C		
PLA	100.0 ± 3.8 ^{a}	100.0 ± 2.1 ^{a}	100.0 ± 1.7 ^{a}	100.0 ± 3.1 ^{a}
PLA/T	57.9 ± 2.2 ^{b}	73.3 ± 1.2 ^{c}	62.9 ± 0.5 ^{b}	71.2 ± 1.4 ^{c}
PLA/D43B2.5	63.8 ± 2.0 ^{b}	84.6 ± 0.9 ^{b}	67.6 ± 0.9 ^{b}	79.4 ± 1.2 ^{b}
PLA/D43B5	64.3 ± 1.5 ^{b}	89.8 ± 2.0 ^{b}	64.8 ± 2.3 ^{b}	71.7 ± 1.2 ^{c}
PLA/T/D43B2.5	44.3 ± 1.4 ^{c}	51.6 ± 0.5 ^{d}	47.2 ± 1.2 ^{c}	49.3 ± 0.5 ^{d}
PLA/T/D43B5	48.5 ± 1.1 ^{c}	59.6 ± 3.2 ^{e}	47.3 ± 1.4 ^{c}	53.2 ± 1.2 ^{d}

Data are expressed as cell viability (%), which corresponds to the percentage of the CFU of bacteria growth in nano-biocomposite films compared to that obtained in PLA (set as 100%). Results are expressed as mean ± standard deviation, $n = 3$. Different superscripts (a, b, c, d, and e) within the same column at a specific temperature indicate statistically significant different values compared at 3 and 24 h for *S. aureus* 8325-4 and *E. coli* at different temperatures.

PLA formulations with only D43B, without the presence of thymol, also showed antibacterial activity against *E.* coli and S. aureus 8325-4 strains at both incubation times and all tested temperatures (4, 24 and 37 °C). PLA/D43B2.5 and PLA/D43B5 showed similar cell viability in most of the tested conditions. The only set of experimental conditions where PLA/D43B2.5 showed significantly higher antimicrobial activity than PLA/D43B5 was at 4 °C and 24 h for S. aureus 8325-4, with values 88.6 ± 0.9 and 94.4 ± 1.7%, respectively. These results are in agreement with those reported by De Azeredo et al. [49], who concluded that organo-modified montmorillonites could also produce the rupture of cell membranes by themselves, resulting in the inactivation of both Gram-positive and Gram-negative bacteria. This effect was attributed to the presence of quaternary ammonium groups able to react with lipids and proteins in the microorganism cell wall. Hong and Rhim [50] proved that organically modified clay powders with a quaternary ammonium salt, such as D43B, possess strong antimicrobial activity against *S. aureus*, *Listeria monocytogenes*, *Salmonella typhimurium*, and *E. coli* O157:H7.

The obtained antibacterial activity in the ternary systems resulted much more efficient and statistically significant with reference to the PLA control film. The percentage of bacteria viability was lower for PLA/T/D43B2.5 and PLA/T/D43B5 incubated with *S. aureus* 8325-4 strains (44.3 ± 1.4% and 48.5 ± 1.1%, respectively) and *E. coli* strains (47.2 ± 1.2% and 47.3 ± 1.4%, respectively) for 3 h at 37 °C (Table 2). Regarding incubation time, the percentage of the surviving fraction of bacteria did not show significant differences ($p > 0.05$) between 3 and 24 h for both bacterial strains, confirming the bacteriostatic action of the PLA-based nanocomposites. Concerning incubation temperature, the percentage of cell viability for both bacterial strains was lower when incubating at 37 °C whereas it was moderately increased at 4 °C. These results are in line with the thymol release data, thus confirming that the temperature may influence the diffusion and swelling properties of the PLA matrix, promoting the diffusion of thymol through the biopolymer structure.

3.3. Disintegrability Under Composting Conditions

Table 3 shows the values of weight loss of each sample and Table 4 shows the visual appearance of nano-biocomposite films at different times under composting conditions. It was observed that after 4 days, the disintegration rate of PLA-based materials increased significantly for binary and ternary systems showing an evident fragmentation, reaching all materials a degree of disintegration exceeding 90% after 35 days. In fact, after 4 days, all samples changed their visual appearance with a general whitening effect, loss of transparency, and evident deformation and size reduction. These results are indicative of the beginning of the hydrolytic degradation process caused by simultaneous changes in the refractive index due to water absorption, with the formation of low molecular weight by-products and the resulting increase in the PLA crystallinity [51].

According to Su et al. [52] this first step, covering the first 5 days of this study where the weight loss was small, corresponds to slow bulk degradation processes with a surface-erosion mechanism that can be mainly caused by hydrolysis, resulting in small molecules (mostly water) that can diffuse through the polymer matrix. The diffusion rate is influenced by several factors, such as crystallinity, cross-linking degree, and other morphological properties. After 5 days, the weight loss increased dramatically reaching values higher than 40% after 7 days and a continuous increase with time up to 35 days when more than 90% of the initial weight was lost (Table 3). In that period, the hydrolysis reaction was still important, and the average molecular weight of PLA decreased continuously forming small fragments easier to disintegrate since the internal chains were increasingly exposed. In that period, water, compost, and the microbiota generated in the reactor can penetrate into the gaps in the PLA structure formed by the hydrolysis reactions, contributing to a clear acceleration of the disintegration process which causes visual modifications by the formation of small particles, fragmentation and irreversible changes in mechanical properties.

Table 3. Disintegrability values (mean ± standard deviation, $n = 3$, %) of PLA and nano-biocomposite films at different times under composting conditions. Different superscripts (a, b, c, and d) within the same row indicate statistically significant different values.

Time (days)	Disintegrability (%)					
	PLA	PLA/T	PLA/T/D43B2.5	PLA/T/D43B5	PLA/D43B2.5	PLA/D43B5
2	0.3 ± 0.1 ^{a}	5.1 ± 0.4 ^{c}	8.4 ± 0.9 ^{d}	7.2 ± 0.3 ^{d}	2.4 ± 0.7 ^{b}	1.2 ± 0.1 ^{a,b}
4	0.4 ± 0.1 ^{a}	3.5 ± 0.2 ^{b,c}	5.2 ± 1.2 ^{c,d}	6.0 ± 0.2 ^{d}	2.3 ± 0.2 ^{a,b}	2.0 ± 0.1 ^{a,b}
7	42.2 ± 3.8 ^{a}	51.3 ± 0.2 ^{b}	57.9 ± 2.3 ^{b}	49.8 ± 0.9 ^{b}	53.0 ± 2.2 ^{b}	54.5 ± 1.8 ^{b}
10	56.3 ± 4.6 ^{a}	72.0 ± 4.3 ^{b}	76.0 ± 1.3 ^{b}	77.3 ± 4.9 ^{b}	67.8 ± 1.5 ^{b}	69.9 ± 4.3 ^{b}
14	72.4 ± 2.8 ^{a}	65.6 ± 3.3 ^{a}	68.9 ± 3.9 ^{a}	65.5 ± 4.5 ^{a}	64.9 ± 6.4 ^{a}	64.0 ± 3.8 ^{a}
21	73.2 ± 6.7 ^{a}	79.5 ± 2.4 ^{a}	82.4 ± 3.4 ^{a}	81.8 ± 2.8 ^{a}	82.2 ± 4.3 ^{a}	78.2 ± 1.6 ^{a}
28	77.3 ± 1.4 ^{a}	81.9 ± 1.6 ^{a}	82.2 ± 0.8 ^{a}	79.7 ± 6.8 ^{a}	76.2 ± 1.6 ^{a}	77.5 ± 4.9 ^{a}
35	95.7 ± 0.7 ^{a}	98.0 ± 0.5 ^{a}	95.5 ± 0.9 ^{a}	97.8 ± 0.5 ^{a}	97.2 ± 1.2 ^{a}	96.5 ± 1.7 ^{a}

Table 4. Visual appearance of PLA and nano-biocomposite films at different times under composting conditions. Different superscripts within the same column indicate statistically significant different values.

Formulation	Time (Days)							
	0	2	4	7	10	14	21	28
PLA	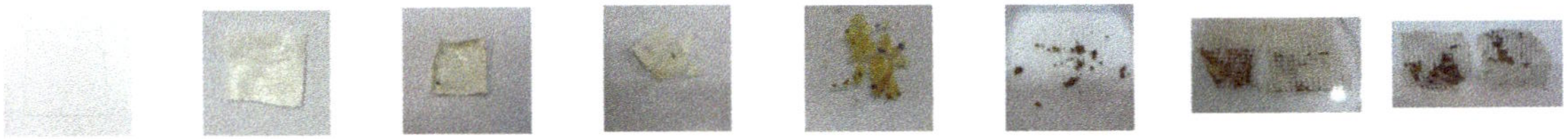							
PLA/T	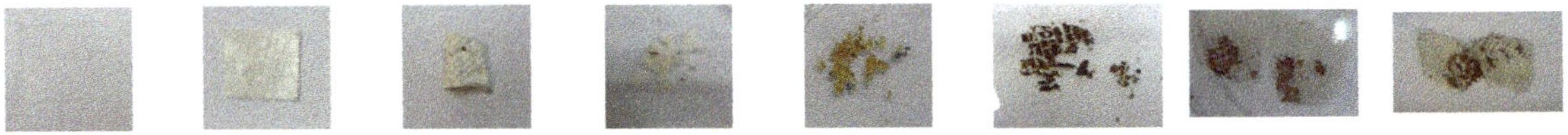							
PLA/T/D43B2.5	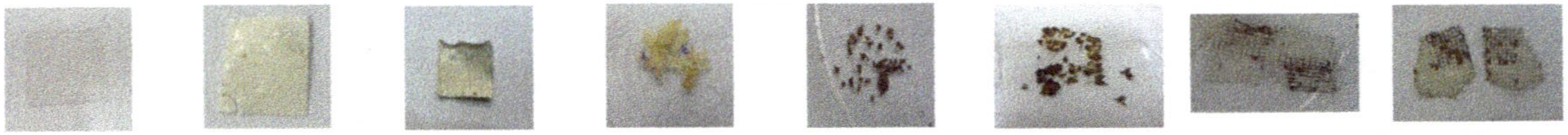							
PLA/T/D43B5	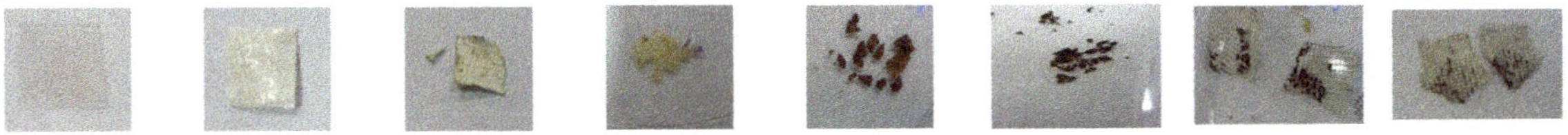							
PLA/D43B2.5	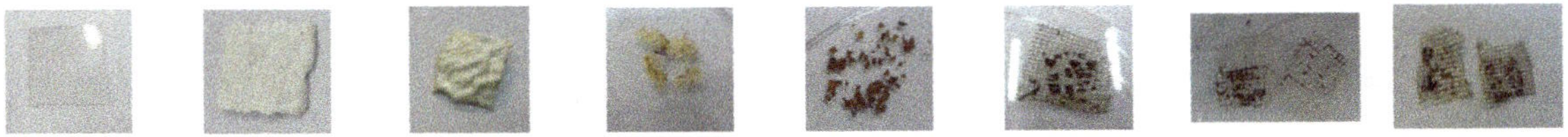							
PLA/D43B5	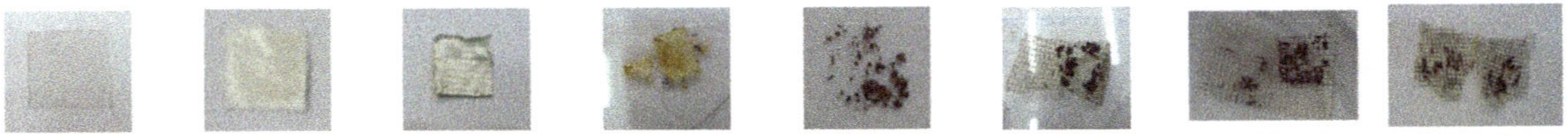							

The use of nanoclays, such as D43B, and thymol as active additive influenced the disintegration rate in compost of PLA since this process is strongly dependent on the hydrophilic/hydrophobic character of the nanocomposite and this parameter changes due to the presence of hydroxyl groups from thymol and the organic modifier of D43B [53]. Hydroxyl groups can contribute to the heterogeneous hydrolysis of PLA by absorbing water from the medium resulting in a noticeable formation of labile bonds in the PLA structure with the consequence of a significantly higher disintegration rate [54]. It was also observed that binary and ternary systems suffered physical breakage with a considerable increase in weight loss (Table 4) after 7 days, showing significant differences in the disintegration profile when comparing neat PLA and nano-biocomposite films. Results at longer times showed that physical degradation progressed with burial time, resulting in the complete disintegration of all the initial samples after 35 days, when the disintegration degree exceeded 90% covering the ISO 20,200 requirements (Table 3).

FTIR analysis of neat PLA and nano-biocomposite films at different times was carried out to evaluate the structural changes produced by the disintegration process in all formulations. Figure 2 shows the FTIR spectra obtained for neat PLA, PLA/T and PLA/T/D43B5 after 0, 7 and 21 days under composting conditions. PLA showed characteristic bands at 1750 cm^{-1} (C = O), 1440 cm^{-1} (CH–CH_3), and 1267 cm^{-1}(C–O–C) as well as three peaks at 1123, 1082 and 1055 cm^{-1} related to the C–C–O groups. After 7 days, the intensity of the three peaks related to the C–C–O groups decreased and after 21 days, these peaks disappeared for all formulations. Similar results were obtained by Fortunati et al. [53] who proposed that the modification of the intensity of peaks related to the C–C–O groups can be associated with the scission of the PLA interchain bonds produced by hydrolysis reactions occurring during disintegration tests. Moreover, the C–O–C stretching vibration at 1267 cm^{-1} was also affected by the depletion of the lactic acid and oligomer molecules caused by microorganisms, leaving highly reactive carboxylate end groups [55]. FTIR results agreed with those achieved for the disintegration weight loss and with the progressive disintegration of all samples with increasing testing time.

Figure 3a shows an example of the DSC thermograms obtained from the first heating scan for PLA, PLA/T and PLA/T/D43B5 films at different composting times. The endothermic peak observed immediately after the T_g at day 0 for all the tested materials corresponds to the enthalpic relaxation process. This effect was related to the aging process of PLA, which was previously observed by other authors [56]. However, the initially amorphous PLA-based materials developed multiple endothermic peaks just after 7 days under composting conditions. In fact, the enthalpic relaxation peak gradually disappeared due to the hydrolysis process at short incubation times. Yang et al. [57] related this behavior to the moisture absorption happening under composting conditions, since water could serve as a plasticizer agent in PLA matrices. These effects could also be related to the well-dispersed nanofiller inside the polymer matrix. Olewnik-Kruszkowska et al. [58] associated the hydrolytic degradation to the decrease in all thermal properties and they confirmed that the highest changes in the T_g value could be related to the dispersion of the nanofiller inside the polymer matrix.

The gradual disintegration suffered when increasing the testing time resulted in the observation of new melting peaks related to the formation of crystalline structures with different perfection degrees in the PLA matrix (Figure 3a). These results were found for all samples and they can be correlated with the observed visual changes, since hydrolysis promotes crystallization in the polymer matrix, resulting in significantly essential changes in the visual appearance of the testing samples and their disintegrability behavior. Similar results were reported by other authors, who suggested that the appearance of multiple melting peaks could be related to the formation of different crystal structures due to the polymer chain scission produced during degradation [59,60].

Figure 3b shows an example of the DSC thermograms recorded during the second heating scan for PLA, PLA/T and PLA/T/D43B5 samples submitted to the disintegration test. It was observed that after 2 days, all PLA-based films showed a significant decrease in T_g. Previous work by our research group showed that this decrease in T_g values was due to the increase in the mobility of the polymer chains as a consequence of the hydrolytic process [55] and the formation of lactic acid oligomers and low molecular weight by-products with a plasticizing effect in the polymer structure and the consequent changes in their visual appearance.

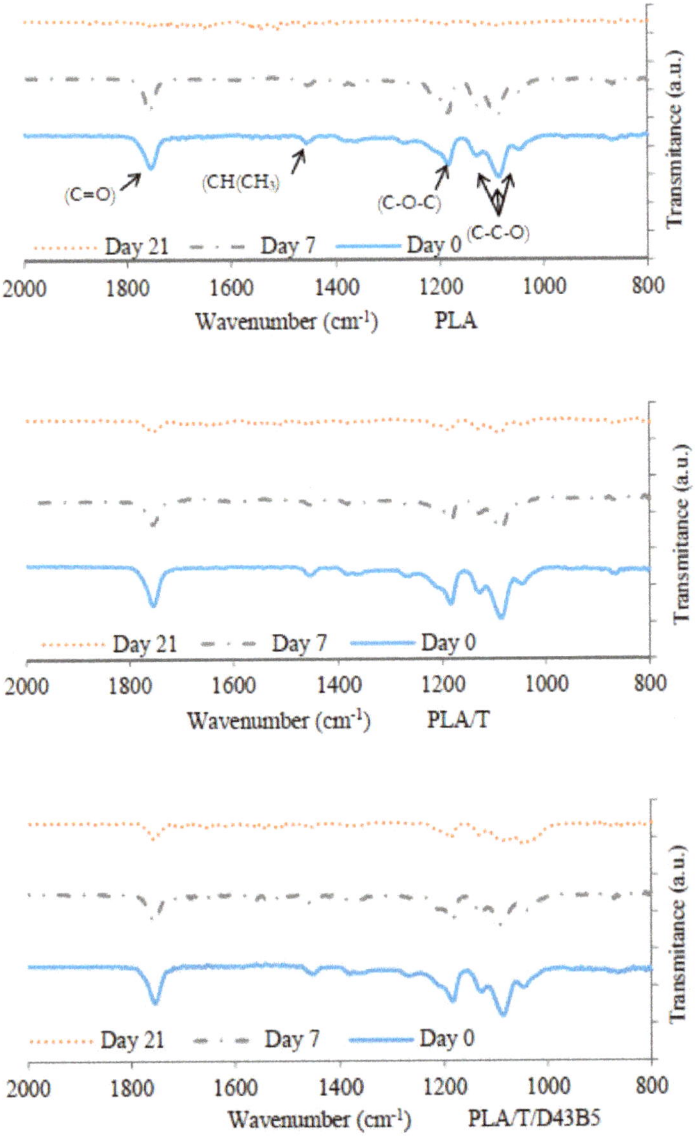

Figure 2. FTIR spectra of PLA, PLA/T, and PLA/T/D43B5 before (0 days) and after different incubation times (7 and 21 days) under composting conditions.

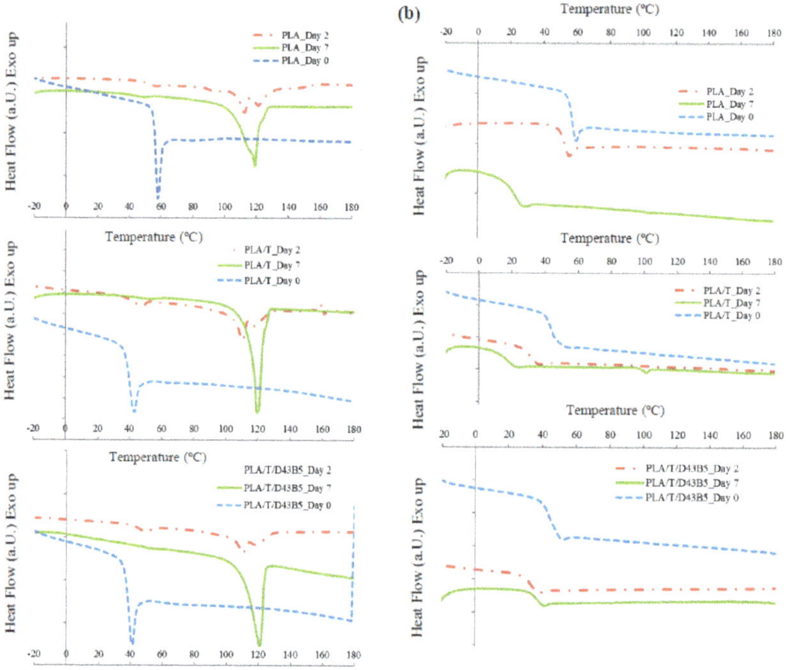

Figure 3. DSC thermograms, 1st heating scan (**a**) and 2nd heating scan (**b**), of PLA-based nano-biocomposite films after different composting times.

4. Conclusions

The incorporation of thymol, as active additive, and D43B, as nano-reinforcing agent, into PLA has shown as an accessible and useful route for the preparation and modification of PLA nano-biocomposite films properties. An improvement in functional properties of PLA-based films was obtained due to the addition of the active additive and the nanoclay resulting in enhanced antimicrobial and antioxidant properties, demonstrating the high potential of the developed formulations for food packaging applications without compromising the inherent biodegradation properties of the PLA matrix. The obtained results suggest the possibility of controlling the release of active additives in the design of active nano-biocomposite films through the incorporation of laminar nanoclays by the decrease in the diffusion of thymol through the polymer matrix by the formation of tortuous paths. The antibacterial activity of these active nanocomposites was proved against two different bacterial strains, showing the PLA/T/D43B2.5 formulation the best results against both *S. aureus 8325-4* (44.3 ± 1.4%) and *E. coli* (47.2 ± 1.2%) at 37 °C and 3 h of incubation time. From a practical point of view, the combination of 8 wt.% of thymol and 2.5 wt.% of D43B added into a commercial PLA matrix showed high potential for the development of new bio-based and biodegradable active packaging films with application in prolonging the shelf-life of fresh food.

Author Contributions: Conceptualization, M.R., M.P., A.J. and M.C.G.; methodology M.R., F.C., L.V. and A.J.M.V.; validation, M.R., E.F., M.C.G. and A.J.; formal analysis, M.R., E.F., A.B., L.V., M.P., A.J.M.V., A.J., J.M.K. and M.C.G.; investigation, M.R., F.C. and A.J.M.V.; resources, M.R., J.M.K., A.J. and M.C.G.; data curation, M.R., A.B., E.F., L.V., A.J.M.V. A.J., and M.C.G.; writing—original draft preparation, M.R.; writing—review and editing, M.R., E.F., M.P., A.B., J.M.K., F.C., L.V., A.J.M.V., A.J. and M.C.G.; supervision, M.R., A.J. and M.C.G. All authors have read and agreed to the published version of the manuscript.

Funding: This research received no external funding.

Acknowledgments: Authors would like to thank Spanish Ministry of Economy and Competitiveness (MAT2017-84909-C2-1-R) and Generalitat Valenciana (IDIFEDER/2018/007) for their support of this research. Marina Ramos would like to thank University of Alicante (Spain) for the UAFPU2011-48539721S predoctoral research grant. AJMV thanks 'Coimbra Chemistry Center and Fundação para a Ciência e Tecnologia (FCT) for the financial support through the programmes UID/QUI/UI0313/2020 and COMPETE.

Conflicts of Interest: The authors declare no conflict of interest.

References

1. Kabir, E.; Kaur, R.; Lee, J.; Kim, K.H.; Kwon, E.E. Prospects of biopolymer technology as an alternative option for non-degradable plastics and sustainable management of plastic wastes. *J. Clen. Prod.* **2020**, *258*, 120536. [CrossRef]
2. Ramos, M.; Beltran, A.; Fortunati, E.; Peltzer, M.; Cristofaro, F.; Visai, L.; Valente, A.J.M.; Jiménez, A.; Kenny, J.M.; Garrigós, M.C. Controlled release of thymol from poly(lactic acid)-based silver nanocomposite films with antibacterial and antioxidant activity. *Antioxidants* **2020**, *9*, 395. [CrossRef] [PubMed]
3. Villegas, C.; Arrieta, M.P.; Rojas, A.; Torres, A.; Faba, S.; Toledo, M.J.; Gutierrez, M.A.; Zavalla, E.; Romero, J.; Galotto, M.J.; et al. PLA/organoclay bionanocomposites impregnated with thymol and cinnamaldehyde by supercritical impregnation for active and sustainable food packaging. *Compos. Part B* **2019**, *176*, 176. [CrossRef]
4. Mayekar, P.C.; Castro-Aguirre, E.; Auras, R.; Selke, S.; Narayan, R. Effect of nano-clay and surfactant on the biodegradation of poly(lactic acid) films. *Polymers* **2020**, *12*, 311. [CrossRef] [PubMed]
5. Souza, P.; Morales, A.; Marin-Morales, M.; Mei, L. PLA and montmorillonite nanocomposites: Properties, biodegradation and potential toxicity. *J. Polym. Environ.* **2013**, *21*, 738–759. [CrossRef]
6. Kalendova, A.; Smotek, J.; Stloukal, P.; Kracalik, M.; Slouf, M.; Laske, S. Transport properties of poly(lactic acid)/clay nanocomposites. *Poly. Eng. Sci.* **2019**, *59*, 2498–2501. [CrossRef]
7. Ramos, M.; Jiménez, A.; Garrigós, M.C. Active nanocomposites in food contact materials. In *Nanoscience in Food and Agriculture 4*; Ranjan, S., Dasgupta, N., Lichtfouse, E., Eds.; Springer International Publishing: Cham, Switzerland, 2017; pp. 1–44.
8. Llanos, J.H.R.; Tadini, C.C. Preparation and characterization of bio-nanocomposite films based on cassava starch or chitosan, reinforced with montmorillonite or bamboo nanofibers. *Int. J. Biol. Macromol.* **2018**, *107*, 371–382. [CrossRef]
9. Dobrucka, R.; Przekop, R. New perspectives in active and intelligent food packaging. *J. Food Process. Preserv.* **2019**, *43*, e14194. [CrossRef]
10. Jahed, E.; Khaledabad, M.A.; Bari, M.R.; Almasi, H. Effect of cellulose and lignocellulose nanofibers on the properties of origanum vulgare ssp. Gracile essential oil-loaded chitosan films. *React. Funct. Polym.* **2017**, *117*, 70–80. [CrossRef]
11. Yildirim, S.; Röcker, B.; Pettersen, M.; Nilsen-Nygaard, J.; Ayhan, Z.; Rutkaite, R.; Radusin, T.; Suminska, P.; Marcos, B.; Coma, V. Active packaging applications for food. *Compr. Rev. Food Sci. Food Saf.* **2017**, *17*, 165–199. [CrossRef]
12. Tornuk, F.; Sagdic, O.; Hancer, M.; Yetim, H. Development of LLDPE based active nanocomposite films with nanoclays impregnated with volatile compounds. *Food Res. Int.* **2018**, *107*, 337–345. [CrossRef] [PubMed]
13. Hammoudi, N.; Ziani Cherif, H.; Borsali, F.; Benmansour, K.; Meghezzi, A. Preparation of active antimicrobial and antifungal alginate-montmorillonite/lemon essential oil nanocomposite films. *Mater. Tech.* **2020**, *35*, 383–394. [CrossRef]
14. Valdes, A.; Mellinas, A.C.; Ramos, M.; Burgos, N.; Jimenez, A.; Garrigos, M.C. Use of herbs, spices and their bioactive compounds in active food packaging. *RSC Adv.* **2015**, *5*, 40324–40335. [CrossRef]
15. Yahyaoui, M.; Gordobil, O.; Herrera Díaz, R.; Abderrabba, M.; Labidi, J. Development of novel antimicrobial films based on poly(lactic acid) and essential oils. *React. Funct. Polym.* **2016**, *109*, 1–8. [CrossRef]
16. Suwanamornlert, P.; Kerddonfag, N.; Sane, A.; Chinsirikul, W.; Zhou, W.; Chonhenchob, V. Poly(lactic acid)/poly(butylene-succinate-co-adipate) (PLA/PBSA) blend films containing thymol as alternative to synthetic preservatives for active packaging of bread. *Food Pack. Shelf Life* **2020**, *25*, 100515. [CrossRef]
17. Cheng, J.; Wang, H.; Kang, S.; Xia, L.; Jiang, S.; Chen, M.; Jiang, S. An active packaging film based on yam starch with eugenol and its application for pork preservation. *Food Hydrocoll.* **2019**, *96*, 546–554. [CrossRef]

18. Li, L.; Wang, H.; Chen, M.; Jiang, S.; Cheng, J.; Li, X.; Zhang, M.; Jiang, S. Gelatin/zein fiber mats encapsulated with resveratrol: Kinetics, antibacterial activity and application for pork preservation. *Food Hydrocoll.* **2020**, *101*, 105577. [CrossRef]
19. Wang, H.; Hao, L.; Wang, P.; Chen, M.; Jiang, S.; Jiang, S. Release kinetics and antibacterial activity of curcumin loaded zein fibers. *Food Hydrocoll.* **2017**, *63*, 437–446. [CrossRef]
20. Liu, J.; Wang, H.; Wang, P.; Guo, M.; Jiang, S.; Li, X.; Jiang, S. Films based on κ-carrageenan incorporated with curcumin for freshness monitoring. *Food Hydrocoll.* **2018**, *83*, 134–142. [CrossRef]
21. Ramos, M.; Jiménez, A.; Peltzer, M.; Garrigós, M.C. Characterization and antimicrobial activity studies of polypropylene films with carvacrol and thymol for active packaging. *J. Food Eng.* **2012**, *109*, 513–519. [CrossRef]
22. Tawakkal, I.S.M.A.; Cran, M.J.; Bigger, S.W. Release of thymol from poly(lactic acid)-based antimicrobial films containing kenaf fibres as natural filler. *LWT Food Sci. Technol.* **2016**, *66*, 629–637. [CrossRef]
23. Galotto, M.J.; López De Dicastillo, C.; Torres, A.; Guarda, A. Chpater 45. Thymol: Use in antimicrobial packaging. In *Antimicrob Food Pack*; Barros-Velázquez, J., Ed.; Academic Press: San Diego, CA, USA, 2016; pp. 553–562.
24. Davoodi, M.; Kavoosi, G.; Shakeri, R. Preparation and characterization of potato starch-thymol dispersion and film as potential antioxidant and antibacterial materials. *Int. J. Biol. Macromol.* **2017**, *104*, 173–179. [CrossRef] [PubMed]
25. Chi, H.; Song, S.; Luo, M.; Zhang, C.; Li, W.; Li, L.; Qin, Y. Effect of PLA nanocomposite films containing bergamot essential oil, TiO_2 nanoparticles, and Ag nanoparticles on shelf life of mangoes. *Sci. Hortic.* **2019**, *249*, 192–198. [CrossRef]
26. Heydari-Majd, M.; Ghanbarzadeh, B.; Shahidi-Noghabi, M.; Najafi, M.A.; Hosseini, M. A new active nanocomposite film based on PLA/ZnO nanoparticle/essential oils for the preservation of refrigerated otolithes ruber fillets. *Food Pack. Shelf Life* **2019**, *19*, 94–103. [CrossRef]
27. Luzi, F.; Dominici, F.; Armentano, I.; Fortunati, E.; Burgos, N.; Fiori, S.; Jiménez, A.; Kenny, J.M.; Torre, L. Combined effect of cellulose nanocrystals, carvacrol and oligomeric lactic acid in PLA_PHB polymeric films. *Carbohydr. Polym.* **2019**, *223*, 115131. [CrossRef] [PubMed]
28. Scaffaro, R.; Maio, A.; Gulino, E.; Morreale, M.; Mantia, F. The effects of nanoclay on the mechanical properties, carvacrol release and degradation of a PLA/PBAT blend. *Materials* **2020**, *13*, 983. [CrossRef]
29. Shebi, A.; Lisa, S. Evaluation of biocompatibility and bactericidal activity of hierarchically porous PLA/TiO_2 nanocomposite films fabricated by breath-figure method. *Mater. Chem. Phys.* **2019**, *230*, 308–318. [CrossRef]
30. Ramos, M.; Jiménez, A.; Peltzer, M.; Garrigós, M.C. Development of novel nano-biocomposite antioxidant films based on poly (lactic acid) and thymol for active packaging. *Food Chem.* **2014**, *162*, 149–155. [CrossRef]
31. Plastic materials and articles intended to come into contact with food. In *Commission Regulation/(EU)/N°-10/2011*; European Commission: Brussels, Belgium, 2011; pp. 2–89.
32. UNE-EN_13130-1. Materials and Articles in Contact with Foodstuffs-Plastics Substances Subject to Limitation-Part 1: Guide to Test Methods for the Specific Migration of Substances from Plastics to Foods and Food Simulants and the Determination of Substances in Plastics and the Selection of Conditions of Exposure to Food Simulants. *Eur. Comm. Stand.* **2005**.
33. Silva, A.S.; Cruz Freire, J.M.; Sendón, R.; Franz, R.; Paseiro Losada, P. Migration and diffusion of diphenylbutadiene from packages into foods. *J. Agric. Food Chem.* **2009**, *57*, 10225–10230. [CrossRef]
34. Crank, J. *The Mathematics of Diffusion*, 2nd ed.; Oxford University Press: Oxford, UK, 1975; p. 414.
35. ISO, UNEEN. Plastics-Determination of the Degree of Disintegration of Plastic Materials under Simulated Composting Conditions in a Laboratory-Scale Test. UNE-EN_20200. 2015. Available online: https://www.iso.org/standard/63367.html (accessed on 19 August 2020).
36. Sanchez-Garcia, M.D.; Ocio, M.J.; Gimenez, E.; Lagaron, J.M. Novel polycaprolactone nanocomposites containing thymol of interest in antimicrobial film and coating applications. *J. Plast. Film Sheeting* **2008**, *24*, 239–251. [CrossRef]
37. Campos-Requena, V.H.; Rivas, B.L.; Pérez, M.A.; Figueroa, C.R.; Sanfuentes, E.A. The synergistic antimicrobial effect of carvacrol and thymol in clay/polymer nanocomposite films over strawberry gray mold. *LWT Food Sci. Technol.* **2015**, *64*, 390–396. [CrossRef]

38. Mastelic, J.; Jerkovic, I.; Blazevic, I.; Poljak-Blazi, M.; Borovic, S.; Ivancic-Bace, I.; Smrecki, V.; Zarkovic, N.; Brcic-Kostic, K.; Vikic-Topic, D.; et al. Comparative study on the antioxidant and biological activities of carvacrol, thymol, and eugenol derivatives. *J. Agric. Food Chem.* **2008**, *56*, 3989–3996. [CrossRef] [PubMed]
39. Quiroga, P.R.; Asensio, C.M.; Nepote, V. Antioxidant effects of the monoterpenes carvacrol, thymol and sabinene hydrate on chemical and sensory stability of roasted sunflower seeds. *J. Sci. Food Agric.* **2015**, *95*, 471–479. [CrossRef] [PubMed]
40. Jamshidian, M.; Tehrany, E.A.; Desobry, S. Release of synthetic phenolic antioxidants from extruded poly lactic acid (PLA) film. *Food Control.* **2012**, *28*, 445–455. [CrossRef]
41. Chung, D.; Papadakis, S.E.; Yam, K.L. Simple models for assessing migration from food-packaging films. *Food Addit. Contam.* **2002**, *19*, 611–617. [CrossRef] [PubMed]
42. Mascheroni, E.; Guillard, V.; Nalin, F.; Mora, L.; Piergiovanni, L. Diffusivity of propolis compounds in polylactic acid polymer for the development of anti-microbial packaging films. *J. Food Eng.* **2010**, *98*, 294–301. [CrossRef]
43. Mhlanga, N.; Ray, S.S. Kinetic models for the release of the anticancer drug doxorubicin from biodegradable polylactide/metal oxide-based hybrids. *Int. J. Biol. Macromol.* **2015**, *72*, 1301–1307. [CrossRef] [PubMed]
44. Samsudin, H.; Soto-Valdez, H.; Auras, R. Poly(lactic acid) film incorporated with marigold flower extract (*tagetes erecta*) intended for fatty-food application. *Food Control.* **2014**, *46*, 55–66. [CrossRef]
45. Ramos, M.; Beltrán, A.; Peltzer, M.; Valente, A.J.M.; Garrigós, M.C. Release and antioxidant activity of carvacrol and thymol from polypropylene active packaging films. *LWT Food Sci. Technol.* **2014**, *58*, 470–477. [CrossRef]
46. Torres, A.; Romero, J.; Macan, A.; Guarda, A.; Galotto, M.J. Near critical and supercritical impregnation and kinetic release of thymol in LLDPE films used for food packaging. *J. Supercrit. Fluids* **2014**, *85*, 41–48. [CrossRef]
47. Liu, D.; Li, H.; Jiang, L.; Chuan, Y.; Yuan, M.; Chen, H. Characterization of active packaging films made from poly(lactic acid)/poly(trimethylene carbonate) incorporated with oregano essential oil. *Molecules* **2016**, *21*, 695. [CrossRef] [PubMed]
48. Rai, M.; Paralikar, P.; Jogee, P.; Agarkar, G.; Ingle, A.P.; Derita, M.; Zacchino, S. Synergistic antimicrobial potential of essential oils in combination with nanoparticles: Emerging trends and future perspectives. *Int. J. Pharm.* **2017**, *519*, 67–78. [CrossRef]
49. de Azeredo, H.M.C. Antimicrobial nanostructures in food packaging. *Trends Food Sci. Technol.* **2013**, *30*, 56–69. [CrossRef]
50. Hong, S.-I.; Rhim, J.-W. Antimicrobial activity of organically modified nano-clays. *J. Nanosci. Nanotechnol.* **2008**, *8*, 5818–5824. [CrossRef]
51. Yang, W.; Fortunati, E.; Dominici, F.; Giovanale, G.; Mazzaglia, A.; Balestra, G.M.; Kenny, J.M.; Puglia, D. Effect of cellulose and lignin on disintegration, antimicrobial and antioxidant properties of PLA active films. *Int. J. Biol. Macromol.* **2016**, *89*, 360–368. [CrossRef] [PubMed]
52. Su, S.; Kopitzky, R.; Tolga, S.; Kabasci, S. Polylactide (PLA) and its blends with poly(butylene succinate) (PBS): A brief review. *Polymers* **2019**, *11*, 1193. [CrossRef]
53. Fortunati, E.; Luzi, F.; Puglia, D.; Dominici, F.; Santulli, C.; Kenny, J.M.; Torre, L. Investigation of thermo-mechanical, chemical and degradative properties of PLA-limonene films reinforced with cellulose nanocrystals extracted from phormium tenax leaves. *Eur. Polym. J.* **2014**, *56*, 77–91. [CrossRef]
54. Proikakis, C.S.; Mamouzelos, N.J.; Tarantili, P.A.; Andreopoulos, A.G. Swelling and hydrolytic degradation of poly(D,L-lactic acid) in aqueous solutions. *Polym. Degrad. Stab.* **2006**, *91*, 614–619. [CrossRef]
55. Ramos, M.; Fortunati, E.; Peltzer, M.; Jimenez, A.; Kenny, J.M.; Garrigós, M.C. Characterization and disintegrability under composting conditions of PLA-based nanocomposite films with thymol and silver nanoparticles. *Polym. Degrad. Stab.* **2016**, *132*, 2–10. [CrossRef]
56. Burgos, N.; Martino, V.P.; Jiménez, A. Characterization and ageing study of poly(lactic acid) films plasticized with oligomeric lactic acid. *Polym. Degrad. Stab.* **2013**, *98*, 651–658. [CrossRef]
57. Yang, W.; Fortunati, E.; Dominici, F.; Kenny, J.M.; Puglia, D. Effect of processing conditions and lignin content on thermal, mechanical and degradative behavior of lignin nanoparticles/polylactic (acid) bionanocomposites prepared by melt extrusion and solvent casting. *Eur. Polym. J.* **2015**, *71*, 126–139. [CrossRef]

58. Olewnik-Kruszkowska, E.; Kasperska, P.; Koter, I. Effect of poly(ε-caprolactone) as plasticizer on the properties of composites based on polylactide during hydrolytic degradation. *React. Funct. Polym.* **2016**, *103*, 99–107. [CrossRef]
59. Gorrasi, G.; Pantani, R. Effect of PLA grades and morphologies on hydrolytic degradation at composting temperature: Assessment of structural modification and kinetic parameters. *Polym. Degrad. Stab.* **2013**, *98*, 1006–1014. [CrossRef]
60. Sedničková, M.; Pekařová, S.; Kucharczyk, P.; Bočkaj, J.; Janigová, I.; Kleinová, A.; Jochec-Mošková, D.; Omaníková, L.; Perd'ochová, D.; Koutný, M.; et al. Changes of physical properties of PLA-based blends during early stage of biodegradation in compost. *Int. J. Biol. Macromol.* **2018**, *113*, 434–442. [CrossRef] [PubMed]

polymers

MDPI

Article

Design and Preparation of Polysulfide Flexible Polymers Based on Cottonseed Oil and Its Derivatives

Yurong Chen [1,2,3], Yanxia Liu [1,2,3], Yidan Chen [2,3], Yagang Zhang [1,2,3,*] and Xingjie Zan [3]

1 School of Materials and Energy, University of Electronic Science and Technology of China, Chengdu 611731, China; chenyurong@cms.net.cn (Y.C.); liuyanxia@ms.xjb.ac.cn (Y.L.)
2 Department of chemical and environmental engineering, Xinjiang Institute of Engineering, Urumqi 830026, China; chenyd2020@ms.xjb.ac.cn
3 Xinjiang Technical Institute of Physics and Chemistry, Chinese Academy of Sciences, Urumqi 830011, China; zanxj@ms.xjb.ac.cn
* Correspondence: ygzhang@ms.xjb.ac.cn or ygzhang@uestc.edu.cn; Tel.: +86-1812-930-7169

Received: 21 July 2020; Accepted: 17 August 2020; Published: 19 August 2020

Abstract: Polysulfide-derived polymers with a controllable density and mechanical strength were designed and prepared successfully using bio-based cottonseed oil (CO) and its derivatives, including fatty acid of cottonseed oil (COF) and sodium soap of cottonseed oil (COS). The reaction features of CO, COF and COS for polysulfide polymers were investigated and compared. Based on the free radical addition mechanism, COF reacts with sulfur to generate serials of polysulfide-derived polymers. COF strongly influences the density and tensile strength of these polymer composites. Whereas COS was not involved in the reaction with sulfur, as a filler, it could increase the density and tensile strength of polysulfide-derived polymers. Moreover, the results showed that these samples had an excellent reprocessability and recyclability. These polysulfide-based polymers, with an adjustable density and mechanical strength based on CO and derivatives, could have potential applications as bio-based functional supplementary additives.

Keywords: polysulfide-derived polymers; cottonseed oil; fatty acid of cottonseed oil; sodium soap of cottonseed oil

1. Introduction

Elemental sulfur derives mainly from industrial process such as petroleum refining. It is currently recognized as a vital basic chemical stock for rubber production, chemical fertilizer, antimicrobial agents and chemical dyes [1–5]. Although elemental sulfur has been applied in many fields, it is extensively stockpiled every year because production far surpasses demand [6]. Since 1,3-diisopropenyl benzene was found to stabilize polysulfide chains in inverse vulcanization and the resulting thermoplastic copolymers exhibited great processability and promising potential for applications in the electrical industry as cathodes [7], various functional sulfur-containing polymers and sulfur–organic copolymers synthesized by using different cross-linkers have been reported [8–10]. These sulfur-containing polymers and sulfur–organic copolymers have drawn considerable attention due to their application potentials for solid electrodes [11,12], camera lenses and medium infrared ranges [13], as well as heavy metal remediation [14]. In order to make the best of mass-produced sulfur and meet the demands of sustainable development, it would be desirable to seek renewable and sustainable bio-based materials to prepare value-added materials with sulfur. Bio-based materials feature many advantages such as their inexpensiveness, accessibility and availability in large quantities. The design and synthesis of bio-based functional materials also fits in with and benefits green chemistry and sustainable engineering, aimed at making the best of natural resources with more environmentally benign and eco-friendly approaches [15–19].

Vegetable oils, such as canola oil, sunflower oil and linseed oil [20], have been explored for the preparation of functional sulfur–organic copolymers with the method of inverse vulcanization. Cotton, as a vital kind of commercial crop, has a wide range of planting areas worldwide and a higher yield compared with other commercial crops. Therefore, cottonseed oil has more advantages in view of yield and price than other vegetable oils. Currently, the explorations for cottonseed oil are concentrated on the development of biodiesel with lower environmental pollution, lower production costs and greater safety [21–24], whereas polysulfide-derived polymers based on cottonseed oil and derivatives are rarely reported on. It is highly desirable to design and prepare value-added functional polymer composites with cottonseed oil and its derivatives.

In our previous work [25], cottonseed oil was used as a renewable cross-linker to react with industrial byproduct sulfur and the resulting sulfur-containing plant rubber polymers were prepared successfully. These plant rubber polymers could remove mercury ions from aqueous solution as bio-based absorbents. Traditional rubber materials could serve as supplementary additives for improving the mechanical properties or densities of polymers [26–29]. It would be interesting and ideal if the mechanical property and rigidity of these sulfur-containing plant rubber polymers were tunable. In the work reported here, we explored the possibility of preparing value-added sulfur-containing functional polymer with cottonseed oil and its derivatives with goal of achieving controllable densities and adjustable mechanical strength.

In this work, though the inverse vulcanization process, mass-produced sulfur, renewable cottonseed oil (CO) and its derivatives were taken advantage of, and a series of novel polysulfide-based polymers with controllable densities and adjustable mechanical strength were successfully prepared. Cottonseed oil derivatives involved in this work included fatty acid of cottonseed oil (COF) and sodium soap of cottonseed oil (COS). These polysulfide-derived polymers could have potential applications as bio-based functional supplementary additives.

2. Materials and Methods

2.1. Materials

Cottonseed oil (CO, food grade), fatty acid of cottonseed oil (COF) and sodium soap of cottonseed oil (COS) from cottonseed oil were obtained from Shihezi Kanglong Oil Industry and Trade Company (Shihezi, China). Sulfur (S, powder, ≥99.5%,) was purchased from Tianjin Baishi Chemical Industry Co. Ltd. (Tianjin, China). CO, COF, COS and their mixture, including CO/COF, CO/COS, as well as CO/COF/COS, is referred to as COX collectively for ease of subsequent discussion. Moreover, except for CO, residual raw materials were also called COY in order to differentiate them from COX.

2.2. Preparation of Polysulfide-Derived Polymers

About 10.0 g of elemental sulfur powder (S) was added into a 100 mL vial equipped with a magnetic stir bar and then melted while stirring at 150 °C. A certain amount of COX was then added to the above molten liquid, while stirring and heating were continued to ensure efficient mixing and reactions between reactants. After a certain period, the mixture was cooled to room temperature and then the resulting polysulfide-derived polymers were obtained. The related details can be found in Table 1.

2.3. Characterization

The morphological analysis of the prepared samples was performed by a field emission scanning electron microscopy (FE-SEM, SUPRA 55vp, ZEISS, Oberkochen, Germany) with an Oxford detector, operating with 2.00 kV electron beams. Infrared (IR) spectra were recorded on a Fourier Transform Infrared Spectrometer (VERTEX-70, Bruker, Karlsruhe, Germany) using the ATR (Attenuated Total Reflectance) method with a wave number ranging from 400 to 4000 cm^{-1}. Gas Chromatography–Mass Spectrometry (GC–MS) was acquired on a Headspace injection gas chromatograph mass spectrometer

(7890B GC/7697A/5977B MSD, KEYSIGHT, Santa Rosa, CA, America) using a Rtx-5MS column (30 m long × 0.25 mm thickness × 0.25 μm ID), with an injection temperature of 130 °C, column temperature of 270 °C, gas flow rate of 24.0 mL·min^{-1}, and electron ionization used to obtained nominal masses. Density analysis was carried out on an electronic densitometer (XF-120MD, Xiongfa, Xiamen, China) with testing samples tailored in the cubic dimension of 10 mm × 10 mm × 10 mm. Mechanical testing was carried out on a microcomputer-controlled electronic tensile testing machine (C43-104, MTS, Rochester, MN, America) according to the national standard GB/T 528-2009 with dumbbell-shaped splines at an elongation rate of 500 mm·min^{-1} and the test length and thickness of splines were 20.0 ± 0.5 mm and 2.0 ± 0.2 mm, respectively.

Table 1. The preparing conditions of the resulting polysulfide-derived polymers.

Samples	Components	Mass Ratio of Reactants	Temperature (°C)	Time [1] (min)
SCO	S:CO	0.5, 0.75, 1, 1.5, 2:1	150	25–30
SCOF	S:COF	0.5, 0.75, 1, 1.5, 2:1	150	25–30
SCOS	S:COS	0.5, 0.75, 1, 1.5, 2:1	150	20–25
SCO/COF	CO:COF S:CO/COF	1:0.5, 0.75, 1.0 0.5, 0.75, 1, 1.5, 2:1	150	15–20
SCO/COS	CO:COS S:CO/COS	1:0.5, 0.75, 1, 1.5, 2 0.5, 0.75, 1, 1.5, 2:1	150	15–20
SCO/COF/COS	COF:COS CO:COF/COS S:CO/COF/COS	0.5, 1, 2:1 1:0.5, 0.75, 1, 1.5, 2 0.5, 0.75, 1, 1.5, 2:1	150	15–20

Time [1] refers to the period from the moment that all reactants were fully mixed together to the moment that the reactions were finished, which does not include the period of cooling samples to room temperature.

2.4. Self-Healing Experiments

To study the reprocessability and recyclability of the related samples, self-healing experiments were performed by cutting samples into pieces and putting the fragments together for remolding under 10 MPa at 160 °C for 10 min using a thermo-compressor (R3212, Qien, Zhengzhou, China). The ratio of the tensile strength of the healed samples to those of the original one was determined as the recovery ratio to measure the reprocessability and recyclability of samples.

3. Results and Discussion

3.1. Synthesis and Characterization

To investigate whether elemental sulfur could react effectively with cottonseed oil derivatives including fatty acid of cottonseed oil (COF) and sodium soap of cottonseed oil (COS), the resulting polysulfide-derived polymers were analyzed. As depicted in Figure 1a–e, when elemental sulfur reacted with CO, COF and COS, respectively, the resulting polysulfide-derived polymers SCO and SCOS appeared as brown elastic bulk and tanned plastic bulk, respectively, while SCOF appeared as a black viscous fluid. Moreover, the tanned plastic SCOS bulk could also be kneaded into a ring-shaped or U-shaped object at room temperature, which implies that SCOS is quite flexible and features excellent processability compared with SCO and SCOF. Figure 1f,g displays the surface morphology of SCO and SCOS under a scanning electron microscope. There were large numbers of fragments on the surface of SCO, whereas the surface of SCOS was occupied by a sticky object conjoined with many small particles. Furthermore, an FT-IR spectrogram (Figure 2) was also used to analyze the structural changes in SCO and SCOF, as well as SCOS. Compared with raw materials CO, COF and COS, the prepared SCO and SCOF both showed that the moderate-strength peak at 3009 cm^{-1} disappeared (with the exception of SCOS). The absence of the peak was attributed to the vanishing of =C–H bond stretching vibration in cottonseed oil and fatty acid of cottonseed oil [4]. Meanwhile, according to Table 2, the relative intensity of the peak at 3009 cm^{-1} in SCO and SCOF samples decreased by about 90% compared with that of raw materials CO and COF, whereas relative intensity of the peak at 3009 cm^{-1} in SCOS samples had no obvious changes compared with that of the raw material COS. The quantitative analysis of the FT-IR spectrogram further indicated that there existed an obvious reaction between raw materials CO and COF and products SCO and SCOF. The macroscopic and microscopic differences between samples as well the as differences in the FT-IR spectrogram implied differences in chemical reactivity between sulfur and different COX. According to the previous work [20], under a high temperature, elemental sulfur could initiate ring-opening polymerization and further react with the unsaturated double bonds in plant oils and their derivatives. The essence of the reaction was that, above the melting point temperature, 120 °C, sulfur was first melted and then heated up further to a higher temperature to generate sulfur free radicals to further react with C=C bonds in plant oils by the free radical addition mechanism.

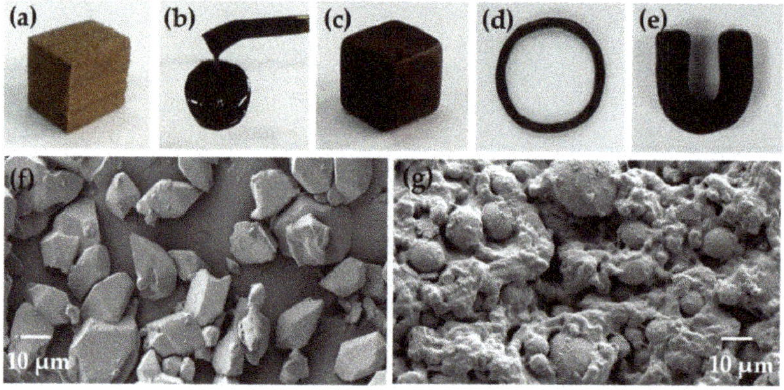

Figure 1. The surface morphology and plasticity of real samples. (**a**,**b**) Digital photos of sulfur cottonseed oil (SCO) and sulfur fatty acid of cottonseed oil (SCOF), respectively; (**c**–**e**) digital images of sulfur–sodium soap of cottonseed oil (SCOS): (**c**) cubic SCOS, (**d**) ring-shaped SCOS and (**e**) U-shaped SCOS; (**f**,**g**) SEM images of SCO and SCOS, respectively (m_S: m_{COX} = 1:1).

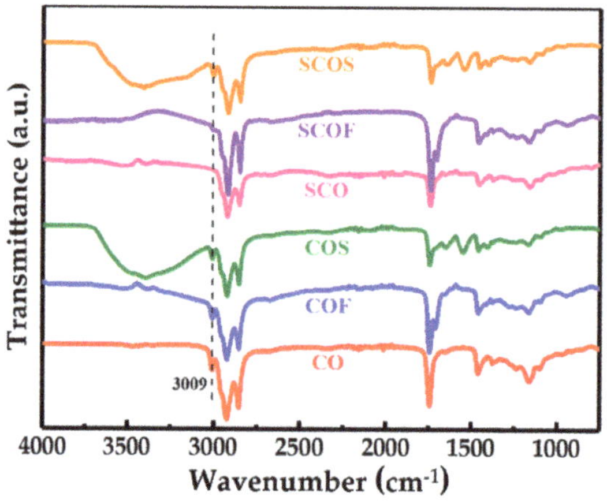

Figure 2. FT-IR spectrogram of the related samples (m_S: m_{COX} = 1:1).

Table 2. Quantitative data for FT-IR spectrogram of the related samples (m_S: m_{COX} = 1:1).

Samples	Peak (cm^{-1})	Relative Intensity (%)
CO	3009	33.82
COF	3009	45.22
COS	3009	47.81
SCO	3009	2.09
SCOF	3009	3.98
SCOS	3009	47.08

GC–MS analysis data (Table 3) show that cottonseed oil contains different fatty acid components, including both saturated and unsaturated fatty acids. There are a large number of C=C bonds derived from unsaturated fatty acid, including linoleic acid and oleic acid. These C=C bonds are the key motifs and act as active cross-linking sites with sulfur to generate cross-linking elastic SCO (Figure 3). This also explains why the =C–H bond stretching vibration in the FT-IR spectra of SCO disappeared compared with raw material CO. The results show that the content of unsaturated fatty acid was about 75% in COF. Theoretically, COF could react with sulfur to generate elastic plant rubber in a similar manner to SCO. However, the resulting SCOF was a viscous fluid instead of presenting solid-state elasticity. This could be due to the fact that the vanishing of glycerinum decreased the cross-linking effect of unsaturated fatty acid in COF, and the obtained SCOF had a low cross-linking degree (Figure 3). This could also explain why the FT-IR spectra of SCO and SCOF were similar but the appearance and character of them were different. Sodium soap based on unsaturated fatty acid was the dominant component of COS, which mainly included sodium linoleate and sodium oleate. In theory, COS could react with sulfur to generate an elasticity similar to SCO. However, the resulting SCOS appeared as a plastic bulk. Moreover, compared with the FT-IR spectrogram of raw material COS, the peak at 3009 cm^{-1} still existed in SCOS, which may imply that there was no reaction between sulfur and COS. This was due to the high melting point of sodium linoleate and sodium oleate, which exceeded 190 °C. Therefore, these sodium soaps were not effectively involved in the reaction with sulfur at 150 °C. Instead, they were mingled with polysulfide fragments as granular padding (Figure 3). The observation of the chemical reactivity with different cottonseed oils and their derivatives was consistent with the above analysis of the SEM images of SCOS (Figure 1f,g).

Table 3. Representative methyl esterification products of cottonseed oil by Gas Chromatography–Mass Spectrometry (GC–MS) analysis.

Fatty Acid Methyl Ester	Relative Content (%)
Methyl tetradecanoate	0.31
Methyl palmitate	16.70
Methyl stearate	1.45
Methyl oleate	15.25
Methyl linoleate	60.61

3.2. *Density Analysis*

As depicted in Figure 4a, when the mass ratio of sulfur to COX was 1.0, the density of SCO/COF as well as SCO/COF/COS serial samples (while the mass ratio of COF to COS was 2.0) decreased with the increase in the mass ratio of COF or COF/COS to CO and was lower than that of SCO samples. These results imply that the introduction of COF could decrease the density of the prepared polymer composites. On the other hand, the density of SCO/COS and SCO/COF/COS serial samples (while the mass ratio of COF to COS was lower than 2.0) increased significantly with the rising mass ratio of COS or COF/COS to CO and, noticeably, the density of these samples were higher than that of SCO serial samples, which indicated that the introduction of COS could increase the density of the resulting polymer composites. Moreover, Figure 4b reveals that the density of samples gradually increased with the increase in the mass ratio of sulfur to COX, which implies that increasing the sulfur content is helpful for elevating the density of samples to some extent.

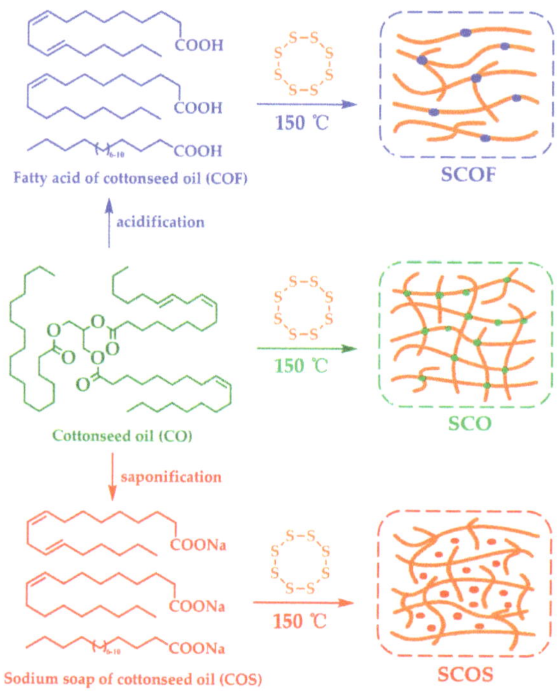

Figure 3. Procedures for the preparation of SCO, SCOF and SCOS, respectively.

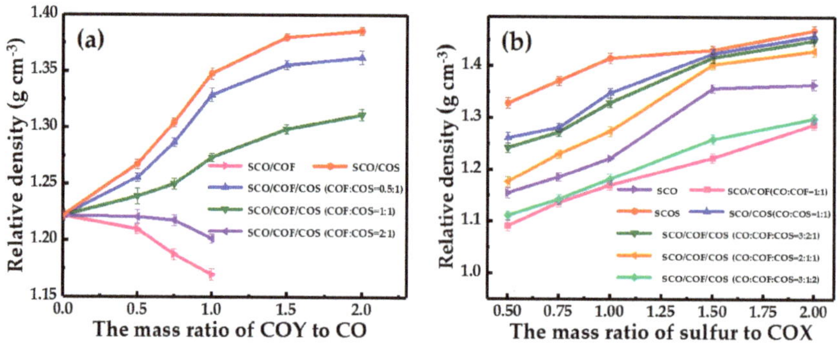

Figure 4. Relative density of the prepared samples. (**a**) Relative density variation in samples versus the mass ratio of COY to CO (m_S/m_{COX} = 1); (**b**) relative density variation in samples versus the mass ratio of sulfur to CO, COF, COS and their mixture, including CO/COF, CO/COS, as well as CO/COF/COS (COX).

3.3. Mechanical Strength Analysis

Figure 5a shows that when the mass ratio of sulfur to COX was 1.0, the tensile strength of SCO/COF serial samples decreased with the increasing mass ratio of COF to CO. Similarly, the tensile strength of SCO/COF/COS serial samples decreased with the increasing mass ratio of COF/COS to CO, while the mass ratio of COF to COS was 2.0. Meanwhile, the tensile strength of both SCO/COF and SCO/COF/COS serial samples were lower than that of SCO serial samples, which implies that the introduction of COF could decrease the mechanical strength of samples. However, when the mass ratio of COY to CO was lower than 1.0, the tensile strength of SCO/COS serial samples increased significantly with the increase in the mass ratio of COS to CO. The tensile strength of SCO/COF/COS serial samples also increased with the increasing mass ratio of COF/COS to CO, while the mass ratio of COF to COS was lower than 2.0. Most importantly, the tensile strength of these samples was higher than that of SCO serial samples, which reveals that the introduction of COS could enhance the mechanical strength of samples effectively. However, when the mass ratio of COY to CO was higher than 1.0, the tensile strength of SCO/COS and SCO/COF/COS serial samples (the mass ratio of COF to COS was lower than 2.0) started to decline, which was due to the decreasing cross-linking density resulting from the lessening of the content of CO. Figure 5b demonstrates that the tensile strength of samples initially increased with the increase in the mass ratio of sulfur to COX and gradually reached the maximum when the mass ratio of sulfur to COX was 1.0 and then decreased dramatically. This could be due to local stress concentration on the surface and inside the samples [30] because of the increase in the content of sulfur.

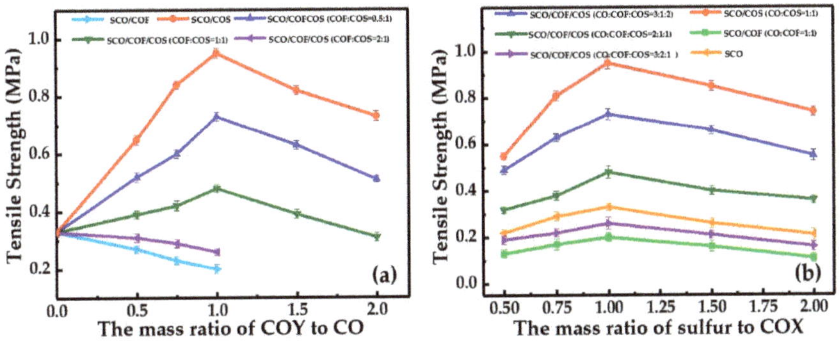

Figure 5. Tensile strength of the prepared samples. (**a**) Tensile strength variation in samples versus the mass ratio of COY to CO (m_S/m_{COX} = 1); (**b**) tensile strength variation in samples versus the mass ratio of sulfur to COX.

3.4. Reprocessability and Recyclability

To further study the reprocessability and recyclability of the prepared samples, SCO/COS serial samples with a higher density and tensile strength were chosen as representative samples. As shown in Figure 6a, SCO/COS serial samples can be remolded into coherent and smooth dumbbell-shaped splines when they are cut into small pieces and after hot pressing. Figure 6b,c show the tensile strength variation in SCO/COS serial samples versus the mass ratio of reactants after multiple reprocesses. The tensile strength of SCO/COS serial samples exhibited a recovery ratio above 90% after first reprocessing and the secondary recovery rate was above 85%, which was due to a large number of reversible disulfide bonds in the samples. These results demonstrate that the reversible cross-linking made the samples capable of reprocessing and recycling [31]. However, the results show that the tensile strength of these materials decreased significantly after more than two cycles. Our hypothesis was that the cross-linked sulfur–sulfur bonds would degrade after more than two cycles and simple hot pressing would not be efficient enough to aid in recovering and rebuilding those damaged bonds. These are actually the disadvantages of this type of material and it is important to find a way to solve these problems.

Figure 6. Reprocessability and recyclability of SCO/COS serial samples. (**a**) Digital photos of thermal reprocessing ability of SCO/COS serial samples; (**b**) tensile strength variation in SCO/COS serial samples versus the mass ratio of COY to CO after multiple reprocesses; (**c**) tensile strength variation in SCO/COS serial samples versus the mass ratio of sulfur to COX after multiple reprocesses.

4. Conclusions

A series of polysulfide-derived polymers with a controllable density and mechanical strength were prepared successfully based on cottonseed oil (CO) and its derivatives, including fatty acid of cottonseed oil (COF) and sodium soap of cottonseed oil (COS). Based on the free radical addition mechanism, which is similar to the reaction mechanism of SCO, COF reacted with sulfur generates serial samples containing COF. COF can decrease the density and tensile strength of polysulfide-based polymers, whereas COS was not effectively involved in the reaction with sulfur due to the high melting point of sodium linoleate and sodium oleate. Even so, COS could act as a padding component, which could increase the density and tensile strength of polysulfide-derived polymers. The results demonstrated that the prepared polymer composites had an excellent reprocessability and recyclability, attributed to the large number of reversible disulfide bonds formed in the formation of plant rubber. These polysulfide-derived polymers with a controllable density and mechanical strength, based on CO and derivatives, could have potential applications as bio-based functional supplementary additives.

Author Contributions: Conceptualization, Y.Z. and Y.C. (Yurong Chen); methodology, Y.C. (Yurong Chen) and Y.L.; software, Y.L. and Y.C. (Yurong Chen); validation, Y.L and Y.C. (Yurong Chen); formal analysis, Y.Z. and X.Z.; investigation, Y.C. (Yurong Chen), L.Y. and Y.C. (Yidan Chen); resources, Y.Z. and X.Z.; data curation, Y.C. (Yurong Chen) and Y.C. (Yidan Chen); writing—original draft preparation, Y.C. (Yurong Chen); writing—review and editing, Y.Z.; visualization, Y.C. (Yurong Chen) and Y.L.; supervision, Y.Z. and X.Z; project admini-stration, Y.Z.; funding acquisition, Y.Z. All authors contributed substantially to the work reported. All authors have read and agreed to the published version of the manuscript.

Funding: This research was financially supported by the National Natural Science Foundation of China (21464015, 21472235), the Xinjiang Tianshan Talents Program (2018xgytsyc 2-3) and UESTC Talents Startup Funds (A1098 5310 2360 1208).

Conflicts of Interest: The authors declare no conflict of interest.

References

1. Trautner, S.; Lackner, J.; Spendelhofer, W.; Huber, N.; Pedarnig, J.D. Quantification of the Vulcanizing System of Rubber in Industrial Tire Rubber Production by Laser-Induced Breakdown Spectroscopy (LIBS). *Anal. Chem.* **2019**, *91*, 5200–5206. [CrossRef]
2. Sun, L.; Xue, Y.; Peng, C.; Xu, C.; Shi, J. Does sulfur fertilizer influence Cu migration and transformation in colloids of soil pore water from the rice (*Oryza sativa* L.) rhizosphere? *Environ. Pollut.* **2018**, *243*, 1119–1125. [CrossRef]
3. Chaudhuri, R.G.; Paria, S. Synthesis of sulfur nanoparticles in aqueous surfactant solutions. *J. Colloid Interface Sci.* **2010**, *343*, 439–446. [CrossRef]
4. Griebel, J.J.; Glass, R.S.; Char, K.; Pyun, J. Polymerizations with elemental sulfur: A novel route to high sulfur content polymers for sustainability, energy and defense. *Prog. Polym. Sci.* **2016**, *58*, 90–125. [CrossRef]
5. Deng, Z.; Hoefling, A.; Théato, P.; Lienkamp, K. Surface Properties and Antimicrobial Activity of Poly (sulfur-co-1,3-diisopropenylbenzene) Copolymers. *Macromol. Chem. Phys.* **2018**, *219*, 1700497. [CrossRef]
6. Worthington, M.J.H.; Kucera, R.L.; Chalker, J.M. Green chemistry and polymers made from sulfur. *Green Chem.* **2017**, *19*, 2748–2761. [CrossRef]
7. Chung, W.J.; Griebel, J.J.; Kim, E.T.; Yoon, H.; Simmonds, A.G.; Ji, H.J.; Dirlam, P.T.; Glass, R.S.; Wie, J.J.; Nguyen, N.A.; et al. The use of elemental sulfur as an alternative feedstock for polymeric materials. *Nat. Chem.* **2013**, *5*, 518–524. [CrossRef]
8. Sun, Z.; Xiao, M.; Wang, S.; Han, D.; Song, S.; Chen, G.; Meng, Y. Sulfur-rich polymeric materials with semi-interpenetrating network structure as a novel lithium–sulfur cathode. *J. Mater. Chem. A* **2014**, *2*, 9280. [CrossRef]
9. Lin, H.K.; Liu, Y.L. Sulfur Radical Transfer and Coupling Reaction to Benzoxazine Groups: A New Reaction Route for Preparation of Polymeric Materials Using Elemental Sulfur as a Feedstock. *Macromol. Rapid Commun.* **2018**, *39*, 1700832. [CrossRef]
10. Gutarowska, B.; Kotynia, R.; Bielinski, D.; Anyszka, R.; Wreczycki, J.; Piotrowska, M.; Koziróg, A.; Berlowska, J.; Dziugan, P. New Sulfur Organic Polymer-Concrete Composites Containing Waste Materials: Mechanical Characteristics and Resistance to Biocorrosion. *Materials* **2019**, *12*, 2602. [CrossRef]
11. Griebel, J.J.; Namnabat, S.; Kim, E.T.; Himmelhuber, R.; Moronta, D.H.; Chung, W.J.; Simmonds, A.G.; Kim, K.J.; van der Laan, J.; Nguyen, N.A.; et al. New infrared transmitting material via inverse vulcanization of elemental sulfur to prepare high refractive index polymers. *Adv. Mater.* **2014**, *26*, 3014–3018. [CrossRef] [PubMed]
12. Dirlam, P.T.; Simmonds, A.G.; Kleine, T.S.; Nguyen, N.A.; Anderson, L.E.; Klever, A.O.; Florian, A.; Costanzo, P.J.; Theato, P.; Mackay, M.E.; et al. Inverse vulcanization of elemental sulfur with 1,4-diphenylbutadiyne for cathode materials in Li–S batteries. *RSC Adv.* **2015**, *5*, 24718–24722. [CrossRef]
13. Griebel, J.J.; Nguyen, N.A.; Namnabat, S.; Anderson, L.E.; Glass, R.S.; Norwood, R.A.; Mackay, M.E.; Char, K.; Pyun, J. Dynamic Covalent Polymers via Inverse Vulcanization of Elemental Sulfur for Healable Infrared Optical Materials. *ACS Macro Lett.* **2015**, *4*, 862–866. [CrossRef]
14. Hasell, T.; Parker, D.J.; Jones, H.A.; McAllister, T.; Howdle, S.M. Porous inverse vulcanised polymers for mercury capture. *Chem. Commun.* **2016**, *52*, 5383–5386. [CrossRef] [PubMed]
15. Hoefling, A.; Lee, Y.J.; Theato, P. Sulfur-Based Polymer Composites from Vegetable Oils and Elemental Sulfur: A Sustainable Active Material for Li-S Batteries. *Macromol. Chem. Phys.* **2017**, *218*, 1600303. [CrossRef]
16. Maraveas, C. Production of Sustainable and Biodegradable Polymers from Agricultural Waste. *Polymers* **2020**, *12*, 1127. [CrossRef] [PubMed]
17. Kim, D.; Kim, I.C.; Kyun, Y.N.; Myung, S. Novel bio-based polymer membranes fabricated from isosorbide-incorporated poly(arylene ether)s for water treatment. *Eur. Polym. J.* **2020**, *136*, 109931. [CrossRef]
18. Jiao, H.W.; Jin, C.H. Role of Bio-Based Polymers on Improving Turbulent Flow Characteristics: Materials and Application. *Polymers* **2017**, *9*, 209. [CrossRef]
19. Tondi, G.; Schnabel, T. Bio-Based Polymers for Engineered Green Materials. *Polymers* **2020**, *12*, 775. [CrossRef]
20. Worthington, M.J.H.; Kucera, R.L.; Albuquerque, I.S.; Gibson, C.T.; Sibley, A.; Slattery, A.D.; Campbell, J.A.; Alboaiji, S.F.K.; Muller, K.A.; Young, J.; et al. Laying Waste to Mercury: Inexpensive Sorbents Made from Sulfur and Recycled Cooking Oils. *Chemistry* **2017**, *23*, 16219–16230. [CrossRef]

21. Moawia, R.M.; Nasef, M.M.; Mohamed, N.H.; Ripin, A.; Zakeri, M. Biopolymer catalyst for biodiesel production by functionalisation of radiation grafted flax fibres with diethylamine under optimised conditions. *Radiat. Phys. Chem.* **2019**, *164*, 108375. [CrossRef]
22. Malhotra, R.; Ali, A. Lithium-doped ceria supported SBA−15 as mesoporous solid reusable and heterogeneous catalyst for biodiesel production via simultaneous esterification and transesterification of waste cottonseed oil. *Renew. Energy* **2018**, *119*, 32–44. [CrossRef]
23. Nantha, G.K.; Ashok, B.; Senthil, K.K.; Thundil, K.R.R.; Denis, A.S.; Varatharajan, V.; Anand, V. Performance analysis and emissions profile of cottonseed oil biodiesel–ethanol blends in a CI engine. *Biofuels* **2017**, *9*, 711–718. [CrossRef]
24. Shrimal, P.; Sanklecha, H.; Patil, P.; Mujumdar, A.; Naik, J. Biodiesel Production in Tubular Microreactor: Optimization by Response Surface Methodology. *Arab. J. Sci. Eng.* **2018**, *43*, 6133–6141. [CrossRef]
25. Chen, Y.; Yasin, A.; Zhang, Y.; Zan, X.; Liu, Y.; Zhang, L. Preparation and Modification of Biomass-Based Functional Rubbers for Removing Mercury(II) from Aqueous Solution. *Materials* **2020**, *13*, 632. [CrossRef]
26. Liang, H.C.; Lu, C.K.; Chang, J.R.; Lee, M.T. Use of waste rubber as concrete additive. *Waste Manag. Res.* **2007**, *25*, 68–76. [CrossRef]
27. Grinys, A.; Sivilevičius, H.; Daukšys, M. Tyre Rubber Additive Effect on Concrete Mixture Strength. *J. Civil. Eng. Manag.* **2012**, *18*, 393–401. [CrossRef]
28. Zarei, M.; Rahmani, Z.; Zahedi, M.; Nasrollahi, M. Technical, Economic, and Environmental Investigation of the Effects of Rubber Powder Additive on Asphalt Mixtures. *J. Transp. Eng. Part. B Pavements* **2020**, *146*, 04019039. [CrossRef]
29. Wulandari, P.S.; Tjandra, D. Use of Crumb Rubber as an Additive in Asphalt Concrete Mixture. *Procedia Eng.* **2017**, *171*, 1384–1389. [CrossRef]
30. Wang, W.; Song, Q.; Liu, Q.; Zheng, H.; Li, C.; Yan, Y.; Zhang, Q. A Novel Reprocessable and Recyclable Acrylonitrile-Butadiene Rubber Based on Dynamic Oxime-Carbamate Bond. *Macromol. Rapid Commun.* **2019**, *40*, 1800733. [CrossRef]
31. Zhang, Q.; Shi, C.Y.; Qu, D.H.; Long, Y.T.; Feringa, B.L.; Tian, H. Exploring a naturally tailored small molecule for stretchable, self-healing, and adhesive supramolecular polymers. *Sci. Adv.* **2018**, *4*, 8192. [CrossRef]

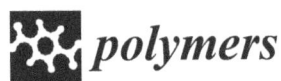

polymers

MDPI

Article

Effect of Iignocellulosic Nanoparticles Extracted from Yerba Mate (*Ilex paraguariensis*) on the Structural, Thermal, Optical and Barrier Properties of Mechanically Recycled Poly(lactic acid)

Freddys R. Beltrán [1,2,*], Marina P. Arrieta [1,2,*], Gerald Gaspar [1], María U. de la Orden [2,3] and Joaquín Martínez Urreaga [1,2]

[1] Dpto. Ingeniería Química Industrial y Medio Ambiente, Universidad Politécnica de Madrid, E.T.S.I. Industriales, 28006 Madrid, Spain; geraldmanuel.gaspar@upm.es (G.G.); joaquin.martinez@upm.es (J.M.U.)
[2] Grupo de Investigación: Polímeros, Caracterización y Aplicaciones (POLCA), 28006 Madrid, Spain; mariula@ucm.es
[3] Dpto. Química Orgánica, Facultad de Óptica y Optometría, Universidad Complutense de Madrid, 28037 Madrid, Spain
* Correspondence: f.beltran@upm.es (F.R.B.); m.arrieta@upm.es (M.P.A.)

Received: 29 June 2020; Accepted: 27 July 2020; Published: 29 July 2020

Abstract: In this work, yerba mate nanoparticles (YMNs) were extracted from *Ilex paraguairiencis* yerba mate wastes and further used to improve the overall performance of mechanically recycled PLA (PLAR). Recycled PLA was obtained by melt reprocessing PLA subjected to an accelerated ageing process, which involved photochemical, thermal and hydrothermal ageing steps, as well as a final demanding washing step. YMNs (1 and 3 wt.%) were added to the PLAR during the melt reprocessing step and further processed into films. The main goal of the development of PLAR-YMNs bionanocomposites was to increase the barrier properties of recycled PLA, while showing good overall performance for food packaging applications. Thus, optical, structural, thermal, mechanical and barrier properties were evaluated. The incorporation of YMNs led to transparent greenish PLAR-based films with an effective blockage of harmful UV radiation. From the backbone FTIR stretching region (bands at 955 and 920 cm^{-1}), it seems that YMNs favor the formation of crystalline domains acting as nucleating agents for PLAR. The morphological investigations revealed the good dispersion of YMNs in PLAR when they are used in the lowest amount of 1 wt.%, leading to bionanocomposites with improved mechanical performance. Although the addition of high hydrophilic YMNs increased the water vapor transmission, the addition of 1 wt.% of YMNs enhanced the oxygen barrier performance of the produced bionanocomposite films. These results show that the synergistic revalorization of post-consumer PLA and nanoparticles obtained from agri-food waste is a potential way for the production of promising packaging materials that meet with the principles of the circular economy.

Keywords: poly(lactic acid); mechanical recycling; yerba mate; bionanocomposites

1. Introduction

The development of bioplastics has raised a fair amount of interest in recent years. This is due to the constant growth of the consumption of fossil-fuel based plastics, which is leading to important environmental and raw materials availability problems. Among the most important bioplastics is poly(lactic acid) (PLA), which is an aliphatic polyester produced from renewable resources. PLA is obtained, on an industrial scale, via the ring-opening polymerization of lactide, the cyclic dimer of lactic acid, which is in turn produced by the fermentation of carbohydrates present in renewable feedstock such as corn, sugar beet or potato [1,2]. Due to its intrinsic biocompatibility and biodegradability,

PLA was initially developed with a focus on biomedical applications. However, the development of new grades with improved thermal, mechanical and optical properties has turned PLA into one of the most important bioplastics on the market, with applications on several industrial sectors, such as the textile, automotive and especially in short-term applications, such as those coming from the food packaging sector [3,4]. This wide variety of applications is leading to a continuous growth on the production of PLA, reaching a global production capacity of 270 kt in 2019 [5].

The use of PLA in applications commonly dominated by fossil-fuel-based plastics, such as packaging, could lead to important advantages from the sustainability point of view. However, it is worth noting that a massive use of PLA might result in some environmental problems. The newer grades of PLA, designed with demanding applications in mind, are very resistant and are only biodegradable at specific industrial conditions (i.e., 58 °C, RH% ≅ 65, pH ≅ 7.5, *C/N* relationship between 20:1 and 40:1) [6,7], which are not available in the environment (i.e., landfill) [8,9]. Hence, an inadequate management of the generated residues could lead to the accumulation of PLA wastes. Furthermore, the transition to a circular economy model has to be considered. In this circular economy model, plastics play a prominent role, as it can be seen from the strategies and directives proposed by the European Commission, including the need to replace single-use plastics by the end of 2020 [10–12]. These policies promote not only the reduction of plastic waste, but also the recovery of such wastes in order to reuse them, retaining their value. Therefore, the incorporation of PLA into the circular economy model constitutes a major challenge, which could be achieved through the mechanical recycling of PLA-based plastic wastes [13].

Although mechanical recycling allows for reducing the consumption of raw materials and the emissions related to the manufacture of PLA, previous studies [14–16] point out that mechanical recycling promotes chain scission reactions in PLA, resulting in a decrease of the molecular weight and of some important properties such as thermal stability and Vickers hardness. Therefore, the development of cost-effective and environmentally friendly methods to recover the properties of mechanically recycled PLA, and thus improve its recyclability, is a key challenge. In this regard, several alternatives have been proposed as valid approaches to increase the overall performance of recycled PLA, such as the use of thermal treatments [17], reactive extrusion with cross-linking agents and chain extender additives [18,19] and the use of inorganic fillers [3]. Furthermore, to guarantee the packaging green nature, another potential alternative is the utilization of reinforcements derived from renewable resources. For instance, in a recent work [20], the addition of small amounts of silk fibroin nanoparticles led to the improvement of thermal, mechanical and gas barrier properties of recycled PLA. It is widely known that nanocomposites show excellent mechanical, thermal, and gas barrier properties compared with the conventional polymeric materials or composites, [21,22]. Thus, the use of natural reinforcements for the development of PLA-based nanocomposites, intended for food packaging applications, represents a good option to improve the overall performance of PLA and recycled PLA (PLAR), without influencing the transparency which is very important for consumers acceptance [3,22,23].

From a circular economy point of view, it would be interesting to evaluate upgrading methods that also allow to valorize other wastes, such as those coming from agri-food or textile industries. In this regard, lignocellulosic residues from agri-food products are mainly considered as waste or low-value by-product [24–26]. Nevertheless, other lignocellulosic biomass derivatives have been recognized as optimal reinforcing fillers for the bioplastic industry due to the fact that they are biobased, light, stiff as well as non-abrasive for the plastic processing machinery [27–29]. In this context, several lignocellulosic nanoparticles have shown their ability to enhance the PLA overall performance, in terms of thermal, mechanical and barrier properties while also providing some anti-oxidant properties, thus increasing its interest in food packaging applications [27,29–31]. Polymer nanocomposites refer to multiphase polymeric systems where at least one of the constituent phases, commonly the nanofiller, has at least one dimension in the nanoscale range (<100 nm) [22]. The nanoparticles dimensions and properties depend on the raw material utilized for the extraction and the chemical process selected for their production [28]. A simple and aqueous extraction procedure to obtain lignocellulosic nanoparticles from yerba mate waste was recently proposed [32]. Yerba mate (*Ilex paraguairiensis*, Saint Hilaire) tree originates from the subtropical region of South America, and naturally grows in a limited zone within Argentina, Brazil and Paraguay. It is generally consumed as infusion due to its good taste and well-known antioxidant properties [32,33]. Yerba mate is composed from about 35% α-cellulose [34], 25% hemicellulose [34] and 25–30% lignin [34,35]. The presence of lignin results in yerba mate containing different amounts of polyphenols (i.e., caffeic and chlorogenic acids) [33], xanthines (i.e., caffeine and theobromine), flavonoids (i.e., catechin, quercetin, kaempferol and rutin) [36,37], amino acids, saponin and tannins as well as some vitamins (i.e., C, B_1, and B_2) [36,38]. Nowadays, Brazil is the largest producer of Mate (around 350 kt annually) [39], followed by Argentina, which produced 270 kt in 2019 [40], and Paraguay (around 100 kt annually). Its high consumption leads to the generation of a high amount of yerba mate wastes, since, after being used as infusion, it is wasted without any kind of revalorization [32,41]. Thus, their use for lignocellulosic nanoreinforcements production could not only provide a sustainable revalorization to such waste, as it was already demonstrated for virgin PLA [32], but it could also potentially help to recover the properties of mechanically recycled PLA by developing bionanocomposites with interest in the food packaging field. In fact, as yerba mate is a rich source of polyphenols, which display an antiradical activity similar to pure gallic acid (20 mg/mL) [38], it has gained interest as sustainable additive that could be used to improve and modulate the properties of biopolymers. For instance, yerba mate extract provided a significant improvement of a starchy polymeric matrix stability in acidic and alkaline media [37]. Moreover, yerba mate extract has been added to starch treated by hydrostatic pressure to increase the loading capacity, obtaining interesting carriers for antioxidants, in which the antioxidant activity was maintained after the high pressure treatment without changing the yerba mate polyphenols profile [33]. Similarly, yerba mate has been encapsulated into electrospun zein fibers, improving the thermal stability and proving antioxidant activity, and thus showing interest as antioxidant releasers for food packaging applications [42]. Lignocellulosic yerba mate nanoparticles (YMNs) has also been added to PLA (in 5 wt.%), showing that the high amount of polyphenols protects the polymeric matrix from the thermal degradation during processing, and yielding bionanocomposites with significantly improved mechanical performance, although they showed a somewhat green tonality [32].

The main objective of the present research is to study the effects of lignocellulosic nanoparticles extracted from yerba mate wastes on the properties of mechanically recycled PLA, aiming to revalorize both yerba mate and PLA wastes by developing high-performance bionanocomposites intended for food packaging applications. Yerba mate nanoparticles (YMNs) were obtained by means of an aqueous extraction procedure, followed by two filtration steps, following a previously optimized recipe [32]. Recycled PLA (PLAR) was obtained by subjecting PLA to an accelerated ageing process previously optimized [14], which involved photochemical, thermal and hydrothermal ageing steps, as well as a final demanding washing step to simulate the washing conditions used on an industrial recycling level. The bionanocomposites were prepared by extrusion followed by a compression molding process. The YMNs were previously freeze dried to obtain a powder. Considering the high amount of –OH on the surface of lignocellulosic nanoparticles, which induces high attraction between them, particularly during the freeze-drying process, the nanoparticles were characterized by means of dynamic light scattering (DLS) and Transmission Electron Microscopy (TEM), before and after the freeze-drying process. The structure of the recycled PLA reinforced with yerba mate nanoparticles was characterized using infrared (FTIR) and UV-visible spectroscopic techniques, Differential Sacanning Calorimetry (DSC), Scanning Electron Microscopy (SEM) and intrinsic viscosity (IV) measurements. The effect of the nanoparticles on the thermal stability was measured using Thermogravimetric analysis (TGA), while the mechanical performance was evaluated by nanoindentation measurements. Finally, regarding the potential application in the packaging field, special attention was given to the gas barrier performance, which is of critical importance in food packaging applications. Thus, the permeability to oxygen gas and water vapor of the obtained materials was measured and compared.

The results show that the yerba mate nanoparticles can significantly enhance the barrier to oxygen in the recycled material.

2. Materials and Methods

2.1. Materials

PLA, under the commercial name Ingeo™ 2003D, was purchased from NatureWorks (Minnetonka, MN, USA). This grade presents a melt mass-flow rate of 6 g/10 min (2.16 kg at 210 °C). The yerba mate (*Ilex paraguariensis*) residue was collected after yerba mate infusion (Taragüi, Argentina) consumption in our own laboratory.

2.2. Nanoparticle Extraction from Yerba Mate Residues

The lignocellulosic-based nanoparticles, named yerba mate nanoparticles (YMNs), were obtained from yerba mate infusion wastes following an already developed recipe [32]. In brief, the residue of yerba mate infusion after its consumption was dried in an oven at 60 °C for 24 h. Then, 6 g of yerba mate infusion residue were mixed with 200 mL of distilled water and heated up to 100 °C under reflux during 60 min with vigorous magnetic stirring (1000 rpm). Next, the solid residue was eliminated by simple filtration, while the obtained mate extract solution was filtered off again (filter paper Whatman Grade 41:20–25 μm particle retention), frozen and further subjected to a freeze-drying process using a Flexi-Dry Freeze Dryer (FTS Systems, Stone Ridge, NY, USA) to obtain a powder as it is schematically represented in Figure 1. The obtained powder of YMNs was stored at 40 °C under vacuum during 24 h to remove any moisture before melt compounding process.

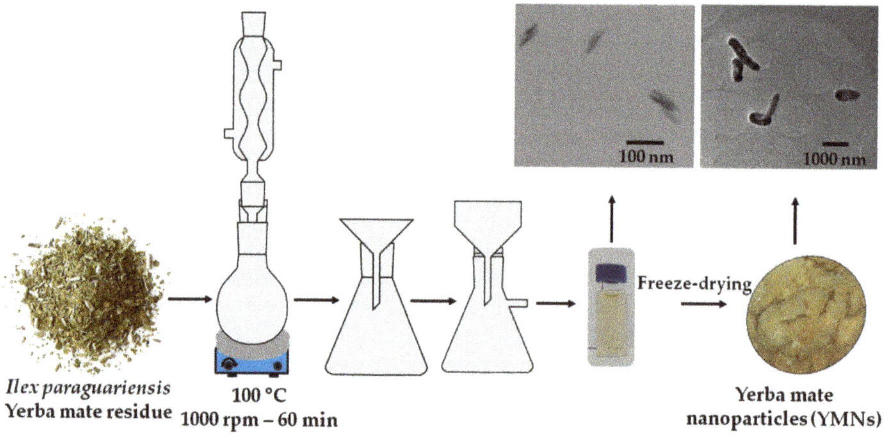

Figure 1. Schematic representation of mate nanoparticles' extraction procedure.

2.3. Preparation of the Samples

The procedure followed for the ageing and subsequent obtainment of recycled PLA based materials is presented on Figure 2. Firstly, Ingeo 2003D pellets were processed by melt extrusion in a Rondol Microlab counter-rotating twin-screw extruder (Microlab, Rondol, France) with an *L*/*D* ratio of 20. The extrusion process was carried out at 60 rpm, using the following temperature profile (from hopper to die): 125, 160, 190, 190, 180 °C. The obtained material was transformed into films (thickness = 200 ± 10 μm) using an IQAP-LAP hot plate press (IQAP Masterbatch Group S.L., Barcelona, Spain) at 190 °C. Secondly, the films (PLAV) were subjected to an accelerated ageing process, consisting of the following stages: (i) 40 h of photochemical degradation using an ATLAS UVCON chamber (Chicago, IL, USA), equipped with eight F40UVB lamps; (ii) 468 h of thermal degradation in an oven at 50 °C and (iii) 240 h of hydrolytic degradation in distilled water at 25 °C. Thirdly, the aged samples were subjected to a demanding washing process, which was used in previous studies [14], using an aqueous solution of NaOH (1.0 wt.%) and Triton X (0.3 wt.%).

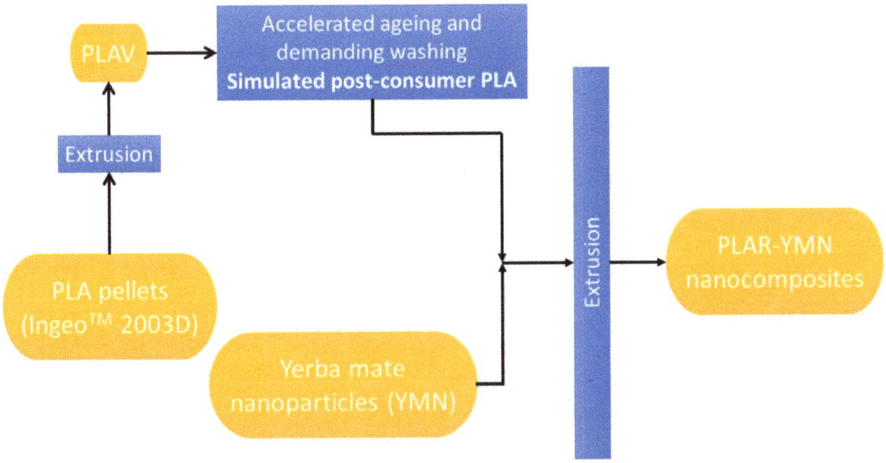

Figure 2. Procedure followed for the obtainment of the PLAR-YMN bionanocomposites.

Lastly, the washed material was ground, and melt compounded together with yerba mate nanoparticles, in different proportions, at the same conditions used for the obtainment of PLAV films. Table 1 summarizes the different materials obtained in this study.

Table 1. Materials obtained after the recycling process.

Sample	Description
PLAV	PLA obtained after the first extrusion and compression molding steps
PLAR	PLA obtained after the accelerated ageing, washing, and melt compounding of PLAV
PLAR-1YMN	PLAR with 1 wt.% of yerba mate nanoparticles
PLAR-3YMN	PLAR with 3 wt.% of yerba mate nanoparticles

2.4. Characterization Techniques

The hydrodynamic size of YMNs were measured by means of a dynamic light scattering (DLS) analyzer. The obtained YMNs, in powder form, were dispersed in water (1 mg mL^{-1}) by sonication and further measured at 20 °C in a Zetasizer Nano series ZS DLS equipment (Malvern Instrument Ltd., Malvern, UK).

YMNs were also observed by Transmission Electron Microscopy (TEM) in a JEOL JEM-1010 operating (JEOL Ltd., Tokyo, Japan) at 100kV. One droplet of YMNs aqueous suspension (1 mg mL^{-1}) was deposited on carbon-coated copper grids and dried at room temperature during 20 min before TEM observation. The nanoparticles' length and width were measured from the TEM images with ImageJ software; the mean and standard deviation of 15 nanoparticle measurements are reported.

Intrinsic viscosity measurements were performed, at 4 different concentrations in chloroform, at 25 °C in an Ubbelohde viscometer. All the solutions were filtered prior to the intrinsic viscosity measurement.

UV-Visible spectroscopy tests were conducted in a Varian Cary 1E UV-Vis spectrophotometer (Varian, Palo Alto, CA, USA) equipped with an integrating sphere and using a scanning speed of 400 nm/min. The overall transmittance in the visible region was then calculated according to the ISO 13468 standard.

Fourier transform infrared (FTIR) spectra of the different materials were recorded in Nicolet iS10 spectrometer (Thermo Fisher Scientific, Waltham, MA, USA), equipped with a diamond Attenuated Total Reflectance (ATR) accessory, using 16 scans and a resolution of 4 cm^{-1}. The surface crystallinity degree (X_c) of each nanocomposite was calculated from the absorbance of the band at 955 cm^{-1}, measured in both the amorphous PLAR (I_0) and the nanocomposite (I_f), using Equation (1) [43]:

$$X_c = \left(\frac{I_0 - I_f}{I_0}\right) \times 100\% \quad (1)$$

The cryo-fractured surface microstructure of the cross section of each bionanocomposite film was observed by field emission scanning electron microscopy (FE-SEM) in a JEOL JSM 7600F microscope (JEOL Ltd., Tokyo, Japan). Films were previously sputtered with a gold layer to make them conductive.

Differential scanning calorimetry (DSC) scans were performed in a TA Instruments Q20 calorimeter (New Castle, DE, USA). Samples of 5 mg were placed in aluminum pans and subjected to the following protocol (under nitrogen atmosphere): (i) heating from 30 to 180 °C at 5 °C/min; (ii) isothermal step at 180 °C for 3 minutes; (iii) cooling from 180 to 0 °C at 5 °C/min; (iv) isothermal step at 0 °C for 1 min and (v) heating from 0 to 180 °C at 5 °C/min.

Thermogravimetric analysis (TGA) was conducted on 10 mg samples using a TA Instruments TGA2050 thermobalance (New Castle, DE, USA). The samples were heated from 40 to 800 °C at 10 °C/min under nitrogen atmosphere. The onset degradation temperature (T_{10}) was calculated at 10% of mass loss, and the maximum degradation temperature (T_{max}) was determined from the peak of the first derivative of the TGA curve (DTG).

The water vapor transmission rate (WVTR) of the materials was measured, three times, by gravimetry according to the ISO 2525 standard. Thin films (9 ± 2 µm) of the samples were prepared by solvent casting from 0.01 g/mL chloroform solutions. The permeability cups were filled with 2 g of dry silica gel, sealed with the sample film and then placed in a desiccator with a saturated KNO_3 solution at 23 °C (approximately 90% RH). The cups were weighed each hour for 6 h. WVTR (g/day cm^2) was determined using Equation (2):

$$WVTR = \frac{240 * (m_t - m_0)}{A * t} \quad (2)$$

where m_t is the mass of the cup at time t, m_0 is the initial mass of the cup and A is the exposed area of the film.

Oxygen permeability tests were conducted at 30 °C in a homemade permeation cell, using a gas pressure of 1 kPa.

Nanoindentation tests were carried out using a Shimadzu DUH-211S dynamic Ultra-Microhardness Tester (Shimadzu Corporation, Kyoto, Japan), equipped with a Berkovich indenter. The measurements were conducted at room temperature (24.5 ± 0.5 °C), using a maximum load of 10 mN and a loading rate of 1.4632 mN/s. Maximum load was held for 5 s, and then it was retired. Each measurement was replicated 6 times.

3. Results and Discussion

3.1. Yerba Mate Nanoparticles' Characterization

The obtained mate extract solution after filtration was analyzed by DLS (Figure 3a), showing a monomodal size distribution from 85 to 103 nm, with an average size of 94 nm. There is a shoulder at higher sizes, around 500 nm, which has been related with the formation of agglomerates [32]. Considering that the DLS technique is designed to calculate the hydrodynamic diameter of spherical particles, the nanosize as well as the morphological aspect of the YMNs were further examined by TEM (Figure 3b). Individualized lignocellulosic 2D YMNs were observed. The yerba mate solution was then freeze-dried to obtain a powder, obtaining a yield of 19% ± 5%, which is in good agreement with previously reported work [32]. The size analysis of the obtained powder of yerba mate nanoparticles was also carried out by DLS (Figure 3c) and it revealed a monomodal size distribution with a dimension ranging from 450 to 545 nm, with an average size of 495 nm. From the TEM image of YMNs powder (Figure 3d), it can be seen that the YMNs tend to agglomerate due to the natural tendency of both, lignin and cellulose, to re-agglomerate and form strong hydrogen bonds as the water sublimate during freeze-drying process [44,45]. Nevertheless, they still showed sub-micron size with dimensions of 525 ± 136 in length and 302 ± 96 nm in width (see zoom in Figure 3d).

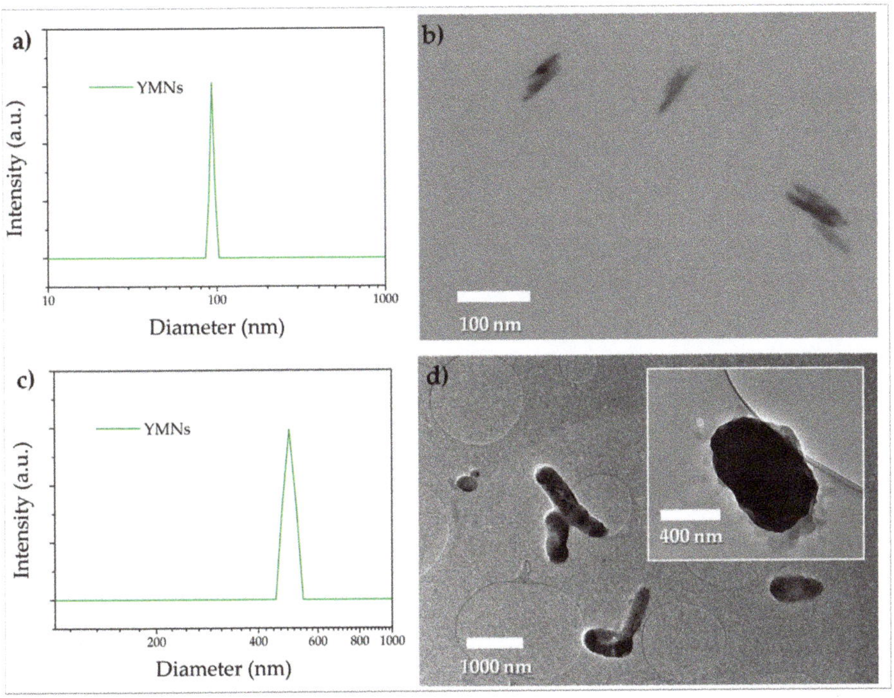

Figure 3. YMNs solution: (**a**) DLS measurements and (**b**) TEM images. YMNs powder: (**c**) DLS measurements and (**d**) TEM images.

3.2. *Structure and Morphology of the PLA-YMN Bionanocomposites*

The FTIR spectra of YMNs, PLAR and PLAR-3YMN are reported in Figure 4. The broad absorption band in the range of 3000–3700 cm^{-1} present in YMNs can be ascribed to the stretching vibration of the –OH groups in lignin as well as in cellulose molecules. The successful hemicellulose removal from yerba mate residue was confirmed by the absence of the band around 1730 cm^{-1} in YMNs [32], which corresponds to the acetyl and ester groups in hemicelluloses [46]. The spectrum that corresponds to PLAR-3YMM bionanocomposite shows a broad band at 3320 cm^{-1} (stretching vibration of the –OH groups) and a shoulder at 2920 cm^{-1} (C–H stretching vibration) (Figure 4a) that confirm the successful incorporation of YMNs in the recycled PLA [32]. The stretching vibration of the carbonyl group (–C=O) of PLA appears at 1750 cm^{-1} (Figure 4b) [47]. Moreover, the FTIR-ATR spectra of the bionanocomposites show very slight changes in the intensity of the bands at 920 and 956 cm^{-1} (Figure 4b). These absorptions have been assigned to skeletal C–C stretching mode coupled with CH_3 rocking one [48–50]; while the band centered at 920 cm^{-1} corresponds to the 10_3 helix chain conformation, characteristic of the crystalline forms, the band at 956 cm^{-1} is assigned to the amorphous phase. In this work, the crystallinity degrees in the surface of the nanocomposites were calculated from the absorbances of the band at 955 cm^{-1} in the different materials, using Equation (1). The values obtained, 16.7% for PLAR-3YMN and 12.5% for PLAR-1YMN indicate that the YMNs act as nucleating agents for recycled PLA.

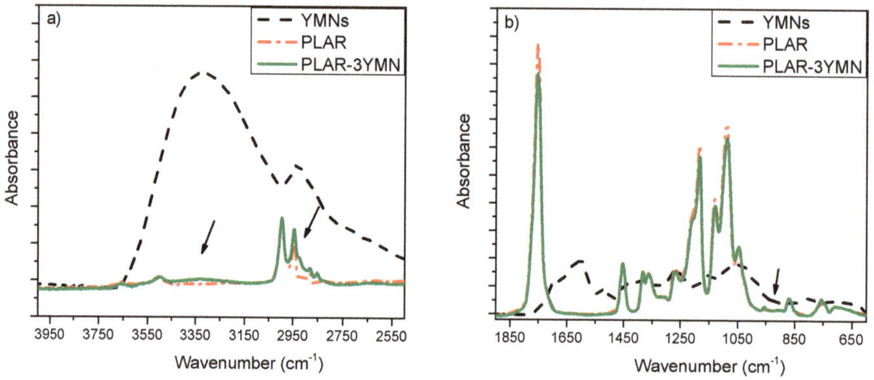

Figure 4. FTIR spectra of YMNs, recycled PLA (PLAR) and PLAR-3YMN bionanocomposite: (**a**) in the 4000–2500 cm^{-1} region and (**b**) in the 1900–600 cm^{-1} region.

The effect of the addition of YMNs on the microstructure of mechanically recycled PLA was studied by means of SEM analysis. Neat virgin PLA (Figure 5a) shows the typical regular and smooth surface of an amorphous polymer. PLAR (Figure 5b) shows a very similar behavior than that of neat PLA (Figure 5a) with a rather more ductile pattern. This more plastic behavior could be ascribed to the already commented chain scission reactions that take place during the accelerated ageing and mechanical recycling, because the shorter polymer chains formed in these degradation processes plasticize the polymeric matrix. Meanwhile, PLAR-YMN bionanocomposites (Figure 5c,d) exhibited a rougher surface due to the YMNs' reinforcing effect, as it has been already observed in virgin PLA blended with lignocellulosic nanoparticles [23,32].

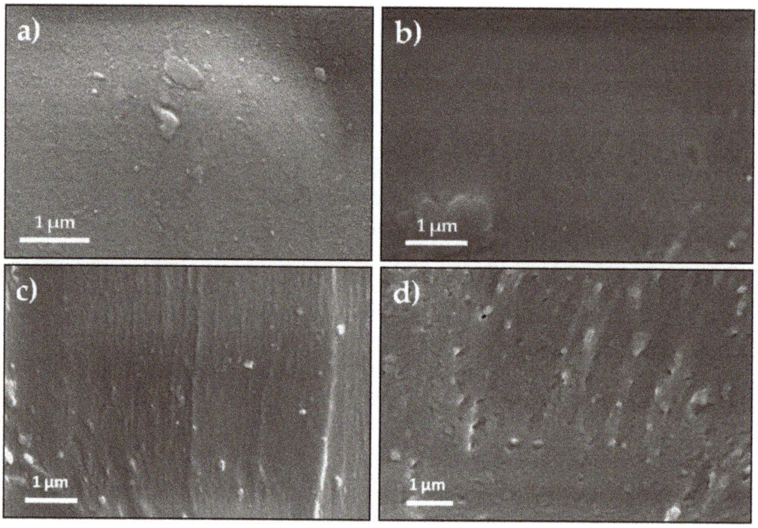

Figure 5. SEM observations of: (**a**) PLAV, (**b**) PLAR, (**c**) PLAR-1YMN and (**d**) PLAR-3YMN. (10,000× magnifications).

The fracture surface depends on the concentration of YMNs (Figure 5c,d). In fact, in PLAR-1YMN bionanocomposite (Figure 5c), YMNs appear uniformly dispersed, with no phase separation between the nanoparticle and the polymeric matrix. However, in PLAR-3YMN (Figure 5d) some micro-holes can be seen, thus suggesting that YMNs in bionanocomposite containing 3 wt.% of YMNs show poor interfacial adhesion with PLAR matrix. Micro-holes have been already observed in virgin PLA reinforced with lignin nanoparticles and have been related to the formation of YMNs' aggregates during bionanocomposite processing [29].

3.3. Properties of the PLA-YMN Bionanocomposites

3.3.1. Effect of the Addition of YMNs on the Intrinsic Viscosity

Intrinsic viscosity is related to the molecular weight of PLA, which plays a very important role in the final thermal and barrier properties of the materials. Furthermore, intrinsic viscosity is important from a processing point of view since industrial forming processes are frequently designed to operate at specific IV values. Thus, in order to get information regarding the effect of YMNs on the processing of PLAR-based bionanocomposites, the values of the intrinsic viscosity (IV) of PLAR in all the samples was determined by dissolving each sample in chloroform, followed by a filtration step to eliminate the YMNs. In accordance with previous works, Figure 6 shows that PLAR has an intrinsic viscosity around 14% lower than PLAV due to the degradation experimented [12,16] during the accelerated ageing, washing and reprocessing steps.

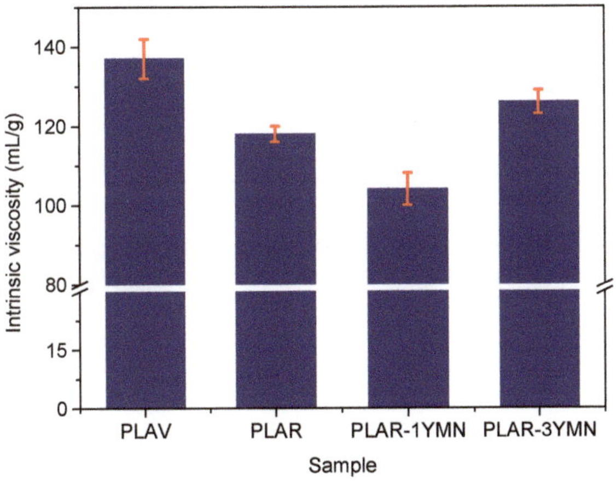

Figure 6. Intrinsic viscosity values of the different samples.

Regarding the effect of the addition of yerba mate nanoparticles, Figure 6 shows that the material with only 1 wt.% of YMNs presents an intrinsic viscosity value 12% lower than PLAR. However, the sample with 3 wt.% of YMNs shows an intrinsic viscosity higher than that of the unfilled recycled material. This behavior might suggest that the addition of the nanoparticles produces two counteracting effects on the intrinsic viscosity of recycled PLA. On the one hand, the high hydrophilicity of the yerba mate nanoparticles might cause the absorption of small amounts of water during processing, which could result in a significant hydrolytic degradation of PLA during melt compounding. A similar behavior was observed in other PLA/lignocellulosic filler composites. For instance, Arrieta et al. [51] observed the reduction in the molecular weight of PLA bionanocomposites in virgin PLA reinforced with cellulose nanocrystals. Similarly, Way et al. [52] reported that PLA filled with lignocellulosic fibers showed a more severe degradation during melt processing than its unfilled counterpart. On the other hand, the antioxidant nature of yerba mate nanoparticles (due to the presence of phenolic compounds) could contribute to reduce the degradation of the polymer during extrusion, as it has been pointed out by Arrieta et al. [32] in a previous work. The results shown on Figure 6 indicate that at lower concentrations of YMNs, the negative effect of the hydrolysis prevails; however, at higher concentrations, the effect of the higher amounts of antioxidant compounds present in YMNs is more important, resulting in materials with higher IV values.

3.3.2. Thermal Properties

The effect of the addition of YMNs on the thermal properties of mechanically recycled PLA was studied by means of DSC and TGA. Figure 7 and Table 2 summarize the DSC results obtained for the different materials. As can be seen in Figure 7, PLAV show the characteristic thermal transitions of PLA: (i) a glass transition (T_g) around 60 °C; (ii) a broad cold crystallization exothermic peak (T_{cc}) above 100 °C and (iii) a melting endotherm (T_m) centered at 150 °C. As for the behavior of PLAR, Figure 7 and Table 2 show that it has, overall, the same thermal transitions as PLAV. However, there are some noteworthy differences. Firstly, PLAR shows a narrower cold crystallization peak, which is also located at temperatures 15 °C lower than the virgin material. This difference could be attributed to the degradation of PLA during mechanical recycling, since the shorter polymer chains have increased mobility and crystallize more easily [14]. This behavior is also reflected in the higher values of the cold crystallization and melting enthalpies (ΔH_{cc} and ΔH_m respectively) of PLAR. Secondly, Figure 7 shows that there are differences in the melting endotherm of the recycled material, since PLAR shows two well-defined melting peaks. This behavior has been reported in previous studies [53], and it has been attributed to the occurrence of a melt recrystallization phenomenon. Such a phenomenon consists of the melting of less ordered crystalline domains at lower temperatures, their rearrangement into crystalline structure as the temperature increases and a final melting of the more ordered crystals at a higher temperature. The fact that PLAR shows this behavior could also be explained by the degradation of the polymer during the recycling process. The shorter polymer chains present in PLAR could rearrange themselves during heating more easily due to their increased mobility and hence form more stable crystalline structure, which melt at higher temperatures.

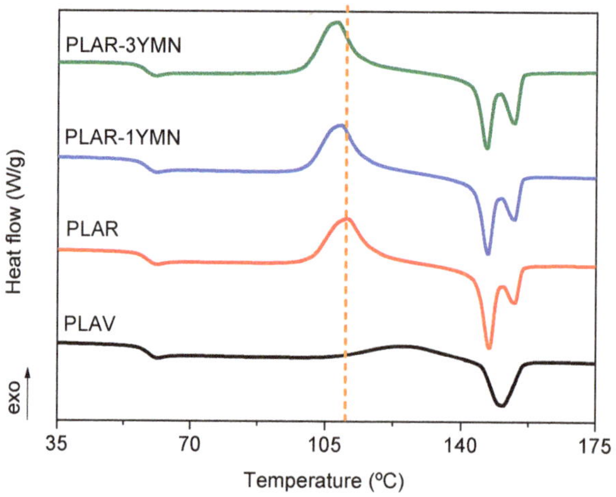

Figure 7. DSC scans corresponding to the second heating of the materials.

Table 2. DSC (second heating) results as well as TGA parameters of the different materials.

Sample	T_g (°C)	T_{cc} (°C)	T_m (°C)	ΔH_{cc} (J/g)	ΔH_m (J/g)	T_{10} (°C)	T_{max} (°C)
PLAV	59.0	125.2	150.8	14.5	15.4	334.2	365.9
PLAR	58.6	110.5	147.3–153.9	27.1	27.7	316.1	355.9
PLAR-1YMN	58.6	108.9	146.9–153.7	26.7	27.8	318.1	354.6
PLAR-3YMN	58.1	107.7	146.5–153.7	28.2	28.3	307.3	349.1

Regarding the effect of the presence of the yerba mate nanoparticles, both Figure 7 and Table 2 show that the thermal behavior of PLAR-1YMN and PLAR-3YMN are very similar to that of mechanically recycled PLA. However, some differences can be seen in the cold crystallization temperature. Both Figure 7 and Table 2 show that the addition of the nanoparticles leads to a slight decrease in T_{cc} values. This behavior suggests that yerba mate nanoparticles act as nucleating agents, promoting the cold crystallization of PLA at lower temperatures, as it was seen by means of FTIR-ATR. The nucleating effect of different organic-based fillers has been previously reported by different authors, such as Fortunati et al. [44], Arrieta et al. [23] and Lizundia et al. [54] in PLA/cellulose nanocrystals bionanocomposites; as well as by Beltrán et al. [3] who studied recycled PLA/silk fibroin nanoparticles nanocomposites. It is also worth noting that PLAR-3YMN presents a melting behavior closer to PLAR than to PLAV, despite its higher IV value. This could also be explained by the nucleating effect of YMNs, since it allows for the occurrence of the melt recrystallization mechanism, despite the limited mobility of the longer polymer chains present in PLAR-3YMN. Nevertheless, the observed differences are rather small, thus indicating that the effect to the yerba mate nanoparticles on the thermal transitions of recycled PLA is limited.

The effect of the recycling process as well as the addition of YMNs on the thermal properties of PLAR was also investigated by dynamic TGA measurements. The weight loss (TGA) and derivative (DTG) curves of virgin PLA (PLAV), recycled PLA (PLAR) and PLAR-YMNs bionanocomposites are reported in Figure 8, while the thermal parameters obtained from these curves are summarized in Table 2. All samples show a one-step degradation processes. While virgin PLA (PLAV) shows the highest maximum onset degradation temperature (T_{10} = 334.2 °C), PLAR-based samples presented a decrease in the thermal stability, as shown by the decrease in the onset degradation temperature, which has been ascribed to the presence of shorter polymer/oligomeric chains with lower thermal stability [14]. These results are in good agreement with the already commented reduction in the molecular weight when discussing the intrinsic viscosity measurements. In this context, Burgos et al. developed different PLA formulations plasticized with oligomeric lactic acid (OLA) and also observed a reduction of the onset degradation temperature, which decrease with increasing amounts of OLA [55]. The incorporation of 1 wt.% of YMNs did not promote significant changes in the thermal behavior of PLAR. While the T_{10} values slightly increased, the T_{max} remained almost invariable. However, with the incorporation of 3 wt.% of YMNs, both the T_{10} and the T_{max} decreased. These results may seem rather surprising considering that the material with 3 wt.% YMNs has a higher IV than PLAR. However, similar findings were observed by Fortunati et al. in virgin PLA reinforced with 1 and 3 wt.% of cellulose nanocrystals. They reported that the thermal stability of PLA decreased as the nanocellulose content increased and ascribed this behavior to the lower thermal stability of the cellulose nanocrystals (maximum degradation rate at about around 291 °C) [30]. The DTG curve of the YMNs used in this work, which has been previously reported and analyzed [32], shows two maxima at 215 and 315 ° C, below the main PLA degradation temperature, so that the presence of 3 wt.% YMNs could explain the decrease in the thermal stability of the nanocomposite.

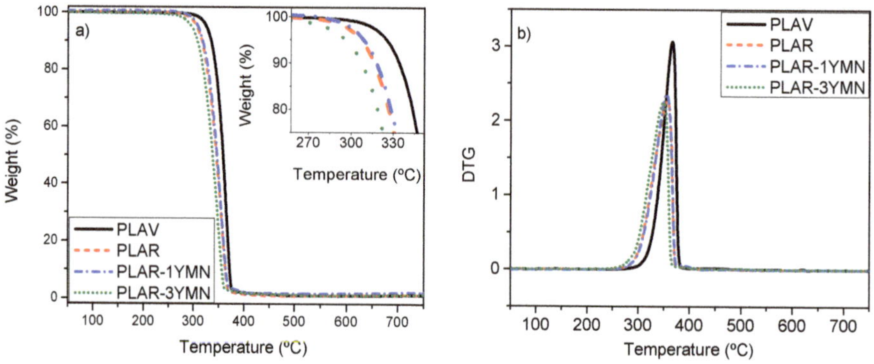

Figure 8. Dynamic (**a**) TGA and (**b**) DTGA curves of binary PLA nanocomposite films.

3.3.3. Optical Properties

The visual appearances of virgin PLA, recycled PLA and YMN-reinforced bionanocomposites are shown in Figure 9a. From the visual appearance of the films, it is possible to observe that the recycled PLA remains transparent, with no apparent differences with virgin PLA. Meanwhile, bionanocomposites presented a somewhat green tonality, which was more evident in the case of the bionanocomposite with a higher amount of YMNs (PLAR-3YMN). In a previous work, virgin PLA has been reinforced with 5 wt.% of similar yerba-mate-based lignocellulosic nanoparticles and the developed films presented a brown tonality [32]. The transmission values in the visible (400–800 nm) and UV region of the spectra were determined by using UV-Vis spectroscopy (Figure 9b,c).

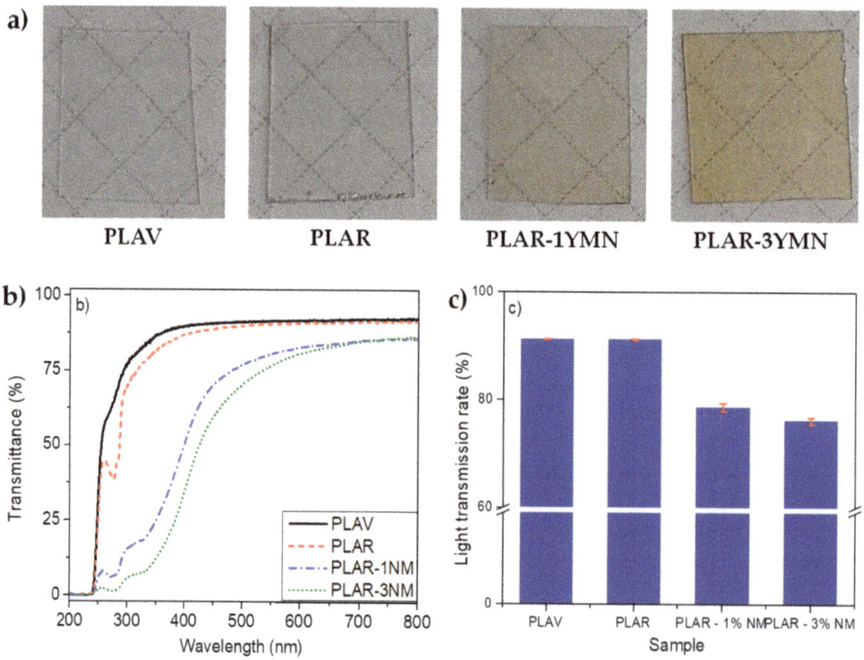

Figure 9. PLAV, PLAR and bionanocomposites (PLAR-1YMN and PLAR-3YMN): (**a**) visual appearance, (**b**) UV-vis spectra and (**c**) visible light transmission rates.

The spectra show that films obtained from PLAV and PLAR are highly transparent in the visible region. In good agreement with the visual appearance of the films, the spectra show that the presence of YMNs leads to significant decreases in the visible light transmission (Figure 9b). The overall transmission rate in this spectral region falls from values higher than 90% in PLAV and PLAR to values clearly below 80% in biocomposites (78.6% and 76.1% of light transmission rates in PLAR-1YMN and PLAR-3YMN, respectively), although these materials remain transparent (Figure 9c). The presence of lignocellulosic aggregates in PLAR-3YMN decreased the visible light transmittance of the PLAR-based film, in good agreement with SEM analysis. Similar results have been observed in PLA/lignin nanoparticles bionanocomposites [31].

It is worth noting that PLAR shows lower UV light transmission than PLAV, with the appearance of a small absorption peak centered at 277 nm. This band is related with the formation of chain-end carboxyl groups, as a consequence of the degradation of the polymer that take place during the recycling process [20,56]. In the case of YMN-reinforced recycled plastics, this region is overlapped with different absorptions due to the polyphenols (i.e., chlorogenic acid, caffeic acid and rutin [33]) present in YMNs.

The above spectra reveal that YMNs produce a strong UV blocking effect in the recycled PLA matrix. Other authors have already reported the UV blocking effect in virgin PLA reinforced with different lignocellulosic nanoparticles, such as in PLA/lignin nanoparticles bionanocomposites [31,57]; PLA/cellulose nanocrystals nanocomposites [51] and also in virgin PLA reinforced with 5 wt.% of similar yerba-mate-based lignocellulosic nanoparticles [32]. In the case of PLAR-1YMN film, the presence of only 1 wt.% of YMNs was able to block around 90% of UV-B and C, and this UV blocking effect was more marked in PLAR-3YMN, as could be expected.

In summary, it can be said that the addition of YMNs nanoparticles to the recycled PLA, in a proportion less than or equal to 3 wt.%, has an overall positive effect on the optical properties of the material. On the one hand, the transparency in the visible region is reduced, but the sheets of these bionanocomposites remain transparent, which is important in many cases of food packaging, because seeing the packed food through the packaging film is one of the most important requirements for consumers' acceptance. On the other hand, the presence of YMNs greatly reduces UV transmission, thus slowing down the degradation of the contents of the container.

3.3.4. Barrier Properties

The barrier properties against different gases are very important in food packaging applications, which is the most important market for PLA. Therefore, the effect of the addition of the YMNs on the gas barrier properties of mechanically recycled PLA was measured; the main results are reported in Figures 10 and 11.

Figure 10 shows the WVTR of the different samples. The obtained values are similar to those reported in previous studies for PLA based materials [58]. It can be seen that mechanical recycling led to a slight increase in the WVTR of PLA. To explain this behavior, one should consider that the gas permeability, and hence the WVTR, of semicrystalline polymers depends on two factors: the diffusion coefficient and the solubility of the gas. These factors are affected by the molecular weight, structure and free volume of the polymer and by the temperature and nature of the gas molecules [12,29]. The observed increase in the WVTR of the mechanically recycled PLA could be related to the generation of terminal carboxyl and hydroxyl groups during the ageing and mechanical recycling, which decreases the hydrophobic character of the polymer, thus facilitating the passage of water vapor through the films. Regarding the effect of the nanoparticles, it can be observed that the nanocomposites show higher WVTR values than both unfilled PLAV and PLAR samples, which could be explained by the hydrophilic nature of the YMNs due to the high amount of –OH groups. In this context, Kim et al. [59] studied the WVTR of PLA reinforced with pristine lignin and acetylated lignin, reporting higher WVTR values for the PLA-lignin composites in comparison with neat PLA. This behavior was ascribed to the hydrophilic nature of pristine lignin. Meanwhile, acetylated lignin-based composites were able to decrease the WVTR values of neat PLA. Similarly, Espino-Perez et al. [60] developed PLA loaded with cellulose nanowhiskers (5, 14 and 30 wt. %), reporting that WVTR increased with the cellulose nanowhiskers content. This behavior was related to the hydrophilic nature of cellulose structures. In this work, PLAR-1YMN, the material with the lower amount of hydrophilic YMNs, shows higher WVTR than PLAR-3YMN, which can be related to the lower viscosity observed in PLAR-1YMN. This low viscosity, due to a stronger degradation, indicates the presence of more hydrophilic terminal groups in the polymer chains, which can explain the higher value of WVTR.

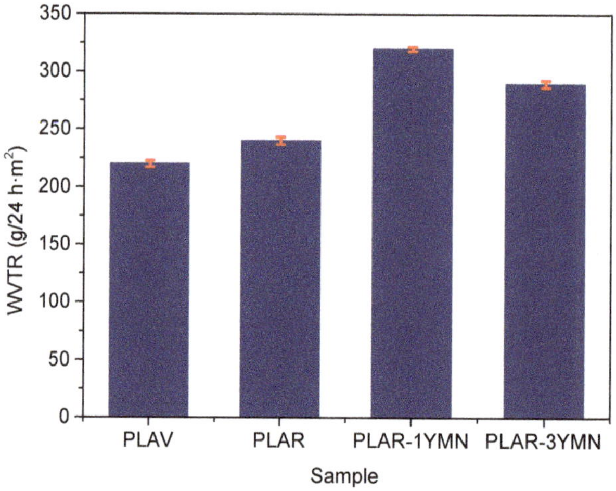

Figure 10. Water vapor transmission rate of the different materials.

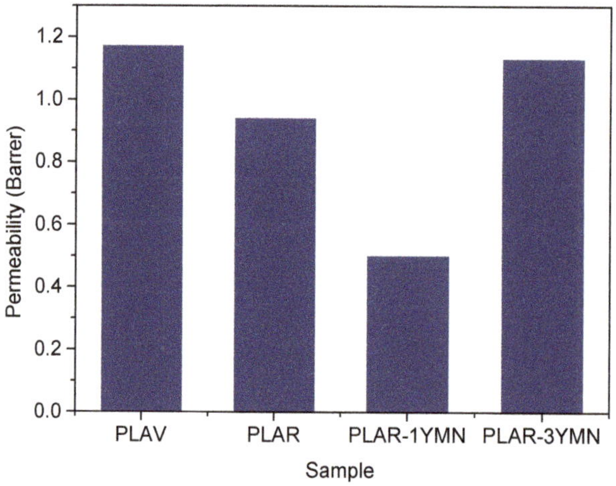

Figure 11. Oxygen permeability of the different materials.

Figure 11 presents the oxygen permeability coefficient, measured in Barrer (1 Barrer = $3.35{\cdot}10^{-16}$ mol m/m^2 s Pa), of the different samples. It can be observed that the ageing and the mechanical recycling cause only a slight decrease in the oxygen permeability of PLA, despite the degradation observed by means of IV measurements. Similar results have been reported in a previous study [14] and have been attributed to the presence of two counteracting effects of the mechanical recycling on the permeability of PLA. On the one hand, the presence of shorter polymer chains might reduce the free volume inside the polymer, due to their better ability to rearrange themselves, reducing the diffusion coefficient. On the other hand, the generation of terminal –COOH and –OH groups during the ageing and recycling lead to an increase in the affinity between the polymer and the gas molecules, increasing the solubility of the gas into the polymer. The concurrence of these counteracting effects leads to the overall small changes observed in the oxygen permeability.

As for the behavior of the bionanocomposites, Figure 11 shows that the oxygen permeability is significantly reduced with the addition of 1 wt.% of YMNs. The reduction of the oxygen permeability due to the incorporation of cellulose nanocrystals have been already observed in virgin PLA/cellulose nanocrystals based bionanocomposites [51,61]. This behavior could be explained by the barrier effect caused by the dispersion of the nanoparticles in the polymer matrix, which leads to an increase in the tortuosity of the diffusion path traveled by the gas going through the polymer film. However, the oxygen permeability increased, reaching values close to those of unfilled PLA, when the amount of YMNs was 3 wt.%. It is well known that the tortuosity of the diffusion path depends on several factors (i.e., shape and aspect ratio of the filler, degree of dispersion or exfoliation, filler loading and orientation, adhesion to the matrix, moisture activity, filler-induced crystallinity, polymer chain immobilization, filler-induced solvent retention and porosity) [61]. Thus, this result could be due to the poor dispersion of the nanoparticles in the PLAR-3YMN sample, as was observed in SEM photographs. The poor dispersion of the nanoparticles might result in the formation of micro-pores in the polymer matrix, which act as low-resistance paths for the gas diffusion through the polymer. Therefore, this result underlines the success of the dispersion of low amounts of YMNs (1 wt.%) during melt-compounding process and its reinforcement effect produced in the final formulation.

3.3.5. Mechanical Properties

Mechanical properties play a very important role in food packaging applications; consequently, nanoindentation tests were conducted to determine the effect of the addition of YMNs on mechanically recycled PLA. Figure 12 shows the indentation hardness and the Young modulus of the different materials. It can be seen that both hardness and modulus values are in good agreement with those found in the literature for PLA samples [62,63]. It could also be seen that mechanical recycling led to a slight decrease in the mechanical properties of PLA, due to the degradation of the polymer during the ageing, washing and reprocessing steps. Similar results have been reported in previous works [14,62], who found that mechanical recycling led to small decrease in the mechanical properties of PLA.

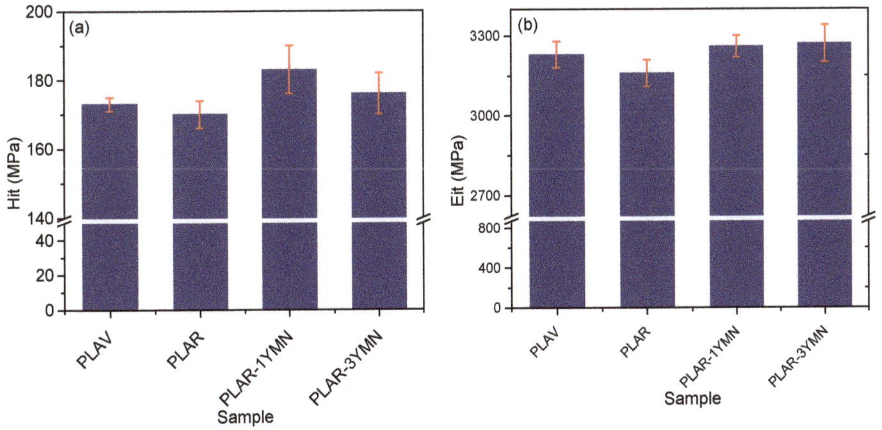

Figure 12. Indentation hardness (**a**) and Young modulus (**b**) of the different samples.

Regarding the effect of the addition of the YMNs, it can be seen that samples with 1 wt.% YMN and 3 wt.% YMN show slightly higher values for hardness and modulus than unfilled PLAR.

This result suggests that the presence of the YMNs nanoparticles has a reinforcing effect on the recycled PLA matrix. Similar trends have been reported in other PLA nanocomposites, for instance, Zaidi et al. [63,64] reported increases in both indentation hardness and the Young modulus with the addition of low amounts of organically modified montmorillonite. It is worth noting that, despite the overall improvement of the mechanical properties of recycled PLA with the addition of YMNs, better results are observed in the material with only 1 wt.% YMN. This behavior agrees with that observed in the oxygen permeability measurements and highlights the relevance of the better dispersion of lower amounts of YMNs.

4. Conclusions

The effect of the addition of lignocellulosic nanoparticles extracted from food waste, specifically yerba mate waste, on the structure, mechanical and barrier properties of mechanically recycled PLA (PLAR) was studied. PLAR was obtained by subjecting a commercial grade of PLA to accelerated ageing followed by mechanical recycling. Lignocellulosic yerba mate nanoparticles (YMNs) were extracted from yerba mate waste in an aqueous extraction process and added to PLAR in the reprocessing step at two levels (1 and 3 wt.%). FTIR and SEM analysis confirmed the successful incorporation of YMNs into the PLAR matrix.

Ageing and mechanical recycling cause the degradation of PLA, leading to a decrease in the molecular weight, thermal stability and barrier performance. The addition of small amounts of YMNs significantly modifies some properties of the material, depending on the YMNs content. The nanoparticles act as nucleating agents, thus facilitating the crystallization of PLAR, without significantly reducing the average molecular weight. Although the nanoparticles slightly reduce the thermal stability of the material, due to their lower thermal stability, the material remains stable under processing conditions. Bionanocomposites with 1 wt.% of YMNs show a good dispersion of the nanoparticles; however, when the YMNs' content rises up to 3 wt.%, although no phase separation was detected, YMNs tend to aggregate, inducing the formation of micro-voids. Thus, the addition of only 1 wt.% YMNs improved the mechanical performance and reduces oxygen permeability, a key property in food packaging materials. However, if the YMNs content rises to 3%, the effect on the oxygen barrier is negative, due to dispersion problems and the formation of micro-voids. In general, the incorporation of YMNs increases the water vapor transmission rate, due to the hydrophilic character of the nanoparticles. As for light transmission, another key property in food packaging, the addition of YMNs slightly reduces transmission in the visible region, but the recycled material remains transparent. However, nanoparticles dramatically reduce transmission in the UV areas of the spectrum, which can help slow down the degradation of the container's content.

Overall, the results obtained indicate that the addition of yerba mate nanoparticles could lead to obtaining recycled PLA with good properties for the intended use and with significant improvements in some key properties, such as the barrier to UV light and oxygen. Considering that these nanoparticles are also obtained from a food residue and using an environmentally friendly extraction process, the use of YMNs could be the basis of a useful and potentially competitive method to improve the recyclability of PLA and other similar polymers.

Author Contributions: Conceptualization, F.R.B., M.P.A. and J.M.U.; methodology, F.R.B., M.P.A., G.G. and M.U.d.l.O.; formal analysis, F.R.B., M.P.A., G.G., M.U.d.l.O. and J.M.U.; investigation, F.R.B., M.P.A. and G.G.; writing—original draft preparation, F.R.B. and M.P.A.; writing—review and editing, F.R.B., M.P.A., M.U.d.l.O. and J.M.U.; supervision, M.U.d.l.O. and J.M.U.; project administration, M.U.d.l.O. and J.M.U.; funding acquisition, M.U.d.l.O. and J.M.U. All authors have read and agreed to the published version of the manuscript.

Funding: This research was founded by European Union's Horizon 2020 research and innovation program under grant agreement No. 860407 BIO-PLASTICS EUROPE; MINECO-SPAIN under project CTM2017-88989-P and Universidad Politécnica de Madrid under project UPM RP 160543006.

Acknowledgments: The authors thank the staff of the ICTS National Center for Electron Microscopy (CNME), UCM, Madrid (Spain) for their assistance with transmission and electron scanning microscopy.

Conflicts of Interest: The authors declare no conflict of interest.

References

1. Castro-Aguirre, E.; Iniguez-Franco, F.; Samsudin, H.; Fang, X.; Auras, R. Poly(lactic acid)—mass production, processing, industrial applications, and end of life. *Adv. Drug Deliver Rev.* **2016**, *107*, 333–366. [CrossRef] [PubMed]
2. Farah, S.; Anderson, D.G.; Langer, R. Physical and mechanical properties of PLA, and their functions in widespread applications—A comprehensive review. *Adv. Drug Deliver Rev.* **2016**, *107*, 367–392. [CrossRef] [PubMed]
3. Beltrán, F.R.; Gaspar, G.; Chomachayi, M.D.; Jalali-Arani, A.; Lozano-Pérez, A.A.; Cenis, J.L.; María, U.; Pérez, E.; Urreaga, J.M.M. Influence of addition of organic fillers on the properties of mechanically recycled PLA. *Environ. Sci. Pollut. Res.* **2020**, 1–14. [CrossRef] [PubMed]
4. Reddy, M.M.; Vivekanandhan, S.; Misra, M.; Bhatia, S.K.; Mohanty, A.K. Biobased plastics and bionanocomposites: Current status and future opportunities. *Prog. Polym. Sci.* **2013**, *38*, 1653–1689. [CrossRef]
5. Chinthapalli, R.; Skoczinski, P.; Carus, M.; Baltus, W.; de Guzman, D.; Käb, H.; Raschka, A.; Ravenstijn, J. Biobased building blocks and polymers—global capacities, production and trends, 2018–2023. *Ind. Biotechnol.* **2019**, *15*, 237–241. [CrossRef]
6. Arrieta, M.P.; Samper, M.D.; Aldas, M.; López, J. On the use of PLA-PHB blends for sustainable food packaging applications. *Materials* **2017**, *10*, 1008. [CrossRef]
7. Kale, G.; Auras, R.; Singh, S.P. Comparison of the degradability of poly (lactide) packages in composting and ambient exposure conditions. *Packag. Technol. Sci.* **2007**, *20*, 49–70. [CrossRef]
8. Haider, T.P.; Völker, C.; Kramm, J.; Landfester, K.; Wurm, F.R. Plastics of the future? The impact of biodegradable polymers on the environment and on society. *Angew. Chem.–Int. Edit.* **2019**, *58*, 50–62. [CrossRef]
9. Niaounakis, M. Recycling of biopolymers–the patent perspective. *Eur. Polym. J.* **2019**, *114*, 464–475. [CrossRef]
10. Bourguignon, D. Plastics in a Circular Economy. European Parliamentary Research Service, 2018. Available online: http://www.europarl.europa.eu/RegData/etudes/ATAG/2018/625163/EPRS_ATA(2018)625163_EN.pdf (accessed on 20 March 2020).
11. European Commission. A European Strategy for Plastics in a Circular Economy. Communication from the Commission to the European Parliament, the Council, the European Economic and Social Committee and the Committee of the Regions. Brussels 2018. Available online: https://eur-lex.europa.eu/legal-content/EN/TXT/HTML/?uri=CELEX:52018DC0028&from=EN (accessed on 20 March 2020).
12. European Commission. Directive (eu) 2019/904 of the european parliament and of the council of 5 june 2019 on the reduction of the impact of certain plastic products on the environment. *Off. J. Eur. Union* **2019**. Available online: https://eur-lex.europa.eu/legal-content/EN/TXT/PDF/?uri=CELEX:32019L0904&from=EN (accessed on 20 March 2020).
13. Payne, J.; McKeown, P.; Jones, M.D. A circular economy approach to plastic waste. *Polym. Degrad. Stabil.* **2019**, *165*, 170–181. [CrossRef]
14. Beltrán, F.; Lorenzo, V.; Acosta, J.; de la Orden, M.; Urreaga, J.M. Effect of simulated mechanical recycling processes on the structure and properties of poly(lactic acid). *J. Environ. Manag.* **2018**, *216*, 25–31. [CrossRef]
15. Botta, L.; Scaffaro, R.; Sutera, F.; Mistretta, M.C. Reprocessing of PLA/graphene nanoplatelets nanocomposites. *Polymers* **2018**, *10*, 18. [CrossRef] [PubMed]
16. Scaffaro, R.; Sutera, F.; Mistretta, M.; Botta, L.; La Mantia, F. Structure-properties relationships in melt reprocessed PLA/hydrotalcites nanocomposites. *Express. Polym. Lett.* **2017**, *11*, 555. [CrossRef]
17. Beltrán, F.R.; Climent-Pascual, E.; de la Orden, M.U.; Martínez Urreaga, J. Effect of solid-state polymerization on the structure and properties of mechanically recycled poly(lactic acid). *Polym. Degrad. Stabil.* **2020**, *171*, 109045. [CrossRef]
18. Beltrán, F.R.; Infante, C.; de la Orden, M.U.; Martínez Urreaga, J. Mechanical recycling of poly(lactic acid): Evaluation of a chain extender and a peroxide as additives for upgrading the recycled plastic. *J. Clean. Prod.* **2019**, *219*, 46–56. [CrossRef]
19. Tuna, B.; Ozkoc, G. Effects of diisocyanate and polymeric epoxidized chain extenders on the properties of recycled poly(lactic acid). *J. Polym. Environ.* **2017**, *25*, 983–993. [CrossRef]
20. Beltrán, F.R.; de la Orden, M.U.; Martínez Urreaga, J. Amino-modified halloysite nanotubes to reduce polymer degradation and improve the performance of mechanically recycled poly(lactic acid). *J. Polym. Environ.* **2018**, *26*, 4046–4055. [CrossRef]

21. Tang, Q.; Wang, F.; Liu, X.; Tang, M.; Zeng, Z.; Liang, J.; Guan, X.; Wang, J.; Mu, X. Surface modified palygorskite nanofibers and their applications as reinforcement phase in *cis*-polybutadiene rubber nanocomposites. *Appl. Clay Sci.* **2016**, *132–133*, 175–181. [CrossRef]
22. Raquez, J.-M.; Habibi, Y.; Murariu, M.; Dubois, P. Polylactide (PLA)-based nanocomposites. *Prog. Polym. Sci.* **2013**, *38*, 1504–1542. [CrossRef]
23. Arrieta, M.P.; Fortunati, E.; Dominici, F.; Rayón, E.; López, J.; Kenny, J.M. Multifunctional PLA-PHB/cellulose nanocrystal films: Processing, structural and thermal properties. *Carbohydr. Polym.* **2014**, *107*, 16–24. [CrossRef] [PubMed]
24. Arrieta, M.P.; Garrido, L.; Faba, S.; Guarda, A.; Galotto, M.J.; Dicastillo, C.L.d. Cucumis metuliferus fruit extract loaded acetate cellulose coatings for antioxidant active packaging. *Polymers* **2020**, *12*, 1248. [CrossRef] [PubMed]
25. Berglund, L.; Noël, M.; Aitomäki, Y.; Öman, T.; Oksman, K. Production potential of cellulose nanofibers from industrial residues: Efficiency and nanofiber characteristics. *Ind. Crop. Prod.* **2016**, *92*, 84–92. [CrossRef]
26. Fajardo, J.; Valarezo, L.; López, L.; Sarmiento, A. Experiencies in obtaining polymeric composites reinforced with natural fiber from ecuador. *Ingenius* **2013**, 28–35. [CrossRef]
27. Arrieta, M.P.; Fortunati, E.; Dominici, F.; López, J.; Kenny, J.M. Bionanocomposite films based on plasticized PLA–PHB/cellulose nanocrystal blends. *Carbohydr. Polym.* **2015**, *121*, 265–275. [CrossRef]
28. Luzi, F.; Fortunati, E.; Jiménez, A.; Puglia, D.; Pezzolla, D.; Gigliotti, G.; Kenny, J.M.; Chiralt, A.; Torre, L. Production and characterization of PLA_PBS biodegradable blends reinforced with cellulose nanocrystals extracted from hemp fibres. *Ind. Crop. Prod.* **2016**, *93*, 276–289. [CrossRef]
29. Yang, W.; Fortunati, E.; Dominici, F.; Kenny, J.M.; Puglia, D. Effect of processing conditions and lignin content on thermal, mechanical and degradative behavior of lignin nanoparticles/polylactic(acid) bionanocomposites prepared by melt extrusion and solvent casting. *Eur. Polym. J.* **2015**, *71*, 126–139. [CrossRef]
30. Fortunati, E.; Luzi, F.; Puglia, D.; Petrucci, R.; Kenny, J.M.; Torre, L. Processing of PLA nanocomposites with cellulose nanocrystals extracted from Posidonia oceanica waste: Innovative reuse of coastal plant. *Ind. Crop. Prod.* **2015**, *67*, 439–447. [CrossRef]
31. Yang, W.; Fortunati, E.; Dominici, F.; Giovanale, G.; Mazzaglia, A.; Balestra, G.M.; Kenny, J.M.; Puglia, D. Synergic effect of cellulose and lignin nanostructures in PLA based systems for food antibacterial packaging. *Eur. Polym. J.* **2016**, *79*, 1–12. [CrossRef]
32. Arrieta, M.P.; Peponi, L.; López, D.; Fernández-García, M. Recovery of yerba mate (*Ilex paraguariensis*) residue for the development of PLA-based bionanocomposite films. *Ind. Crop. Prod.* **2018**, *111*, 317–328. [CrossRef]
33. Deladino, L.; Teixeira, A.S.; Navarro, A.S.; Alvarez, I.; Molina-García, A.D.; Martino, M. Corn starch systems as carriers for yerba mate (*Ilex paraguariensis*) antioxidants. *Food Bioprod. Process.* **2015**, *94*, 463–472. [CrossRef]
34. Dahlem, M.A.; Borsoi, C.; Hansen, B.; Catto, A.L. Evaluation of different methods for extraction of nanocellulose from yerba mate residues. *Carbohydr. Polym.* **2019**, *218*, 78–86. [CrossRef] [PubMed]
35. Pagliosa, C.M.; de Simas, K.N.; Amboni, R.D.; Murakami, A.N.N.; Petkowicz, C.L.; de Deus Medeiros, J.; Rodrigues, A.C.; Amante, E.R. Characterization of the bark from residues from mate tree harvesting (*Ilex paraguariensis* st. Hil.). *Ind. Crop. Prod.* **2010**, *32*, 428–433. [CrossRef]
36. Burris, K.P.; Harte, F.M.; Davidson, P.M.; Stewart Jr, C.N.; Zivanovic, S. Composition and bioactive properties of yerba mate (*Ilex paraguariensis* a. St.-hil.): A review. *Chil. J. Agricl. Res.* **2012**, *72*, 268. [CrossRef]
37. Medina Jaramillo, C.; Gutiérrez, T.J.; Goyanes, S.; Bernal, C.; Famá, L. Biodegradability and plasticizing effect of yerba mate extract on cassava starch edible films. *Carbohydr. Polym.* **2016**, *151*, 150–159. [CrossRef] [PubMed]
38. González de Mejía, E.; Song, Y.S.; Heck, C.I.; Ramírez-Mares, M. Yerba mate tea (*Ilex paraguariensis*): Phenolics, antioxidant capacity and in vitro inhibition of colon cancer cell proliferation. *J. Funct. Food* **2010**, *2*, 23–34. [CrossRef]
39. Instituto Brasileiro de Geografia e Estatística—IBGE. 2018. Available online: Https://sidra.Ibge.Gov.Br/tabela/289#resultado (accessed on 20 March 2020).
40. Instituto Nacional de Yerba Mate (INYM). 2020. Available online: Https://www.Inym.Org.Ar/aumentaron-la-produccion-y-el-consumo-de-yerba-mate-en-la-argentina/ (accessed on 20 March 2020).
41. Nunes Ferraz Junior, A.D.; Etchelet, M.I.; Braga, A.F.M.; Clavijo, L.; Loaces, I.; Noya, F.; Etchebehere, C. Alkaline pretreatment of yerba mate (*Ilex paraguariensis*) waste for unlocking low-cost cellulosic biofuel. *Fuel* **2020**, *266*, 117068. [CrossRef]

42. Pinheiro Bruni, G.; dos Santos Acunha, T.; de Oliveira, J.P.; Martins Fonseca, L.; Tavares da Silva, F.; Martins Guimarães, V.; da Rosa Zavareze, E. Electrospun protein fibers loaded with yerba mate extract for bioactive release in food packaging. *J. Sci. Food Agric.* **2020**, *100*, 3341–3350. [CrossRef]
43. Meaurio, E.; López-Rodríguez, N.; Sarasua, J.R. Infrared spectrum of poly(l-lactide): Application to crystallinity studies. *Macromolecules* **2006**, *39*, 9291–9301. [CrossRef]
44. Fortunati, E.; Armentano, I.; Zhou, Q.; Iannoni, A.; Saino, E.; Visai, L.; Berglund, L.A.; Kenny, J.M. Multifunctional bionanocomposite films of poly(lactic acid), cellulose nanocrystals and silver nanoparticles. *Carbohydr. Polym.* **2012**, *87*, 1596–1605. [CrossRef]
45. Yang, W.; Owczarek, J.; Fortunati, E.; Kozanecki, M.; Mazzaglia, A.; Balestra, G.; Kenny, J.; Torre, L.; Puglia, D. Antioxidant and antibacterial lignin nanoparticles in polyvinyl alcohol/chitosan films for active packaging. *Ind. Crop. Prod.* **2016**, *94*, 800–811. [CrossRef]
46. Mondragon, G.; Fernandes, S.; Retegi, A.; Peña, C.; Algar, I.; Eceiza, A.; Arbelaiz, A. A common strategy to extracting cellulose nanoentities from different plants. *Ind. Crop. Prod.* **2014**, *55*, 140–148. [CrossRef]
47. Auras, R.; Harte, B.; Selke, S. An overview of polylactides as packaging materials. *Macromol. Biosci.* **2004**, *4*, 864. [CrossRef]
48. Badia, J.; Santonja-Blasco, L.; Martínez-Felipe, A.; Ribes-Greus, A. Hygrothermal ageing of reprocessed polylactide. *Polym. Degrad. Stabil.* **2012**, *97*, 1881–1890. [CrossRef]
49. Beltrán, F.; de la Orden, M.; Lorenzo, V.; Pérez, E.; Cerrada, M.; Urreaga, J.M. Water-induced structural changes in poly(lactic acid) and PLLA-clay nanocomposites. *Polymer* **2016**, *107*, 211–222. [CrossRef]
50. Chen, X.; Han, L.; Zhang, T.; Zhang, J. Influence of crystal polymorphism on crystallinity calculation of poly (l-lactic acid) by infrared spectroscopy. *Vib. Spectrosc.* **2014**, *70*, 1–5. [CrossRef]
51. Arrieta, M.P.; Fortunati, E.; Dominici, F.; Rayón, E.; López, J.; Kenny, J.M. PLA-PHB/cellulose based films: Mechanical, barrier and disintegration properties. *Polym. Degrad. Stabil.* **2014**, *107*, 139–149. [CrossRef]
52. Way, C.; Dean, K.; Wu, D.Y.; Palombo, E. Biodegradation of sequentially surface treated lignocellulose reinforced polylactic acid composites: Carbon dioxide evolution and morphology. *Polym. Degrad. Stabil.* **2012**, *97*, 430–438. [CrossRef]
53. Di Lorenzo, M.L. Calorimetric analysis of the multiple melting behavior of poly (l-lactic acid). *J. App Polym. Sci.* **2006**, *100*, 3145–3151. [CrossRef]
54. Lizundia, E.; Fortunati, E.; Dominici, F.; Vilas, J.L.; León, L.M.; Armentano, I.; Torre, L.; Kenny, J.M. PLLA-grafted cellulose nanocrystals: Role of the CNC content and grafting on the PLA bionanocomposite film properties. *Carbohydr. Polym.* **2016**, *142*, 105–113. [CrossRef]
55. Burgos, N.; Martino, V.P.; Jiménez, A. Characterization and ageing study of poly(lactic acid) films plasticized with oligomeric lactic acid. *Polym. Degrad. Stabil.* **2013**, *98*, 651–658. [CrossRef]
56. Chariyachotilert, C.; Joshi, S.; Selke, S.E.; Auras, R. Assessment of the properties of poly (L-lactic acid) sheets produced with differing amounts of postconsumer recycled poly(L-lactic acid). *J. Plast. Film Sheeting* **2012**, *28*, 314–335. [CrossRef]
57. Yang, W.; Weng, Y.; Puglia, D.; Qi, G.; Dong, W.; Kenny, J.M.; Ma, P. Poly(lactic acid)/lignin films with enhanced toughness and anti-oxidation performance for active food packaging. *Int. J. Biol. Macromol.* **2020**, *144*, 102–110. [CrossRef]
58. Shogren, R. Water vapor permeability of biodegradable polymers. *J. Environ. Polym. Degrad.* **1997**, *5*, 91–95. [CrossRef]
59. Kim, Y.; Suhr, J.; Seo, H.-W.; Sun, H.; Kim, S.; Park, I.-K.; Kim, S.-H.; Lee, Y.; Kim, K.-J.; Nam, J.-D. All biomass and UV protective composite composed of compatibilized lignin and poly(lactic-acid). *Sci. Rep.* **2017**, *7*, 43596. [CrossRef]
60. Espino-Pérez, E.; Bras, J.; Ducruet, V.; Guinault, A.; Dufresne, A.; Domenek, S. Influence of chemical surface modification of cellulose nanowhiskers on thermal, mechanical, and barrier properties of poly(lactide) based bionanocomposites. *Eur. Polym. J.* **2013**, *49*, 3144–3154. [CrossRef]
61. Fortunati, E.; Peltzer, M.; Armentano, I.; Torre, L.; Jiménez, A.; Kenny, J.M. Effects of modified cellulose nanocrystals on the barrier and migration properties of pla nano-biocomposites. *Carbohyd. Polym.* **2012**, *90*, 948–956. [CrossRef]
62. Pillin, I.; Montrelay, N.; Bourmaud, A.; Grohens, Y. Effect of thermo-mechanical cycles on the physico-chemical properties of poly (lactic acid). *Polym. Degrad. Stabil.* **2008**, *93*, 321–328. [CrossRef]

63. Zaidi, L.; Bruzaud, S.; Bourmaud, A.; Médéric, P.; Kaci, M.; Grohens, Y. Relationship between structure and rheological, mechanical and thermal properties of polylactide/cloisite 30b nanocomposites. *J. Appl. Polym. Sci.* **2010**, *116*, 1357–1365. [CrossRef]
64. Zaidi, L.; Kaci, M.; Bruzaud, S.; Bourmaud, A.; Grohens, Y. Effect of natural weather on the structure and properties of polylactide/cloisite 30b nanocomposites. *Polym. Degrad. Stabil.* **2010**, *95*, 1751–1758. [CrossRef]

polymers

MDPI

Article

Cucumis metuliferus Fruit Extract Loaded Acetate Cellulose Coatings for Antioxidant Active Packaging

Marina Patricia Arrieta [1,*], Luan Garrido [2], Simón Faba [2], Abel Guarda [2], María José Galotto [2] and Carol López de Dicastillo [2,*]

1 Departamento de Química Orgánica, Facultad de Óptica y Optometría, Universidad Complutense de Madrid (UCM), Arcos de Jalón 118, 28037 Madrid, Spain

2 Center of Innovation in Packaging (Laben), Department of Science and Food Technology, Faculty of Technology, Center for the Development of Nanoscience and Nanotechnology (CEDENNA), Universidad de Santiago de Chile (USACH), 9170201 Santiago, Chile; luan.garrido@usach.cl (L.G.); simon.riverosf@usach.cl (S.F.); abel.guarda@usach.cl (A.G.); maria.galotto@usach.cl (M.J.G.)

* Correspondence: marrie06@ucm.es (M.P.A.); analopez.dediscastillo@usach.cl (C.L.d.D.); Tel.: +34-913-946-885 (M.P.A.); +56-2-2718-4510 (C.L.d.D.)

Received: 12 May 2020; Accepted: 28 May 2020; Published: 29 May 2020

Abstract: A new active coating was developed by using *Cucumis metuliferus* fruit extract as antioxidant additive with the aim of obtaining an easy way to functionalize low-density polyethylene (LDPE) films for food packaging applications. Thus, an extraction protocol was first optimized to determine the total phenolic compounds and the antioxidant activity of CM. The aqueous CM antioxidant extract was then incorporated into cellulose acetate (CA) film-forming solution in different concentrations (1, 3 and 5 wt.%) to be further coated in corona-treated LDPE to obtain LDPE/CA-CM bilayer systems. CA and CA-CM film-forming solutions were successfully coated onto the surface of LDPE, showing good adhesion in the final bilayer structure. The optical, microstructural, thermal, mechanical and oxygen barrier performance, as well as the antioxidant activity, were evaluated. The active coating casted onto the LDPE film did not affect the high transparency of LDPE and improved the oxygen barrier performance. The antioxidant effectiveness of bilayer packaging was confirmed by release studies of *Cucumis metuliferus* from the cellulose acetate layer to a fatty food simulant. Finally, the LDPE/CA-CM active materials were also tested for their application in minimally processed fruits, and they demonstrated their ability to reduce the oxidation process of fresh cut apples. Thus, the obtained results suggest that CA-CM-based coating can be used to easily introduce active functionality to typically used LDPE at industrial level and enhance its oxygen barrier, without affecting the high transparency, revealing their potential application in the active food packaging sector to extend the shelf-life of packaged food by prevention of lipid oxidation of fatty food or by prevention fruit browning.

Keywords: *Cucumis metuliferus*; extraction; antioxidant activity; coating; cellulose acetate; LDPE; bilayer packaging; active packaging

1. Introduction

Increasing ecological concern aimed towards a reduction of the environmental impact of short-term plastics (i.e., packaging, disposable cutlery, agricultural mulch films, etc.) is contributing to a move towards a circular economy model, in which a more sustainable plastic industry is continuously developing. The deliberate introduction of bio-based and biodegradable plastics in the field of packaging will make it possible to reduce the consumption of non-renewable petrochemical-based resources, as well as to prevent the accumulation of plastics waste in the environment (i.e., landfills, oceans, etc.) within the frame of the circular economy [1,2].

Cellulose acetate (CA) is a thermoplastic biodegradable polymer extensively studied for packaging applications owing to its excellent optical transparency and high toughness, and because it has the advantage of being produced primarily through the esterification of the renewable and most abundant polymer in nature: cellulose [3–5]. In fact, among cellulosic derivates, CA is extremely attractive for the packaging field mainly because of its low price, good biodegradability and non-toxicity, as well as due to its having a better processability than cellulose, as it can be processed either by solvent-casting, melt-blending or electrospinning approaches [6–8]. Moreover, CA has been widely used for the development of active packaging materials by means of the incorporation of active compounds (i.e., antioxidant and antimicrobial substances) into the CA polymeric matrix [9–11]. Active food packaging has advantages over direct addition of active compounds into the foodstuff, such as the use of additives in lower concentrations, as well as extension of shelf life due to the controlled release of active compounds during storage [12,13]. Moreover, it has an important effect on the reduction of deterioration reactions that begin at the food surface due to a more significant interaction with the surface of the packed food [11,14,15]. Additionally, there is a growing trend in the food packaging industry to replace synthetic additives with natural antioxidants in both petrochemical-based [16–19] and bio-based and biodegradable polymers [10,20–22], particularly with tocopherol, catechols, essential oils, and plant extracts [16,20,23].

In this regard, coatings are progressively becoming more widely recognized as a powerful tool for extending the shelf life of food by improving many properties of plastic materials as well as a simple method for providing specific active functions (e.g., antioxidant, antimicrobial, etc.) to the final food packaging [24]. In fact, the addition of active compounds into food coatings or packaging coatings has some advantages, as they act only at the surface level, and can be applied at any stage of the food supply chain [13,21]. In the food packaging sector, coating technology represents the most efficient and affordable solution for attaining high barrier properties against oxygen for light packaging, particularly in the case of polyolefins (i.e., polypropylene (PP) and polyethylene (PE)) [25]. In this regard, the application of CA-based active coatings on a typical film packaging material (e.g., low-density polyethylene (LDPE)) is an advantageous alternative for easily providing the final material with specific active performance. In fact, LDPE is one of the most widely consumed polymers in the food packaging field; it is extensively used in film to cover foodstuff due to its low cost, high resistance to tearing, low heat seal temperature, and high water barrier, as well as its high production efficiency [2,19,26]. However, it presents high oxygen permeability, which is a crucial property for plastic food packaging films [27]. Several strategies have been explored to improve the LDPE barrier performance for food packaging applications through blending [28,29], development of nanocomposites [30,31], or by using multi- or bi-layer approaches (i.e., surface coatings, sandwich structures, electrospun deposition, etc.) [25,32–35]. Coating approaches are of high interest since they make it possible to obtain a bi-layer structure using an easy, scalable and cost-effective method at an industrial level. Thus, applying a CA-CM layer to commercially available LDPE films by a simple coating process has the potential to reduce the oxidation process of packed food, providing the CA with better intrinsic oxygen barrier performance than LDPE, as well as offering the additional advantage of giving active packaging technology to the final formulation by simply incorporating antioxidant compounds into the CA-based film-forming solution.

Cucumis metuliferus (Cucurbitaceae) is an annual climber plant, native to Africa, that grows specifically from South Africa to tropical Africa [36]. It is known for its potential benefits to human health, and it has been suggested that it possesses antifungal, antimicrobial, antiviral and antioxidant effects, as well as chelating power [37,38]. It is called African homed melon, jelly melon and kiwano. The commercial culture of *C. metuliferus* began in New Zealand, where it was commercialized as an exotic fruit during the eighties. The main commercial advantages of *C. metuliferus* are that it grows rapidly and remains in good condition for around 6 months without cold storage [39,40]. For this reason, the commercialization was extended, and nowadays it grows in New Zealand, Australia, Chile, Argentina, Venezuela, Spain, Portugal, Germany, Italy, Israel and California [37,40]. Young *C. metuliferus* fruit is dark green with mottled light green spots, while as it ripens it becomes bright orange with very sharp spines [36]. In the interior is a mass of green, translucent, slightly mucilaginous juice-sacs enclosing many tightly packed, flat seeds [39]. Although the antioxidant ability of *C. metuliferus* has been determined [36,38], to the best of our knowledge its use in the development of antioxidant active packaging coating has not yet been proposed.

The main goal of the present research work was to assess the potential production of antioxidant active coatings for food packaging proposes based on cellulose acetate loaded with *C. metuliferus* fruit extract (CM). Initially, the extraction of antioxidant agents from *C. metuliferus* fruit was optimized by evaluating extraction procedures using different solvents: water, ethanol and ethanol 50%. Then, the obtained extract was incorporated into a cellulose acetate solution in different proportions (1, 3 and 5 wt.%). While coatings require substrates with high surface energy, LDPE is known to possess low surface energy as a consequence of its non-polar nature. Hence, LDPE is frequently surface treated to promote good adhesion between the polyolefin and the coating [33,41]. Thus, the obtained CM-functionalized CA film-forming solutions were coated onto commercial LDPE films. The obtained bi-layer structures were fully characterized considering the intended use in the active food packaging field. Thus, the correct adhesion of CA coating into corona-treated LDPE film was corroborated by scanning electron microscopy (SEM). The effect of the CA-based coating on the optical properties of LDPE was investigated by UV-visible measurements and the determination of colorimetric properties in the CIELab space. The mechanical and barrier performances were also evaluated with the aim of assessing their suitability for the food packaging sector. Finally, since these materials are intended for active food packaging applications, the release ability of the antioxidant compounds of *C. metuliferus* fruit from bilayer materials was analyzed in a fatty food simulant, as well as in direct contact with fresh-cut apples in order to get information of the possible application of these sustainable materials at an industrial level intended for both fatty food and fresh fruit protection.

2. Materials and Methods

2.1. Materials

Cellulose acetate with Mn = 30,000 and 39.8 wt.% acetyl content (CA degree of substitution = 2.5 [42]) was supplied by Sigma-Aldrich (Santiago, Chile). A commercial corona-treated low-density polyethylene (LDPE) film was kindly supplied by EDELPA (Santiago, Chile). The *Cucumis metuliferus* fruits were obtained at a local market in Santiago de Chile. 2,2-azinobis(3-ethylbenzothiazoline-6-sulphonate) (ABTS), Folin Ciocalteu phenol reagent, anhydrous sodium carbonate, gallic acid and 6-hydroxy-2,5,7,8-tetramethylchroman-2-carboxylic acid (Trolox) were purchased from Sigma Aldrich. Acetone (99.9% HPLC grade) and absolute ethanol (99.9% HPLC grade) were supplied by Merck (Santiago, Chile).

2.2. Methods

2.2.1. *Cucumis metuliferus* Extraction Optimization

The *C. metuliferus* fruits were cut into slices and dried at 40 °C for 48 h. They were mechanically grinded to obtain powder by means of a cutter grinder. With the aim of obtaining the CM extract with highest antioxidant performance, it was extracted from *C. metuliferus* fruit powder using three different solvent systems: water, ethanol and 50% aqueous ethanol (v/v, EtOH 50%). Approximately 500 mg of *C. metuliferus* fruit powder was dispersed in 30 mL of each solvent and vigorously stirred for 180 min at 40 °C. The obtained viscous extracts were then filtered twice and used for the determination of radical scavenging activity and the measurement of total phenolic content (PC). Figure 1 show the schematic representation of *C. metuliferus* extract (CM) extraction procedure from *Cucumis metuliferus* fruit.

The total phenolic content (PC) of the *C. metuliferus* extract was colorimetrically determined by means of the Folin-Ciocalteu method according to the methodology adapted in previous work by Lopez de Dicastillo et al. [10]. In brief, 0.2 mL of Folin-Ciocalteu reagent and 3.1 mL of distilled water were mixed with 0.1 mL of each CM extract and kept in darkness for 5 min. Then, 0.6 mL of anhydrous Na_2CO_3 20% (w/v) was added, shaken and then kept in the dark for 2 h. The PC was determined by means of the absorbance at 765 nm in a UV-vis spectrophotometer. The measurements were performed in triplicate and the results were expressed as mg gallic acid equivalent (GAE) per 100 g of dried sample.

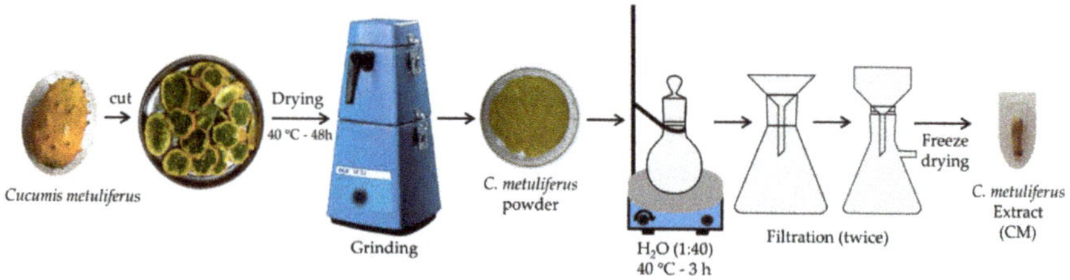

Figure 1. Schematic representation of antioxidants extraction procedure from *Cucumis metuliferus* fruit.

The evaluation of the antioxidant ability of *C. metuliferus* fruit extract in the three different solvents assayed was determined by means of Ferric Reducing Antioxidant Power (FRAP), as well as radical 2,2′-azinobis(3-ethylbenzothiazoline-6-sulphonate ($ABTS^{\cdot+}$) methods, since they are simple, inexpensive and robust techniques [10,43]. FRAP follows a Single Electron Transfer (SET) method, and thus detects the ability of the antioxidant to transfer one electron to deactivate the reactive functional group of ferric 2,4,6-tripyridyl-s-triazine (TPTZ) [12]. In fact, FRAP measures the reduction to a blue ferrous product, which is absorbed at 593 nm [12,44]. Meanwhile, the ABTS method monitors the inhibition of oxidation of a suitable substrate, which may be neutralized either by direct reduction via SET or by radical quenching via Hydrogen Atom Transfer (HAT) [12,44]. HAT-based methods measure the antioxidant ability to quench free radicals by hydrogen donation [44]. The inhibition of the cationic radical $ABTS^{\cdot+}$ due to the presence of antioxidant compounds from *C. metuliferus* fruit extract was followed by the reduction of the characteristic wavelength absorption spectrum at 715 nm [16,45]. FRAP and ABTS methods were performed in triplicate, and results were expressed as Trolox equivalents per 100 g of fruit sample.

2.2.2. Preparation of *Cucumis metuliferus*-Loaded Cellulose Acetate-Coated LDPE Films

Cellulose acetate (CA) coating was prepared by dissolving 6 g of CA in 40 mL of acetone (0.15 g/mL) at 50 °C under stirring. The antioxidant cellulose acetate coating (CA-CM) was prepared by adding 1, 3 and 5 wt.% of the *C. metuliferus* antioxidant extract (CM) respect to CA weight (Figure 2a).

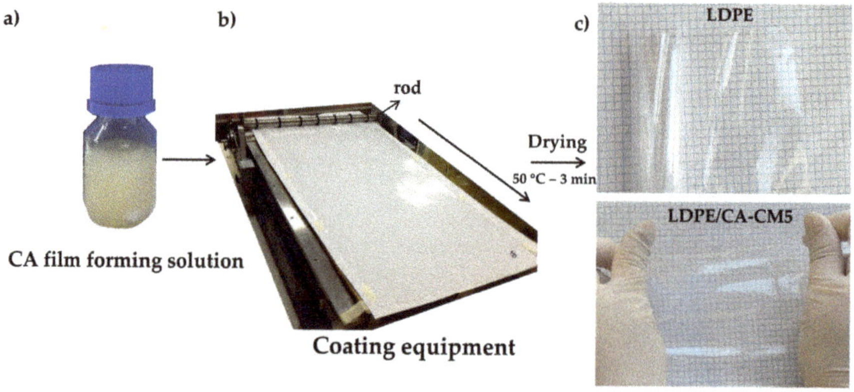

Figure 2. Schematic representation of the bi-layer film-forming process: (**a**) film coating solution, (**b**) coating application over LDPE film followed by drying process, and (**c**) visual appearance of the CA-CM coated LDPE films.

Each acetate film-forming solution was poured onto the corona-treated side of LDPE films with a lab-scale automatic applicator (Multicoater RK Printcoat, model K303, Royston, UK) equipped with a steel horizontal rod to obtain a homogeneous wet coating material of 80 μm (speed of 5 m/min) and at room temperature (Figure 2b). The coated LDPE films were immediately dried at 50 °C for 3 min and transparent films were obtained (Figure 2c). CA-CM films were also prepared for comparison. Thus, the film-forming solutions were casted on a 50 mm-diameter Petri dish and dried at 50 °C for 3 h.

Figure 2 schematically summarizes the bi-layer structure film preparation procedure, starting from the film-forming solution (Figure 2a), its application onto the LDPE film surface with the automatic applicator (Figure 2b), and the resulting bilayer LDPE film coated with CA-CM (Figure 2c).

2.2.3. Film Characterization

Characterization of *Cucumis metuliferus*-Loaded Cellulose Acetate-Coated LDPE Films

The thickness of the obtained bilayer films was measured with a Digimatic Micrometer Series 293 MDC-Lite (Mitutoyo, Tokyo, Japan) at ten random positions over the film surface.

The absorption spectra in the 700–250 nm region of bilayer films were obtained using a Perkin Elmer (Lambda 35, Waltham, MA, USA) UV-VIS spectrophotometer.

The color properties of the films were measured in the CIELab space in a Minolta colorimeter CR-410 Chroma Meter (Minolta Series, Tokyo, Japan). The colorimeter was calibrated with a white standard tile. Measurements were carried out in quintuplicate at random positions over the CA-based coating surface layer of the LDPE films and average values for these five tests were used to calculate the total color differences (ΔE) induced by the presence of CA and CA-CM coatings into LDPE by means of Equation (1):

$$\Delta E = \sqrt{(\Delta a^*)^2 + (\Delta b^*)^2 + (\Delta L)^2} \quad (1)$$

where a*, b* and L are the color coordinates L (lightness), a* (red-green) and b* (yellow-blue).

The cross cryo-fractured surface microstructure of the cross-section of the bi-layer structures was observed by scanning electron microscopy (SEM) using a JEOL F-6335 microscope. Samples were previously sputtered with a gold layer to make them conductive.

Thermogravimetric measurements were carried out in a Mettler Toledo Gas Controller GC20 Stare System TGA/DCS thermal analyzer (Schwerzenbach, Switzerland). The experiments were conducted under dynamic mode and under nitrogen atmosphere (flow rate of 50 mL/min). Film samples were heated from room temperature to 700 °C at 10 °C/min. The initial degradation temperatures (T_0) were determined at 5% mass loss. Meanwhile, the temperatures at the maximum degradation rate (T_{max}) were calculated from the first derivative of the TGA curves (DTG) for CA (T_{maxCA}), as well as for LDPE ($T_{maxLDPE}$).

The mechanical properties of the LDPE and LDPE/CA bilayer films were determined by tensile test measurements at room temperature with an IBERTEST ELIB 30 (S.A.E. Ibertest, Madrid, Spain) machine equipped with a 100 N load cell. Tests were performed in rectangular strips (dimensions: 100 × 10 mm^2), initial grip separation of 50 mm and crosshead speed of 2 mm/min. Five different samples were tested, and average values of tensile strength and elongation at break were reported.

The oxygen permeation rates of the LDPE and LDPE/CA films were determined at 23 °C and 0% relative humidity (RH) by means of an OXTRAN MOCON model 2/21 ML (Lippke, Neuwied, Germany). Films were previously purged with nitrogen for a minimum of 16 h prior to exposure to an oxygen flow of 10 mL/min. The oxygen permeability coefficient (OP) is proportional to oxygen transmission rate per thickness, OTR*e (e = thickness, mm), and thus, the OTR*e values were used to compare the oxygen barrier properties of the films.

Release studies of the active compounds from the CA-CM coated LDPE films were conducted by immersion of the films into a fatty food simulant (simulant D1 = solution of 50% ethanol) at 40 °C for 10 days [46]. Double sided, total immersion migration tests were carried out by total immersion of

3 cm^2 piece of each film in 5 mL of food simulant (area-to-volume ratio = 6 dm^2/L) contained in a glass vial. Since $ABTS^{\cdot+}$ is an indicator radical that can be neutralized either by direct reduction via SET or by radical quenching via HAT [12], the antioxidant performance of the developed film formulations was measured by means of ABTS method. Therefore, the antioxidant activity of the *C. metuliferus* fruit extract released in the fatty food simulant was regularly analyzed by the scavenging activity of stable free $ABTS^{\cdot+}$ radicals, expressed as Trolox equivalents per film area.

The obtained CA-CM coated LDPE films were also tested as fresh fruit browning prevention systems. Thus, LDPE/CA and LDPE/CA-CM films were used to pack fresh-cut apples. Apples were previously washed with tap water, peeled and sliced with a clean knife and packed with the developed bilayer materials in direct contact with the CA-CM layer. The browning of apples was indirectly measured by colorimetrical measurements in the CIELab space at 30 °C for 92 h. The packed sliced apples color changes were measured in a Minolta colorimeter CR-410 Chroma Meter (Minolta Series, Tokyo, Japan). The colorimeter was calibrated with a white standard tile. Measurements were carried out in quintuplicate at random positions over the packed apple surfaces and average values for these five tests were used to calculate the total color differences (ΔE) by Equation (1).

Significant differences in the determination of PC as well as in the assessment of antioxidant activity of *C. metuliferus* fruit extract (FRAP and ABTS methods) were statistically calculated by one-way analysis of variance (ANOVA) with OriginPro 8 software using Tukey's test with a 95% confidence level. Similarly, for bilayer films the colorimetric coordinates determinations, tensile test measurements, the release studies of CM-CA-coated LDPE films, as well as the color changes measurements in packed sliced apples, were also statistically calculated by one-way analysis of variance (ANOVA) with OriginPro 8 software using Tukey's test with a 95% confidence level.

3. Results and Discussion

3.1. Antioxidant Activity of Fruit Extracts

The antioxidant ability of natural extracts is highly dependent on the chemical structure of the active compounds, as well as on the mechanisms used (SET and/or HAT) [12]. Therefore, there is no general standardized method for the extraction of antioxidant agents from heterogeneous systems, such as foods and crops [10]. With the aim of evaluating the most effective extraction process, various solvent systems were assayed based on previous works [43,47]. Table 1 reports the polyphenolic content (PC), as well as the antioxidant activities measured by two methods: FRAP (which operates by the SET mechanism) and ABTS (which operates by both the HAT and SET mechanisms) of the resulting fruit extracts. The lowest polyphenolic extraction efficiency was obtained for pure ethanol, while PC and antioxidant power values of aqueous and aqueous/ethanol extractive solutions did not present significant differences. Matsusaka et al. studied the PC of edible (pulp) and non-edible (seed and peel) parts of *C. metuliferus* from Japan, extracted in EtOH 50%, and similar values were obtained [38]. The peel and seeds showed higher phenolic content than the edible pulp [38].

Table 1. Total phenolic content (PC) and antioxidant measurements (ABTS and FRAP) of *C. metuliferus* extract in different solvents.

C. metuliferus Extract	PC (mg GAE/100g Fruit)	FRAP (mg Trolox/100g Fruit)	ABTS (mg Trolox/100g Fruit)
H_2O	89.0 ± 5.1^{a}	238.6 ± 8.6^{a}	8.0 ± 0.2^{a}
EtOH	47.2 ± 7.3^{b}	161.3 ± 2.0^{b}	9.6 ± 0.8^{a}
EtOH 50%	101.5 ± 1.9^{a}	241.1 ± 3.2^{a}	9.0 ± 0.1^{a}

$^{a-b}$ Different superscripts within the same column indicate significant differences between formulations ($p < 0.05$).

With respect to antioxidant ability, it is known that the FRAP method is more specific for hydrophilic antioxidants, while ABTS is a good method for evaluating both lipophilic as well as hydrophilic antioxidants [48]. The FRAP and ABTS methods (Table 1) indicated that FRAP values were higher for aqueous and 50% ethanolic extractive solutions (without significant differences, $p > 0.05$) than for ethanol ($p < 0.05$). Matsusaka et al. also determined the radical scavenging activity using the ABTS method, obtaining around 200 µmol Trolox/g of whole fruit (edible and non-edible parts), which is approximately 5 mg GAE/100g fruit, which is lower than the results obtained here. Motlhanka studied the antioxidant performance of aqueous, methanolic and under chloroform *C. metuliferus* extracts (pulp and skin) using the DPPH method, and the results indicated that aqueous extract exhibited the strongest antioxidant response, while methanolic extract possessed moderate antioxidant response and low activity in chloroform [36]. Both works were in accordance in confirming that principal phenolic compounds were mainly extracted by using distilled water. Although the chemical composition of *C. metuliferus* has aroused little scientific interest, it is known that the pulp contains beta carotene and vitamins A (retinol), B (B_1–B_3, B_5, B_6 and B_9) and C [37]. Meanwhile, the seeds are rich in linoleic acid [49], α-tocopherol and γ-tocopherol [37,49], lipases, lipoxygenases enzymes [49] and inorganic ions, such as potassium, calcium, iron, magnesium, phosphorus and zinc [37,49]. It has been reported that the fruit also comprises alkaloids, carbohydrates, cardiac glycosides, flavonoids (i.e., rutin, miricetin and quercetin), saponins, tannins, steroids and terpenoids [37,40]. Although PC values manifested clear differences between extracts, the results concluded that extracts were rich in both hydrophilic antioxidants (as determined by FRAP and ABTS), and lipophilic antioxidants (as determined by ABTS). In fact, the antioxidant activity determined by FRAP showed the lowest values for pure ethanol ($p > 0.05$) and higher values for water and aqueous/ethanol extractive solutions, without significant differences between the water and aqueous/ethanol extractive solutions ($p > 0.05$), in accordance with the PC results. This fact was probably because flavonoids, which are generally more soluble in ethanol, can be bonded with saccharide groups, which are more water soluble, as has already been reported in previous work [43]. The correlation between PC and FRAP values between extracts occurred principally because the PC method is based on the oxidation of phenols by a molybdotungstophosphoric reagent through single electron transfer [43]. On the other hand, the ABTS values of extracts did not present significant differences. Non-glycoside phenolic compounds such as flavanol and flavones generally present better solubility in alcoholic extractive solutions. These PCs were probably molecules with higher chemical composition where a simple molecule is able to scavenge several radical molecules and whose antioxidant activities through the HAT mechanism were also taken into account [43,44,50–52]. Due to the obtained results (Table 1), and considering environmental aspects in evaluating the extractive effectiveness, water was selected for the extraction procedure, which is schematically represented in Figure 1. In brief, dried *C. metuliferus* fruit was mechanically grinded to obtain a powder and dispersed in distilled water, which was then heated (40 °C, 180 min at 500 rpm). The resulting solution was filtered twice (double ring qualitative filter paper GE, Grade fast 101), frozen and then the extract was concentrated to dryness by means of a freeze drying process. The obtained viscous extract was then used for the preparation of antioxidant coatings.

3.2. Coating Process for Bilayer Film Forming

C. metuliferus extract (CM) was incorporated at three different concentrations (1, 3 and 5 wt.%) into cellulose acetate solution for the development of active coating film-forming solutions. It is widely known that a polymer should be soluble in a solvent with a similar solubility parameter (δ), and thus δ represents an important parameter when working with polymeric solutions [8,14,42]. Thus, good solubility of CA into acetone is ascribed to their close solubility parameters, which are between 19.6 and 25.1 $MPa^{1/2}$ [8,42,53] for CA and between 19.9 and 20.1 $MPa^{1/2}$ [8,53] for acetone. Concerning the solubility parameters of the main components of CM which are 18.9 $MPa^{1/2}$ for betacarotene [54], 18.7 $MPa^{1/2}$ for retinol [55], 20.2 $MPa^{1/2}$ for α-tocopherol and 20.3 $MPa^{1/2}$ for γ-tocopherol [54], good miscibility between CA and CM should be expected.

It is known that the viscosity of a polymeric solution greatly depends on the polymer concentration [56], thus it can be regulated by simple varying the polymer concentration in the film-forming solution. Therefore, 6 g of CA powder was firstly dispersed in 40 mL of acetone under continuous stirring at 50 °C, until complete dissolution [42]. Since in this work CM was obtained by means of an aqueous extraction procedure, it should be taken into account that low amounts of residual water can act as a non-solvent, which can potentially compete against the interactions between CA and acetone [42]. In fact, the solubility parameter of water is 47.9 $MPa^{1/2}$ [8,42], and thus the solubility parameter of solvent will increase as the presence of water increases in the acetone:water system. From semi-dilute to concentrated polymeric solution, the polymer dimensions decrease until critical concentration (c^+) is reached, at which point they shrink to their unperturbed dimensions and remain constant [57]. Necula et al. studied several polymeric solutions of CA in acetone 95% *v/v* at different concentrations up to 0.4 g/mL, at different temperatures (from 20 to 50 °C). CA critical concentration at which the polymer coils begin to overlap with each other (0.013 > c^* > 0.018 g/mL), as well as the critical concentration for reaching the unperturbed state, c^+ (0.098 > c^+ > 0.142 g/mL) ($c^+ \cong 8\ c^*$), were determined [58]. Thus, in the present work, in order to ensure that the cellulose acetate coils in acetone (or acetone with low amounts of water as non-solvent, i.e., acetone > 90%) were able to contract toward the unperturbed size state, the selected concentration of CA and/or CA-CM film-forming solution was 0.15 g/mL.

Due to the non-polar nature of LDPE for coating applications and for effectively formation of adhesion joints, it needs a previous surface treatment [41]. Thus, a commercial corona-treated LDPE was selected as substrate in order to increase the poor adhesion properties of LDPE. It should also be taken into account that mass and/or heat transfer takes place, and the polymeric systems become thermodynamically unstable during the solvent coating process, and therefore, phase separation can take place [42]. The molecules of CA in acetone (boiling point 56 °C) are characterized by high chain rigidity, but the chain stiffness decreases as temperature increases and, as a result, their flexibility increase [59]. Thus, in order to select the coating drying conditions, the temperature was increasingly varied within the temperature range from 45 °C to 50 °C and the time was varied between 2 and 3 min through trial-and-error practice until bilayer films with good-quality visual appearance were obtained. That is, the coating parameters were adjusted until a homogeneous solution coating completely covered the LDPE film without apparent phase separation. The processing drying temperature and time of CA and/or CA-CM coated onto the LDPE films were 50 °C and 3 min, since these processing parameters made it possible to obtain transparent films without visual defects (Figure 2c). The obtained film thicknesses ranged from 50 ± 2 µm to 63 ± 3 µm, confirming the low thickness of the CA coating in the final bilayer formulation. All LDPE/CA-coated film formulations were transparent without affecting the high transparency of the LDPE (see upper image in Figure 2c), even at the highest *C. metuliferus* fruit extract concentration of 5 wt.% (see lower image of LDPE/CA-CM5 in Figure 2c).

3.3. Optical and Morphological Properties

The processing conditions used here made it possible to obtain transparent and thin bilayer films (see thickness in Table 2). It should be highlighted that transparency is one of the most important characteristics of the polymeric films for food packaging. Thus, these results were confirmed by means of the determination of their optical properties (Figure 3). No significant differences were observed on the light transmission along the visible region of the spectra (400–700 nm), suggesting that CM was homogeneously dispersed over the CA matrix. The transparency of the LDPE/CA films was measured in the range 540–560 nm (see zoom in Figure 3). The addition of CA had practically no effect on the high transparency of the LDPE. Similarly, the incorporation of CM into the CA matrix had practically no effect on the high transparency of the LDPE/CA, particularly when it was added at low amounts such as 1 wt.% and 3 wt.% (LDPE/CA-CM1 and LDPE/CA-CM3). Meanwhile, the incorporation of the highest amount of CM (5 wt.%) produced a slight reduction in the transparency of the LDPE/CA (see zoom Figure 3), but high transparency was still observed, as can be seen in the visual appearance of this bilayer film (see as example the lower image of LDPE/CA-CM5 in Figure 2c). With respect to the UV spectra region (250–400 nm), the LDPE/CA film showed a reduction of the transmittance of

LDPE due to the fact that CA absorb light in the region below 250 nm. This absorption was slightly reduced with increasing amounts of CM due to the decreasing CA content in the formulation.

The color parameters of films were measured in the CIELab space (Table 2). All materials showed high lightness values. The CA and CA-CM coating application did not produce significant changes in *L* values, which is in good accordance with the high transparency observed for the visual appearance of the films (Figure 2c). The negative values obtained for the a* coordinate are indicative of deviation towards green color, but these values were very close to zero. This coordinate decreased particularly in the LDPE/CA-CM5 film with the highest amount of CM, showing significant ($p > 0.05$) differences with respect to LDPE/CA-CM materials with lower amounts of CM (LDPE/CA-CM1 and LDPE/CA-CM3 films). Meanwhile, no significant differences in the b* coordinate were observed between LDPE/CA-based films with respect to the LDPE film, with the exception of LDPE/CA-CM5 film, which showed significant differences ($p > 0.05$) towards positive values, which are indicative of deviation towards yellow color. Similarly, the highest color differences with respect to uncoated LDPE film were observed for LDPE/CA-CM5. Nevertheless, it should be highlighted that all formulations showed lower ΔE values than 0.3, being considerably lower than ΔE of ± 2.0, which is the value typically considered to be the threshold of perceptible color difference for the human eye [13], and even lower than ΔE of ± 0.5, which is the total color difference able to be recognized by a sensorial panel [60].

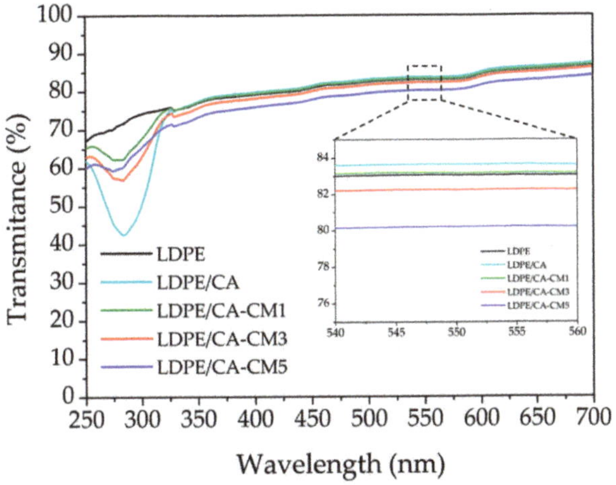

Figure 3. UV-vis spectra of LDPE and LDPE/CA-coated films.

Table 2. Color properties of CA-CM-coated LDPE films.

Formulations	Thickness (µm)	L	a*	b*	ΔE
LDPE	50 ± 2	98.7 ± 0.1 ^{a}	−0.03 ± 0.02 ^{a,b}	2.02 ± 0.03 ^{a}	- ^{a}
LDPE/CA	55 ± 3	98.6 ± 0.1 ^{a}	−0.02 ± 0.01 ^{a,b}	2.05 ± 0.06 ^{a}	0.05 ± 0.03 ^{a,b}
LDPE/CA-CM1	60 ± 1	98.8 ± 0.3 ^{a}	−0.01 ± 0.02 ^{a}	2.04 ± 0.06 ^{a}	0.10 ± 0.10 ^{b}
LDPE/CA-CM3	60 ± 4	98.5 ± 0.2 ^{a}	−0.04 ± 0.02 ^{a}	2.09 ± 0.03 ^{a}	0.18 ± 0.09 ^{a,b,}
LDPE/CA-CM5	63 ± 3	98.5 ± 0.1 ^{a}	−0.06 ± 0.01 ^{b}	2.19 ± 0.02 ^{b}	0.27 ± 0.03 ^{b}

$^{a–b}$ Different superscripts within the same column indicate significant differences between formulations ($p < 0.05$).

The morphological structure of polymeric films is an essential characteristic, since it directly affects the mechanical and barrier performance of the final materials, particularly important in the packaging sector, where it can ultimately influence the commercial success. The adhesion between polymeric layers in multilayer systems is frequently evaluated by observing the microstructure of the materials using microscopic techniques [21,61]. Figure 4 shows the micrograph of the cross-section surfaces of LDPE-, LDPE/CA- and LDPE/CA-CM-based films analyzed by SEM. The SEM analysis was carried out to evaluate the morphological investigation of the bilayer structures, as well as to evaluate the effect of active films on the microstructure at the different concentrations of CM (1, 3, and 5 wt.%) with respect to the CA polymeric matrix used to produce the different LDPE/CA-CM-based formulations. In the SEM micrographs, both polymeric layers can be clearly distinguished (see arrows), showing very good adhesion, with no detachment being observed, revealing that cellulose acetate had been successfully coated onto the surface of corona-treated LDPE (Figure 4a). The LDPE layer presented the typical smooth surface of LDPE in all bilayer formulations [60]. In LDPE/CA film, the CA layer presented a homogenous structure without the presence of pores on the coating structure, suggesting that no pores were formed during the process as a consequence of the acetone evaporation, as can occur in CA-based film processed by solvent casting [62]. This result confirmed the success of the coating process developed here in which the CA/acetone ratio used, as well as the drying conditions (50 °C during 3 min), are crucial. The addition of CM into CA coating film-forming solutions did not affect the adhesion of either polymeric layer (Figure 4b–d). However, some compact rougher structures were observed with increasing amounts of CM in the CA layer, which was particularly evident in the LDPE/CA-CM5 film (Figure 4d). This behavior can be ascribed to interaction among active components that tend to agglomerate at high concentrations.

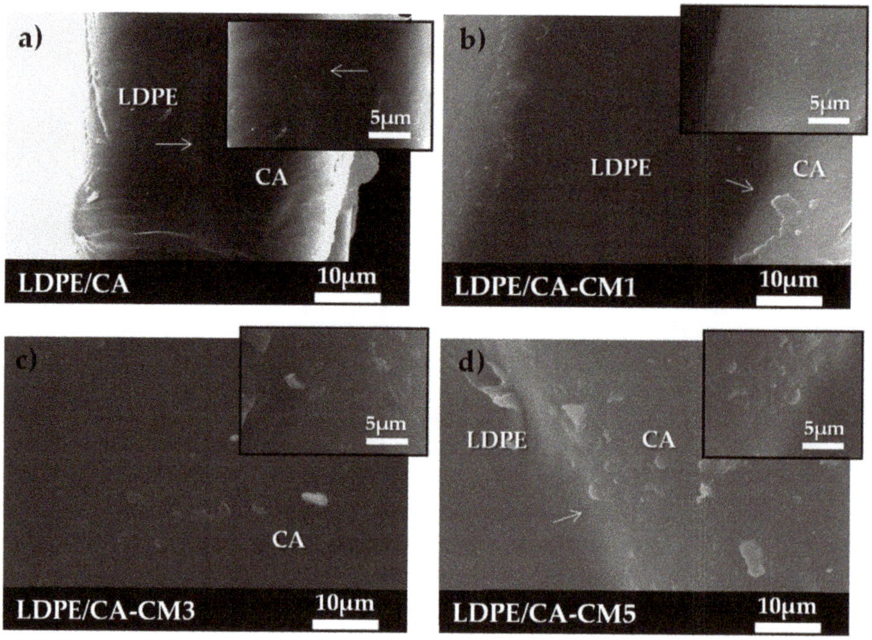

Figure 4. SEM of cross-fracture surface: (**a**) LDPE/CA, (**b**) LDPE/CA-CM1, (**c**) LDPE/CA-CM3, and (**d**) LDPE/CA-CM5. 2000× (inset figures 5000×).

Although further studies should be performed to ensure the good adhesion between CA-CM coating and corona-treated LDPE substrate (e.g., sealability, as well as friction and scratch resistance), optical and morphological SEM analysis of bilayer systems revealed that good adhesion had been achieved through the applied drying process parameters (time and temperature) immediately after coating CA-CM film forming solution onto LDPE film. In fact, on one hand, CA-CM coating layer had practically no effect on the LDPE substrate transparency, while on the other hand the absence of porous structures and/or phase separation in SEM images suggests good adhesion between both layers.

3.4. Thermal Properties

With the aim of studying the thermal degradation of each layer, CA and CA-CM-based formulations were prepared by solvent casting, and the TGA parameters, as well as the residue at 700 °C, are summarized in Table 3. Meanwhile, the TGA and DTG curves are shown in Figure 5a,b, respectively.

Table 3. TGA thermal properties of CA-CM-based films and LDPE/CA-CM-based films.

Formulations	T_0 (°C)	T_{maxCA} (°C)	$T_{maxLDPE}$ (°C)	Residue at 700 °C (%)
CA	281	364	-	9.7
CA-CM1	265	360	-	11.2
CA-CM3	254	353	-	12.2
CA-CM5	247	351	-	13.5
LDPE	429	-	476	0.1
LDPE/CA	347	363	474	0.9
LDPE/CA-CM1	340	356	474	0.4
LDPE/CA-CM3	339	355	474	0.5
LDPE/CA-CM5	319	355	474	0.1

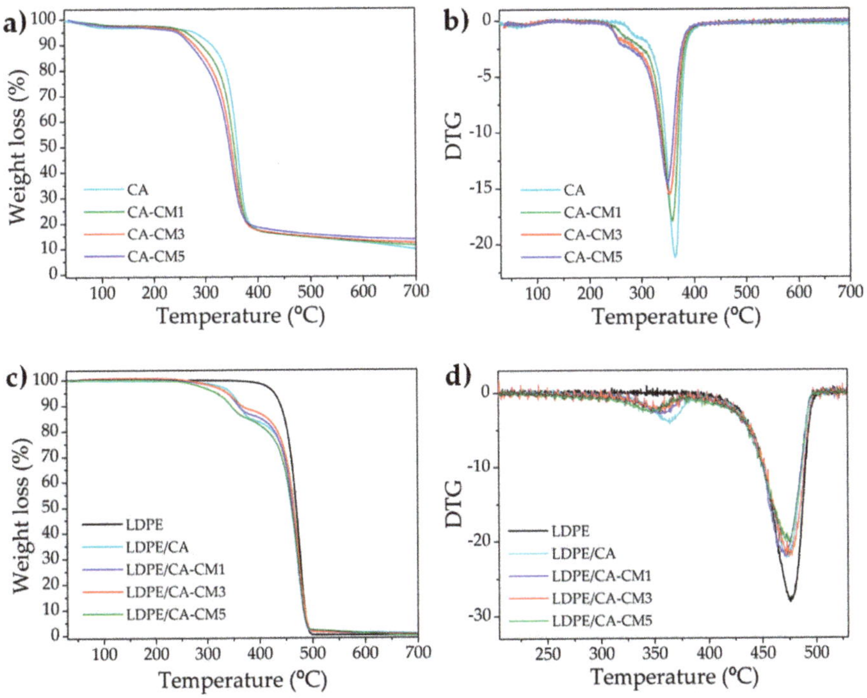

Figure 5. Thermogravimetric analysis: (**a**) TGA of CA-based layer, (**b**) DTG of CA-based layer, (**c**) TGA of LDPE/CA-based bilayer, and (**d**) DTG of LDPE/CA-based bilayer.

A small mass loss below 130 °C belonging to the volatilization of the volatile matter and/or to the evaporation of absorbed and bound water was seen in all CA-based films [6,60]. Subsequently, there was a thermal degradation (from around 180 to 300 °C) related to the loss of acetyl groups, followed by acetic acid volatilization, which could further catalyze the decomposition of cellulose [63]. Next, the two typical thermal degradation processes of cellulose acetate were also observed, corresponding to the fragmentation of macromolecular structure of the cellulosic chain (T_{maxCA} = 364 °C), followed by the last thermal degradation step, which starts at around 450 °C, belonging to the carbonization of products (≈550 °C) to ash [6,60]. Neat CA film still yielded small residual ashes after degradation at 700 °C (less than 10%), since CA requires higher temperatures in order to achieve practically no residue (790 °C) [63]. The incorporation of *C. metuliferus* extract reduced the thermal stability of cellulose acetate matrix by reducing the onset degradation and maximum degradation temperatures to lower values. After 700 °C, the residual ashes for the CA-CM samples were somewhat higher, probably consisting of: (i) positive interaction between CM components and cellulosic structures formed during degradation (i.e., hydrogen-bonding interaction) that delayed the end of the main degradation step of the cellulose structure, (ii) lignocellulosic structures extracted from CM (lignin degradation takes place in a wide range of temperature, from 100 to 900 °C [64]), and/or (iii) inorganic components of CM.

The effect of the addition of the CA coating onto LDPE film was also investigated by TGA (Figure 5c,d). The addition of the cellulose acetate coating layer reduced the high thermal stability of LDPE, since CA presented lower thermal stability than LDPE. Thus, the onset degradation temperature was shifted approximately 80 °C toward lower values, from 429 °C in LDPE to 347 °C in the LDPE/CA film, while the T_{max} of LDPE was slightly reduced or largely maintained.

The effect of the addition of CA-CM coating produced a similar behavior, and both the T_0 and T_{max} of CA shifted to lower values, following the same tendency as that of the CA films (Table 3). Meanwhile, the T_{max} of LDPE was not affected by the presence of the CA-CM-based coatings. After 700 °C, the bilayer films presented practically no residue (less than 1%). Nevertheless, it should be highlighted that no significant degradation took place in the temperature region from room temperature to 200 °C, which is a considerably higher temperature than that at which the films are intended to be used during the food packing process, as well as during storage.

3.5. Mechanical and Oxygen Barrier Properties

Films for food packaging are required to maintain their integrity with the aim of withstand the stress that occurs during shipping, handling and storage [1,2]. Thus, the mechanical properties of LDPE-, LDPE/CA- and LDPE/CA-CM-based films were studied by mean of tensile test measurements. Based on the tensile test results (Table 4), it seems that the mechanical properties of the coated LDPE films (LDPE/CA- and LDPE/CA-CM-based films) were controlled by the polyethylene layer. Nevertheless, it should be mentioned that multilayer films of plastic combined with biopolymers generally possess poor mechanical properties due to the poor mechanical strength of the biopolymers [65]. For instance, Shin et al. studied corn zein-coated LDPE, and their mechanical strength could not be measured due to the high brittleness of the corn zein layer, since it broke before the LDPE in the bilayer system [65]. In the present work, the LDPE/CA-based bilayer films exhibited a somewhat higher tensile strength, probably due to the composite structure and the higher tensile strength of CA polymeric matrix [5] with respect to that of LDPE [32], although without significance differences ($p < 0.05$). However, CA possessed very low elongation at break [5] and thus, it is probable that the very thin CA-based coating broke before the LDPE in the bilayer structure during the tensile test measurements. However, this was undetectable from the stress-strain curve (not shown) due to the very thin character of the CA layer. In fact, as Table 4 shows, it seems that the high flexibility of LDPE was not affected by the presence of the CA-CM coating ($p > 0.05$) in bilayer formulations. Moreover, comparing the LDPE/CA-CM-based films with respect to the LDPE/CA formulation, it seems that CM did not affect the mechanical performance of the LDPE/CA film, confirming the well dispersion of the *C. metuliferus* extract in the CA polymeric matrix, as was noted in SEM analysis (see Figure 4). Similar findings on the mechanical performance of LDPE coated with methylcellulose containing murta leaf (*Ugni molinae* Turcz) extract were observed in a previous work by Hauser et al. (2016). In that case, although the elongation at break of neat methylcellulose did not exceed 15%, high elongation at break (higher than 160%) in the bilayer structures was observed [32]. They ascribed this behavior to homogenous methylcellulose coating formation with good adhesion to the corona-treated LDPE [32].

One of the major challenges for coatings intended for LDPE is to increase the low oxygen barrier performance of this polymer. CA is recognized to have a higher barrier performance (OTR values around 650 cm^3/m^2 day [6]) than LDPE (LDPE film = 4750 650 cm^3/m^2 day, thickness = 0.05 mm). Thus, the application of a CA coating onto LDPE film drastically reduces the oxygen permeability, reducing the OTR*e values by between 19% and 31% (Table 4). The LDPE/CA-CM5 film presented slightly higher oxygen transmission values, probably due to its having the lowest homogeneity as a result of its high CM extract concentration. Although the oxygen barrier performance obtained here did not provide the final packaging material with a strong oxygen barrier performance, such as those provided by other polymeric matrices with well-known oxygen barrier performance (i.e., poly(ethylene terphthalate) (PET) with OTR*e < 3 cm^3 mm/m^2 day [4,66], EVOH which exhibits low OTR values under dry conditions with OTR*e < 4 cm^3 mm/m^2 day [67], or calcium and sodium caseinates with OTR*e < 7 cm^3 mm/m^2 day [13]), it showed the effectiveness of CA and CA-CM coating to improve these properties due to the good adhesion onto the corona-treated LDPE substrate. The improvement of the LDPE films' barrier performance by coating it with different biopolymers such as whey protein [35], gelatin [66], chitosan or corn zein [65] has been already observed.

Table 4. Tensile test and oxygen barrier properties of LDPE/CA-CM-based films.

Formulations	Tensile Strength (MPa)	Elongation at Break (%)	OTR*e (cm^3 mm/m^2 day)
LDPE	8.5 ± 1.5 [a]	455 ± 35 [a]	237.7
LDPE/CA	9.7 ± 2.4 [a]	480 ± 45 [a]	170.7
LDPE/CA-CM1	9.5 ± 0.2 [a]	455 ± 10 [a]	164.2
LDPE/CA-CM3	8.1 ± 3.1 [a]	420 ± 35 [a]	171.6
LDPE/CA-CM5	8.7 ± 1.0 [a]	475 ± 70 [a]	193.2

[a] Different superscripts within the same column indicate significant differences between formulations ($p < 0.05$).

3.6. Antioxidant Activity of Active Bilayer Systems

Considering that lipids are one of the main targets of oxidative reactions and, thus, lipid oxidation process responsible of a major problem in both natural and processed foodstuff [68], the antioxidant developed bilayer films were studied in direct contact with a fatty food simulant (simulant D1 = solution of 50% ethanol) [46]. Meanwhile, considering the complexity of different compounds in *C. metuliferus* fruit extract, the release studies of the active agents from the CA-CM coating layer of the LDPE films were indirectly measured through the determination of the total antioxidant activity into the food simulant, due to the fact that it is proportional to antioxidant release kinetics [10]. To evaluate both, lipophilic as well as hydrophilic antioxidants released, ABTS method was used and the measurements were performed after 1, 3 and 10 days in contact with. Non containing *C. metuliferus* fruit extract LDPE/CA film was also analyzed as control material and, as expected, did not show any ABTS radical scavenging activity (not shown). The antioxidant release kinetics indicated that more than 50% of active compounds were released during the first day (Figure 6). Subsequently, the release capacity moved on towards an equilibrium value on the third day in contact with the food simulant. The release kinetic of the active compounds of CM followed the second Ficks' law of diffusion with an exponential growth to a maximum, in accordance with already reported works of active cellulose acetate films (i.e., CA loaded with ascorbic acid [69], L-tyrosine [69], thymol [70] and red onion extract [10]). As it was expected, the antioxidant activity increased with increasing amount of CM in the formulations. Thus, the higher antioxidant effectiveness was for the film with the higher amount of CM (LDPE/CA-CM5 film). However, higher antioxidant effect was observed in other cellulose acetate films such as CA loaded with 5 wt% or red onion extract which showed around 1 mg Trolox/dm^2 film [10]. This result can be related with the lower PC and antioxidant performance of the *C. metuliferus* extract with respect to that of red onion. Nevertheless, it should be taken into account that the use of an active internal CA-based layer coated in an external LDPE layer may contribute to the effectiveness of antioxidant performance by slowing down the release rates and extending their action due to the interaction between both polymeric layers and the less exposition to the food simulant [61]. The major antioxidant ability of CM has been attributed to non-edible parts of the fruit (seed and peal) [38]. Thus, CM can result interesting not only for the development of antioxidant packaging materials, but also towards the use of this fruit waste from agri-food industry as a valorization resource of bioactive compounds giving an added value to the non-edible waste.

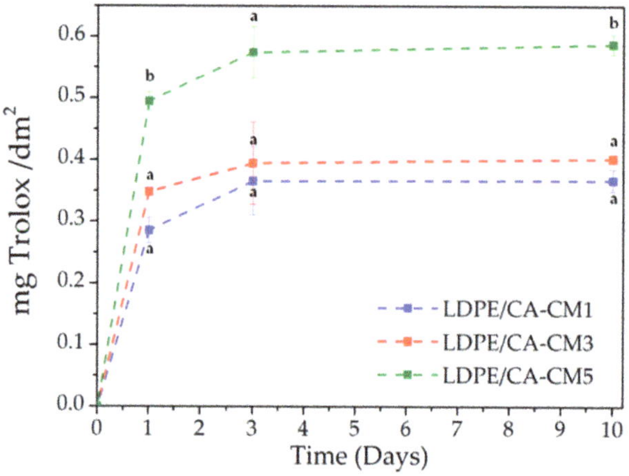

Figure 6. The *C. metuliferus* fruit extract antioxidant activity measured by ABTS method. [a–b] Different superscripts within the same day indicate significant differences between formulations ($p < 0.05$).

3.7. Anti-Browning Effect on Packaged Fresh-Cut Apple

Another promising field of coating technology application is in the fresh and minimally processed fruit sector, which are highly perishable products [21]. Thus, the obtained active-coated LDPE films were also tested in the prevention of browning in fresh-cut apples (Figure 7).

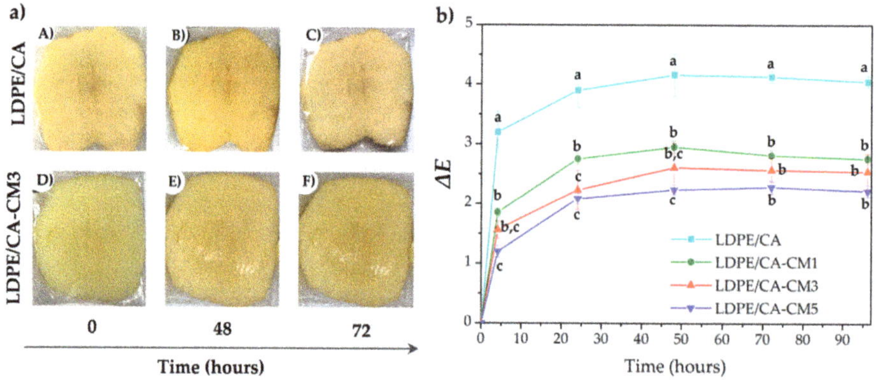

Figure 7. (**a**) Visual appearance of apples packed with LDPE/CA film: (**a**)/(**A**) immediately packed, (**a**)/(**B**) after 48 h and (**a**)/(**C**) after 72 h; and LDPE/CA-CM3 film: (**a**)/(**D**) immediately packed, (**a**)/(**E**) after 48 h, and (**a**)/(**F**) after 72 h packed; (**b**) Total color change evolution of fresh-cut apples packed with LDPE/CA and LDPE/CA-CM bilayer films over 96 h. [a–c] Different superscripts within the same day indicate significant differences between formulations ($p < 0.05$).

It is well known that fresh fruit browning is caused by enzymatic oxidation of phenolic compounds mediated by polyphenol oxidase activity, and two strategies for inhibiting this process are through the reduction of oxygen and addition of antioxidants [71]. Figure 7 shows the visual appearance of apple slices packed with non-active bilayer film (LDPE/CA, upper images in Figure 7b), as well as with active LDPE/CA-CM films, using LDPE/CA-CM3 as example (down images in Figure 7a) stored at 30 °C to simulate the worst foreseeable conditions. As can be seen, apples packed in LDPE/CA without *C. metuliferus* extract clearly exhibited a browning effect after 48 h (Figure 7a(B)) and 72 h (Figure 7a(C)). Meanwhile, this effect was less pronounced in LDPE/CA-CM-based formulations (Figure 7a(D–F)). These findings were corroborated by the determination of the evolution of color differences (ΔE) of packed apples with all LDPE/CA-based bilayer formulations over 96 h of storage (Figure 7b). As expected, the highest color differences were observed in packed apples with un-functionalized LDPE/CA film Figure 7b, which was in good accordance with the visual browning observed in Figure 7a(B,C). Meanwhile, those packed fresh-cut apples containing CM in the coating layer showed less color difference, which decreased with increasing amounts of CM in the formulations (Figure 7b). In fact, LDPE/CA-CM5 formulation was able to reduce the ΔE value by around 50% with respect to the LDPE/CA film, reaching values around $\Delta E = 2$.

4. Conclusions

Antioxidant compounds of *Cucumis metuliferus* (CM) fruit were successfully extracted by means of an aqueous extraction process and further incorporated into cellulose acetate (CA) matrix to develop antioxidant active coatings. CA was dissolved in acetone and CM was further added in concentrations of 1, 3 and 5 wt.%. The good miscibility of the film-forming solution was directly related to the fact that the main components of *C. metuliferus* show solubility parameters close to those of CA, as well as to acetone. The CA-CM-based film-forming solutions were successfully coated onto corona-treated LDPE films through a simple process. Very thin CA-CM layers were obtained, since the films' thickness varied from 50 μm in the case of LDPE to thicknesses between 60 and 65 μm in the case of the bilayer LDPE/CA-CM films. SEM observations confirmed the proper adhesion of the CA coating onto the LDPE film for intended use as bilayer packaging materials. CA and CA-CM-based coatings induced a decrease in the thermal stability of LDPE, but exhibited enough thermal stability ($T_0 > 300$ °C) for the intended use (i.e., during food packing or storage). The CA-CM-based layer provided improved oxygen barrier to LDPE film and did not affect its high transparency or colorlessness. Meanwhile, the CA layer containing different amounts of CM extract (1, 3 and 5 wt.%) showed its effectiveness as an antioxidant carrier, since CM either underwent a sustained release into a fatty food simulant, exerting free radical scavenging activity, or reduced the browning of fresh-cut apples in direct contact. Since the coating process proposed here is simple, and extremely flexible and low-cost, it is expected that the transfer of these active coatings from laboratory scale to industrial production will be easily feasible.

Author Contributions: Conceptualization, M.P.A., C.L.d.D., A.G. and M.J.G.; methodology, M.P.A., C.L.d.D., L.G., S.F., A.G. and M.J.G.; formal analysis, M.P.A., C.L.d.D., L.G. and S.F.; investigation, M.P.A., C.L.d.D., L.G., S.F., A.G. and M.J.G.; resources, A.G. and M.J.G.; data curation, M.P.A., C.L.d.D., L.G. and S.F.; writing—original draft preparation, M.P.A., C.L.d.D.; writing—review and editing, M.P.A., C.L.d.D., L.G., S.F., A.G. and M.J.G.; visualization, M.P.A., C.L.d.D., L.G., S.F., A.G. and M.J.G.; supervision, A.G. and M.J.G.; project administration, M.P.A. and C.L.d.D.; funding acquisition, M.P.A., C.L.d.D., A.G. and M.J.G. All authors have read and agreed to the published version of the manuscript.

Funding: This research was funded by Santander-UCM (PR87/19-22628) and FONDECYT 1200766 projects. Marina Patricia Arrieta thanks MINECO for her postdoctoral contract: Juan de la Cierva-Incorporación (FJCI-2017-33536).

Acknowledgments: The authors thank the staff of the ICTS National Center for Electron Microscopy (CNME), UCM, Madrid (Spain) for their assistance with transmission and electron scanning microscopy. The authors thank EDELPA who kindly provide corona-treated low-density polyethylene (LDPE) film.

Conflicts of Interest: The authors declare no conflict of interest.

References

1. Arrieta, M.P.; Samper, M.D.; Aldas, M.; López-Martínez, J. On the Use of PLA-PHB Blends for Sustainable Food Packaging Applications. *Materials* **2017**, *10*, 1008. [CrossRef] [PubMed]
2. Luzi, F.; Torre, L.; Kenny, J.M.; Puglia, D. Bio- and Fossil-Based Polymeric Blends and Nanocomposites for Packaging: Structure–Property Relationship. *Materials* **2019**, *12*, 471. [CrossRef] [PubMed]

3. De Moraes, A.C.M.; Andrade, P.F.; Faria, A.F.; Simões, M.B.; Salomão, F.C.C.S.; Barros, E.B.; Gonçalves, M.D.C.; Alves, O.L. Fabrication of transparent and ultraviolet shielding composite films based on graphene oxide and cellulose acetate. *Carbohydr. Polym.* **2015**, *123*, 217–227. [CrossRef] [PubMed]
4. Arrieta, M.P.; Fortunati, E.; Burgos, N.; Peltzer, M.; López, J.; Peponi, L. Nanocellulose-Based Polymeric Blends for Food Packaging Applications. In *Multifunctional Polymeric Nanocomposites Based on Cellulosic Reinforcements*; Puglia, D., Fortunati, E., Kenny, J.M., Eds.; William Andrew Publishing: Norwich, NY, USA, 2016; pp. 205–252.
5. Rodríguez, F.; Cortés, L.A.; Guarda, A.; Galotto, M.J.; Bruna, J.E. Characterization of cetylpyridinium bromide-modified montmorillonite incorporated cellulose acetate nanocomposite films. *J. Mater. Sci.* **2015**, *50*, 3772–3780. [CrossRef]
6. Rodríguez, F.; Galotto, M.J.; Guarda, A.; Bruna, J.E. Modification of cellulose acetate films using nanofillers based on organoclays. *J. Food Eng.* **2012**, *110*, 262–268. [CrossRef]
7. Wu, L.; Liu, M. Preparation and characterization of cellulose acetate-coated compound fertilizer with controlled-release and water-retention. *Polym. Adv. Technol.* **2008**, *19*, 785–792. [CrossRef]
8. Ghorani, B.; Russell, S.J.; Goswami, P. Controlled Morphology and Mechanical Characterisation of Electrospun Cellulose Acetate Fibre Webs. *Int. J. Polym. Sci.* **2013**, *2013*, 256161. [CrossRef]
9. Rodríguez, F.; Abarca, R.L.; Bruna, J.E.; Moya, P.E.; Galotto, M.J.; Guarda, A.; Padula, M. Effect of organoclay and preparation method on properties of antimicrobial cellulose acetate films. *Polym. Compos.* **2018**, *40*, 2311–2319. [CrossRef]
10. De Dicastillo, C.L.; Navarro, R.; Guarda, A.; Galotto, M.J. Development of Biocomposites with Antioxidant Activity Based on Red Onion Extract and Acetate Cellulose. *Antioxidants* **2015**, *4*, 533–547. [CrossRef]
11. Assis, R.Q.; Rios, P.D.; Rios, A.D.O.; Olivera, F.C. Biodegradable packaging of cellulose acetate incorporated with norbixin, lycopene or zeaxanthin. *Ind. Crops Prod.* **2020**, *147*, 112212. [CrossRef]
12. Gómez-Estaca, J.; De Dicastillo, C.L.; Hernández-Muñoz, P.; Catalá, R.; Gavara, R. Advances in antioxidant active food packaging. *Trends Food Sci. Technol.* **2014**, *35*, 42–51. [CrossRef]
13. Arrieta, M.P.; Peltzer, M.; López-Martínez, J.; Garrigós, M.C.; Valente, A.J.M.; Jiménez, A. Functional properties of sodium and calcium caseinate antimicrobial active films containing carvacrol. *J. Food Eng.* **2014**, *121*, 94–101. [CrossRef]
14. Arrieta, M.P.; García, A.D.; López, D.; Fiori, S.; Peponi, L. Antioxidant Bilayers Based on PHBV and Plasticized Electrospun PLA-PHB Fibers Encapsulating Catechin. *Nanomaterials* **2019**, *9*, 346. [CrossRef] [PubMed]
15. Del-Valle, V.; Hernández-Muñoz, P.; Guarda, A.; Galotto, M. Development of a cactus-mucilage edible coating (*Opuntia ficus indica*) and its application to extend strawberry (*Fragaria ananassa*) shelf-life. *Food Chem.* **2005**, *91*, 751–756. [CrossRef]
16. De Dicastillo, C.L.; Ares-Pernas, A.; López, M.D.M.C.; López-Vilariño, J.; Rodríguez, M.V.G. Enhancing the Release of the Antioxidant Tocopherol from Polypropylene Films by Incorporating the Natural Plasticizers Lecithin, Olive Oil, or Sunflower Oil. *J. Agric. Food Chem.* **2013**, *61*, 11848–11857. [CrossRef] [PubMed]
17. Joaquin, H.-F.; Rayón, E.; López-Martínez, J.; Arrieta, M.P. Enhancing the Thermal Stability of Polypropylene by Blending with Low Amounts of Natural Antioxidants. *Macromol. Mater. Eng.* **2019**, *304*. [CrossRef]
18. Peltzer, M.; Jiménez, A.; Matisová-Rychlá, L.; Rychlý, J. Use of isothermal and nonisothermal chemiluminescence measurements for comparison of stabilizing efficiency of hydroxytyrosol (3,4-dihydroxy-phenylethanol), α-tocopherol and irganox®1076 in polypropylene. *J. Appl. Polym. Sci.* **2011**, *121*, 3393–3399. [CrossRef]
19. Torres, A.; Romero, J.; Macan, A.; Guarda, A.; Galotto, M.J. Near critical and supercritical impregnation and kinetic release of thymol in LLDPE films used for food packaging. *J. Supercrit. Fluids* **2014**, *85*, 41–48. [CrossRef]
20. Arrieta, M.P.; Sessini, V.; Peponi, L. Biodegradable poly(ester-urethane) incorporated with catechin with shape memory and antioxidant activity for food packaging. *Eur. Polym. J.* **2017**, *94*, 111–124. [CrossRef]
21. Cano, A.; Andres, M.; Chiralt, A.; González-Martínez, C. Use of tannins to enhance the functional properties of protein based films. *Food Hydrocoll.* **2020**, *100*, 105443. [CrossRef]
22. Sessini, V.; Arrieta, M.P.; Fernández-Torres, A.; Peponi, L. Humidity-activated shape memory effect on plasticized starch-based biomaterials. *Carbohydr. Polym.* **2018**, *179*, 93–99. [CrossRef] [PubMed]
23. Arrieta, M.P.; Peponi, L.; López, D.; Fernández-García, M. Recovery of yerba mate (*Ilex paraguariensis*) residue for the development of PLA-based bionanocomposite films. *Ind. Crops Prod.* **2018**, *111*, 317–328. [CrossRef]

24. Farris, S.; Piergiovanni, L. Emerging coating technologies for food and beverage packaging materials. In *Emerging Food Packaging Technologies*; Elsevier: Amsterdam, The Netherlands, 2012; pp. 274–302. [CrossRef]
25. Rovera, C.; Ghaani, M.; Farris, S. Nano-inspired oxygen barrier coatings for food packaging applications: An overview. *Trends Food Sci. Technol.* **2020**, *97*, 210–220. [CrossRef]
26. Panrong, T.; Karbowiak, T.; Harnkarnsujarit, N. Effects of acetylated and octenyl-succinated starch on properties and release of green tea compounded starch/LLDPE blend films. *J. Food Eng.* **2020**, *284*, 110057. [CrossRef]
27. Ayuso, C.F.; Agüero, A.A.; Hernández, J.A.P.; Santoyo, A.B.; Gómez, E.G. High Oxygen Barrier Polyethylene Films. *Polym. Polym. Compos.* **2017**, *25*, 571–582. [CrossRef]
28. Datta, D.; Halder, G. Blending of phthalated starch and surface functionalized rice husk extracted nanosilica with LDPE towards developing an efficient packaging substitute. *Environ. Sci. Pollut. Res.* **2019**, *27*, 1533–1557. [CrossRef] [PubMed]
29. Ol'khov, A.A.; Iordanskii, A.L.; Zaikov, G.E.; Shibryaeva, L.S.; Litwinov, I.A.; Vlasov, S.V. Morphological features of poly(3-hydroxybutyrate)/low density polyethylene blends. *Int. J. Polym. Mater. Polym. Biomater.* **2000**, *47*, 457–468. [CrossRef]
30. Paul, D.R.; Robeson, L. Polymer nanotechnology: Nanocomposites. *Polymer* **2008**, *49*, 3187–3204. [CrossRef]
31. Rojas, A.; Torres, A.; Martínez, F.; Salazar, L.; Villegas, C.; Galotto, M.J.; Guarda, A.; Romero, J. Assessment of kinetic release of thymol from LDPE nanocomposites obtained by supercritical impregnation: Effect of depressurization rate and nanoclay content. *Eur. Polym. J.* **2017**, *93*, 294–306. [CrossRef]
32. Hauser, C.; Peñaloza, A.; Guarda, A.; Galotto, M.J.; Bruna, J.E.; Rodríguez, F. Development of an Active Packaging Film Based on a Methylcellulose Coating Containing Murta (*Ugni molinae Turcz*) Leaf Extract. *Food Bioprocess Technol.* **2015**, *9*, 298–307. [CrossRef]
33. Lasprilla-Botero, J.; Torres-Giner, S.; Pardo-Figuerez, M.; Alvarez-Lainez, M.; Lagaron, J.M. Superhydrophobic Bilayer Coating Based on Annealed Electrospun Ultrathin Poly(ε-caprolactone) Fibers and Electrosprayed Nanostructured Silica Microparticles for Easy Emptying Packaging Applications. *Coatings* **2018**, *8*, 173. [CrossRef]
34. Quiles-Carrillo, L.; Montanes, N.; Lagaron, J.M.; Balart, R.; Torres-Giner, S. Bioactive Multilayer Polylactide Films with Controlled Release Capacity of Gallic Acid Accomplished by Incorporating Electrospun Nanostructured Coatings and Interlayers. *Appl. Sci.* **2019**, *9*, 533. [CrossRef]
35. Schmid, M.; Dallmann, K.; Bugnicourt, E.; Cordoni, D.; Wild, F.; Lazzeri, A.; Noller, K. Properties of Whey-Protein-Coated Films and Laminates as Novel Recyclable Food Packaging Materials with Excellent Barrier Properties. *Int. J. Polym. Sci.* **2012**, *2012*, 562381. [CrossRef]
36. Motlhanka, D. Free radical scavenging activity of selected medicinal plants of Eastern Botswana. *Pak. J. Boil. Sci.* **2008**, *11*, 805–808. [CrossRef] [PubMed]
37. Usman, J.; Sodipo, O.; Kwaghe, A.; Sandabe, U. Uses of cucumis metuliferus: A review. *Cancer Biol.* **2015**, *5*, 24.
38. Matsusaka, Y.; Kawabata, J. Evaluation of Antioxidant Capacity of Non-Edible Parts of Some Selected Tropical Fruits. *Food Sci. Technol. Res.* **2010**, *16*, 467–472. [CrossRef]
39. Morton, J.F. The horned cucumber, alias "kiwano" (*Cucumis metuliferus*, cucurbitaceae). *Econ. Bot.* **1987**, *41*, 325–327.
40. Ferrara, L. A fruit to discover: Cucumis metuliferus E.Mey Ex Naudin(Kiwano). *Clin. Nutr. Metab.* **2018**, *1*, 1–2. [CrossRef]
41. Fombuena, V.; Balart, J.; Boronat, T.; Sánchez, L.; Garcia-Sanoguera, D. Improving mechanical performance of thermoplastic adhesion joints by atmospheric plasma. *Mater. Des.* **2013**, *47*, 49–56. [CrossRef]
42. Ferrarezi, M.M.F.; Rodrigues, G.V.; Felisberti, M.I.; Gonçalves, M.D.C. Investigation of cellulose acetate viscoelastic properties in different solvents and microstructure. *Eur. Polym. J.* **2013**, *49*, 2730–2737. [CrossRef]
43. Arrieta, M.P.; De Dicastillo, C.L.; Garrido, L.; Roa, K.; Galotto, M.J. Electrospun PVA fibers loaded with antioxidant fillers extracted from Durvillaea antarctica algae and their effect on plasticized PLA bionanocomposites. *Eur. Polym. J.* **2018**, *103*, 145–157. [CrossRef]
44. Prior, R.L.; Wu, X.; Schaich, K. Standardized Methods for the Determination of Antioxidant Capacity and Phenolics in Foods and Dietary Supplements. *J. Agric. Food Chem.* **2005**, *53*, 4290–4302. [CrossRef] [PubMed]

45. De Dicastillo, C.L.; Jordá, M.; Catalá, R.; Gavara, R.; Hernández-Muñoz, P. Development of Active Polyvinyl Alcohol/β-Cyclodextrin Composites to Scavenge Undesirable Food Components. *J. Agric. Food Chem.* **2011**, *59*, 11026–11033. [CrossRef] [PubMed]
46. European Commission. *No. 10/2011 of 14 January 2011. On Plastic Materials and Articles Intended to Come into Contact with Food*; European Commission: Brussels, Belgium, 2011.
47. De Dicastillo, C.L.; Rodríguez, F.; Guarda, A.; Galotto, M.J. Antioxidant films based on cross-linked methyl cellulose and native Chilean berry for food packaging applications. *Carbohydr. Polym.* **2016**, *136*, 1052–1060. [CrossRef] [PubMed]
48. Rivero-Pérez, M.D.; Muñiz, P.; González-Sanjosé, M.L. Antioxidant Profile of Red Wines Evaluated by Total Antioxidant Capacity, Scavenger Activity, and Biomarkers of Oxidative Stress Methodologies. *J. Agric. Food Chem.* **2007**, *55*, 5476–5483. [CrossRef] [PubMed]
49. Sadou, H.; Sabo, H.; Saadou, M.; Leger, C.-L.; Alma, M.M. Chemical Content of The Seeds And Physico-Chemical Characteristic of The Seed Oils from Citrullus Colocynthis, Coccinia Grandis, Cucumis Metuliferus and Cucumis Prophetarum of Niger. *Bull. Chem. Soc. Ethiop.* **2007**, *21*, 323–330. [CrossRef]
50. De Dicastillo, C.L.; Bustos, F.; Valenzuela, X.; López-Carballo, G.; Vilariño, J.M.; Galotto, M.J. Chilean berry *Ugni molinae* Turcz. fruit and leaves extracts with interesting antioxidant, antimicrobial and tyrosinase inhibitory properties. *Food Res. Int.* **2017**, *102*, 119–128. [CrossRef]
51. De Dicastillo, C.L.; López-Carballo, G.; Gavara, R.; Galet, V.M.; Guarda, A.; Galotto, M.J. Improving polyphenolic thermal stability of Aristotelia Chilensis fruit extract by encapsulation within electrospun cyclodextrin capsules. *J. Food Process. Preserv.* **2019**, *43*, e14044. [CrossRef]
52. Sendra, J.M.; Sentandreu, E.; Navarro, J.L. Kinetic Model for the Antiradical Activity of the Isolatedp-Catechol Group in Flavanone Type Structures Using the Free Stable Radical 2,2-Diphenyl-1-picrylhydrazyl as the Antiradical Probe. *J. Agric. Food Chem.* **2007**, *55*, 5512–5522. [CrossRef]
53. Nguyen, T.P.N.; Yun, E.-T.; Kim, I.-C.; Kwon, Y.-N. Preparation of cellulose triacetate/cellulose acetate (CTA/CA)-based membranes for forward osmosis. *J. Membr. Sci.* **2013**, *433*, 49–59. [CrossRef]
54. Kagliwal, L.; Patil, S.C.; Pol, A.S.; Singhal, R.S.; Patravale, V. Separation of bioactives from seabuckthorn seeds by supercritical carbon dioxide extraction methodology through solubility parameter approach. *Sep. Purif. Technol.* **2011**, *80*, 533–540. [CrossRef]
55. Laredj-Bourezg, F.; Bolzinger, M.-A.; Pelletier, J.; Valour, J.-P.; Rovère, M.-R.; Smatti, B.; Chevalier, Y. Skin delivery by block copolymer nanoparticles (block copolymer micelles). *Int. J. Pharm.* **2015**, *496*, 1034–1046. [CrossRef] [PubMed]
56. Liu, H.; Hsieh, Y.-L. Ultrafine fibrous cellulose membranes from electrospinning of cellulose acetate. *J. Polym. Sci. Part B Polym. Phys.* **2002**, *40*, 2119–2129. [CrossRef]
57. Bercea, M.; Ioan, C.; Ioan, S.; Simionescu, B.; Simionescu, C. Ultrahigh molecular weight polymers in dilute solutions. *Prog. Polym. Sci.* **1999**, *24*, 379–424. [CrossRef]
58. Necula, A.M.; Olaru, N.; Olaru, L.; Ioan, S. Influence of the Substitution Degree on the Dilute Solution Properties of Cellulose Acetate. *J. Macromol. Sci. Part B* **2008**, *47*, 913–928. [CrossRef]
59. Johnston, H.K.; Sourirajan, S. Viscosity–temperature relationships for cellulose acetate–acetone solutions. *J. Appl. Polym. Sci.* **1973**, *17*, 3717–3726. [CrossRef]
60. Bruna, J.; Peñaloza, A.; Guarda, A.; Rodríguez, F.; Galotto, M. Development of MtCu2+/LDPE nanocomposites with antimicrobial activity for potential use in food packaging. *Appl. Clay Sci.* **2012**, *58*, 79–87. [CrossRef]
61. Sogut, E.; Seydim, A.C.; Seydim, A.C. Development of Chitosan and Polycaprolactone based active bilayer films enhanced with nanocellulose and grape seed extract. *Carbohydr. Polym.* **2018**, *195*, 180–188. [CrossRef]
62. Gemili, S.; Yemenicioğlu, A.; Altınkaya, S.A.; Altinkaya, S.A. Development of cellulose acetate based antimicrobial food packaging materials for controlled release of lysozyme. *J. Food Eng.* **2009**, *90*, 453–462. [CrossRef]
63. Junior, A.R.D.O.; Ferrarezi, M.M.F.; Yoshida, I.V.P.; Gonçalves, M.D.C. Cellulose acetate/polysilsesquioxane composites: Thermal properties and morphological characterization by electron spectroscopy imaging. *J. Appl. Polym. Sci.* **2011**, *123*, 2027–2035. [CrossRef]
64. Rayón, E.; Ferrándiz, S.; Rico, M.I.; Martínez, J.L.; Arrieta, M.P. Microstructure, Mechanical, and Thermogravimetric Characterization of Cellulosic By-Products Obtained from Biomass Seeds. *Int. J. Food Prop.* **2014**, *18*, 1211–1222. [CrossRef]

65. Shin, G.H.; Lee, Y.H.; Lee, J.S.; Kim, Y.S.; Choi, W.S.; Park, H.J. Preparation of Plastic and Biopolymer Multilayer Films by Plasma Source Ion Implantation. *J. Agric. Food Chem.* **2002**, *50*, 4608–4614. [CrossRef] [PubMed]
66. Farris, S.; Introzzi, L.; Piergiovanni, L. Evaluation of a bio-coating as a solution to improve barrier, friction and optical properties of plastic films. *Packag. Technol. Sci.* **2009**, *22*, 69–83. [CrossRef]
67. Lagaron, J.; Nunez, E.; Onishi, H. Novel evoh compound with enhanced moisture resistance for high oxygen barrier packaging. In *18th Iapri World Packaging Conference*; DEStech Publications, Inc.: Lancaster, PA, USA, 2012; pp. 225–228.
68. Tian, F.; Decker, E.A.; Goddard, J.M. Controlling lipid oxidation of food by active packaging technologies. *Food Funct.* **2013**, *4*, 669. [CrossRef] [PubMed]
69. Gemili, S.; Yemenicioğlu, A.; Altinkaya, S.A. Development of antioxidant food packaging materials with controlled release properties. *J. Food Eng.* **2010**, *96*, 325–332. [CrossRef]
70. Rodríguez, F.; Torres, A.; Peñaloza, Á.; Sepulveda, H.; Galotto, M.J.; Guarda, A.; Bruna, J. Development of an antimicrobial material based on a nanocomposite cellulose acetate film for active food packaging. *Food Addit. Contam. Part A* **2014**, *31*, 342–353. [CrossRef]
71. Putnik, P.; Kovačević, D.B.; Herceg, K.; Levaj, B. Influence of antibrowning solutions, air exposure, and ultrasound on color changes in fresh-cut apples during storage. *J. Food Process. Preserv.* **2017**, *41*, e13288. [CrossRef]

polymers

MDPI

Article

Manufacturing and Properties of Binary Blend from Bacterial Polyester Poly(3-hydroxybutyrate-*co*-3-hydroxyhexanoate) and Poly(caprolactone) with Improved Toughness

Juan Ivorra-Martinez *, Isabel Verdu, Octavio Fenollar, Lourdes Sanchez-Nacher, Rafael Balart and Luis Quiles-Carrillo

Technological Institute of Materials (ITM), Universitat Politècnica de València (UPV), Plaza Ferrándiz y Carbonell 1, 03801 Alcoy, Spain; isvergar@epsa.upv.es (I.V.); ocfegi@epsa.upv.es (O.F.); lsanchez@mcm.upv.es (L.S.-N.); rbalart@mcm.upv.es (R.B.); luiquic1@epsa.upv.es (L.Q.-C.)

* Correspondence: juaivmar@doctor.upv.es; Tel.: +34-966-528-421

Received: 29 April 2020; Accepted: 12 May 2020; Published: 14 May 2020

Abstract: Polyhydroxyalkanoates (PHAs) represent a promising group of bacterial polyesters for new applications. Poly(3-hydroxybutyrate-*co*-3-hydroxyhexanoate) (PHBH) is a very promising bacterial polyester with potential uses in the packaging industry; nevertheless, as with many (almost all) bacterial polyesters, PHBH undergoes secondary crystallization (aging) which leads to an embrittlement. To overcome or minimize this, in the present work a flexible petroleum-derived polyester, namely poly(ε-caprolactone), was used to obtain PHBH/PCL blends with different compositions (from 0 to 40 PCL wt %) using extrusion followed by injection moulding. The thermal analysis of the binary blends was studied by means of differential scanning calorimetry (DSC) and thermogravimetry (TGA). Both TGA and DSC revealed immiscibility between PHBH and PCL. Mechanical dynamic thermal analysis (DMTA) allowed a precise determination of the glass transition temperatures (T_g) as a function of the blend composition. By means of field emission scanning electron microscopy (FESEM), an internal structure formed by two phases was observed, with a PHBH-rich matrix phase and a finely dispersed PCL-rich phase. These results confirmed the immiscibility between these two biopolymers. However, the mechanical properties obtained through tensile and Charpy tests, indicated that the addition of PCL to PHBH considerably improved toughness. PHBH/PCL blends containing 40 PCL wt % offered an impact resistance double that of neat PHBH. PCL addition also contributed to a decrease in brittleness and an improvement in toughness and some other ductile properties. As expected, an increase in ductile properties resulted in a decrease in some mechanical resistant properties, e.g., the modulus and the strength (in tensile and flexural conditions) decreased with increasing wt % PCL in PHBH/PCL blends.

Keywords: bacterial polyesters; poly(3-hydroxybutyrate-*co*-3hydroxyhexanoate)—PHBH; poly(ε-caprolactone)—PCL; binary blends; improved toughness; mechanical and thermal characterization

---

1. Introduction

Nowadays, awareness of environmental protection, sustainable development, and the use of renewable energies has become a priority for our society. The high volume of wastes generated that are harmful for the environment, oceans, ecosystems, and so on, has become a major problem to be solved. Furthermore, the waste generated in a consumer society, such as the present one, comes mainly from the packaging sector. This need has favoured the development of new environmentally friendly materials [1]. For this reason, the use of the so-called biopolymers is increasing in the packaging sector. Most of these materials are obtained from renewable resources and they are, in many cases, biodegradable (or compostable in controlled compost soil). Therefore, they positively contribute to minimizing plastic wastes, thus reducing the carbon footprint and also contributing to circular economies by upgrading industrial wastes [2] and/or by-products [3].

In this area, researchers have successfully developed new polymeric materials from renewable and/or biodegradable sources. These important research works have allowed the optimization of interesting biopolymers to be scaled in the industry such as poly(lactic acid)—PLA [4], poly(hydroxyalkanoates)—PHA [5], thermoplastic starch—TPS [6], poly(ε-caprolactone)—PCL [7], and poly(butylene succinate)—PBS [8], among others. Biopolymers can perfectly replace some petroleum-derived polymers [9], since they offer similar performance to most commodities and some engineering plastic. Biopolyesters are an interesting group which includes petroleum-based polymers such as poly(glycolic acid)—PGA, poly(butylene succinate)—PBS, poly(butylene adipate-*co*-terephthalate)—PBAT, and PCL, among others. However, biopolyesters also include bacterial polyesters (PHAs) and some starch-derived polymers such as PLA. The main advantage of polyesters (from both natural or fossil resources) is that they can undergo biodegradation (disintegration in controlled conditions with special compost soil), through the action of microorganisms. This makes composting an important and simple sustainable option for the management of these wastes [10].

Nowadays, biopolymers produced by bacterial fermentation, such as polyhydroxyalkanoates (PHAs), are becoming very promising as there are more than 300 potential PHAs and copolymers. Despite this wide variety, the most commonly used and commercially available PHAs are poly(3-hydroxybutyrate)—P3HB—and poly(3-hydroxybutyrate-*co*-3-hydroxyvalerate)—PHBV [11,12]. PHAs are biologically synthesized polyesters by controlled fermentation with bacteria, such as *Gram-negative* bacteria (*Azobacter*, *Bacillus* and *Pseudomonas*) and *Gram-positive* bacteria (*Rhodococcus*, *Nocardia* and *Streptomyces*). These bacteria, under food stress, can produce energy reserves as intracellular food in the form of granules. These granules are stored in the form of PHAs [13].

Arrieta et al. [10] reported that these bacteria, under feeding conditions of limited macro-elements (such as phosphorus, nitrogen, trace elements or oxygen) and in the presence of an abundant source of carbon (e.g., glucose or sucrose) and/or lipids (e.g., vegetable oils or glycerin), are capable of accumulating up to 60–80 wt % in the form of PHAs. In this way, they can subsist under conditions of food restriction [14,15]. Similar to plants that store energy in the form starch polymer, some bacteria are able to accumulate energy reserves in the form of PHAs [16].

It should be noted that these bacterial polyesters are high-molecular-weight, semi-crystalline, biocompatible thermoplastic polymers. They have very good biodegradability even under environmental conditions. They tend to exhibit rigid behaviour, due to high crystallinity, low thermal stability, and small temperature windows for conventional processing [14,17].

These limitations have been improved by the bacterial synthesis of different copolymers. In this way, a wide range of physical and thermal properties can be tailored, depending on the chemical structure of the used comonomers. More than 150 types of monomers have been successfully synthesized by selecting different raw or modified bacteria and/or the fermentation conditions [13]. The work of Alata et al. [18] reported the effect of medium-length side groups of 3-hydroxyhexanoate (3-HH) units from 5 to 18 mol % on properties of poly(3-hydroxybutyrate-*co*-3-hydroxyhexanoate)—P(3HB-*co*-3HH) or simply PHBH. They observed a remarkable decrease in crystallinity (χ_c) from 41.6% to 25.1% for copolymers containing 5 mol % and 18 mol % 3-HH, respectively. In addition, due to the reduced crystallinity, secondary crystallization is very low for high 3-HH (above 10 mol %) content in P(3-HB-*co*-3-HH). Other studies have also reported similar results, together with an interesting decrease in the melting temperature of P(3HB-*co*-3HH) [15].

PHBH consists of a random copolymer of 3-HB and 3-HH (see Scheme 1). 3-HH medium-length chains act as short branches of the main 3-HB chains; therefore, stereoregularity is lost, and subsequently, crystallinity is remarkably reduced. Besides this, the presence of aleatory 3-HH chains, broadens the melt peak, but the storage modulus and the overall strength is reduced [17,19]. Nevertheless,

PHBH copolymers with low 3-HH content undergo physical aging with time (increase in modulus and strength and reduction of ductile properties such as elongation at break, toughness, and impact strength) [20], which is ascribed to secondary crystallization above the glass transition temperature, T_g [21]. It is worthy to note that typical values of T_g for P3HB are −5 to 5 °C, and this interval is remarkably reduced to values as low as −38 °C for medium-to-long alkanoate chains, e.g., the T_g of poly(3-hydroxyhexanoate) is close to −28 °C [22].

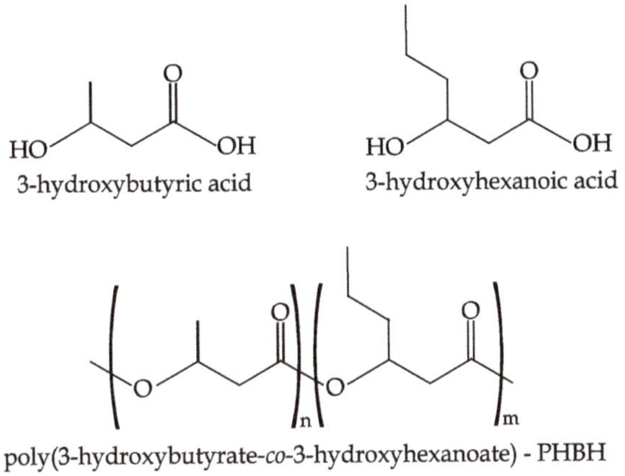

Scheme 1. Chemical structure of 3-hydroxyalkanoic acids used to synthesize poly(3-hydroxybutyrate-*co*-3-hydroxyhexanoate)—PHBH.

Another key issue in the massive use of PHAs at industrial scale is their "relatively" low cost due to the use of renewable resources such as coconut oil, sugarcane, beet, molasses, other vegetable oils, and, most importantly, carbon-rich industrial wastes such as those obtained from agro-food industry or even sludge coming from sewage treatment plants, by selecting the appropriate bacteria or using bacterial engineering to tailor the desired behaviour of a particular bacteria strain [23,24].

These properties make PHBH an environmentally efficient biopolymer suitable for applications in different packaging applications such as disposable plastic bags, food packaging, catering, agricultural mulch films, and so on [25]. However, as mentioned above, most PHAs undergo secondary crystallization or aging that makes them fragile, thus limiting their possible applications [26]. Xu et al. [27] suggested that one disadvantage of PHBH is that the secondary crystallization process is very slow (for low mol % 3-HH), due to the irregularity of its polymer chain. Large spherulites and secondary crystallization give them poor mechanical properties.

Plasticization of PHAs has been studied with an improvement of ductile properties [28]. Since plasticizers are based on low-molecular-weight compounds, they usually show potential migration problems [29]. An interesting approach to overcome this drawback is blending PHBH with another ductile polymer. However, it is important to bear in mind that the selected polymer for the blend must not compromise biodegradation or disintegration in controlled compost soil. Some researchers have already used poly(ε-caprolactone)—PCL—in blends with PHAs. PCL is a semi-crystalline biodegradable polyester with a very low T_g of about −60 °C, which gives an overall ductile behaviour with high elongation at break [26,30]. The addition of PCL decreases the fragility of the PHAs, reducing the elastic modulus and improving the blend processability. However, its low melting temperature (around 50–60 °C) means the obtained blends should not be used at temperatures above 50–60 °C since dimensional stability could be compromised. Garcia-Garcia et al. [31] reported a noteworthy improvement in the impact behaviour of P3HB by blending it with PCL. In addition, P3HB/PCL blends improved the flexibility and ductility.

The aim of this work is to overcome the intrinsic fragility of a bacterial copolyester, namely poly(3-hydroxybutyrate-*co*-3-hydroxyhexanoate)—PHBH—, by blending with a flexible polyester, namely poly(ε-caprolactone)—PCL. The effect of the incorporation of different amounts of PCL is evaluated by means of mechanical, thermal, thermo-mechanical, and morphological characterization. The evaluation of the results allows the optimum PHBH/PCL blends to be established for applications in the packaging sector that do not compromise the environment at the end-of-life cycle. In this way, it contributes to the reduction of the current serious problem of eliminating the large volume of plastic waste generated by the packaging sector. In addition, these developed formulations could be used in medical applications, as improved toughness is expected with PCL addition and both are resorbable biopolyesters.

2. Materials and Methods

2.1. Materials

Poly(3-hydroxybutyrate-*co*-3-hydroxyhexanoate)—PHBH—commercial grade ErcrosBio® PH 110 was supplied in pellet form by Ercros S.A. (Barcelona, Spain). This has a density of 1.2 g cm^{-3} and a melt flow index of 1.0 g/10 min, measured at 160 °C. As indicated by the supplier, it is suitable for injection moulded parts, and it can be melt-blended with other polyesters to obtain tailored properties. Regarding poly(ε-caprolactone)—PCL, commercial grade Capa™ 6800, in pellet form, with a mean molecular weight of 80,000 Da, was supplied by Perstorp (Cheshire, UK). This PCL grade has an MFI (Melt Flow Index) of 3 g/10 min at 160 °C.

2.2. Manufacturing of PHBH/PCL Binary Blends

PHBH pellets were dried for 8 h at 80 °C, while PCL pellets were dried at 45 °C for 24 h, in an air-circulating oven CARBOLITE Eurotherm 2416 CG (Hope Valley, UK). As has been reported in other works, the typical weight content (wt %) of flexible polymer blended with brittle polymers to improve toughness is comprised in the 20–40 wt % range. In this work, we selected a maximum PCL content of 40 wt %, since at this composition, PCL is still the minor component in the blend [32–34]. Garcia et al. [31] studied the whole PHB/PCL system and revealed, as expected, that with above 50 wt % PCL, it is PCL which defines the properties of the blend. Ferry et al. [35] also confirmed a maximum loading of 30 wt % PCL to improve the high brittleness of neat PLA. Then, different wt % of PHBH and PCL (see Table 1) were mechanically mixed in a zipper bag to provide initial homogenization. After that, all compositions were extruded using a twin-screw corotating extruder manufactured by DUPRA S.L. (Alicante, Spain) with a temperature profile (four barrels, from the hopper to the extrusion die) of 110, 120, 130, and 140 °C respectively and a screw speed of 20 rpm. The extruded material was cooled in air and subsequently pelletized for further processing. After pelletizing, the different blends were subjected to injection moulding in a Meteor 270/75 from Mateu & Solé (Barcelona, Spain). The temperature profile in the injection moulding process was 150 °C (hopper), 140, 130, and 120 °C (nozzle) in a heated mould at 60 °C as recommended by the supplier, since this PHBH has a very low melt strength. The filling time was set to 3 s while the cooling time was 60 s. Standard samples (rectangular and dog-bone shape) were obtained for further characterization. After processing, the specimens were stored at room temperature in a vacuum desiccator for 15 days before characterization, due to the secondary crystallization process or aging that PHBH undergoes with time at 25 °C [21,26,27].

Table 1. Code and composition (wt %) of binary blends of poly(3-hydroxybutyrate-*co*-3-hydroxyhexanoate)/poly(ε-caprolactone) (PHBH/PCL) blends.

Code	PHBH (wt %)	PCL (wt %)
100PHBH-0PCL	100	0
90PHBH-10PCL	90	10
80PHBH-20PCL	80	20
70PHBH-30PCL	70	30
60PHBH-40PCL	60	40
0PHBH-100PCL	0	100

2.3. *Characterizations Techniques*

2.3.1. Thermal and Thermomechanical Characterization

The thermal transitions of PHBH/PCL binary blends were analyzed using differential scanning calorimetry (DSC) in a Q2000 DSC from TA Instruments (New Castle, DE, USA). The temperature program was scheduled in three different stages: 1st heating, 1st cooling, and 2nd heating. The first heating was scheduled from −50 to 200 °C. The second stage consisted of a cooling program from 200 °C down to −50 °C (this stage is interesting to remove the thermal history and allow crystallization); finally, a 2nd heating cycle identical to the first one (−50 to 200 °C) was launched. The heating/cooling rates were all set to 10 °C·min^{-1}. All the DSC runs were performed in an inert nitrogen atmosphere with a flow rate of 66 mL·min^{-1}. In addition to parameters such as the melt peak temperature (T_{m}), the cold crystallization peak temperatures (T_{cc}), enthalpies related to the melting (ΔH_{m}), and cold crystallization (ΔH_{cc}) processes, the degree of crystallinity, χ_{c} (%), was calculated for each polymer in the blend as

$$\chi_{\mathrm{c}}(\%) = \left[\frac{\Delta H_{\mathrm{m}} - \Delta H_{\mathrm{cc}}}{\Delta H_{\mathrm{m}}^{0}\cdot(1-w)}\right]\cdot 100 \quad (1)$$

The normalized enthalpy values ($\Delta H_{\mathrm{m}}^{0}$) for a theoretical 100% crystalline PHBH and PCL were taken as 146 and 156 J·g^{-1}, respectively, as reported in the literature [19,36,37]. Finally, the term $(1-w)$ stands for the actual weight of the polymer whose crystallinity is being evaluated.

To study the thermal degradation, thermogravimetry (TGA) was carried out in a Mettler-Toledo Inc. TGA 851-E thermobalance (Schwerzenbach, Switzerland). The thermal program used in this case was a unique dynamic ramp from 30 °C up to 700 °C at 20 °C·min^{-1} in an N_2 inert atmosphere with a flow rate of 66 mL·min^{-1}. In this analysis, the onset degradation temperature was taken as the temperature related to a mass loss of 2 wt % and was denoted as $T_{2\%}$.

Dynamic-mechanical thermal analysis, or DMTA, was carried out in a Mettler-Toledo dynamic analyzer DMA1 (Columbus, OH, USA) in single cantilever mode on rectangular samples sized 40 × 10 × 4 mm^3. Samples were subjected to slightly different temperature ramps since PCL melts at 58–60 °C. The maximum dynamic deflection was 10 µm and the frequency for the sinusoidal stress wave was set to 1 Hz. Thus, for neat PCL the heating range was from −70 to 50 °C to avoid melting. In the case of neat PHBH, the heating ramp was set from −70 to 100 °C, and finally, PHBH/PCL blends were subjected to a heating program from −70 to 70 °C. The heating rate was the same, 2 °C·min^{-1}, for all the different scheduled temperature programs

To evaluate the dimensional stability, neat PHBH and PCL, as well as PHBH/PCL blends, were tested in a thermomechanical analyzer (TMA) Q400 from TA Instruments (New Castle, DE, USA) on rectangular samples sized 10 × 10 × 4 mm^3. The temperature sweep was from −70 to 70 °C, with a constant heating rate of 2 °C min^{-1} and a constant load of 20 mN. The coefficient of linear thermal expansion (CLTE) of all specimens was determined as the slope for the linear correlation between the expansion and temperature, both below and above T_{g}.

2.3.2. Mechanical Properties

The mechanical characterization of the PHBH/PCL blends was studied by means of tensile, flexural, impact, and hardness tests on five standardized specimens for each test. The tensile and flexure tests were carried out according to ISO 527 and ISO 178 respectively, in an ELIB 30 universal machine from S.A.E. Ibertest (Madrid, Spain). The load cell was 5 kN for both tests. The crosshead rate was 5 mm·min^{-1} for flexural tests and 20 mm·min^{-1} for the tensile tests.

The impact resistance (absorbed-energy during impact conditions, per unit area) was determined according to ISO 179, using a Charpy pendulum from Metrotec S.A. (San Sebastian, Spain) with an energy of 1-J. A standardized "V-type" notch was produced on standard rectangular samples.

The hardness properties were measured according to ISO 868. The equipment used was a Shore D hardness tester model 673-D from J. Bot, S.A. (Barcelona, Spain).

2.3.3. Morphology Characterization

The surface analysis of the fractured specimens from impact tests was performed with a field emission scanning electron microscope (FESEM) model ZEISS ULTRA55 (Oxford Instruments, Abingdon, UK). The working accelerating voltage was 2 kV. Prior to this analysis, the samples were metallized with a gold-palladium alloy in an EMITECH mod. SC7620 sputter coater from Quorum Technologies Ltd. (East Sussex, UK). In a second analysis, the samples were subjected to a selective PCL extraction in acetone at room temperature for 24 h. In this way, PCL can be extracted, and therefore, it is possible to observe more accurately the phase distribution in the developed binary blends [38].

3. Results and Discussion

3.1. Thermal Properties of PHBH/PCL Blends

Thermal analysis, using DSC of PHBH/PCL binary blends with different amounts of PCL (wt %), and neat PHBH and PCL, was done from the first heating cycle (Figure 1a) to obtain the thermal parameters of the starting material, just after 15 days from its processing, thus allowing secondary crystallization. In addition, the second heating cycle after a slow cooling (Figure 1b) allowed the thermal history of the material to be removed and the main thermal characterization parameters to be obtained.

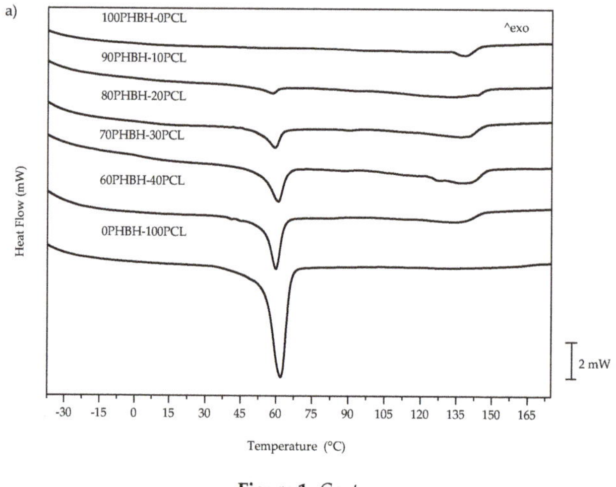

Figure 1. *Cont.*

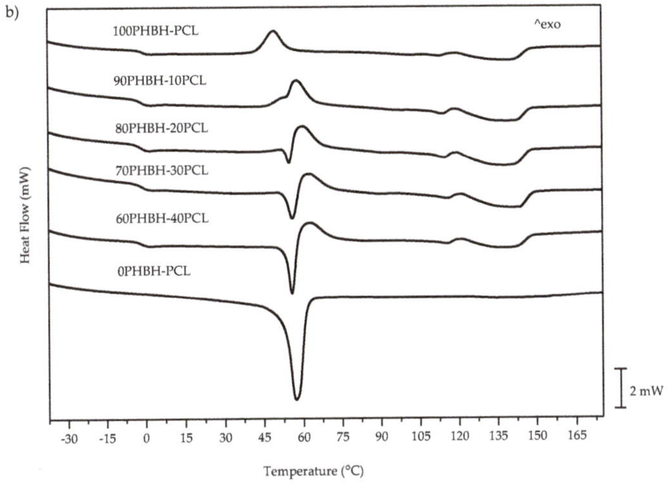

Figure 1. Comparative plot of the differential scanning calorimetry (DSC) thermograms of PHBH, PCL and PHBH/PCL blends with different PCL wt %: (**a**) first heating cycle after processing and aging for 15 days, (**b**) 2nd heating after cooling (**a**) at a controlled rate.

Table 2 summarizes the main thermal parameters corresponding to the DSC thermograms of the first heating cycle plus an additional aging time of 15 days, obtained using DSC runs from Figure 1a. The thermogram of neat PCL showed a single endothermic peak at around 62 °C (T_{m_PCL}), which was attributed to the melting of packed crystallites of PCL embedded in an amorphous PCL fraction. Since DSC tests were run from −50 °C, the T_{g_PCL} could not be clearly observed. This was because the T_{g_PCL} is very low, with values below −50 °C, so that, with this thermal program, it could not be accurately determined. On the other hand, the DSC thermogram of neat PHBH did not allow its T_{g_PHBH} to be identified either. This is because the crystalline fraction of PHBH remarkably increased after aging for 15 days; therefore, the remaining amorphous fraction was noticeably reduced, and then, the step in the baseline (around 0 °C), attributed to the T_{g_PHBH}, could not be clearly seen. As one can see in Figure 1a, only the melt process of the crystalline fraction in PHBH could be observed with a peak located at 138 °C (T_{m_PHBH}). Regarding binary PHBH/PCL blends, the two above-mentioned endothermal peaks could be seen in all the developed compositions: a first peak at around 60 °C, corresponding to melting of PCL, and a second peak at around 135 °C, related to the melting of PHBH. One can see that as the PCL wt % increased in the PHBH/PCL blends, the melt peak of PCL became larger, while inversely, the melt peak of PHBH was slightly diluted. These two independent peaks suggest some lack of miscibility between the two biopolyesters, as each polymer melted at its corresponding temperature [26,31].

Table 2. Thermal properties of PHBH, PCL, and PHBH/PCL binary blends with different PCL wt % obtained during the 1st heating cycle after processing plus 15 aging days to complete secondary crystallization.

Code	T_{m_PCL} (°C)	ΔH_{m_PCL} (J·g^{-1})	ΔH_{m_PCL} * (J·g^{-1})	χ_{c_PCL} (%)	T_{m_PHBH} (°C)	ΔH_{m_PHBH} (J·g^{-1})	ΔH_{m_PHBH} * (J·g^{-1})	χ_{c_PHBH} (%)
100PHBH-0PCL	-	-	-	-	138.3	18.92	18.92	13.0
90PHBH-10PCL	58.0	6.05	60.52	38.6	134.0	19.72	21.91	15.0
80PHBH-20PCL	59.5	13.41	67.00	42.7	134.0	19.70	24.62	16.8
70PHBH-30PCL	60.8	17.34	57.70	36.8	135.0	18.69	26.68	18.2
60PHBH-40PCL	60.1	25.27	63.21	40.3	135.0	15.91	26.51	18.1
0PHBH-100PCL	62.0	72.23	72.23	46.0	-	-	-	-

* Standardized enthalpies based on the actual weight of the polymer present in the samples.

The results in Table 2 indicate that neat PHBH is characterized by a small degree of crystallinity, χ_{c_PHBH} around 13%, even after the 15-day aging process at room temperature. According to the results of Xu et al. [27], this low χ_c was due to the irregularities in the structure of the polymer chain of the copolymer, which hinders the formation of crystallites, with increasing mol % 3-HH. The addition of PCL wt % slightly increases the crystallinity values, from 15% for the sample with 10 PCL wt % to 18% for 40 PCL wt %. Garcia-Garcia et al. [31] found similar results in the PHB/PCL system with an increase in the degree of crystallization χ_c from 55.1% to 58.2% with 25 wt % PCL. They attributed this to the fact that PCL can affect the crystallization kinetics of neat PHB. As expected, PHBH showed lower χ_c due to the hindering effects of 3-HH, as mentioned above. Contrary to this, Antunes et al. [39] reported a decrease in the degree of crystallinity of PHB by increasing PCL wt % up to 20 wt %, while an increase was observed for 30 wt % PCL. The thermograms in Figure 1a, are interesting as they clearly indicate the aging process after 15 days has been able to complete the secondary crystallization. Moreover, the results gathered in Table 2 corroborated the absence of secondary crystallization after 15 aging days.

With respect to the DSC thermograms obtained in a second heating cycle, shown in Figure 1b (after the cooling of the first heating cycle), it is worthy to note that these showed a clear change. In these DSC thermograms a step in the base line at about 0 °C could be clearly observed, which was attributable to the T_{g_PHBH} [38]. In this case, as the thermal history is completely different, the results regarding crystallinity were somewhat variable. Przybysz et al. [40] reported a remarkable decrease in PHB/PCL blends due to the addition of different peroxide-based compatibilizers, while Oyama et al. [26] showed completely different results for a PHBH/PCL system with peroxide-based compatibilizers, which showed an increase in χ_c of PHBH. Nevertheless, Antunes et al. [39] reported a decrease in χ_c of PHBH without any compatibilizer, which was attributed to changes in crystallization kinetics. In this work, we obtained somewhat varying effects of PCL wt % on the degree of crystallinity of PHBH, as its complex structure (hindering crystallization due 3-HH units) and the additional effects of PCL on crystallization kinetics could overlap with some simultaneous processes and lead to these changes. Obviously, PCL did not show its corresponding step change in the baseline as its T_{g_PCL} is lower than −50 °C. The compatibility of a polymer blend can be assessed by changes in T_g as Garcia et al. reported [41]. In this case, the addition of different PCL wt % to PHBH/PCL blends did not affect the T_{g_PHBH} values obtained, as shown in Table 3.

Table 3. Thermal properties of PHBH/PCL binary blends obtained during the 2nd heating cycle after a heating-cooling process to remove thermal history.

Code	T_{g_PHBH} (°C)	T_{m_PCL} (°C)	ΔH_{m_PCL} (°C)	ΔH_{m_PCL} * (°C)	T_{m1_PHBH} (°C)	T_{m2_PHBH} (°C)	ΔH_{m_PHBH} (°C)	ΔH_{m_PHBH} * (°C)
100PHBH-0PCL	0.46	-	-	-	112.8	137.8	24.9	24.9
90PHBH-10PCL	0.12	54.6	**	**	114.5	138.0	28.4	31.5
80PHBH-20PCL	−0.17	55.0	**	**	113.5	139.0	26.2	32.7
70PHBH-30PCL	0.59	56.0	**	**	112.4	141.3	20.3	28.9
60PHBH-40PCL	−0.46	55.6	**	**	114.8	139.4	15.4	25.7
0PHBH-100PCL	-	57.0	45.7	45.7	-	-	-	-

* Standardised enthalpies based on the actual weight of the polymer present in the samples. ** Melting enthalpies of PCL on PHBH/PCL blends could not be obtained by the overlapping with cold crystallization process in PHBH.

The thermogram corresponding to 100% PHBH showed an exothermic peak at around 49 °C which stood for the peak temperature, T_{cc_PHBH}, of the cold crystallization, with a crystallization enthalpy (ΔH_{cc_PHBH}) of 21.14 J·g^{-1} [19]. At temperatures close to 113 °C, PHBH showed a first and small endothermic peak that corresponded to the melting of PHBH crystalline fraction (T_{m1_PHBH}) and a second melt peak located at 120 °C (T_{m2_PHBH}). These two endothermic peaks could be due to two effects. The first is based on the polymorphism presented by some copolymers such as PHBH as observed in other aliphatic polyesters. Due to the heterogeneous composition of the copolymer itself, different crystalline morphologies can be formed, with different thermal stability. The second effect is that crystallization produces primary crystals with low degree of perfection; these may melt and recrystallize to produce crystals of greater perfection or greater thickness, and this could be the explanation for the presence of two overlapped melting peaks [21,23,25,42].

On the other hand, the 100 PCL wt % sample only showed a very marked endothermic peak, at 55 °C, which corresponded to its melting temperature, T_{m_PCL}, as mentioned above. Due to the similarity between the cold crystallization process of PHBH and the melting of the crystalline fraction of PCL, the binary PHBH/PCL blends showed two overlapped endothermal (PCL melting)-exothermal (PHBH cold crystallization) peaks in the DSC thermograms (Figure 1b). This overlapping did not allow either the melting enthalpy of the PCL (ΔH_{m_PCL}) or the cold crystallization enthalpy of the PHBH (ΔH_{cc_PHBH}) to be quantified accurately. This effect did not allow the correct calculation of χ_c in the PHBH/PCL system blends in this 2nd heating cycle as they were overlapped. Nevertheless, these 2nd heating DSC runs were interesting as the samples had undergone a thermal heating-cooling cycle to remove the thermal history, and subsequently, all the thermal transitions could be detected in a clearer way. In particular, the secondary crystallization of PHBH, which disappeared after the aging process (Figure 1a) and could not be detected, was clearly seen in the 2nd heating cycle.

On the other hand, the thermograms of the binary PHBH/PCL blends showed the characteristic melting peak corresponding to PCL and the two melting peaks of PHBH. It should be noted that as the PCL wt % in the PHBH/PCL blends increased, it had virtually no influence on the melting peak temperatures of PHBH (T_{m1_PHBH} and T_{m2_PHBH}), which was the major component in the developed PHBH/PCL blends. The immiscibility between PHBH and PCL suggests that they form two separate phases with almost independent thermal parameters [30].

Regarding the enthalpies of the thermal transitions related to melting, the results obtained are shown in Table 3. The addition of small amounts of PCL (10 and 20 wt %) slightly increased the values of ΔH_{m_PHBH} *, which indicated a higher amount of energy to melt the crystalline fraction due to there was a slight increase in crystallinity. Higher amounts of PCL (30 and 40 wt %) offer the opposite effect leading to a decrease in crystallinity, probably due to changes in the crystallization kinetics due to the intrinsic structural complexity of PHBH (with 3-HH units which hinder crystallization) and PCL, which could affect crystallization, as has been described previously. When comparing these results with those obtained from samples aged for 15 days, the values of the melting enthalpies were slightly higher. Slow cooling favoured the phenomenon of cold crystallization of PHBH, so that the final melting enthalpy of the crystalline fraction was higher in samples with no previous thermal history.

The thermal stability of binary PHBH/PCL blends was analyzed using thermogravimetric analysis (TGA). Figure 2 presents a comparative plot of the TGA curves for neat PHBH and PCL and PHBH/PCL blends with different PCL wt %.

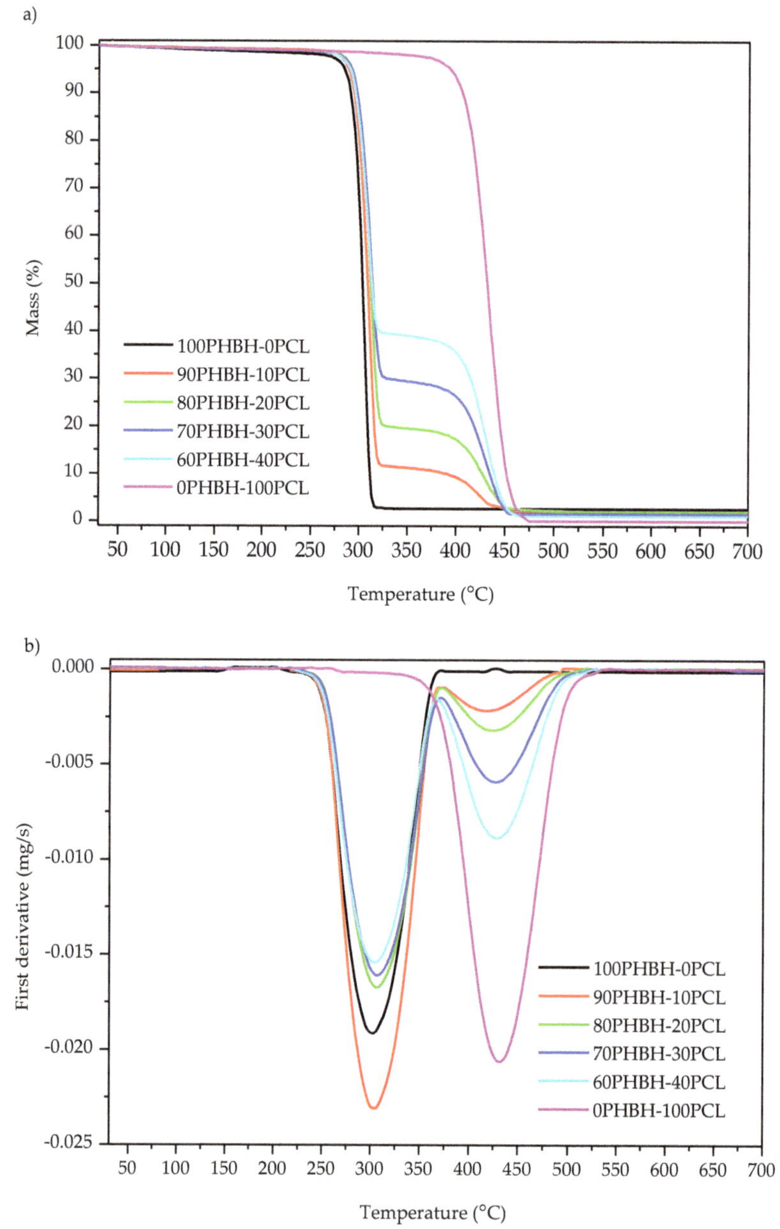

Figure 2. Comparative plot of the thermal degradation of PHBH, PCL, and PHBH/PCL with different PCL wt %. (**a**) TGA degradation curves in terms of mass loss and (**b**) first derivative of TGA thermograms.

The TGA curve of neat PHBH shows a single-step thermal degradation process. The onset degradation temperature ($T_{2\%}$) for neat PHB was 264.6 °C. Once the thermal degradation/decomposition had started, it proceeded very fast with a maximum degradation rate temperature (T_{deg}) of 301.3 °C. These results for individual PHBH were in accordance with those reported by Hosoda et al. [43] and Mahmood et al. [19]. Almost all the mass was thermally decomposed very quickly with an endset degradation temperature of 321 °C. A small residual char of 2.72 wt % was obtained for neat PHBH. It is worth mentioning that the thermal degradation of PHBH occurred in a very narrow temperature range of 57 °C. On the other hand, despite its crystalline fraction melting at relatively low temperature (58–60 °C), it is important to note that PCL was much more thermally stable since its onset degradation temperature ($T_{2\%}$) was 358 °C. In fact, PCL is one of the most thermally stable polyesters. Once the onset degradation temperature was reached, the thermal degradation proceeded in a single-step degradation process, with a maximum degradation rate of 430.9 °C (T_{deg}), and also occurred in a very narrow range up to 474.8 °C, with a residual mass of 0.12 wt %.

The immiscibility between PHBH and PCL was reflected in the TGA curves of the blends. In all of them, two separated degradation steps could be observed: a first step at around 300–310 °C, which corresponded to the degradation of the PHBH phase, and a second step, at a higher temperature occurring in the 390–440 °C range, which corresponded to the degradation of PCL in the blend, as reported by Garcia-Garcia et al. [31] in P3HB-PCL binary blends. The immiscibility was clearly detected by observing the TGA curves. By observing the TGA curve of the 60PHBH-40PCL sample, the mass loss after the first degradation step was almost 40 wt %, which corresponded exactly to the PCL wt % in the blend. The same effect was observed in all other binary blends, which corroborated the lack of miscibility between PHBH and PCL. On the other hand, the blends also presented low residual char formation, between 2 and 1.13 wt %. Comparatively, there was a very slight shift of TGA thermograms to the right with increasing PCL wt % content.

The first derivative of the thermogravimetric curves (DTG), Figure 2b, shows two clearly differentiated thermal degradation processes without almost any overlapping, which corroborates the above-mentioned immiscibility of the PHBH/PCL binary blends. The first peak observed at temperatures between 301–307 °C corresponded to the maximum degradation rate temperature (T_{deg}) of PHBH in all blends. On the other hand, the second peak, located between 420–430 °C, was attributed to the maximum degradation rate temperature of PCL in all the blends. There were very slight changes in the peak maximum values for each process, thus corroborating this immiscibility.

3.2. *Thermomechanical Properties of PHBH/PCL Blends*

In addition to the thermal stability, it is important to evaluate the effect of temperature on dimensional stability. To this end, samples of the different PHBH/PCL blends were subjected to a thermomechanical characterization, which allowed the coefficients of linear thermal expansion (CLTE) were obtained below and above the T_{g_PHBH} (Table 4).

Table 4. Coefficient of linear thermal expansion (CLTE) of PHBH/PCL blends with different PCL wt %, below and above T_{g_PHBH}, obtained using thermomechanical analysis (TMA).

Code	*CLTE* below T_{g_PHBH} (μm·m^{-1}·°C^{-1})	*CLTE* above T_{g_PHBH} (μm·m^{-1}·°C^{-1})
100PHBH-0PCL	68.0 ± 1.2	172.5 ± 2.8
90PHBH-10PCL	70.7 ± 1.5	175.6 ± 2.7
80PHBH-20PCL	93.4 ± 1.1	178.6 ± 2.2
70PHBH-30PCL	104.6 ± 1.0	196.6 ± 3.5
60PHBH-40PCL	106.9 ± 0.5	198.9 ± 3.0

As expected, at temperatures below T_{g_PHBH}, the CLTE values were much lower than those obtained above T_{g_PHBH}. The dimensional expansion of the material was lower below the characteristic T_g, since the material offers a rigid, brittle, glassy behaviour. On the other hand, above T_g, the polymer material changed its behaviour to a plastic, rubbery-like behaviour, and subsequently, the dimensional expansion was allowed. Neat PHBH showed a CLTE of 68 μm·m^{-1}·°C^{-1} below its T_g. As PCL is a much more flexible polymer, its addition to the PHBH/PCL blends provided increased flexibility, and subsequently, the CLTE increased according to PCL wt % contained in the blends. This increase in CLTE was proportional and increased up to 106.9 μm·m^{-1}·°C^{-1} for the blend with 40 PCL wt %. This increase in the CLTE, which was directly proportional to the PCL wt %, was representative for a somewhat loss of fragility and an improvement in the ductile behaviour of PHBH at low temperatures. The same effect was observed for the CLTE values obtained at temperatures above T_{g_PHBH} in the blends. In this case, neat PHBH offered a remarkable increase in its CLTE to 172.5 μm·m^{-1}·°C^{-1}, which was remarkably higher than the CLTE below its T_g (68 μm·m^{-1}·°C^{-1}). The increasing tendency for the CLTE was similar to that mentioned above, and the PHBH/PCL blend with 40 PCL wt % showed the maximum CLTE of about 198.9 μm·m^{-1}·°C^{-1}.

The thermal dynamic mechanical analysis of the PHBH/PCL blends allowed the variation of the storage modulus, E', and the damping factor (tan δ) to be obtained. The damping factor was directly related to the phase angle (δ), which is representative for the delay between the applied dynamic stress (σ_d) and the obtained dynamic elongation (ε_d). The DMTA technique is much more sensitive to the glass transition temperature, T_g, detection since this technique measures changes in mechanical properties as a function of temperature, which includes the definition of T_g with a change from a glassy state to a rubber-like behaviour [30,31,34,37,38]. The dependence of E' on temperature is shown graphically in Figure 3a.

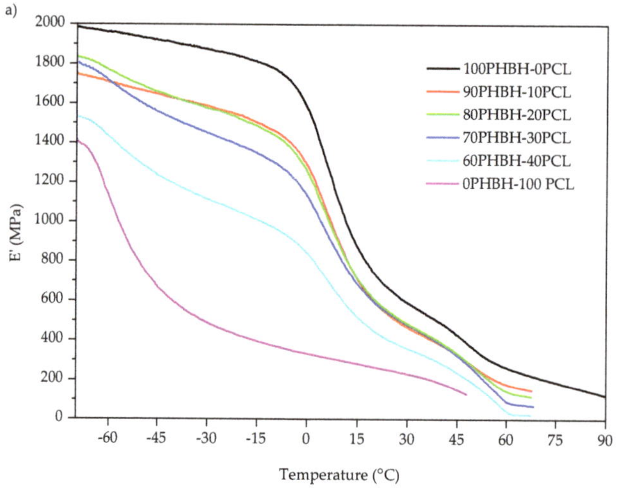

Figure 3. *Cont.*

b)

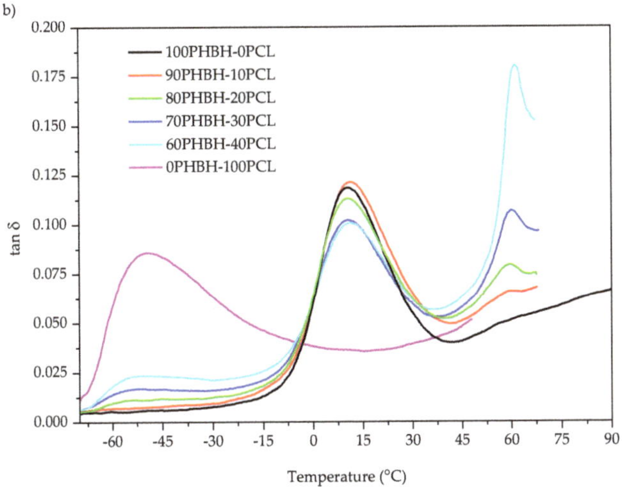

Figure 3. Comparative plot of the dynamic mechanical thermal analysis (DMTA) properties of PHBH, PCL, and PHBH/PCL blends with different PCL wt %, as a function of temperature, (**a**) storage modulus, E', and (**b**) dynamic damping factor (tan δ).

The values of E' obtained for neat PHBH at very low temperatures (e.g., –70 °C) were high, about 2 GPa, since this temperature zone corresponds to an elastic-glassy behaviour of the material, which means it was far (below) from its T_g. For example, at −40 °C, the E' of neat PHBH was 1903.4 MPa. For this same temperature, the E' for neat PCL was much lower, with a value of 596.3 MPa. In this case, the material showed a visco-elastic behaviour, due to it was above T_g (T_{g_PCL} about −60 °C). Thus, adding different PCL wt % to the PHBH led to blends with a clear decreasing tendency of E' which was proportional to the PCL wt %. As the temperature increased, a remarkable decrease in E' values was observed, which represented the change from a glassy state (rigid with high E' values) to a viscous, rubber-like behaviour (viscoplastic with low E' values). Obviously, this was directly related to the glass transition temperature range from −10 to 20 °C. By taking the T_g criterium corresponding to the peak maximum of the dynamic damping factor, T_{g_PHBH} was 9 °C. Since the T_g of both PHBH and PCL is relatively low, at room temperature both polymers are in the rubbery-plateau zone, with a relatively flexible viscoelastic behaviour, as shown in the respective E' plots. This is the typical behaviour of polymers above their T_g, as Burgos et al. and Avolio et al. reported [44,45]. At 25 °C, the value of E' was 653.2 MPa, almost three times lower than neat PHBH below its T_g. The same ranges of variation were maintained for the studied blends, with values between [539 MPa, 398 MPa] at 25 °C vs. [1666 MPa, 1289 MPa] at −40 °C. At higher temperatures, E' tended to have very low values (close to 0 MPa) for blends with high PCL wt %. This effect was because PCL melted at about 60 °C, and once this temperature was overpassed, PCL changed from a rubber-like state to a melt state with extremely low elastic properties. Nevertheless, in blends with 10 and 20 wt % PCL (it is important to bear in mind that PHBH content in these blends was still very high and is not highly affected by PCL addition in terms of dynamic-mechanical behaviour at high temperatures) the effect was less pronounced so that at 60 °C, E' was 178.5 and 142.0 MPa respectively. These values are typical of a rubber-like material and decrease as the wt % PCL increases.

Figure 3b shows the variation of the dynamic damping factor (tan δ) as a function of temperature for neat PHBH and PCL and for their blends with different PCL wt %. With respect to the neat PCL curve, a marked and broad peak was observed with a peak maximum located at –47 °C, corresponding to the glass transition temperature T_g of PCL [30,31] With regard to neat PHBH, a narrow peak (compared to that of PCL) can be seen, with its maximum peak value located at 9 °C. As mentioned above, all the applied techniques suggested poor (or even lack) of miscibility between these two polymers. Therefore, the changes in the respective T_g values of neat PCL (−47 °C) and neat PHBH (9 °C) were not changed in a significant way, therefore corroborating the lack of miscibility of PHBH/PCL blends [38]. In addition, close to 60 °C, PHBH/PCL blends show a third peak which was related to the partial melting of one of the components, i.e., PCL. This third peak was much more pronounced in blends with high PCL wt % and shows proportionality to the PCL wt %, but this third peak appears always at the same temperature of about 60 °C.

3.3. Mechanical Properties and Morphology of PHBH/PCL Blends

Table 5 presents a summary of the mechanical properties obtained from tensile, flexural, and hardness (Shore D) tests, corresponding to neat PHBH and PCL and PHBH/PCL binary blends with different PCL wt %. All the blends of this binary system offered a noticeable decrease in the tensile modulus (E_t) with an increase in the PCL wt % content compared to neat PHBH. PHBH/PCL blends with 40 PCL wt % showed an E_t value of 722 MPa, which was remarkably lower than that of neat PHBH at about 1022 MPa. This represented a % decrease of almost 30%. The addition of PCL resulted in lowering the overall stiffness of the PHBH/PCL blends. Similar results were reported by Hinüber et al. in PHBH/PCL blends [46]. It should be noted that neat PCL is a biopolymer characterized by a very low tensile strength (σ_t) of 12.2 MPa (typical of a rubber-like polymer), high elongation at break (ε_b) (no break means more than 600% as this is the maximum elongation in the used machine), and a low tensile modulus (E_t) of 386 MPa. All these properties positively contributed to improving the ductility of neat PHBH by blending with PCL. The same tendency could be observed for the tensile strength of the PHBH/PCL blends. The addition of different PCL wt % progressively decreased the values of σ_t, from 16 MPa for neat PHBH to [13.4, 13.9] MPa with 20 PCL wt % and 40 PCL wt %, respectively. This decrease in the resistance parameters was due to the two-phase structure of the blends, resulting from the lack of (or very poor) miscibility, as mentioned above. The immiscibility between PHBH and PCL forms a dispersed PCL phase that interrupts the continuity of the PHBH-rich matrix phase, making the stress transfer difficult and, subsequently, decreasing the E_t and σ_t of the PHBH/PCL blends with increasing PCL wt %.

Table 5. Summary of the mechanical properties from tensile, flexural, and hardness tests of neat PHBH and PCL and PHBH/PCL blends with different PCL wt %. σ_t and σ_f represent the tensile and flexural strength, respectively. E_t and E_f are the respective values for the tensile and flexural modulus.

Code	σ_t (MPa)	E_t (MPa)	ε_b(%)	σ_f (MPa)	E_f (MPa)	Shore D Hardness
100PHBH-0PCL	16.0 ± 0.9	1022 ± 412	13.9 ± 1.3	29.5 ± 0.6	1029 ± 31	61.0 ± 0.8
90PHBH-10PCL	14.4 ± 0.9	966 ± 22	19.4 ± 0.8	30.2 ± 1.7	966 ± 36	59.0 ± 0.8
80PHBH-20PCL	13.4 ± 0.7	837 ± 29	67.9 ± 4.1	29.3 ± 1.5	946 ± 47	58.4 ± 1.1
70PHBH-30PCL	13.3 ± 1.2	817 ± 29	308.3 ± 3.6	29.2 ± 1.0	813 ± 20	58.3 ± 0.6
60PHBH-40PCL	14.0 ± 0.5	722 ± 52	461.0 ± 4.1	28.3 ± 1.1	802 ± 64	58.0 ± 0.1
0PHBH-100PCL	12.2 ± 0.9	386 ± 22	No break	22.3 ± 0.3	354 ± 26	55.0 ± 2.0

With respect to the elongation at break, ε_b(%), the effect was opposite and very positive. The increase in ε_b(%) in the PHBH/PCL blends also increased ductility with PCL wt %. PHBH is a rather fragile polymer (after aging or secondary crystallization) with only 13.9% elongation at break. Some researchers attributed this fragility to the secondary crystallization or aging on PHBH, which reduced the amorphous fraction [18,19]. With the addition of only 20 PCL wt %, the ε_b(%) increased up to 67.8% (which represented a percentage increase of almost 387%). Even more, the ε_b(%) for the blend with the highest PCL wt % considered in this study, i.e., 40 wt %, showed an ε_b(%) of 461% (this was 33 times higher than neat PHBH).

Table 5 also offers the flexural parameters of neat PHBH and PCL and the PHBH/PCL blends. As in tensile conditions, PHBH/PCL blends became less rigid with PCL addition. This can be confirmed by a clear decrease in the flexural modulus (E_f). Neat PHBH had an E_f value of 1029 MPa, and this decreased progressively to 801.7 MPa for the PHBH/PCL blend containing 40 PCL wt %. In this case, the decrease in the flexural strength (σ_f) was not so pronounced as that observed in tensile conditions.

In addition, the Shore D hardness, as it is a mechanical resistant property, it followed the same tendency as that observed for both modulus and strength. PHBH showed a Shore D of 61, and the addition of the flexible PCL to PHBH decreased the Shore D values down to 55 for the blend with 40 PCL wt %. Despite this, all Shore D values were close to 58 with very slight variations.

Another important property of polymers is toughness. Figure 4 shows the variation of the absorbed energy per unit area (impact resistance) obtained using a Charpy's test. PHBH is rather brittle, and consequently, it offers low toughness. Graphically, an interesting increase in the impact resistance of PHBH/PCL blends could be observed as the PCL wt % increased. It is important to bear in mind that the energy absorption capacity under impact conditions is directly related to the plastic deformation capacity of the material before the breakage occurs, and the supported stress, too [47]. The presence of a biphasic structure (as reported in morphology analysis) could contribute to improving the toughness, as Ferri et al. reported [48]. Thus, the results obtained corroborated those previously analyzed for tensile and flexural characterizations. Neat PHBH showed a low impact resistance of 5.56 kJ·m^{-2}; with the addition of only 10 PCL wt %, it was increased to 8.7 kJ·m^{-2} (which represented a % increase of 56%). This same trend was proportionally maintained as the PCL wt % increased. It is worth noting that the impact energy for the PHBH/PCL blend with 40 PCL wt % was around 10.5 kJ·m^{-2}, almost twice as much as neat PHBH. These results were consistent with the above-mentioned decrease in the intrinsic brittleness of neat PHBH by blending with PCL [37].

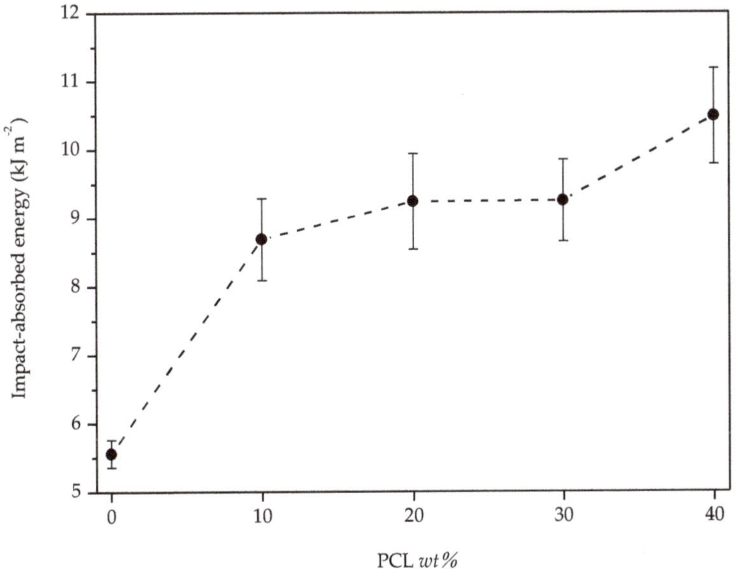

Figure 4. Plot evolution of the impact-absorbed energy of neat PHBH and PHBH/PCL blends with increasing PCL wt %.

Figure 5 shows the FESEM images of the impact fracture surfaces of the PHBH/PCL binary blends with different PCL wt %. All the images show a structure with a continuous and homogeneous matrix phase, which corresponded to the PHBH, and a scattered phase of special morphology, which corresponded to the PCL. This two-phase structure confirmed the lack of miscibility between PHBH and PCL as already concluded with previous thermal analyses. Similar findings were proposed by Quiles-Carrillo et al. [49]. Nevertheless, the typical drop-like structure could not be observed in this system. In addition, the special morphology of the dispersed phase of the PCL, forming small, thin "sheets or flakes" homogeneously distributed with very regular sizes, must be emphasized. FESEM images also revealed that the higher the PCL wt %, the greater the amount of dispersed phase that could be observed.

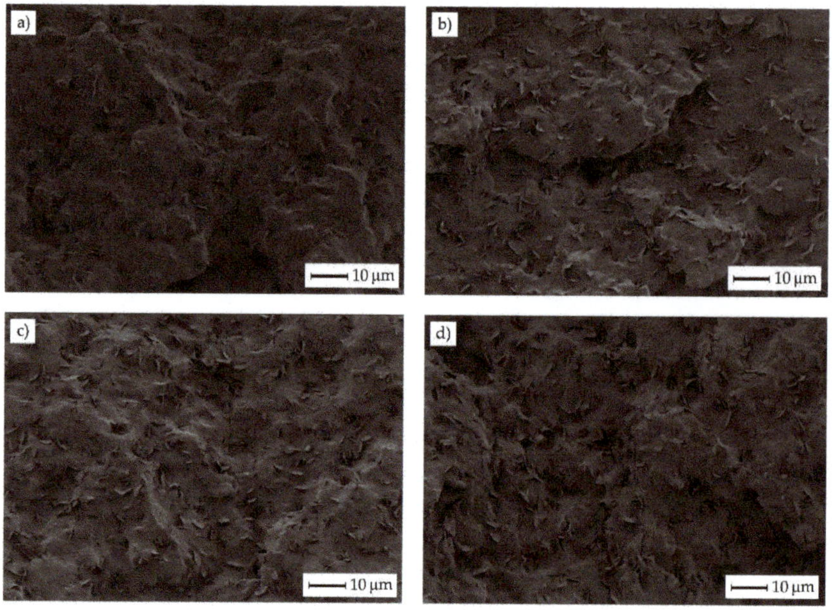

Figure 5. Field emission scanning electron microscopy (FESEM) images at 1000× of the impact fracture surface morphologies of PHBH/PCL binary blends with different PCL wt %, (**a**) 10, (**b**) 20, (**c**) 30, and (**d**) 40.

To check that this dispersed phase corresponded to PCL present in the binary blends, a selective PCL extraction with acetone was performed for 24 h [38]. Figure 6 shows the FESEM images obtained after this selective extraction. In all of them, the dispersed phase with small flake-like shapes was no longer observed. Instead, small and thin empty voids appeared, which corresponded exactly to the geometric shape of the PCL phase before the selective attack (Figure 5). This observation allowed us to conclude that the dispersed phase indeed corresponded to the PCL present in the binary blend. The lack of miscibility between the biopolymers of the PHBH/PCL system was responsible for the internal biphasic structure formed in the blends. In addition, FESEM images showed how the dispersed PCL-rich phase interrupted the continuity of the PHBH matrix, so that the stress transmission inside the material when subjected to external stresses was not adequate [38]. This reduced the mechanical resistant parameters, which corroborated the results obtained in the mechanical characterization of the PBHB/PCL binary blends.

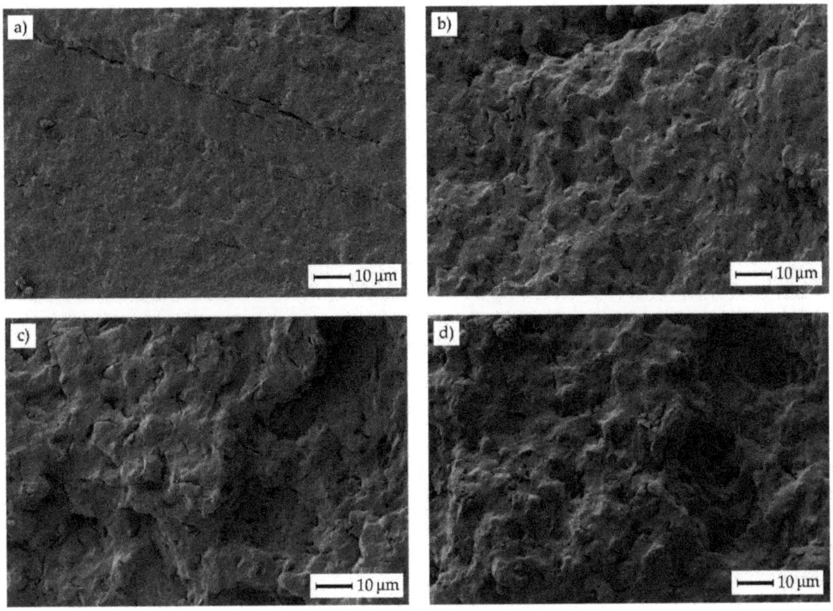

Figure 6. Field emission scanning electron microscopy (FESEM) images at 1000× of the impact fracture surface morphologies of PHBH/PCL binary blends, subjected to PCL selective extraction, with different PCL wt %, (**a**) 10, (**b**) 20, (**c**) 30, and (**d**) 40.

4. Conclusions

The processing and obtaining of PHBH/PCL binary blends allowed a new environmentally friendly material to be obtained with improved toughness and ductile properties, suitable for industrial use in the packaging sector and for medical applications too.

Considering the intrinsic fragility of PHBH, the addition of different PCL wt %, allowed its toughness, ductility, and, above all, impact resistance to be improved. Blends with a 10 PCL wt % offered a percentage increase in the impact resistance of about 56%. The impact resistance was even improved up to double the initial value of neat PHBH by adding 40 PCL wt %. This effect of increasing the ductility of PHBH by increasing the PCL content in the PHBH/PCL blends had an opposite effect on mechanical resistant properties such as modulus and strength (tensile and flexural). In contrast, the elongation at break was remarkably improved, from 13.9% for neat PHBH up to 461% for the PHBH/PCL blend with 40 PCL wt %.

On the other hand, thermal analysis suggested high immiscibility between PHBH and PCL, since the main thermal parameters of neat PHBH and PCL remained unchanged in blends, which is characteristic of very poor or lack of miscibility. Only a slight increase in the thermal stability of neat PHBH was obtained by adding different PCL wt %, since PCL is much more thermally stable than most biopolyesters. The dynamic mechanical thermal analysis (DMTA) allowed accurate values of T_g to be obtained, with two clear and unchanged values located at −47 °C (T_g of PCL) and at 9 °C (T_g of PHBH). These characteristic T_g values remained unchanged in the PHBH/PCL blends, thus suggesting lack of miscibility. Regarding the morphology of the PHBH/PCL blends, they did not show the typical drop-like PCL phase embedded in a PHBH matrix. PCL appeared in the form of small flakes which could exert a positive effect on ductile properties.

Author Contributions: Conceptualization, R.B., L.Q.-C., and J.I.-M.; methodology, L.S.-N., J.I.-M., and O.F.; validation, L.S.-N. and I.V.; formal analysis, L.Q.-C. and R.B.; investigation, I.V. and L.S.-N.; data curation, L.Q.-C., I.V., and O.F.; writing—original draft preparation, L.S.-N.; writing—review and editing, R.B. and J.I.-M.; supervision, R.B. and L.S.-N.; project administration, R.B. All authors have read and agreed to the published version of the manuscript.

Funding: This research work was funded by the Spanish Ministry of Science, Innovation, and Universities (MICIU), project numbers MAT2017-84909-C2-2-R. This work was supported by the POLISABIO program, grant number (2019-A02).

Acknowledgments: Juan Ivorra-Martinez is the recipient of an FPI grant from Universitat Politècnica de València (PAID-2019-SP20190011). Luis Quiles-Carrillo wants to thank GVA for his FPI grant (ACIF/2016/182) and MECD for his FPU grant (FPU15/03812). Microscopy services at UPV are acknowledged for their help in collecting and analyzing FESEM images.

Conflicts of Interest: The authors declare no conflict of interest.

References

1. Fombuena, V.; Samper, M.D. Study of the properties of thermoset materials derived from epoxidized soybean oil and protein fillers. *J. Am. Oil Chem. Soc.* **2013**, *90*, 449–457. [CrossRef]
2. Carbonell-Verdu, A.; Bernardi, L.; Garcia-Garcia, D.; Sanchez-Nacher, L.; Balart, R. Development of environmentally friendly composite matrices from epoxidized cottonseed oil. *Eur. Polym. J.* **2015**, *63*, 1–10. [CrossRef]
3. Ferrero, B.; Boronat, T.; Moriana, R.; Fenollar, O.; Balart, R. Green composites based on wheat gluten matrix and posidonia oceanica waste fibers as reinforcements. *Polym. Compos.* **2013**, *34*, 1663–1669. [CrossRef]
4. Quiles-Carrillo, L.; Montanes, N.; Sammon, C.; Balart, R.; Torres-Giner, S. Compatibilization of highly sustainable polylactide/almond shell flour composites by reactive extrusion with maleinized linseed oil. *Ind. Crop. Prod.* **2018**, *111*, 878–888. [CrossRef]
5. Wong, J.X.; Ogura, K.; Chen, S.; Rehm, B.H. Bioengineered Polyhydroxyalkanoates as Immobilized Enzyme Scaffolds for Industrial Applications. *Front. Bioeng. Biotechnol.* **2020**, *8*, 156. [CrossRef]
6. Dang, K.M.; Yoksan, R. Development of thermoplastic starch blown film by incorporating plasticized chitosan. *Carbohydr. Polym.* **2015**, *115*, 575–581. [CrossRef]
7. Carter, P.; Rahman, S.M.; Bhattarai, N. Facile fabrication of aloe vera containing PCL nanofibers for barrier membrane application. *J. Biomater. Sci. Polym. Ed.* **2016**, *27*, 692–708. [CrossRef]
8. Zhang, W.; Xiang, Y.; Fan, H.; Wang, L.; Xie, Y.; Zhao, G.; Liu, Y. Biodegradable urea-formaldehyde/PBS and its ternary nanocomposite prepared by a novel and scalable reactive extrusion process for slow-release applications in agriculture. *J. Agric. Food Chem.* **2020**, *68*, 4595–4606. [CrossRef]
9. Boronat, T.; Fombuena, V.; Garcia-Sanoguera, D.; Sanchez-Nacher, L.; Balart, R. Development of a biocomposite based on green polyethylene biopolymer and eggshell. *Mater. Des.* **2015**, *68*, 177–185. [CrossRef]
10. Arrieta, M.P.; Samper, M.D.; Aldas, M.; López, J. On the use of PLA-PHB blends for sustainable food packaging applications. *Materials* **2017**, *10*, 1008. [CrossRef]
11. Mukherjee, T.; Kao, N. PLA based biopolymer reinforced with natural fibre: A review. *J. Polym. Environ.* **2011**, *19*, 714–725. [CrossRef]
12. Averous, L. Biodegradable multiphase systems based on plasticized starch: A review. *J. Macromol. Sci. Part C* **2004**, *44*, 231–274. [CrossRef]
13. Pilania, G.; Iverson, C.N.; Lookman, T.; Marrone, B.L. Machine-Learning-Based Predictive Modeling of Glass Transition Temperatures: A Case of Polyhydroxyalkanoate Homopolymers and Copolymers. *J. Chem. Inf. Model.* **2019**, *59*, 5013–5025. [CrossRef] [PubMed]
14. Asrar, J.; Valentin, H.E.; Berger, P.A.; Tran, M.; Padgette, S.R.; Garbow, J.R. Biosynthesis and properties of poly(3-hydroxybutyrate-*co*-3-hydroxyhexanoate) polymers. *Biomacromolecules* **2002**, *3*, 1006–1012. [CrossRef]
15. Misra, S.K.; Valappil, S.P.; Roy, I.; Boccaccini, A.R. Polyhydroxyalkanoate (PHA)/inorganic phase composites for tissue engineering applications. *Biomacromolecules* **2006**, *7*, 2249–2258. [CrossRef]
16. Torres-Giner, S.; Montanes, N.; Boronat, T.; Quiles-Carrillo, L.; Balart, R. Melt grafting of sepiolite nanoclay onto poly(3-hydroxybutyrate-*co*-4-hydroxybutyrate) by reactive extrusion with multi-functional epoxy-based styrene-acrylic oligomer. *Eur. Polym. J.* **2016**, *84*, 693–707. [CrossRef]

17. Corre, Y.-M.; Bruzaud, S.; Audic, J.-L.; Grohens, Y. Morphology and functional properties of commercial polyhydroxyalkanoates: A comprehensive and comparative study. *Polym. Test.* **2012**, *31*, 226–235. [CrossRef]
18. Alata, H.; Aoyama, T.; Inoue, Y. Effect of aging on the mechanical properties of poly(3-hydroxybutyrate-*co*-3-hydroxyhexanoate). *Macromolecules* **2007**, *40*, 4546–4551. [CrossRef]
19. Mahmood, H.; Pegoretti, A.; Brusa, R.S.; Ceccato, R.; Penasa, L.; Tarter, S.; Checchetto, R. Molecular transport through 3-hydroxybutyrate co-3-hydroxyhexanoate biopolymer films with dispersed graphene oxide nanoparticles: Gas barrier, structural and mechanical properties. *Polym. Test.* **2020**, *81*, 106181. [CrossRef]
20. Garcia-Garcia, D.; Fenollar, O.; Fombuena, V.; Lopez-Martinez, J.; Balart, R. Improvement of Mechanical Ductile Properties of Poly(3-hydroxybutyrate) by Using Vegetable Oil Derivatives. *Macromol. Mater. Eng.* **2017**, *302*, 1600330. [CrossRef]
21. Hu, Y.; Zhang, J.; Sato, H.; Noda, I.; Ozaki, Y. Multiple melting behavior of poly(3-hydroxybutyrate-*co*-3-hydroxyhexanoate) investigated by differential scanning calorimetry and infrared spectroscopy. *Polymer* **2007**, *48*, 4777–4785. [CrossRef]
22. Sharma, P.K.; Munir, R.I.; Blunt, W.; Dartiailh, C.; Cheng, J.; Charles, T.C.; Levin, D.B. Synthesis and physical properties of polyhydroxyalkanoate polymers with different monomer compositions by recombinant Pseudomonas putida LS46 expressing a novel PHA synthase (PhaC116) enzyme. *Appl. Sci.* **2017**, *7*, 242. [CrossRef]
23. Watanabe, T.; He, Y.; Fukuchi, T.; Inoue, Y. Comonomer compositional distribution and thermal characteristics of bacterially synthesized poly(3-hydroxybutyrate-*co*-3-hydroxyhexanoate) s. *Macromol. Biosci.* **2001**, *1*, 75–83. [CrossRef]
24. Morgan-Sagastume, F.; Valentino, F.; Hjort, M.; Cirne, D.; Karabegovic, L.; Gerardin, F.; Johansson, P.; Karlsson, A.; Magnusson, P.; Alexandersson, T. Polyhydroxyalkanoate (PHA) production from sludge and municipal wastewater treatment. *Water Sci. Technol.* **2014**, *69*, 177–184. [CrossRef] [PubMed]
25. Sato, H.; Nakamura, M.; Padermshoke, A.; Yamaguchi, H.; Terauchi, H.; Ekgasit, S.; Noda, I.; Ozaki, Y. Thermal behavior and molecular interaction of poly(3-hydroxybutyrate-*co*-3-hydroxyhexanoate) studied by wide-angle X-ray diffraction. *Macromolecules* **2004**, *37*, 3763–3769. [CrossRef]
26. Oyama, T.; Kobayashi, S.; Okura, T.; Sato, S.; Tajima, K.; Isono, T.; Satoh, T. Biodegradable Compatibilizers for Poly(hydroxyalkanoate)/Poly(ε-caprolactone) Blends through Click Reactions with End-Functionalized Microbial Poly(hydroxyalkanoate) s. *ACS Sustain. Chem. Eng.* **2019**, *7*, 7969–7978. [CrossRef]
27. Xu, P.; Cao, Y.; Lv, P.; Ma, P.; Dong, W.; Bai, H.; Wang, W.; Du, M.; Chen, M. Enhanced crystallization kinetics of bacterially synthesized poly(3-hydroxybutyrate-*co*-3-hydroxyhexanate) with structural optimization of oxalamide compounds as nucleators. *Polym. Degrad. Stab.* **2018**, *154*, 170–176. [CrossRef]
28. Gamba, A.; Fonseca, J.S.; Mendez, D.; Viloria, A.; Fajardo, D.; Moreno, N.; Rojas, I.C. Assessment of Different Plasticizer–Polyhydroxyalkanoate Mixtures to Obtain Biodegradable Polymeric Films. *Chem. Eng. Trans.* **2017**, *57*, 1363–1368.
29. Fenollar, O.; Sanchez-Nacher, L.; Garcia-Sanoguera, D.; López, J.; Balart, R. The effect of the curing time and temperature on final properties of flexible PVC with an epoxidized fatty acid ester as natural-based plasticizer. *J. Mater. Sci.* **2009**, *44*, 3702–3711. [CrossRef]
30. Gassner, F.; Owen, A. Physical properties of poly(β-hydroxybutyrate)-poly(ε-caprolactone) blends. *Polymer* **1994**, *35*, 2233–2236. [CrossRef]
31. Garcia-Garcia, D.; Ferri, J.; Boronat, T.; López-Martínez, J.; Balart, R. Processing and characterization of binary poly(hydroxybutyrate)(PHB) and poly(caprolactone)(PCL) blends with improved impact properties. *Polym. Bull.* **2016**, *73*, 3333–3350. [CrossRef]
32. Garcia-Garcia, D.; Garcia-Sanoguera, D.; Fombuena, V.; Lopez-Martinez, J.; Balart, R. Improvement of mechanical and thermal properties of poly(3-hydroxybutyrate)(PHB) blends with surface-modified halloysite nanotubes (HNT). *Appl. Clay Sci.* **2018**, *162*, 487–498. [CrossRef]
33. Garcia-Garcia, D.; Lopez-Martinez, J.; Balart, R.; Strömberg, E.; Moriana, R. Reinforcing capability of cellulose nanocrystals obtained from pine cones in a biodegradable poly(3-hydroxybutyrate)/poly(ε-caprolactone)(PHB/PCL) thermoplastic blend. *Eur. Polym. J.* **2018**, *104*, 10–18. [CrossRef]
34. Garcia-Campo, M.J.; Quiles-Carrillo, L.; Masia, J.; Reig-Pérez, M.J.; Montanes, N.; Balart, R. Environmentally friendly compatibilizers from soybean oil for ternary blends of poly(lactic acid)-PLA, poly(ε-caprolactone)-PCL and poly(3-hydroxybutyrate)-PHB. *Materials* **2017**, *10*, 1339. [CrossRef]

35. Ferri, J.M.; Fenollar, O.; Jorda-Vilaplana, A.; García-Sanoguera, D.; Balart, R. Effect of miscibility on mechanical and thermal properties of poly(lactic acid)/polycaprolactone blends. *Polym. Int.* **2016**, *65*, 453–463. [CrossRef]
36. Arifin, W.; Kuboki, T. Effects of thermoplastic elastomers on mechanical and thermal properties of glass fiber reinforced poly(3-hydroxybutyrate-*co*-3-hydroxyhexanoate) composites. *Polym. Compos.* **2018**, *39*, E1331–E1345. [CrossRef]
37. Simões, C.; Viana, J.; Cunha, A. Mechanical properties of poly(ε-caprolactone) and poly(lactic acid) blends. *J. Appl. Polym. Sci.* **2009**, *112*, 345–352. [CrossRef]
38. Katsumata, K.; Saito, T.; Yu, F.; Nakamura, N.; Inoue, Y. The toughening effect of a small amount of poly(ε-caprolactone) on the mechanical properties of the poly(3-hydroxybutyrate-*co*-3-hydroxyhexanoate)/PCL blend. *Polym. J.* **2011**, *43*, 484–492. [CrossRef]
39. Antunes, M.C.M.; Felisberti, M.I. Blends of poly(hydroxybutyrate) and poly(epsilon-caprolactone) obtained from melting mixture. *Polímeros* **2005**, *15*, 134–138. [CrossRef]
40. Przybysz, M.; Marć, M.; Klein, M.; Saeb, M.R.; Formela, K. Structural, mechanical and thermal behavior assessments of PCL/PHB blends reactively compatibilized with organic peroxides. *Polym. Test.* **2018**, *67*, 513–521. [CrossRef]
41. Garcia, D.; Balart, R.; Sanchez, L.; Lopez, J. Compatibility of recycled PVC/ABS blends. Effect of previous degradation. *Polym. Eng. Sci.* **2007**, *47*, 789–796. [CrossRef]
42. Ding, C.; Cheng, B.; Wu, Q. DSC analysis of isothermally melt-crystallized bacterial poly(3-hydroxybutyrate-*co*-3-hydroxyhexanoate) films. *J. Therm. Anal. Calorim.* **2011**, *103*, 1001–1006. [CrossRef]
43. Hosoda, N.; Tsujimoto, T.; Uyama, H. Green composite of poly(3-hydroxybutyrate-*co*-3-hydroxyhexanoate) reinforced with porous cellulose. *ACS Sustain. Chem. Eng.* **2014**, *2*, 248–253. [CrossRef]
44. Burgos, N.; Tolaguera, D.; Fiori, S.; Jiménez, A. Synthesis and characterization of lactic acid oligomers: Evaluation of performance as poly(lactic acid) plasticizers. *J. Polym. Environ.* **2014**, *22*, 227–235. [CrossRef]
45. Avolio, R.; Castaldo, R.; Gentile, G.; Ambrogi, V.; Fiori, S.; Avella, M.; Cocca, M.; Errico, M.E. Plasticization of poly(lactic acid) through blending with oligomers of lactic acid: Effect of the physical aging on properties. *Eur. Polym. J.* **2015**, *66*, 533–542. [CrossRef]
46. Hinüber, C.; Häussler, L.; Vogel, R.; Brünig, H.; Heinrich, G.; Werner, C. Hollow fibers made from a poly(3-hydroxybutyrate)/poly-ε-caprolactone blend. *Express Polym. Lett.* **2011**, *5*, 643–652. [CrossRef]
47. Quiles-Carrillo, L.; Duart, S.; Montanes, N.; Torres-Giner, S.; Balart, R. Enhancement of the mechanical and thermal properties of injection-molded polylactide parts by the addition of acrylated epoxidized soybean oil. *Mater. Des.* **2018**, *140*, 54–63. [CrossRef]
48. Ferri, J.; Garcia-Garcia, D.; Carbonell-Verdu, A.; Fenollar, O.; Balart, R. Poly(lactic acid) formulations with improved toughness by physical blending with thermoplastic starch. *J. Appl. Polym. Sci.* **2018**, *135*, 45751. [CrossRef]
49. Quiles-Carrillo, L.; Blanes-Martínez, M.; Montanes, N.; Fenollar, O.; Torres-Giner, S.; Balart, R. Reactive toughening of injection-molded polylactide pieces using maleinized hemp seed oil. *Eur. Polym. J.* **2018**, *98*, 402–410. [CrossRef]

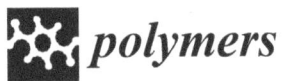

polymers

MDPI

Article

Development and Characterization of Sustainable Composites from Bacterial Polyester Poly(3-Hydroxybutyrate-*co*-3-hydroxyhexanoate) and Almond Shell Flour by Reactive Extrusion with Oligomers of Lactic Acid

Juan Ivorra-Martinez *, Jose Manuel-Mañogil, Teodomiro Boronat, Lourdes Sanchez-Nacher, Rafael Balart and Luis Quiles-Carrillo

Technological Institute of Materials (ITM), Universitat Politècnica de València (UPV), Plaza Ferrándiz y Carbonell 1, 03801 Alcoy, Spain; jomaoam@epsa.upv.es (J.M.-M.); tboronat@dimm.upv.es (T.B.); lsanchez@mcm.upv.es (L.S.-N.); rbalart@mcm.upv.es (R.B.); luiquic1@epsa.upv.es (L.Q.-C.)

* Correspondence: juaivmar@doctor.upv.es; Tel.: +34-966-528-421

Received: 9 April 2020; Accepted: 8 May 2020; Published: 11 May 2020

Abstract: Eco-efficient Wood Plastic Composites (WPCs) have been obtained using poly(hydroxybutyrate-co-hexanoate) (PHBH) as the polymer matrix, and almond shell flour (ASF), a by-product from the agro-food industry, as filler/reinforcement. These WPCs were prepared with different amounts of lignocellulosic fillers (wt %), namely 10, 20 and 30. The mechanical characterization of these WPCs showed an important increase in their stiffness with increasing the wt % ASF content. In addition, lower tensile strength and impact strength were obtained. The field emission scanning electron microscopy (FESEM) study revealed the lack of continuity and poor adhesion among the PHBH-ASF interface. Even with the only addition of 10 wt % ASF, these green composites become highly brittle. Nevertheless, for real applications, the WPC with 30 wt % ASF is the most attracting material since it contributes to lowering the overall cost of the WPC and can be manufactured by injection moulding, but its properties are really compromised due to the lack of compatibility between the hydrophobic PHBH matrix and the hydrophilic lignocellulosic filler. To minimize this phenomenon, 10 and 20 *phr* (weight parts of OLA-Oligomeric Lactic Acid per one hundred weight parts of PHBH) were added to PHBH/ASF (30 wt % ASF) composites. Differential scanning calorimetry (DSC) suggested poor plasticization effect of OLA on PHBH-ASF composites. Nevertheless, the most important property OLA can provide to PHBH/ASF composites is somewhat compatibilization since some mechanical ductile properties are improved with OLA addition. The study by thermomechanical analysis (TMA), confirmed the increase of the coefficient of linear thermal expansion (CLTE) with increasing OLA content. The dynamic mechanical characterization (DTMA), revealed higher storage modulus, E′, with increasing ASF. Moreover, DTMA results confirmed poor plasticization of OLA on PHBH-ASF (30 wt % ASF) composites, but interesting compatibilization effects.

Keywords: PHBH; almond shell flour; mechanical properties; thermal characterization; WPCs

1. Introduction

The current problem related to the negative environmental impact of large volumes of wastes [1] in a consumer society has promoted a significant awareness and sensitiveness about this problem. Some governments are facing this through legislation that protects environment and minimizes the harmful impact on nature. This, in part, has led to the extensive development of new eco-efficient materials, from the point of view of their renewable origin, low carbon footprint, possibility of composting, biodegradability, and so on [2]. An interesting family of these new eco-efficient materials are the so-called Wood Plastic Composites, WPC. These composites consist on a polymeric matrix in which wood (or whatever lignocellulosic subproduct of the food industry or agroforestry) particles (from 10 to 60 wt %, depending on the manufacturing process) are embedded, leading to an appearance and surface finishing similar to natural wood. As many times, the lignocellulosic fillers are by-products from other sectors, they are cheap and do not increase the cost of the WPC; in addition, they come from natural resources and, subsequently, they represent a sustainable source for use in new and environmentally friendly materials [3–5].

These WPCs are already replacing the use of traditional woods in some applications, which is an important protection of forest resources. WPCs formulations have been optimized in sectors as important as automotive, outdoor furniture, interior design, railings, floors, coatings, decks, fences, pergolas, decking, and so on [3–8].

The fact that they are composed of a polymeric matrix, gives them better behaviour against water or in humid environments. According to Singh et al., WPCs have gained a significant share of the consumer market, becoming the fastest growing segment of the plastics industry [4]. Within the wide range of possibilities that these eco-efficient materials offer as substitutes for wood, those that use thermoplastic polymers as the matrix are of particular interest, precisely because of the ease and versatility of manufacturing processes. Poly(ethylene) (PE), poly(styrene) (PS), poly(vinyl chloride) (PVC) and poly(propylene) (PP) are some of the most widely used polymers in WPCs. Nevertheless, these polymer matrices are petroleum-derived polymers.

Due to the need to protect the environment, the possibility of using biopolymers as matrices in WPCs is currently being studied. The use of a biodegradable thermoplastic polymer (actually, a compostable polymer that disintegrates in controlled compost soil) from natural resources, together with a natural lignocellulosic filler from industrial wastes or by-products, allows the obtaining of totally biodegradable and eco-efficient WPCs [6]. These new green composites represent the new generation of biobased, sustainable, low environmental impact WPCs. Nowadays, there are already a large number of natural biopolymers on a commercial level, among which three main families stand out. The first one, consists on polymers from biomass which include polysaccharides such as starch (and starch-derived polymers such as polylactide), cellulose, chitosan, chitin, and proteins such as casein, keratin, collagen, and so on. The second group includes conventional polymers such as poly(ethylene), poly(urethanes), poly(amides), that are partially or fully obtained from natural resources but they show identical (or very similar) properties to their petroleum-derived counterparts. Finally, a new family of very promising polymers is that of bacterial polyesters which are generally referred to as polyhydroxyalkanoates PHAs. PHAs include more than 300 different polyesters and copolymers such as poly(3-hydroxybutyrate) (P3HB or just PHB), poly(3-hydroxybutyrate-co-3-hydroxyvalerate) (PHBV), among others [5,9].

One of the most interesting biopolymers, obtained by bacterial fermentation, is poly(3-hydroxybutyrate-*co*-3-hydroxyhexanoate), PHBH. This copolymer is obtained by incorporating into the polyhydroxybutyrate chain, PHB, 3-hydroxyhexanoate units with medium-length side groups, [P(3HB-co-3HH)]. Mahmood and Corre identify a structure formed by branches of short 3HH units, on the main 3HB chain, thus reducing regularity. Yang and Liao compare the formation of these units by dielectric spectroscopy and melt viscosity [10,11]. Moreover, the addition of the 3HH units extends the temperature range for processing of this copolymer, but the storage modulus and the strength is reduced [12–15]. Watanabe and Oyama, synthesized PHBH from cheap natural resources such as coconut oil, biomass, beet, sugar cane, molasses and vegetable oils [16,17]. These characteristics allow it to be used as a substitute for traditional petroleum-derived polymers in some applications, such as disposable plastic bags, food packaging, catering, agricultural mulch film, and so on [18].

The main objective of this work is to obtain fully biobased WPCs. For this purpose, PHBH was chosen as the thermoplastic matrix; this matrix was reinforced with almond shell flour (ASF). The almond shell flour, ASF, is a waste of the agro-food industry. It is very cheap, fully biobased and biodegradable. By incorporating ASF into the PHBH matrix it gives a wood-like appearance. In this work, the effect of the amount of ASF on the mechanical, thermal, thermomechanical and water absorption properties of PHBH-ASF composites is investigated. In addition, the optimization of the behaviour of these composites is used by the addition of an oligomer of lactic acid (OLA), to provide some plasticization and to increase toughness. Due to the lack of compatibility between the different elements, reactive extrusion (REX) has been proposed as a strategy to improve the properties of the mixtures. This process will improve the chemical bonding of the biopolymer chains to the surface of the lignocellulosic fillers by the action of reactive molecules with at least two functional sites.

2. Experimental Section

2.1. Materials

The PHBH commercial grade (ErcrosBio® PH 110) used in this study was supplied in pellet by Ercros S.A. (Barcelona, Spain). This polymer has a density of 1.2 g cm^{-3} and a melt flow index (MFI) of 1 (g/10 min^{-1}) measured at 160 °C. Even with this low MFI, this is suitable for injection moulding as it has very low melt strength, so requires an appropriate temperature profile for extrusion and injection moulding. Almond shell powder/flour (ASF) was purchased from Jesol Materias Primas (Valencia, Spain). This powder was sieved in a vibrational sieve RP09 CISA® (Barcelona, Spain) to obtain a maximum particle size of 150 μm. Figure 1 shows the irregular particle size of ASF with average size below 150 μm (the average size is 75 μm). As plasticizer/impact modifier, an oligomer of lactic acid (OLA), commercial grade Glyplast OLA 8 was kindly provided by Condensia Química S.A. (Barcelona, Spain). Glyplast OLA 8 is a liquid polyester (with an ester content above 99%) with a viscosity of 22.5 mPa s measured at 100 °C. Its density is 1.11 g cm^{-3}; it has a maximum acid index of 1.5 mg KOH g^{-1} and a maximum moisture content of 0.1%.

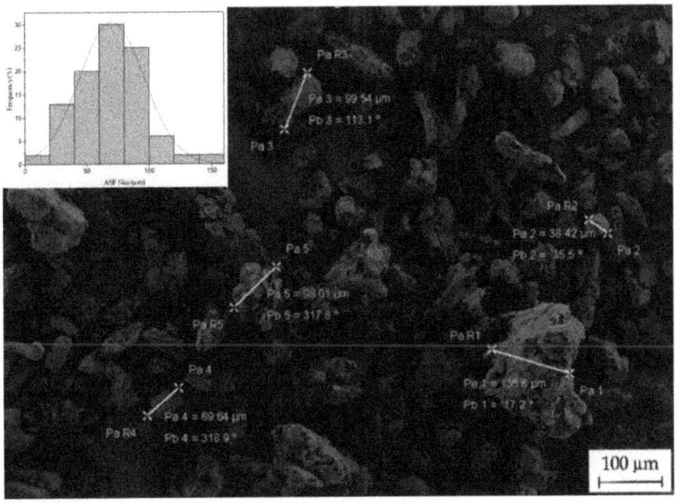

Figure 1. Visual aspect of almond shell flour particles obtained by field emission scanning electron microscopy (FESEM) at 100× and a histogram of their size distribution.

2.2. Manufacturing of PHBH-ASF Composites

Before further processing of composites, PHBH pellets and almond shell flour were dried for 6 h at 80 °C, in a dehumidifier model MDEO, supplied by Industrial Marsé (Barcelona, Spain). Then, different amounts (see Table 1) of PHBH, ASF (in wt %) and OLA (in *phr*–weight parts of OLA per one hundred weight parts of PHBH) were mechanically pre-mixed in a zipper bag to obtain pre-homogenization. These six materials were then extruded in a twin-screw corotating extruder from DUPRA S.L. (Alicante, Spain). The four temperature barrels were programmed to the following temperature program: 110 °C (hopper), 120 °C, 130 °C and 140 °C (extrusion die) and the screw speed was maintained in the 20–25 rpm range. The extruded material was cooled down to room temperature and then, pelletized for further processing by injection moulding. The injection moulding process was carried out in a Sprinter 11 injection machine from Erinca S.L. (Barcelona, Spain) to obtain standard samples for further characterization. As PHBH has low melt strength, it needs some particular processing conditions. The injection temperature profile was set to 150 °C (hopper), 140 °C, 130 °C and 120 °C (nozzle). In addition, it requires a tempered mould at 60 °C. The filling and cooling times were set to 1 s and 20 s, respectively. It is well known that bacterial polyesters undergo secondary crystallization or recrystallization with time (sometimes designed as physical aging since this leads to an embrittlement), especially at temperatures above T_g. Recrystallization rate is directly related to temperature; therefore, samples have been subjected to a recrystallization process at 25 °C for 15 days since it has been reported that almost all recrystallization takes place after two weeks from the processing [17,19,20]. To avoid potential hydrolysis of the polyester surface, samples were stored in a vacuum desiccator with constant moisture.

Table 1. Summary of sample compositions according to the weight content (wt %) of PHBH (Poly(3-hydroxybutyrate-*co*-3-hydroxyhexanoate)), and ASF (Almond Shell Flour) and the addition of OLA (Oligomeric Lactic Acid) as parts per hundred resin (*phr*) of PHBH-ASF composite.

Code	PHBH (wt %)	ASF (wt %)	OLA (*phr*)
PHBH	100	-	-
PHBH-10ASF	90	10	-
PHBH-20ASF	80	20	-
PHBH-30ASF	70	30	-
PHBH-30ASF/10OLA	70	30	10
PHBH-30ASF/20OLA	70	30	20

2.3. *Mechanical Properties of PHBH-ASF/OLA Composites*

The mechanical characterization of PHBH-ASF/OLA composites was carried out by means of tensile tests according to ISO 527-2:2012 in a universal testing machine, model ELIB-50 from Ibertest (Madrid, Spain). A 5 kN load cell was used and the crosshead rate was set to 10 mm min^{-1}. The standardized specimens corresponded to the designation A12 from ISO 20753:2018. Impact resistance was quantified by means of a Charpy test, with a 1-J pendulum from Metrotec S.A. (San Sebastian, Spain), on specimens with a standardized "V" notch, according to ISO 179-1:2010. In addition, the hardness of PHBH-ASF/OLA composites, was obtained using a Shore-D hardness tester model 673-D from J. Bot Instruments, S.A. (Barcelona, Spain) according to ISO 868:2003. All mechanical tests were performed on 5 specimens of each composition.

2.4. *Morphology of PHBH-ASF/OLA Composites*

The morphology study of the impact fractured specimens from impact tests was carried out by field emission scanning electron microscopy (FESEM) in a ZEISS ULTRA 55 microscope from Oxford Instruments (Abingdon, Oxfordshire, UK). The accelerating voltage was 2 kV. Prior to this analysis, the samples were metallized with platinum in a sputtering metallizer EMITECH mod. SC7620 from Quorum Technologies Ltd. (East Sussex, UK).

2.5. *Thermal Characterization of PHBH-ASF/OLA Composites*

The thermal characterization of PHBH-ASF/OLA composites, by means of differential scanning calorimetry (DSC), was performed in a TA Instruments calorimeter mod. Q2000 (New Castle, DE, USA). For the thermal study, a dynamic temperature cycle was scheduled with the following sequence:

1st cycle: −50 °C to 200 °C at a constant heating rate of 10 °C min^{-1}, 2nd cycle: 200 °C to −50 °C at a constant cooling rate of 10 °C min^{-1}; this step was scheduled to remove the thermal history. Finally, a 3rd cycle from −50 °C up to 200 °C at 10 °C min^{-1} was programmed. The DSC analysis was performed in an inert nitrogen atmosphere with a flow rate of 50 mL min^{-1}, with samples between (5–10 mg), in standard 40 µL aluminium crucibles. The degree of crystallinity (X_c) was calculated by using Equation (1) where ΔH_m and ΔH_{cc} (J g^{-1}) are melt enthalpy and cold crystallization enthalpy respectively. ΔH_m^0 (J g^{-1}) is the theoretical value that corresponds to fully crystalline PHBH; this was taken as 146 (J g^{-1}) as reported in Reference [14]. Finally, w is the fraction weight of PHBH.

$$X_c = \left[\frac{\Delta H_m - \Delta H_{cc}}{\Delta H_m^0 \times w}\right] \times 100 \quad (1)$$

Thermogravimetric analysis (TGA) was carried out in a Mettler-Toledo TGA/SDTA 851 thermobalance (Schwerzenbach, Switzerland). Samples consisted on small pieces with a total weight of 5–7 mg. These samples were placed in standard alumina pans (70 µL), and then subjected to a heating ramp from 30 to 700 °C at a constant heating rate of 20 °C min^{-1} in nitrogen atmosphere. All the thermal tests were done in triplicate.

2.6. *Thermomechanical Characterization of PHBH-ASF/OLA Composites*

The dynamic-mechanical-thermal analysis, DMTA, was done in a Mettler-Toledo dynamic analyzer (Columbus, OH, USA), on rectangular samples of 40 × 10 × 4 mm^3. Heating was programmed from −70 °C to 70 °C at a constant rate of 2 °C min^{-1}; samples were subjected to a single cantilever test in dynamic conditions with a maximum deflection of 10 µm and a frequency of 1 Hz. The coefficient of linear thermal expansion (CLTE) of the PHBH-ASF/OLA composites was determined using a TA Instruments mod. Q400 (New Castle, DE, USA). The heating program was set from −70 °C to 70 °C, using a constant heating rate of 2 °C min^{-1}. Rectangular samples with dimensions 10 × 10 × 4 mm^3 were subjected to a constant force of 0.02 N.

2.7. *Water Uptake of PHBH-ASF/OLA Composites*

The water absorption study was carried out according to the method described in ISO 62:2008, with distilled water at 23 ± 1 °C, for 9 weeks. The specimens had rectangular dimensions of 80 × 10 × 4 mm^3. Before starting the immersion, samples were dried at 60 °C for 24 h in an air circulating oven, model 2001245 DIGIHEAT-TFT from J.P. Selecta, S.A. (Barcelona, Spain).

Samples were extracted periodically from the water every planned period. They were dried to remove any remaining surface moisture and weighed on a precision analytical balance model AG245, from Mettler-Toledo Inc. (Schwerzenbach, Switzerland). After this measurement, they were re-immersed in the distilled water bath.

The amount of absorbed water during the water uptake process can be calculated following this expression:

$$\Delta m_t(\%) = \left(\frac{W_t - W_0}{W_0}\right) \times 100 \quad (2)$$

where w_t stands for the sample weight after an immersion time of t; w_0 corresponds to the initial weight of the dried, before the immersion.

ISO 62:2008 establishes the application of first Fick's Law to determine the diffusion coefficient, D, from the collected data regarding the increase of mass by immersion, by means of the expression (2). The calculation of D can be done in the linear zone of the water absorption plot. In this initial stage, w_t/w_s is a linear function Δm_t = f($\sqrt{t}$), that allows to determine D from the slope, θ [21–24].

$$\frac{W_t}{W_S} = \frac{4}{d}\left(\frac{D\,t}{\pi}\right)^{\frac{1}{2}} \quad (3)$$

where D represents the coefficient of diffusion, d stands for the initial thickness of the specimen and w_s stands for the saturation mass in the linear zone. If we plot w_t/w_s against $\sqrt{t}$, it is possible to obtain the slope (θ) as this condition is met, w_t/w_s ($\leq$0.5), then the D value can be calculated by following the above-mentioned expression [25].

A correction (Stefan's approximation) is applied to this calculation for the exact calculation of the D according to the dimensions of the specimens:

$$D_c = D\left(1 + \frac{d}{h} + \frac{d}{w}\right)^{-2} \tag{4}$$

where D_c is the geometrically corrected diffusion coefficient, h is the length, w is the width of the sample and d is the thickness. This equation is based on the assumption that the diffusion velocities are the same in all directions [23–25].

3. Results and Discussion

3.1. Mechanical Properties of PHBH-ASF/OLA Composites

Table 2 shows the results obtained from mechanical characterization (tensile test, hardness and impact Charpy) of PHBH-ASF/OLA composites. The addition of ASF to the PHBH matrix resulted in composites with greater stiffness with increasing ASF wt %. With only 10 wt % ASF, the elastic modulus in tensile test (E_t) increases to 1310 MPa from 1065 MPa (neat PHBH without lignocellulosic filler). This means an increase of 23%. This % increase is, obviously higher, for PHBH composites containing 30 wt % ASF. Regarding the maximum tensile strength (σ_{max}), the incorporation of natural fillers to the polymeric PHBH matrix, promotes a noticeable decrease. Neat PHBH offers a tensile strength of 20 MPa, which decreases to 16 MPa with only 10 wt % ASF and to 12 MPa with 30 wt % ASF. Singh et al. [4] established that the decrease of tensile strength results from stress concentration at the polymer/filler interfaces. There is a lack of interface interactions between the polymeric matrix (highly hydrophobic) and the lignocellulosic particles (highly hydrophilic), which gets more pronounced with increasing particle content [4]. The mechanical behaviour of these composites highly depends on the potential interactions between the polymer matrix and the surrounding lignocellulosic filler/particle. The lack of (or poor) adhesion leads to formation of microscopic gaps that are responsible for a discontinuous material with the subsequent stress concentration phenomenon [26].

Table 2. Summary of the mechanical properties of the PHBH-ASF/OLA composites with different compositions, in terms of the tensile modulus (E_t), maximum tensile strength (σ_{max}), elongation at break (ε_b), Shore-D hardness and impact strength.

Code	Tensile Strength (MPa)	Elastic Modulus (MPa)	Elongation at Break (%)	Hardness Shore-D	Impact Strength (kJ m^{-2})
PHBH	20 ± 1	1065 ± 23	8.1 ± 0.7	60.2 ± 0.2	4.3 ± 0.3
PHBH-10ASF	16 ± 1	1310 ± 35	5.2 ± 0.4	63.5 ± 0.4	1.8 ± 0.2
PHBH-20ASF	14 ± 1	1543 ± 23	4.0 ± 0.4	64.7 ± 0.6	1.7 ± 0.2
PHBH-30ASF	12 ± 1	1744 ± 31	3.5 ± 0.3	66.2 ± 0.6	1.6 ± 0.3
PHBH-30ASF/10OLA	10 ± 1	1158 ± 23	6.2 ± 0.2	58.6 ± 0.5	2.4 ± 0.4
PHBH-30ASF/20OLA	8 ± 1	735 ± 28	9.7 ± 0.8	50.0 ± 0.4	2.9 ± 0.3

Furthermore, as usual in composites with lignocellulosic fillers/reinforcements [27–34], the plastic deformation capacity of WPCs decreases in a dramatic way. If we focus on the elongation at break (ε_b), the intrinsic very low ε_b values for neat PHBH (around 8% after an aging time of 15 days), are reduced to half with 20 wt % ASF. Composites with 30 wt % ASF, has a ε_b of only 3.5%. This means a much more fragile and less resistant behaviour of PHBH-ASF composites with increasing wt % ASF without any other component. This mechanical behaviour is like those obtained in other thermoplastic matrix composite systems with natural fillers [3,26,35–37]. Nevertheless, the addition of OLA leads to an improvement of the elongation at break due to a compatibilization between PHBH/ASF by the interaction of compatibilizer with terminal groups of PHBH and lignocellulosic particles. Additionally, a plasticization effect on the matrix can be expected by OLA acting as lubricant inside de polymer chain. Both effects were reported by Quiles-Carillo et al. with different biobased and petroleum-derived compatibilizers on PLA/ASF [38].

With respect to the hardness values, the increase in the wt % ASF content favours a slight increase in Shore-D hardness as expected since tensile characterization suggested increased stiffness. In fact, the Shore-D hardness increases from 60.2 (neat PHBH) to 66.2 (composite containing 30 wt % ASF) [30]. On the other hand, the impact resistance is one of the properties with the greatest decrease in uncompatibilized PHBH-ASF composites. First, it should be noted that PHBH is a thermoplastic with an intrinsically low impact resistance. This fragile behaviour of PHBH was greatly affected by the addition of ASF (even with low wt % ASF content. The results show how the addition of 10 wt % ASF, decreases the impact strength to values lower to the half (1.8 kJ m^{-2}) of neat PHBH (4.3 kJ m^{-2}).

It is worthy to note that WPCs are widely used in applications that include fencing, garden objects, furniture, decking, and so on. The technical requirements will depend on the final application. Considering the mechanical results, these WPCs offer relatively low tensile strength and low elongation at break even without ASF filler. The addition of ASF up to 30 wt % and OLA as compatibilizer, gives interesting materials with a wood-like appearance but they cannot be used for medium technological applications as mechanical properties are low. Moreover, addition of 30 wt % ASF leads to a cost-effective material as PHBH matrix is still an expensive bacterial polyester.

The impact resistance values dropped down to similar values with increasing wt % ASF content. For 30 wt % ASF the absorbed-energy per unit area is around 1.6 kJ m^{-2}, which represents a loss of almost 63% of the capacity to absorb energy during impact conditions, which is representative for the overall toughness. These results showed a clear embrittlement and loss of toughness on PHBH-ASF composites as the wt % ASF content increases. The small lignocellulosic ASF particles form a dispersed phase in the thermoplastic matrix (due to the high hydrophilicity of ASF particles, it is possible to form aggregates which lead to worse properties). This dispersed phase interrupts the continuity of the PHBH matrix; in these conditions, the stress transfer between the particle and the matrix is not allowed. In addition, as observed in Figure 1, ASF particles are not spherical; their shape is very irregular (with angular shapes) and could act as micro-notches that promote formation of microcracks and subsequently affect the crack growth. This phenomenon justifies the decrease in toughness in PHBH-ASF composites [3,4,29,37,39]. Furthermore, since the polymeric matrix is non-polar (hydrophobic) and the ASF particles are highly polar (highly hydrophilic due to its lignocellulosic composition), there is no (or very poor) matrix–particle interaction along the interface. This lack of interface causes a fragilizing effect by concentrating the stresses and decreasing the potential plastic deformation capacity in PHBH-ASF composites [26,35,40–43].

Figure 2 shows in a detailed way the lack of interface between PHBH and ASF particles. Figure 2a,b show a clear micro-gap surrounding the ASF particle. This gap sizes range from 1 μm to 3–4 μm (see white arrows), and the gaps are responsible for lack of interactions in the polymer-particle interface. This gaps are responsible for interrupting the continuity in uncompatibilized PHBH-ASF composites and do not allow stress transfer. This suggests ASF particles do not act as reinforcing material; furthermore, they promote stress concentration leading to a brittle material. Despite this, addition of an oligomer of lactic acid (OLA), could potentially provide improved toughness as observed in Figure 2c,d, which corresponds to the compatibilized composite with 10 *phr* OLA. At higher magnifications ASF particles are completely embedded by the PHBH-rich matrix (Figure 2d, see white arrow with a circle end). This situation is similar to that obtained in composites with 20 *phr* OLA (Figure 2e,f) which shows absence of gap between the ASF particles and the surrounding matrix. This oligomer has carboxylic acid and hydroxyl terminal groups that can readily react (interact) with hydroxyl terminal groups in PHBH and, obviously, with the hydroxyl groups in ASF (mainly, cellulose, lignin and hemicelluloses). As can be seen, the gap is remarkably reduced (white arrow) and this could contribute to improve toughness and stress transfer [44,45]. Despite high polarity ester groups can establish somewhat interactions with polar groups in cellulose, the main compatibilizing effects are obtained with high reactive groups such as maleic anhydride, carboxyl acids, end-chain hydroxyl groups, glycidyl methacrylate as reported by Pracela et al. [46] by using functionalized copolymers to provide increased interface interactions between a polymer matrix and cellulose particles. Some interactions between ester groups and cellulose particles have been described by Chabros et al. [47] in thermosetting unsaturated polyester resins with cellulose fillers; in particular they describe some interactions between the polar ester groups and hydroxyl groups in cellulose by hydrogen bonding. These small range interactions can also occur in PHBH/ASF composites, but their intensity is lower than that provided by the reaction of carboxylic acid and hydroxyl terminal groups in OLA with both hydroxyl groups in cellulose and PHBH through condensation or esterification reactions. As described by Mokhena et al. [48] the ester groups in PLA are not enough to provide intense interactions between the polyester-type matrix and the cellulose filler. They report the need of different treatments on cellulose such as acetylation, glyoxalization, silylation, treatment with glycidyl methacrylate (GMA), among others to improve polymer-matrix interactions. This could be related not only to the polarity but also with the hydrophilic nature of ASF and the hydrophobic nature of PHBH.

Figure 2. Field emission scanning electron microscopy (FESEM) images at 1000× (left side) and 2500× (right side) corresponding to PHBH-ASF composite with 30 wt % ASF with different OLA content, (**a**) & (**b**) 0 *phr* OLA, (**c**) & (**d**) 10 *phr* OLA, (**e**) & (**f**) 20 *phr* OLA.

The PHBH-ASF composite with 30 wt % ASF, showed the worst mechanical properties in terms of ductility and toughness. This was taken as a reference material to improve its properties by the addition of a compatibilizer/plasticizer. An oligomer of lactic acid ester (OLA) was added in different proportions (10 and 20 *phr*) to provide compatibilization and some plasticization [49,50]. In Table 2, the increase in impact resistance for the reference uncompatibilized composite is observed with the addition of 10 *phr* and 20 *phr* of OLA. The addition of small amounts (10 *phr* $\approx$ 0.1 wt %) of this oligomer significantly improves the impact resistance of the composite [51]. It changes from 1.6 kJ m^{-2}

(uncompatibilized PHBH-ASF with 30 wt % ASF) to 2.4 kJ m^{-2} for the same composite with 20 *phr* OLA, which represents a % increase of 33%. Considering that this increase in toughness is related to an improvement in ductility, subsequently, Shore-D hardness values decreased, changing from 66.2 Shore-D to 58.6 and 50.0 Shore-D for 10 *phr* and 20 *phr* OLA content, respectively.

This improvement on toughness is corroborated by the capacity of deformation observed in compatibilized composites. By adding only 10 *phr* OLA to the reference uncompatibilized PHBH-ASF composite, its elongation at break is almost doubled. The compatibilization effect reported in FESEM images (gap reduction) provided a more efficient load transfer between PHBH and ASF leading to an improvement of elongation at break as Quiles-Carrillo et al. reported [38]. With 20 *phr* OLA, the ε_b, increase up to 9.7%, which means an increase of 177% compared to the same composite without OLA. However, the most striking thing is that the ε_b with 20 *phr* OLA is even higher than that of neat PHBH which is a very positive feature, mainly in this highly brittle material. The composite with 30 wt % ASF and 20 *phr* OLA shows a ε_b value of almost 20% higher than neat PHBH without any filler. These results indicate a marked plasticizing effect of this OLA oligomer, which is corroborated by the values of the tensile strength (σ_t) and the elastic modulus (E_t). The incorporation of short-chain oligomers OLA increases the free volume of the polymer chains in PHBH, which leads to a reduction in the stiffness and an increase in the ductility of composites. The improvement in ductility these PHBH-ASF composites with the addition of OLA, produces a decrease in σ_t to 8 MPa with 20 *phr* OLA, which is slightly lower than the σ_t compared to the same composite without OLA (12 MPa). On the other hand, the decrease in the E_t observed by OLA addition, indicated that the compatibilized composites are not as rigid as uncompatibilized materials. The obtained E_t values for 10 and 20 *phr* OLA are 1158 MPa and 735 MPa respectively. This represents a decrease of 33% and 58%, with regard to the reference uncompatibilized composite with an E_t value of 1744 MPa [51].

3.2. Thermal Properties of PHBH-ASF/OLA Composites

A comparative plot of the DSC thermograms is represented in Figure 3 and the main thermal parameters are summarized in Table 3. All thermograms are characterized by a first change at very low temperature (around 0 °C) in the corresponding baseline, which is attributable to the corresponding glass transition temperature (T_g). Neat PHBH is a thermoplastic with low T_g, close to 0 °C; similar values to this have been reported in several studies with PHBH [14]. In a first analysis, it was determined that the addition of lignocellulosic filler, ASF, to PHBH, slightly decreases the T_g to values comprised in the −0.5–1.9 °C range for all uncompatibilized composites. Considering that T_g are not a unique temperature, but a temperature range in which the material undergoes a change from a glassy state to a rubbery state, it can be assessed that these slight variations in T_g are not significative. The marked exothermic peak observed in the DSC thermogram of neat PHBH corresponds to the cold crystallization phenomenon, and its peak (corresponding to the temperature in which the crystallization rate is maximum) is located at 54.6 °C (T_{cc}). The addition of lignocellulosic ASF particles does not influence the T_{cc} of PHBH in the developed composites. It can only be observed a dilution effect (the cold crystallization peak height is smaller in PHBH/ASF composites, compared to neat PHBH; this is because the PHBH/ASF composite contains 30 wt % ASF which has no thermal transition in this temperature range, and consequently, the intensity of the peak is lower). At higher temperatures, the DSC thermograms show three small and broad endothermic peaks corresponding to the melting process of PHBH [52]. As already indicated in other studies [16,18,19,53–55], due to the polymorphism of the PHBH crystals during the crystallization process, its melting occurs at different temperatures. This situation of PHBH, is identical to other polyhydroxyalkanoates (PHAs) which present three melting peaks at different temperatures T_{m1}, T_{m2} and T_{m3}, too. Neat PHBH used in this work, show three melting peak temperatures located at 111 °C, 130 °C and 162 °C, respectively. The addition of lignocellulosic ASF particles does not produce significant changes in melting temperatures as observed in other studies [36]. The analysis of the enthalpies corresponding to the cold crystallization process (ΔH_{cc}), indicated that the highest enthalpy corresponded to neat PHBH during the second heating cycle (ΔH_{cc2} = 26.7 J g^{-1}). These values decreased gradually with the addition of ASF particles, down to values of 3.7 J g^{-1} for the sample with 30 wt % ASF. The dilution effect (which means considering the actual PHBH content without taking into account the wt % ASF for the cold crystallization enthalpy calculation), would give a theoretical diluted enthalpy of 18.69 J g^{-1}, which is remarkably higher than the actual obtained value of 3.7 J g^{-1}. These differences are not so pronounced for 10 and 20 wt % ASF.

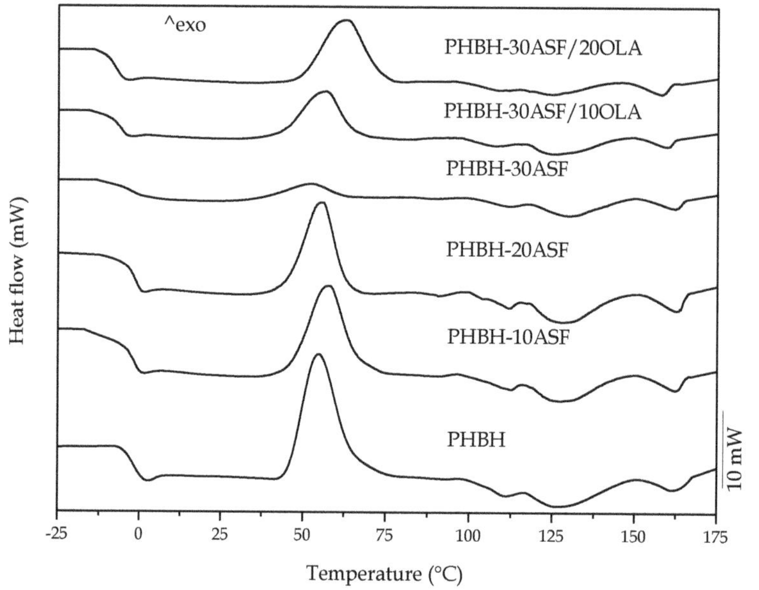

Figure 3. Comparative plot of the second heating curves obtained by dynamic differential scanning calorimetry (DSC) of the different PHBH-ASF/OLA composites with different compositions.

Table 3. Main thermal parameters of the PHBH-ASF/OLA composites with different compositions, obtained by differential scanning calorimetry (DSC).

Code	T_g (°C)	T_{cc} (°C)	T_{m1} (°C)	T_{m2} (°C)	T_{m3} (°C)	ΔH_{m1}* (J g^{-1})	ΔH_{cc2} (J g^{-1})	ΔH_{m2} (J g^{-1})	X_{c1}* (%)	X_{c2} (%)
PHBH	0.3 ± 0.1	54.6 ± 1.1	111.5 ± 1.9	130.8 ± 2.0	162.5 ± 1.2	20.3 ± 0.5	26.7 ± 0.8	33.6 ± 1.3	13.9 ± 1.1	4.7 ± 0.3
PHBH-10ASF	−0.5 ± 0.1	57.5 ± 1.8	112.1 ± 1.8	129.7 ± 1.7	162.3 ± 1.8	19.6 ± 0.4	17.4 ± 0.4	29.0 ± 2.2	14.9 ± 1.1	8.8 ± 0.4
PHBH-20ASF	−1.9 ± 0.2	55.5 ± 1.9	111.5 ± 2.0	129.6 ± 2.1	163.4 ± 1.4	18.1 ± 0.3	15.0 ± 0.1	26.4 ± 2.1	15.5 ± 0.8	9.8 ± 0.7
PHBH-30ASF	−1.1 ± 0.1	51.8 ± 1.2	111.3 ± 1.3	130.6 ± 1.9	164.1 ± 1.2	8.5 ± 0.2	3.7 ± 0.3	21.4 ± 0.9	8.3 ± 0.7	17.3 ± 0.9
PHBH-30ASF/10OLA	−5.2 ± 0.2	56.3 ± 1.4	107.6 ± 2.1	127.0 ± 1.8	159.4 ± 1.6	7.4 ± 0.1	11.4 ± 0.1	26.1 ± 1.3	8.0 ± 0.8	14.7 ± 0.8
PHBH-30ASF/20OLA	−5.6 ± 0.3	62.6 ± 2.0	109.9 ± 2.3	127.5 ± 2.0	157.9 ± 1.5	6.4 ± 0.3	15.9 ± 0.4	22.6 ± 1.5	7.5 ± 1.2	6.7 ± 0.4

* ΔH_{m1} and X_{c1} correspond to the first heating scan.

In a second stage, an oligomer of lactic acid (OLA) was added to improve toughness of the PHBH-AS composite with worst toughness, i.e., composite with 30 wt % ASF. The DSC thermograms, (Figure 3), show a similar thermal behaviour to the OLA-free composites. The addition of 10 *phr* and 20 *phr* OLA to this composite, shows a decrease of the PHBH T_g down to values around −5 °C, with respect to −1 °C for the same composite without OLA. The low molecular weight OLA chains offer slightly increased mobility in comparison with the polymer chains of neat PHBH; similar results were reported by Quiles-Carrillo et al. [56]. This phenomenon increases the free volume in the polymeric structure and leads to a poor plasticizing effect. It is true that addition of oligomers of lactic acid to PLA, usually leads to a remarkable decrease in PLA's T_g as reported by Burgos et al. [57] from 66 °C (neat PLA) down to −10 °C. Nevertheless, the decrease in T_g, is directly related to the PLA and OLA structure. D. Lascano et al. [51] reported a decrease of PLA T_g from 63.3 °C (neat PLA; different grade) down to 50.8 °C with 20% OLA (different commercial grade of OLA). Armentano et al. [58] reported a dual plasticization effect of PHB and OLA on PLA, but the decrease in T_g was not as important as the above-mentioned by Burgos et al. Moreover, Amor et al. [59], reported a slight plasticization effect on PLA/PHBH blends by using OLA in pellet form with a dual plasticization effect of PHBH (which provided a 3 °C decrease in T_g of PLA with a PHBH loading of 10 wt %) and OLA (which provided a decrease of 1 °C with a load of 1 wt %). All these results show the disparity in plasticization of PLA with OLA even they share the same chemical structure. Therefore, the plasticization effect on PHAs is even more complex and has not been studied previously independently of blends with PLA. In this work, this decrease in T_g values are very low, but we must bear in mind that PHBH structure contains medium chain hydroxyalkanoates and these, contribute to lowering crystallinity compared to P3HB. Therefore, this slight decrease could be representative of somewhat plasticization effect provided by OLA. Besides this, it corroborates the mechanical results analyzed previously [31,40,51]. T_{cc} values increased from 52 °C for the sample without OLA, to 62 °C for the sample with 20 *phr* OLA. This is not the typical effect of a plasticizer which increases chain mobility and, therefore, the cold crystallization process is shifted to lower temperatures as reported by Lascano et al. [51] and Ferri et al. [60]. Nevertheless, some additives such as maleinized linseed oil (MLO), maleinized cottonseed oil (MCSO), or even epoxidized vegetable oils, promote an overlapping of several phenomena such as slight plasticization, chain extension, branching and, in some cases, potential crosslinking due to the multifunctionality of these compounds. Some of these vegetable-oil derivatives, produce the same behaviours as observed in this study, i.e., a shift of the cold crystallization process to higher temperatures, due to the disrupted overall structure they provide with branches, chain extension, and so on, as reported by Garcia-Campo et al. [61]. Limiñana et al. [62] reported the potential of these modified vegetable oils as compatibilizers for PBS and lignocellulosic fillers, due to reaction of oxirane, maleic anhydride groups with hydroxyl groups in almond shell flour. In this study, it seems OLA has a similar effect to those modified vegetable oils, since changes in T_g are very low (which is representative for very slight plasticization effect, almost inexistent, since the technique itself has some uncertainty in the obtained values depending on the sample size, geometry, surface contact, and so on), and the cold crystallization is shifted to higher temperatures, thus indicating that other phenomena could be occurring, such as compatibilization and/or chain extension.

The ΔHcc_2 values of PHBH in OLA-compatibilized composites is 11.4 and 15.9.6 J g^{-1}, for 10 *phr* and 20 *phr* OLA respectively. It is a striking fact that the same composite without OLA shows a much lower ΔH_{cc2}, 3.7 J g^{-1}. With respect to T_{m1}, T_{m2} and T_{m3} it was observed that these thermal transitions were slightly decreased with the addition OLA, which could be related to more or less perfect crystals [51].

On the first heating scan the polymer has been recrystallized at 25 °C for 15 days. Consequently, during the first heating scan the cold crystallization peak did not appear which means that the polymer structure was not able to form crystallites. Under this conditions PHBH reaches a X_{c1} value of 13.9%. This increases with the amount of ASF filler up to 15.5% with 20 wt % ASF which suggests that ASF (mainly crystalline cellulose fractions) acts as a nucleant agent [62]. Furthermore, with the addition of 30 wt % ASF, X_{c1} decreases to 8.3% due to the decrease of free volume necessary for nucleation of polymer as Thomas et al. reported [63]. Mechanical characterization shows no correlation between the degree of crystallinity, while elastic modulus increases up to 30 wt % ASF, the degree of crystallinity is saturated with only 20 wt %. The second scan was performed after a controlled cooling process of 10 °C min^{-1}, as a result in the second scan the polymer was able to form crystallites due to a cold crystallization process. Under this condition the degree of crystallinity could increase until 30 wt %

ASF. The compatibilizing effect of OLA in both conditions decreased the degree of crystallinity by reducing the gaps between the filler and the matrix as it is reported in FESEM analysis and Gong et al. proposed [64].

Thermogravimetric analysis, TGA, allowed to analyze the thermal stability of PHBH-ASF composites. The TGA curves of the studied materials are gathered in Figure 4, and the main thermal degradation parameters are summarized in Table 4. The thermal degradation process of neat PHBH occurs in a single step. PHBH shows good thermal stability up to 266.8 °C (the onset degradation temperature was taken as the temperature for a weight loss of 2%, and it is denoted as $T_{2\%}$) of PHBH. Above this temperature, thermal degradation starts, with a very fast weight loss and a temperature of maximum degradation rate, T_{max}, of 308.9 °C, obtained from peak corresponding to the first derivative of its TGA curve or first DTG (Derivative thermogravimetry) (Figure 4b). The results obtained by Singh for PHBV indicated that the degradation process involves the breaking of polymer chains and hydrolysis. Since PHBH presents a similar structure, the mechanism of degradation should be similar [4]. Reaching the endset of the degradation process, located at 371 °C, PHBH generates a small residue or ash of 2.4 wt % of its initial weight. These results are in accordance to those obtained in other works [14,65].

Table 4. Summary of the main thermal degradation parameters of PHBH-ASF/OLA composites with different compositions, in terms of onset degradation temperature ($T_{2\%}$), temperature of maximum degradation (T_{max}), and residual mass at 700 °C.

Code	$T_{2\%}$ (°C)	T_{deg} (°C)	Residual Mass (wt %)
ASF	101.4 *	300.6/460.7	1.5 ± 0.2
PHBH	286.8	308.9	2.4 ± 0.3
PHBH-10ASF	253.2	288.1	2.3 ± 0.2
PHBH-20ASF	250.5	284.3	2.1 ± 0.1
PHBH-30ASF	223.6	279.1	2.0 ± 0.3
PHBH-30ASF/10OLA	258.4	292.0	2.0 ± 0.2
PHBH-30ASF/20OLA	226.3	283.5	2.0 ± 0.2

* Initial weight loss in ASF due to residual water evaporation.

The thermogram obtained for ASF particles shows different degradation processes corresponding to three different sections [31,35,37]. Since it is an agro-food waste of lignocellulosic nature, it shows a first weight loss around 100 °C, which corresponds to the loss of remaining water in ASF, specifically 6.3 wt %. During the dynamic degradation process, when temperature reaches 213 °C, a rapid weight loss is observed in two main steps. The first step of weight loss, of about 44.2 wt %, corresponds to the degradation of the cellulose and hemicellulose contained in ASF particles. The first component to start degradation is hemicellulose, followed by cellulose and lignin. Lignin shows a slower (in a wide temperature range) degradation process, so the third section of the TGA curve shows a lower slope, starting at 357.3 °C (temperature change of slope in the thermogram) up to 500 °C, with a loss of 47.6 wt %. Complete degradation occurs around 500 °C, leaving a final carbonaceous residue or ash of 1.5 wt %, mainly from lignin. In Figure 4b, it can be seen how the temperature corresponding to the maximum degradation rate of hemicellulose-cellulose fraction is located at 300.6 °C, while the lignin fraction maximum degradation rate is close to 460.7 °C. Perinovic determined that degradation of polysaccharides, hemicellulose and cellulose starts between 220–290 °C, while lignin degradation range is comprised between 200 °C and 500 °C [32,33,35,37,41,66].

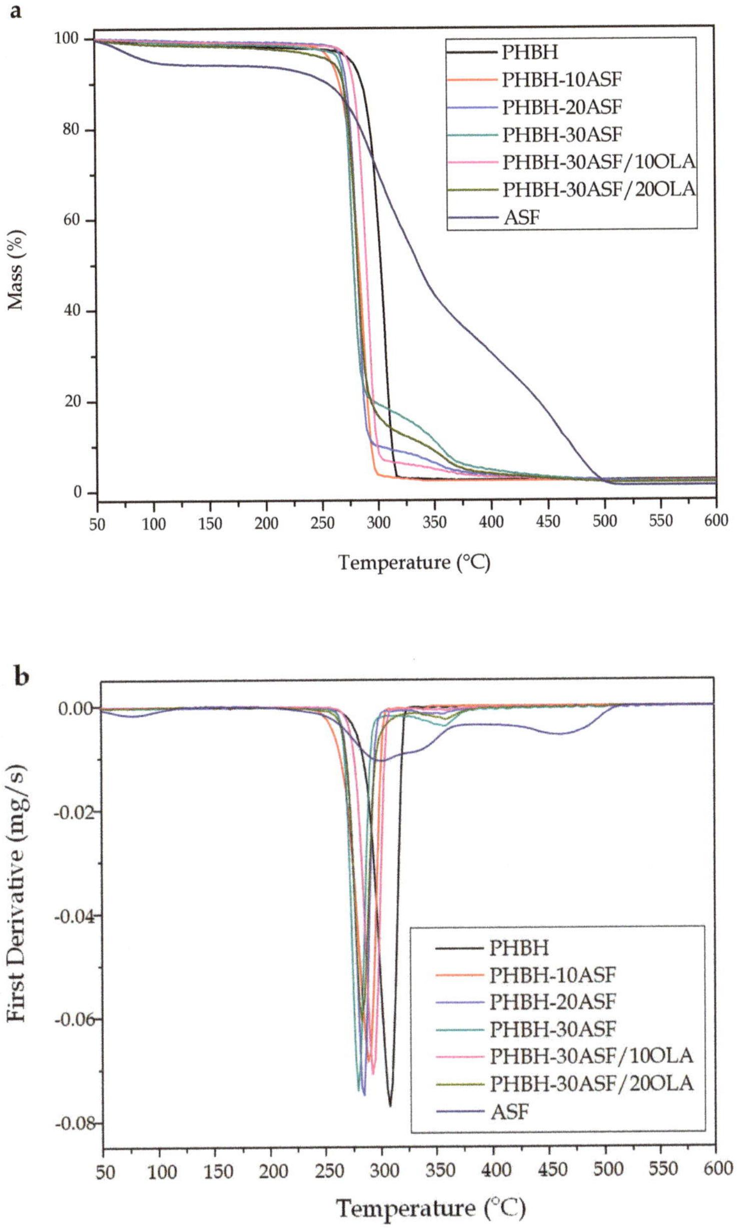

Figure 4. Comparative plot of (**a**) thermogravimetric analysis (TGA) curves and (**b**) first derivative (DTG) of the PHBH-ASF/OLA composites with different compositions.

The TGA curves of PHBH-ASF composites indicated that the thermal degradation process is a linear combination of the two individual degradation phenomena observed in PHBH and ASF, the first part of the curve is identical to neat PHBH, and a small hump at the end of the curve can be detected, which is attributable to residual degradation of ASF which changes with the ASF wt % [4,31]. For any wt % content in almond shells, the TGAs are characterized by presenting degradation start temperatures

($T_{2\%}$) slightly lower than neat PHBH (with a $T_{2\%}$ of 286.8 °C), in the 220–250 °C temperature range. The addition of lignocellulosic fillers leads to slightly lower thermal stability, since ASF particles degrade separately from PHBH, and the overall effect is the PHBH-ASF composites has reduced its thermal stability. This factor is due to the initial degradation of low molecular weight components on almond shell flour such as hemicelluloses. Quiles-Carrillo et al. [38] reported similar results with PLA composites with almond shell flour.

TGA curves of composites are very similar to those of the PHBH (as it is the main component), with practically only one step degradation stage, but with a small hump at higher temperatures, corresponding to lignin degradation. As the ASF content increases, this hump becomes more pronounced. This process ends at temperatures around 500 °C, generating small amounts of residue close to 2 wt %. The TGA curves of PHBH-ASF composites indicated that from a thermal point of view, the incorporation of lignocellulosic fillers such as ASF, slightly reduces the stability, but even in this case, the processing window is not compromised since all the onset degradation temperatures are above 250 °C, and the recommended processing temperature for this polymer is 140–150 °C.

On the other hand, composites with 10 *phr* and 20 *phr* OLA, show a very similar thermal degradation behaviour to the reference uncompatibilized composite (PHBH-ASF with 30 wt % ASF) without OLA. The addition of 10 *phr* OLA seems to slightly improve the thermal stability of the developed composites. The characteristic thermal degradation values are delayed by 34 °C (onset degradation temperature, $T_{2\%)}$ and by 10 °C (for the maximum degradation rate, T_{deg}) compared to the unmodified reference composite without OLA. The biggest improvement in thermal stability is observed for the composite with 10 wt % OLA. This improvement is due to the chemical interaction of the compatibilizer with both components of the composite as above-mentioned. The complex structure formed after reaction of OLA with both PHBH and ASF, can act as a physical barrier that obstructs the removal of volatile products produced during decomposition [67].

3.3. *Thermomechanical Properties of PHBH-ASF/OLA Composites*

Figure 5a shows the variation of the flexural single cantilever) storage modulus, E', with respect to temperature, obtained by DMTA analysis. It can be seen graphically how E' decreases with increasing temperature in all the developed composites, as expected due to the softening of the polymeric PHBH matrix. At low temperatures, E' values are high in all composites, since this temperature range corresponds to the elastic-glassy behaviour of the PHBH matrix. In this first zone, E' for neat PHBH is 1869 MPa at −40 °C, which is lower than E' values of any of the uncompatibilized PHBH-ASF composites (e.g., E' is 2019 MPa for the PHBH-ASF composite with 30 wt % ASF at −40 °C). These results are in accordance with those obtained by mechanical characterization which suggested a stiffening as the wt % ASF increased [37]. Table 5 shows the numerical comparison of the variation of E' as a function of ASF wt % and OLA *phr*, at two different temperatures.

Table 5. Main dynamic-mechanical thermal parameters of PHBH-ASF/OLA composites with different compositions: flexural storage modulus (E') measured at −40 °C and 25 °C and glass transition temperature (T_g), obtained by dynamic-mechanical thermal analysis (DMTA).

Code	T_g (°C)	E' at −40 °C (MPa)	E' at 25 °C (MPa)
PHBH	10.6 ± 0.9	1869 ± 42	1345 ± 28
PHBH-10ASF	14.3 ± 0.8	1910 ± 49	1431 ± 40
PHBH-20ASF	12.0 ± 0.7	1948 ± 30	1512 ± 20
PHBH-30ASF	11.4 ± 0.9	2019 ± 52	1604 ± 45
PHBH-30ASF/10OLA	9.7 ± 0.7	1601 ± 36	1352 ± 29
PHBH-30ASF/10OLA	9.3 ± 0.6	853 ± 25	767 ± 23

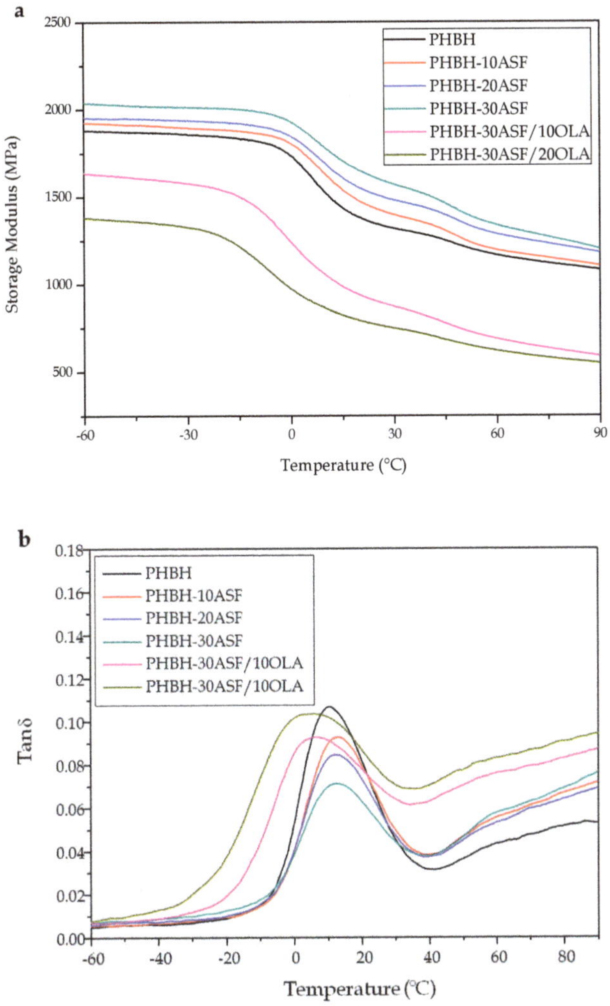

Figure 5. Comparative plot of dynamic-mechanical thermal analysis (DMTA) curves of PHBH-ASF/OLA composites with different compositions: (**a**) flexural storage modulus (E') and (**b**) dynamic damping factor (*tan* δ).

As the temperature increases, E' decreases rapidly as it acquires a rubbery state behaviour. This is related to the α-relaxation process or the glass transition temperature (T_g). Table 5 shows how at 25 °C, the E' value for neat PHBH has decreased to 1345 MPa from 1869 MPa at −40 °C and, subsequently, the stiff-elastic behaviour changes to a rubber-like behaviour. The same trend can be observed for uncompatibilized PHBH-ASF composites. However, when compared to the neat PHBH matrix, at the same temperature, the higher the wt % of ASF, the stiffer the composite becomes. With only 10 wt % ASF, E' at 25 °C increases 6.4% with respect to PHBH at the same temperature. Accordingly, the composite with 30 wt % ASF offers higher stiffness (a percentage increase of 19% regarding neat PHBH). The presence of ASF particles finely dispersed in the PHBH matrix restricts the mobility of the polymer chains, thus decreasing their viscous behaviour, which causes an increase in the E' value as the ASF loading increases [4,27–29,39]. In addition, at this temperature range, lignocellulosic components are below its T_g, which means they show a glassy behaviour that promotes increased stiffness. Nevertheless, it is important to bear in mind that the main component in PHBH/ASF composites is PHBH and the dynamic behaviour is highly influenced by PHBH behaviour since conditions are not as aggressive as a conventional tensile test up to fracture. These results corroborate those obtained in the mechanical characterization of PHBH-ASF composites.

As in previous analyses, by adding small amounts of OLA to the PHBH-ASF composite with the highest ASF loading, which shows the worst ductile/toughness properties, the DMTA graphs in Figure 5a show the lowest E' in the temperature range analyzed. At −40 °C, E' decreases from 2019 MPa without OLA addition, to 1601 MPa and 853 MPa for the addition of 10 *phr* and 20 *phr* OLA, respectively. The trend is the same at room temperature (25 °C). With an addition of only 10 *phr* OLA the E' is decreased by 16%, and with 20 *phr* OLA E' is reduced even more, thus decreasing the rigidity of PHBH-ASF composites and, subsequently their E' [40]. It is worth noting the extremely small changes in T_g obtained by DMTA which suggests, as DSC, very poor plasticization effect, suggesting compatibilization is one of the most representative effects of OLA in this PHBH/ASF system.

Figure 5b shows the variation of the dynamic damping factor ($tan\ \delta$) as a function of temperature, for neat PHBH, uncompatibilized PHBH-ASF and PHBH-ASF/OLA composites. Despite there are several criteria to obtain the T_g from DMTA graphs, the most used is the peak maximum of $tan\ \delta$. The T_g values obtained for all developed materials are gathered in Table 5. It can be observed that $tan\ \delta$ peaks are slightly moved towards higher temperatures in uncompatibilized PHBH-ASF composites, compared to neat PHBH (a maximum shift of 3–4 °C). This change is not significant as observed in other techniques such as DSC. On the other hand, the addition of OLA leads to slightly lower T_g values by DMTA, but once again, these changes are not high enough to give a clear evidence of the chain mobility restriction by ASF particles or increased chain mobility by OLA.

For potential structural/engineering/conventional applications of WPCs, it is very important to know their dimensional stability with temperature. This can be assessed by thermomechanical analysis (TMA) which allows us to obtain the coefficient of linear thermal expansion (*CLTE*). It must be stated that a good dimensional stability involves low *CLTE* values. Table 6 summarizes the CLTE values for the developed composites, at temperatures below and above their corresponding T_g. In general, at temperatures below T_g the *CLTE* values are much lower than above their T_g. The dimensional expansion of the material is lower at low temperatures because the material is more rigid, which is the typical glassy behaviour below T_g. Above T_g, the behaviour is viscous or rubber-like and so that, the dimensional expansion is favoured, with higher *CLTE* values.

Table 6. Summary of the main thermo mechanical properties of neat PHBH and PHBH-ASF/OLA with different compositions, regarding the thermal expansion, obtained by thermomechanical analysis (TMA).

Code	T_g (°C)	CLTE (μm m^{-1} °C^{-1})	
		Below T_g	Above T_g
PHBH	−0.3 ± 0.1	77.1 ± 2.2	160.7 ± 2.3
PHBH-10ASF	0.2 ± 0.1	76.9 ± 2.1	157.0 ± 1.3
PHBH-20ASF	−0.4 ± 0.1	75.6 ± 2.1	157.4 ± 2.9
PHBH-30ASF	1.4 ± 0.2	66.8 ± 0.8	140.3 ± 2.6
PHBH-30ASF/10OLA	−1.3 ± 0.1	72.0 ± 0.9	169.1 ± 3.8
PHBH-30ASF/20OLA	−1.4 ± 0.2	90.7 ± 4.1	194.3 ± 2.83

First, the analysis of the *CLTE* values at temperatures below T_g shows that increasing wt % ASF gives more dimensional stability, which is in accordance with previous mechanical results that suggested a clear stiffening with ASF addition. Pure PHBH has an initial value of 77.1 μm m^{-1} °C^{-1}, which decreases to 66.8 μm m^{-1} °C^{-1} with 30 wt % ASF particles (which represents a % decrease of 13%), which involves improved dimensional stability. However, OLA addition significantly increases the *CLTE* values, as typical plasticizers do, thus leading to slightly lower dimensional stability. With 10 *phr* and 20 *phr* OLA, *CLTE* is 72.0 and 90.7 μm m^{-1} °C^{-1}, respectively.

Secondly, the results obtained in the study at temperatures above T_g, show the same tendency as the results discussed in the previous paragraph. Neat PHBH shows an initial *CLTE* of 160.7 μm m^{-1} °C^{-1} (much higher than below T_g), which decreases to 140.3 μm m^{-1} °C^{-1} with 30 wt % ASF particles. Identically as observed previously, *CLTE* becomes greater again with the addition of OLA, reaching values of 194.3 μm m^{-1} °C^{-1} with 20 *phr* OLA. In general, as the ASF content increases, composites show improved dimensional stability [4,29,68]. However, these dimensional expansions are higher than those of neat PHBH in PHBH-ASF composites (30 wt % ASF) with 10 *phr* and 20 *phr* OLA, due to the plasticizing effect [51].

3.4. Evolution of the Water Uptake and Water Diffusion Process in PHBH-ASF/OLA Composites

The water absorption capacity of WPCs is an important feature in some applications due to the lignocellulosic component. This creates a three-dimensional path inside the polymer matrix that allows water entering (for example when the composite is subjected to high relative humidity environments) and this causes an expansion. It is possible that after this initial stage, this WPC could be subjected to drying at sun with low humidity; then this 3D-path allows water/moisture removal, promoting a contraction. This situation is quite usual in WPCs such as those used in fences, decking, and so on. This repeated expansion–contraction cycles could lead to formation of microcracks. Figure 6 shows the mass increase (wt %) with respect to immersion time in water for the PHBH-ASF/OLA composites. It can be seen graphically that during the first week of immersion, the developed composites show a rapid increase in mass by water absorption (Δ*mass*). As immersion time increases, the mass increase is slower. Some samples even show an asymptotic behaviour, which indicates that saturation has been reached (Δ*mass*∞). This type of behaviour corresponds to that indicated by the first Fick's Law.

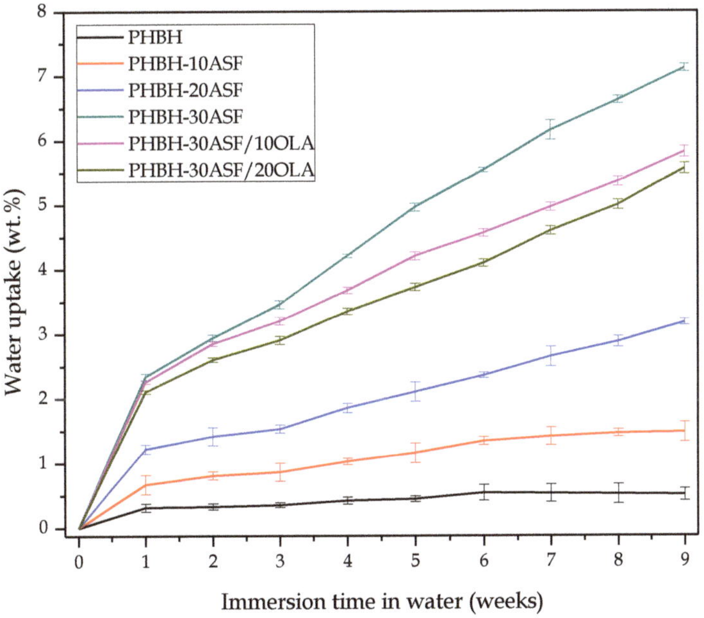

Figure 6. Water uptake of PHBH-ASF/OLA composites with different compositions. Evolution of the water uptake for a period of nine weeks.

Obviously, due to the hydrophobicity of neat PHBH, it shows the lowest water absorption for nine weeks. After 35 days of immersion it reaches a constant saturation mass, Δ*mass*∞ of 0.53%, which is maintained practically until 63 days of immersion. As mentioned above, the X_c of neat PHBH is 13.9%.

The addition of ASF (highly hydrophilic particles due to its composition: cellulose, hemicellulose and lignin as the main components) considerably increases water absorption, but the effects of PHBH crystallinity can also be observed in this behaviour. The composite with 10 wt % ASF reaches a water saturation mass, $\Delta mass\infty$ of 1.46 wt % after 42 days (X_c of PHBH in this composite is 14.9% and this prevents from water entering). In a similar way, uncompatibilized composites containing 20 wt % show a relatively low $\Delta mass\infty$ of 3.1 wt %. This is almost double the previous value (with 10 wt % ASF). This value is expected since the X_c of PHBH in this composite is close to 15.5%. Nevertheless, the composite containing 30 wt % ASF, shows a remarkable increase in $\Delta mass\infty$ up to values of 7.1 wt % after nine weeks, which represents almost 14 times the value for neat PHBH, representing a typical result in most WPCs [23,36,43]. The result is higher than the extrapolation from the ASF wt % (which should be of about three times the value of the composite with 10 wt % ASF, i.e., 4.5 wt %); despite this, we have to bear in mind that the X_c of PHBH in this composite has decreased to 8.3% and this has a negative effect on water absorption as amorphous regions allow water entering [69,70]. Therefore, the increase in water absorption is not only related to the number of lignocellulosic components, but also with the degree of crystallinity of the polymer matrix. Cellulose promotes water absorption due to hydroxyl (–OH) groups that interact with water molecules [21–23,25,41,71].

On the other hand, the addition of OLA to the reference composite (PHBH with 30 wt % ASF), shows an unexpected behaviour. As one can see in Figure 6, the water absorption curve with OLA, moves to lower wt % absorbed water. The values of the mass increase after nine weeks of immersion reach values of 5.8 wt % and 5.5 wt % for OLA contents of 10 *phr* and 20 *phr*, respectively. This means a decrease in water absorption of 18% and 22.6% respectively when OLA is added to the PHBH-ASF system. Typical plasticizers, increase the water absorption as they are responsible for an increase in the free volume, thus allowing water molecules to enter. Nevertheless, OLA not only provides plasticization effects, but also improved polymer-particle interaction among the interface due to the interaction between the hydroxyl groups in OLA and the hydroxyl groups of both PHBH (terminal groups), and cellulose/hemicellulose/lignin in ASF particles. Therefore, in addition to a plasticization phenomenon, we could think on an additional compatibilization effect which was also evidenced in Figure 2 (FESEM characterization), in which the gap size was higher on OLA-free composites than composites with OLA.

Table 7 shows the values of the diffusion coefficient (D) or diffusivity of water into the developed composites, by applying the first Fick's Law. The lowest corrected diffusion coefficient, D_c value is offered by neat PHBH, as expected due to its hydrophobicity. The only addition of 10 wt % ASF, leads to a D_c value, almost triple compared to neat PHBH. Obviously, ASF is responsible for water entering the composite structure; therefore, uncompatibilized composites with 30 wt % ASF, shows an increase of two orders of magnitude. Due to the hydrophilic nature of ASF particles, and possibly accentuated by the capillarity of the micro-gaps between PHBH and the embedded ASF particles, water molecules can easily enter into the composite structure as in most WPCs [51,71]. Finally, the D_c values PHBH-ASF/OLA composites remain with similar values to those of the same composite without OLA. Thus, it was deduced that the amount of wt % ASF is the parameter with the greatest influence on the water diffusion process in PHBH-ASF/OLA composites, together with the degree of crystallinity as previously discussed with the relationship of the water absorption and the wt % ASF loading and the degree of crystallinity of the PHBH matrix. Since the D_c values for the composites with 10 and 20 *phr* OLA are similar to that of the same composite with 30 wt % ASF, it is possible to conclude the poor plasticization effect of this OLA, suggesting, once again, that compatibilization is the main acting mechanisms of OLA on PHBH/ASF composites.

Table 7. Values of the diffusion coefficient (D) and the corrected diffusion coefficient (D_c) for PHBH and the PHBH-ASF composites processed with OLA.

Code	$D \times 10^{-9}$ (cm^2 s^{-1})	$Dc \times 10^{-9}$ (cm^2 s^{-1})
PHBH	0.14 ± 0.03	0.07 ± 0.01
PHBH-10ASF	0.54 ± 0.05	0.25 ± 0.02
PHBH-20ASF	1.56 ± 0.07	0.74 ± 0.04
PHBH-30ASF	6.08 ± 0.08	2.89 ± 0.05
PHBH-30ASF/10OLA	6.66 ± 0.09	3.17 ± 0.07
PHBH-30ASF/20OLA	7.08 ± 0.09	3.37 ± 0.03

4. Conclusions

The results obtained in this study indicate that the analyzed system of poly(3-hydroxybutyrate-co-3-hydroxyhexanoate) (PHBH) and almond shell flour (ASF), is suitable for the manufacture of fully biobased and environmentally friendly Wood Plastic Composites (WPCs). PHBH-ASF composites present a very interesting set of properties for technical applications as wood substitute materials. The characterization of PHBH-ASF composites showed that the addition of lignocellulosic particles of ASF leads to an embrittlement and reduced toughness. These effects are much more evident with increasing the wt % ASF. To overcome or minimize these negative properties, an oligomer of lactic acid, OLA was added to give PHBH-ASF/OLA composites with improved ductile properties and, subsequently, improved toughness. It is worth noting a remarkable increase in impact strength with 20 *phr* OLA. A higher mobility of the PHBH polymer chains by the addition of OLA is the reason for the improvement on toughness, even on composites with 30 wt % ASF.

The study of the PLA-ASF/OLA composites allowed us to obtain good balanced properties and, therefore, these materials can be used in the WPC industry as they are suitable for technical applications that require certain stiffness and thermal stability, in an interesting range of properties depending on the wt % ASF content. Furthermore, the addition of OLA oligomer decreases the water absorption capacity of PHBH-ASF/OLA, thus broadening potential uses in high humidity environments. Finally, its thermoplastic nature allows it to be easily processed by conventional extrusion–injection moulding, and overall, these composites contribute to a sustainable development and a reduction of the carbon footprint as all the used materials are bio-sourced.

Author Contributions: Conceptualization, R.B. and L.Q.-C.; methodology, L.S.-N., J.I.-M. and J.M.-M.; validation, J.M.-M., J.I.-M. and L.Q.-C.; formal analysis, L.Q.-C. and J.I.-M.; investigation, J.M.-M., T.B., L.Q.-C. and J.I.-M.; data curation, L.Q.-C. and J.I.-M.; writing-original draft preparation, L.Q.-C. and T.B.; writing-review and editing, R.B., J.I.-M.; supervision, R.B., L.Q.-C. and T.B.; project administration, R.B. and T.B. All authors have read and agreed to the published version of the manuscript.

Funding: This research work was funded by the Spanish Ministry of Science, Innovation, and Universities (MICIU) project number MAT2017-84909-C2-2-R. This work was supported by the POLISABIO program grant number (2019-A02).

Acknowledgments: J. Ivorra-Martinez is the recipient of an FPI grant from Universitat Politècnica de València (PAID-2019). L. Quiles-Carrillo wants to thank GV for his FPI grant (ACIF/2016/182) and MECD for his FPU grant (FPU15/03812). Microscopy services at UPV are acknowledged for their help in collecting and analyzing FESEM images.

Conflicts of Interest: The authors declare no conflicts of interest.

References

1. Carbonell-Verdu, A.; Bernardi, L.; Garcia-Garcia, D.; Sanchez-Nacher, L.; Balart, R. Development of environmentally friendly composite matrices from epoxidized cottonseed oil. *Eur. Polym. J.* **2015**, *63*, 1–10. [CrossRef]
2. España, J.; Samper, M.; Fages, E.; Sánchez-Nácher, L.; Balart, R. Investigation of the effect of different silane coupling agents on mechanical performance of basalt fiber composite laminates with biobased epoxy matrices. *Polym. Compos.* **2013**, *34*, 376–381. [CrossRef]

3. Basalp, D.; Tihminlioglu, F.; Sofuoglu, S.C.; Inal, F.; Sofuoglu, A. Utilization of Municipal Plastic and Wood Waste in Industrial Manufacturing of Wood Plastic Composites. *Waste Biomass Valorization* **2020**. [CrossRef]
4. Singh, S.; Mohanty, A.K. Wood fiber reinforced bacterial bioplastic composites: Fabrication and performance evaluation. *Compos. Sci. Technol.* **2007**, *67*, 1753–1763. [CrossRef]
5. Mukheem, A.; Hossain, M.; Shahabuddin, S.; Muthoosamy, K.; Manickam, S.; Sudesh, K.; Saidur, R.; Sridewi, N.; Campus, N.M. Bioplastic Polyhydroxyalkanoate (PHA): Recent Advances in Modification and Medical Applications. *Prepr. Org.* **2018**. [CrossRef]
6. Mohanty, A.K.; Misra, M.; Drzal, L.T. Sustainable bio-composites from renewable resources: Opportunities and challenges in the green materials world. *Abstr. Pap. Am. Chem. Soc.* **2002**, *223*, D70.
7. Petchwattana, N.; Covavisaruch, S. Mechanical and Morphological Properties of Wood Plastic Biocomposites Prepared from Toughened Poly(lactic acid) and Rubber Wood Sawdust (Hevea brasiliensis). *J. Bionic Eng.* **2014**, *11*, 630–637. [CrossRef]
8. Summerscales, J.; Dissanayake, N.; Virk, A.; Hall, W. A review of bast fibres and their composites. Part 2-Composites. *Compos. Part A Appl. Sci. Manuf.* **2010**, *41*, 1336–1344. [CrossRef]
9. Averous, L. Biodegradable multiphase systems based on plasticized starch: A review. *J. Macromol. Sci. Polym. Rev.* **2004**, *C44*, 231–274. [CrossRef]
10. Yang, Y.; Ke, S.; Ren, L.; Wang, Y.; Li, Y.; Huang, H. Dielectric spectroscopy of biodegradable poly (3-hydroxybutyrate-co-3-hydroxyhexanoate) films. *Eur. Polym. J.* **2012**, *48*, 79–85. [CrossRef]
11. Liao, Q.; Noda, I.; Frank, C.W. Melt viscoelasticity of biodegradable poly (3-hydroxybutyrate-co-3-hydroxyhexanoate) copolymers. *Polymer* **2009**, *50*, 6139–6148. [CrossRef]
12. Alata, H.; Aoyama, T.; Inoue, Y. Effect of aging on the mechanical properties of poly(3-hydroxybutyrate-co-3-hydroxyhexanoate). *Macromolecules* **2007**, *40*, 4546–4551. [CrossRef]
13. Misra, S.K.; Valappil, S.P.; Roy, I.; Boccaccini, A.R. Polyhydroxyalkanoate (PHA)/inorganic phase composites for tissue engineering applications. *Biomacromolecules* **2006**, *7*, 2249–2258. [CrossRef] [PubMed]
14. Mahmood, H.; Pegoretti, A.; Brusa, R.S.; Ceccato, R.; Penasa, L.; Tarter, S.; Checchetto, R. Molecular transport through 3-hydroxybutyrate co-3-hydroxyhexanoate biopolymer films with dispersed graphene oxide nanoparticles: Gas barrier, structural and mechanical properties. *Polym. Test.* **2019**, *81*, 106181. [CrossRef]
15. Corre, Y.-M.; Bruzaud, S.; Audic, J.-L.; Grohens, Y. Morphology and functional properties of commercial polyhydroxyalkanoates: A comprehensive and comparative study. *Polym. Test.* **2012**, *31*, 226–235. [CrossRef]
16. Watanabe, T.; He, Y.; Fukuchi, T.; Inoue, Y. Comonomer compositional distribution and thermal characteristics of bacterially synthesized poly(3-hydroxybutyrate-co-3-hydroxyhexanoate)s. *Macromol. Biosci.* **2001**, *1*, 75–83. [CrossRef]
17. Oyama, T.; Kobayashi, S.; Okura, T.; Sato, S.; Tajima, K.; Isono, T.; Satoh, T. Biodegradable Compatibilizers for Poly(hydroxyalkanoate)/Poly(epsilon-caprolactone) Blends through Click Reactions with End-Functionalized Microbial Poly(hydroxyalkanoate)s. *ACS Sustain. Chem. Eng.* **2019**, *7*, 7969. [CrossRef]
18. Sato, H.; Nakamura, M.; Padermshoke, A.; Yamaguchi, H.; Terauchi, H.; Ekgasit, S.; Noda, I.; Ozaki, Y. Thermal behavior and molecular interaction of poly(3-hydroxybutyrate-co-3-hydroxyhexanoate) studied by wide-angle X-ray diffraction. *Macromolecules* **2004**, *37*, 3763–3769. [CrossRef]
19. Hu, Y.; Zhang, J.; Sato, H.; Noda, I.; Ozaki, Y. Multiple melting behavior of poly (3-hydroxybutyrate-co-3-hydroxyhexanoate) investigated by differential scanning calorimetry and infrared spectroscopy. *Polymer* **2007**, *48*, 4777–4785. [CrossRef]
20. Xu, P.; Cao, Y.; Lv, P.; Ma, P.; Dong, W.; Bai, H.; Wang, W.; Du, M.; Chen, M. Enhanced crystallization kinetics of bacterially synthesized poly(3-hydroxybutyrate-co-3-hydroxyhexanate) with structural optimization of oxalamide compounds as nucleators. *Polym. Degrad. Stab.* **2018**, *154*, 170–176. [CrossRef]
21. Tham, W.L.; Ishak, Z.A.M.; Chow, W.S. Water Absorption and Hygrothermal Aging Behaviors of SEBS-g-MAH Toughened Poly(lactic acid)/Halloysite Nanocomposites. *Polym. Plast. Technol. Eng.* **2014**, *53*, 472–480. [CrossRef]
22. Tham, W.L.; Poh, B.T.; Ishak, Z.A.M.; Chow, W.S. Water Absorption Kinetics and Hygrothermal Aging of Poly(lactic acid) Containing Halloysite Nanoclay and Maleated Rubber. *J. Polym. Environ.* **2015**, *23*, 242–250. [CrossRef]

23. Arbelaiz, A.; Fernandez, B.; Ramos, J.A.; Retegi, A.; Llano-Ponte, R.; Mondragon, I. Mechanical properties of short flax fibre bundle/polypropylene composites: Influence of matrix/fibre modification, fibre content, water uptake and recycling. *Compos. Sci. Technol.* **2005**, *65*, 1582–1592. [CrossRef]
24. Deroine, M.; Le Duigou, A.; Corre, Y.-M.; Le Gac, P.-Y.; Davies, P.; Cesar, G.; Bruzaud, S. Accelerated ageing of polylactide in aqueous environments: Comparative study between distilled water and seawater. *Polym. Degrad. Stab.* **2014**, *108*, 319–329. [CrossRef]
25. Gil-Castell, O.; Badia, J.D.; Kittikorn, T.; Stromberg, E.; Martinez-Felipe, A.; Ek, M.; Karlsson, S.; Ribes-Greus, A. Hydrothermal ageing of polylactide/sisal biocomposites. Studies of water absorption behaviour and Physico-Chemical performance. *Polym. Degrad. Stab.* **2014**, *108*, 212–222. [CrossRef]
26. Petinakis, E.; Yu, L.; Edward, G.; Dean, K.; Liu, H.; Scully, A.D. Effect of Matrix-Particle Interfacial Adhesion on the Mechanical Properties of Poly(lactic acid)/Wood-Flour Micro-Composites. *J. Polym. Environ.* **2009**, *17*, 83–94. [CrossRef]
27. Pilla, S.; Gong, S.; O'Neill, E.; Rowell, R.M.; Krzysik, A.M. Polylactide-pine wood flour composites. *Polym. Eng. Sci.* **2008**, *48*, 578–587. [CrossRef]
28. Shah, B.L.; Selke, S.E.; Walters, M.B.; Heiden, P.A. Effects of wood flour and chitosan on mechanical, chemical, and thermal properties of polylactide. *Polym. Compos.* **2008**, *29*, 655–663. [CrossRef]
29. Balart, J.F.; Garcia-Sanoguera, D.; Balart, R.; Boronat, T.; Sanchez-Nacher, L. Manufacturing and properties of biobased thermoplastic composites from poly(lactid acid) and hazelnut shell wastes. *Polym. Compos.* **2018**, *39*, 848–857. [CrossRef]
30. Kumar, S.; Vedrtnam, A.; Pawar, S.J. Effect of wood dust type on mechanical properties, wear behavior, biodegradability, and resistance to natural weathering of wood-plastic composites. *Front. Struct. Civ. Eng.* **2019**, *13*, 1446–1462. [CrossRef]
31. Ling, S.L.; Koay, S.C.; Chan, M.Y.; Tshai, K.Y.; Chantara, T.R.; Pang, M.M. Wood Plastic Composites Produced from Postconsumer Recycled Polystyrene and Coconut Shell: Effect of Coupling Agent and Processing Aid on Tensile, Thermal, and Morphological Properties. *Polym. Eng. Sci.* **2020**, *60*, 202–210. [CrossRef]
32. Quitadamo, A.; Massardier, V.; Valente, M. Eco-Friendly Approach and Potential Biodegradable Polymer Matrix for WPC Composite Materials in Outdoor Application. *Int. J. Polym. Sci.* **2019**. [CrossRef]
33. Salasinska, K.; Polka, M.; Gloc, M.; Ryszkowska, J. Natural fiber composites: The effect of the kind and content of filler on the dimensional and fire stability of polyolefin-based composites. *Polimery* **2016**, *61*, 255–265. [CrossRef]
34. Wang, X.; Yu, Z.; McDonald, A.G. Effect of Different Reinforcing Fillers on Properties, Interfacial Compatibility and Weatherability of Wood-plastic Composites. *J. Bionic Eng.* **2019**, *16*, 337–353. [CrossRef]
35. Yussuf, A.A.; Massoumi, I.; Hassan, A. Comparison of Polylactic Acid/Kenaf and Polylactic Acid/Rise Husk Composites: The Influence of the Natural Fibers on the Mechanical, Thermal and Biodegradability Properties. *J. Polym. Environ.* **2010**, *18*, 422–429. [CrossRef]
36. Kuciel, S.; Jakubowska, P.; Kuzniar, P. A study on the mechanical properties and the influence of water uptake and temperature on biocomposites based on polyethylene from renewable sources. *Compos. Part B-Eng.* **2014**, *64*, 72–77. [CrossRef]
37. Liminana, P.; Quiles-Carrillo, L.; Boronat, T.; Balart, R.; Montanes, N. The Effect of Varying Almond Shell Flour (ASF) Loading in Composites with Poly(Butylene Succinate (PBS) Matrix Compatibilized with Maleinized Linseed Oil (MLO). *Materials* **2018**, *11*, 2179. [CrossRef]
38. Quiles-Carrillo, L.; Montanes, N.; Garcia-Garcia, D.; Carbonell-Verdu, A.; Balart, R.; Torres-Giner, S. Effect of different compatibilizers on injection-molded green composite pieces based on polylactide filled with almond shell flour. *Compos. Part B Eng.* **2018**, *147*, 76–85. [CrossRef]
39. Mathew, A.P.; Oksman, K.; Sain, M. Mechanical properties of biodegradable composites from poly lactic acid (PLA) and microcrystalline cellulose (MCC). *J. Appl. Polym. Sci.* **2005**, *97*, 2014–2025. [CrossRef]
40. Ghaffar, S.H.; Madyan, O.A.; Fan, M.; Corker, J. The Influence of Additives on the Interfacial Bonding Mechanisms between Natural Fibre and Biopolymer Composites. *Macromol. Res.* **2018**, *26*, 851–863. [CrossRef]
41. Tserki, V.; Matzinos, P.; Kokkou, S.; Panayiotou, C. Novel biodegradable composites based on treated lignocellulosic waste flour as filler. Part I. Surface chemical modification and characterization of waste flour. *Compos. Part A Appl. Sci. Manuf.* **2005**, *36*, 965–974. [CrossRef]

42. Niaraki, P.R.; Krause, A. Correlation between physical bonding and mechanical properties of wood plastic composites: Part 1: Interaction of chemical and mechanical treatments on physical properties. *J. Adhes. Sci. Technol.* **2020**, *34*, 744–755. [CrossRef]
43. Akesson, D.; Fazelinejad, S.; Skrifvars, V.-V.; Skrifvars, M. Mechanical recycling of polylactic acid composites reinforced with wood fibres by multiple extrusion and hydrothermal ageing. *J. Reinf. Plast. Compos.* **2016**, *35*, 1248–1259. [CrossRef]
44. Torres-Giner, S.; Montanes, N.; Fenollar, O.; García-Sanoguera, D.; Balart, R. Development and optimization of renewable vinyl plastisol/wood flour composites exposed to ultraviolet radiation. *Mater. Des.* **2016**, *108*, 648–658. [CrossRef]
45. Juárez, D.; Ferrand, S.; Fenollar, O.; Fombuena, V.; Balart, R. Improvement of thermal inertia of styrene–ethylene/butylene–styrene (SEBS) polymers by addition of microencapsulated phase change materials (PCMs). *Eur. Polym. J.* **2011**, *47*, 153–161. [CrossRef]
46. Pracella, M.; Haque, M.; Alvarez, V. Functionalization, compatibilization and properties of polyolefin composites with natural fibers. *Polymers* **2010**, *2*, 554–574. [CrossRef]
47. Chabros, A.; Gawdzik, B.; Podkościelna, B.; Goliszek, M.; Pączkowski, P. Composites of Unsaturated Polyester Resins with Microcrystalline Cellulose and Its Derivatives. *Materials* **2020**, *13*, 62. [CrossRef]
48. Mokhena, T.; Sefadi, J.; Sadiku, E.; John, M.; Mochane, M.; Mtibe, A. Thermoplastic processing of PLA/cellulose nanomaterials composites. *Polymers* **2018**, *10*, 1363. [CrossRef]
49. Patwa, R.; Saha, N.; Saha, P.; Katiyar, V. Biocomposites of poly(lactic acid) and lactic acid oligomer-grafted bacterial cellulose: It's preparation and characterization. *J. Appl. Polym. Sci.* **2019**, *136*. [CrossRef]
50. Tripathi, N.; Katiyar, V. Lactic acid oligomer (OLLA) grafted gum arabic based green adhesive for structural applications. *Int. J. Biol. Macromol.* **2018**, *120*, 711–720. [CrossRef]
51. Lascano, D.; Moraga, G.; Ivorra-Martinez, J.; Rojas-Lema, S.; Torres-Giner, S.; Balart, R.; Boronat, T.; Quiles-Carrillo, L. Development of Injection-Molded Polylactide Pieces with High Toughness by the Addition of Lactic Acid Oligomer and Characterization of Their Shape Memory Behavior. *Polymers* **2019**, *11*, 2099. [CrossRef] [PubMed]
52. Zhou, Y.-x.; Huang, Z.-g.; Diao, X.-q.; Weng, Y.-x.; Wang, Y.-Z. Characterization of the effect of REC on the compatibility of PHBH and PLA. *Polym. Test.* **2015**, *42*, 17–25. [CrossRef]
53. Asrar, J.; Valentin, H.E.; Berger, P.A.; Tran, M.; Padgette, S.R.; Garbow, J.R. Blosynthesis and properties of poly(3-hydroxybutyrate-co-3-hydroxyhexanoate) polymers. *Biomacromolecules* **2002**, *3*, 1006–1012. [CrossRef] [PubMed]
54. Ding, C.; Cheng, B.; Wu, Q. DSC analysis of isothermally melt-crystallized bacterial poly(3-hydroxybutyrate-co-3-hydroxyhexanoate) films. *J. Therm. Anal. Calorim.* **2011**, *103*, 1001–1006. [CrossRef]
55. Jacquel, N.; Tajima, K.; Nakamura, N.; Miyagawa, T.; Pan, P.; Inoue, Y. Effect of Orotic Acid as a Nucleating Agent on the Crystallization of Bacterial Poly(3-hydroxybutyrate-co-3-hydroxyhexanoate) Copolymers. *J. Appl. Polym. Sci.* **2009**, *114*, 1287–1294. [CrossRef]
56. Quiles-Carrillo, L.; Duart, S.; Montanes, N.; Torres-Giner, S.; Balart, R. Enhancement of the mechanical and thermal properties of injection-molded polylactide parts by the addition of acrylated epoxidized soybean oil. *Mater. Des.* **2018**, *140*, 54–63. [CrossRef]
57. Burgos, N.; Martino, V.P.; Jiménez, A. Characterization and ageing study of poly (lactic acid) films plasticized with oligomeric lactic acid. *Polym. Degrad. Stab.* **2013**, *98*, 651–658. [CrossRef]
58. Armentano, I.; Fortunati, E.; Burgos, N.; Dominici, F.; Luzi, F.; Fiori, S.; Jiménez, A.; Yoon, K.; Ahn, J.; Kang, S. Processing and characterization of plasticized PLA/PHB blends for biodegradable multiphase systems. *Expresss Polym. Lett.* **2015**, *9*, 583–596. [CrossRef]
59. Miquelard, G.; Guinault, A.; Sollogoub, C.; Gervais, M. Combined compatibilization and plasticization effect of low molecular weight poly (lactic acid) in poly (lactic acid)/poly (3-hydroxybutyrate-co-3-hydroxyvalerate) blends. *Expresss Polym. Lett.* **2018**, *12*, 114–125. [CrossRef]
60. Ferri, J.M.; Garcia-Garcia, D.; Montanes, N.; Fenollar, O.; Balart, R. The effect of maleinized linseed oil as biobased plasticizer in poly (lactic acid)-based formulations. *Polym. Int.* **2017**, *66*, 882–891. [CrossRef]
61. Garcia-Campo, M.J.; Quiles-Carrillo, L.; Masia, J.; Reig-Pérez, M.J.; Montanes, N.; Balart, R. Environmentally friendly compatibilizers from soybean oil for ternary blends of poly (lactic acid)-PLA, poly (ε-caprolactone)-PCL and poly (3-hydroxybutyrate)-PHB. *Materials* **2017**, *10*, 1339. [CrossRef] [PubMed]

62. Liminana, P.; Garcia-Sanoguera, D.; Quiles-Carrillo, L.; Balart, R.; Montanes, N. Development and characterization of environmentally friendly composites from poly (butylene succinate)(PBS) and almond shell flour with different compatibilizers. *Compos. Part B Eng.* **2018**, *144*, 153–162. [CrossRef]
63. Thomas, S.; Shumilova, A.; Kiselev, E.; Baranovsky, S.; Vasiliev, A.; Nemtsev, I.; Kuzmin, A.P.; Sukovatyi, A.; Avinash, R.P.; Volova, T. Thermal, mechanical and biodegradation studies of biofiller based poly-3-hydroxybutyrate biocomposites. *Int. J. Biol. Macromol.* **2019**. [CrossRef] [PubMed]
64. Gong, X.; Gao, X.; Tang, C.Y.; Law, W.C.; Chen, L.; Hu, T.; Wu, C.; Tsui, C.P.; Rao, N. Compatibilization of poly (lactic acid)/high impact polystyrene interface using copolymer poly (stylene-ran-methyl acrylate). *J. Appl. Polym. Sci.* **2018**, *135*, 45799. [CrossRef]
65. Hosoda, N.; Tsujimoto, T.; Uyama, H. Green Composite of Poly(3-hydroxybutyrate-co-3-hydroxyhexanoate) Reinforced with Porous Cellulose. *Acs Sustain. Chem. Eng.* **2014**, *2*, 248–253. [CrossRef]
66. Perinovic, S.; Andricic, B.; Erceg, M. Thermal properties of poly(L-lactide)/olive stone flour composites. *Thermochim. Acta* **2010**, *510*, 97–102. [CrossRef]
67. Quiles-Carrillo, L.; Montanes, N.; Sammon, C.; Balart, R.; Torres-Giner, S. Compatibilization of highly sustainable polylactide/almond shell flour composites by reactive extrusion with maleinized linseed oil. *Ind. Crop. Prod.* **2018**, *111*, 878–888. [CrossRef]
68. Liminana, P.; Garcia-Sanoguera, D.; Quiles-Carrillo, L.; Balart, R.; Montanes, N. Optimization of Maleinized Linseed Oil Loading as a Biobased Compatibilizer in Poly(Butylene Succinate) Composites with Almond Shell Flour. *Materials* **2019**, *12*, 685. [CrossRef]
69. Yin, C.; Wang, Z.; Luo, Y.; Li, J.; Zhou, Y.; Zhang, X.; Zhang, H.; Fang, P.; He, C. Thermal annealing on free volumes, crystallinity and proton conductivity of Nafion membranes. *J. Phys. Chem. Solids* **2018**, *120*, 71–78. [CrossRef]
70. Oliver-Ortega, H.; Méndez, J.A.; Espinach, F.X.; Tarrés, Q.; Ardanuy, M.; Mutjé, P. Impact strength and water uptake behaviors of fully bio-based PA11-SGW composites. *Polymers* **2018**, *10*, 717. [CrossRef]
71. Pfister, D.P.; Larock, R.C. Thermophysical properties of conjugated soybean oil/corn stover biocomposites. *Bioresour. Technol.* **2010**, *101*, 6200–6206. [CrossRef] [PubMed]

MDPI
St. Alban-Anlage 66
4052 Basel
Switzerland
Tel. +41 61 683 77 34
Fax +41 61 302 89 18
www.mdpi.com

Polymers Editorial Office
E-mail: polymers@mdpi.com
www.mdpi.com/journal/polymers

www.ingramcontent.com/pod-product-compliance
Lightning Source LLC
LaVergne TN
LVHW070122100526
838202LV00016B/2214